Lecture Notes in Computer Science 2442

Edited by G. Goos, J. Hartmanis, and J. van Leeuwen

T0191515

Springer
Berlin
Heidelberg
New York
Barcelona
Hong Kong
London
Milan
Paris
Tokyo

Springer
Berlin
Heidelberg
New York
Barcelona
Hong Kong
London
Milan
Paris
Tokyo

Moti Yung (Ed.)

Advances in Cryptology – CRYPTO 2002

22nd Annual International Cryptology Conference
Santa Barbara, California, USA, August 18-22, 2002
Proceedings

 Springer

Series Editors

Gerhard Goos, Karlsruhe University, Germany
Juris Hartmanis, Cornell University, NY, USA
Jan van Leeuwen, Utrecht University, The Netherlands

Volume Editor

Moti Yung
Columbia University
Department of Computer Science, 450 Computer Science Building
1214 Amsterdam Ave., New York, N.Y. 10027, USA
E-mail: moti@cs.columbia.edu

Cataloging-in-Publication Data applied for

Die Deutsche Bibliothek - CIP-Einheitsaufnahme

Advances in cryptology : proceedings / CRYPTO 2002, 22nd Annual
International Cryptology Conference, Santa Barbara, California, USA, August
18 - 22, 2002. Moti Yung (ed.). [IACR]. - Berlin ; Heidelberg ; New York ;
Barcelona ; Hong Kong ; London ; Milan ; Paris ; Tokyo : Springer, 2002
  (Lecture notes in computer science ; Vol. 2442)
  ISBN 3-540-44050-X

CR Subject Classification (1998): E.3, G.2.1, F.2.1-2, D.4.6, K.6.5, C.2, J.1

ISSN 0302-9743
ISBN 3-540-44050-X Springer-Verlag Berlin Heidelberg New York

Springer-Verlag Berlin Heidelberg New York
a member of BertelsmannSpringer Science+Business Media GmbH

http://www.springer.de

© Springer-Verlag Berlin Heidelberg 2002
Printed in Germany

Typesetting: Camera-ready by author, data conversion by Olgun Computergrafik
Printed on acid-free paper   SPIN 10871005   06/3142   5 4 3 2 1 0

Preface

Crypto 2002, the 22nd Annual Crypto Conference, was sponsored by IACR, the International Association for Cryptologic Research, in cooperation with the IEEE Computer Society Technical Committee on Security and Privacy and the Computer Science Department of the University of California at Santa Barbara. It is published as Vol. 2442 of the Lecture Notes in Computer Science (LNCS) of Springer Verlag. Note that 2002, 22 and 2442 are all palindromes... (Don't nod!)

The conference received 175 submissions, of which 40 were accepted; two submissions were merged into a single paper, yielding the total of 39 papers accepted for presentation in the technical program of the conference. In this proceedings volume you will find the revised versions of the 39 papers that were presented at the conference. The submissions represent the current state of work in the cryptographic community worldwide, covering all areas of cryptologic research. In fact, many high-quality works (that surely will be published elsewhere) could not be accepted. This is due to the competitive nature of the conference and the challenging task of selecting a program. I wish to thank the authors of all submitted papers. Indeed, it is the authors of all papers who have made this conference possible, regardless of whether or not their papers were accepted.

The conference program was also immensely benefited by two plenary talks. The first invited talk was by Andrew Chi-Chih Yao, who spoke on "New Directions in Quantum Cryptographic Protocols." In the second talk, David Chaum gave the 2002 IACR Distinguished Lecture, entitled "Privacy Technology: A Survey of Security without Identification."

My deepest thanks go to the program committee members. Serving on a program committee seems, at times, like a thankless job. When a paper is accepted certain people may believe it is due to the paper's intrinsic quality, whereas when a paper is rejected it is attributed to the misjudgment of committee members. The demanding nature of the task of careful evaluation and selection is, at times, easily forgotten. In reality, the reviewing process for this conference was a huge challenge that demanded from committee members top-level scientific capabilities, combined with a lot of time-consuming hard work. Each paper was reviewed by at least three members, and some papers (including those submitted by committee members) were reviewed by as many as six reviewers. The process followed the review directives of IACR. We reached our decisions via electronic discussions and in a meeting of the program committee; this was a tough job, the successful completion of which should be credited to each and every committee member. We were assisted by the program committee's advisory members, as well as by an army of external reviewers whose expertise and help is highly appreciated. Their names are given in a separate list. (I apologize for any possible omission.)

The conference was run by Rebecca Wright, who served as the general chair. I thank her for all her work, and in particular for her continuous assistance to the program committee and the program chair. Some of the committee members as well as other members of the community served as session chairs during the conference, and I thank them for their help in running the program. The conference program also included the traditional Rump Session, chaired by Stuart Haber, featuring short informal talks on recently completed research and work in progress.

The committee task was an international effort (as befits the IACR, where the "I" stands for "International"). We had members from all over the world, a chair in the USA, a program committee meeting in The Netherlands and a web server in Belgium. We utilized Internet technology as much as we could. This was possible due to efforts by a number of individuals. I thank Berry Schoenmakers for making all the necessary local arrangements for the Program Committee meeting in Amsterdam (just before Eurocrypt 2002). I thank Bart Preneel, and his great team at K.U. Leuven, Thomas Herlea and Wim Moreau, who administered the submission and web-review software. Their support has been instrumental. I thank my Ph.D. student Aggelos Kiayias, who served as a technical assistant to the chairs and helped me with the various technical and technological aspects of running the committee and preparing the conference proceedings. Further thanks are due to Bart Preneel, Wim Moreau and Joris Claessens for authoring the web-review software that was used in the refereeing process, and to Chanathip Namprempre, Sam Rebelsky and SIGACT's Electronic Publishing Board, for authoring the software for the electronic submissions. Thanks are also due to the publisher, Springer-Verlag.

To summarize, I benefited greatly from the pleasant and effective working relationships that I enjoyed with the many individuals I had to collaborate with in order to make the program possible, and it was a real learning experience. Indeed, the making of a program for a conference such as Crypto 2002 is an effort that requires a lot of work from a lot of individuals. Fortunately, the IACR and the cryptographic community at large form the active, strong, vibrant, and relevant community that supports our successful conferences. Long live Crypto!

June 2002   Moti Yung

CRYPTO 2002

August 18–22, 2002, Santa Barbara, California, USA

Sponsored by the
International Association for Cryptologic Research (IACR)

In cooperation with
IEEE Computer Society Technical Committee on Security and Privacy,
Computer Science Department, University of California, Santa Barbara

General Chair
Rebecca N. Wright, Stevens Institute of Technology, NJ, USA

Program Chair
Moti Yung, Columbia University, NY, USA

Program Committee

Tom Berson .... Anagram Laboratories, USA
Don Coppersmith .... IBM Research, USA
Giovanni Di Crescenzo .... Telcordia, USA
Hans Dobbertin .... University of Bochum, Germany
Matt Franklin .... UC Davis, USA
Juan Garay .... Bell Labs, USA
Stuart Haber .... Surety, Inc., USA
Johan Håstad .... Royal Institute of Technology, Sweden
Kwangjo Kim .... ICU, Korea
Alfred Menezes .... University of Waterloo, Canada
David Naccache .... Gemplus, France
Tatsuaki Okamoto .... NTT Labs, Japan
Rafail Ostrovsky .... Telcordia, USA
Erez Petrank .... Technion, Israel
Bart Preneel .... K.U. Leuven, Belgium
Ron Rivest .... Massachusetts Institute of Technology, USA
Rei Safavi-Naini .... University of Wollongong, Australia
Dan Simon .... Microsoft Research, USA
Nigel Smart .... University of Bristol, UK
Markus Stadler .... Crypto AG, Switzerland
Eric Verheul .... PricewaterhouseCoopers, The Netherlands
Yiqun Lisa Yin .... NTT MCL, USA

Advisory Members

Joe Kilian (Crypto 2001, Program Chair) .... NEC, USA
Dan Boneh (Crypto 2003, Program Chair) ... Stanford University, USA

External Reviewers

Yonathan Aumann
Dirk Balfanz
Mihir Bellare
Josh Benaloh
Alex Biryukov
John Black
Simon Blackburn
Dan Boneh
Antoon Bosselaers
Thomas Breuel
Eric Brier
Dan Brown
Joe Buhler
Christian Cachin
Jan Camenisch
Ran Canetti
Christophe De Cannière
Sungtaek Chee
Lily Chen
Jung Hee Cheon
Christophe Clavier
Scott Contini
Jean-Sébastien Coron
Ronald Cramer
Anand Desai
Glenn Durfee
Andreas Enge
Lars Engebretsen
Uri Feige
Marc Fischlin
Yair Frankel
Atsushi Fujioka
Eiichiro Fujisaki
Steven Galbraith
Clemente Galdi
Rosario Gennaro
Craig Gentry
Virgil Gligor
Mikael Goldmann
Jovan Golić
Guang Gong
Daniel Gottesman
Louis Goubin
Louis Granboulan

Rich Graveman
Shai Halevi
Helena Handschuh
Darrel Hankerson
Gustav Hast
Jon Herzog
Florian Hess
Martin Hirt
Susan Hohenberger
Jonas Holmerin
Yuval Ishai
Markus Jakobsson
Stanislaw Jarecki
Thomas Johansson
Antoine Joux
Marc Joye
Charanjit Jutla
Jonathan Katz
Aggelos Kiayias
Joe Kilian
Seung Joo Kim
Lars Knudsen
Neil Koblitz
Hugo Krawczyk
Hartono Kurnio
Eyal Kushilevitz
Tanja Lange
Alan Lauder
Arjen Lenstra
Matt Lepinski
Yehuda Lindell
Moses Liskov
Anna Lysyanskaya
Phil MacKenzie
Tal Malkin
John Malone-Lee
Renato Menicocci
Daniele Micciancio
Miodrag Mihaljevic
Tal Mor
Shiho Moriai
Christophe Mourtel
Yi Mu
Jen Mulligan

Mats Näslund
Kenny Nguyen
Svetla Nikova
Kazuo Ohta
Pascal Paillier
Rafael Pass
Christopher Peikert
Benny Pinkas
Michaël Quisquater
Tal Rabin
Raj Rajagopalan
Anna Redz
Omer Reingold
Vincent Rijmen
Phil Rogaway
Tomas Sander
Werner Schindler
Jasper Scholten
Stefaan Seys
Alice Silverberg
Diana Smetters
Adam Smith
David Soldera
Jessica Staddon

Martijn Stam
Koutarou Suzuki
Edlyn Teske
Prasad Tetali
Dong To
Yuki Tokunaga
Marten Trolin
Yiannis Tsiounis
Christophe Tymen
Ugo Vaccaro
Serge Vaudenay
Frederik Vercauteren
Huaxiong Wang
Yejing Wang
John Watrous
Steve Weis
Michael Wiener
Peter Winkler
Douglas Wikstrom
Duncan Wong
Hao-Chi Wong
Yoav Yerushalmi
Xian-Mo Zhang
Yuliang Zheng

Table of Contents

Secure Multiparty Computation

Public-Key Encryption

Information Theory and Secret Sharing

Cipher Design and Analysis

Elliptic Curves and Abelian Varieties

Password-Based Authentication

Distributed Cryptosystems

Pseudorandomness and Applications

Variations on Signatures and Authentication

Stream Ciphers and Boolean Functions

Commitment Schemes

Signature Schemes

Author Index

Essential Algebraic Structure within the AES

Sean Murphy and Matthew J.B. Robshaw

Information Security Group,
Royal Holloway, University of London,
Egham, Surrey, TW20 0EX, UK
{s.murphy,m.robshaw}@rhul.ac.uk, mrobshaw@supanet.com

Abstract. One difficulty in the cryptanalysis of the Advanced Encryption Standard AES is the tension between operations in the two fields $GF(2^8)$ and $GF(2)$. This paper outlines a new approach that avoids this conflict. We define a new block cipher, the BES, that uses only simple algebraic operations in $GF(2^8)$. Yet the AES can be regarded as being identical to the BES with a restricted message space and key space, thus enabling the AES to be realised solely using simple algebraic operations in one field $GF(2^8)$. This permits the exploration of the AES within a broad and rich setting. One consequence is that AES encryption can be described by an extremely sparse overdetermined multivariate quadratic system over $GF(2^8)$, whose solution would recover an AES key.

Keywords: Advanced Encryption Standard, AES, Rijndael, BES, Algebraic Structure, (Finite) Galois Field, (Field) Conjugate, Multivariate Quadratic (MQ) Equations.

1 Introduction

Rijndael [7,8] was chosen as the Advanced Encryption Standard (AES) and published as FIPS 197 [21] on 26 November 2001. The AES is carefully designed to resist standard block cipher attacks [1,18]. Here we move our attention to a cipher that is an extension of AES, but which offers one particular advantage. All of the operations in this new cipher, the BES, are entirely described using very simple operations in $GF(2^8)$. Thus while the AES is embedded within the BES, and while the BES fully respects encryption with the AES, there are no $GF(2)^8$ operations.

The properties of this new cipher are intimately related to the properties of the AES, as the AES is essentially the BES with a restricted message and key space. The AES is, in essence, woven into the fabric of the BES. Yet, in many ways, the new cipher is easier to analyse. It is certainly easier to describe; one round of the cipher consists exclusively of inversion in $GF(2^8)$, matrix multiplication in $GF(2^8)$, and key addition in $GF(2^8)$.

By recasting the AES in this way we highlight some important structural features of the AES. We illustrate this with a differential-type effect in the BES that seems surprising given the design principles of the AES. Furthermore, we show that the AES preserves algebraic curves and that it can be expressed as a

M. Yung (Ed.): CRYPTO 2002, LNCS 2442, pp. 1–16, 2002.
© Springer-Verlag Berlin Heidelberg 2002

very simple system of multivariate quadratic equations over $GF(2^8)$. It is entirely possible that such a new approach might offer significant improvements to the cryptanalysis of the AES.

2   Previous Work and Notation

Throughout the AES process, Rijndael (the eventual AES) received considerable cryptanalytic attention [10, 12, 17]. The simplicity of Rijndael was emphasized by its designers [7, 8], and much work has concentrated on the structural properties of the cipher [9, 11, 15, 19, 20, 23, 24].

In this paper we introduce a new technique which further simplifies analysis of the AES. While the AES encryption process is typically described using operations on an array of bytes, we represent the data as column vectors, so matrix multiplication of such a column vector occurs on the left. We regard a byte as an element of the binary field defined by the irreducible "Rijndael" polynomial $X^8 + X^4 + X^3 + X + 1$. We denote this field by $\mathbf{F}$ and a root of this polynomial by θ, so

$$\mathbf{F} = GF(2^8) = \frac{GF(2)[X]}{(X^8 + X^4 + X^3 + X + 1)} = GF(2)(\theta).$$

Each byte therefore represents a polynomial in θ and we adopt the convention that the most significant bit in a byte (the θ^7 term) is represented by the leftmost, and most significant, bit of the hexadecimal representation of a byte.

The version of the AES we consider has a 128-bit or 16-byte message and key space, though our comments are more generally applicable. The new cipher BES has a 128-byte message and key space. We later define a restriction of the BES spaces to a subset of size 2^{128} that corresponds to the AES. We denote these three sets by $\mathbf{A}$, $\mathbf{B}$ and $\mathbf{B_A}$ respectively, so

$\mathbf{A}$	State space of the AES	Vector space $\mathbf{F}^{16}$
$\mathbf{B}$	State space of the BES	Vector space $\mathbf{F}^{128}$
$\mathbf{B_A}$	Subset of $\mathbf{B}$ corresponding to $\mathbf{A}$	Subset of $\mathbf{F}^{128}$.

3   The Basic Structure of the AES

We refer to FIPS 197 [21] for a full description of the cipher, but we list the significant steps here. We concentrate our attentions on a typical round; the first and last rounds have a different (but related) form that is easily assimilated. We consider the basic version of the AES, which encrypts a 16-byte block using a 16-byte key with 10 encryption rounds.

The input to the AES round function can be viewed as a rectangular array of bytes or, equivalently, as a column vector of bytes. Throughout the encryption process this byte-structure is fully respected. The AES specification defines a round in terms of the following three transformations.

1. **The AES S-Box.** The value of each byte in the array is substituted according to a table look-up. This table look up $S[\cdot]$ is the combination of three transformations.

   (a) The input w is mapped to $x = w^{(-1)}$ where $w^{(-1)}$ is defined by

$$w^{(-1)} = w^{254} = \begin{cases} w^{-1} & w \neq 0 \\ 0 & w = 0 \end{cases}$$

   Thus "AES inversion" is identical to standard field inversion in $\mathbf{F}$ for non-zero field elements with $0^{(-1)} = 0$.

   (b) The intermediate value x is regarded as a $GF(2)$-vector of dimension 8 and transformed using an (8×8) $GF(2)$-matrix L_A. The transformed vector $L_A \cdot x$ is then regarded in the natural way as an element of $\mathbf{F}$.

   (c) The output of the AES S-Box is $(L_A \cdot x) + 63$, where addition is with respect to $GF(2)$.

2. **The AES linear diffusion (mixing) layer.**

   (a) Each row of the array is rotated by a certain number of byte positions.

   (b) Each column of the array is considered to be an $\mathbf{F}$-vector, and a column $\mathbf{y}$ is transformed to the column $C \cdot \mathbf{y}$, where C is a (4×4) $\mathbf{F}$-matrix.

3. **The AES subkey addition.** Each byte of the array is added (with respect to $GF(2)$) to a byte from the corresponding array of round subkeys.

The additive constant (63) in the AES S-box can be removed by incorporating it within a (slightly) modified key schedule [19]. For simplicity, we use this description of the AES in this paper.

4   The Big Encryption System (BES)

We introduce a new iterated block cipher, the *Big Encryption System (BES)*, which operates on 128-byte blocks with a 16-byte key. Both the AES and the BES are defined in terms of bytes and we now describe the common mathematical framework for both ciphers.

Both the AES and the BES use a *state vector* of bytes, which is transformed by the basic operations within a round. In both cases, the plaintext is the input state vector while the ciphertext is the output state vector. As described in Section 2, the state spaces of the AES and the BES are the vector spaces $\mathbf{A} = \mathbf{F}^{16}$ and $\mathbf{B} = \mathbf{F}^{128}$ respectively. We now describe the basic techniques required to establish the relationship between the AES and the BES.

Inversion. The inversion operation is easily described. For $a \in \mathbf{F}$, it is identical to standard field inversion for non-zero field elements with $0^{(-1)} = 0$. For an n-dimensional vector $\mathbf{a} = (a_0, \ldots, a_{n-1}) \in \mathbf{F}^n$, we view inversion as a componentwise operation and set

$$\mathbf{a}^{(-1)} = (a_0^{(-1)}, \ldots, a_{n-1}^{(-1)}).$$

Vector conjugates. For any element $a \in \mathbf{F}$ we can define the *vector conjugate* of a, $\tilde{\mathbf{a}}$, as the vector of the eight $GF(2)$-conjugates of a, so

$$\tilde{\mathbf{a}} = \left(a^{2^0}, a^{2^1}, a^{2^2}, a^{2^3}, a^{2^4}, a^{2^5}, a^{2^6}, a^{2^7}\right).$$

We use a *vector conjugate mapping* ϕ from $\mathbf{F}^n$ to a subset of $\mathbf{F}^{8n}$. For $n = 1$ and $a \in \mathbf{F}$, we have

$$\tilde{\mathbf{a}} = \phi(a) = \left(a^{2^0}, a^{2^1}, a^{2^2}, a^{2^3}, a^{2^4}, a^{2^5}, a^{2^6}, a^{2^7}\right).$$

This definition extends in the obvious way to a vector conjugate mapping ϕ from $\mathbf{F}^n$ to a subset of $\mathbf{F}^{8n}$. The n-dimensional vector $\mathbf{a} = (a_0, \ldots, a_{n-1}) \in \mathbf{F}^n$ is mapped to

$$\tilde{\mathbf{a}} = \phi(\mathbf{a}) = (\phi(a_0), \ldots, \phi(a_{n-1})).$$

The vector conjugate mapping ϕ has desirable algebraic properties, namely that it is additive and preserves inverses, so

$$\phi(\mathbf{a} + \mathbf{a}') = \phi(\mathbf{a}) + \phi(\mathbf{a}') \text{ and}$$
$$\phi(\mathbf{a}^{-1}) = \phi(\mathbf{a})^{-1}.$$

When each successive set of eight components in $\mathbf{a} \in \mathbf{F}^{8n}$ form an ordered set of $GF(2)$-conjugates, we say that $\mathbf{a}$ has the *conjugacy property*. Such vectors lie in $Im(\phi)$, and we can consider $\phi^{-1} : Im(\phi) \to \mathbf{F}^n$ as an *extraction mapping* which recovers the basic vector from a vector conjugate.

Embedding the AES state space in the BES state space. Any plaintext, ciphertext, intermediate text, or subkey for the AES is an element of the state space $\mathbf{A}$. Similarly, any plaintext, ciphertext, intermediate text, or subkey for the BES is an element of the state space $\mathbf{B}$.

We can use the vector conjugate map ϕ to embed any element of the AES state space $\mathbf{A}$ into the BES state space $\mathbf{B}$. We define

$$\mathbf{B_A} = \phi(\mathbf{A}) \subset \mathbf{B} \text{ to be the } AES \text{ subset of } BES,$$

that is the embedded image of the AES state space in the BES state space. Elements of $\mathbf{B_A}$, that is embedded images of AES states or subkeys, have the vector conjugacy property. Furthermore, $\mathbf{B_A}$ is an additively closed set that also preserves inverses.

In the following sections we describe the cipher BES. This is done in such a way that the "commuting" diagram in Figure 1 is fully respected.

4.1   AES and BES

As previously described, we regard a state vector of the AES to be an element $\mathbf{a} \in \mathbf{A}$. We further regard each round subkey as an element $\mathbf{k}_i \in \mathbf{A}$. We do

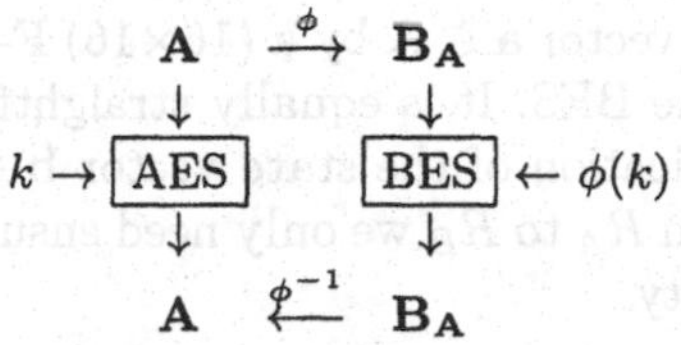

$$A \xrightarrow{\phi} B_A$$
$$\downarrow \qquad \downarrow$$
$$k \to \boxed{AES} \qquad \boxed{BES} \leftarrow \phi(k)$$
$$\downarrow \qquad \downarrow$$
$$A \xleftarrow{\phi^{-1}} B_A$$

Fig. 1. The relationship between the AES and the BES. The important feature of the BES is that it is defined exclusively using simple operations in one field, $GF(2^8)$.

not use the standard AES way of representing an element $\mathbf{a}$ as a square array. Instead we view the state vector $\mathbf{a}$ as a column vector, where

$$\mathbf{a} = \begin{array}{|c|c|c|c|} \hline a_{00} & a_{01} & a_{02} & a_{03} \\ \hline a_{10} & a_{11} & a_{12} & a_{13} \\ \hline a_{20} & a_{21} & a_{22} & a_{23} \\ \hline a_{30} & a_{31} & a_{32} & a_{33} \\ \hline \end{array} = (a_{00}, \dots, a_{30}, a_{01}, \dots, a_{31}, \dots, a_{33})^T.$$

For the BES, we also view the state vector $\mathbf{b} \in \mathbf{B}$ as a column vector where

$$\mathbf{b} = (b_{000}, \dots, b_{007}, b_{100}, \dots, b_{107}, \dots, \dots, b_{330}, \dots, b_{337})^T.$$

It should be obvious how we intend to use the embedding mapping ϕ. We set

$$\phi(a_{ij}) = (b_{ij0}, \dots, b_{ij7}).$$

Each basic operation in a round of the AES describes a bijective mapping on $\mathbf{A}$. These can be readily replaced with similar operations in the BES. Our aim in doing this is to ensure that every operation (including the $GF(2)$-linear map from the AES S-box) is expressed using simple algebraic operations over $\mathbf{F}$.

Subkey addition. This is obvious for both the AES and the BES. For the AES we combine the state vector $\mathbf{a} \in \mathbf{A}$ with an AES subkey $(\mathbf{k}_A)_i \in \mathbf{A}$ by $\mathbf{a} \mapsto \mathbf{a} + (\mathbf{k}_A)_i$. We do exactly the same in BES and we combine the state vector $\mathbf{b} \in \mathbf{B}$ with a subkey $(\mathbf{k}_B)_i \in \mathbf{B}$ by $\mathbf{b} \mapsto \mathbf{b} + (\mathbf{k}_B)_i$. We consider the generation of the BES subkeys below.

S-box inversion. As inversion operates componentwise on bytes, it is just as easy to describe in the BES as the AES. In the AES, inversion can be viewed as a componentwise vector inversion of the state vector $\mathbf{a} \in \mathbf{A}$. Thus the AES inversion operation is given by $\mathbf{a} \mapsto \mathbf{a}^{(-1)}$. This can be translated in the obvious manner, and for $\mathbf{b} \in \mathbf{B}$, inversion in the BES is given by $\mathbf{b} \mapsto \mathbf{b}^{(-1)}$.

Row operation. The AES RowShift operation permutes the bytes in the array. Clearly this process can be considered as a transformation of the components of a column vector $\mathbf{a} \in \mathbf{A}$. It is straightforward to represent this transformation as multiplication of the state vector $\mathbf{a} \in \mathbf{A}$ by a (16×16) F-matrix R_A. Consider the equivalent operation in the BES. It is equally straightforward to represent this transformation as multiplication of the state vector $\mathbf{b} \in \mathbf{B}$ by a (128×128) F-matrix R_B. In moving from R_A to R_B we only need ensure that vector conjugates are moved as a single entity.

Column operation. The AES MixColumn operation is defined using a (4×4) F-matrix C_A. A column $\mathbf{y} \in \mathbf{F}^4$ of the conceptual state array is transformed into a replacement column $\mathbf{z} \in \mathbf{F}^4$ by

$$\mathbf{z} = C_A \cdot \mathbf{y} = \begin{pmatrix} \theta & (\theta+1) & 1 & 1 \\ 1 & \theta & (\theta+1) & 1 \\ 1 & 1 & \theta & (\theta+1) \\ (\theta+1) & 1 & 1 & \theta \end{pmatrix} \cdot \mathbf{y} = \begin{pmatrix} 02\ 03\ 01\ 01 \\ 01\ 02\ 03\ 01 \\ 01\ 01\ 02\ 03 \\ 03\ 01\ 01\ 02 \end{pmatrix} \cdot \mathbf{y}.$$

We can readily view this as a transformation of the AES state space $\mathbf{A}$ by the (16×16) F-matrix transformation Mix_A, where Mix_A is a block diagonal matrix with 4 identical blocks C_A, so $\mathrm{Mix}_A = Diag_4(C_A)$. Consider now the equivalent transformation within the BES. Our aim is to replicate the actions of the AES, but to maintain the condition that each byte in the AES is represented by a conjugate vector in the BES. To do this we consider eight versions of the matrix C_A. These versions are denoted by $C_B^{(k)}$ and they are defined as

$$C_B^{(k)} = \begin{pmatrix} \theta^{2^k} & (\theta+1)^{2^k} & 1 & 1 \\ 1 & \theta^{2^k} & (\theta+1)^{2^k} & 1 \\ 1 & 1 & \theta^{2^k} & (\theta+1)^{2^k} \\ (\theta+1)^{2^k} & 1 & 1 & \theta^{2^k} \end{pmatrix} \quad \text{for } k = 0, \ldots, 7,$$

so $C_B^{(0)} = C_A$. We note that $C_B^{(k)}$ is an MDS matrix, thereby offering certain diffusion properties [7, 8], and that if

$$(z_0, z_1, z_2, z_3)^T = C_A \cdot (y_0, y_1, y_2, y_3)^T \text{ then}$$

$$(z_0^{2^k}, z_1^{2^k}, z_2^{2^k}, z_3^{2^k})^T = C_B^{(k)} \cdot \left(y_0^{2^k}, y_1^{2^k}, y_2^{2^k}, y_3^{2^k} \right)^T.$$

This provides a way of preserving the conjugacy property through the MixColumn transformation in the BES. The matrices $C_B^{(k)}$ can be used to define the (128×128) F-matrix Mix_B that respects the vector conjugate embedding mapping $\phi : \mathbf{A} \to \mathbf{B}_A$, so the action of MixColumn on bytes in the AES is replicated by the action of Mix_B on vector conjugates in the BES. Under a simple basis re-ordering, Mix_B is a block diagonal matrix comprising 32 (4×4) MDS matrices.

The S-box $GF(2)$-linear operation. In the AES, there is no easy way to represent this transformation of the state space $\mathbf{A}$ as a matrix multiplication However, in the BES there is a simple matrix representation of this operation.

The AES $GF(2)$-linear operation $\sigma_A : \mathbf{F}^{16} \to \mathbf{F}^{16}$ is defined using a function $f : \mathbf{F} \to \mathbf{F}$ that operates on each component of the state vector $\mathbf{a}$, so

$$\mathbf{a} = (a_{00}, \ldots, a_{33}) \mapsto \sigma_A(\mathbf{a}) = (f(a_{00}), \ldots, f(a_{33})).$$

In the AES specification, f is defined by considering $\mathbf{F} = GF(2^8)$ as the vector space $GF(2)^8$. The transformation f is then represented by the action of an (8×8) $GF(2)$ matrix L_A where

$$L_A = \begin{pmatrix} 1 & 0 & 0 & 0 & 1 & 1 & 1 & 1 \\ 1 & 1 & 0 & 0 & 0 & 1 & 1 & 1 \\ 1 & 1 & 1 & 0 & 0 & 0 & 1 & 1 \\ 1 & 1 & 1 & 1 & 0 & 0 & 0 & 1 \\ 1 & 1 & 1 & 1 & 1 & 0 & 0 & 0 \\ 0 & 1 & 1 & 1 & 1 & 1 & 0 & 0 \\ 0 & 0 & 1 & 1 & 1 & 1 & 1 & 0 \\ 0 & 0 & 0 & 1 & 1 & 1 & 1 & 1 \end{pmatrix}.$$

To accomplish the change from $GF(2^8)$ to $GF(2)^8$, the natural mapping $\psi : GF(2^8) \to GF(2)^8$ is used in the AES. The componentwise AES $GF(2)$-linear operation $f : \mathbf{F} \to \mathbf{F}$ is then defined by $f(a) = \psi^{-1}(L_A(\psi(a)))$ for $a \in \mathbf{F}$. It is the need for the maps ψ and ψ^{-1} that complicates analysis of the AES.

However, there exists a polynomial with co-efficients in $\mathbf{F}$ which interpolates $f : \mathbf{F} \to \mathbf{F}$. This polynomial may be regarded as an equivalent definition of f. Further, since f is an *additive* or *linearized* polynomial [16] on $\mathbf{F}$, it is necessarily described by a linear combination of conjugates. Thus we obtain

$$f(a) = \sum_{k=0}^{7} \lambda_k a^{2^k} \text{ for } a \in \mathbf{F},$$

where $(\lambda_0, \lambda_1, \lambda_2, \lambda_3, \lambda_4, \lambda_5, \lambda_6, \lambda_7) = (05, 09, \mathtt{f9}, 25, \mathtt{f4}, 01, \mathtt{b5}, \mathtt{8f})$.

This polynomial is essentially given in [8] as part of the derivation of the related S-Box interpolation polynomial [7, 8]. However, our interest is in separating out the $\mathbf{F}$-inversions from the rest of the $\mathbf{F}$-linear round function, since this separation seems algebraically the most natural.

The $GF(2)$-linear operation from the AES S-box can now be defined in the BES using an (8×8) $\mathbf{F}$-matrix. This matrix replicates the (AES) action of the $GF(2)$-linear map on the first byte of a vector conjugate set while ensuring that the property of *vector conjugacy* is preserved on the remaining bytes. We set

$$L_B = \begin{pmatrix} (\lambda_0)^{2^0} & (\lambda_1)^{2^0} & (\lambda_2)^{2^0} & (\lambda_3)^{2^0} & (\lambda_4)^{2^0} & (\lambda_5)^{2^0} & (\lambda_6)^{2^0} & (\lambda_7)^{2^0} \\ (\lambda_7)^{2^1} & (\lambda_0)^{2^1} & (\lambda_1)^{2^1} & (\lambda_2)^{2^1} & (\lambda_3)^{2^1} & (\lambda_4)^{2^1} & (\lambda_5)^{2^1} & (\lambda_6)^{2^1} \\ (\lambda_6)^{2^2} & (\lambda_7)^{2^2} & (\lambda_0)^{2^2} & (\lambda_1)^{2^2} & (\lambda_2)^{2^2} & (\lambda_3)^{2^2} & (\lambda_4)^{2^2} & (\lambda_5)^{2^2} \\ (\lambda_5)^{2^3} & (\lambda_6)^{2^3} & (\lambda_7)^{2^3} & (\lambda_0)^{2^3} & (\lambda_1)^{2^3} & (\lambda_2)^{2^3} & (\lambda_3)^{2^3} & (\lambda_4)^{2^3} \\ (\lambda_4)^{2^4} & (\lambda_5)^{2^4} & (\lambda_6)^{2^4} & (\lambda_7)^{2^4} & (\lambda_0)^{2^4} & (\lambda_1)^{2^4} & (\lambda_2)^{2^4} & (\lambda_3)^{2^4} \\ (\lambda_3)^{2^5} & (\lambda_4)^{2^5} & (\lambda_5)^{2^5} & (\lambda_6)^{2^5} & (\lambda_7)^{2^5} & (\lambda_0)^{2^5} & (\lambda_1)^{2^5} & (\lambda_2)^{2^5} \\ (\lambda_2)^{2^6} & (\lambda_3)^{2^6} & (\lambda_4)^{2^6} & (\lambda_5)^{2^6} & (\lambda_6)^{2^6} & (\lambda_7)^{2^6} & (\lambda_0)^{2^6} & (\lambda_1)^{2^6} \\ (\lambda_1)^{2^7} & (\lambda_2)^{2^7} & (\lambda_3)^{2^7} & (\lambda_4)^{2^7} & (\lambda_5)^{2^7} & (\lambda_6)^{2^7} & (\lambda_7)^{2^7} & (\lambda_0)^{2^7} \end{pmatrix}$$

We can now represent the entire set of $GF(2)$-linear operations in the AES with a (128×128) **F**-matrix in the BES, Lin_B. Thus Lin_B is a block diagonal matrix with 16 identical blocks L_B, so $\mathrm{Lin}_B = Diag_{16}(L_B)$.

Key schedule. We can use the techniques from previous sections to describe the key schedule for the BES. This effectively replicates the actions of the AES key schedule, in which a 16-byte AES key $\mathbf{k}_A$ provides eleven subkeys, each in **A**. In the BES, a 128-byte BES key $\mathbf{k}_B$ provides eleven subkeys, each in **B**.

The key schedule in the AES uses the same operations as the AES encryption process, namely the $GF(2)$-linear map, componentwise inversion, byte rotation, and addition. Thus the key schedule can also be described using the same simple algebraic operations over **F**. Whenever a constant is required in the AES, we use the embedded image of that constant in the BES. Whenever a byte in the AES has to be moved to a different position, we ensure that the corresponding vector conjugate is moved as a single entity in the BES. In this way, we ensure that if a BES key the conjugacy property, then so do all its derived subkeys. If the embedded image of the AES key $\mathbf{k}_A$ is the BES key $\mathbf{k}_B = \phi(\mathbf{k}_A)$, then $(\mathbf{k}_B)_i = \phi((\mathbf{k}_A)_i)$ for every round subkey, so the embedded images of an AES subkey sequence form a BES subkey sequence.

Round function of BES. We have now completely described a round of BES. If the inputs to the BES round function are $\mathbf{b} \in \mathbf{B}$ and subkey $(\mathbf{k}_B)_i \in \mathbf{B}$, then the BES round function is given by

$$\mathrm{Round}_B(\mathbf{b}, (\mathbf{k}_B)_i) = \mathrm{Mix}_B \left(R_B \left(\mathrm{Lin}_B \left(\mathbf{b}^{(-1)} \right) \right) \right) + (\mathbf{k}_B)_i$$
$$= M_B \cdot (\mathbf{b}^{(-1)}) + (\mathbf{k}_B)_i,$$

where M_B is a (128×128) **F**-matrix performing linear diffusion within the BES. Furthermore, if the inputs to the AES round function are $\mathbf{a} \in \mathbf{A}$ and subkey $(\mathbf{k}_A)_i \in \mathbf{A}$, then we have

$$\mathrm{Round}_A(\mathbf{a}, (\mathbf{k}_A)_i) = \phi^{-1} \left(\mathrm{Round}_B \left(\phi(\mathbf{a}), \phi((\mathbf{k}_A)_i) \right) \right).$$

4.2   The Relationship between the AES and the BES

The BES is a 128-byte block cipher, which consists entirely of simple algebraic operations over **F**. It has the property that $\mathbf{B_A}$, the set of embedded images of AES vectors, or equivalently the set of all BES inputs with the conjugacy property, is closed under the action of the BES round function. Furthermore, encryption in the BES fully respects encryption in the AES and the commuting diagram given in Figure 1 holds. Thus the BES restricted to $\mathbf{B_A}$ provides an alternative description of the AES and analysis of the BES may well provide additional insight into the AES.

5   Algebraic Observations on the BES

The round function of the BES, and hence essentially the AES, is given by

$$\mathbf{b} \mapsto M_B \cdot \mathbf{b}^{(-1)} + (\mathbf{k}_B)_i.$$

Thus a round of the AES is simply componentwise inversion and an affine transformation with respect to the *same* field $\mathbf{F} = GF(2^8)$. This suggests many possible areas for future investigation. We offer some preliminary observations.

5.1   Linear Diffusion in BES

The linear diffusion $\mathbf{F}$-matrix M_B of the BES is a sparse matrix and can be analysed using similar techniques to those used in [19]. These were originally used to analyse the related linear diffusion $GF(2)$-matrix (denoted by M in [19]). However, this linear diffusion matrix (M) and the AES inversion are with respect to *different* fields.

The minimum polynomial of M_B is $(X + 1)^{15}$, effectively the same as the minimum polynomial of M. In some sense, the BES is structurally no more complicated than the AES. Following [19], we find an $\mathbf{F}$-matrix P_B such that

$$R_B = P_B^{-1} \cdot M_B \cdot P_B,$$

where R_B is essentially the *Jordan* form of M_B. The matrix R_B has 112 rows with two ones and 16 rows with a single one while all other entries are zero. It is effectively the simple matrix R given in [19], and has similar interesting properties. The significant change here is that the properties of M_B are properties in $\mathbf{F}$ and not $GF(2)$. Such properties have the potential to interact directly with the inversion operation. Many of these properties involve *linear functionals* or *parity equations*. A *parity equation* is a row vector $\mathbf{e}^T$, and the parity of a vector $\mathbf{b} \in \mathbf{B}$ is $\mathbf{e}^T \cdot \mathbf{b} = \sum_{i=0}^{127} e_i \cdot b_i$. We note a few interesting properties.

- M_B has order 16.
- The columns of P_B form a basis for $\mathbf{B}$. In this basis, the action of the linear diffusion layer is given by the very simple matrix M_B.
- In particular, M_B fixes a subspace of $\mathbf{B}$ of dimension 16. The intersection with $\mathbf{B_A}$, the embedded AES state space, has 2^{16} elements.
- The rows of P_B^{-1} form linear functionals or parity equations (defined above) that always evaluate to 0 or 1 on $\mathbf{B_A}$ (by considering dual spaces).
- The set of parity equations whose value is fixed by M_B form a 16-dimensional vector subspace over $\mathbf{F}$.

These observations may seem somewhat abstract, but they do have important consequences. We discuss an example below in which these observations can be used to illustrate certain differential properties of the BES.

5.2   Related Encryptions in the BES

As noted in Section 5.1, it is possible to find parity equations whose values are fixed by M_B, the linear diffusion layer of the BES. One example is

$$\overbrace{\text{e} = (\text{b4}, \text{fd}, 17, 0\text{e}, 54, \text{a0}, \text{f6}, 52, \ldots)^T,}^{\text{repeat 16 times}}$$

for which $\text{e}^T = \text{e}^T \cdot M_B$, so $\text{e}^T \cdot \text{b} = \text{e}^T \cdot (M_B \cdot \text{b})$. We now describe some interesting properties relating two plaintext-ciphertext pairs generated under related subkey sequences. These properties hold with probability one and so they can be appropriately extended to any number of rounds.

Suppose p has parity $p_e = \text{e}^T \cdot \text{p}$ under parity equation e^T, so $t \cdot \text{p}$ has parity tp_e for any $t \in \text{F}$. Consider two state and subkey pairs $\text{p}, \text{k}_i \in \text{B}$ and $t\text{p}, t^{(-1)}\text{k}_i \in \text{B}$ ($t \neq 0, 1$). A typical BES round function is given by

$$\text{p} \mapsto M_B \cdot \text{p}^{(-1)} + \text{k}_i, \text{ and}$$
$$t \cdot \text{p} \mapsto t^{-1} M_B \cdot \text{p}^{(-1)} + t^{-1}\text{k}_i.$$

When we consider the effect of the BES round function on the parities, we obtain

$$p_e = (\text{e}^T \cdot \text{p}) \mapsto \text{e}^T \cdot M_B \cdot \text{p}^{(-1)} + \text{e}^T \cdot \text{k}_i \qquad = \text{e}^T \cdot \text{p}^{(-1)} + \text{e}^T \cdot \text{k}_i$$
$$= \text{e}^T \cdot (\text{p}^{(-1)} + \text{k}_i),$$
$$tp_e = (\text{e}^T \cdot t\text{p}) \mapsto \text{e}^T \cdot t^{-1} M_B \cdot \text{p}^{(-1)} + \text{e}^T \cdot t^{-1}\text{k}_i = \text{e}^T \cdot t^{-1}\text{p}^{(-1)} + \text{e}^T \cdot t^{-1}\text{k}_i$$
$$= t^{-1}\text{e}^T \cdot (\text{p}^{(-1)} + \text{k}_i).$$

Thus if (p_e, tp_e) are the parities under e^T, then after one round using subkeys k_i and $t^{-1}\text{k}_i$ respectively, the respective parities are $(p'_e, t^{-1}p'_e)$ for some p'_e. Hence if we encrypt two plaintexts $(\text{p}, t\text{p})$ under different sets of related subkey sequences as detailed below, then we obtain two ciphertexts that are related by their parities c_e and tc_e.

Plaintext Parity	Subkey sequence	Ciphertext Parity
p_e	$\text{k}_0, \text{k}_1, \text{k}_2, \text{k}_3 \cdots, \text{k}_9\text{k}_{10}$	c_e
tp_e	$t\text{k}_0, t^{-1}\text{k}_1, t\text{k}_2, t^{-1}\text{k}_3, \cdots, t^{-1}\text{k}_9, t\text{k}_{10}$	tc_e

Differential-type effect in BES. We can increase the sophistication slightly and consider two pairs of plaintext $\text{p}_0, \text{p}_1 \in \text{B}$ and $t\text{p}_0, t\text{p}_1 \in \text{B}$. The difference in the first pair is $\text{p}_0 + \text{p}_1$ with parity $p_e = \text{e}^T \cdot (\text{p}_0 + \text{p}_1)$, and similarly the parity of the difference in the second pair is tp_e.

Plaintext Difference Parity	Subkey sequence	Ciphertext Difference Parity
$\text{e}^T \cdot (\text{p}_0 + \text{p}_1) = p_e$	$\text{k}_0, \text{k}_1, \cdots, \text{k}_{10}$	c_e
$\text{e}^T \cdot (t\text{p}_0 + t\text{p}_1) = tp_e$	$t\text{k}_0, t^{-1}\text{k}_1, \cdots, t\text{k}_{10}$	tc_e

Suppose we encrypt the two pairs of plaintexts under two sets of related subkey sequences as detailed in the above table, then the plaintext and ciphertext difference parities have the same relationship, as shown in the above table. This relationship holds with probability one, so would be applicable for any number of rounds. Thus there exists a probability one differential effect under related subkey sequences in the BES in which every S-Box is active.

Relevance of these BES observations to the AES. These preliminary observations do not apply when specific details of the key schedule are considered. Even if they did, they would not apply directly to the AES for a rather subtle reason. If $(\mathbf{p}, t\mathbf{p}) \in \mathbf{B} \times \mathbf{B}$, then $(\mathbf{p}, t\mathbf{p}) \notin \mathbf{B_A} \times \mathbf{B_A}$; that is if $\mathbf{p}$ has the conjugacy property, then $t\mathbf{p}$ cannot have the conjugacy property ($t \neq 0, 1$). Thus, if $\mathbf{p}$ is an embedded AES plaintext, then $t\mathbf{p}$ cannot be an embedded AES plaintext.

However, these observations are very interesting for the light they shed on the AES design philosophy [7, 8]. As far as linear and differential cryptanalysis are concerned, the BES would be expected to have similar properties to the AES. In particular, the diffusion in both has the same reliance on MDS matrices. However in the BES, which is intricately entwined with the AES, we have exhibited a differential-like property that occurs with certainty even though every S-Box is active.

5.3   Preservation of Algebraic Curves

Each of the BES operations, namely "inversions" (ignoring 0-inversion for the moment) and affine transformations over $\mathbf{F}$, are simple algebraic transformations of $\mathbf{B}$. Thus each BES operation maps an algebraic curve defined on $\mathbf{B} = \mathbf{F}^{128}$ to an isomorphic algebraic curve. For a given key 128-bit key $\mathbf{k}$, more than half (about 53%) of AES plaintexts are encrypted without "inverting" 0 (since 160 inversions are performed). Let $\mathbf{A_k} \subset \mathbf{A}$ denote this set of AES plaintexts for key $\mathbf{k}$. If embedded plaintexts from $\mathbf{A_k}$ lie on a curve, then the corresponding embedded ciphertexts lie on an isomorphic curve over $\mathbf{F}$. Thus, the AES and the BES can be considered to preserve algebraically simple curves over $\mathbf{F}$ with a reasonable probability. In particular, the inversion and the affine transformation of the BES round function map quadratic forms over $\mathbf{F}$ to quadratic forms over $\mathbf{F}$, so the AES can be described using a very simple system of multivariate quadratic equations over $\mathbf{F}$. We consider the consequences of this observation below.

6   Multivariate Quadratic Equations

We now demonstrate that recovering an AES key is equivalent to solving particular systems of extremely sparse multivariate quadratic equations by expressing a BES (and hence an AES) encryption as such a system. The problem of solving such systems of equations lies at the heart of several public key cryptosystems [3, 22], and there has been some progress in providing solutions to such problems [4, 5, 14]. Recently, Courtois and Pieprzyk [6] have suggested the use of a system of multivariate quadratic equations over $GF(2)$ to analyse the AES. However, such a $GF(2)$-system derived directly from the AES is far more complicated than the $\mathbf{F}$-system derived from the BES.

6.1   A Simple Multivariate Quadratic System for the AES

We first establish the notation that we need. We denote the plaintext and ciphertext by $\mathbf{p} \in \mathbf{B}$ and $\mathbf{c} \in \mathbf{B}$ respectively, and the state vectors before and after the ith invocation of the inversion layer by $\mathbf{w}_i \in \mathbf{B}$ and $\mathbf{x}_i \in \mathbf{B}$ ($0 \leq i \leq 9$) respectively. A BES encryption is then described by the following system of equations:

$$
\begin{aligned}
\mathbf{w}_0 &= \mathbf{p} + \mathbf{k}_0, \\
\mathbf{x}_i &= \mathbf{w}_i^{(-1)} & \text{for } i = 0, \ldots, 9, \\
\mathbf{w}_i &= M_B \mathbf{x}_{i-1} + \mathbf{k}_i & \text{for } i = 1, \ldots, 9, \\
\mathbf{c} &= M_B^* \mathbf{x}_9 + \mathbf{k}_{10},
\end{aligned}
$$

where $M_B^* = R_B \cdot Lin_B = Mix_B^{-1} \cdot M_B$, since the final round in the BES (equivalently the AES) does not use the MixColumn operation.

We now consider these equations componentwise. We first denote the matrix M_B by (α) and the matrix M_B^* by (β). We represent the $(8j+m)^{\text{th}}$ component of $\mathbf{x}_i$, $\mathbf{w}_i$ and $\mathbf{k}_i$ by $x_{i,(j,m)}$, $w_{i,(j,m)}$ and $k_{i,(j,m)}$ respectively. We can now express the previous set of equations in the following way:

$$
\begin{aligned}
w_{0,(j,m)} &= p_{(j,m)} + k_{0,(j,m)}, \\
x_{i,(j,m)} &= w_{i,(j,m)}^{(-1)} & \text{for } i = 0, \ldots, 9, \\
w_{i,(j,m)} &= (M_B \mathbf{x}_{i-1})_{(j,m)} + k_{i,(j,m)} & \text{for } i = 1, \ldots, 9, \\
c_{(j,m)} &= (M_B^* \mathbf{x}_9)_{(j,m)} + k_{10,(j,m)}.
\end{aligned}
$$

We assume that 0-inversion does not occur as part of the encryption or the key schedule. This assumption is true for 53% of encryptions and 85% of 128-bit keys, and even if the assumption is invalid, only a very few of the following equations are incorrect. Under the stated assumption, the system of equations can be written as:

$$
\begin{aligned}
0 &= w_{0,(j,m)} + p_{(j,m)} + k_{0,(j,m)}, \\
0 &= x_{i,(j,m)} w_{i,(j,m)} + 1 & \text{for } i = 0, \ldots, 9, \\
0 &= w_{i,(j,m)} + (M_B \mathbf{x}_{i-1})_{(j,m)} + k_{i,(j,m)} & \text{for } i = 1, \ldots, 9, \\
0 &= c_{(j,m)} + (M_B^* \mathbf{x}_9)_{(j,m)} + k_{10,(j,m)}.
\end{aligned}
$$

We thus obtain a collection of simultaneous multivariate quadratic equations which fully describe a BES encryption. These are given for $j = 0, \ldots, 15$ and $m = 0, \ldots, 7$ by:

$$
\begin{aligned}
0 &= w_{0,(j,m)} + p_{(j,m)} + k_{0,(j,m)}, \\
0 &= x_{i,(j,m)} w_{i,(j,m)} + 1 & \text{for } i = 0, \ldots, 9, \\
0 &= w_{i,(j,m)} + k_{i,(j,m)} + \sum_{(j',m')} \alpha_{(j,m),(j',m')} x_{i-1,(j',m')} & \text{for } i = 1, \ldots, 9, \\
0 &= c_{(j,m)} + k_{10,(j,m)} + \sum_{(j',m')} \beta_{(j,m),(j',m')} x_{9,(j',m')}.
\end{aligned}
$$

A BES encryption can therefore be described as a multivariate quadratic system using 2688 equations over $\mathbf{F}$, of which 1280 are (extremely sparse) quadratic equations and 1408 are linear (diffusion) equations. These equations comprise 5248 terms, made from 2560 state variables and 1408 key variables.

When we consider an AES encryption embedded in the BES framework, we obtain more multivariate quadratic equations because the embedded state variables of an AES encryption are in $\mathbf{B_A}$ and possess the conjugacy property. We thus obtain the following very simple multivariate quadratic equations for $j = 0, \ldots, 15$ and $m = 0, \ldots, 7$ (where $m+1$ is interpreted modulo 8). We divide these equations into linear equations and multivariate quadratic equations.

$$0 = w_{0,(j,m)} + p_{(j,m)} + k_{0,(j,m)};$$
$$0 = w_{i,(j,m)} + k_{i,(j,m)} + \sum_{(j',m')} \alpha_{(j,m),(j',m')} x_{i-1,(j',m')} \qquad \text{for } i = 1, \ldots, 9,$$
$$0 = c_{(j,m)} + k_{10,(j,m)} + \sum_{(j',m')} \beta_{(j,m),(j',m')} x_{9,(j',m')}.$$

$$0 = x_{i,(j,m)} w_{i,(j,m)} + 1 \qquad \text{for } i = 0, \ldots, 9,$$
$$0 = x_{i,(j,m)}^2 + x_{i,(j,m+1)} \qquad \text{for } i = 0, \ldots, 9,$$
$$0 = w_{i,(j,m)}^2 + w_{i,(j,m+1)} \qquad \text{for } i = 0, \ldots, 9.$$

An AES encryption can therefore be described as an overdetermined multivariate quadratic system using 5248 equations over $\mathbf{F}$, of which 3840 are (extremely sparse) quadratic equations and 1408 are linear equations. These encryption equations comprise 7808 terms, made from 2560 state variables and 1408 key variables. Furthermore, the AES key schedule can be expressed as a similar multivariate quadratic system. In its most sparse form, the key schedule system uses 2560 equations over $\mathbf{F}$, of which 960 are (extremely sparse) quadratic equations and 1600 are linear equations. These key schedule equations comprise 2368 terms made from the 2048 variables, of which 1408 are basic key variables and 640 are auxiliary variables. We can, of course, immediately reduce the sizes of these multivariate quadratic systems by using the linear equations to substitute for state and key variables, though the resulting system is slightly less sparse.

6.2 Potential Attack Techniques

It is clear that an efficient method for the solution of this type of multivariate quadratic system would give a cryptanalysis of the AES with potentially very few plaintext-ciphertext pairs. While there is some connection to work on interpolation attacks [13], techniques such as *relinearisation* [14] or the *extended linearisation* or *XL* algorithm [5] have been specifically developed for the solution of such systems. A simple overview of these techniques is given below.

- Generate equations of higher degree from the original equations by multiplying the original equations by certain other terms or equations.
- Regard the generated system of equations of higher degree as linear combinations of formal terms.
- If there are more linearly independent equations than terms, solve the linear system.

The recently proposed *extended sparse linearisation* or *XSL* algorithm [6] is a modification of the XL algorithm that attempts to solve the types of multivariate quadratic systems that can occur in iterated block ciphers. A discussion of the use of the XSL algorithm on the AES multivariate quadratic $GF(2)$-system is given in [6]. The AES **F**-system derived from the BES is far simpler, which would suggest that the XSL algorithm would solve this **F**-system far faster (2^{100} AES encryptions) than the $GF(2)$-system. However, the estimate given for the number of linearly independent equations generated by the XSL technique [6] appears to be inaccurate [2].

It is obvious that much urgent research is required on the solution of AES multivariate quadratic systems over **F** to see what new cryptanalytic approaches and attacks are possible. In particular, refinements to XL-type techniques and the applicability of sparse matrix techniques seem to be important topics for future work. It is certainly important to know the degree and size of linearly soluble systems generated from the AES multivariate quadratic systems. If the degree and size of such a generated system is too small, then attacks on the AES might be possible. We note that the BES representation of the AES gives other simple quadratic equations over **F**, such as $x_{i,(j,m+1)} w_{i,(j,m)} = x_{i,(j,m)}$ or $x_{i,(j,m+2)} w_{i,(j,m)} = x_{i,(j,m+1)} x_{i,(j,m)}$. These can be used to build other simple multivariate quadratic systems over **F** for the AES. Indeed, the first of these equations is essentially used to construct the $GF(2)$ system for the AES given in [6]. We can also use simple higher degree equations over **F** to build other simple multivariate systems for the AES. It is clear from this brief discussion that many aspects of the AES representation over **F** remain to be investigated.

6.3   Implications for the AES

The cryptanalysis of the AES is equivalent to the solution of some particular system of extremely sparse multivariate quadratic equations over **F**. The analysis of the AES as a complicated multivariate quadratic system over $GF(2)$ by Courtois and Pieprzyk [6] is related to the problem of finding such a solution. Most of the other published security results on the AES are concerned with demonstrating that bit-level linear and differential techniques do not compromise the AES. However, from an algebraic viewpoint, such techniques are trace ($\mathbf{F} \to GF(2)$) function techniques, and trace function techniques are not normally employed in the solution of multivariate systems. It is arguable that an important aspect of the security of the AES, namely the solubility of an extremely sparse multivariate quadratic system over **F**, is yet to be explored.

7   Conclusions

In this paper we have introduced a novel interpretation of the AES as being embedded in a new cipher, the BES. However, the BES does not necessarily inherit security properties we might have expected from the AES. Furthermore, the BES has a simple algebraic round function consisting solely of a componentwise inversion and and a highly structured affine transformation over the same field $GF(2^8)$. Indeed, this alternative description of the AES is mathematically much simpler than the original specification. One consequence is that the security of the AES is equivalent to the solubility of certain extremely sparse multivariate quadratic systems over $GF(2^8)$.

Acknowledgements

We would like to thank Fred Piper, Simon Blackburn and Don Coppersmith for some interesting and useful discussions about this paper.

References

1. E. Biham and A. Shamir. *Differential Cryptanalysis of the Data Encryption Standard*. Springer–Verlag, New York, 1993.
2. D. Coppersmith. Personal communication, 30 April 2002.
3. N. Courtois, L. Goubin, and J. Patarin. Quartz, 128-bit long digital signatures. In D. Naccache, editor, Proceedings of *Cryptographers' Track RSA Conference 2001*, LNCS 2020, pages 282–297, Springer–Verlag, 2001.
4. N. Courtois, L. Goubin, W. Meier, and J. Tacier. Solving underdefined systems of multivariate quadratic equations. In D. Paillier, editor, Proceedings of *Public Key Cryptography 2002*, LNCS 2274, pages 211-227, Springer-Verlag, 2002.
5. N. Courtois, A. Klimov, J. Patarin, and A. Shamir. Efficient algorithms for solving overdefined systems of multivariate polynomial equations. In B. Preneel, editor, Proceedings of *Eurocrypt 2000*, LNCS 1807, pages 392–407, Springer-Verlag, 2000.
6. N. Courtois and J. Pieprzyk. Cryptanalysis of block ciphers with overdefined systems of equations. IACR eprint server www.iacr.org, April 2002.
7. J. Daemen and V. Rijmen. AES Proposal: Rijndael (Version 2). NIST AES website csrc.nist.gov/encryption/aes, 1999.
8. J. Daemen and V. Rijmen. *The Design of Rijndael: AES—The Advanced Encryption Standard*. Springer–Verlag, 2002.
9. J. Daemen and V. Rijmen. Answers to "New Observations on Rijndael". NIST AES website csrc.nist.gov/encryption/aes, August 2000.
10. N. Ferguson, J. Kelsey, B. Schneier, M. Stay, D. Wagner, and D. Whiting. Improved cryptanalysis of Rijndael. In B. Schneier, editor, Proceedings of *Fast Software Encryption 2000*, LNCS , pages 213–230, Springer-Verlag, 2000.
11. N. Ferguson, R. Shroeppel, and D. Whiting. A simple algebraic representation of Rijndael. In S. Vaudenay and A. Youssef, editors, Proceedings of *Selected Areas in Cryptography*, LNCS, pages 103–111, Springer-Verlag, 2001.
12. H. Gilbert and M. Minier. A collision attack on seven rounds of Rijndael. *Third AES Conference*, NIST AES website csrc.nist.gov/encryption/aes, April 2000.
13. T. Jakobsen and L.R. Knudsen. The interpolation attack on block ciphers. In E. Biham, editor, Proceedings of *Fast Software Encryption 1997*, LNCS 1267, pages 28–40, Springer-Verlag, 1997.
14. A. Kipnis and A. Shamir. Cryptanalysis of the HFE Public Key Cryptosystem be Relinearization. In M.Wiener, editor, Proceedings of *Crypto '99*, LNCS 1666, pages 19–30, Springer-Verlag, 1999.
15. L. Knudsen and H. Raddum. Recommendation to NIST for the AES. *NIST second round comment*, NIST AES website csrc.nist.gov/encryption/aes/, 2000.
16. R. Lidl and H. Niederreiter. *Introduction to Finite Fields and Their Applications*. Cambridge University Press, 1984.

17. S. Lucks. Attacking seven rounds of Rijndael under 192-bit and 256-bit keys. In Proceedings of *Third AES Conference* and also via NIST AES website csrc.nist.gov/encryption/aes, April 2000.
18. M. Matsui. Linear cryptanalysis method for DES cipher. In T. Helleseth, editor, Proceedings of *Eurocrypt '93*, LNCS 765, pages 386–397, Springer-Verlag, 1994.
19. S. Murphy and M.J.B. Robshaw. New observations on Rijndael. NIST AES website csrc.nist.gov/encryption/aes, August 2000.
20. S. Murphy and M.J.B. Robshaw. Further comments on the structure of Rijndael. NIST AES website csrc.nist.gov/encryption/aes, August 2000.
21. National Institute of Standards and Technology. Advanced Encryption Standard. FIPS 197. 26 November 2001.
22. J. Patarin. Hidden field equations (HFE) and isomorphisms of polynomials (IP): Two new families of asymmetric algorithms. In U. Maurer, editor, Proceedings of *Eurocrypt '96*, LNCS 1070, pages 33–48, Springer-Verlag, 1996.
23. R. Schroeppel. Second round comments to NIST. *NIST second round comment*, NIST AES website csrc.nist.gov/encryption/aes/, 2000.
24. R. Wernsdorf. The round functions of Rijndael generate the alternating group. In V. Rijmen, editor, Proceedings of *Fast Software Encryption*, LNCS, Springer-Verlag, to appear.

Blockwise-Adaptive Attackers

Revisiting the (In)Security of Some Provably Secure Encryption Modes: CBC, GEM, IACBC

Antoine Joux, Gwenaëlle Martineti, and Frédéric Valette

DCSSI Crypto Lab
18, rue du Docteur Zamenhof
92131 Issy-les-Moulineaux
Antoine.Joux@m4x.fr
Gwenaelle.Martinet@ens.fr
fred.valette@wanadoo.fr

Abstract. In this paper, we show that the natural and most common way of implementing modes of operation for cryptographic primitives often leads to insecure implementations. We illustrate this problem by attacking several modes of operation that were proved to be semantically secure against either chosen plaintext or chosen ciphertext attacks.

The problem stems from the simple following fact: in the definition and proofs of semantic security, messages are considered as atomic objects that cannot be split; however, in most practical implementations, messages are subdivided into smaller chunks than can be easily manipulated. Depending on the implementation, each chunk may consist of one or several blocks of the underlying primitive. The key point here is that upon reception of a processed chunk, the attacker can now adapt his choice for the next chunk. Since the possibility of adapting within a single message is not taken into account in the current security models, this leaves room for unexpected attacks.

We illustrate this new paradigm by attacking three symmetric and hybrid encryption schemes based on the chaining mode in spite of their security proofs.

1 Introduction

Currently, the strongest definition of security for an encryption scheme captures the idea that an attacker can adapt his queries according to the previously received answers. A scheme is said to be secure, if no attacker is able to distinguish between different scenarios. These definitions exist in several flavors, depending on the allowed scenarios, e.g. Find-Then-Guess (FTG) security, Left-or-Right (LOR) security, Real-or-Random (ROR) security ([2]). Moreover, the attacker can be given access to an encryption oracle only, when considering chosen-plaintext attacks (CPA), or to an encryption/decryption oracle when considering chosen-ciphertext attacks (CCA). The case of chosen ciphertext secure modes of operation has been specially studied in [11].

M. Yung (Ed.): CRYPTO 2002, LNCS 2442, pp. 17–30, 2002.

However, all these definitions consider messages as atomic objects that cannot be split into smaller pieces. While very convenient from a theoretical point of view, this approach does not really model the reality of many cryptographic implementations. Indeed, in real-life implementation, encryption is usually performed "on the fly", *i.e.* ciphertext chunks are computed and sent as soon as possible. The potential attacks induced by such implementations have already been taken into account in some cryptographic constructions, such as in [7] (signed digital streams) and in [6] (pseudorandom number generators). However, we are not aware of any work that takes advantage of this to attack practical implementations of previously existing schemes. The first example that comes to mind is the case of encryption with a smart card. Usually, the host computer sends blocks of plaintext one at a time to the smart card and immediately receives the corresponding ciphertext block. Thus an hostile host can adapt his next plaintext block as a function of the previously received ciphertext block. Even when the encryption is performed by a general purpose computer, messages are divided into smaller chunks. For example, in SSH ([14]), the plaintext to be encrypted is stored in a buffer of finite size. Whenever the buffer is full[1], it is encrypted and sent. Moreover, as described in [14], "initialization vectors should be passed from the end of one packet to the beginning of the next packet". As a consequence, even though attackers cannot be adaptive within a buffer, they can adapt from one buffer to the next, within a single message. Finally, even if several blocks are stored in the cryptographic component and if buffers are longer than one block, the attacker can force a dependency between the last block of a buffer and the first block of the next one.

In the rest of paper, we show how to take advantage of this extra degree of freedom to attack some modes of operations that were previously thought (and proven) secure. These cryptanalysis are presented in the Find-Then-Guess model, described in appendix A. For the sake of simplicity, we will allow the attacker to be adaptive from one block to the next within a single message. This mimics the behavior of smart card implementations. Throughout the paper, this kind of attacker is said to be blockwise-adaptive.

The first and simplest cryptanalysis we present is the attack on CBC mode of operation. The attacker adapts directly the plaintext block according to the previous ciphertext block. The proposed attack is very efficient, it uses a small constant number of queries to the encryption oracle and always succeeds.

The second attack is against an hybrid scheme, called GEM, that was proposed by Coron *et al.* in [5]. It is an academic attack of higher complexity in time and memory. However, the attack beats the bound of the security proof. The main idea is to feed the challenge oracle with message blocks until a collision appears on the ciphertexts blocks. Then with a single query to a decryption oracle the attacker can distinguish which message has been encrypted.

Finally we attack the IACBC encryption mode proposed by Jutla in [10] and proved secure by Halevi [8] in a slightly modified variant. Here the attack

---

[1] To avoid useless waits, the SSH layer can encrypt and send incomplete buffers. However, this detail is irrelevant at this point.

exploits some relations between the values used to mask the ciphertext blocks. As for CBC, the attack is very efficient since the attacker just needs to feed the encryption oracle with a constant number of queries and always succeeds. Furthermore, we show in appendix B that this weakness was already present in the initial proposal of Jutla.

2   Attack on the CBC Mode of Operation

The CBC (Cipher Block Chaining) mode of encryption security has been ana- lyzed in [2]. It was proved to be secure in the LOR-CPA sense, assuming that the underlying block cipher is a family of PRP. The definition of this security notion is standard and can be found in [2]. It is also briefly described in appendix A. In this section, we briefly recall the CBC encryption mode and then we describe how it can be attacked when allowing the attacker to be adaptive from one block to the next within a single message.

Let E_K be a block cipher with secret key K and block-size n bits and let M be the (padded) message to encrypt. M is divided into ℓ n–bit blocks denoted by $(M[1], M[2], \ldots, M[\ell])$. A random n–bit initial value IV is generated by the encryption box. The CBC mode of encryption with random initial value is a stateless symmetric encryption scheme $CBC(E_K)$. The ciphertext blocks $C[i]$ are computed as follows:

$$C[0] = IV,$$
$$C[i] = E_K(C[i-1] \oplus M[i])$$

The transmitted ciphertext is $(C[0], C[1], \ldots, C[\ell])$.

The crux of the security proof of [2], is that since each message block is randomized by xoring it with a block cipher output, each new call to E_K is independent from the previous ones and no attacker can succeed unless a ran- dom collision occurs. However, if $C[i-1]$ is known when choosing $M[i]$, the independence is clearly lost and the proof fails.

It turns out that this can be illustrated by a very simple attack in the (block- wise) FTG-CPA sense. The attack proceeds as follows:

Step 1 The attacker chooses its FTG challenge. This challenge consists of two- blocks messages M_0 and M_1, such that $M_0[2] \neq M_1[2]$.
Step 2 The black-box computes the encryption of either M_0 and M_1, accord- ing to the value of a random bit b. It transmits $(C_b[0], C_b[1], C_b[2])$ to the attacker. The goal of the attacker is now to guess the value of b.
Step 3 The attacker starts the encryption of a test message M'. It first sends the first block $M'[1]$ chosen uniformly at random.
Step 4 The attacker receives the beginning of the encryption of M', namely $(C'[0], C'[1])$. It sends the second block $M'[2] = M_0[2] \oplus C_b[1] \oplus C'[1]$.
Step 5 The attacker receives $C'[2]$.
Step 6 If $C'[2] = C_b[2]$ the attacker guesses that $b = 0$, otherwise it guesses $b = 1$.

We claim that the guess of the attacker is always correct. Indeed, when $b = 0$ we can check that:

$$C'[2] = E_K(M_0[2] \oplus C_b[1] \oplus C'[1] \oplus C'[1])$$
$$= E_K(M_0[2] \oplus C_b[1])$$
$$= C_b[2]$$

Moreover, when $b = 1$ we can check that:

$$C'[2] = E_K(M_0[2] \oplus C_b[1] \oplus C'[1] \oplus C'[1])$$
$$= E_K(M_0[2] \oplus C_b[1])$$
$$\neq E_K(M_1[2] \oplus C_b[1])$$

and thus $C'[2] \neq C_b[2]$. As a consequence, the attacker can easily find which of the two challenge messages was encrypted.

One can remark that the proposed attack could be even more efficient. Indeed, if the attacker can be adaptive *during* the challenge phase itself (and not only after, as described above), the test message is no longer necessary and the adversary can guess the bit b by just seeing the challenge ciphertext.

A simple and efficient countermeasure to this attack could be considered. The encryption process $\mathcal{E}$ can delay the outputs by one block. That is, when receiving the kth plaintext block, $\mathcal{E}$ encrypts and stores it, and returns the $(k-1)$th block of the ciphertext. In this case, an adversary against this scheme cannot adapt each plaintext block according to the previous ciphertext block during the encryption process, and the above attack fails. This scheme will be called the Delayed Cipher-Block Chaining and will be denoted by DCBC.

Remark 1. The same cryptanalysis can also be mounted against the ABC encryption mode (Accumulated Block Chaining) proposed by Knudsen in [12]. However we do not explicitly describe the attack which is related to the proof by Bellare *et al.* in [1] that ABC mode of operation with public or secret initial value is not a secure OPRP. We just remark here that this attack is possible since each plaintext block is masked with the previous ciphertext block and with a value issued from a function h evaluated at the previous plaintext block. As the h function is not kept secret, the attacker can predict the mask values and adapt each message block accordingly.

Remark 2. In many cases, encrypted messages are also authenticated using a message authentication code (MAC). It is known from recent papers [3] that the right way of doing this is the Encrypt-Then-MAC paradigm. When encryption and authentication are correctly combined, the complete system was shown to be CCA secure in the current security model. However, it is easily remarked that adding authenticity does not prevent the above attack.

3   An Hybrid Example: The GEM Schemes

Two chosen ciphertext secure asymmetric encryption schemes for messages of arbitrary length, GEM-1 and GEM-2, have recently been presented in [5]. In fact they are based on an hybrid construction using an asymmetric encryption scheme and a block cipher. The security proof is made in the random oracle model with a very weak assumption on the underlying block cipher: any fixed-length indistinguishable secure symmetric scheme can be used.

In this section we show how to cryptanalyze these schemes with help of our new kind of attacks. In order to simplify the analysis of the attack, we assume that the underlying symmetric encryption scheme is the XOR, as proposed in the original paper. We mount a chosen ciphertext attack in the sense of the indistinguishability of the encryption. This proposed attack is blockwise-adaptive in a stronger sense than the attack against CBC encryption. Indeed, in the case of GEM, the attacker needs to be adaptive during the challenge transmission phase. We will focus on the first scheme GEM–1, but the same attack can be mounted against the second, GEM–2.

3.1   Overview of GEM–1

Let us briefly describe the GEM–1 scheme according to [5]. The system makes use of several cryptographic primitives, a trapdoor one way function $\mathcal{E}_{pk}$ (such as RSA) and a weak symmetric encryption scheme E_K. In fact, using the XOR function is proposed by the authors. The scheme also makes use of a family of hash functions H_i and of an additional hash function F which are modeled as random oracles. For practical instantiations, it is proposed to use SHA-1 together with a counter, i.e. $H_i(.) = \text{SHA-1}(\ . \parallel i)$. The additional hash function F can be defined similarly using a special value for i, e.g. $F = H_0$.

Given the public key pk, one can encrypt a message M formed of n l–bit blocks, $(M[1], M[2], \ldots, M[n])$ by randomly choosing w and u and by computing the ciphertext $(T_1, C[1], C[2], \ldots, C[n], T_2)$ as follows:

$$T_1 = \mathcal{E}_{pk}(w, u)$$
$$k_1 = H_1(w, T_1)$$
$$C[1] = E_{k_1}(M[1])$$
$$k_i = H_i(k_{i-1}, M[i-1], w)$$
$$C[i] = E_{k_i}(M[i])$$
$$T_2 = F(k_n, M[n], w)$$

This is summarized in figure 1.

3.2   Attack on GEM–1

The security of GEM–1 is proved in [5] in the random oracle model, assuming that $\mathcal{E}_{pk}$ is "reasonably" secure, even when E_K is quite weak (a simple XOR suffices). Without writing down the explicit security bound given for GEM–1, let us remark that the advantage of a CCA adversary in the usual security model is linear in the size of the processed data. As a consequence, square-root attacks are ruled out by the security proof.

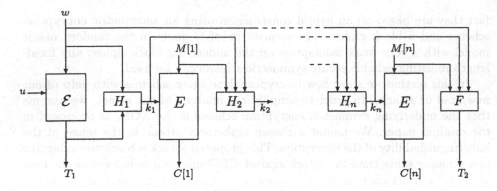

Fig. 1. The GEM–1 algorithm.

In this section, we show that this is no longer the case when using a blockwise-adaptive attacker and give an explicit square-root attack using such an attacker. Note that the proposed attacker is blockwise-adaptive during the challenge phase itself.

The attacker needs to transmit a challenge of size $\mathcal{O}(2^{n/2})$ where n is the size in bits of the C_i values. In other words, this can be described as a square-root attack.

For the sake of simplicity, we assume that the XOR function is used as symmetric encryption. The important property of the XOR function for our purpose is that for a given pair consisting of one plaintext block and its related ciphertext block, the encryption key is uniquely determined. With a different block cipher algorithm, several keys could be possible. However, when the block size and the key size are both equal to n, the number of possible keys is always small. As a consequence, the proposed attack would still have a good probability of success with an ordinary block cipher.

After constructing its challenge message and getting the corresponding ciphertext, the attacker asks for the decryption of a different (but of course related) message. This decryption message tells him which of the two challenge messages was encrypted with probability 1.

The attack goes as follows:

Step 1 The attacker chooses and transmits the first block of each challenge message, $M_0[1]$ and $M_1[1]$, such that $M_0[1] \neq M_1[1]$. At this point in time, the attacker has not yet decided the length of the challenge messages.

Step 2 The encryption box computes the tag T_1, picks a bit b in $\{0,1\}$, encrypts $M_b[1]$ and returns $(T_1, C_b[1])$.

Step 3 The attacker now sends the second block of each challenge message $M_0[2] = M_0[1]$ and $M_1[2] = M_1[1]$.

Step 4 The encryption box encrypts $M_b[2]$ and returns $C_b[2]$.

Step 5 The attacker continues to send the challenge messages one block at a time, with $M_0[i] = M_0[1]$ and $M_1[i] = M_1[1]$. It receives the encrypted blocks $C_b[i]$ and waits for a collision among these encrypted blocks.

Step 6 When a collision occurs, namely when the attacker receives a ciphertext block $C_b[i]$ such that there exists $j < i$ with $C_b[i] = C_b[j]$, the attacker tells the encryption box that the challenge messages are complete.

Step 7 The encryption box computes and returns the tag T_2.

Step 8 The attacker now requests the decryption of the truncated ciphertext $(T_1, C_b[1], \dots, C_b[j], T_2)$. This decryption is either a truncation of M_0 or a truncation of M_1. The attacker guesses b accordingly.

In order to check that the attacker always succeeds, it suffices to verify the validity of the tag T_2 for the truncated message. For the original message, T_2 was computed as $F(k_i, M_b[i], w)$. When decrypting the truncated message, w is the same (since T_1 has not changed), and $M_b[j] = M_b[i]$ by choice of the challenge messages. Moreover, since $C_b[j] = C_b[i]$ thanks to the collision check performed by the attacker, we have $k_j = k_i$. As a consequence, $T_2 = F(k_j, M_b[j], w)$ is a valid tag for the truncated message and the truncated plaintext is indeed returned by the decryption box.

In order to determine the complexity of the attack, we must evaluate the expected length of challenge messages needed before a collision occurs. Thanks to the birthday paradox, since the keys and ciphertext blocks are coded on n bits, collisions are expected after $\mathcal{O}(2^{n/2})$ blocks. According to the security proof given in [5], no attack in the usual (non blockwise-adaptive) model can be that efficient.

4   Jutla's IACBC

In [10, 9], Jutla proposes two new encryption modes that provide confidentiality and integrity in a single pass. One of these modes, IACBC, is a CBC encryption of the plaintext where the encrypted blocks are hidden by xoring them with a sequence of masks $(S_0, \dots, S_{\ell-1})$. For the scheme to be secure, these masks need to be pairwise independent within each encryption, however full pseudorandomness is not necessary. Furthermore, in [8] Halevi proposed a slight modification where he generates the mask values using a non cryptographic process. Halevi then proves the security of the modified scheme in the ROR-CCA sense. The proof is based on the pairwise independence of the masks. However in the sequel we use the fact that the masks are not truly independent to attack the scheme using a blockwise adaptive attacker.

4.1   Overview of IACBC

The IACBC mode works as follows: let E_K be a block cipher with block size n and key length k, along with secret key K_1. Let r be a random initial vector of n bits used to generate ℓ mask values $S_0, S_1, \dots, S_{\ell-1}$, where $\ell - 1$ is the size in blocks of the message to encrypt. In Jutla's paper [10], these masks are generated from $t = \lceil \log(\ell + 1) \rceil$ random and independent vectors W_i, computed from $r+1, \dots, r+t$ using the block cipher with another secret key K_0. To speed up

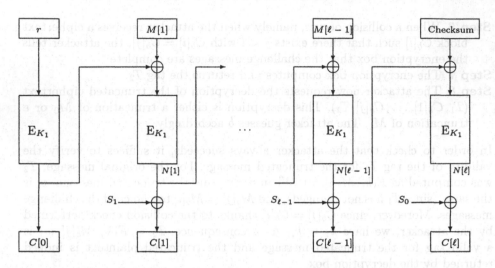

Fig. 2. The IACBC encryption mode.

the generation of the masks, Gray codes are used. With this generation technique the masks are pairwise independent within a single encryption. Moreover the sequences of masks are independent between encryptions. In [8] Halevi proves that second property is not necessary to prove the security of the scheme. Then he proposes a new method to generate the masks values: let r a random initial vector of n bits, and M a secret random boolean matrix of dimension $n \times (\log \bar{L}+1+n)$, where $\bar{L}$ is an upper bound on the ciphertext length. For $j = 1$ to $\ell - 1$ we have $S_j = M \cdot (< 2j >, < r >)$, where $(< 2j >, < r >)$ is the boolean vector of length $\log \bar{L} + 1 + n$ composed with the binary representation of $2j$ on $\log \bar{L} + 1$ bits and the binary representation of r on n bits. Furthermore $S_0 = M \cdot (< 2L+1 >, < r >)$, where L is the ciphertext length.

Then the ciphertext is generated as follows: the message is divided into $\ell - 1$ blocks $M[1], \ldots, M[\ell - 1]$, of n bits each. The ciphertext is defined by:

$C[0] = \mathrm{E}_{K_1}(r)$
$N[0] = C[0]$
for $i = 1$ to $\ell - 1$ do
$\qquad N[i] = \mathrm{E}_{K_1}(M[i] \oplus N[i - 1])$
$\qquad C[i] = N[i] \oplus S_i$
end for
$C[\ell] = \mathrm{E}_{K_1}(\text{checksum} \oplus N[l - 1]) \oplus S_0$, where $\text{checksum} = \bigoplus_{i=1}^{l-1} M[i]$.

This is summarized in figure 2.

To decrypt a ciphertext C, the receiver parses it into $\ell + 1$ blocks denoted by $(C[0], C[1], \ldots, C[\ell])$ and computes $r = \mathrm{D}_{K_1}(C[0])$. He can then recover the mask values $(S_0, \ldots, S_{\ell-1})$ with the help of the secret boolean matrix M. Each plaintext block $M[i]$ is computed as $M[i] = \mathrm{D}_{K_1}(C[i] \oplus S_i) \oplus C[i - 1] \oplus S_{i-1}$. The message integrity is verified by checking the correctness of the Checksum.

4.2   Blockwise Adaptive Cryptanalysis

In this section we exhibit a cryptanalysis of the scheme, in the blockwise adaptive adversarial model. The main idea is to used deterministic relations verified by the masks. Indeed, even though the values used for different blocks are pairwise independent, by construction they satisfy some relations. For every set of masks $S = (S_0, S_1, \ldots, S_{\ell-1})$ and every pair of indices (i, j), $S_i \oplus S_j$ is a constant. To prove this claim we have to look at the mask generation. We have:

$$S_i = M \times (< 2i >, < r >) \text{ and } S_j = M \times (< 2j >, < r >)$$

Thus we get:

$$S_i \oplus S_j = M \times (< 2i > \oplus < 2j >, < r > \oplus < r >)$$
$$= M \times (< 2i > \oplus < 2j >, < 0 >)$$

Then the vector $S_i \oplus S_j$ is independent of r and only depends on some columns of the secret matrix M. Thus, for every set of masks and every pair of indices, $S_i \oplus S_j$ is constant. In the attack we will use this fact for $S_1 \oplus S_2$.

The proposed blockwise adaptive attacker is adaptive during the encryption query but not during the challenge phase itself. However the encryption box has to send the initial ciphertext block $C[0]$ before it receives the first plaintext block.

Here is the scenario of the attack:

Step 1 The attacker chooses at random two messages of two blocks M_0 and M_1 at random and such that $M_0[1] = M_1[1]$ and $M_0[2] \neq M_1[2]$.

Step 2 The challenge box generates the masks values (S_0, S_1, S_2) from a random initial value r. It then picks at random a bit b, encrypts r and M_b under the secret key and transmits $C_b[0] \| C_b[1] \| C_b[2] \| C_b[3]$. The aim of the attacker is to guess the bit b.

Step 3 The attacker now queries the encryption box for one message of two blocks. It first receives $C'[0]$ and sends $M[1] = C'[0] \oplus M_0[1] \oplus C_b[0]$.

Step 4 After receiving $C'[1]$ the attacker outputs $M[2] = M_0[2]$. Then it receives $C'[2]$ and ends the query. The encryption box finally outputs $C'[3]$.

Step 5 if the equality $C_b[1] \oplus C_b[2] = C'[1] \oplus C'[2]$ holds, the attacker guesses the bit $b' = 0$, else he guesses $b' = 1$.

We claim that the attacker always guesses correctly the bit b. Indeed, suppose that message M_0 has been encrypted, meaning that $b = 0$. Then we get:

$$C_b[1] \oplus C_b[2] = E_K(M_0[1] \oplus C_b[0]) \oplus S_1$$
$$\oplus E_K(M_0[2] \oplus E_K(M_0[1] \oplus C_b[0])) \oplus S_2$$

Furthermore, we have:

$$C'[1] \oplus C'[2] = E_K(M[1] \oplus C'[0]) \oplus S'_1$$
$$\oplus E_K(M[2] \oplus E_K(M[1] \oplus C'[0]) \oplus S'_2$$
$$= E_K(C'[0] \oplus M_0[1] \oplus C_b[0] \oplus C'[0]) \oplus S'_1$$
$$\oplus E_K(M[2] \oplus E_K(C'[0] \oplus M_0[1] \oplus C_b[0] \oplus C'[0]) \oplus S'_2$$
$$= E_K(M_0[1] \oplus C_b[0]) \oplus S'_1$$
$$\oplus E_K(M_0[2] \oplus E_K(M_0[1] \oplus C_b[0])) \oplus S'_2$$

Now, we have proved above that $S_1 \oplus S_2 = S'_1 \oplus S'_2$. Consequently, if $b = 0$, we always have $C_b[1] \oplus C_b[2] = C[1] \oplus C[2]$.

Moreover, if $b = 1$ this equality never holds. Indeed, challenge messages M_0 and M_1 have been chosen such that $M_0[1] = M_1[1]$ and $M_0[2] \neq M_1[2]$, and the test message is such that $M[1] = C'[0] \oplus M_0[1] \oplus C_b[0]$. Then it is easy to check that in this case $C_b[1] \oplus C_b[2]$ never equals $C'[1] \oplus C'[2]$. Indeed, we have:

$$S_1 \oplus S_2 = S'_1 \oplus S'_2$$
$$M_1[1] \oplus C_b[0] = M[1] \oplus C'[0]$$
$$M_1[2] \oplus E_K(M_1[1] \oplus C_b[0])) \neq M[2] \oplus E_K(M[1] \oplus C'[0]))$$

and as a consequence $C'[1] \oplus C'[2] \neq C_b[1] \oplus C_b[2]$. Thus the attacker's guess of b is always correct.

The crucial step in this attack is the encryption query made by the adversary and the way in which the oracle returns the ciphertext blocks. Indeed, if the initial value is not sent before the beginning of the encryption, the adversary cannot adapt the next plaintext blocks and the attack fails. Thus, if correctly implemented, IACBC encryption scheme is not subject to such an attack.

Remark 3. Note that the initial IACBC scheme proposed by Jutla in [10] can be attacked in a similar way. Indeed, even when sequences of masks are independent between encryptions, it is however possible to find non trivial relations within a single encryption. This property can be used to cryptanalyze the scheme in the blockwise adaptive adversarial model. See appendix B for more details.

5   Conclusion

In this paper, we proposed a new class of attacks against modes of operation. These attacks, called blockwise adaptive, take advantage of the properties of most practical implementations to allow cryptanalysis of some modes that were previously thought (and proven) secure. Some other modes of operation do not seem to be vulnerable to such attacks, especially when there is no chaining (as in OCB, [13]), or when secret masks are used to randomized inputs and outputs of the block cipher (as XCBC, [4], and HPCBC, [1]). Furthermore, although the impact of this attack on the CBC is huge, this can be simply avoided by using the Delayed CBC (DCBC) that consists in delaying the outputs by one block.

We believe that dealing with blockwise adaptive attacks is the next step towards secure implementations of cryptographic modes of operation.

References

1. M. Bellare, A. Boldyreva, L. Knudsen, and C. Namprempre. On-Line Ciphers and the Hash-CBC Construction. In J. Kilian, editor, *Advances in Cryptology – Crypto'01*, volume 2139 of *Lecture Notes in Computer Science*, pages 292 – 309. Springer-Verlag, Berlin, 2001.
2. M. Bellare, A. Desai, E. Jokipii, and P. Rogaway. A Concrete Security Treatment of Symmetric Encryption. In *Proceedings of the 38th Symposium of Fundations of Computer Science*. IEEE, 1997.
3. M. Bellare and C. Namprempre. Authenticated Encryption: Relations among notions and analysis of the generic composition paradigm. In T. Okamoto, editor, *Advances in Cryptology – Asiacrypt'00*, volume 1976 of *Lecture Notes in Computer Science*. Springer-Verlag, Berlin, 2000.
4. V.D. Gligor and P. Donescu. Fast Encryption and Authentication: XCBC and XECB Authentication Modes. In *Fast Software Encryption*, Lecture Notes in Computer Science. Springer-Verlag, Berlin, 2001.
5. J.S Coron, H. Handshuh, M. Joye, P. Paillier, D. Pointcheval, and C. Tymen. Real-life Chosen-Ciphertext Secure Encryption of Arbitrary-Length Messages. In D. Naccache, editor, *PKC'2002*, volume 2274 of *Lecture Notes in Computer Science*, pages 17 – 33. Springer-Verlag, Berlin, 2002.
6. A. Desai, A. Hevia, and Y.L Yin. A Practice-Oriented Treatment of Pseudorandom Number Generators. In L. Knudsen, editor, *Advances in Cryptology – Eurocrypt 2002*, volume 2332 of *Lecture Notes in Computer Science*. Springer-Verlag, Berlin, 2002.
7. R. Gennaro and P. Rohatgi. How to Sign Digital Streams. In Burt Kaliski, editor, *Advances in Cryptology – Crypto'97*, volume 1294 of *Lecture Notes in Computer Science*, pages 180 – 197. Springer-Verlag, Berlin, 1997.
8. S. Halevi. An Observation regarding Jutla's modes of operation. Crytology ePrint archive, Report 2001/015, available at http://eprint.iacr.org.
9. C. Jutla. Encryption modes with almost free message integrity. Cryptology ePrint archive, Report 2000/039, available at http://eprint.iacr.org.
10. C. Jutla. Encryption modes with almost free message integrity. In B. Pfitzmann, editor, *Advances in Cryptology – Eurocrypt'01*, volume 2045 of *Lecture Notes in Computer Science*. Springer-Verlag, Berlin, 2001.
11. J. Katz and M. Yung. Unforgeable Encryption and Chosen Ciphertext Secure Modes of Operation. In Bruce Schneier, editor, *Fast Software Encryption*, volume 1978 of *Lectures Notes in Computer Science*. Springer-Verlag Berlin, 2000.
12. L. Knudsen. Block chaining modes of operation. Technical report, Department of Informatics, University of Bergen, 2000.
13. P. Rogaway, M. Bellare, J. Black, and T. Krovetz. OCB: A Block-Cipher Mode of Operation for Efficient Authenticated Encryption. In *Eighth ACM conference on Computer and Communications Security*. ACM Press, 2001.
14. T. Ylonen, T. Kivinen, M. Saarinen, T. Rinne, and S. Lehtinen. SSH Transport Layer Protocol, Network Working Group. January 2002. Internet-Draft available at http://www.ietf.org/internet-drafts/draft-ietf-secsh-transport-12.txt.

A   Security Notions

In the standard model, privacy of an encryption scheme is viewed as ciphertext indistinguishability. In [2] the authors have defined different security notions and proved that the strongest one is the LOR ("Left or Right"). However, we focus here on the Find-Then-Guess (FTG) model. We can modelize this notion through a "Find-Then-Guess" game. In this setting the adversary is first given access to an encryption oracle $\mathcal{E}$ that he can feed with plaintexts of his choice. At the end of the first phase (the "Find" phase) the adversary returns two plaintexts M_0 and M_1 of equal length. The encryption oracle flips a bit b, encrypts M_b and returns the challenge ciphertext C_b. The adversary's goal is to guess with non negligible advantage the bit b. In the "Guess" phase, he is again given access to the encryption oracle, he can feed with plaintexts of his choice At the end of the game, the adversary returns a bit b' representing his guess. This attack is called a *Chosen Plaintext Attack* (CPA). However the adversary can also performed *Chosen Ciphertext Attacks* (CCA). In this setting, he also has access to a decryption oracle he can feed with queries of his choice, except with the challenge ciphertext C_b itself.

A symmetric encryption scheme is said to be FTG-CPA secure (respectively FTG-CCA secure), if no polynomial time adversary can guess the bit b in the respective games, with non negligible advantage.

B   Cryptanalysis of the Original Jutla's IACBC

In the original Jutla's proposal in [10], the mask generation is slightly different. The random value r is expanded into $t = \log(\ell + 1)$ random and independent vectors $W_1, \ldots, W_t$ such that $W_i = E_{K_0}(r+i)$, where K_0 is another secret key for the block cipher. Then ℓ pairwise independent and differentially uniform mask values $(S_0, S_1, \ldots, S_{\ell-1})$ are generated from the W_i, with a Gray Code or with the following method, proposed in [10]:

```
input:  Wᵢ, for 1 ≤ i ≤ t
output: S₀, S₁,..., Sₗ₋₁
For i = 0 to ℓ − 1 do
     Let < aᵢ[1], aᵢ[2],..., aᵢ[t] > be the binary
          representation of i + 1
     Sᵢ = ⊕ⱼ₌₁ʲ⁼ᵗ aᵢ[j] · Wⱼ
end for
```

In [10], Jutla claims the security of IACBC in the sense of the message integrity and in the Find-Then-Guess model. However no security proof is given for this claim. In the sequel we show how to attack the scheme in the blockwise adaptive adversarial model. The attack is similar to the one described section 4: some relations between the masks values are exploited. Indeed, each mask is defined with the following relation:

$$S_{i-1} = \bigoplus_{j=1}^{j=t} a_{i-1}[j] \cdot W_j$$

for all $0 \leq i \leq l-1$ and where $< a_{i-1}[1], \ldots, a_{i-1}[t] >$ is the binary representation of i. Then in particular, we have:

$$S_1 = W_2$$
$$S_2 = W_2 \oplus W_1$$
$$S_3 = W_3$$
$$S_4 = W_3 \oplus W_1$$

Then, for every set of mask, we have $S_1 \oplus S_2 \oplus S_3 \oplus S_4 = 0$.

During the proposed attack the attacker has access to an encryption box to mount a chosen plaintext blockwise adaptive attack. For this attack a single query to the encryption box allows to always guess correctly the message encrypted. Let us present the attacker's algorithm:

Step 1 The attacker chooses uniformly two messages of four blocks each M_0 and M_1 such that $M_0[1] = M_1[1]$, $M_0[2] = M_1[2]$, $M_0[3] = M_1[3]$ and $M_0[4] \neq M_1[4]$.

Step 2 The masks values $(S_0, \ldots, S_4)$ are generated from the random r and the vectors W_i. The encryption box encrypts r, randomly chooses a bit b and encrypts message M_b under the secret key K and transmits the ciphertext $C_b[0] \parallel C_b[1] \parallel C_b[2] \parallel C_b[3] \parallel C_b[4] \parallel C_b[5]$.

Step 3 The attacker then queries the encryption box with a message of four blocks. It first receives $C'[0]$ and outputs $M[1] = M_0[1] \oplus C'[0] \oplus C_b[0]$.

Step 4 The oracle encrypts $M[1]$ and returns $C'[1]$.

Step 5 The query continues with plaintext blocks defined by: $M[2] = M_0[2]$, $M[3] = M_0[3]$, and $M[4] = M_0[4]$.

Step 6 After having received $C'[1]$, $C'[2]$, $C'[3]$ and $C'[4]$, the adversary ends the game, receives $C'[5]$ and sends the bit $b' = 0$ if

$$C[1] \oplus C[2] \oplus C[3] \oplus C[4] = C_b[1] \oplus C_b[2] \oplus C_b[3] \oplus C_b[4] \quad (1)$$

and $b' = 1$ otherwise.

Let us look at the equality checked by the adversary. We see that if $b = 0$ we have:

$$\begin{aligned}
C_b[1] \oplus C_b[2] \oplus C_b[3] \oplus C_b[4] = {} & E_K(C_b[0] \oplus M_0[1]) \oplus S_1 \\
& \oplus E_K(M_0[2] \oplus N_b[1]) \oplus S_2 \\
& \oplus E_K(M_0[3] \oplus N_b[2]) \oplus S_3 \\
& \oplus E_K(M_0[0] \oplus N_b[3]) \oplus S_4
\end{aligned}$$

where $N_b[i]$ denotes $E_K(M_0[i] \oplus N_b[i-1])$ for $1 \leq i \leq 3$. Due to the choice of the test message, we also have:

$$C'[1] \oplus C'[2] \oplus C'[3] \oplus C'[4] = E_K(C_b[0] \oplus M_0[1]) \oplus S_1'$$
$$\oplus E_K(M_0[2] \oplus N[1]) \oplus S_2'$$
$$\oplus E_K(M_0[3] \oplus N[2]) \oplus S_3'$$
$$\oplus E_K(M_0[4] \oplus N[3]) \oplus S_4'$$

Then if $b = 0$, since we have $S_1 \oplus S_2 \oplus S_3 \oplus S_4 = S_1' \oplus S_2' \oplus S_3' \oplus S_4' = 0$ and $N[i] = N_b[i]$ for $1 \leq i \leq 3$, equality (1) always holds.

Moreover if $b = 1$ equality (1) is never satisfied. Indeed we have $N[1] = N_b[1]$, $N[2] = N_b[2]$ and $N[3] \neq N_b[3]$ due to the special choice of the challenge messages.

Thus the attacker always guesses correctly the bit b.

Tweakable Block Ciphers

Moses Liskov[1], Ronald L. Rivest[1], and David Wagner[2]

[1] Laboratory for Computer Science
Massachusetts Institute of Technology
Cambridge, MA 02139, USA
mliskov@theory.lcs.mit.edu, rivest@mit.edu
[2] University of California Berkeley
Soda Hall
Berkeley, CA 94720, USA
daw@cs.berkeley.edu

Abstract. We propose a new cryptographic primitive, the *"tweakable block cipher."* Such a cipher has not only the usual inputs – message and cryptographic key – but also a third input, the "tweak." The tweak serves much the same purpose that an initialization vector does for CBC mode or that a nonce does for OCB mode. Our proposal thus brings this feature down to the primitive block-cipher level, instead of incorporating it only at the higher modes-of-operation levels. We suggest that (1) tweakable block ciphers are easy to design, (2) the extra cost of making a block cipher "tweakable" is small, and (3) it is easier to design and prove modes of operation based on tweakable block ciphers.

Keywords: block ciphers, tweakable block ciphers, initialization vector, modes of operation

1 Introduction

A conventional block cipher takes two inputs – a *key* $K \in \{0,1\}^k$ and a *message* (or *plaintext*) $M \in \{0,1\}^n$ – and produces a single output – a *ciphertext* $C \in \{0,1\}^n$. The signature for a block cipher is thus (see Figure 1(a)):

$$E : \{0,1\}^k \times \{0,1\}^n \to \{0,1\}^n \ . \tag{1}$$

On the other hand, the corresponding operators for variable-length encryption have a different signature. These operators are usually defined as "modes of operation" for a block cipher, but they may also be viewed abstractly as another set of encryption operators. They take as input a *key* $K \in \{0,1\}^k$, an *initialization vector* (or *nonce*) $V \in \{0,1\}^v$, and a *message* $M \in \{0,1\}^*$ of arbitrary length, and produce as output a *ciphertext* $C \in \{0,1\}^*$. The signature for a typical encryption mode is thus:

$$\mathcal{E} : \{0,1\}^k \times \{0,1\}^v \times \{0,1\}^* \to \{0,1\}^* \ .$$

Block ciphers (pseudorandom permutations) are inherently deterministic: every encryption of a given message with a given key will be the same. Many modes

M. Yung (Ed.): CRYPTO 2002, LNCS 2442, pp. 31–46, 2002.

of operation and other applications using block ciphers have nonetheless a requirement for "essentially different" instances of the block cipher in order to
prevent attacks that operate by, say, permuting blocks of the input. Attempts to
resolve the conflict between keeping the same key for efficiency and yet achieving
variability often results in a design that uses a fixed key, but which attempts
to achieve variability by manipulating the input before encryption, the output
after encryption, or both. Such designs seem inelegant – they are attempting to
solve a problem with a primitive (a basic block cipher) that is not well suited
for the problem at hand. Better to rethink what primitives are really wanted for
such a problem.

This paper proposes to revise the signature of a block cipher so that it contains a notion of variability as well. The revised primitive operation, which we
call a *tweakable block cipher*, has the signature:

$$\widetilde{E} : \{0,1\}^k \times \{0,1\}^t \times \{0,1\}^n \to \{0,1\}^n . \tag{2}$$

For this operator, we call the new (second) input a "tweak" rather than a "nonce"
or "initialization vector," but the intent is similar. A tweakable block cipher thus
takes three inputs – a *key* $K \in \{0,1\}^k$, a *tweak* $T \in \{0,1\}^t$, and a *message* (or
plaintext) $M \in \{0,1\}^n$ – and produces as output a *ciphertext* $C \in \{0,1\}^n$ (see
Figure 1(b)).

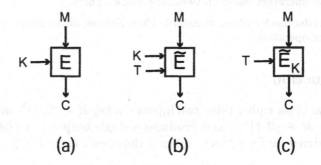

Fig. 1. (a) *Standard block cipher* encrypts a message M under control of a key K to
yield a ciphertext C. (b) *Tweakable block cipher* encrypts a message M under control
of not only a key K but also a "tweak" T to yield a ciphertext C. The "tweak" can be
changed quickly, and can even be public. (c) Another way of representing a tweakable
block cipher; here the key K shown inside the box.

In designing a tweakable block cipher, we have certain goals. First of all,
obviously, we want any tweakable block ciphers we design to be as efficient as
possible (just as with any scheme). Specifically, a tweakable block cipher should
have the property that changing the tweak should be less costly than changing
the key. Many block ciphers have the property that changing the encryption key

is relatively expensive, since a "key setup" operation needs to be performed. In contrast, changing the tweak should be cheaper[1].

A tweakable block cipher should also be secure, meaning that even if an adversary has control of the tweak input, we want the tweakable block cipher to remain secure. We'll define what this means more precisely later on. But intuitively, *each fixed setting of the tweak gives rise to a different, apparently independent, family of standard block cipher encryption operators.* We wish to carefully distinguish between the function of the *key*, which is to provide uncertainty to the adversary, and the role of the *tweak*, which is to provide variability. The tweak is not intended to provide additional uncertainty to an adversary. Keeping the tweak secret need not provide any greater cryptographic strength.

The point of this paper is to suggest that by cleanly separating the roles of cryptographic key (which provides uncertainty to the adversary) from that of tweak (which provides independent variability) we may have just the right tool for many cryptographic purposes.

1.1 Related Work

One motivating example for this introduction of tweakable block ciphers is the DESX construction introduced by Rivest (unpublished). The reason for introducing DESX was to cheaply provide additional key information for DES. The security of DESX has been analyzed by Kilian and Rogaway [10]; they show that DESX with n-bit inputs (and tweaks) and k-bit keys has an effective key-length of $k + n - 1 - \lg m$ where the adversary is limited to m oracle calls. In the DESX construction secret pre- and post-whitening values were added as additional key information.

Even and Mansour [8] have also investigated a similar construction where the inner encryption operator F is fixed and public, and encryption is performed by $E_{K_1 K_2}(M) = K_2 \oplus F(K_1 \oplus M)$. They show (see also Daemen[7]) that the effective key length here is $n - \lg l - \lg m$ where the adversary is allowed to make l calls to the encryption/decryption oracles and m calls to an oracle for F or F^{-1}.

Similarly, if one looks at the internals of the recently proposed "offset codebook mode" (OCB mode) of Rogaway et al. [12], one sees DESX-like modules that may also be viewed as instances of a tweakable block ciphers. That is, the pre- and post-whitening operations are essentially there to provide distinct families of encryption operators, i.e. they are "tweaked."

In a similar vein, Biham and Biryukov [4] suggest strengthening DES against exhaustive search by (among other things) applying a DESX-like construction to each of DES's S-boxes.

---

[1] Some cryptographic modes of operation such as the Davies-Meyer hash function (see Menezes et al. [11, Section 9.40]) have fallen into disfavor because they have a feedback path into the key input of the block cipher. Since for many block ciphers it is relatively expensive to change the key, these modes of operation are relatively inefficient compared to similar modes that use the same key throughout. See, for example, the discussion by Rogaway et al. [12] explaining the design rationale for the OCB mode of operation, which uses the same cryptographic key throughout.

Finally, two block cipher proposals, the Hasty Pudding Cipher (HPC) [14] and the Mercy cipher [6] include an extra input for variability, called in their design specifications a "spice," a "randomiser," or a "diversification parameter." These proposals include a basic notion of what kind of security is needed for a block cipher with this extra input, but no formal notions or proofs are given.

1.2 Outline of This Paper

In Section 2 we then discuss and formalize the notion of security for tweakable block ciphers. In Section 3 we suggest several ways of constructing tweakable block ciphers from existing block ciphers, and prove that the existence of tweakable block ciphers is equivalent to the existence of block ciphers. Then in Section 4 we suggest several new modes of operation utilizing tweakable block ciphers, and give simple proofs for some of them. Section 5 concludes with some discussion and open problems.

2 Definitions

The security of a block cipher E (e.g. parameterized as in equation (1)) can be quantified as $\mathrm{Sec}_E(q, t)$–the maximum advantage that an adversary can obtain when trying to distinguish $E(K, \cdot)$ (with a randomly chosen key K) from a random permutation $\Pi(\cdot)$, when allowed q queries to an unknown oracle (which is either $E(K, \cdot)$ or $\Pi(\cdot)$) and when allowed computation time t. This advantage is defined as the difference between the probability the adversary outputs 1 when given oracle access to E and the probability the same adversary outputs 1 when given oracle access to Π. A block cipher may be considered secure when $\mathrm{Sec}_E(q, t)$ is sufficiently small.

We may measure the security of a tweakable block cipher $\widetilde{E}$ (parameterized as in equation (2)) in a similar manner as the maximum advantage $\mathrm{Sec}_{\widetilde{E}}(q, t)$ an adversary can obtain when trying to distinguish $\widetilde{E}(\cdot, \cdot)$ from a "tweakable random permutation" $\widetilde{\Pi}(\cdot, \cdot)$ where $\widetilde{\Pi}$ is just a family of independent random permutations parametrized by T. That is, for each T, we have that $\widetilde{\Pi}(T, \cdot)$ is an independent randomly chosen permutation of the message space. Note that the adversary is allowed to choose both the message and tweak for each oracle call. A tweakable block cipher $\widetilde{E}$ may be considered secure when $\mathrm{Sec}_{\widetilde{E}}(q, t)$ is sufficiently small.

A tweakable block cipher should also be efficient: both encryption $\widetilde{E}_K(\cdot, \cdot)$ and decryption $\widetilde{D}_K(\cdot, \cdot)$ should be easy to compute.

2.1 Strong Tweakable Block Ciphers

A stronger definition for a block cipher, $\mathrm{Sec}'_E(q, t)$, can be defined as the maximum advantage than an adversary can obtain when trying to distinguish the pair of oracles $E(K, \cdot), D(K, \cdot)$ from the pair Π, Π^{-1}, when allowed q queries and computation time t. This advantage is defined as the difference between the probability the adversary outputs 1 when given oracle access to E, D and the probability the same adversary outputs 1 when given oracle access to Π, Π^{-1}. A block cipher is considered "chosen-ciphertext" secure when $\mathrm{Sec}'_E(q, t)$ is sufficiently small.

Similarly, we define $\mathrm{Sec}'_{\widetilde{E}}(q, t)$ as the maximum advantage an adversary can obtain when trying to distinguish $\widetilde{E}_K(\cdot, \cdot), \widetilde{D}_K(\cdot, \cdot)$ from $\widetilde{\Pi}, \widetilde{\Pi}^{-1}$, when given q queries and t time. We say a tweakable block cipher is chosen-ciphertext secure when $\mathrm{Sec}'_{\widetilde{E}}(q, t)$ is sufficiently small, and we call such a secure tweakable block cipher a "strong tweakable block cipher."

3 Constructions

In this section we show that the existence of block ciphers and the existence of tweakable block ciphers are equivalent. One direction is easy: if we let $E_K(M) = \widetilde{E}_K(0^t, M)$, it is easy to see that if $\widetilde{E}$ is a secure tweakable block cipher then E must be a secure block cipher.

The other direction is more difficult. Some simple attempts to construct a tweakable block cipher from a block cipher fail.

For example, the DESX analogue:

$$\widetilde{E}_K((T_1, T_2), M) = E_K(M \oplus T_1) \oplus T_2$$

fails because an adversary can notice that flipping the same bits in both T_1 and m has no net effect.

Similarly, taking an ordinary block cipher and splitting its key into a key for the tweakable cipher and a tweak:

$$\widetilde{E}_K(T, M) = E_{K\|T}(M)$$

or xoring the tweak into the key:

$$\widetilde{E}_K(T, M) = E_{K \oplus T}(M)$$

need not yield secure tweakable block ciphers, since a block cipher need not depend on every bit of its key. (Biham's related-key attacks of Biham [3] would be relevant to this sort of design.)

The following theorem gives a construction that works.

Theorem 1. *Let*
$$\widetilde{E}_K(T, M) = E_K(T \oplus E_K(M)).$$

$\widetilde{E}$ *is a secure tweakable block cipher. More precisely,*

$$\mathrm{Sec}_{\widetilde{E}}(q, t) < \mathrm{Sec}_E(q, t) + \Theta(q^2/2^n) .$$

Proof. We assume that E has security function $\text{Sec}_E(q,t)$ and assume that an adversary $A^?$ exists that achieves and advantage $\text{Sec}_{\tilde{E}}(q,t)$ when distinguishing $\tilde{E}$ from a tweakable random permutation.

We have the following cases.

Case i: A can distinguish between $\tilde{E}_K$ and H^1, where $H^1(T,M) = \Pi(T \oplus \Pi(M))$. If this is the case, we can use A to distinguish E from Π.

Case ii: A can distinguish between H^1 and H^2 where $H^2(T,M) = R(T \oplus R(M))$, where R is a random function. It is easy to see that the advantage in distinguishing a random function from a random permutation is $\Theta(q^2/2^n)$.

Case iii: A can distinguish H^2 from H^3, where $H^3(T,M) = R_2(T \oplus R_1(M))$, where R_1 and R_2 are random functions. Suppose $(T_1, M_1), \ldots, (T_q, M_q)$ are all the queries A makes to the oracle, and suppose no collisions of the following type happen: $T_i \oplus R(M_i) = M_j$. With no such collisions, H^2 cannot be distinguished from H^3 as the outer application of R takes place on a set of inputs disjoint from the inputs to the inner application of R, and so the outer outputs are independently random, just as the outputs of R_2 would be.

Furthermore, the probability of any such collisions occuring is $\Theta(q^2)/2^n$. What is the probability that (T_i, M_i) collides with any previous pair? If M_i is a new value then it is easy to see that the probability is at most $(i-1)/2^n$. What if M_i is not new? In this case, either (T_i, M_i) will collide or it won't, since all the random decisions have been made. However it is important to note that if no collisions have happened before, then every oracle response the adversary gets is just a new random value. Thus, the values the adversary gets are independent from the T's the adversary produces. Conversely, T_i must be independent from the distribution of R. Thus, even though T_i is not necessarily chosen randomly, no matter how the adversary picks T_i, it has a probability of at most $(i-1)/2^n$ of being one that causes a collision. Adding all these probabilities up, we see that the probability that any collision occurs is $\Theta(q^2)/2^n$.

Case iv: A can distinguish H^3 from H^4, where $H^4(T,M) = R(T,M)$, where R is a random function.

In order for there to be a difference between H^3 and H^4, the output of R must be constrained for two different input pairs. Thus, there must be a pair i,j such that $T_i \oplus R_1(M_i) = T_j \oplus R_1(M_j)$ for $i \neq j$. What is the probability that this happens for any given j? Well, if M_j is a new M, this will only happen with probability $(j-1)/2^n$. Now suppose that up through the jth query there have been no collisions. The adversary then receives purely random values back in response. Thus, since the outputs the adversary sees are independent of the queries, the queries must be independent of the values $T_i \oplus R_1(M_i)$.

If M_j is not a new value, but T_j has never been asked with M_j before, then the probability of a collision is at most $(j-2)/2^n$, since the only possible values to collide with are those where $M_i \neq M_j$. This critically relies on the observation that the adversary's queries are independent of the values so long as there have been no collisions. Thus, we can bound the total probability of collisions in the same way. The probability of distinguishing can be bounded by $q^2/2^n$ where q is the number of queries.

Case v: A can distinguish between H^4 and $\widetilde{\Pi}$. Note that $R(T, M)$ differs from $\widetilde{\Pi}(T, M)$ only in that for any given T, one will be a random function and the other will be a random permutation. Since random functions and random permutations are indistinguishable, this is impossible: we use a simple hybrid argument, providing permutations for more and more T.

Thus, we see that this construction only "degrades" $\mathrm{Sec}_E q, t$ by $\Theta(q^2/2^n)$ to obtain $\mathrm{Sec}_{\widetilde{E}}(q, t)$.   □

Note that this construction has the nice property that changing the tweak is easy (no "key setup" required). Furthermore, we do not require a longer key than the block cipher did for the same level of security. However, the construction has an overall cost (running time) that is twice that of the underlying block cipher.

This completes our proof that the existence of (secure) tweakable block ciphers is equivalent to the existence of (secure) block ciphers. We leave it as an open problem to devise a construction with a tighter bound than Theorem 1.

3.1 Another Construction

We can do better than this, however. We now give a construction that is more efficient, and is also a strong tweakable block cipher.

First, we need a definition. A set $\mathcal{H}$ of functions with signature $\{0, 1\}^t \rightarrow \{0, 1\}^n$ is said to be ϵ-almost 2-xor-universal (ϵ-AXU$_2$, for short) if $\Pr_h[h(x) \oplus h(y) = z] \leq \epsilon$ holds for all x, y, z, where the probability is taken over h chosen uniformly at random from $\mathcal{H}$.

With these definitions, we prove

Theorem 2. *Let $\widetilde{E}_{K,h}(T, M) = E_K(M \oplus h(T)) \oplus h(T)$, and let $\mathcal{H}$ be an ϵ-AXU$_2$ family with $\epsilon \geq 1/2^n$. Then $\widetilde{E}$ is a strong tweakable block cipher. Specifically,*

$$\mathrm{Sec}'_{\widetilde{E}}(q, t) \leq \mathrm{Sec}'_E(q, t) + 3\epsilon q^2.$$

We give the proof in Appendix A.

As there are plenty of known constructions of AXU$_2$ hash families with $\epsilon \approx 1/2^n$, the security theorem shows that we can obtain a construction with good security against adaptive chosen-ciphertext attacks for up to the birthday bound, i.e., for $q \ll 2^{n/2}$.

Moreover, we expect that our construction will be reasonably fast. For instance, for $t = n = 128$, a generalized division hash runs in something like 300 cycles [15], UMAC/UHASH runs in about 200 cycles [5], hash127 runs in about 150 cycles [2] and a DFC-style decorrelation module should run in about 200 cycles [9] (all speeds on a Pentium II class machine, and are rough estimates). If we compare to AES, which runs in about 230–300 cycles [1], we expect that a version of AES tweaked in this way will run about 50–80% slower than the plain AES. Though this is likely to be faster than the previous construction, it does require a longer key.

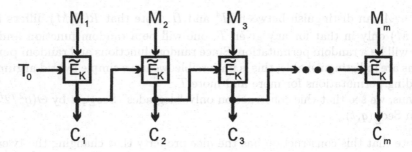

Fig. 2. Tweak block chaining: a chaining mode for a tweakable block cipher. Each ciphertext becomes the tweak for the next encryption.

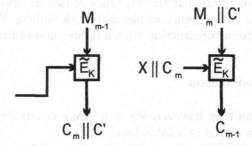

Fig. 3. Ciphertext stealing for tweak block chaining handles messages whose length is at least n bits long but not a multiple of n. Let r denote the length of the last (short) block M_m of the message. Then $|C_m| = |M_m| = r$ and $|C'| = n - r$. Here X denotes the rightmost $n - r$ bits of C_{m-2} (or of T_0 if $m = 2$).

4　Tweakable Modes of Operation

The new "tweak" input of a tweakable block ciphers enables a multitude of new modes of operation. Indeed, these new modes may really be the "payoff" for introducing tweakable block ciphers. In this section we sketch three such possible modes, and leave the remainder to your imagination. We just describe the first two, and prove secure the third, which is the most interesting of the three (it is an analogue to OCB mode for authenticated encryption).

4.1　Tweak Block Chaining (TBC)

Tweak block chaining (TBC) is similar to cipher block chaining (CBC). An initial tweak T_0 plays the role of the initialization vector (IV) for CBC. Each successive message block M_i is encrypted under control of the encryption key K and a tweak T_{i-1}, where $T_i = C_i$ for $i > 0$. See Figure 2.

To handle messages whose length is greater than n but not a multiple of n, a variant of ciphertext-stealing [13] can be used; see Figure 3.

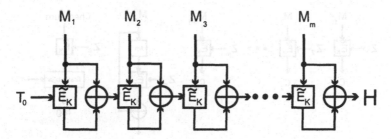

Fig. 4. The tweak chain hash (TCH). Here T_0 is a fixed initialization vector. A fixed public key K is used in the tweakable block cipher. The message M is padded in some fixed reversible manner, such as by appending a 1 and then enough 0's to make the length a multiple of n. The value H is the output of the hash function.

One can also adapt the TBC construction to make a TBC-MAC in the same manner that one can use the CBC construction to make a CBC-MAC, though these constructions still need a security analysis.

4.2   Tweak Chain Hash (TCH)

To make a hash function, one can adapt the Matyas-Meyer-Oseas construction (see Menezes et al. [11, Section 9.40]). See Figure 4, using a fixed public key K in the tweakable block cipher, and chaining through the tweak input.

We don't know if this construction is secure. With a strong additional property on the tweakable block cipher, namely that for a fixed known key and fixed unknown tweak, we still get a pseudorandom permutation, we could adapt the proof of the Davies-Meyer hash function. However, as we noted in section 2, this is not the case for all tweakable block ciphers[2].

4.3   Tweakable Authenticated Encryption (TAE)

In this section we suggest an authenticated mode of encryption (TAE) based on the use of a tweakable block cipher. This mode can be viewed as a paraphrase or restatement of the architecture of the OCB (offset codebook) mode proposed by Rogaway et al. [12] to utilize tweakable block ciphers rather than DESX-like modules. The result is shown in Figure 5. (The reader may need to consult the OCB paper to follow the rather terse description given here.)

The OCB paper goes to considerable effort to analyze the probability that various encryption blocks all have distinct inputs. We feel that an authenticated encryption mode such as TAE should be much simpler to analyze, since the use of tweaks obviate this concern.

We will in fact give a fairly easy proof that a tweakable block cipher used in TAE mode gives all the security properties claimed for OCB mode. Rogaway et al claim that OCB mode is:

---

[2] One tweakable block cipher construction that does have this property is $\widetilde{E}_K(T, M) = E_K(E_T(E_K(M)))$, but this is not as desirable a construction as it is not easy to change the tweak.

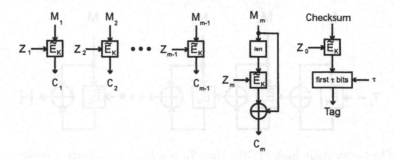

Fig. 5. Authenticated encryption based on a tweakable block cipher. This mode takes as input an $n/2$-bit nonce N. The tweak Z_i for $i > 0$ is defined as the concatenation of the nonce N, an $n/2-1$-bit representation of the integer i, and a zero bit 0: $Z_i = N\|i\|0$. The tweak Z_0 is defined as the concatentation of the nonce N, an $n/2-1$-bit representation of the integer b, where b is the bit-length of the message M, and a one bit 1: $Z_0 = N\|b\|1$. The message M is divided into $m-1$ blocks $M_1, \ldots, M_{m-1}$ of length n and one last block M_m of length r for $0 < r \leq n$ (except that if $|M| = 0$ then the last (and only) block has length 0). Each ciphertext block C_i has same length as M_i. The function len(M_m) produces an n-bit binary representation of the length r of the last message block. The last message block M_m is padded with zeros if necessary to make it length n before xoring. The checksum is $(M_1 \oplus \cdots \oplus M_{m-1} \oplus (M_m\|0^*))$. The parameter τ, $0 \leq \tau \leq n$ specifies the desired length of the authentication tag.

– *Unforgeable.* Any nonce-respecting[3] adversary can forge a new valid encryption with probability at worst negligibly greater than $2^{-\tau}$.
– *Pseudorandom.* To any nonce-respecting adversary, the output of OCB mode is pseudorandom. In other words, no adversary can distinguish between an OCB mode oracle and a random function oracle. [12]

We now prove that TAE mode satisfies these properties.

Theorem 3. *If $\widetilde{E}$ is a secure tweakable block cipher, then $\widetilde{E}$ used in TAE mode will be unforgeable and pseudorandom.*

Proof. To prove that TAE mode is pseudorandom, we note that no tweak is ever repeated when the adversary is nonce-respecting. Now, if an adversary A were able to distinguish between a random function oracle and a TAE mode oracle, then we could distinguish the tweakable block cipher $\widetilde{E}$ from $\widetilde{\Pi}$ as follows. Given an oracle $\mathcal{O}$, we simply run A, and answer A's oracle queries by simulating TAE mode with $\mathcal{O}$ instead of $\widetilde{E}$. Now, if $\mathcal{O} = \widetilde{E}$ then we are in fact providing A with a TAE mode oracle. However, if $\mathcal{O} = \widetilde{\Pi}$ we are providing a random oracle. To

---

[3] By "nonce-respecting," it is meant that while the adversary has oracle access and control of the nonce, the adversary may never ask that a nonce be used more than once; the idea is that any oracle a real adversary would have access to would still not repeat nonces, even if manipulated in order to accept nonces from elsewhere.

see this, note that since no tweak is ever repeated, every part of every output is an independent random value. Thus, if we just give the answer A gives, we are correct whenever A is correct, and thus we defeat the security of $\widetilde{E}$.

To prove that TAE mode is unforgeable, we do the same thing. Suppose some adversary A can forge encryptions in TAE mode. We will break $\widetilde{E}$ as follows. Given an oracle $\mathcal{O}$ we just run A and answer A's oracle queries by simulating TAE mode with $\mathcal{O}$. When A gives an answer, we check to see if the answer is a successful forgery. If it is, we guess that $\mathcal{O} = \widetilde{E}$ and if not, we guess that $\mathcal{O} = \widetilde{\Pi}$. Since A is a successful adversary, if $\mathcal{O} = \widetilde{E}$, it forges successfully with probability nonnegligibly greater than $2^{-\tau}$. We will now show that if $\mathcal{O} = \widetilde{\Pi}$ then A forges with probability at most $2^{-\tau}$. Once we prove this we'll be done, since this reduction will be correct non-negligibly more often than it is incorrect.

Suppose now that $\mathcal{O} = \widetilde{\Pi}$. First we note that if A returns an answer with a new nonce, then Z_0 for the answer will be new and thus the correct answer will be a totally random τ-bit string. In other words, A will be correct with probability exactly $2^{-\tau}$. Secondly, if A returns an answer with an old nonce, then there are several cases. In the first case, all the ciphertext blocks are the same ciphertext blocks that were returned when that nonce was used previously. In this case, the forgery cannot possibly be correct since either it will be wrong or it will not be new. In the second case, the message is a different length than the message this nonce was queried with before. In this case, the forgery is correct with probability $2^{-\tau}$ since Z_0 will be different from before. In the final case, the message is the same length but there is at least one new ciphertext block. Now, the preimage of every block which is different is a random new value. Thus, the checksum is a random value, so the input to the $\widetilde{\Pi}$ that computes the tag is a new value. Thus, with probability exactly $2^{-\tau}$, the forgery is correct. Thus, the probability that the forgery is correct is at most $2^{-\tau}$, which concludes the proof. $\square$

It is interesting to note that the construction *loses nothing* in terms of its advantage compared to the advantage of the tweakable block cipher! This is somewhat remarkable, and helps to emphasize our main point that tweakable block ciphers may be the most natural and useful construct for designing higher-level modes of operation. What's more, we note that if we use $\widetilde{E}$ from section 3.1, TAE mode is very similar to OCB mode. One critical difference is that OCB mode (essentially) derives its choice of h from the key K whereas our construction would require h to be additional key information. Also, OCB mode uses a Gray code to fine-tune efficiency, which we do not. However, this proof is significantly shorter and simpler than the proof for OCB mode, which further strengthens our point that tweakable block ciphers are the right primitive for this kind of task.

5   Conclusions and Open Problems

By introducing tweakable block ciphers, we have "re-partitioned" the design problem into two (new) parts: designing good tweakable block ciphers, and designing good modes of operation based on tweakable block ciphers. We feel that this re-partitioning is likely to be more useful and fruitful than the usual structure, since certain issues (e.g. having to do with collisions, say) can be handled once and for all at the lower level, and can then be ignored at the higher levels, instead of having to be dealt with repeatedly at the higher levels.

We feel that the notions of a *tweakable block cipher* and *tweakable modes of operation* (that is, modes of operation based on tweakable block ciphers) are interesting and worthy of further study.

One advantage of this framework is the new division of issues between design and analysis of the underlying primitive and the design and analysis of the higher-level modes of operation. We feel that the new primitive may result in a more fruitful partition.

Some interesting open problems are:

- What is the security of TAES (AES with our proposed "standard tweak")?
- Design efficient and secure tweakable block ciphers directly.
- Improve the construction of Theorem 1 to achieve a tighter bound.
- Analyze the security of the tweak-block-chaining mode of encryption.
- Analyze the security of the tweak chain hash.
- Devise and analyze the security of other modes of operation based on tweakable block ciphers.
- Define, devise and analyze the security of tweakable stream ciphers.

Acknowledgments

We would like to thank Rogaway, Bellare, Black, and Krovetz for inspiring this line of research with their proposed OCB mode and Mihir Bellare, Burt Kaliski, Zulfikar Ramzan, and Matthew Robshaw for very helpful discussions. Moses Liskov would like to acknowledge support from NTT grant #6762700.

References

1. Kazumaro Aoki and Helger Lipmaa. Fast implementations of AES candidates. In *Third AES Candidate Conference*, April 2000.
2. D.J. Bernstein. Floating-point arithmetic and message authentication, March 2000.
3. Eli Biham. New types of cryptanalytic attacks using related keys. *Journal of Cryptology*, 7(4):229–246, Fall 1994.
   Also available at: citeseer.nj.nec.com/biham94new.html.
4. Eli Biham and Alex Biryukov. How to strengthen DES using existing hardware. In *Proceedings ASIACRYPT '94*, volume 917 of *Lecture Notes in Computer Science*, pages 398–412. Springer-Verlag, 1994.
   Also available at: citeseer.nj.nec.com/biham94how.html.
5. John Black, Shai Halevi, Hugo Krawczyk, Ted Krovetz, and Phillip Rogaway. UMAC: Fast and secure message authentication. In *Proceedings CRYPTO '99*, volume 1666 of *Lecture Notes in Computer Science*, pages 216–233. Springer-Verlag, 1999.

6. Paul Crowley. Mercy: A fast large block cipher for disk sector encryption. In *Fast Software Encryption: 7th International Workshop*, volume 1978 of *Lecture Notes in Computer Science*, pages 49–63. Springer-Verlag, 2000. Also available at: www.ciphergoth.org/crypto/mercy.
7. Joan Daemen. Limitations of the Even-Mansour construction. In *Proceedings ASIACRYPT '91*, volume 739 of *Lecture Notes in Computer Science*, pages 495–499. LNCS, Springer-Verlag, 1991.
   Also available at: citeseer.nj.nec.com/daemen92limitation.html.
8. Shimon Even and Yishay Mansour. A construction of a cipher from a single pseudorandom permutation. *Journal of Cryptology*, 10(3):151–161, Summer 1997. Also available at: citeseer.nj.nec.com/even91construction.html.
9. L. Granboulan, P. Nguyen, F. Noilhan, and S. Vaudenay. DFCv2. In *Selected Areas in Cryptography*, volume 2012 of *Lecture Notes in Computer Science*, pages 57–71. Springer-Verlag, 2001.
10. Joe Kilian and Phillip Rogaway. How to protect DES against exhaustive search (an analysis of DESX). In *Proceedings CRYPTO '96*, volume 1109 of *Lecture Notes in Computer Science*, pages 252–267. Springer, 1996. See http://www.cs.ucdavis.edu/~rogaway/papers/desx.ps for an updated version.
11. Alfred J. Menezes, Paul C. van Oorschot, and Scott A. Vanstone. *Handbook of Applied Cryptography*. CRC Press, 1997.
12. Phillip Rogaway, Mihir Bellare, John Black, and Ted Krovetz. A block-cipher mode of operation for efficient authenticated encryption. In *Eighth ACM Conference on Computer and Communications Security (CCS-8)*, pages 196–205. ACM Press, Aug 16 2001. See http://www.cs.ucdavis.edu/~rogaway/ocb/ocb-doc.htm.
13. Bruce Schneier. *Applied Cryptography, Second Edition: Protocols, Algorithms, and Source Code in C*. John Wiley & Sons, New York, 1996.
14. Rich Schroeppel. The hasty pudding cipher.
   Available at http://www.cs.arizona.edu/~rcs/hpc/., 1999.
15. Victor Shoup. On fast and provably secure message authentication based on universal hashing. In *Proceedings CRYPTO '96*, volume 1109 of *Lecture Notes in Computer Science*, pages 313–328. Springer, 1996.
16. Serge Vaudenay. Provable security for block ciphers by decorrelation. In *Proceedings STACS '98*, volume 1373 of *Lecture Notes in Computer Science*, pages 249–275. Springer-Verlag, 1998.

A   Proof of Theorem 2

In this section, we give a proof of Theorem 2. First, though, we establish some notation. We use $\Pr_0[\cdot]$ to represent the probability measure in the case where A interacts with $\widetilde{E}_{K,h}$, where the probability is taken over the choice of $K \in \{0,1\}^k$ and $h \in \mathcal{H}$ uniformly and independently at random. Also, we let $\Pr_1[\cdot]$ denote the measure where A interacts with $\widetilde{\Pi}$. In either case, we write $\mathcal{O}$ for A's oracle, so in the former case $\mathcal{O} = \widetilde{E}_K$, and in the latter case $\mathcal{O} = \widetilde{\Pi}$.

We let the random variable T_i denote the tweak input on A's i-th oracle call, and we let M_i and C_i denote the plaintext and ciphertext corresponding to this call, so that $\mathcal{O}(T_i, M_i) = C_i$. In other words, if A's i-th oracle query is an encryption query (to $\mathcal{O}$), then (T_i, M_i) denotes the input and C_i the return value, whereas if A's i-th oracle query is a decryption query (to $\mathcal{O}^{-1}$), then the input is (T_i, C_i) and the result of the query is M_i. Moreover, we define the random variables N_i, B_i by $N_i = M_i \oplus h(T_i)$ and $B_i = C_i \oplus h(T_i)$. Note that if $\mathcal{O} = \widetilde{E}_K$, then $E_K(N_i) = B_i$. We define the random variable τ_n by $\tau_n = \langle (T_1, M_1, C_1), \ldots, (T_n, M_n, C_n) \rangle$, and we use $\tau = \tau_q$ to represent the full transcript of interaction with the oracle.

We fix an adversary A, and we assume without loss of generality that A does not make any repeated or redundant queries to its oracle. As a consequence of this assumption, the pairs (T_i, M_i) are all distinct, or in other words, for all $i \neq j$, we have $(T_i, M_i) \neq (T_j, M_j)$. Similarly, the pairs (T_i, C_i) are also distinct, as are the (T_i, N_i)'s and the (T_i, B_i)'s. Also, the output of A can be viewed as a function of the transcript τ, so we sometimes write the output of A as $A(\tau)$.

Our proof is separated into two parts. In the information-theoretic part, we let $E_K = \Pi$ denote a permutation chosen uniformly at random, we set $\widetilde{E'}(T, M) = \Pi(M \oplus h(T)) \oplus h(T)$, and we show that $\widetilde{E'}$ is a secure tweakable block cipher. Then, in the computational part, we let E be arbitrary, and we show that if E_K and Π are computationally indistinguishable, then $\widetilde{E}$ will also be a secure tweakable block cipher.

The information-theoretic part of the proof uses the following strategy. We define a bad event Bad. We show that when conditioning on the complement event, the probability measures $\mathrm{Pr}_0[\cdot|\overline{\mathsf{Bad}}]$ and $\mathrm{Pr}_1[\cdot|\overline{\mathsf{Bad}}]$ are in fact identical. Then, we show that $\mathrm{Pr}_0[\mathsf{Bad}]$ and $\mathrm{Pr}_1[\mathsf{Bad}]$ are both small. The result will then follow using standard arguments.

In our arguments, we define Bad_n to be the event that, for some $1 \leq i < j \leq n$, either $N_i = N_j$ or $B_i = B_j$. Also, we let $\mathsf{Bad} = \mathsf{Bad}_q$.

Lemma 1. *For every possible transcript t, if $E_K = \Pi$, then $\mathrm{Pr}_0[\tau = t|\overline{\mathsf{Bad}}] = \mathrm{Pr}_1[\tau = t|\overline{\mathsf{Bad}}]$.*

Proof. We show this by induction on the length of the transcript, q. Consider the q-th oracle query: it is either an encryption or a decryption query. Suppose first that the q-th oracle query is an encryption query, with inputs (T_q, M_q). By the inductive hypothesis, we can assume that the distribution of τ_{q-1} is the same for both the $\mathrm{Pr}_0[\cdot|\overline{\mathsf{Bad}}_{q-1}]$ and $\mathrm{Pr}_1[\cdot|\overline{\mathsf{Bad}}_{q-1}]$ probability measures, hence the same is true of the distribution of (τ_{q-1}, T_q, M_q).

Now fix any h such that $N_q \notin \{N_1, \ldots, N_{q-1}\}$, so that the only remaining random choice is over Π or $\widetilde{\Pi}$. When $\mathcal{O} = \widetilde{\Pi}$, $C_q = \widetilde{\Pi}(T_q, M_q)$ is uniformly distributed on the set $S = \{0,1\}^n \setminus \{C_i : T_i = T_q \text{ and } 1 \leq i < q\}$, if we condition on $\overline{\mathsf{Bad}}_{q-1}$ (but before conditioning on $\overline{\mathsf{Bad}}_q$). When $\mathcal{O} = \widetilde{E'}$, we find something slightly different: $B_q = \Pi(N_q)$ is uniformly distributed on the set $\{0,1\}^n \setminus \{B_1, \ldots, B_{q-1}\}$ (conditioned on $\overline{\mathsf{Bad}}_{q-1}$, but before conditioning on $\overline{\mathsf{Bad}}_q$), hence $C_q = B_q \oplus h(T_q)$ is uniformly distributed on the set $S' = \{0,1\}^n \setminus \{C_i \oplus h(T_i) \oplus h(T_q) : i = 1, \ldots, q-1\}$. In both cases, the probabilities are independent of the choice of h. Also, note that $S' \subseteq S$, since when $T_i = T_q$ we have $C_i \oplus h(T_i) \oplus h(T_q) = C_i$. Adding the condition $\overline{\mathsf{Bad}}_q$ amounts to adding the condition that $B_q \in \{0,1\}^n \setminus \{B_1, \ldots, B_{q-1}\}$, i.e., that $C_q \in S'$. Thus, after conditioning on $\overline{\mathsf{Bad}}_q$, we see that C_q is uniformly distributed on S' and independent of the rest of the transcript, and hence the distribution of τ is the same for both the $\mathrm{Pr}_0[\cdot|\overline{\mathrm{Bad}}_q]$ and $\mathrm{Pr}_1[\cdot|\overline{\mathrm{Bad}}_q]$ probability measures. (Here we have used the following simple fact: if the random variable X is uniform on a set S, and if S' is some subset of S, then after conditioning on the event $X \in S'$ we find that the resulting r.v. is uniform on S'.)

This covers the case where the q-th query is a chosen-plaintext query. The other case, where the q-th query is a chosen-ciphertext query, is treated similarly. This concludes the proof of Lemma 1.

Lemma 2. *If $\mathcal{H}$ is ϵ-AXU_2, then $\mathrm{Pr}_1[\mathrm{Bad}_q] \leq \epsilon q(q-1)$.*

Proof. Note that, when $\mathcal{O} = \widetilde{\Pi}$, h is independent of the transcript τ. Hence, we can defer the choice of h until after A completes all q of its queries and the values of T_i, M_i, C_i are fixed. Then, we find $\mathrm{Pr}_1[\mathrm{Bad}_q] = \mathrm{Pr}_h[\exists i,j.N_i = N_j \vee B_i = B_j]$(by the definition of Bad_q) $\leq \sum_{1 \leq i < j \leq q} \mathrm{Pr}_h[N_i = N_j] + \mathrm{Pr}_h[B_i = B_j]$(by a union bound) $= \sum_{1 \leq i < j \leq q} \mathrm{Pr}_h[h(T_i) \oplus h(T_j) = M_i \oplus M_j] \mathrm{Pr}_h[h(T_i) \oplus h(T_j) = C_i \oplus C_j]$(by the definition of N_i, B_i) $\leq \sum_{1 \leq i < j \leq q} 2\epsilon = \epsilon q(q-1)$.(since $\mathcal{H}$ is ϵ-AXU_2)

Lemma 3. *If $\mathcal{H}$ is ϵ-AXU_2 for $\epsilon \geq 1/2^n$, and if $E_K = \Pi$, then $\mathrm{Pr}_0[\mathrm{Bad}_q] \leq 1.5\epsilon q(q-1)$.*

Proof. We will prove $\mathrm{Pr}_0[\mathrm{Bad}_q] \leq 1.5\epsilon q(q-1)$ by induction on q. Let E denote that event that, for some i, we have $N_i = N_q$, and let E' denote the event that, for some i, we have $B_i = B_q$. Note that $\mathrm{Pr}_0[\mathrm{Bad}_q] = \mathrm{Pr}_0[\mathrm{Bad}_{q-1}] + \mathrm{Pr}_0[\mathrm{Bad}_q|\overline{\mathrm{Bad}}_{q-1}] \mathrm{Pr}_0[\overline{\mathrm{Bad}}_{q-1}]$. By the inductive hypothesis, $\mathrm{Pr}_0[\mathrm{Bad}_{q-1}] \leq 1.5\epsilon(q-1)(q-2)$. Also, $\mathrm{Pr}_0[\overline{\mathrm{Bad}}_{q-1}] \leq 1$. Hence all that remains is to bound the term $\mathrm{Pr}_0[\mathrm{Bad}_q|\overline{\mathrm{Bad}}_{q-1}]$.

Applying a union bound shows $\mathrm{Pr}_0[\mathrm{Bad}_q|\overline{\mathrm{Bad}}_{q-1}] \leq \mathrm{Pr}_0[\mathrm{E}|\overline{\mathrm{Bad}}_{q-1}] + \mathrm{Pr}_0[\mathrm{E}' | \overline{\mathrm{E}} \wedge \overline{\mathrm{Bad}}_{q-1}]$. We next bound each of these two terms in turn. By Lemma 1, and since $\mathcal{H}$ is ϵ-AXU_2, we see $\mathrm{Pr}_0[\mathrm{E}|\overline{\mathrm{Bad}}_{q-1}] = \mathrm{Pr}_1[\mathrm{E}|\overline{\mathrm{Bad}}_{q-1}] \leq \epsilon(q-1)$. Moreover, $\mathrm{Pr}_0[\mathrm{E}'|\overline{\mathrm{E}} \wedge \overline{\mathrm{Bad}}_{q-1}] = \mathrm{Pr}_0[\Pi(N_q) \in \{B_1, \ldots, B_{q-1}\}|\overline{\mathrm{E}} \wedge \overline{\mathrm{Bad}}_{q-1}] \leq (q-1)/(2^n - q + 1) \leq 2(q-1)/2^n \leq 2\epsilon(q-1)$, since $\Pi(N_q)$ is uniformly distributed on a set of size at least $2^n - q + 1$ and since $\epsilon \geq 1/2^n$. Finally, $1.5\epsilon(q-1)(q-2) + \epsilon(q-1) + 2\epsilon(q-1) \leq 1.5\epsilon q(q-1)$. The statement of the lemma now follows.

We are now ready to prove the security theorem.

Proof (of Theorem 2). First, we do a simple calculation:

$$\mathrm{Sec}'_{\widetilde{E'}}(q,t) = \max_A |\mathrm{Pr}_0[A(\tau) = 1] - \mathrm{Pr}_1[A(\tau) = 1]| \qquad \text{(by definition)}$$

$$= \max_A |\mathrm{Pr}_0[A(\tau) = 1|\overline{\mathrm{Bad}}] \mathrm{Pr}_0[\overline{\mathrm{Bad}}] + \mathrm{Pr}_0[A(\tau) = 1|\mathrm{Bad}] \mathrm{Pr}_0[\mathrm{Bad}]$$

$$- \mathrm{Pr}_1[A(\tau) = 1|\overline{\mathrm{Bad}}] \mathrm{Pr}_1[\overline{\mathrm{Bad}}] - \mathrm{Pr}_1[A(\tau) = 1|\mathrm{Bad}] \mathrm{Pr}_1[\mathrm{Bad}]|$$
$$\text{(by conditional probabilities)}$$

$$\leq \max_A | \Pr_0[A(\tau) = 1|\overline{\mathsf{Bad}}] \Pr_0[\overline{\mathsf{Bad}}] - \Pr_1[A(\tau) = 1|\overline{\mathsf{Bad}}] \Pr_1[\overline{\mathsf{Bad}}]|$$

$$+ | \Pr_0[A(\tau) = 1|\mathsf{Bad}] \Pr_0[\mathsf{Bad}] - \Pr_1[A(\tau) = 1|\mathsf{Bad}] \Pr_1[\mathsf{Bad}]|$$

(by the triangle inequality)

$$\leq \max_A | \Pr_0[A(\tau) = 1|\overline{\mathsf{Bad}}] \Pr_0[\overline{\mathsf{Bad}}] - \Pr_1[A(\tau) = 1|\overline{\mathsf{Bad}}] \Pr_1[\overline{\mathsf{Bad}}]|$$

$$+ 1.5\epsilon q(q - 1) \qquad\qquad \text{(since } \Pr[\mathsf{Bad}] \leq 1.5\epsilon q(q - 1))$$

$$\leq \max_A \Pr[A(\tau) = 1|\overline{\mathsf{Bad}}] \cdot | \Pr_0[\overline{\mathsf{Bad}}] - \Pr_1[\overline{\mathsf{Bad}}]| + 1.5\epsilon q(q - 1)$$

(by Lemma 1)

$$\leq 3\epsilon q(q - 1) \qquad\qquad \text{(since } 1 - 1.5\epsilon q(q - 1) \leq \Pr[\overline{\mathsf{Bad}}] \leq 1)$$

The result then follows from the triangle inequality: $\mathrm{Sec}'_{\widetilde{E}}(q, t) = \max_A | \Pr_0[A^{\widetilde{E}_K, \widetilde{E}_K^{-1}} = 1] - \Pr_1[A^{\widetilde{\Pi}, \widetilde{\Pi}^{-1}} = 1]| \leq \max_A | \Pr_0[A^{\widetilde{E}_K, \widetilde{E}_K^{-1}} = 1] - \Pr_0[A^{\widetilde{E}'_K, \widetilde{E}'_K^{-1}} = 1]| + | \Pr_0[A^{\widetilde{E}'_K, \widetilde{E}'_K^{-1}} = 1] - \Pr_1[A^{\widetilde{\Pi}, \widetilde{\Pi}^{-1}} = 1]| \leq \mathrm{Sec}'_E(q, t) + \mathrm{Sec}'_{\widetilde{E}'}(q, t) \leq \mathrm{Sec}'_E(q, t) + 3\epsilon q(q - 1).$

The LSD Broadcast Encryption Scheme

Dani Halevy and Adi Shamir

Applied Math. Dept.
The Weizmann Institute of Science
Rehovot 76100, Israel
{danih,shamir}@wisdom.weizmann.ac.il

Abstract. Broadcast Encryption schemes enable a center to broadcast encrypted programs so that only designated subsets of users can decrypt each program. The stateless variant of this problem provides each user with a fixed set of keys which is never updated. The best scheme published so far for this problem is the "subset difference" (SD) technique of Naor Naor and Lotspiech, in which each one of the n users is initially given $O(\log^2(n))$ symmetric encryption keys. This allows the broadcaster to define at a later stage any subset of up to r users as "revoked", and to make the program accessible only to their complement by sending $O(r)$ short messages before the encrypted program, and asking each user to perform an $O(log(n))$ computation. In this paper we describe the "Layered Subset Difference" (LSD) technique, which achieves the same goal with $O(\log^{1+\epsilon}(n))$ keys, $O(r)$ messages, and $O(\log(n))$ computation. This reduces the number of keys given to each user by almost a square root factor without affecting the other parameters. In addition, we show how to use the same LSD keys in order to address any subset defined by a nested combination of inclusion and exclusion conditions with a number of messages which is proportional to the complexity of the description rather than to the size of the subset. The LSD scheme is truly practical, and makes it possible to broadcast an unlimited number of programs to 256,000,000 possible customers by giving each new customer a smart card with one kilobyte of tamper-resistant memory. It is then possible to address any subset defined by t nested inclusion and exclusion conditions by sending less than $4t$ short messages, and the scheme remains secure even if all the other users form an adversarial coalition.

1 Introduction

Broadcast Encryption schemes enable a center to deliver encrypted data to a large set of users so that only a particular subset of *privileged users* can decrypt it. Such schemes are useful in pay-TV systems, the distribution of copyrighted material on encrypted CD/DVD disks, internet multicasting of video music and magazines, the distribution of commercial catalogs and price lists on a need-to-know basis, etc.

This basic problem has many variants. For example, the privileged sets can be arbitrary, size-limited, or with tree-like structure. They can be fixed, slowly

M. Yung (Ed.): CRYPTO 2002, LNCS 2442, pp. 47–60, 2002.

changing, or rapidly changing. The scheme can be used to support a single, bounded, or unbounded number of broadcasts. The keys stored by each user can be fixed, time-dependent, or modifiable by previous transmissions. User revocation can be permanent or limited to a single program. The scheme can be resistant to random or arbitrary adversarial coalitions of various sizes. And so on.

The most interesting (and arguably the hardest) variant of broadcast encryption deals with *stateless receivers* and has the following requirements:

- Each user is initially given a collection of symmetric encryption keys.
- The keys can be used to access any number of broadcasts.
- The keys can be used to define any subset of users as privileged.
- The keys are not affected by the user's viewing history.
- The keys do not change when other users join or leave the system.
- Consecutive broadcasts can address unrelated privileged subsets.
- Each privileged user can decrypt the broadcast by himself.
- Even a coalition of all the non-privileged users cannot decrypt the broadcast.

Early papers on the topic concentrated on general key management issues in star shaped networks. In 1991, S. Berkovits published an article, "How to Broadcast a secret"[1], in which he presented several broadcast schemes based on secret sharing (see [9]). However, his matrix based schemes were impractical for large sets of users and insecure under repeated use.

In 1994, Moni Naor and Amos Fiat[5] formalized the basic definitions and paradigms of this field. In particular, they presented schemes in which each user has a fixed reusable set of keys. However, the complexity of their schemes was strongly dependent on the size k of the adversarial coalition, and their best result required storage of $o(k\log(k)\log(n))$ keys per user and transmision of $o(k^2(\log^2(k))\log(n))$ messages by the broadcaster. These complexities are too high in applications such as pay-TV in which thousands of smart cards can be obtained and analysed by commercial pirates.

A simple solution to the broadcast encryption problem is to give each user u a unique symmetric key K_u which is known only to the broadcaster and the user. The program is broadcast multiple times, encrypted under the key of each privileged user. To save bandwidth, we can use the broadcast encryption scheme only in order to predeliver a short program key K to the privileged subset. The actual program is broadcast only once, encrypted by this K. This scheme requires $O(1)$ storage and processing per user, but the transmission length is $O(s)$ where s is the size of the privileged set. Consequently, this scheme is practical only when the privileged sets are small (or slowly expanding if the broadcaster reuses old program keys and the privileged users memorize them).

A folklore extension of this technique is to consider the n users as the leaves of a balanced binary tree of height $\log(n)$. A unique key is assigned to each vertex in this tree, and each user knows the $\log(n)$ keys of all its tree ancestors. This makes it possible to use a single message in order to address a complete subtree of privileged users, and the total number of messages in the general case is proportional to the number of subtrees required to cover the privileged subset. To minimize the average number of subtrees, the broadcaster should assign users to leaves based on their similarity rather than in random order.

This tree representation makes it possible to revoke any single user u by describing all the other users as a union of the $\log(n)$ subtrees that "hang off" the tree path from the root to u. To address them, the broadcaster sends $\log(n)$ short messages which contain the program key encrypted under the unique key of each one of the roots of these subtrees. An extension of this technique (which is called *the complete subtree method* in [6]) makes it possible to simultaneously revoke any r users with $r\log(n/r)$ messages, and the scheme is secure against any coalition of revoked users.

A major improvement of this idea was the "subset difference" (SD) technique developed in Naor Naor and Lotspiech (NNL[6]), which is the current state of the art in stateless broadcast schemes. In the SD algorithm, the number of keys given to each user is $O(\log^2(n))$ (which depends only on the total number of users), and the number of messages is $O(r)$ (which depends only on the number of revoked users). Each privileged user has to perform $O(log(n))$ cryptographic operations in order to retrieve the program key, and the scheme can be used an arbitrary number of times with arbitrary subsets of revoked users.

In this paper we improve and generalize this technique. We introduce the "Layered Subset Difference" (LSD) scheme, and show that for any $\epsilon > 0$ we can create a stateless broadcast encryption scheme with $O(\log^{1+\epsilon}(n))$ keys, $O(r)$ messages, and $O(log(n))$ cryptographic operations. The improved scheme is truly practical: By using less than one kilobyte of storage in each user's smart card, the broadcaster can revoke any r out of $2^{28} \approx 256,000,000$ possible users by sending at most $4r$ messages. The actual number of messages is typically smaller than this worst case bound, and initial experiments indicate that for random subsets of r revoked users, the average number of messages is approximately $2r$.

In the last part of this paper, we generalize the scheme by considering more complicated types of privileged sets defined by nested inclusion and exclusion conditions. Such a representation is common in legal documents (which initially define the general rule, then list the exceptions, then list exceptions to the exceptions, and so on). As a typical example, consider a satellite TV operator that wants to broadcast a baseball event which is held in Boston. In general, he wants to make the program accessible to all its customers in New England who subscribe to the sport channel, except for those who live in Boston (in order to encourage local ticket sales). However, within Boston the game should be available to sport bars (which have special subscriptions) except for Joe's Bar and Moe's Bar (who stopped paying their fees last month). If the customers are organized in a tree-like structure based on their geographic location and type of subscription, the broadcaster can describe this complex set with a small number of nested inclusion and exclusion conditions on the vertices of the tree. We show that we can use the same LSD keys in order to address any set defined by t conditions with an essentially optimal number of $O(t)$ messages. Since the number of messages in our scheme depends only on the difficulty of describing the privileged set and not on its size or the total number of customers, it can be orders of magnitude more efficient than previous algorithms which try to individually list or revoke millions of users.

2 Improvement: The Layered Subset Difference Scheme

The basic idea in all the stateless broadcast encryption schemes is to represent any privileged set as the union of s subsets of users of a particular form. A different key is associated with each one of these sets, and a user knows a key if and only if he belongs to the corresponding set. The broadcaster encrypts the program key s times under all the keys associated with the sets in the cover. Consequently, each privileged user can easily access the program, but even a coalition of all the non-privileged users cannot find the program key.

The simplest implementation of this idea is to cover the privileged set with singleton sets. A better solution is to associate the users with the leaves of a binary tree, and to cover the privileged set of leaves with a collection of subtrees. However, these covering strategies are inefficient when the privileged set is the complement of a small number of revoked users.

The improved performance of the SD algorithm is primarily due to its more sophisticated choice of covering sets:

Definition 1. *Let i be any vertex in the tree and let j be any descendant of i. Then $S_{i,j}$ is the subset of leaves which are descendants of i but are not descendants of j.*

Note that $S_{i,j}$ is empty if $i = j$. Otherwise, $S_{i,j}$ looks like a tree with a smaller subtree cut out. An alternative view of this set is as a collection of subtrees which are hanging off the tree path from i to j.

The SD scheme covers any privileged set P defined as the complement of r revoked users by the union of $O(r)$ of these $S_{i,j}$ sets. What we show in this paper is that this collection of sets can be drastically pruned: A small subcollection of the $S_{i,j}$ sets suffices to represent any such P as the union of $O(r)$ of the remaining sets, with a slightly larger constant. Since there are fewer possible sets, we can reduce the number of initial keys given to each user. We first show that if we allow the number of sets in the cover to grow by a factor of two, we can reduce the number of keys from $O(\log^2(n))$ to $O(\log^{3/2}(n))$, and then we extend the technique and show how to reduce the number of keys to $O(\log^{1+\epsilon}(n))$ for any fixed $\epsilon > 0$.

2.1 The Original SD Scheme

In this section we describe only those parts of the Subset Difference scheme which are required in order to understand our improvement. Further details and discussions can be found in the NNL paper.

During the initialization stage, the broadcaster defines a binary tree with n leaves associated with the users. To simplify our notation, we assume that the tree is balanced and that $log_2(n)$ is an integer. The broadcaster associates with each non-leaf i an m bit label which is chosen randomly and independently. A pseudo random generator G extends an m bit input x into a $3m$ bit output, which is used to define new labels and keys recursively for all the nodes j in the subtree whose root is i in the following way: Given the label x of vertex j, use the three parts of $G(x)$ to define the key $K_{i,j}$ associated with the set $S_{i,j}$, the label of the left child of j, and the label of the right child of j. Note that each vertex j gets multiple labels derived from the different starting points i above it, and each one of them yields a different key. The crucial property of this derivation technique is that given the label of vertex j derived from starting point i, it is easy to compute the key of any set $S_{i,j'}$ in which j' is a descendant of j, but it is infeasible to compute any other keys (whose sets either have a different i or the same i but a j' which is not a descendant of j).

The number of nonempty sets $S_{i,j}$ that contain a particular leaf u is $O(n)$. User u cannot store so many independent keys in a large system with millions of users, but due to their interdependence he can derive all of them from the smaller set of labels in which i is an ancestor of u and j is just off the tree path from i to u (at distance 1). The number of these labels is $O(\log^2(n))$, and u is entitled to know all of them since he belongs to all these sets $S_{i,j}$.

To revoke r users, the broadcaster uses the following covering strategy: As long as there are at least two revoked users in the tree, choose a smallest subtree with this property, and denote its root by v. Add to the cover the sets $S_{k,l}$ and $S_{p,q}$ where k is the left child of v, l is the revoked leaf in the subtree rooted at k, p is the right child of v and q is the revoked leaf in the subtree rooted at p. (Note that if $k = l$ or $p = q$ the corresponding set is empty, and there is no need to add it to the cover). Delete from the tree all the vertices beneath v, mark it as revoked, and repeat the process. If we end with the root marked as revoked we are done, otherwise we add to the cover the set $S_{r,v}$ where r is the root and v is the single remaining revoked vertex, and stop.

The number of sets in the resultant cover is at most 2r-1 since at each step we pick at most 2 subsets and decrease the number of revoked leaves by 1. So, before the last step we pick at most 2r-2 subsets, and in the last step we pick at most one additional subset. Most of the splitting is expected to happen at the top of the tree, and thus many subsets from that region are likely to be empty. For a random choice of r revoked leaves, the average number of sets in the cover was experimentally found to be approximately 1.25r.

Figure 1 demonstrates the seven subsets actually picked by the SD algorithm in a tree with 4 revoked users v_6, v_9, v_{13}, v_{14}. It suffices to consider their Steiner tree, which is the minimal subtree which contains all of them and the root. A thick edge represents the "descendant of" relation (e.g., $v2$ is a descendant but not necessarily a direct child of $v1$), whereas a thin edge represents the "child of" relation (e.g., $v3$ is a child of $v2$). The sets picked by the algorithm described above are actually all the subsets $S_{i,j}$ where $parent(i)$ is either undefined (when i is the root) or a splitting point in the Steiner tree, and j is the nearest splitting point underneath i. These sets are denoted by the double pointed arcs, and each

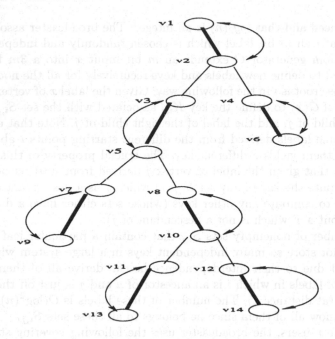

Fig. 1. A Steiner tree

one of them contains all the leaves of the original tree which hang off the chain
of edges in the Steiner tree represented by the thick line.

2.2   The Basic LSD Scheme

In this section we describe the simplest version of the Layered Subset Difference
scheme. Our main observation is the following Lemma:

Lemma 1. *If i, k, j are vertices which occur in this order on some root-to-leaf
path in the tree, then $S_{i,j}$ can be described as the disjoint union $S_{i,j} = S_{i,k} \cup S_{k,j}$*

Proof: $S_{i,k}$ is defined as a set difference $A \setminus B$ and $S_{k,j}$ is defined as a set
difference $B \setminus C$ for some sets of leaves that satisfy $A \supseteq B \supseteq C$. The sets $S_{i,k}$
and $S_{k,j}$ are clearly disjoint, and their union is $A \setminus C$ which is the definition
of $S_{i,j}$. Note that if k is not on the path from i to j, and we use the natural
extension of the definition of these sets, the lemma is incorrect. For example, if
k is the root and i, j are distinct leaves then the set of leaves below i but not
below j is $\{i\}$, the set of leaves below i but not below k is empty, the set of
leaves below k but not below j contains all the leaves except j, and these sets
do not satisfy the desired relationship. $\diamond$

The basic idea of the LSD scheme is to retain only a small subcollection of
the $S_{i,j}$ sets used by the SD scheme. Whenever the broadcaster wants to use
a discarded set in his set cover, he replaces it by the union of two smaller sets
which are in the subcollection.

The subcollection of sets $S_{i,j}$ in the LSD scheme is defined by restricting the levels in which the vertices i and j can occur in the tree. We define some of the $\log(n)$ levels as "special". The root is considered to be at a special level, and in addition we consider every level of depth $k \cdot \sqrt{\log(n)}$ for $k = 1..\sqrt{\log(n)}$ as special (wlog, we assume that these numbers are integers). There are thus $\sqrt{\log(n)}$ special levels which are equally spaced at a distance of $\sqrt{\log(n)}$ from each other. The collection of levels between (and including) adjacent special levels is defined as a "layer".

The subcollection of sets in the LSD scheme is defined in the following way:

Definition 2. *$S_{i,j}$ is a useful set if it is not empty, and at least one of the following conditions is true: both i and j belong to the same layer, or i is at a special layer.*

Lemma 2. *Any nonempty set $S_{i,j}$ is either a useful set or the disjoint union of two useful sets.*

Proof: Since $S_{i,j}$ is nonempty, j is a strict descendant of i. If they belong to the same layer, then $S_{i,j}$ is a useful set. Otherwise, define k as the first vertex on the path from i to j which is in a special level (possibly i itself). Since i and k are in the same layer, $S_{i,k}$ is a useful set (unless it is empty). Since k is in a special level, $S_{k,j}$ is useful (unless it is empty) even if k and j belong to different layers. Consequently, $S_{i,j}$ is the disjoint union of the two useful sets $S_{i,k}$ and $S_{k,j}$, or equal to one of them if the other is empty. ◇

Since the number of covering sets in the SD scheme is bounded by $2r - 1$ and each one of them is replaced by at most two useful sets during the actual broadcast, the number of messages sent by the broadcaster in this scheme is at most $4r - 2$. The $1.25r$ average complexity for r randomly chosen revoked users in the SD scheme suggests an $2.5r$ average complexity in the modified scheme, but in fact there is an additional saving since many of the sets do not get split. Actual experiments on trees with 200,000 users indicate that the average complexity is closer to $2r$, which is only 1.6 times larger than in the original SD scheme.

What we gain from this slightly increased message complexity is a significant reduction in the size of the tamper resistant memory in each user's smart card:

Lemma 3. *The number of labels memorized by each user u in the basic LSD scheme is $O(\log^{3/2}(n))$.*

Proof: The keys associated with the broadcast sets are generated by the user in the same recursive way as in the SD algorithm. The only labels a user should memorize are those that correspond to sets $S_{i,j}$ in which i is an ancestor of u, j is just hanging off the path from i to u, and the levels of i and j are those specified in the definition of useful sets. Note that at each level in the tree there can be at most one vertex which can serve as i and one vertex which can serve as j wrt u, and these two vertices are siblings.

To count the number of memorized labels, consider the two possible cases of useful sets:

– Local sets: Each layer contains $\sqrt{\log(n)}$ levels, and thus the number of i and j pairs which belong to that layer is $O((\sqrt{\log(n)})^2) = O(\log(n))$. Since there are $\sqrt{\log(n)}$ possible layers, the number of i and j pairs of this type is $O(\log^{3/2}(n))$.

– Special sets: Each i in a special level can be associated with a j in any one of the $O(\log(n))$ levels underneath it. Since there are $\sqrt{\log(n)}$ possible i's, the number of labels of this type is also $O(\log^{3/2}(n))$.

Consequently, the total number of labels u has to know is reduced from $O(\log^2(n))$ to $O(\log^{3/2}(n))$. It is easy to show that the choice of $\sqrt{\log(n)}$ as the distance between consecutive special levels is optimal among all the equidistant partitions. For any choice of distance s between consecutive special levels, the number of keys each user has to remember is $O(s^2 \times (\log(n)/s) + (\log(n) \times \log(n)/s))$. The first derivative of this expression(as a function of s) is $\log(n) - \log^2(n)/s^2$ which is equal to 0 for $s = \sqrt{\log(n)}$. Since the second derivative is positive, choosing this value of s minimizes the storage complexity of this scheme. ◇

2.3   The General LSD Scheme

In this section we show how to further reduce the memory requirements of the user revocation scheme, by solving an interesting graph theoretic problem.

The basic LSD algorithm represents each $S_{i,j}$ as the disjoint union of two sets from a smaller subcollection. It is easy to generalize this observation and represent $S_{i,j}$ as the disjoint union of d sets:

Lemma 4. *Let $i, k_1, k_2, \ldots, k_{d-1}, j$ be any sequence of vertices which occur in this order (but not necessarily consecutively) along some root-to-leaf path in the tree. Then*

$$S_{i,j} = S_{i,k_1} \cup S_{k_1,k_2} \cup \cdots \cup S_{k_{d-1},j}$$

Proof: This is just a telescoping formula of set differences for any descending chain of sets. ◇

Any root to leaf path can be viewed as a line graph of length $\log(n)$ with directed edges between adjacent vertices. Broadcasting the set $S_{i,j}$ corresponds to walking from vertex i to vertex j, and addressing all the subtrees that hang off this segment. The original line graph has very few edges (whose labels require very little memory) but these edges provide only a slow way of walking from i to j (with many messages). Since for each original edge the corresponding set is a single subtree, this covering technique is equivalent to the Complete Subtree method of [6].

The Subset Difference technique adds to the line graph all the edges in its directed transitive closure. We can now jump from any i to any descendant j in a single jump (and thus address all the users in $S_{i,j}$ with a single message), but each user has to memorize the labels of $O(\log^2(n))$ edges.

The basic LSD scheme shows that we only have to use $O(\log^{3/2}(n))$ edges (labels) in order to get from any i to any descendant j in two steps (messages). The general LSD scheme considers the following graph theoretic problem: What is the smallest number of edges we have to add to the line graph in order to guarantee the existence of a directed path of length at most d from any i to any descendant j?

To make the graph construction applicable to our user revocation problem, we have to add two additional constraints:

- Monotonicity: We can only add an edge from i to one of its descendants.
- Shrinkage: If we add an edge from i to j, we also have to add all the edges from i to vertices j' between i and j.

Note that without the monotonicity condition, we can get in two steps from any i to any j by adding the $2\log(n)$ directed edges from each vertex to the root and from the root to each vertex. However, such a nonmonotonic path does not correspond to a legal set partition. The shrinkage condition is required in order to guarantee that the shrunk versions of sets provided to users (in which we stop at the first j' on the path from i to j that hangs off the path from the root to the user) also belong to the subcollection. To see the importance of this condition, consider the basic LSD construction. Our original definition of useful sets was closed under shrinkage. However, we could have used the following alternative definition: $S_{i,j}$ is useful if i and j belong to the same layer, or j is in a special layer. We can still get from any i to any descendant j in two steps (through the last special vertex k on the path from i to j), and the number of sets of this type is still $O(\log^{3/2}(n))$. However, this collection of sets is not closed under shrinkage, and thus we have to give each user the labels of all the $O(\log^2(n))$ possible $S_{i,j'}$ sets (where neither i nor j' are special), since each one of them can be the shrunk version of some useful set $S_{i,j}$ that will be used by the broadcaster.

We describe an efficient solution to this graph theoretic problem by representing each vertex on the line graph by its distance from the root, expressed as a d digit number in base $b = O(\log^{1/d}(n))$. The root is represented by $0\ldots00$, its child is represented by $0\ldots01$, etc.

Our goal is to define a small subcollection of *useful transformations* between pairs of numbers which satisfy the monotonicity and shrinkage condition, and allow us to change any i to any larger j with a sequence of at most d useful transformations. Consider for example the problem of changing $i = 825917$ to $j = 864563$ in standard decimal notation. The simplest solution is to allow arbitrary single-digit transformations such as

$$825917 \longrightarrow 865917 \longrightarrow 864917 \longrightarrow 864517 \longrightarrow 864567 \longrightarrow 864563$$

However, these transformations do not satisfy either the monotonicity or the shrinkage condition. Consequently, we have to use a more complicated sequence of transformations:

Definition 3. *Let i be represented as a d digit number in base b by $\vec{x}a\vec{0}$ where a is the rightmost nonzero digit, $\vec{x}$ is a sequence of arbitrary digits, and $\vec{0}$*

is a sequence of zeroes. The transformation of i to j is called useful if j is represented either by $\overrightarrow{x+10}\,\overrightarrow{0}$ or by any number $\overrightarrow{x}a'\overrightarrow{y}$ in which $a' \geq a$ and $\overrightarrow{y}$ is an arbitrary sequence of digits of the same length as $\overrightarrow{0}$.

The basic LSD scheme can be viewed as a special case of this definition for $d = 2$, where two digit numbers ending with 0 are considered to be special. In the general LSD scheme the number of trailing zeroes in the representation of i determines how special it is and how big is the layer within which it is allowed to jump (j can be any destination between $i + 1$ and the first vertex which is even more special than i, inclusive). In our previous example, these useful transformations allow the broadcaster to split the $S_{i,j}$ set into the following segments:

$$825917 \longrightarrow 825920 \longrightarrow 826000 \longrightarrow 830000 \longrightarrow 864563$$

Note that from the point of view of the broadcaster, the first three transitions are of the first type (which jumps to the end of the layer and increases the specialty level of the vertex), and the last transition is of the second type (which jumps to the middle of the layer). However, from the point of view of the user he knows the label associated with at most one of these transitions, and its memorized shrunk version is likely to be of the second type even if the broadcast transition was of the first type.

To count the number of useful transformations, consider any pair of i and j linked by a single useful transformation. We can choose the location of the digit a within i in d ways, and for each location we can choose the $d - 1$ digits in the sequences $\overrightarrow{x}$ and $\overrightarrow{y}$ in b^{d-1} ways, and the two digits $a \leq a'$ in $b^2/2$ ways. Since $b = O(\log^{1/d}(n))$, we get a total number of useful transformations of $O(d \cdot b^{d+1}) = O(d \cdot \log^{(d+1)/d}(n))$. This is an upper bound on the number of labels each user has to memorize, and the corresponding bound on the number of broadcast messages is $d(2r-1)$. Note that our construction of useful transformations is clearly optimal for regular graphs: To reach every vertex within d steps in a regular graph of size e and degree f, the condition $f^d \geq e$ must be clearly satisfied. For our parameters the degree f must satisfy $f \geq O(\log^{1/d}(n)))$, and our construction has such an indegree for any j. Noga Alon(private communication) has recently proved that this lower bound applies to all graphs(regular and nonregular) and thus our broadcast encryption scheme cannot be further improved by analyzing this graph theoretic problem. To find the asymptotic complexity of this general LSD scheme, choose d as a large enough constant. The message complexity remains $O(r)$, while the number of memorized labels is reduced to $O(\log^{1+\epsilon}(n))$ for any $\epsilon > 0$. Alternatively, if we use binary numbers with $d = \log\log(n)$ bits to represent each one of the $log(n)$ levels in the tree, we get another interesting tradeoff point of $O(r \cdot \log\log(n))$ messages and $O(\log(n) \cdot \log\log(n))$ storage per user.

The reader should be warned that these extended LSD schemes are mostly of theoretical interest. The basic LSD algorithm is significantly better than the original SD algorithm for practical values of n, but the additional improvements are noticable only for an astronomical number of users.

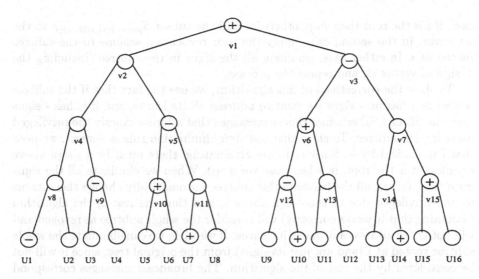

Fig. 2. A typical IE-tree

3   Generalization: Addressing IE-Trees

Standard broadcast encryption schemes deal with the problem of either addressing or avoiding a given list of users, and their complexity depends on the size
of this list. In this section we consider a new way of defining the privileged sets,
which allows the broadcaster to define large complex sets in a natural way by
nesting multiple inclusion and exclusion conditions on the vertices of the user
tree. The corresponding data structure is called an IE-tree (where IE stands for
Inclusion Exclusion), and a typical example of such a tree appears in Fig. 2. Each
vertex in the tree is either unmarked, marked with +, or marked with -. We assume that the root is marked, and thus for any leaf u there is a well defined closest
(i.e., lowest, or most specific) marked vertex on its path to the root. The mark of
this vertex determines whether u is in the privileged set or not. For example, in
Fig. 2 the privileged set consists of the leaves $u_2, u_5, u_6, u_7, u_{10}, u_{11}, u_{12}, u_{14}, u_{15}$
and u_{16}, whereas the non-privileged set consists of u_1, u_3, u_4, u_8, u_9 and u_{13}.

An IE-tree is a very compact way of representing interesting subsets of users,
provided we assign users to leaves in a meaningful order (e.g., by geographic
location, type of subscription, type of smart card, profession, etc). Fortunately,
the SD and LSD schemes can efficiently deal with such a generalized definition
of privileged sets, with only minor changes.

Given an IE-tree, we can assume WLOG that the root is marked, and all
the marked vertices along any root to leaf path in the tree contain alternating
marks (if two vertices along such a path are marked in the same way and there
is no opposite mark between them, the lower mark is redundant and can be
eliminated). We then apply the following iterative procedure:

If there are any vertices marked by +, find a lowest such vertex i. The subtree
below i is either completely unmarked, or it contains some - signs. In the first

case, if i is the root then stop, otherwise add the subset $S_{parent(i),sibling(i)}$ to the set cover. In the second case, apply the user revocation scheme to the subtree rooted at i. In either case, eliminate all the signs in the subtree (including the + sign of vertex i), and repeat the process.

To show the correctness of this algorithm, we use the fact that if the subtree rooted at i has no - signs we want to address all its leaves, and if it has - signs then the SD or LSD schemes choose messages that address exactly the privileged users in this subtree. To show that the sign elimination rule is correct, we note that i is marked by +. Since signs are alternating, there must be a - sign above i (unless it is the root, in which case we stop). When we eliminate all the signs from the subtree, all the leaves in this subtree automatically change their status to non-privileged due to the - sign above i, and thus the rest of the algorithm (assuming that it works correctly) will consider the whole subtree as revoked and will not try to address any one of its leaves. If we wish, we can eliminate the whole subtree rooted at i (and not just its signs) from the original tree, since it will not be considered by the rest of the algorithm. The broadcast messages correspond to all the sets $S_{i,j}$ chosen during any one of the phases of the algorithm.

To count the number of messages broadcast by an algorithm which uses the SD revocation scheme, we note that for every vertex i marked by + we either pick one subset (if i is the only marked vertex in its tree) or at most $2r_i$ subsets where r_i is the number of vertices marked by '-' in this tree, and then we eliminate all these marks. Hence, the number of picked subsets is bounded by $2M + P$ where M is the number of vertices marked by - and P is the number of vertices marked by + in the original IE-tree. When we use the basic LSD revocation scheme, we get a slightly larger message bound of $4M + P$, but each user has to memorize a considerably smaller number of labels.

Since there can be several meaningful ways in which users should be assigned to the leaves of the tree, and the broadcaster may want to use different orders on different occasions, he can use as his data structure the extended notion of an *IE-multitree*, defined as a collection of h independent trees which share their leaves (in possibly different orders). A simple example of a multitree consisting of two trees with four common leaves is described in Fig 3. Each user gets h sets of labels (one per tree), and thus the broadcaster can address users purely by geographic location in one program, and purely by their type of subscription in a second program. While it is possible to use different levels in a single tree to partition users by different types of criteria, the number of marked vertices (and thus the message complexity) for typical privileged sets can be much higher than in the IE-multitree representation.

4   Final Comments

4.1   Security Analysis

Since in all the variations of the LSD scheme each user stores only a subset of the labels he stores in the original SD algorithm, the power of each coalition (in terms of the keys they know) can only diminish, and thus we can use exactly

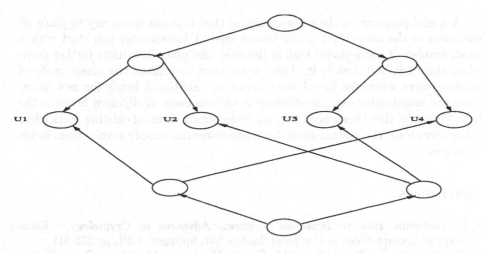

Fig. 3. An IE-multitree

the same argument as in the original NNL paper to show that broadcasts remain inaccessible even to a coalition of all the non-privileged users, under the assumption that the generator G and the symmetric encryption scheme are cryptographically secure. Note that if we don't need this strong notion of security and only want to protect the broadcast from a single non-privileged user, there are simpler and more efficient constructions.

4.2   A Practical Example

Consider a large broadcasting operation with $2^{28} \approx 256,000,000$ users (which is comparable to the number of TV or Internet users in the US). These users can be represented by a complete binary tree with 28 levels, and we split them in the basic LSD scheme into five layers of sizes 6, 6, 6, 5 and 5. Each user has to memorize 81 labels associated with special i's and 65 labels associated with non-special i's. The total number of labels is thus 146, and if each label is a 7-byte DES key they all fit into less than one kilobyte of memory. In the original SD scheme, each user has to memorize $28 * 29/2 = 406$ keys which require almost three kilobytes of memory. This difference is particularly significant if the broadcaster wants to use IE-multitrees with several sets of memorized labels which correspond to several selection criteria (one tree based on geographic location, another based on type of subscription, etc.). With the same memory size, smart cards using the LSD scheme can use three times more useful criteria than smart cards using the SD scheme. Note that the actual algorithm executed by the user is exactly the same in the SD and LSD schemes: check whether for one of the broadcast (i, j) pairs you are a descendant of i but not of j (there can be at most one such pair), and if so, derive the corresponding key by using the generator G at most 27 times (which requires negligible time). The only difference between the SD and LSD schemes is in the broadcaster's algorithms for choosing the user's keys and the subset covers.

60   Dani Halevy and Adi Shamir

A useful property of these algorithms is that it is not necessary to place all the users at the same level in the binary tree. A broadcaster can start with a small number of users placed high in the tree, and place new users further down when their numbers justify it. There is no need to replace the smart cards of existing users when the broadcaster opens up additional levels for new users, since the broadcaster can simultaneously address users at different levels in the tree. Similarly, there is no need to replace the smart cards of existing users when other users leave the system since the broadcaster can simply revoke them in his messages.

References

1. S. Berkovits, How to Broadcast a secret, Advances in Cryptology - Eurocrypt'91,Lecture Notes in Computer Science 547, Springer, 1991, pp.536-541.
2. Ran Canetti, Juan Garay, Gene Itkis, Daniele Micciancio, Moni Naor, Benny Pinkas, Multicast Security: A Taxonomy and Some Efficient Constructions.
3. E. Gafni, J.Staddon and Y.L. Yin, Efficient methods for integrating traceability and broadcast encryption,Proc. Advances in Cryptology - Crypto '99, LNCS 1666, Springer, 1999, 372-387.
4. J.A. Garay, J. Staddon and A. Wool, Long-Lived Broadcast Encryption. Advances in Cryptology - CRYPTO'2000, Lecture Notes in Computer Science, vol 1880, pp. 333-352, 2000.
5. M. Naor, A. Fiat, Broadcast Encryption, Advances in Cryptology - Crypto 93',Lecture Notes in Computer Science 773, Springer, 1994, pp. 480-491.
6. D. Naor., M. Naor, J. Lotspiech, Revocation and Tracing Schemes for Stateless Receivers. February, 2001.
7. M. Naor, B. Pinkas, Threshold Traitor Tracing, Crypto 98.
8. M. Naor, B. Pinkas, Efficient Trace and Revoke Schemes, FC'2000.
9. Shamir, A., "How to Share a Secret", Communications of the ACM, vol. 22, NO. 11, November 1979, pp. 612-613.

Dynamic Accumulators and Application to Efficient Revocation of Anonymous Credentials

Jan Camenisch[1] and Anna Lysyanskaya[2]

[1] IBM Research
Zurich Research Laboratory
CH–8803 Rüschlikon
jca@zurich.ibm.com
[2] MIT LCS
545 Technology Square
Cambridge, MA 02139 USA
anna@theory.lcs.mit.edu

Abstract. We introduce the notion of a *dynamic accumulator*. An accumulator scheme allows one to hash a large set of inputs into one short value, such that there is a short proof that a given input was incorporated into this value. A dynamic accumulator allows one to dynamically add and delete a value, such that the cost of an add or delete is independent of the number of accumulated values. We provide a construction of a dynamic accumulator and an efficient zero-knowledge proof of knowledge of an accumulated value. We prove their security under the strong RSA assumption. We then show that our construction of dynamic accumulators enables efficient revocation of anonymous credentials, and membership revocation for recent group signature and identity escrow schemes.

Keywords: Dynamic accumulators, anonymity, certificate revocation, group signatures, credential systems, identity escrow.

1 Introduction

Suppose a set of users is granted access to a resource. This set changes over time: some users are added, and for some, the access to the resource is revoked. When a user is trying to access the resource, some verifier must check that the user is in this set. The immediate solution is to have the verifier look up the user in some database to make sure that the user is still allowed access to the resource in question. This solution is expensive in terms of communication. Another approach is of certificate revocation chains, where every day eligible users get a fresh certificate of eligibility. This is somewhat better because the communication burden is now shifted from the verifier to the user, but still suffers the drawback of high communication costs, as well as the computation costs needed to reissue certificates. Moreover, it disallows revocation at arbitrary time as need arises. A satisfactory solution to this problem has been an interesting question for some time, especially in a situation where the users in the system are anonymous.

M. Yung (Ed.): CRYPTO 2002, LNCS 2442, pp. 61–76, 2002.

Accumulators were introduced by Benaloh and de Mare [4] as a way to combine a set of values into one short accumulator, such that there is a short witness that a given value was incorporated into the accumulator. At the same time, it is infeasible to find a witness for a value that was not accumulated. Extending the ideas due to Benaloh and de Mare [4], Barić and Pfitzmann [3] give an efficient construction of so-called collision-resistant accumulators, based on the strong RSA assumption.

We propose a variant of the cited construction with the additional advantage that, using additional trapdoor information, the work of deleting a value from an accumulator can be made independent of the number of accumulated values, at unit cost. Better still, once the accumulator is updated, updating the witness that a given value is in the accumulator (provided that this value has not been revoked, of course!) can be done without the trapdoor information at unit cost. Accumulators with these properties are called *dynamic*. Dynamic accumulators are attractive for the application of granting and revoking privileges.

In the anonymous access setting, where a user can prove eligibility without revealing his identity, revocation appeared impossible to achieve, because if a verifier can tell whether a user is eligible or ineligible, he seems to gain some information about the user's identity. However, it turns out that this intuition was wrong! Indeed, using accumulators in combination with zero-knowledge proofs allows one to prove that a committed value is in the accumulator. We show that this can be done efficiently (i.e., *not* by reducing to an NP-complete problem and then using the fact that $NP \subseteq ZK$ [20] and *not* by using cut-and-choose for the Barić and Pfitzmann's [3] construction).

From the above, we obtain an efficient mechanism for revoking group membership for the Ateniese et al. identity escrow/group signature scheme [1] (the most efficient secure identity escrow/group signature scheme known to date) and a credential revocation mechanism for Camenisch and Lysyanskaya's [9] credential system. The construction can be applied to other such schemes as well. The idea is to incorporate the public key for an accumulator scheme into the group manager's (resp., organization's) public key, and the secret trapdoor of the accumulator scheme into the corresponding secret key. Each time a user joins the group (resp., obtains a credential), the group manager (resp., organization) gives her a membership certificate (resp., credential certificate). An integral part of this certificate is a prime number e. This will be the value added to the accumulator when the user is added, and deleted from the accumulator if the user's privileges have to be revoked. This provably secure mechanism does not add any significant communication or computation overhead to the underlying schemes (at most a factor of 2). We note that both our dynamic accumulator scheme and the ACJT identity escrow/group signature scheme rely on the strong RSA assumption. While one could add membership revocation using our dynamic accumulator also to other group signature and identity escrow schemes, such a combination would not make much sense as one would get a less efficient scheme and might even require additional cryptographic assumption. We therefore do not discuss the detail involved here.

Related Work. For the class of group signature schemes [15, 7] where the group's public key contains a list of the public keys of all the group members, excluding a member is straightforward: the group manager only needs to remove the affected member's key from the list. These schemes, however, have the drawback that the complexity of proving and verifying membership is linear in the number of current members and therefore becomes inefficient for large groups. This drawback is overcome by schemes where the size of the group's public key as well as the complexity of proving and verifying membership is independent of the number of members [13, 21, 12, 1]. The idea underlying these schemes is that the group public key contains the group manager's public key of a suitable signature scheme. To become a group member, a user chooses a membership public key which the group manager signs. Thus, to prove membership, a user has to prove possession of membership public key, of the corresponding secret key and of a group manager's signature on a membership public key.

The problem of excluding group members within such a framework without incurring big costs has been considered, but until now no solution was satisfactory. One approach is to change the group's public key and reissue all the membership certificates (cf. [2]). Clearly, this puts quite a burden on the group manager, especially for large groups. Another approach is to incorporate a list of revoked certificates and their corresponding membership keys into the group's public key [6]. In this solution, when proving membership, a user has to prove that his or her membership public key does not appear on the list. Hence, the size of the public key as well as the complexity of proving and verifying signatures are linear in the number of excluded members. In particular, this means that the size of a group signature grows with the number of excluded members.

Song [27] presents an alternative approach in conjunction with a construction that yields forward secure group signature schemes based on the ACJT group signature scheme [1]. While here the size of a group signature is independent of the number of excluded members, the verification task remains computationally intensive, and is linear in the number of excluded group members. Moreover, her approach does not work for ordinary group signature schemes as it relies heavily on the different time periods peculiar to forward secure signatures. Ateniese and Tsudik [2] adapt this approach to the ACJT group signature/identity escrow scheme. Their solution retains the property that the verification task is linear in the number of excluded group members. Moreover, it uses so-called double discrete logarithms which results in the complexity of proving/signing and verifying to be rather high compared to underlying scheme (about a factor of 90 for reasonable security parameters).

Finally, we point out that the proposal by Kim et al. [22] is broken, i.e., excluded group members can still prove membership (after the group manager changed the group's key, excluded members can update their membership information in the very same way as non-excluded members).

Thus, until now, all schemes have a linear dependency either on the number of current members, or on the total number of deleted members. As we have noted above, this linear dependency comes in three flavors: (1) the burden being on the group manager to re-issue certificates in every time period; (2) the burden being on the group member to prove that his certificate is different from any of those that have been revoked and on the verifier to check this; or (3) the burden being on the verifier to perform a computational test on the message received from the user for each item in the list of revoked certificates.

In contrast, in our solution no operation is linearly dependent on the number of current or total deleted members. Its only overhead over a scheme without revocation is the following: We require some public archive that stores information on added and deleted users. Then, the public key (whose size depends only on the security parameter) needs to be updated each time a user is added or deleted. The work to do so dependents only on the number of users added and deleted. Each user must read the public key from time to time (e.g., prior to proving his membership), and if the public key has changed since the last time he looked, he must read the news in the public archive and then perform a local computation. The amount of data to read and the local computation are linear in the number of changes that have taken place in the meantime, but *not* in the total number of changes. Furthermore, the data to read is the *same* for all users. The additional burden on the verifier is simply that he should look at the public key frequently (which seems unavoidable); the verifier need not read the archive.

We note that Sander, Ta-Shma, and Yung [25] also provide a zero-knowledge proof of member knowledge for the Barić-Pfitzmann accumulator. Their proof uses commitments for each of the bits of value contained in the accumulator. In contrast, the proof we provide is a factor of $\mathcal{O}(n)$ more efficient, where n is the number of bits of the used RSA modulus.

2 Preliminaries

Let $A(\cdot)$ be an algorithm. By $y \leftarrow A(x)$ we denote that y was obtained by running A on input x. In case A is deterministic, then this y is unique; if A is probabilistic, then y is a random variable.

Let A and B be interactive Turing machines. By $(a \leftarrow A(\cdot) \leftrightarrow B(\cdot) \rightarrow b)$, we denote that a and b are random variables that correspond to the outputs of A and B as a result of their joint computation.

Let b be a boolean function. By $(y \leftarrow A(x) : b(y))$ we denote the event that $b(y)$ is true after y was generated by running A on input x. The statement

$$\Pr[x_1 \leftarrow A_1(y_1);\ x_2 \leftarrow A_2(y_2);\ \ldots\ ;x_n \leftarrow A_n(y_n)\ :\ b(x_n)] = \alpha$$

means that the probability that $b(x_n)$ is TRUE after the value x_n was obtained by running algorithms $A_1, \ldots, A_n$ on inputs $y_1, \ldots, y_n$, is α.

We say that $\nu(k)$ is a negligible function, if for all polynomials $p(k)$, for all sufficiently large k, $\nu(k) < 1/p(k)$.

Let a be a real number. We denote by $\lfloor a \rfloor$ the largest integer $b \leq a$, by $\lceil a \rceil$ the smallest integer $b \geq a$, and by $\lceil a \rfloor$ the largest integer $b \leq a + 1/2$. Let q be a positive integer. Sometime we need to do modular arithmetic centered around 0; in these cases we use 'rem' as the operator for modular reduction rather than 'mod', i.e., $(c \text{ rem } q) = c - \lceil c/q \rfloor q$.

The *flexible* RSA problem is the following. Given an RSA modulus n and a random element $v \in Z_n^*$ find $e > 1$ and u such that $z = u^e$. The *strong* RSA assumption states that this problem is hard to solve. The strong RSA assumption [3, 18] is a common number-theoretic assumption that, in particular, is the basis for several cryptographic schemes (e.g., [1, 11, 16, 19]). By QR_n we denote the group of quadratic residues modulo n.

We use notation introduced by Camenisch and Stadler [13] for the various zero-knowledge proofs of knowledge of discrete logarithms and proofs of the validity of statements about discrete logarithms. For instance,

$$PK\{(\alpha, \beta, \gamma) : y = g^\alpha h^\beta \ \wedge \ \mathfrak{y} = \mathfrak{g}^\alpha \mathfrak{h}^\gamma \ \wedge \ (u \le \alpha \le v)\}$$

denotes a *"zero-knowledge Proof of Knowledge of integers α, β, and γ such that $y = g^\alpha h^\beta$ and $\mathfrak{y} = \mathfrak{g}^\alpha \mathfrak{h}^\gamma$ holds, where $v < \alpha < u$,"* where $y, g, h, \mathfrak{y}, \mathfrak{g}$, and $\mathfrak{h}$ are elements of some groups $G = \langle g \rangle = \langle h \rangle$ and $\mathfrak{G} = \langle \mathfrak{g} \rangle = \langle \mathfrak{h} \rangle$. The convention is Greek letters denote quantities the knowledge of which is being proved, while all other parameters are known to the verifier. Using this notation, a proof-protocol can be described by just pointing out its aim while hiding all details.

3 Dynamic Accumulators

3.1 Definition

Definition 1. *A secure accumulator for a family of inputs $\{X_k\}$ is a family of families of functions $\mathcal{G} = \{\mathcal{F}_k\}$ with the following properties:*

Efficient generation: *There is an efficient probabilistic algorithm G that on input 1^k produces a random element f of $\mathcal{F}_k$. Moreover, along with f, G also outputs some auxiliary information about f, denoted t_f.*

Efficient evaluation: *$f \in \mathcal{F}_k$ is a polynomial-size circuit that, on input $(u, x) \in \mathcal{U}_f \times X_k$, outputs a value $v \in \mathcal{U}_f$, where $\mathcal{U}_f$ is an efficiently-samplable input domain for the function f; and X_k is the intended input domain whose elements are to be accumulated.*

Quasi-commutative: *For all k, for all $f \in \mathcal{F}_k$, for all $u \in \mathcal{U}_f$, for all $x_1, x_2 \in X_k$, $f(f(u, x_1), x_2) = f(f(u, x_2), x_1)$. If $X = \{x_1, \dots, x_m\} \subset X_k$, then by $f(u, X)$ we denote $f(f(\dots(u, x_1), \dots), x_m)$.*

Witnesses: *Let $v \in \mathcal{U}_f$ and $x \in X_k$. A value $w \in \mathcal{U}_f$ is called a witness for x in v under f if $v = f(w, x)$.*

Security: *Let $\mathcal{U}_f' \times X_k'$ denote the domains for which the computational procedure for function $f \in \mathcal{F}_k$ is defined (thus $U_f \subseteq \mathcal{U}_f'$, $X_k \subseteq X_k'$). For all probabilistic polynomial-time adversaries A_k,*

$$\Pr[f \leftarrow G(1^k); u \leftarrow \mathcal{U}_f; (x, w, X) \leftarrow A_k(f, U_f, u) :$$
$$X \subset X_k; w \in \mathcal{U}_f'; x \in X_k'; x \notin X; f(w, x) = f(u, X)] = \text{neg}(k)$$

Note that only the legitimate accumulated values, $(x_1, \dots, x_m)$, must belong to X_k; the forged value x can belong to a possibly larger set X_k'.

(This definition is essentially the one of Barić and Pfitzmann, with the difference that they do not require that the accumulator be quasi-commutative; as a consequence they need to introduce two further algorithms, one for generation and one for verification of a witness value.)

The above definition is seemingly tailored for a static use of the accumulator. In this paper, however, we are interested in a dynamic use where there is a manager controlling the accumulator, and several users. First, let us show that dynamic addition of a value is done at unit cost in this setting.

Lemma 1. *Let* $f \in \mathcal{F}_k$. *Let* $v = f(u, X)$ *be the accumulator so far. Let* $v' = f(v, x') = f(u, X')$ *be the value of the accumulator when* x' *is added to the accumulated set,* $X' = X \cup \{x'\}$. *Let* $x \in X$ *and* w *be the witness for* x *in* v. *The computation of* w' *which is the witness for* x *in* v', *is independent on the size of* X.

Proof. w' is computed as follows: $w' = f(w, x')$. Let us show correctness using the quasi-commutative property: $f(w', x) = f(f(w, x'), x) = f(f(w, x), x') = f(v, x') = v'$.

We must also be able to handle dynamic deletions of a value from the accumulator. It is clear how to do that using computations that are linear in the size of the accumulated set X. Here, we restrict the definition so as to make the complexity of this operation independent of the size of X:

Definition 2. *A secure accumulator is* dynamic *if it has the following property:*

Efficient deletion: *there exist efficient algorithms* D *and* W *such that, if* $v = f(u, X)$, $x, x' \in X$, *and* $f(w, x) = v$, *then*

1. $D(t_f, v, x') = v'$ *such that* $v' = f(u, X \setminus \{x'\})$; *and*

2. $W(f, v, v', x, x', w) = w'$ *such that* $f(w', x) = v'$.

Note that D is given the trap-door information t_f while W is not.

Now, we show that a dynamic accumulator is secure against an adaptive adversary, in the following scenario: An accumulator manager sets up the function f and the value u and hides the trapdoor information t_f. The adversary adaptively modifies the set X. When a value is added to it, the manager updates the accumulator value accordingly. When a value $x \in X$ is deleted, the manager algorithm D and publishes the result. In the end, the adversary attempts to produce a witness that $x' \notin X$ is in the current accumulator v.

Theorem 1. *Let* $\mathcal{G}$ *be a dynamic accumulator algorithm. Let* M *be an interactive Turing machine set up as follows: It receives input* (f, t_f, u), *where* $f \in \mathcal{F}_k$ *and* $u \in \mathcal{U}_f$. *It maintains a list of values* X *which is initially empty, and the current accumulator value,* v, *which is initially* u. *It responds to two types of messages: in response to the ("add",* x) *message, it checks that* $x \in \mathcal{X}_k$, *and if so, adds* x *to the list* X *and modifies* v *by evaluating* f, *it then sends back this updated value; similarly, in response to the ("delete",* x) *message, it checks that*

$x \in X$, and if so, deletes it from the list and updates v by running D and sends back the updated value. In the end of the computation, M outputs the current values for X and v. Let $\mathcal{U}'_f \times \mathcal{X}'_k$ denote the domains for which the computational procedure for function $f \in \mathcal{F}_k$ is defined. For all probabilistic polynomial-time adversaries $\mathcal{A}_k$,

$$\Pr[((f,t_f) \leftarrow G(1^k); u \leftarrow \mathcal{U}_f; (x,w) \leftarrow \mathcal{A}_k(f,\mathcal{U}_f,u) \leftrightarrow M(f,t_f,u) \rightarrow (X,v):$$
$$w \in \mathcal{U}'_f; x \in \mathcal{X}'_k; x \notin X; f(w,x) = f(u,X)] = \text{neg}(k)$$

Proof. Let us exhibit a reduction from the adversary that violates the theorem to the adversary that breaks the collision-resistance property of a secure accumulator. The reduction will proceed in the following (straightforward) manner: On input $(f, \mathcal{U}_f, u)$, feed these values to the adversary. To respond to an ("add", x) query, simply update X and compute $v = f(u, X)$. To respond to a ("delete",x) query, compute $v = f(u, X \setminus \{x\})$, and then update X. The success of the adversary directly corresponds to the success of our reduction.

Finally, in the application we have in mind we require that the accumulator allows for an efficient proof that a secret value given by some commitment is contained in a given accumulator value. That is, we require that the accumulator be *efficiently provable* with respect to some commitment scheme (*Commit*).

Zero-knowledge proof of member knowledge: There exists an efficient zero-knowledge proof of knowledge system where the common inputs are c (where $c = Commit(x,r)$ with a r being a randomly chosen string), the accumulating function f and $v \in U_f$, and the prover's inputs are $(r, x \in \mathcal{X}_k, u \in U_f)$ for proving knowledge of x, w, r such that $c = Commit(x,r)$ and $v = f(w,x)$.

If by "efficient" we mean "polynomial-time," then any accumulator satisfies this property. However we consider a proof system efficient if it compares well with, for example, a proof of knowledge of a discrete logarithm.

3.2   Construction

The construction due to Barić and Pfitzmann [3] is the basis for our construction below. The differences from the cited construction are that (1) the domain of the accumulated values consists of prime numbers only; (2) we give a method for deleting values from the accumulator, i.e., we construct a *dynamic* accumulator; (3) we give efficient algorithms for deleting a user and updating a witness; and (4) we provide an efficient zero-knowledge proof of membership knowledge.

- $\mathcal{F}_k$ is the family of functions that correspond to exponentiating modulo safe-prime products drawn from the integers of length k. Choosing $f \in \mathcal{F}_k$ amounts to choosing a random modulus $n = pq$ of length k, where $p = 2p'+1$, $q = 2q' + 1$, and p,p',q,q' are all prime. We will denote f corresponding to modulus n and domain $\mathcal{X}_{A,B}$ by $f_{n,A,B}$. We denote $f_{n,A,B}$ by f_n and by f when it does not cause confusion.

- $\mathcal{X}_{A,B}$ is the $\{e \in \text{primes} : e \neq p', q' \land A \leq e \leq B\}$, where A and B can be chosen with arbitrary polynomial dependence on the security parameter k, as long as $2 < A$ and $B < A^2$. $\mathcal{X}'_{A,B}$ is (any subset of) of the set of integer from $[2, A^2 - 1]$ such that $\mathcal{X}_{A,B} \subseteq \mathcal{X}'_{A,B}$.
- For $f = f_n$, the auxiliary information t_f is the factorization of n.
- For $f = f_n$, $\mathcal{U}_f = \{u \in QR_n : u \neq 1\}$ and $\mathcal{U}'_f = \mathbb{Z}^*_n$.
- For $f = f_n$, $f(u, x) = u^x \bmod n$. Note that $f(f(u, x_1), x_2) = f(u, \{x_1, x_2\}) = u^{x_1 x_2} \bmod n$
- Update of the accumulator value. As mentioned earlier, adding a value $\tilde{x}$ to the accumulator value v can be done as $v' = f(v, \tilde{x}) = v^{\tilde{x}} \bmod n$. Deleting a value $\tilde{x}$ from the accumulator is as follows. $D((p, q), v, \tilde{x}) = v^{\tilde{x}^{-1} \bmod (p-1)(q-1)} \bmod n$.
- Update of witness: As mentioned, updating the witness u after $\tilde{x}$ has been added can be done by $u' = f(u, \tilde{x}) = u^{\tilde{x}}$. In case, $\tilde{x} \neq x \in \mathcal{X}_k$ has be deleted from the accumulator, the witness u can be updated as follows. By the extended GCD algorithm, one can compute the integers a, b such that $ax + b\tilde{x} = 1$ and then $u' = W(u, x, \tilde{x}, v, v') = u^b v'^a$. Let us verify that $f(u', x) = u'^x \bmod n = v'$:

$$(u^b v'^a)^x \overset{(1)}{=} ((u^b v'^a)^{x\tilde{x}})^{1/\tilde{x}} = ((u^x)^{b\tilde{x}} (v'^{\tilde{x}})^{ax})^{1/\tilde{x}} = (v^{b\tilde{x}} v^{ax})^{1/\tilde{x}} = v^{1/\tilde{x}} = v'$$

Equation (1) is correct because $\tilde{x}$ is relatively prime to $\varphi(n)$.

We note that adding or deleting several values at once can be done simply by letting x' be the product of the added or deleted values. This also holds with respect to updating the witness. More precisely, let π_a be the product the x's to add to and π_d be the ones to delete from the accumulator value v. Then, the new accumulator value $v' := v^{\pi_a \pi_d^{-1} \bmod (p-1)(q-1)} \bmod n$. If u was the witness that x was contained in v and x was not removed from the accumulator, i.e., $x \nmid \pi_d$, then $u' u^{a\pi_a} v'^b \bmod n$ is a witness that x is contained in v', where a and b satisfy $ax + b\pi_d = 1$ and are computed using the extended GCD algorithm.

Theorem 2. *Under the strong RSA assumption, the above construction is a secure dynamic accumulator.*

Proof. Everything besides security is immediate. By Theorem 1, it is sufficient to show that the construction satisfies security as defined in Definition 1. Our proof is similar to the one given by Barić-Pfitzmann for their construction (the difference being that we do not require x' to be prime). The proof by Barić-Pfitzmann is actually the same as one given by Shamir [26].

Suppose we are given an adversary $\mathcal{A}$ that, on input n and $u \in_R QR_n$, outputs m primes $x_1, \ldots, x_m \in \mathcal{X}_{A,B}$ and $u' \in \mathbb{Z}^*_n$, $x' \in \mathcal{X}'_{A,B}$ such that $(u')^{x'} = u^{\prod x_i}$. Let us use $\mathcal{A}$ to break the strong RSA assumption.

Suppose $n = pq$ that is a product of two safe primes, $p = 2p' + 1$ and $q = 2q' + 1$, is given. Suppose the value $u \in QR_n$ is given as well. To break the strong RSA assumption, we must output a value $e > 1$, y such that $y^e = u$.

We shall proceed as follows: Give (n, u) to the adversary. Suppose the adversary comes up with a forgery $(u', x', (x_1, \ldots, x_m))$. Let $x = \prod_{i=1}^{m} x_i$. Thus we have $u'^{x'} = u^x$.

Claim. Let $d = \gcd(x, x')$. Then either $d = 1$ or $d = x_j$ for some $1 \leq j \leq m$.

Proof of claim: Suppose $d|x$ and $d \neq 1$. Then, as $x_1, \ldots, x_m$ are primes, it follows that d is the product of a subset of primes. Suppose for some x_i and x_j we have $x_i x_j | d$. Then $x_i x_j | x'$. But this is a contradiction as $x_i x_j > x'$ must hold due to the definitions of $\mathcal{X}_{A,B}$ and $\mathcal{X}'_{A,B}$: Because $x' \in \mathcal{X}'_{A,B}$ we have $x' < A^2$. For any $x_i, x_j \in \mathcal{X}_{A,B}$, $x_i x_j \geq A^2 > x'$, as $x_1, x_2 \geq A$.

Back to the proof of the theorem: Suppose that $d \neq 1$ is not relatively prime to $\phi(n)$. Then, by the claim, for some j, $d = x_j$. Because $d = x_j \in \mathcal{X}_{A,B}$, $d > 2$ and d is prime. $\phi(n) = 4p'q'$, therefore $d = p'$ or $d = q'$. Then $2d + 1$ is a non-trivial divisor of n, so in this case we can factor n.

Suppose d is relatively prime to $\phi(n)$. Then, because $(u^{x/d})^d = ((u')^{x'/d})^d$, it follows that $u^{x/d} = (u')^{x'/d}$. Let $\tilde{x} = x/d$, and $\tilde{x}' = x'/d$. Because $\gcd(x, x') = d$, the equation $\gcd(\tilde{x}, \tilde{x}') = 1$ holds and thus one can compute $\mathfrak{a}, \mathfrak{b}$ such that $\mathfrak{a}\tilde{x} + \mathfrak{b}\tilde{x}' = 1$ by extended GCD algorithm. Output $(y := \tilde{u}^{\mathfrak{a}} u^{\mathfrak{b}}, \tilde{x}')$. Note that $y^{\tilde{x}'} = (y^{\tilde{x}\tilde{x}'})^{1/\tilde{x}} ((\tilde{u}^{\tilde{x}'})^{\mathfrak{a}\tilde{x}} (u^{\tilde{x}})^{\mathfrak{b}\tilde{x}'})^{1/\tilde{x}} = ((u^{\tilde{x}})^{\mathfrak{a}\tilde{x}+\mathfrak{b}\tilde{x}'})^{1/\tilde{x}} u$ and thus y and $\tilde{x}'$ are a solution to the instance (n, u) of the flexible RSA problem.

3.3 Efficient Proof That a Committed Value Was Accumulated

Here we show that the accumulator exhibited above is efficiently provable with respect to the Pedersen commitment scheme. Suppose that the parameters of the commitment scheme are a group $\mathfrak{G}_q$, and two generators $\mathfrak{g}$ and $\mathfrak{h}$ such that $\log_{\mathfrak{h}} \mathfrak{g}$ is unknown. Recall that to commit to a value $x \in \mathbb{Z}_q$, one picks a random $r \in_R \mathbb{Z}_q$ and outputs $Commit(x, r) := \mathfrak{g}^x \mathfrak{h}^r$. This information-theoretically hiding commitment scheme is binding under the discrete-logarithm assumption.

For the definitions of $\mathcal{X}_{A,B}$ and the choice of q, we require that $B2^{k'+k''+2} < A^2 - 1 < q/2$ holds, where k' and k'' are security parameters, i.e., k' is the bit length of challenges in the PK protocol below and k'' determines the statistical zero-knowledge property of the same protocol. We set $\mathcal{X}'_{A,B}$ the largest possible set, i.e., to $[2, A^2 - 1]$.

Finally, we require that two elements g and h of QR_n are available such that $\log_g h$ is not known to the prover, where n is the public key of the accumulator.

To prove that a given commitment $\mathfrak{C}_e$ and a given accumulator v contain the same (secret) value e, the following protocol is carried out. The common inputs to the protocol are the values $\mathfrak{C}_e, \mathfrak{g}, \mathfrak{h}, n, g, h$, and v. The prover's additional inputs are the values e, u, and r such that $u^e = v \bmod n$ and $\mathfrak{C}_e = \mathfrak{g}^e \mathfrak{h}^r$.

The prover will form a commitment C_u to u and prove that this commitment corresponds to the e-th root of the value v. This is carried out as follows:

1. The prover chooses $r_1, r_2, r_3 \in_R \mathbb{Z}_{\lfloor n/4 \rfloor}$, computes $C_e := g^e h^{r_1}$, $C_u := u h^{r_2}$, $C_r := g^{r_2} h^{r_3}$, and sends $\mathfrak{C}_e, C_e, C_u$, and C_r to the verifier.

2. The prover and verifier carry out the following proof of knowledge:

$$PK\{(\alpha, \beta, \gamma, \delta, \varepsilon, \zeta, \varphi, \psi, \eta, \varsigma, \xi) :$$

$$\mathfrak{C}_e = \mathfrak{g}^\alpha \mathfrak{h}^\varphi \ \wedge \ \mathfrak{g} = (\frac{\mathfrak{C}_e}{\mathfrak{g}})^\gamma \mathfrak{h}^\psi \ \wedge \ \mathfrak{g} = (\mathfrak{g}\mathfrak{C}_e)^\varsigma \mathfrak{h}^\xi \ \wedge$$

$$C_r = h^\varepsilon g^\varsigma \ \wedge \ C_e = h^\alpha g^\eta \ \wedge \ v = C_u^\alpha (\frac{1}{h})^\beta \ \wedge \ 1 = C_r^\alpha (\frac{1}{h})^\delta (\frac{1}{g})^\beta \ \wedge$$

$$\alpha \in [-B2^{k'+k''+2}, B2^{k'+k''+2}] \ \} \ .$$

The details of this protocol can be found in full version of this paper [8].

Theorem 3. *Under the strong RSA assumption the PK protocol in step 2 is a proof of knowledge of two integers $e \in \mathcal{X}'_{A,B} = [2, A^2 - 1]$ and u such that $v \equiv u^e$ (mod n) and $\mathfrak{C}_e$ is a commitment to e.*

Proof. Showing that the protocol is statistical zero-knowledge is standard. Also, it is easy to see that $\mathfrak{C}_e$, C_e, C_u, and C_r are statistically independent from u and e.

It remains to show that if the verifier accepts, then value e committed to in $\mathfrak{C}_e$ and a witness w that e is in v can be extracted from the prover. Using standard rewinding techniques, the knowledge extractor can get answers $(s_\alpha, s_\beta, s_\gamma, s_\delta, s_\varepsilon, s_\zeta, s_\eta, s_\varphi, s_\psi)$ and $(s'_\alpha, s'_\beta, s'_\gamma, s'_\delta, s'_\varepsilon, s'_\eta, s'_\zeta, s'_\varphi, s'_\psi)$ for the two different challenges c and c'. Let $\Delta\alpha = s_\alpha - s'_\alpha$, $\Delta\beta = s_\beta - s'_\beta$, $\Delta\gamma = s_\gamma - s'_\gamma$, $\Delta\delta = s_\delta - s'_\delta$, $\Delta\varepsilon = s_\varepsilon - s'_\varepsilon$, $\Delta\zeta = s_\zeta - s'_\zeta$, $\Delta\eta = s_\eta - s'_\eta$, $\Delta\varphi = s_\varphi - s'_\varphi \mod q$, $\Delta\psi = s_\psi - s'_\psi$, $\Delta\varsigma = s_\varsigma - s'_\varsigma$, $\Delta\xi = s_\xi - s'_\xi$, and $\Delta c = c' - c$. Then we have

$$\mathfrak{C}_e^{\Delta c} = \mathfrak{g}^{\Delta\alpha}\mathfrak{h}^{\Delta\varphi} , \qquad \mathfrak{g}^{\Delta c} = (\frac{\mathfrak{C}_e}{\mathfrak{g}})^{\Delta\gamma}\mathfrak{h}^{\Delta\psi} , \qquad \mathfrak{g}^{\Delta c} = (\mathfrak{g}\mathfrak{C}_e)^{\Delta\varsigma}\mathfrak{h}^{\Delta\xi} \quad (1)$$

$$C_r^{\Delta c} = h^{\Delta\varepsilon}g^{\Delta\zeta} , \qquad C_e^{\Delta c} = h^{\Delta\alpha}g^{\Delta\eta} , \qquad (2)$$

$$v^{\Delta c} = C_u^{\Delta\alpha}(\frac{1}{h})^{\Delta\beta} , \qquad 1 = C_r^{\Delta\alpha}(\frac{1}{h})^{\Delta\delta}(\frac{1}{g})^{\Delta\beta} . \qquad (3)$$

We first show that $\mathfrak{C}_e$ commits to an integer different from 1 and consider the first two equations (1). Let $\tilde{\alpha} := \Delta\alpha\Delta c^{-1} \mod q$, $\tilde{\gamma} := \Delta\gamma\Delta c^{-1} \mod q$, $\tilde{\varphi} := \Delta\varphi\Delta c^{-1} \mod q$, and $\tilde{\psi} := \Delta\psi\Delta c^{-1} \mod q$. Then we have

$$\mathfrak{C}_e = \mathfrak{g}^{\tilde{\alpha}}\mathfrak{h}^{\tilde{\varphi}} \quad \text{and} \quad \mathfrak{g} = (\frac{\mathfrak{C}_e}{\mathfrak{g}})^{\tilde{\gamma}}\mathfrak{h}^{\tilde{\psi}} = \mathfrak{g}^{(\tilde{\alpha}-1)\tilde{\gamma}}\mathfrak{h}^{\tilde{\varphi}\tilde{\gamma}}\mathfrak{h}^{\tilde{\psi}} .$$

Under the hardness of computing discrete logarithms, $1 \equiv (\tilde{\alpha}-1)\tilde{\gamma}$ (mod q) must hold and therefore $\tilde{\alpha} \neq 1$ (mod q) as otherwise $\tilde{\gamma}$ would not exists. Similarly, from the first and third equation of (1) one can conclude that $\tilde{\alpha} \neq -1$ (mod q).

We next show that $\tilde{\alpha}$ is accumulated in v. From the next two equations (2) one can derive that Δc divides $\Delta\alpha$, $\Delta\eta$, $\Delta\varepsilon$, and $\Delta\zeta$ provided the strong RSA assumption. (While we do not investigate this claim here, one can show that if Δc does not divide $\Delta\alpha$, $\Delta\eta$, $\Delta\varepsilon$, and $\Delta\zeta$, then from the Equations (2) one can compute a non-trivial root of g with probability at least $1/2$. This, however, is not feasible under the strong RSA assumption. We refer to, e.g., [17] for the details of such a reduction.) Let $\hat{\alpha} = \Delta\alpha/\Delta c$, $\hat{\eta} = \Delta\eta/\Delta c$, $\hat{\varepsilon} = \Delta\varepsilon/\Delta c$ and $\hat{\zeta} = \Delta\zeta/\Delta c$. Because $|c|, |c'| < p', q'$, we get $C_r = ah^{\hat{\varepsilon}}g^{\hat{\zeta}}$ for some a such that $a^2 = 1$. Moreover, the value a must be either 1 or -1 as otherwise $1 < \gcd(a - 1, n) < n$ and we could factor n. Plugging C_r into the second equation of (3) we get $1 = a^{\Delta\alpha}h^{\Delta\alpha\hat{\varepsilon}}g^{\Delta\alpha\hat{\zeta}}(\frac{1}{h})^{\Delta\delta}(\frac{1}{g})^{\Delta\beta}$, where $a^{\Delta\alpha}$ must be 1 as 1, g, and h are in QR_n and $a^2 = 1$ otherwise. Under the hardness of computing discrete logarithms we can conclude that $\Delta\alpha\hat{\zeta} \equiv \hat{\beta} \pmod{\mathrm{ord}(g)}$ and hence we get $v^{\Delta c} = (\frac{C_v}{h^{\hat{\zeta}}})^{\Delta\alpha}$ and $v = b(\frac{C_v}{h^{\hat{\zeta}}})^{\hat{\alpha}}$ with some b such that $b^2 = 1$. Again $b = \pm 1$ as otherwise $1 < \gcd(b \pm 1, n) < n$ and we could factor n. Let $s = -1$ if $\hat{\alpha} < 0$ and $s = 1$. Thus we have $v = u^{|\hat{\alpha}|}$,

$$ u = \begin{cases} (b\frac{C_v}{h^{\hat{\zeta}}})^s & \text{if } \hat{\alpha} \text{ is odd} \\ (\frac{C_v}{h^{\hat{\zeta}}})^s & \text{if } \hat{\alpha} \text{ is even.} \end{cases} $$

The latter holds because $v \in QR_n$ and $-1 \notin QR_n$ and therefore $b = -1$ is not possible. Also note that $\hat{\alpha} \neq 0$ as $v \neq 1$. Because $s_\alpha, s'_\alpha \in [-B2^{k'+k''+1}, -B2^{k'+k''+1}]$ we have $\Delta\alpha, \hat{\alpha} \in [-B2^{k'+k''+2}, -B2^{k'+k''+2}]$. Because $B2^{k'+k''+2} < q/2$ it follows that $\hat{\alpha} = (\Delta\alpha\hat{c} \bmod q)(\tilde{\alpha} \bmod q)$, and hence that the absolute value committed to by $\mathfrak{C}_e$ is indeed accumulated in v. As $B2^{k'+k''+2} < A^2 - 1$, $\hat{\alpha} \neq \pm 1$ mod q and $\hat{\alpha} \neq 0$ we can conclude that $|\hat{\alpha}| \in \mathcal{X}'_{A,B}$. Therefore, due to Theorem 2, we can conclude that $|\hat{\alpha}|$ must be contained in the accumulator value v.

4 Application to Revocation of Anonymous Credentials

In this section we describe how dynamic accumulators can be used to add a revocation capability to anonymous credentials. We then show how this can be done efficiently for the ACJT identity escrow [1] and describe how to adapt this solution to related group signature schemes and credential systems [1, 9].

4.1 Revocation of Anonymous Credentials

We first note that dynamic accumulators can be used for revocation of ordinary credentials (and certificates): First, one adds a unique value to each credential. Then, the accumulator values of the unique values of all valid credentials is published authentically. Now, a user can convince a verifier that the credential is still valid by providing the witness for the unique value contained in her credential. Thus, to check a credential, the verifier has to check the issuer's signature, to get the current accumulator value and to verify whether the unique value contained in the credential is contained in the accumulator value using the witness provided by the user.

In case of anonymous credentials the same approach can be used. Now, however, the witness and the value contained in the accumulator can no longer be revealed to the verifier as this would compromise anonymity completely. Instead, the user can apply zero-knowledge proofs to convince the verifier of that the value contained in her credential is also contained in the accumulator. In the previous section we exhibited an efficient zero-knowledge protocol to prove that a value contained in a commitment is contained in an accumulator. One can thus get efficient revocation for any anonymous credential scheme if one finds an efficient protocol to prove that a value contained in a commitment is also contained in the credential.

In the remainder of this section we provide such a protocol for the ACJT identity escrow scheme and the Camenisch-Lysyanskaya credential system [1, 9]. However, it is not hard to see how to add revocation for other schemes and systems that use some form of anonymous credentials (e.g., [5,11,12,10,13,21, 23]).

4.2   The ACJT Identity Escrow Scheme and Its Friends

An identity escrow scheme involves a *membership manager*, who is responsible for adding and deleting members, an *anonymity revocation manager*, who can identify the user who provided an anonymous membership-proof to a verifier, and finally *users* that can become members. There are the following procedures: a *setup* algorithm, that allows the manager to choose their secret and public keys; *join* protocol that a user runs the membership manager if she want to become a member; and a *prove membership* protocol with which a user can anonymously convince a verifier that she is a member. We refer to [1, 8] for details of these procedures and for the security properties.

Recall the ACJT [1] identity escrow scheme. (Recall that the ACJT group signature scheme is obtained from the ACJT identity escrow by applying the Fiat-Shamir heuristic to the protocol for proving membership.) The group manager has a public key PK, consisting of a number $\mathfrak{n}$, which is a product of two safe primes, the values $\mathfrak{a}$, $\mathfrak{b}$, $\mathfrak{g}$, $\mathfrak{h}$, and $\mathfrak{y}$ which are quadratic residues modulo $\mathfrak{n}$, and two intervals Γ and Δ. The value $z = \log_{\mathfrak{g}} \mathfrak{y}$ is a secret key of the group manager used for revocation. A user U_i's membership certificate consists of a user's secret x_i selected jointly by the user and the group manager (it is selected in a secure manner that ensures that the group manager obtains no information about this value) from an appropriate integer range, i.e., Δ, and the values $\mathfrak{v}_i$ and e_i, where e_i is a prime number selected from another appropriate range, i.e., Γ, and $\mathfrak{v}_i^{e_i} = \mathfrak{a}^{x_i} \mathfrak{b} \bmod \mathfrak{n}$. The value $\mathfrak{a}^{x_i}$ is user U_i's public key. When U_i proves membership in a group, he effectively proves knowledge of a membership certificate $(x, \mathfrak{v}, e)$. This proof is as follows. The group member chooses $r_1', r_2' \in_R \mathbb{Z}_{\lfloor \mathfrak{n}/4 \rfloor}$ and computes $\mathfrak{T}_1 := \mathfrak{v} \mathfrak{y}^{r_1'}$, $\mathfrak{T}_2 := \mathfrak{g}^{r_1'}$, and $\mathfrak{T}_3 := \mathfrak{g}^e \mathfrak{h}^{r_2'}$. The group member sends $\mathfrak{T}_1$, $\mathfrak{T}_2$, and $\mathfrak{T}_3$ to the verifier and carries out with the verifier the protocol denoted

$$PK\{(\alpha, \beta, \gamma, \delta, \varepsilon) : \mathfrak{b} = \mathfrak{T}_1^{\alpha} \left(\frac{1}{\mathfrak{a}}\right)^{\beta} \left(\frac{1}{\mathfrak{y}}\right)^{\gamma} \wedge 1 = \mathfrak{T}_2^{\alpha} \left(\frac{1}{\mathfrak{g}}\right)^{\gamma} \wedge \mathfrak{T}_2 = \mathfrak{g}^{\delta} \wedge \mathfrak{T}_3 = \mathfrak{g}^{\alpha} \mathfrak{h}^{\varepsilon} \wedge$$
$$\alpha \in \Gamma \wedge \beta \in \Delta\} \ .$$

As with all group signature and identity escrow schemes, only the group manager can assert a signature/protocol transcript to a group member, i.e., knowing z, the group manager can compute the value $\hat{v} = \mathcal{T}_1/\mathcal{T}_2{}^z$ that identifies the user.

The Camenisch and Lysyanskaya [9] credential system has a similar construction. An organization's public key consists of a number $\mathfrak{n}$, which is a product of two safe primes, and the values $\mathfrak{a}$, $\mathfrak{b}$, $\mathfrak{c}$, $\mathfrak{g}$ and $\mathfrak{h}$ which are all quadratic residues modulo $\mathfrak{n}$. A user U_i's secret key x_i, selected from an appropriate integer range, is incorporated into all of U_i's credentials. A credential tuple for user U_i consists of his secret key x_i, a secret value s_i selected jointly by the U_i and the organization (via a secure computation which ensures secrecy for the user) from an appropriate integer range, and the values $\mathfrak{v}_i$ and e_i such that e_i is a prime number selected by the organization from an appropriate integer interval, and $\mathfrak{v}_i$ is such that $\mathfrak{v}_i^{e_i} = \mathfrak{a}^x \mathfrak{b}^s \mathfrak{c} \bmod N$. Proving possession of a credential is effectively a proof of knowledge of a credential tuple.

Variations of these schemes incorporate such features as anonymity revocation, non-transferability, one-show credentials, expiration dates, and appointed verifiers. For all these variations, an integral part of a group membership certificate and of a credential, is the prime number e_i. Using one-way accumulators, we can accumulate e_i's into a single public value u. Proof of group membership will now have to include proof of knowledge of a witness to the fact that e_i was accumulated into u.

In the sequel, we will talk about augmenting the ACJT identity escrow scheme with the membership revocation property; however, all our results and discussion apply immediately to the credential scheme and group signature discussed above.

4.3 Incorporating Revocation into the ACJT Identity Escrow Scheme

To make certificate revocation possible, the additions outlined below have to be made to the usual operations the ACJT identity escrow scheme.

Modifications to the group manager's operations are as follows:

Setup: In addition to setting up the identity escrow scheme, the group manager creates the public modulus n for the accumulator, chooses a random $u, g, h \in QR_n$ and publishes (n, u, g, h). She sets up (empty for now) public archives E_{add} for storing values that correspond to added users and E_{delete} for storing values that correspond to deleted users. Set $\mathcal{X}'_{A,B} = \Gamma$ and $\mathcal{X}_{A,B}$ to the interval from which the group manager chooses e in the ACJT scheme ($\mathcal{X}_{A,B} \subseteq \mathcal{X}'_{A,B} \subseteq [2, A^2 - 1]$ will be satisfied).

Join: Issue the user's membership certificate, as in the identity escrow scheme. Add the current u to the user's membership certificate. (Denote it by u_i.) Let e_i be the prime number used in this certificate. Update u in the public key: $u := f_n(u, e_i)$. Update E_{add}: store e_i there.

Revoke membership: Retrieve e_i which is the prime number corresponding to the user's membership certificate. Update u in the public key: $u := D(\varphi(n), u, e_i)$. Update E_{delete}: store e_i there.

We stress that the archives are E_{add} and E_{delete} are *not* part of the group's public key, i.e., the verifier is not required to read them for any verification purposes. Also, note that is it not necessary to restrict read access to these archives only to group members.

A user U_i must augment the ACJT protocol as follows:

Join: Store the value u_i along with the rest of the membership certificate. Verify that $f_n(u_i, e_i) = u_i^{e_i} = u$.

Update membership: An entry in the archive is called "new" if it was entered after the last time U_i performed an update.

1. Let y denote the old value of u.
2. For all new $e_j \in E_{\text{add}}$, $u_i := f(u_i, \prod e_j) = u_i^{\prod e_j}$ and $y := y^{\prod e_j}$.
3. For all new $e_j \in E_{\text{delete}}$, $u_i := W(u_i, e_i, \prod e_j, y, u)$.

(Note that as a result $u = f(u_i, e_i)$.)

Prove membership: Proving membership is augmented with the step of proving that a committed value is part of the accumulated value u (contained in the current public key). That is, in addition to $\mathfrak{T}_1, \mathfrak{T}_2$, and $\mathfrak{T}_3$ the group member computes the values $C_e := g^e h^{r_1}$, $C_{u_i} := u_i h^{r_2}$, and $C_r := g^{r_2} h^{r_3}$ and sends them to verifier, with random choices $r_1, r_2, r_3 \in_R \mathbb{Z}_{\lfloor n/4 \rfloor}$. Then the verifier and the group member engage in the protocol denoted

$$PK\{(\alpha, \beta, \gamma, \delta, \varepsilon, \xi, \zeta, \varphi, \psi, \eta):$$
$$\mathfrak{w} \equiv \mathfrak{T}_1^\alpha \left(\frac{1}{\mathfrak{a}}\right)^\beta \left(\frac{1}{\mathfrak{h}}\right)^\gamma \ \wedge \ 1 \equiv \mathfrak{T}_2^\alpha \left(\frac{1}{\mathfrak{g}}\right)^\gamma \ \wedge \ \mathfrak{T}_2 \equiv \mathfrak{g}^\delta \ \wedge \ \mathfrak{T}_3 \equiv \mathfrak{g}^\alpha \mathfrak{h}^\varepsilon \ \wedge$$
$$C_r = h^\xi g^\zeta \ \wedge \ C_e = h^\alpha g^\eta \ \wedge \ u = C_{u_i}^\alpha \left(\frac{1}{h}\right)^\varphi \ \wedge \ 1 = C_r^\alpha \left(\frac{1}{h}\right)^\psi \left(\frac{1}{g}\right)^\varphi \ \wedge$$
$$\alpha \in \Gamma \ \wedge \ \beta \in \Delta \} \ .$$

This protocol is already an optimized union of the PK protocol given in the previous section and the ACJT PK protocol for proving group membership. That is, different from the previous section, we do not require the group $\mathfrak{G}_q$ for the commitment scheme because here the value $\mathfrak{T}_3$ acts as commitment to the value whose membership in the accumulator is claimed. Furthermore, as $-1, 0, 1 \notin \Gamma$, we need not show that $\alpha \neq -1, 0, 1$.

The complexity of this augmented proof is about twice that of the original one. The definition of Γ is compatible with the accumulator and the proof that a committed value is contained in the accumulator as presented in the previous section. Also, Γ excludes 1 and hence it is not required to explicitly prove that the committed value is not 1.

Remark 1. Updates after a user joined the group can be avoided: the group manager knows the factorization of n and therefore can always compute a witness $u_i := u^{1/e_i}$ for the e_i of the added user, where u is the old *and* new accumulator value. It's easy to see that this modifided construction retains security.

Lemma 2. *Under the strong RSA assumption the above is a secure identity escrow scheme with membership revocation.*

Proof (sketch). It is not hard to show the security of this lemma in a formal model given the security proofs of the ACJT scheme and the proof of Theorem 3. Let us provide an informal argument here.

First of all, note that all the properties of the original ACJT scheme are retained as the amount of information revealed by C_e, C_u, and C_r about the group member's certificate is negligible (i.e., C_e, C_u, and C_r are statistically hiding commitments and the PK-protocol is statistical zero-knowledge). It remains to argue that excluded group members can no longer prove group membership even if they collude in an adaptive attack against the group manager. Similarly as in the proof of Theorem 3, one can show that the above of a protocol is a proof of knowledge of a quadruple $(\hat{x}, \hat{v}, \hat{e}, \hat{u})$ such that $a^{\hat{x}}b = \hat{v}^{\hat{e}}$ and $\hat{u}^{\hat{e}} = u$ hold, i.e., such that $(\hat{x}, \hat{v}, \hat{e})$ is valid group membership certificate and $\hat{e}$ is contained in the accumulator value u. In [1], Ateniese et al. show that under the strong RSA assumption an adaptive adversary controlling all users cannot find a triple $(\tilde{x}, \tilde{v}, \tilde{e})$ that is different from the ones legitimately obtained through the *join* protocol. On other words, the values a^{x_i} and e_i are tightly linked. Therefore, the user with public key a^{x_i} is no longer able to prove membership of the group once an e_i is removed from the accumulator value as the accumulator is secure against an adaptive adversary (Theorem 1). We note that all these arguments hold in spite of the fact that all members' (current and past one) e_i's are public. It follows that anonymity and unlinkability is retained for actions past members made prior to their exclusion from the group.

References

1. G. Ateniese, J. Camenisch, M. Joye, and G. Tsudik. A practical and provably secure coalition-resistant group signature scheme. In *Advances in Cryptology – CRYPTO 2000*, vol. 1880 of *LNCS*, pp. 255–270. Springer Verlag, 2000.
2. G. Ateniese and G. Tsudik. Quasi-efficient revocation of group signatures. http://eprint.iacr.org/2001/101, 2001.
3. N. Barić and B. Pfitzmann. Collision-free accumulators and fail-stop signature schemes without trees. In *EUROCRYPT '97*, vol. 1233 of *LNCS*, pp. 480–494.
4. J. Benaloh and M. de Mare. One-way accumulators: A decentralized alternative to digital signatures. In *EUROCRYPT '93*, vol. 765 of *LNCS*, pp. 274–285.
5. S. Brands. *Rethinking Public Key Infrastructure and Digital Certificates – Building in Privacy*. PhD thesis, Eindhoven Institute of Technology, Eindhoven, The Netherlands, 1999.
6. E. Bresson and J. Stern. Group signatures with efficient revocation. In *Proceedings of PKC2001*, vol. 1992 of *LNCS*, pp. 190–206. Springer, 2001.
7. J. Camenisch. Efficient and generalized group signatures. In *Advances in Cryptology – EUROCRYPT '97*, vol. 1233 of *LNCS*, pp. 465–479.
8. J. Camenisch and A. Lysyanskaya. Dynamic accumulators and application to efficient revocation of anonymous credentials. http://eprint.iacr.org/2001, 2001.

9. J. Camenisch and A. Lysyanskaya. Efficient non-transferable anonymous multi-show credential system with optional anonymity revocation. In *Advances in Cryptology – EUROCRYPT 2001*, vol. 2045 of *LNCS*, pp. 93–118.

10. J. Camenisch and A. Lysyanskaya. An identity escrow scheme with appointed verifiers. In *CRYPTO 2001*, vol. 2139 of *LNCS*, pp. 388–407.

11. J. Camenisch and M. Michels. A group signature scheme with improved efficiency. In *Advances in Cryptology – ASIACRYPT '98*, vol. 1514 of *LNCS*, pp. 160–174.

12. J. Camenisch and M. Michels. Separability and efficiency for generic group signature schemes. In *CRYPTO '99*, vol. 1666 of *LNCS*, pp. 413–430.

13. J. Camenisch and M. Stadler. Efficient group signature schemes for large groups. In *Advances in Cryptology – CRYPTO '97*, vol. 1296 of *LNCS*, pp. 410–424, 1997.

14. R. Canetti. Security and composition of multi-party cryptographic protocols. *Journal of Cryptology*, 13(1):143–202, 2000.

15. L. Chen and T. P. Pedersen. New group signature schemes. In *Advances in Cryptology – EUROCRYPT '94*, vol. 950 of *LNCS*, pp. 171–181, 1995.

16. R. Cramer and V. Shoup. A practical public key cryptosystem provably secure against adaptive chosen ciphertext attack. In *Advances in Cryptology – CRYPTO '98*, vol. 1642 of *LNCS*, pp. 13–25, Berlin, 1998. Springer Verlag.

17. I. Damgård and E. Fujisaki. An integer commitment scheme based on groups with hidden order. http://eprint.iacr.org/2001/064, 2001.

18. E. Fujisaki and T. Okamoto. Statistical zero knowledge protocols to prove modular polynomial relations. In *CRYPTO '97*, vol. 1294 of *LNCS*, pp. 16–30.

19. R. Gennaro, S. Halevi, and T. Rabin. Secure hash-and-sign signatures without the random oracle. In *EUROCRYPT '99*, vol. 1592 of *LNCS*, pp. 123–139.

20. O. Goldreich, S. Micali, and A. Wigderson. How to prove all NP statements in zero-knowledge and a methodology of cryptographic protocol design. In *Advances in Cryptology – CRYPTO '86*, vol. 263 of *LNCS*, pp. 171–185, 1987.

21. J. Kilian and E. Petrank. Identity escrow. In *Advances in Cryptology – CRYPTO '98*, vol. 1642 of *LNCS*, pp. 169–185, Berlin, 1998. Springer Verlag.

22. H.-J. Kim, J. I. Lim, and D. H. Lee. Efficient and secure member deletion in group signature schemes. In *ICISC 2000*, number 2015 in LNCS, pp. 150–161, 2001.

23. A. Lysyanskaya, R. Rivest, A. Sahai, and S. Wolf. Pseudonym systems. In *Selected Areas in Cryptography*, vol. 1758 of *LNCS*. Springer Verlag, 1999.

24. B. Pfitzmann and M. Waidner. Composition and integrity preservation of secure reactive systems. In *Proc. 7th ACM CCS*, pp. 245–254. ACM press, nov 2000.

25. T. Sander, A. Ta-Shma, and M. Yung. Blind, auditable membership proofs. In *Financial Cryptography '00*, vol. 1962 of *LNCS*, pp. 53–71, 2000.

26. A. Shamir. On the generation of cryptographically strong pseudorandom sequences. In *ACM Transaction on Computer Systems*, vol.1, pp. 38–44, 1983.

27. D. X. Song. Practical forward secure group signature schemes. In *Proc. 8th ACM CCS*, pp. 225–234. ACM press, nov 2001.

Provably Secure Steganography

(Extended Abstract)

Nicholas J. Hopper, John Langford, and Luis von Ahn

Computer Science Department, Carnegie Mellon University,
Pittsburgh PA 15213, USA
{hopper,jcl,biglou}@cs.cmu.edu

Abstract. Informally, *steganography* is the process of sending a secret message from Alice to Bob in such a way that an eavesdropper (who listens to all communications) cannot even tell that a secret message is being sent. In this work, we initiate the study of steganography from a complexity-theoretic point of view. We introduce definitions based on computational indistinguishability and we prove that the existence of one-way functions implies the existence of secure steganographic protocols.

1 Introduction

The scientific study of steganography began in 1983 when Simmons [13] stated the problem in terms of communication in a prison. In his formulation, two inmates, Alice and Bob, are trying to hatch an escape plan. The only way they can communicate with each other is through a public channel, which is carefully monitored by the warden of the prison, Ward. If Ward detects any encrypted messages or codes, he will throw both Alice and Bob into solitary confinement. The problem of steganography is, then: how can Alice and Bob cook up an escape plan by communicating over the public channel in such a way that Ward doesn't suspect anything fishy is going on. (Notice how steganography is different from classical cryptography, which is about hiding the *content* of secret messages: steganography is about hiding the very existence of the secret messages.)

Steganographic "protocols" have a long and intriguing history that goes back to antiquity. There are stories of secret messages written in invisible ink or hidden in love letters (the first character of each sentence can be used to spell a secret, for instance). More recently, steganography was used by prisoners and soldiers during World War II because all mail in Europe was carefully inspected at the time [8]. Postal censors crossed out anything that looked like sensitive information (e.g. long strings of digits), and they prosecuted individuals whose mail seemed suspicious. In many cases, censors even randomly deleted innocent-looking sentences or entire paragraphs in order to prevent secret messages from going through. Over the last few years, steganography has been studied in the framework of computer science, and several algorithms have been developed to hide secret messages in innocent looking data.

M. Yung (Ed.): CRYPTO 2002, LNCS 2442, pp. 77–92, 2002.
© Springer-Verlag Berlin Heidelberg 2002

The main goal of this paper is to put steganography on a solid complexity-theoretic foundation. We define steganographic secrecy in terms of computational indistinguishability, and we define steganographic robustness, which deals with the case of active wardens (ones that cross out innocent-looking sentences or modify the messages just to prevent secrets from going through). Our main result is a positive one: secret and robust steganographic protocols exist within our model, given that one-way functions exist.

Related Work

There has been considerable work on digital steganography. The first International Workshop on Information Hiding occurred in 1996, with five subsequent workshops, and even books have been published about the subject [9]. Surprisingly, though, very little work has attempted to formalize steganography, and most of the literature consists of heuristic approaches: steganography using digital images [9], steganography using video systems [9], etc. A few papers have given information theoretic models for steganography [3, 10, 11, 14], but these are limited in the same way that information theoretic cryptography is limited. We believe complexity theory is the right framework in which to view steganography and, to the best of our knowledge, this is the first paper to treat steganography from a complexity-theoretic point of view (and to achieve provably positive results).

Organization of the Paper

In section 2 we define the basic cryptographic quantities used throughout the paper, as well as the notions of a cover *channel* and a *stegosystem*. In section 3 we define steganographic secrecy and state protocols which are steganographically secret assuming the existence of one-way functions. In section 4 we define robust steganographic secrecy for adversaries with bounded power to perturb stegotext messages and state protocols which satisfy this definition. Section 5 closes the paper with a discussion of implications.

2   Definitions

2.1   Preliminaries

A function $\mu : \mathbb{N} \rightarrow (0, 1)$ is said to be *negligible* if for every $c > 0$, for all sufficiently large n, $\mu(n) < 1/n^c$. The concatenation of string s_1 and string s_2 will be denoted by $s_1 || s_2$, and when we write "Parse s as $s_1^t || s_2^t || \cdots || s_l^t$" we mean to separate s into strings $s_1, \ldots s_l$ where each $|s_i| = t$, $l = \lceil |s|/t \rceil$, and $s = s_1 || s_2 || \cdots || s_l$. We will let $U(k)$ denote the uniform distribution on k bit strings, and $U(L, l)$ denote the uniform distribution on functions from L bit strings to l bit strings. If X is finite a set, we let $U(X)$ denote the uniform distribution on X.

2.2   Cryptographic Notions

Let $F : \{0,1\}^k \times \{0,1\}^L \to \{0,1\}^l$ denote a family of functions. Let A be an oracle probabilistic adversary. Define the *prf-advantage of A over F* as

$$\mathbf{Adv}_F^{\mathrm{prf}}(A) = \left| \Pr_{K \leftarrow U(k), r \leftarrow \{0,1\}^*}[A_r^{F_K(\cdot)} = 1] - \Pr_{g \leftarrow U(L,l), r \leftarrow \{0,1\}^*}[A_r^g = 1] \right|.$$

where r is the string of random bits used by adversary A. Define the insecurity of F as

$$\mathbf{InSec}_F^{\mathrm{prf}}(t,q) = \max_{A \in \mathcal{A}(t,q)} \left\{ \mathbf{Adv}_F^{\mathrm{prf}}(A) \right\}$$

where $\mathcal{A}(t,q)$ denotes the set of adversaries taking at most t steps and making at most q oracle queries. Then F is a (t,q,ϵ)-*pseudorandom function* if $\mathbf{InSec}_F^{\mathrm{prf}}(t,q) \leq \epsilon$. Suppose that $l(k)$ and $L(k)$ are polynomials. A sequence $\{F_k\}_{k \in \mathbb{N}}$ of families $F_k : \{0,1\}^k \times \{0,1\}^{L(k)} \to \{0,1\}^{l(k)}$ is called *pseudorandom* if for all polynomially bounded adversaries A, $\mathbf{Adv}_{F_k}^{\mathrm{prf}}(A)$ is negligible in k. We will sometimes write $F_k(K, \cdot)$ as $F_K(\cdot)$.

Let $E : \mathcal{K} \times \mathcal{R} \times \mathcal{P} \to \mathcal{C}$ be a probabilistic private key encryption scheme, which maps a random number and an $|m|$-bit plaintext to a ciphertext. Consider a game in which an adversary A is given access to an oracle which is either:

- E_K for $K \leftarrow U(\mathcal{K})$; that is, an oracle which given a message m, uniformly selects random bits R and returns $E_K(R,m)$; or
- $g(\cdot) = U(|E_K(\cdot)|)$; that is, an oracle which on any query ignores its input and returns a uniformly selected output of the appropriate length.

Let $\mathcal{A}(t,q,l)$ be the set of adversaries A which make q queries to the oracle of at most l bits and run for t time steps. Define the CPA advantage of A against E as

$$\mathbf{Adv}_E^{\mathrm{cpa}}(A) = \left| \Pr_{K \leftarrow U(\mathcal{K}), s, r \leftarrow \{0,1\}^*}[A_r^{E_{K,s}} = 1] - \Pr_{g, r \leftarrow \{0,1\}^*}[A_r^g = 1] \right|$$

where $E_{K,s}$ denotes E_K with random bit source s. Define the insecurity of E as

$$\mathbf{InSec}_E^{\mathrm{cpa}}(t,q,l) = \max_{A \in \mathcal{A}(t,q,l)} \left\{ \mathbf{Adv}_E^{\mathrm{cpa}}(A) \right\}.$$

Then E is (t,q,l,ϵ)-*indistinguishable from random bits under chosen plaintext attack* if $\mathbf{InSec}_E^{\mathrm{cpa}}(t,q,l) \leq \epsilon$. A sequence of cryptosystems $\{E_k\}_{k \in \mathbb{N}}$ is called *indistinguishable from random bits under chosen plaintext attack* (IND\$-CPA) if for every PPTM A, $\mathbf{Adv}_{E_k}^{\mathrm{cpa}}(A)$ is negligible in k.

Let $\mathcal{C}$ be a distribution with finite support X. Define the *minimum entropy* of $\mathcal{C}$, $H_\infty(\mathcal{C})$, as

$$H_\infty(\mathcal{C}) = \min_{x \in X} \left\{ \log_2 \frac{1}{\Pr_{\mathcal{C}}[x]} \right\}.$$

2.3 Steganography

Steganography will be thought of as a game between the warden, Ward, and the inmate, Alice. The goal of Alice is to pass a secret message to Bob over a communication channel (known to Ward). The goal of Ward is to detect whether a secret message is being passed. In this and the following sections we will formalize this game. We start by defining a communication channel.

Definition. A *channel* is a distribution on bit sequences where each bit is also timestamped with monotonically increasing time value. Formally, a channel is a distribution with support $(\{0,1\}, t_1), (\{0,1\}, t_2), ...,$ where $\forall i > 0 : t_{i+1} \geq t_i$.

This definition of a channel is sufficiently general to encompass nearly any form of communication. It is important to note that our protocols may depend upon the timing information as well as the actual bits sent on a channel. For example, it may be possible to do steganography over email using only the timing of the emails rather than the contents of the message. It may also be possible for an enemy to detect steganography via timing analysis.

Anyone communicating on a channel can be regarded as implicitly drawing from the channel, so we will assume the existence of an oracle capable of drawing from the channel. In fact, we will assume something stronger: an oracle that can *partially* draw from the channel a (finite, fixed length) sequence of bits. This oracle can draw from the channel in steps and at any point the draw is conditioned on what has been drawn so far. We let C_h be the channel distribution conditional on the history h of already drawn timestamped bits. We also let C_h^b be the marginal channel distribution over the next block of b timestamped bits conditional on the history h. Intuitively, C_h^b is a distribution on the next b timestamped bits conditioned on the history h.

Fix b. We assume the existence of an oracle which can draw from C_h^b. We will call such a partial draw a "block". We will require that the channel satisfy a minimum entropy constraint for all blocks:

$$\forall h \text{ drawn from } C : \quad H_\infty(C_h^b) > 1$$

This partial draw will be conditional on all past draws and so we can regard a sequence of partial draws as a draw from the channel. This notion of randomness is similar to Martingale theory where random variable draws are conditional on previous random variable draws (and we use Martingale theory in our analysis).

It is important to note that a "block" might (for example) contain timestamped bits which span multiple emails. We will overload the definition of the concatenation operator $||$ for sequences of timestamped bits. Thus $c_1||c_2$ will consist of the timestamped bits of c_1 followed by the timestamped bits of c_2.

One example of a channel might be electronic mail. We can map an email system allowing communication from Alice to Bob to a channel by considering all of the bits used to encode a *particular* email as a sequence of channel bits, each with the same timestamp. The timestamp of emailed bits would be the time of transmission. The complete channel consists of a distribution over sequences of emails.

Remark. In the remainder of this paper, we will assume that cryptographic primitives remain secure with respect to an oracle which draws from a channel distribution C_h^b. Thus channels which can be used to solve the hard problems that standard primitives are based on must be ruled out. In practice this is of little concern, since the existence of such channels would have previously led to the conclusion that the primitive in question was insecure.

Definition 1. (Stegosystem) A steganographic protocol, or stegosystem, is a pair of probabilistic algorithms $S = (SE, SD)$. SE takes a key $K \in \{0,1\}^k$, a string $m \in \{0,1\}^*$ (the *hiddentext*), a message history h, and an oracle $M(h)$ which samples blocks according to a channel distribution C_h^b. $SE^M(K, m, h)$ returns a sequence of blocks $c_1 \| c_2 \| \dots \| c_l$ (the *stegotext*) from the support of C_h^{l*b}. SD takes a key K, a sequence of blocks $c_1 \| c_2 \| \dots \| c_l$, a message history h, and an oracle $M(h)$, and returns a hiddentext m. There must be a polynomial $p(k) > k$ such that SE^M and SD^M also satisfy the relationship:

$$\forall m, |m| < p(k): \quad \Pr(SD^M(K, SE^M(K, m, h), h) = m) \geq \frac{2}{3}$$

where the randomization is over any coin tosses of SE^M, SD^M, and M. (In the rest of the paper we will use (SE, SD) instead of (SE^M, SD^M).)

Note that we choose a probability of failure for the stegosystem of $1/3$ in order to include a wide range of possible stegosystems. In general, given a protocol with any reasonable probability of failure, we can boost the system to a very low probability of failure using error-correcting codes.

Although all of our oracle-based protocols will work with the oracle $M(h)$, we will always use it in a particular way. Consequently, it will be convenient for us to define the rejection sampling function $RS^{M,F} : \{0,1\}^* \times \mathbb{N} \to \{0,1\}$.

> **Procedure RS^{M,F}:**
> **Input:** target x, iteration *count*
> $i = 0$
> repeat: $c \leftarrow M$; increment i
> until $F(c) = x$ or $i = count$
> **Output:** c

The function RS simply samples from the distribution provided by the sample oracle M until $F(M) = x$. The function will return c satisfying $F(c) = x$ or the *count*-th sample from M. Note that we use an iteration count to bound the worst case running time of RS and that RS may fail to return a c satisfying $F(c) = x$.

Comment. We have taken the approach of assuming a channel which can be drawn from freely by the stegosystem; most current proposals for stegosystems act on a single sample from the channel (one exception is [3]). While it may be possible to define a stegosystem which is steganographically secret or robust and works in this style, this is equivalent to a system in our model which merely makes a single draw on the channel distribution. Further, we believe that the lack of reference to the channel distribution may be one of the reasons for the failure of many such proposals in the literature.

It is also worth noting that we assume that a stegosystem has very little knowledge of the channel distribution—SE and SD may only *sample* from an oracle according to the distribution. This is because in many cases the full distribution of the channel has never been characterized; for example, the oracle may be a human being, or a video camera focused on some complex scene. However, our definitions do not rule out encoding procedures which have more detailed knowledge of the channel distribution.

Sampling from C_h^b might not be trivial. In some cases $M(h)$ is a human, and in others a simple randomized program. We stress that it is important to minimize the use of such an oracle, because oracle queries can be extremely expensive. In practice, this oracle is also the weakest point of all our constructions. We assume the existence of a *perfect* oracle: one that can perform independent draws, one that can be rewound, etc. This assumption can be justified in some cases, but not in others. If the oracle is a human, the human may not be able to perform independent draws from the channel as is required by the function RS. A real world Warden would use this to his advantage. We therefore stress the following cautionary remark: *our protocols will be shown to be secure under the assumption that the oracle is perfect.*

Finally, we mention that our decoding algorithm, SD, is defined to have access to the oracle $M(h)$. This is a general definition, and there are cases in which this access will not be necessary. Protocols in which SD needs no access to $M(h)$ are clearly preferred.

3 Steganographic Secrecy

A *passive* warden, W, is an adversary which plays the following game:

1. W is given access to an oracle $M(h)$ which samples blocks (one at a time) from the distribution C_h^b, for past histories h drawn from the channel. W makes as many draws from $M(h)$ as it likes.
2. W is given access to a second oracle which is either $SE(K, \cdot, \cdot)$ or $O(\cdot, \cdot)$ defined by $O(m, h) \leftarrow C_h^{|SE(K, m, h)|}$. Ward W makes at most q queries totaling l bits (of hiddentext) to this oracle.
3. W outputs a bit.

We define W's advantage against a stegosystem S by

$$\mathbf{Adv}_{S,C}^{ss}(W) = \left| \Pr_{K,r,M,SE}[W_r^{M,SE(K,\cdot,\cdot)} = 1] - \Pr_{r,M,O}[W_r^{M,O(\cdot,\cdot)} = 1] \right| ,$$

where the warden uses random bits r. Define the insecurity of S by

$$\mathbf{InSec}_{S,C}^{ss}(t, q, l) = \max_{W \in \mathcal{W}(t,q,l)} \left\{ \mathbf{Adv}_{S,C}^{ss}(W) \right\} ,$$

where $\mathcal{W}(t, q, l)$ denotes the set of all adversaries which make at most q queries totaling at most l bits (of hiddentext) and running in time at most t.

Definition 2. (Steganographic secrecy) A Stegosystem $S = (SE, SD)$ is called (t, q, l, ϵ) *steganographically secret against chosen hiddentext attack* for the channel $\mathcal{C}$ ((t, q, l, ϵ)-SS-CHA-$\mathcal{C}$) if $\mathbf{InSec}^{SS}_{S,\mathcal{C}}(t, q, l) \leq \epsilon$.

Definition 3. (Universal Steganographic Secrecy) A Stegosystem S is called (t, q, l, ϵ)-*universally steganographically secret against chosen hiddentext attack* ((t, q, l, ϵ)-USS-CHA) if it is (t, q, l, ϵ)-SS-CHA-$\mathcal{C}$ for every channel distribution $\mathcal{C}$ that satisfies $H_\infty(\mathcal{C}^b_h) > 1$ for all h drawn from $\mathcal{C}$. A sequence of stegosystems $\{S_k\}_{k \in \mathbb{N}}$ is called *universally steganographically secret* if for every channel distribution $\mathcal{C}$ and for every PPTM W, $\mathbf{Adv}^{SS}_{S_k,\mathcal{C}}(W)$ is negligible in k.

Note that steganographic secrecy can be thought of roughly as encryption which is indistinguishable from *arbitrary* distributions $\mathcal{C}$.

3.1   Construction: Stateful Universal Steganographic Secrecy

We now show a stateful protocol which is universally steganographically secret. This protocol is not new and can be found (up to small differences) in [1], for instance. The novel aspect is the proof of steganographic secrecy within our model.

Assume Alice and Bob share a channel and let $\mathcal{C}$ denote the channel distribution. We let $M(h)$ be an oracle which draws blocks (one at a time) from the conditional distribution $\mathcal{C}^b_h$. Let $F_K(\cdot, \cdot)$ denote a pseudorandom function family indexed by $k = |K|$ key bits which maps a pair consisting of a d-bit number and b timestamped bits to the set $\{0, 1\}$, i.e., $F : \{0, 1\}^k \times \{0, 1\}^d \times (\{0, 1\}, t)^b \to \{0, 1\}$. Let Alice and Bob share a secret key $K \in \{0, 1\}^k$ and a synchronized d bit counter N. Let Enc and Dec be the encoding and decoding procedures, respectively, for an error-correcting code which corrects up to a $\mu(k)$ fraction of errors (where $\mu(k)$ is defined below) and which has stretch function $\ell(\cdot)$. The following procedures allow Alice and Bob to encode and decode a hiddentext and to send it via their shared channel.

Construction 1. (Steganographic Secrecy)

Procedure S1.Encode:
Input: key K, hiddentext m', history h
Let $m = Enc(m')$
Parse m as $m^1_1 || m^1_2 || \cdots || m^1_l$
for $i = 1 \ldots l$ do
    $c_i \leftarrow RS^{M(h), F_K(N, \cdot)}(m_i, |K|)$
    set $h = h || c_i$
    increment N
Output: $c_1 || c_2 || \ldots || c_l$

Procedure S1.Decode:
Input: key K, Stegotext c
Parse c as $c^b_1 || c^b_2 || \ldots || c^b_l$
for $i = 1 \ldots l$ do
    set $m_i = F_K(N, c_i)$
    increment N
let $m = m_1 || m_2 || \cdots || m_l$
Output: $Dec(m)$

The idea behind this construction is simple. First, the encoding algorithm splits the hiddentext m into single bits. Then it draws blocks from the channel until it finds c_1 such that $F_K(N, c_1) = m_1$. After that it finds c_2 such that $F_K(N +$

$1, c_2) = m_2$, and c_3, and so on. This continues until all bits of m have been encoded. The decoding algorithm simply applies $F_K(N + i - 1, \cdot)$ to all the c_i's.

Note that the encoding procedure actually has a small probability of failure per bit. In particular, RS might not return a c_i such that $F_K(N + i - 1, c_i) = m_i$, because RS "gives up" after $|K|$ timesteps. The probability of failure is dependent on the sampling oracle M and the evaluation oracle F.

Lemma 1. *If* C_h^b *has minimum entropy at least* H *and* $|K| = k$, *the probability of failure of RS is bounded above by* $fail(k, C_h^b) + \mathbf{InSec}_F^{prf}(O(kl), kl)$, *where*

$$fail(k, C_h^b) \leq \sum_{j=0}^{2^H} \frac{\binom{2^H}{j}}{2^{2^H}} \left(\frac{j}{2^H}\right)^k$$

Proof. See the full version of the paper [6]. ∎

It can be verified that $fail(k, C_h^b)$ is exponentially small in $|K|$ and H. We will choose our error-correcting code to correct a $\mu(k) = (1 + \epsilon) * fail(k, U(1))$ fraction of errors (note that $U(1)$ has minimum entropy 1). A Chernoff bound will then suffice to show that the probability of decoding failure is exponentially small in l. For a channel C_h^b with minimum entropy 1, we have $fail(k, C_h^b) \approx \frac{1}{4}$. Consequently, we will at worst require a code with a stretch function $\ell(n)$ approximately $2n$. We will assume for simplicity that the running times of Enc and Dec are linear.

Theorem 1. *Let* $k = |K|$. *For any* $l \leq 2^d$:

$$\mathbf{InSec}_{S1,C}^{ss}(t, q, l) \leq \mathbf{InSec}_F^{prf}(t + O(k\ell(l)), k\ell(l))$$

Proof. For any warden, W, running in time t and making q queries totaling l bits, we construct a corresponding PRF adversary A, where

$$\mathbf{Adv}_{S1,C}^{ss}(W) = \mathbf{Adv}_F^{prf}(A)$$

The running time of A is the running time of warden W plus the time of rejection sampling (RS): $O(k\ell(l))$ in the worst case. The number of calls to the sampling oracle, $M(h)$, is at most $k\ell(l)$.

A^f simply runs W, emulating the encoding procedure $S1.\text{Encode}$ using the function oracle f in place of $F_K(\cdot, \cdot)$. Note that when f is a uniformly chosen random function, the output of $RS^{M(h),f}(\cdot, k)$ is distributed identically to the channel distribution C_h^b. Similarly, when f is chosen from $F_K(\cdot, \cdot)$, the output of $RS^{M(h),f}(\cdot, k)$ is distributed identically to the output of Construction 1, by the definition of the construction. So the advantage is:

$$\mathbf{Adv}_F^{prf}(A) = \left| \Pr_{K \leftarrow U(k), r \leftarrow \{0,1\}^*} [A_r^{F_K(\cdot,\cdot)} = 1] - \Pr_{g, r \leftarrow \{0,1\}^*} [A_r^g = 1] \right|$$

$$= \left| \Pr_{K,r,M,SE} [W_r^{M,SE(K,\cdot,\cdot)} = 1] - \Pr_{r,M,O} [W_r^{M,O(\cdot,\cdot)} = 1] \right|$$

$$= \mathbf{Adv}_{S1,C}^{ss}(W)$$

And the theorem follows by the definition of insecurity. ∎

Corollary 1. *If $F_K(\cdot, \cdot)$ is pseudorandom then construction 1 is universally steganographically secret against chosen-hiddentext attacks.*

3.2   Stateless Universal Steganographic Secrecy

The following protocol also satisfies our definition for universal steganographic secrecy. The proofs are omitted due to space considerations, but are similar to the proof of Theorem 1 in the previous section, and can be found in the full version of the paper Once again, this protocol (up to small differences) is not new and can be found in [1]; an information theoretic version of the protocol can also be found in [3].

We will say that a function $f : D \to R$ is an *unbiased function on a distribution* C if for all $r \in R, h$, $\Pr_{d \leftarrow C_h^b}[f(d) = r] = \frac{1}{|R|}$. Let $f : (\{0,1\}, t)^b \to \{0,1\}$ be a public function which is unbiased on C. Let $E_K(\cdot, \cdot)$ and $D_K(\cdot)$ denote the encryption and decryption functions for a cryptosystem which is indistinguishable from random bits under chosen plaintext attack (i.e., IND\$-CPA) [12]. Suppose Alice and Bob share a key $K \in \{0,1\}^k$. The following procedures allow encoding and decoding of messages in a manner which is steganographically secret under chosen hiddentext attack for the channel distribution C.

Construction 2. (Stateless Steganographic Secrecy)

Procedure S2.Encode:
Input: key K, hiddentext m, history h
Let $s = E_K(m)$
Parse s as $s_1^1 || s_2^1 || \cdots || s_l^1$
for $i = 1 \ldots l$ do
$\quad c_i \leftarrow RS^{M(h), f(\cdot)}(s_i, |K|)$
$\quad$ set $h = h || c_i$
Output: $c_1 || c_2 || \ldots || c_l$

Procedure S2.Decode:
Input: key K, Stegotext c
Parse c as $c_1^b || c_2^b || \cdots || c_l^b$
for $i = 1 \ldots l$ do
$\quad$ set $s_i = f(c_i)$
$\quad$ set $s = s_1 || s_2 || \cdots || s_l$.
Output: $D(K, s)$

Note that the execution time for SD is essentially linear in l and the execution time for SE is an expected $O(l)$. Also note that our assumption of an unbiased function implies that an error correcting code as in construction 1 is unnecessary: the probability of failure here is $l/2^{|K|}$. Clearly, the assumption of an unbiased f is quite strong. In practice, perfect unbiasedness may be hard to meet and stronger assumptions may be required. For example, when f is a pseudorandom function, under the assumption of a large minimum entropy in the channel, f will be nearly unbiased.

Lemma 2. *For any warden $W \in \mathcal{W}(t, q, l)$, we can construct a corresponding adversary A where*

$$\mathbf{Adv}_{S2,C}^{ss}(W) = \mathbf{Adv}_E^{cpa}(A) .$$

The running time of A is at most $t + O(kl)$, and A makes q encryption queries for a total of l bits of plaintext.

Theorem 2. $\mathbf{InSec}_{S2,C}^{ss}(t, q, l) \leq \mathbf{InSec}_E^{cpa}(t + O(kl), q, l).$

Generalization. The assumption that the balanced function, f, is unbiased can be weakened to the assumption of an ϵ-biased function where the probability of any value is within ϵ of uniform. The same proofs work with the insecurity increased by at most ϵ (however, error correcting codes might be necessary in this case).

A few easy corollaries follow from Theorem 2. If E is indistinguishable from random bits under chosen plaintext attack then construction 2 is SS-CHA-$\mathcal{C}$ secure. Additionally, if E is replaced by a public key cryptosystem which is indistinguishable from random bits under chosen plaintext attack, then construction 2 is a public key stegosystem which is steganographically secret under chosen hiddentext attack (under an appropriate generalization of our definitions to a public-key scenario).

4   Robust Steganography

4.1   Definitions for Robust Steganography

Robust steganography will be modelled as a game between Alice and Ward in which Ward is allowed to make some alterations to Alice's messages. Alice wins if she can pass a message with high probability, even when Ward alters her message. For example, if Alice passes a single bit per channel message and Ward is unable to change the bit with probability at least $\frac{1}{2}$, Alice can use error correcting codes to reliably transmit her message. It will be important to state the limitations we impose on Ward, since otherwise he can replace all messages with a new draw from the channel distribution, effectively destroying any hidden information. In this section we give a formal definition of robust steganography with respect to a limited adversary.

We will model Ward's power as defined by a relation R which is constrained to not corrupt the channel too much. This general notion of constraint is sufficient to include many simpler notions such as (for example) "only alter at most 1% of the bits".

Let $\mathcal{D}$ be a finite distribution with support X and let R be a relation between the set X and the set Y such that for every $x \in X$, there exists a $y \in Y$ where $(x, y) \in R$. Consider a single-player game of chance which is played as follows:

1. The player draws x according to $\mathcal{D}$.
2. The player chooses an arbitrary y such that $(x, y) \in R$.
3. The player makes an independent draw x' from $\mathcal{D}$.

The player wins if $(x', y) \in R$. Define the *obfuscation probability* of R for $\mathcal{D}$ by

$$\mathcal{O}(R, \mathcal{D}) = \max_y \sum_{(x', y) \in R} \Pr_{\mathcal{D}}[x'] \ .$$

This function represents an upper bound on the player's winning probability. In particular, for any y the player chooses in step 2, $\mathcal{O}(R, \mathcal{D})$ bounds the probability $\sum_{(x', y) \in R} \Pr_{\mathcal{D}}[x']$ of winning. Note that the $\log_2 \mathcal{O}(R, \mathcal{D})$ gives the minimum amount of conditional information retained about draws from $\mathcal{D}$ when they are substituted arbitrarily amongst possibilities which satisfy R. The obfuscation probability is therefore a worst-case conditional entropy (just as minimum entropy is a worst-case entropy), except that logarithms have been removed.

Now let R be an efficiently computable relation on blocks and let $R(x) = \{y : (x, y) \in R\}$. We say that the pair $(R, \mathcal{C}_h^b)$ is δ-admissible if $\mathcal{O}(R, \mathcal{C}_h^b) \leq \delta$ and a pair $(R, \mathcal{C})$ is δ-admissible if $\forall h$ $(R, \mathcal{C}_h^b)$ is δ-admissible. An R-bounded active warden W can be thought of as an adversary which plays the following game against a stegosystem $S = (SE, SD)$:

1. W is given oracle access to the channel distribution $\mathcal{C}$ and makes as many draws as it likes.
2. W is given oracle access to $SE(K, \cdot, \cdot)$, and makes at most q queries totaling at most l_1 bits to SE.
3. W presents an arbitrary message $m \in \{0, 1\}^{l_2}$ and history h.
4. W is then given a sequence of blocks $c = c_1 || c_2 || \ldots || c_u$ from the support of $\mathcal{C}_h^{(u*b)}$, and returns a sequence $c' = c_1' || c_2' || \ldots || c_u'$ where $c_i' \in R(c_i)$ for each $1 \leq i \leq u$. Here u is the number of blocks of stegotext output by $SE(K, m, h)$.

Define the success of W against S by

$$\mathbf{Succ}_S^R(W) = \Pr_{K \leftarrow U(k), r \leftarrow \{0,1\}^*, o \leftarrow \{0,1\}^*} [SD_o(K, W_r(SE_o(K, m, h)), h) \neq m]$$

Here, r and o are the random bits used by Ward and the protocol, respectively. Define the failure rate of S by

$$\mathbf{Fail}_S^R(t, q, l) = \max_{W \in \mathcal{W}(R, t, q, l)} \left\{ \mathbf{Succ}_S^R(W) \right\},$$

where $\mathcal{W}(R, t, q, l)$ denotes the set of all R-bounded active wardens that submit at most q queries of total length at most l_1, produce a plaintext of length at most $l_2 = l - l_1$ and run in time at most t.

Definition 4. (Robust Steganography) A stegosystem $S = (SE, SD)$ is called $(t, q, l, \epsilon, \delta)$ *steganographically robust against R-bounded adversaries* for the distribution $\mathcal{C}$ (denoted $(t, q, l, \epsilon, \delta)$-SR-CHA-$(\mathcal{C}, R)$) if the following conditions hold:

- (Secrecy): S is (t, q, l, ϵ)-SS-CHA-$\mathcal{C}$.
- (Robustness): $\mathbf{Fail}_S^R(t, q, l) \leq \delta$.

A stegosystem is called $(t, q, l, \epsilon, \delta)$ *steganographically robust* (SR-CHA) if it is $(t, q, l, \epsilon, \delta)$-SR-CHA-$(\mathcal{C}, R)$ for every δ-admissible pair $(\mathcal{C}, R)$.

Definition 5. (Universal Robust Steganography) A sequence of stegosystems $\{S_k\}_{k \in \mathbb{N}}$ is called *universally steganographically robust* if it is universally steganographically secret and there exists a polynomial $q(\cdot)$ and a constant $\delta \in [0, \frac{1}{2})$ such that for every PPTM W, every δ-admissible $(R, \mathcal{C})$, and all sufficiently large k, $\mathbf{Succ}_{S_k}^R(W) < 1/q(k)$.

4.2   Universally Robust Stegosystem

In this section we give a stegosystem which is Steganographically robust against any bounding relation R, under a slightly modified assumption on the channel oracles, and assuming that Alice and Bob know some efficiently evaluable, δ-admissible relation R' such that R' is a superset of R. For several reasons, this stegosystem appears impractical but it serves as a proof that robust steganography is possible for any admissible relation.

Suppose that the channel distribution C is efficiently sampleable, that is, there is an efficient algorithm M which, given a uniformly chosen string $s \in \{0,1\}^m$ and history h produces a block distributed according to C_h^b (or statistically close to C_h^b). We will assume that Alice, Bob, and Ward all have access to this algorithm. Furthermore, we assume Alice and Bob share a key K to a pseudorandom function; and have a synchronized counter N. Let n be a robustness parameter.

Construction 3. (Universally Robust Steganography)

Procedure S3.Encode:	**Procedure S3.Decode:**
Input: K, m, h	**Input:** key K, stegotext c, history h
Parse m as $m_1^1 \|\| m_2^1 \|\| \cdots \|\| m_l^1$	Parse c as $c_1^b \|\| c_2^b \|\| \ldots \|\| c_{ln}^b$
for $i = 1 \ldots l$ do	for $i = 1 \ldots l$ do
$\quad$ for $j = 1 \ldots n$ do	$\quad$ set $h_0 = h_1 = h$
$\quad\quad$ set $c_{i,j} = M(F_K(N, m_i), h)$	$\quad$ for $j = 1 \ldots n$ do
$\quad\quad$ increment N	$\quad\quad$ for $\sigma \in \{0,1\}$ do
$\quad\quad$ set $h = h \|\| c_{i,j}$	$\quad\quad\quad$ set $m_\sigma = M(F_K(N, \sigma), h_\sigma)$
Output: $c_{1,1} \|\| c_{1,2} \|\| \ldots \|\| c_{l,n}$	$\quad\quad\quad$ set $h_\sigma = h_\sigma \|\| m_\sigma$
	$\quad\quad$ increment N
	$\quad$ if $(\forall j.(h_{0,j}, c_{i,j}) \in R')$
	$\quad\quad$ then $p_i = 0$; else $p_i = 1$
	$\quad$ set $h = h_{p_i}$
	set $p = p_1 \|\| p_2 \|\| \cdots \|\| p_l$.
	Output: p

Suppose that instead of sharing a key to a pseudorandom function F, Alice and Bob shared two secret blocks b_0, b_1 drawn independently from C_h^b. Then Alice could send Bob the message bit σ by sending block b_σ, and Bob could recover σ by checking to see if the block he received was related (by R') to b_0 or b_1. Since the adversary is R bounded and (C, R') is δ-admissible, the probability of a decoding error — caused either by the adversary, or by accidental draw of b_0, b_1 — would be at most δ. Intuitively, Construction 3 simply extends this notion to multiple bits by replacing the b_0, b_1 by draws from $M(\cdot)$ with shared pseudorandom inputs; and reduces the probability of decoding error to δ^n by encoding each hiddentext bit n times.

Lemma 3. $\mathbf{InSec}_{S3,C}^{ss}(t, q, l) \leq \mathbf{InSec}_F^{prf}(t + O(nl), nl)$.

Lemma 4. $\mathbf{Fail}_{S3}^{R}(t, q, l_1, l_2) \leq \mathbf{InSec}_F^{prf}(t + O(nl), nl) + l_2 \delta^n$.

Proof. Let W be an active R-bounded (t, q, l_1, l_2) warden. We construct a PRF adversary A which runs in time $t + O(nl)$, makes at most nl PRF queries, and satisfies $\mathbf{Adv}_F^{\mathrm{prf}}(A) \geq \mathbf{Succ}_S^R(W) - l_2 \delta^n$. A^f works by running W, using its function oracle f in place of $F_K(\cdot, \cdot)$ to emulate Construction 3 in responding to the queries of W. Let m, c' be the hiddentext and the stegotext sequence returned by W, respectively. Then A^f returns 1 iff $SD(K, c', h) \neq m$. Consider the following two cases for f:

- f is chosen uniformly from all appropriate functions. Then, for each i, j, the stegotexts $c_{i,j} = M(f(N_i + j, p_i), h_{i,j})$ are distributed independently according to $C_{h_{i,j}}^b$. Consider the sequence of "alternative stegotexts" $d_{i,j} = M(f(N_i + j, 1 - p_i), h_{i,j})$; each of these is also distributed independently according to $C_{h_{i,j}}^b$; and since W is never given access to the $d_{i,j}$, the $c_{i,j}'$ are independent of the $d_{i,j}$. Now SD will fail (causing A^f to output 1) only if the event $\forall j.(d_{i,j}, c_{i,j}') \in R'$ occurs for some i. Because the d_i are independent of the actions of W, and because (C, R') is δ-admissible, each event $(d_{i,j}, c_{i,j}') \in R'$ happens independently with probability at most δ. So for any fixed i, the probability of failure is at most δ^n. The union bound then gives

$$\Pr_{f \leftarrow U(b,n)}[A^f = 1] \leq l_2 \delta^n.$$

- f is chosen uniformly from $F_K(\cdot, \cdot)$. Then A^F outputs 1 exactly when W succeeds against S, by the definition of S:

$$\Pr_{K \leftarrow U(k), r \leftarrow \{0,1\}^*}[A_r^{F_K} = 1] = \mathbf{Succ}_S^R(W).$$

Taking the difference of these probabilities, we get:

$$\mathbf{Adv}_F^{\mathrm{prf}}(A) = \Pr_{K \leftarrow U(k), r \leftarrow \{0,1\}^*}[A_r^{F_K} = 1] - \Pr_{f \leftarrow U(b,n), r \leftarrow \{0,1\}^*}[A_r^f = 1]$$
$$\geq \mathbf{Succ}_S^R(W) - l_2 \delta^n.$$

Theorem 3. *If F is $(t + O(nl), nl, \epsilon)$-pseudorandom then Construction 3 is $(t, l_1, l_2, \epsilon, \epsilon + l_2 \delta^n)$-SR-CHA.*

4.3 Robust Steganography for Text-Like Channels

We provide a protocol which is steganographically robust against R-bounded adversaries for a *particular* class of admissible relations R on *particular* channels. For some channel distributions this class of relations may provide an accurate model of the limitations of real wardens; in particular it seems reasonable to suppose that a predominantly text-based channel might limit the power of an active warden by such a relation.

A text based channel (such as email) might consist of "documents" built out of many bits with the same timestamp (= sending time). Fix a symbol alphabet $\Sigma = \{0, 1\}^s$, and represent every document d in the support of C by a string of symbols $d = d_0 d_1 \cdots d_l$ and a single timestamp. The n-shingling of d, denoted by $ss_n(d)$ is the set $\{d_0 d_1 \cdots d_{n-1}, d_1 d_2 \cdots d_n, d_2 d_3 \cdots d_{n+1}, \ldots, d_{l-n} d_{l-n+1} \cdots d_l\}$. Define the family of relations R_δ^n for $0 \le \delta \le 1$ by

$$(x, y) \in R_\delta^n \Leftrightarrow \frac{|ss_n(x) \cap ss_n(y)|}{|ss_n(x) \cup ss_n(y)|} \ge 1 - \delta.$$

Let $F_K(\cdot, \cdot)$ denote a pseudorandom function family indexed by $|K|$ bits which maps n-shingles to the set $\{0, 1\}^p$. Let E_K be a length-preserving, stateful (rather than randomized), IND\$-CPA encryption scheme which works by producing a pseudorandom stream s and outputs $E_K(m) = s \oplus m$, (such a scheme can be produced using a pseudorandom function in (stateful) CTR mode, for example) and let D be the corresponding decryption function. Let Enc and Dec be the encoding and decoding algorithms for an error-correcting code which can correct up to a $\delta + \xi$ fraction of symbol errors; and let $\ell(\cdot)$ be the stretch function of this code. We will assume Enc and Dec work in linear time for simplicity. Define the function $\max_f(d)$ as the shingle in d which maximizes f, that is

$$\max_f(d) = \text{argmax}_{s \in ss_n(d)} \{f(s)\} .$$

Let $g : \{0, 1\}^* \to \{0, 1\}$ be unbiased for the distribution on shingles induced by $\max_{F_K}(C)$. Let Alice and Bob share secret keys K_f, K_e. The following procedures allow Alice and Bob to robustly encode and decode a message.

Construction 4. (Robust Steganographic Protocol)

Procedure S4.Encode:	**Procedure S4.Decode:**
Input: m, h, keys K_f, K_e	**Input:** stegotext c, keys K_f, K_e
Let $p = E_{K_e}(Enc(m))$	Parse c as $c_1^b \| c_2^b \| \ldots \| c_l^b$
Parse p as $p_1^1 \| p_2^1 \| \cdots \| p_l^1$	for $i = 1 \ldots l$ do
for $i = 1 \ldots l$ do	set $s_i = \max_{F_{K_f}(\cdot)}(c_i)$
set $c_i = RS^{M(h), \text{gomax}_{F_f}}(p_i)$	set $p_i = g(s_i)$
set $h = h \| c_i$	set $p = p_1 \| p_2 \| \cdots \| p_l$.
Output: $c_1 \| c_2 \| \ldots \| c_l$	**Output:** $Dec(D_{K_e}(p))$

Note that it is important that encryption and bit errors commute here which holds for only some encryption techniques.

In the following, Let ℓ_q be the maximum size of q encoded strings with total length l_1 plus $\ell(l_2)$.

Lemma 5. $\text{InSec}_{S4}^{ss}(t, q, l) \le \text{InSec}_E^{cpa}(t + O(k\ell_q), q, \ell_q)$.

Lemma 6. $\text{Fail}_{S4}^{R_\delta^n}(t, q, l_1, l_2) \le 2e^{-\xi^2 \ell(l_2)/2} + \text{InSec}_F^{prf}(t + O(k\ell_q), k\ell_q)$.

Theorem 4. *If F is $(t + O(k\ell_q), k\ell_q, \epsilon)$-pseudorandom and E is $(t + \ell_q, q, \ell_q, \mu)$ - IND\$-CPA, then Construction 4 is $(t, l_1, l_2, \epsilon + \mu, 2e^{-\xi^2 \ell(l_2)/2} + \epsilon)$ - SR-CHA against R_δ^n-bounded adversaries.*

5   Discussion

5.1   Alternative Security Conditions

There are several conceivable alternatives to our security conditions; we will briefly examine these alternatives and justify our choices.

Find-Then-Guess: This is the standard model in which an attacker submits two plaintexts p_0 and p_1, receives $SE(p_b)$, and attempts to guess b. Security in our attack model implies find-then-guess security; moreover the essence of steganographic secrecy is not merely the inability to distinguish between messages (as in the find-then-guess model) but the inability to detect a message.

Fixed History: In this model the adversary may not submit alternate histories to the encryption model. Security under a chosen-history attacks implies security against a fixed-history attacks. This notion may be of interest however, especially because in many situations a chosen-history attack may not be physically realizable. Our attacks can be considered chosen-history attacks.

Integrity of Hiddentexts. Intuitively, Integrity of Hiddentexts requires that an active warden is unable to create a sequence of covertexts which decodes to a valid, new hiddentext. Suppose we amend the description of a stego system to allow the decoding algorithm to output the "fail" symbol $\perp$. Then suppose we give the adversary oracle access to SE and allow the adversary to make at most q queries $p_0, \ldots, p_q$ to $SE(K, \cdot, \cdot)$ totaling l bits. The adversary then produces a sequence of covertexts $c = c_1 || \ldots || c_m$. Denote the advantage of A against S by

$$\mathbf{Adv}_{S,C}^{\text{int}}(A) = \Pr\left[SD(K, c, h) \neq \perp \wedge \forall i. SD(K, c, h) \neq p_i\right],$$

and denote the integrity failure of a stegosystem by

$$\mathbf{Fail}_{S,C}^{int}(t, q, l) = \max_{A \in \mathcal{A}(t,q,l)} \left\{\mathbf{Adv}_{S,C}^{\text{int}}(A)\right\}.$$

A stegosystem has (t, q, l, ϵ) integrity of hiddentexts if $\mathbf{Fail}_{S,C}^{\text{int}}(t, q, l) \leq \epsilon$.

Note that in practice this notion by itself is too weak because it allows the possibility for the warden to disrupt the communication between Alice and Bob. Finally, we note that if the symmetric encryption scheme E is INT-CTXT secure as defined by Bellare and Namprempre [2], then construction 2 also provides integrity of hiddentexts.

5.2   Complexity Theoretic Ramifications

Construction 1 gives a stegosystem which is steganographically secret for any channel distribution C which has minimum entropy greater than 1, assuming the existence of a pseudorandom function family. Goldreich *et al* [4] show how to construct a pseudorandom function from a pseudorandom generator, which in turn can be constructed from any one-way function, as demonstrated by Hastad *et al* [5]. Thus in an asymptotic sense, our constructions show that one-way functions are sufficient for steganography. Conversely, it is easy to see that a stegosystem which is steganographically secret for some C is a secure weak private key encryption protocol in the sense of Impagliazzo and Luby [7]; and they prove that the existence of such a protocol implies the existence of a one-way function. Thus the existence of secure steganography is equivalent to the existence of one-way functions.

Acknowledgments

We are greatful to Manuel Blum for his suggestions and his unconditional encouragment. We also thank Steven Rudich and the anonymous CRYPTO reviewers for helpful discussions and comments. This work was partially supported by the National Science Foundation (NSF) grants CCR-0122581 and CCR-0085982 (The Aladdin Center). Nicholas Hopper is also partially supported by an NSF graduate research fellowship.

References

1. Ross J. Anderson and Fabien A. P. Petitcolas. *On The Limits of Steganography*. IEEE Journal of Selected Areas in Communications, 16(4). May 1998.
2. Mihir Bellare and Chanathip Namprempre. *Authenticated Encryption: Relations among notions and analysis of the generic composition paradigm*. In: *Advances in Cryptology – Asiacrypt '00*. December 2000.
3. C. Cachin. *An Information-Theoretic Model for Steganography*. In: *Information Hiding – Second International Workshop, Preproceedings*. April 1998.
4. Oded Goldreich, Shafi Goldwasser and Silvio Micali. *How to Construct Random Functions*. Journal of the ACM, 33(4):792 – 807, 1986.
5. Johan Hastad, Russell Impagliazzo, Leonid A. Levin, and Michael Luby. *A pseudorandom generator from any one-way function*. SIAM Journal on Computing, 28(4):1364–1396, 1999.
6. Nicholas J. Hopper, John Langford, and Luis von Ahn. *Provably Secure Steganography*. CMU Tech Report CMU-CS-TR-02-149, 2002.
7. Russell Impagliazzo and Michael Luby. *One-way Functions are Essential for Complexity Based Cryptography*. In: 30th FOCS, November 1989.
8. D. Kahn. *The Code Breakers*. Macmillan 1967.
9. Stefan Katzenbeisser and Fabien A. P. Petitcolas. *Information hiding techniques for steganography and digital watermarking*. Artech House Books, 1999.
10. T. Mittelholzer. *An Information-Theoretic Approach to Steganography and Watermarking* In: *Information Hiding – Third International Workshop*. 2000.
11. J. A. O'Sullivan, P. Moulin, and J. M. Ettinger *Information theoretic analysis of Steganography*. In: *Proceedings ISIT '98*. 1998.
12. Phillip Rogaway, Mihir Bellare, John Black and Ted Krovetz. *OCB: A Block-Cipher Mode of Operation for Efficient Authenticated Encryption*. In: *Proceedings of the Eight ACM Conference on Computer and Communications Security (CCS-8)*. November 2001.
13. G.J. Simmons. *The Prisoner's Problem and the Subliminal Channel*. In: *Proceedings of CRYPTO '83*. 1984.
14. J Zollner, H.Federrath, H.Klimant, A.Pftizmann, R. Piotraschke, A.Westfield, G.Wicke, G.Wolf. *Modeling the security of steganographic systems*. In: *Information Hiding – Second International Workshop, Preproceedings*. April 1998.

Flaws in Applying Proof Methodologies to Signature Schemes

Jacques Stern[1,*], David Pointcheval[1], John Malone-Lee[2], and Nigel P. Smart[2]

[1] Dépt d'Informatique, ENS – CNRS, 45 rue d'Ulm, 75230 Paris Cedex 05, France
{Jacques.Stern,David.Pointcheval}@ens.fr
http://www.di.ens.fr/~{stern,pointche}
[2] Computer Science Dept, Woodland Road, University of Bristol, BS8 1UB, UK
{malone,nigel}@cs.bris.ac.uk
http://www.cs.bris.ac.uk/~{malone,nigel}

Abstract. Methods from *provable security*, developed over the last twenty years, have been recently extensively used to support emerging standards. However, the fact that proofs also need time to be validated through public discussion was somehow overlooked. This became clear when Shoup found that there was a gap in the widely believed security proof of OAEP against adaptive chosen-ciphertext attacks. We give more examples, showing that provable security is more subtle than it at first appears. Our examples are in the area of signature schemes: one is related to the security proof of ESIGN and the other two to the security proof of ECDSA. We found that the ESIGN proof does not hold in the usual model of security, but in a more restricted one. Concerning ECDSA, both examples are based on the concept of duplication: one shows how to manufacture ECDSA keys that allow for two distinct messages with identical signatures, a *duplicate signature*; the other shows that from any message-signature pair, one can derive a second signature of the same message, the *malleability*. The security proof provided by Brown [7] does not account for our first example while it surprisingly rules out malleability, thus offering a proof of a property, non-malleability, that the actual scheme does not possess.

1 Introduction

In the last twenty years *provable security* has dramatically developed, as a means to validate the design of cryptographic schemes. Today, emerging standards only receive widespread acceptance if they are supported by some form of provable argument. Of course, cryptography ultimately relies on the P vs. $\mathcal{NP}$ question and actual proofs are out of reach. However, various security models and assumptions allow us to interpret newly proposed schemes in terms of related mathematical results, so as to gain confidence that their underlying design is not flawed. There is however a risk that should not be underestimated: the use of provable security is more subtle than it appears, and flaws in security proofs themselves might have

---

* The first and last examples in this paper are based on the result of an evaluation requested by the Japanese Cryptrec program and performed by this author.

M. Yung (Ed.): CRYPTO 2002, LNCS 2442, pp. 93–110, 2002.
© Springer-Verlag Berlin Heidelberg 2002

a devastating effect on the trustworthiness of cryptography. By flaws, we do not mean plain mathematical errors but rather ambiguities or misconceptions in the security model. The first such example appeared recently, when Victor Shoup noted in [29] that there was a gap in the widely believed security proof of OAEP against adaptive chosen-ciphertext attacks. By means of a nice counter-example in a relativized model of computation, he showed that, presumably, OAEP could not be proven secure from the one-wayness of the underlying trapdoor permutation. A closer look at the literature, notably [4, 2], showed that the security proof was actually valid in a weaker security model, namely against indifferent chosen-ciphertext attacks (IND-CCA1), also called lunchtime attacks [18], and *not* in the full (IND-CCA2) adaptive setting [24]. This came as a shock, even though Fujisaki, Okamoto, Pointcheval and Stern [12] were quickly able to establish that the security of RSA–OAEP could actually be proven under the RSA assumption alone, in the random oracle model. Since the more general result could not hold, a different argument based on specific properties of the RSA function had to be used.

Goldwasser, Micali and Rivest [14] introduced the notion of *existential forgery against adaptive chosen-message attacks* for public key signature schemes. This notion has become the *de facto* security definition for digital signature algorithms, against which all new signature algorithms are measured. The definition involves a game in which the adversary is given a target user's public key and is asked to produce a valid message/signature pair with respect to this public key. The adversary is given access to an oracle which will produce signatures on messages of his choice. However, the above definition does not directly deal with the most important property of a digital signature, namely *non-repudiation*: the signer should be unable to repudiate his signature. One should not that an adversary against the non-repudiation property of a signature scheme would be the legitimate signer himself. Hence, such an adversary has access to the private key, and may even control the key generation process.

The present paper gives further examples of flaws in security proofs, related to signature schemes. Two of them stem from a subtle point that has apparently been somehow overlooked: in non deterministic signature schemes, several signatures may correspond to a given message. Accordingly, the security model should unambiguously decide whether an adaptive attacker is allowed to query several signatures of the same message. Similarly, it should make clear whether obtaining a second signature of a given message, different from a previously obtained signature of the same message, is a forgery or not, and namely an existential forgery.

The first example that we give is related to the security proof offered in [22] for the ESIGN signature scheme. Crosschecking the proof, with the above observations in mind, it can be seen that it *implicitly* assumes that the attacker is not allowed to query the same message twice. Thus, the security proof does not provide security against existential forgeries under adaptive chosen-message attacks. It only applies to a more restricted class, which may be termed *single-occurrence* chosen-message attacks.

The two other examples are related to the elliptic curve digital signature algorithm ECDSA [1]. In [7], Brown uses the so-called *generic group model* to prove the security of the generic DSA, a natural analog of DSA and ECDSA in this setting. This result is viewed as supporting the security of the actual ECDSA: in the generic model, ECDSA prevents existential forgeries under adaptive chosen-message attacks. But as already remarked, this security notion does not deal with the important non-repudiation property that signature schemes should guarantee. The obvious definition is that it should be hard for a legitimate signer to produce two messages which have the same signature with respect to the same public key. If a signature scheme did not have this property then a user could publish the signature on one message and then claim it was actually the signature on another. Such a signature we shall call a *duplicate signature*, since it is the signature on two messages. This shows an inadequacy between the classical security notions and the practical requirements. Furthermore, we show that with ECDSA a signer which controls the key generation process can easily manufacture duplicate signatures, without finding a collision in the hash function. Luckily, however, our construction of duplicate signatures means that, as soon as the signer reveals the second message, the signer's private key is revealed. Concerning the generic group model, which was the sole assumption on which relies the security result provided in [7], carefully crosschecking the proof, with the above observations in mind, we see that it actually prevents a forgery which creates a different signature to a previously obtained signature of the same message. Hence, the proof implies the scheme produces non-malleable signatures. Unfortunately, ECDSA does *not* withstand such forgeries. What goes wrong here is the adequacy of the model. The proof is correct but the underlying model is flawed, since it disallows production of malleable signatures.

Note that we have not broken any of the two schemes. In particular, there are some easy ways of revising ESIGN so that it satisfies the classical security notions (see e.g. [15]).

2   Digital Signature Schemes and Security Proofs

2.1   Formal Framework

In modern terms (see [14]), a digital signature scheme consists of three algorithms $(\mathcal{K}, \Sigma, V)$:

- A *key generation algorithm* $\mathcal{K}$, which, on input 1^k, where k is the security parameter, outputs a pair $(\mathsf{pk}, \mathsf{sk})$ of matching public and private keys. Algorithm $\mathcal{K}$ is probabilistic.
- A *signing algorithm* Σ, which receives a message m and the private key sk, and outputs a signature $\sigma = \Sigma_{\mathsf{sk}}(m)$. The signing algorithm might be probabilistic.
- A *verification algorithm* V which receives a candidate signature σ, a message m and a public key pk, and returns an answer $V_{\mathsf{pk}}(m, \sigma)$ as to whether or not σ is a valid signature of m with respect to pk. In general, the verification algorithm need not be probabilistic.

Attacks against signature schemes can be classified according to the goals of the adversary and to the resources that it can use. The goals are diverse:

- Disclosing the private key of the signer. This is the most drastic attack. It is termed *total break*.
- Constructing an efficient algorithm which is able to sign any message with significant probability of success. This is called *universal forgery*.
- Providing a single message/signature pair. This is called *existential forgery*.

In terms of resources, the setting can also vary. We focus on two specific attacks against signature schemes: the *no-message attacks* and the *known-message attacks*. In the first scenario, the attacker only knows the public key of the signer. In the second, the attacker has access to a list of valid message/signature pairs. Again, many sub-cases appear, depending on how the adversary gains knowledge. The strongest is the *adaptive chosen-message attack* (CMA), where the attacker can require the signer to sign any message of its choice, where the queries are based upon previously obtained answers. When signature generation is not deterministic, there may be several signatures corresponding to a given message. A slightly weaker security model, which we call *single-occurrence adaptive chosen-message attack* (SO-CMA), allows the adversary at most one signature query for each message. In other words the adversary cannot submit the same message twice for signature.

In known-message attacks, one should point out that existential forgery becomes the ability to forge a fresh message/signature pair that has not been obtained during the attack. Again there is a subtle point here, related to the context where several signatures may correspond to a given message. We actually adopt the stronger rule that the attacker needs to forge the signature of message, whose signature was not queried. The more liberal rule, which makes the attacker successful, when it outputs a second signature of a given message, different from a previously obtained signature of the same message, will be called *malleability*.

Conversely, the *non-repudiation* property means the impossibility to produce two messages with the same signature, which will be called a *duplicate signature*. However, one should note that the adversary for such a forgery is the signer himself, who may furthermore have control on the key generation process. Such a security notion is not covered by the usual notions, and should be studied independently.

2.2   The Random Oracle Model

Ideally, one would like to obtain provable security for a signature scheme, based on the sole assumption that some underlying computational problem is hard. Unfortunately, very few schemes are currently known that allow such a proof.

The next step is to hope for a proof in a non-standard computational model, as proposed by Bellare and Rogaway [3], following an earlier suggestion by Fiat and Shamir [11]. In this model, called the random oracle model, concrete objects such as hash functions are treated as random objects. This allows one to carry through the usual reduction arguments to the context of relativized computations, where the hash function is treated as an oracle returning a random answer for each new query. A reduction still uses an adversary as a subroutine of a program that contradicts a mathematical assumption, such as the assumption that RSA is one-way [25]. However, probabilities are taken not only over coin tosses but also over the random oracle.

Of course, the significance of proofs carried in the random oracle is debatable. Hash functions are deterministic and therefore do not return random answers. Along those lines, Canetti *et al.* [8] gave an example of a signature scheme which is secure in the random oracle model, but insecure under any instantiation of the random oracle. Despite these restrictions, the random oracle model has proved extremely useful to analyze many encryption and signature schemes. It clearly provides an overall guarantee that a scheme is not flawed, based on the intuition that an attacker would be forced to use the hash function in a non generic way.

2.3   Generic Algorithms

Recently, several authors have proposed to use yet another model to argue in favor of the security of cryptographic schemes, that could not be tackled by the random oracle model. This is the so-called *black-box* group model, or *generic* model [27, 7, 17]. In particular, paper [7] considered the security of ECDSA in this model. Generic algorithms had been earlier introduced by Nechaev and Shoup [19, 28] to encompass group algorithms that do not exploit any special property of the encodings of group elements other than the property that each group element is encoded by a unique string. Typically, algorithms like Pollard's ρ algorithm [23] fall under the scope of this formalism, while index-calculus methods do not.

We will now go into a bit more detail of proofs in this generic model, because in one of our examples, this model is the origin of the apparent paradox. More precisely, we will focus on groups which are isomorphic to $(\mathbb{Z}_q, +)$, where q is a prime. Such groups will be called *standard cyclic groups*. An encoding of a standard cyclic group Γ is an injective map from Γ into a set of bit-strings S. We give an example: consider a subgroup of prime order of the group of points of a non-singular elliptic curve E over a finite field $\mathbb{F}$. Given a generator $\mathbf{g}$ of E, an encoding is obtained by computing $\sigma(x) = x \cdot \mathbf{g}$, where $x \cdot \mathbf{g}$ denotes the scalar multiplication of $\mathbf{g}$ by the integer x and providing coordinates for $\sigma(x)$. Note that the encoding set appears much larger than the group size, but compact encodings using only one coordinate and a sign bit ± 1 exist and, for such encodings, the image of σ is included in the binary expansions of integers $< tq$ for some small integer t, provided that q is close enough to the size of the underlying field $\mathbb{F}$. This is exactly what is recommended for cryptographic applications [16, 9].

A *generic* algorithm $\mathcal{A}$ over a standard cyclic group Γ is a probabilistic algorithm that takes as input an *encoding list* $\mathcal{L} = \{\sigma(x_1), \ldots, \sigma(x_k)\}$, where each x_i is in Γ. While it executes, the algorithm may consult an oracle for further encodings. Oracle calls consist of triples $\{i, j, \epsilon\}$, where i and j are indices of the encoding list $\mathcal{L}$ and ϵ is $\pm$. The oracle returns the string $\sigma(x_i \pm x_j)$, according to the value of ϵ and this bit-string is appended to the list $\mathcal{L}$, unless it was already present. In other words, $\mathcal{A}$ cannot access an element of Γ directly but only through its name $\sigma(x)$ and the oracle provides names for the sum or difference of two elements addressed by their respective names. Note however that $\mathcal{A}$ may access the list $\mathcal{L}$ at any time. In many cases, $\mathcal{A}$ takes as input a pair $\{\sigma(1), \sigma(x)\}$. Probabilities related to such algorithms are computed with respect to the internal coin tosses of $\mathcal{A}$ as well as the random choices of σ and x.

In [7], the adversary is furthermore allowed to include additional elements z_i' in the encoding list $\mathcal{L}$, without calling the oracle. This is consistent with the fact that one may detect whether an element is in the group or not (e.g. whether the coordinates of a point satisfy the equation which defines the elliptic curve.) However, this definitely enlarges the class of generic algorithm, compared to [19, 28]. One can keep the number of additional elements smaller than twice the number of queries, since additional elements not appearing in a further query can be deleted and since each query involves at most two additional elements. Some useful results about the generic model are provided in Appendix A.1.

Again, from a methodological point of view, proofs in the generic model have to be handled with care. A specific group is not generic and specific encodings may further contradict genericity. If it happens, the exact meaning of a security proof may become highly questionable.

3   The Provable Security of ESIGN

3.1   Description of ESIGN

We follow [22], where a specification of ESIGN appears. The key generation algorithm of ESIGN chooses two large primes p, q of equal size k and computes the modulus $n = p^2 q$. The sizes of p, q are set in such a way that the binary length $|n|$ of n equals $3k$. Additionally, an exponent $e > 4$ prime to $\varphi(n)$ is chosen.

Signature generation is performed as follows, using a hash function $\mathcal{H}$, outputting strings of length $k - 1$.

1. Pick at random r in $\mathbb{Z}_{pq}^*$.
2. Convert $(0\|\mathcal{H}(m)\|0^{2k})$ into an integer y and compute $z = (y - r^e) \bmod n$.
3. Compute $w_0 = \lceil z/pq \rceil$ and $w_1 = w_0.pq - z$. If $w_1 \geq 2^{2k-1}$, return to step 1.
4. Set $u = w_0 \cdot (er^{e-1})^{-1} \bmod p$ and $s = r + upq$.
5. Output s as the signature of m.

The basic paradigm of ESIGN is that the arithmetical progression $r^e \bmod n + tpq$ consists of e-th powers of easily computed integers: one adjusts t so as to fall into a prescribed interval of length 2^{2k-1}.

Signature verification converts integer $s^e \bmod n$ into a bit string S of length $3k$ and checks that $[S]^k = 0\|\mathcal{H}(m)$, where $[S]^k$ denotes the k leading bits of S.

3.2 The Approximate e-th Root Problem

As noted in the previous section, RSA moduli of the from $p^2 q$ offer a very efficient way to solve the following problem, having knowledge of the factorization of n: given n and y in $\mathbb{Z}_n^\star$, find x such that $x^e \bmod n$ lies in the interval $[y, y + 2^{2k-1})$, where the bit-size of n is $3k$ and $[y, y + 2^{2k-1})$ denotes $\{u | y \leq u < y + 2^{2k-1}\}$.

It is conjectured that the above problem, called the approximate e-th root problem (AERP) in [22], is hard to solve. More precisely, denote by $\mathrm{Succ}^{\mathrm{aerp}}(\tau, k)$ the probability for any adversary $\mathcal{A}$ to find an element whose e-th power lies in the prescribed interval, within time τ. In symbols, it reads

$$\Pr[(n, e) \leftarrow \mathcal{K}(1^k), y \leftarrow \mathbb{Z}_n, x \leftarrow \mathcal{A}(n, e, y) : (x^e \bmod n) \in [y, y + 2^{2k-1})],$$

then, for large enough moduli, this probability is extremely small. Variants of the above can be considered, where the length of the interval is replaced by 2^{2k} or 2^{2k+1}.

Of course, the factorization of n allows to solve the AERP problem. It is unknown whether the converse is true, i.e. whether AERP and inverting RSA are computationally equivalent. Various attacks against AERP are known for $e = 2, 3$ (see [5, 30]). However, it is fair to say that there is no known attack against AERP when e is greater or equal than 4.

3.3 The Security Proof

For this signature scheme, one can prove, in the random oracle model, the following security result, where $T_{exp}(k)$ denotes the computing time of modular exponentiation modulo a $3k$-bit integer.

Theorem 1. Let $\mathcal{A}$ be a SO-CMA-adversary against the ESIGN signature scheme that produces an existential forgery, with success probability c, within time τ, making q_H queries to the hash function and q_s distinct requests to the signing oracle respectively. Then, AERP can be solved with probability ε', and within time τ', where

$$\varepsilon' \geq \frac{\varepsilon}{q_H} - (q_H + q_s) \times (3/4)^k - \frac{1}{2^{k-1}} \quad and \quad \tau' \leq \tau + k(q_s + q_H) \cdot T_{exp}(k).$$

Our method of proof is inspired by Shoup [29] and differs from [22]: we define a sequence of Game_1, Game_2, etc of modified attack games starting from the actual game Game_0. Each of the games operates on the same underlying probability space, only the rules defining how the view is computed differ from game to game.

Proof. (of Theorem 1). We consider an adversary $\mathcal{A}$ outputting an existential forgery (m, s), with probability ε, within time τ. We denote by q_H and q_s respectively the number of queries from the random oracle $\mathcal{H}$ and from the signing oracle. As explained, we start by playing the game coming from the actual adversary, and modify it step by step, until we reach a final game, whose success probability has an upper-bound obviously related to solving AERP.

Game_0: The key generation algorithm $\mathcal{K}(1^k)$ is run and produces a pair of keys $(\mathsf{pk}, \mathsf{sk})$. The adversary $\mathcal{A}$ is fed with pk and, querying the random oracle $\mathcal{H}$ and the signing oracle Σ_{sk}, it outputs a pair (m, s). We denote by S_0 the event that $V_{\mathsf{pk}}(m, s) = 1$. We use a similar notation S_i in any Game_i below. By definition, we have $\Pr[S_0] = \varepsilon$.

Game_1: In this game, we discard executions, which end up outputting a valid message/signature pair (m, s), such that m has not been queried from $\mathcal{H}$. This means restricting to the event AskH that m has been queried from $\mathcal{H}$. Unwinding the ESIGN format, we write: $s^e = 0 \| w \| \star \bmod n$. If AskH does not hold, $\mathcal{H}(m)$ is undefined, and the probability that $\mathcal{H}(m) = w$ holds is $1/2^{k-1}$: $\Pr[S_0 \,|\, \neg\mathsf{AskH}] \leq 2^{-k+1}$. Thus, $\Pr[S_1] = \Pr[S_0 \wedge \mathsf{AskH}] \geq \Pr[S_0] - 2^{-k+1}$.

Game_2: In this game, we choose at random an index κ between 1 and q_H. We let m_κ be the κ-th message queried to $\mathcal{H}$. We then discard executions which output a valid message/signature pair (m, s), such that $m \neq m_\kappa$. Since the additional random value κ is chosen independently of the execution of Game_1, $\Pr[S_2] = \Pr[S_1]/q_H$.

Game_3: In this game, we immediately abort if a signing query involves message m_κ. By the definition of existential forgery, this only eliminates executions outside S_2. Thus: $\Pr[S_3] = \Pr[S_2]$.

Game_4: We now simulate the random oracle $\mathcal{H}$, by maintaining an appropriate list, which we denote by $\mathsf{H\text{-}List}$. For any fresh query m, we pick at random $u \in \mathbb{Z}_n$ and compute $z = u^e \bmod n$, until the most significant bit of z is 0. We next parse z as $0 \| w \| \star$, where w is of length $k - 1$ and check whether $z - w.2^{2k}$ is less than 2^{2k-1}. If this is true, we store (m, u, w) in $\mathsf{H\text{-}List}$ and returns w as the answer to the oracle call. Otherwise we restart the simulation of the current query. However, we stop and abort the game after k trials. This game differs from the previous one if z remains undefined after k attempts: $|\Pr[S_4] - \Pr[S_3]| \leq (q_H + q_s) \times (3/4)^k$.

Game_5: We modify the simulation by replacing $\mathcal{H}(m_\kappa)$ by v, where v is a bit string of length $k - 1$, which serves as an additional input. The distribution of $\mathcal{H}$-outputs is unchanged: $\Pr[S_5] = \Pr[S_4]$.

Game_6: We finally simulate the signing oracle: for any m, whose signature is queried, we know that $m \neq m_\kappa$ cannot hold, since corresponding executions have been aborted. Thus $\mathsf{H\text{-}List}$ includes a triple (m, u, w), such that $u^e \bmod n$ has its k leading bits of the form $0\|\mathcal{H}(m)$. Accordingly, u provides a valid signature of m. Therefore, $\Pr[S_6] = \Pr[S_5]$.

Summing up the above inequalities, we obtain

$$\Pr[S_6] \geq \Pr[S_3] - (q_H + q_s) \times (3/4)^k \geq \frac{\varepsilon}{q_H} - (q_H + q_s) \times (3/4)^k - \frac{1}{2^{k-1}}.$$

When Game_6 terminates outputting a valid message/signature pair (m, s), we unwind the ESIGN format and get $s^e = (0 \| v \| \star) \bmod n$, with $v = \mathcal{H}(m)$. If S_6 holds, we know that $m = m_\kappa$ and $\mathcal{H}(m) = v$. This leads to an element whose e-th power lies in the interval $[v2^{2k}, v2^{2k} + 2^{2k})$, thus solving an instance of AERP.

We finally have: $\Pr[S_6] \leq \text{Succ}^{\text{aerp}}(\tau', k)$, where τ' denotes the running time of Game_6. This is the requested bound. Observe that τ' is the sum of the time for the original attack, plus the time required for simulations, which amounts to at most $k(q_s + q_H)$ modular exponentiations. We get $\tau' \leq \tau + k(q_s + q_H) \cdot T_{exp}(k)$.
□

3.4 Comments on the Security Model

We definitely had to use the SO-CMA model. If the adversary was allowed to submit the same message twice to the signing oracle, the simulation would fail at the second call, since there is a single signature available. Thus, contrarily to what is claimed in [22], the result only applies to single-occurrence adaptive chosen-message attacks. We do not know how to extend the proof to deal with the stronger CMA model.

4   Duplicates in ECDSA

Let us now turn to the ECDSA signature scheme, on which we give two more examples.

4.1   Description of ECDSA

The ElGamal signature scheme [10] appeared in 1985 as the first DL-based signature scheme. In 1989, using the Fiat and Shamir heuristic [11] based on fair zero-knowledge [13], Schnorr provided a zero-knowledge identification scheme [26], together with the corresponding signature scheme. In 1994, a digital signature standard DSA [20] was proposed, whose flavor was a mixture of ElGamal and Schnorr. The standard was later adapted to the elliptic curve setting under the name ECDSA [1,20]. Following [6,7], we propose the description of a generic DSA (see Figure 1), which operates in any cyclic group $\mathcal{G}$ of prime order q, thanks to a reduction function. This reduction function f applies to any element of the group $\mathcal{G}$, into $\mathbb{Z}_q$. In the DSA, f takes as input an integer modulo p and outputs $f(r) = r \bmod q$. In the elliptic curve version [1,20,9], the function is

Initialization	Σ: Signature of $m \to (r,s)$
$\mathcal{G}$ a cyclic group of prime order q	k randomly chosen $0 < k < q$
$\mathbf{g}$ a generator of $\mathcal{G}$	$\mathbf{r} = k \cdot \mathbf{g}$ $r = f(\mathbf{r})$
$H : \{0,1\}^* \to \{0,1\}^h$ a hash function	if $r = 0$ abort and start again
$f : \mathcal{G} \to \mathbb{Z}_q$ a reduction function	$e = H(m)$ $s = k^{-1}(e + xr) \bmod q$
$\mathcal{K}$: **Key Generation** $\to (\mathbf{y}, x)$	if $s = 0$ abort and start again
private key $0 < x < q$	V: **Verification of** $(m, r, s) \to$ valid ?
public key $\mathbf{y} = x \cdot \mathbf{g}$	check whether $0 < r, s < q$ and $r = f(\mathbf{r}')$
	where $e = H(m)$ and $\mathbf{r}' = es^{-1} \cdot \mathbf{g} + rs^{-1} \cdot \mathbf{y}$

Fig. 1. The Generic DSA

defined in a more intricate manner, which we now describe. An elliptic curve point $\mathbf{r}$ is given by two coordinates (x, y), which take values in the base field. For elliptic curves over prime fields, one simply sets $f(\mathbf{r}) = x \bmod q$. For curves over $\mathbb{F}_{2^m}$, x is a sequence of m bits and $f(\mathbf{r})$ is obtained by first turning x into an integer less than 2^m, by a standard conversion routine. Anyway, one just has to keep in mind that in ECDSA the function f depends on the x-coordinate only, and thus $f(-\mathbf{r}) = f(\mathbf{r})$.

Before we review the security results proven about ECDSA, namely in [7], let us show some surprising properties of the scheme due to the above choice of reduction function f.

4.2 Duplicate Signatures

Let us first describe how to produce duplicate signatures for ECDSA. Recall we have two messages m_1 and m_2 and we wish to produce a signature which is valid for both messages, with a possible control on the key generation process. We will do this by "concocting" a public/private key pair, hence we see that our method assumes that the two target messages are known to the signer before he generates his public/private key pair. We note that the special key pair is still valid and the user is still able to sign other messages as usual.

We first compute $h_1 = H(m_1)$ and $h_2 = H(m_2)$. We generate a random $k \in \{1, \ldots q - 1\}$, compute $r = f(k \cdot \mathbf{g})$, and then set the private key to be

$$x = -\left(\frac{h_1 + h_2}{2r}\right) \bmod q,$$

with the public key being given by $\mathbf{y} = x \cdot \mathbf{g}$. To generate our duplicate signature on m_1 and m_2 we compute $s = k^{-1}(h_1 + xr) \bmod q$.

That (r, s) is a valid signature on m_1 follows from the definition of ECDSA, we only need to show that (r, s) is also a valid signature on m_2. We evaluate the $\mathbf{r}'$ in the verification algorithm for the signature (r, s) on the message m_2, noting that $rx = -(h_1 + h_2)/2 \bmod q$,

$$\mathbf{r}' = (h_2/s)\mathbf{g} + (r/s)\mathbf{y} = \left(\frac{h_2 + rx}{s}\right)\mathbf{g} = k\left(\frac{h_2 - h_1}{h_1 - h_2}\right)\mathbf{g} = -k \cdot \mathbf{g} = -\mathbf{r}.$$

Hence, $f(\mathbf{r}') = f(-\mathbf{r}) = f(\mathbf{r}) = r$ and the signature verifies.

Example. As an example we use one of the recommended curves from X9.62 [1]. The curve is defined over $\mathbb{F}_p$ where $p = 2^{192} - 2^{64} - 1$, and is given by equation $y^2 = x^3 - 3x + b$, where

$$b = \texttt{0x64210519E59C80E70FA7E9AB72243049FEB8DEECC146B9B1}.$$

This curve has prime group order given by

$$q = 6277101735386680763835789423176059013767194773182842284081,$$

and a base point is given by $\mathbf{g} = (X, Y)$ where

$X = $ 0x188DA80EB03090F67CBF20EB43A18800F4FF0AFD82FF1012,

$Y = $ 0x07192B95FFC8DA78631011ED6B24CDD573F977A11E794811.

Suppose we have a public key given by $\mathbf{y} = (X', Y')$

$X' = $ 0xA284DB03CAC23298DF9FD9C60560B16292FBE5C7E2C26C25,

$Y' = $ 0x3F9EABD65A25DA6E72285670AA3D639B381952AFDDECEBAA.

Consider the two, hundred byte messages $m_1 = [0, 1, 2, 3, \ldots, 99]$ and $m_2 = [10, 11, 12, 13, \ldots, 109]$, with hash values, computed via SHA-1 [21],

$h_1 = $ SHA-1$(m_1) = $ 0x1E6634BFAEBC0348298105923D0F26E47AA33FF5,

$h_2 = $ SHA-1$(m_2) = $ 0x71DDBA9666E28406506F839DAA4ECAF8D03D2440.

A duplicate signature on both m_1 and m_2 is provided by (r, s), with

$r = $ 0x7B3281ED9C01372E09271667D88F840BEB888F43AF4A7783,

$s = $ 0xAFC81CEC549C77F00B4790160A584FD636BB049FD9D9E0BD.

Note that, as soon as one publishes the duplicate signature, a third party can recover the signer's private key and so is able to forge messages. Hence, this example of duplicate signatures should not be considered a security weakness. However, one does not know that no other duplicate signature exists, for this or any other signature scheme, which do not arise from collisions in the hash function.

4.3   Malleability

Still using the above specific property of f, that is $f(-\mathbf{r}) = f(\mathbf{r})$, ECDSA is easily malleable. Indeed, from a signature (r, s) of a message m, whatever the keys are, one can derive a second signature, namely $(r, -s)$. Referring to Figure 1, we see that the values of $\mathbf{r}'$ that appear in the verification of both signatures are symmetric, so that their image by f is the same.

4.4   Comments on the Security Results

Let us now see whether the above security notions have been appropriately dealt with or not in the security analyses which appeared in the literature. For the reader's convenience, we include in Appendix A.3 our own version of the theorem and its proof (it is highly based on [7]). The original theorem [7] claims that the generic DSA withstands existential forgeries against adaptive chosen-message attacks, in the generic model, under some assumptions, namely the collision-resistance of the hash function and the *almost-invertibility* of the reduction function (see more details in Appendix A.2). This latter property is not satisfied for DSA, but is clearly satisfied with the reduction function used in ECDSA: given an x-value, if it does not correspond to the x-coordinate of a point on the curve, g outputs Fail, otherwise it randomly outputs one of the (two) corresponding points. Hasse's theorem ensures that g is an almost inverse of f. It furthermore helps to say that the statistical distance between $\mathcal{D}_g$ and $\mathcal{U}$ is less than $5/q$. Therefore, f is $(5/q, t)$-almost-invertible for any t.

Going through the proof, the reader can check that it actually establishes that, in the *generic model*, ECDSA is non-malleable under the collision-resistance of the hash function only. The question now becomes: what is the meaning of a proof supporting a scheme by means of an ideal model where the scheme has a property (non-malleability) that it does not have in reality? The flaw here comes from the encoding which is not generic because of the automorphism. Notice that Koblitz curves, as advocated in some standards, are even "less" generic since they have more automorphisms.

About the duplicate signatures, the proof does not deal with the problem at all, since as already remarked, for non-repudiation the adversary is the signer himself. The methodological lesson is that in some scenarios non-repudiation does not necessarily follows from resistance to existential forgeries. In other words, the security model does not properly account on a possible collusion between the key generation algorithm and the signing algorithm. Whilst our example of duplicate signatures is not a security concern, there may be others. Hence, the proof methodology and security model should allow for this.

5   Conclusion

We have shown that the version of the ESIGN cryptosystem described in the P1363 submission [22] withstands existential forgery against single-occurrence adaptive chosen-message attacks, based on the hardness of AERP. However, the proof does not extend to the usual CMA scenario. We have also considered a new kind of attack, independent of existential forgeries, since the attacker may be the signer himself. We have illustrated it on ECDSA. It shows that non-repudiation is not totally encompassed by usual security analyses. Finally, we have proved the non-malleability of the generic DSA under adaptive chosen-message attacks. This is in contrast with the actual malleability of ECDSA and puts some doubts on the significance of the generic model.

In conclusion, we give the warning to practitioners, that security proofs need some time to be discussed, accepted, and interpreted within the research community.

Acknowledgments

We thank Tatsuaki Okamoto for fruitful discussions. It should be emphasized that the first and the last examples are based on the result of an evaluation requested by the Japanese Cryptrec program and performed by the first named author. This author wishes to thank Cryptrec.

References

1. American National Standards Institute. Public Key Cryptography for the Financial Services Industry: The Elliptic Curve Digital Signature Algorithm. ANSI X9.62-1998, January 1999.

2. M. Bellare, A. Desai, D. Pointcheval, and P. Rogaway. Relations among Notions of Security for Public-Key Encryption Schemes. In *Crypto '98*, LNCS 1462, pages 26–45, Springer-Verlag, 1998.
3. M. Bellare and P. Rogaway. Random Oracles Are Practical: a Paradigm for Designing Efficient Protocols. In *Proc. of the 1st CCS*, pages 62–73, ACM Press, 1993.
4. M. Bellare and P. Rogaway. Optimal Asymmetric Encryption – How to Encrypt with RSA. In *Eurocrypt '94*, LNCS 950, pages 92–111, Springer-Verlag, 1995.
5. E. Brickell and J. M. DeLaurentis. An Attack on a Signature Scheme proposed by Okamoto and Shiraishi. In *Crypto '85*, LNCS 218, pages 28–32, Springer-Verlag, 1986.
6. E. Brickell, D. Pointcheval, S. Vaudenay, and M. Yung. Design Validations for Discrete Logarithm Based Signature Schemes. In *PKC '2000*, LNCS 1751, pages 276–292, Springer-Verlag, 2000.
7. D. R. L. Brown. The Exact Security of ECDSA, January 2001. IEEE 1363 [16].
8. R. Canetti, O. Goldreich, and S. Halevi. The Random Oracles Methodology, Revisited. In *Proc. of the 30th STOC*, pages 209–218, ACM Press, 1998.
9. Certicom. Standards for efficient cryptography, September 2000.
10. T. ElGamal. A Public Key Cryptosystem and a Signature Scheme Based on Discrete Logarithms. *IEEE Transactions on Information Theory*, IT–31(4):469–472, July 1985.
11. A. Fiat and A. Shamir. How to Prove Yourself: Practical Solutions of Identification and Signature Problems. In *Crypto '86*, LNCS 263, pages 186–194, Springer-Verlag, 1987.
12. E. Fujisaki, T. Okamoto, D. Pointcheval, and J. Stern. RSA–OAEP is Secure under the RSA Assumption. In *Crypto '2001*, LNCS 2139, pages 260–274, Springer-Verlag, 2001.
13. S. Goldwasser, S. Micali, and C. Rackoff. The Knowledge Complexity of Interactive Proof Systems. In *Proc. of the 17th STOC*, pages 291–304, ACM Press, 1985.
14. S. Goldwasser, S. Micali, and R. Rivest. A Digital Signature Scheme Secure Against Adaptive Chosen-Message Attacks. *SIAM Journal of Computing*, 17(2):281–308, April 1988.
15. L. Granboulan. How to repair ESIGN. NESSIE internal document, may 2002. See http://www.cryptonessie.org/. Document NES/DOC/ENS/WP5/019.
16. IEEE P1363. Standard Specifications for Public Key Cryptography, August 1998. See http://grouper.ieee.org/groups/1363/.
17. D. Naccache, D. Pointcheval, and J. Stern. Twin Signatures: an Alternative to the Hash-and-Sign Paradigm. In *Proc. of the 8th CCS*, ACM Press, 2001.
18. M. Naor and M. Yung. Public-Key Cryptosystems Provably Secure against Chosen Ciphertext Attacks. In *Proc. of the 22nd STOC*, pages 427–437. ACM Press, 1990.
19. V. I. Nechaev. Complexity of a Determinate Algorithm for the Discrete Logarithm. *Mathematical Notes*, 55(2):165–172, 1994.
20. NIST. Digital Signature Standard (DSS). Federal Information Processing Standards Publication 186, November 1994. Revision (To include ECDSA) : 186-2, January 2000.
21. NIST. Secure Hash Standard (SHS). Federal Information Processing Standards Publication 180–1, April 1995.
22. T. Okamoto, E. Fujisaki and H. Morita. TSH-ESIGN: Efficient Digital Signature Scheme Using Trisection Size Hash, 1998. IEEE 1363 [16].
23. J. M. Pollard. Monte Carlo Methods for Index Computation (mod p). *Mathematics of Computation*, 32(143):918–924, July 1978.

24. C. Rackoff and D. R. Simon. Non-Interactive Zero-Knowledge Proof of Knowledge and Chosen Ciphertext Attack. In *Crypto '91*, LNCS 576, pages 433–444. Springer-Verlag, 1992.
25. R. Rivest, A. Shamir, and L. Adleman. A Method for Obtaining Digital Signatures and Public Key Cryptosystems. *Communications of the ACM*, 21(2):120–126, February 1978.
26. C. P. Schnorr. Efficient Signature Generation by Smart Cards. *Journal of Cryptology*, 4(3):161–174, 1991.
27. C. P. Schnorr and M. Jakobsson. Security of Signed ElGamal Encryption. In *Asiacrypt '2000*, LNCS 1976, pages 458–469, Springer-Verlag, 2000.
28. V. Shoup. Lower Bounds for Discrete Logarithms and Related Problems. In *Eurocrypt '97*, LNCS 1233, pages 256–266, Springer-Verlag, 1997.
29. V. Shoup. OAEP Reconsidered. In *Crypto '2001*, LNCS 2139, pages 239–259, Springer-Verlag, 2001.
30. B. Vallée, M. Girault and P. Toffin. How to break Okamoto's Cryptosystem by Reducing Lattice Bases. In *Eurocrypt '88*, LNCS 330, pages 281–292, Springer-Verlag, 1988.

A   The Security Proof of ECDSA

A.1   Proofs in the Generic Model

With the proofs in the generic model, we identify the underlying probabilistic space with the space $S^{n+2} \times \Gamma \times \Gamma^{2n}$, where S is the set of bit-string encodings. Given a tuple $\{z_1, \ldots, z_{n+2}, x, x_1, \ldots, x_{2n}\}$ in this space, z_1 and z_2 are used as $\sigma(1)$ and $\sigma(x)$, the successive z_i are used in sequence to answer the n oracle queries and the $x_i \in \Gamma$ serve as pre-images of the additional elements z_i' (in the group) included by the adversary into the encoding list $\mathcal{L}$. However, this interpretation may yield inconsistencies as it does not take care of possible collisions.

We give another interpretation of the encoding σ. This interpretation is based on defining from the tuple $\{z_1, \ldots, z_{n+2}\}$, a sequence of polynomials $F_i(X, X_1, \ldots, X_{2n})$, with coefficients modulo q, depending on the execution of $\mathcal{A}$:

- Polynomials F_1 and F_2 are set to $F_1 = 1$ and $F_2 = X$, respectively. Thus $\mathcal{L} = \{F_1, F_2\}$.
- When the adversary puts an additional k-th element z_k' in the encoding list, polynomial F_{n+k+2} is defined as X_k, and added to $\mathcal{L}$.
- At the ℓ-th query $\{i, j, \epsilon\}$, polynomial F_ℓ is defined as $F_i \pm F_j$, where the sign $\pm$ is chosen according to ϵ. If F_ℓ is already listed as a previous polynomial $F_h \in \mathcal{L}$, then F_ℓ is marked and $\mathcal{A}$ is fed with the answer corresponding to h. Otherwise, z_ℓ is returned by the oracle and F_ℓ is added to $\mathcal{L}$.

Observe that all F_i polynomials are *affine*, i.e. of the form $a_0 + \sum_{i=1}^{j} a_i X_i$.

Once $\mathcal{A}$ has come to a stop, variable X is set to x, and the X_ks are set to x_k. In other words, σ is set at random, subject to the conditions $z_\ell = \sigma(F_\ell(x, x_1, \ldots, x_{2n}))$, $\ell = 1, \ldots, n+2$ and $z_k' = \sigma(x_k)$, $k = 1, \ldots, 2n$. It is easy to check that the behavior of the algorithm that is driven by the polynomials F_i is exactly similar to the behavior of the regular algorithm, granted that elements in the sequence $(z_1, \ldots, z_{n+2})$ are all distinct, and that no polynomial $F_i - F_j$ vanishes at $(x, x_1, \ldots, x_{2n})$, where i, j range over the $3n + 2$ indices of polynomials in $\mathcal{L}$. We call a sequence $\{z_1, \ldots, z_{n+2}, x, x_1, \ldots, x_{2n}\}$ which satisfies both requirements a *safe* sequence. As explained, an encoding σ can be defined from a safe sequence, such that:

$$\sigma(F_i(x, x_1, \ldots, x_{2n})) = z_i, \text{ for all unmarked } F_i, \text{ and } 1 \leq i \leq n+2,$$
$$\sigma(x_k) = z'_k, \text{ for } k = 1, \ldots, 2n.$$

This correspondence preserves probabilities. However, it does not completely cover the sample space $\{\sigma, x\}$ since executions such that $F_i(x, x_1, \ldots, x_{2n}) = F_j(x, x_1, \ldots, x_{2n})$, for some indices i, j, such that F_i and F_j are not identical are omitted. The following lemmas allow to bound the probability of unsafe sequences.

Lemma 1. *Let P be a non-zero affine polynomial in $\mathbb{Z}_q[X_1, \ldots, X_j]$, then*

$$\Pr_{x_1, \ldots, x_j \in \mathbb{Z}_q} [P(x_1, \ldots, x_j) = 0] \leq \frac{1}{q}.$$

Lemma 2. *Assume $n^2 < q$. The probability of unsafe sequences is at upper-bounded by $5(n + 1)^2/q$.*

Proof. We first observe that sequences of random elements $\{z_1, z_2, \ldots, z_{n+2}\}$, which are not all distinct appear with probability

$$1 - \prod_{k=1}^{n+1} \left(1 - \frac{k}{q}\right) \leq 1 - \left(1 - \sum_{k=1}^{n+1} \frac{k}{q}\right) \leq \frac{(n+1)(n+2)}{2q}.$$

Next, using Lemma 1, we can bound the probability that $F_i - F_j$ vanishes at $(x, x_1, \ldots, x_{2n})$ by $1/q$. Since there are at most $\binom{3n+2}{2}$ such polynomials, we infer that, once $\{z_1, \ldots, z_{n+2}\}$ have been set and are distinct, the set of $(x, x_1, \ldots, x_{2n})$ such that $\{z_1, \ldots, z_{n+2}, x, x_1, \ldots, x_{2n}\}$ is not safe has probability bounded by $\binom{3n+2}{2}/q = (3n + 2)(3n + 1)/2q$. One easily completes the proof. □

A.2 Preliminaries

Let f be a reduction function $f : \mathcal{G} \to \mathbb{Z}_q$. An almost-inverse g of f is a probabilistic algorithm g, possibly outputting Fail, such that

$$(i) \Pr_{b \in_R \mathbb{Z}_q} [g(b) \in \mathcal{G} \wedge f(g(b)) = b] \geq 1/3$$

Function f is (δ, t)-*almost-invertible*, with almost-inverse g, if furthermore:

$$(ii) \ \mathcal{D}_g \approx_\delta \mathcal{U}, \text{ where } \begin{cases} \mathcal{D}_g = \{g(b) \,|\, b \in_R \mathbb{Z}_q \wedge g(b) \in \mathcal{G}\} \\ \mathcal{U} = \{a \,|\, a \in_R \mathcal{G}\}. \end{cases}$$

In the second item, notation $\mathcal{D}_g \approx_\delta \mathcal{U}$ means that no distinguisher with running time bounded by t can get an advantage greater than δ.

A.3   The Security Proof

We now prove the security of the generic DSA in the generic model. We follow [7], but we adopt a different style of proof, inspired by Shoup [29]. Referring to Figure 1, we clarify our use of encodings. The base point **g** of the group $\mathcal{G}$ is identified with the canonical generator 1 of $\mathbb{Z}_q$ and therefore labeled by $\sigma(1)$. Similarly, the public key **y** is labeled by $\sigma(x)$, where x is the private key. When an element **r** is requested, at signature generation, it is obtained as $\sigma(k)$, where k is randomly chosen. Finally, the reduction function f directly operates on the set of encodings S. Contrary to the earlier approach of [27], we do not model the hash function as a random oracle. Rather, along the lines first investigated in [6], we use specific properties of the hash function, such as one-wayness or collision resistance.

A couple of lemmas will be needed. We first show how one can perfectly simulate the distribution of valid signatures. We define a simulator S. The simulator, picks elements $u \in_R S$, and $s \in_R \mathbb{Z}_q$, and outputs the pair (r, s), with $r = f(u)$.

Lemma 3. *For any message m, the output distribution of S is perfectly indistinguishable from the output distribution of $\Sigma_{\mathsf{sk}}(m)$.*

We also state an easy lemma from elementary probability theory.

Lemma 4. *Let $\mathbf{S}$ be a binomial distribution, which is the sum of $k = 5 \ln n$ Bernoulli trials with probability for success $\geq 1/3$. Then, the probability that $\mathbf{S} = 0$ is at most $1/n^2$.*

We finally state the security result.

Theorem 2. *Let Γ be a standard cyclic group of prime order q. Let S be a set of bit-string encodings. Let $H : \{0,1\}^* \to \{0,1\}^h$ be a hash function and $f : S \to \mathbb{Z}_q$ be a reduction function with almost-inverse g. Let $\mathcal{A}$ be a generic algorithm over Γ, that makes at most q_s queries to the signing oracle and n queries to the group-oracle, respectively. Assume that $\mathcal{A}$, on input $\{\sigma(1), \sigma(x)\}$, returns a message m and a valid generic DSA signature (r, s) of m, achieving malleability with probability $\varepsilon = \mathsf{Succ}^{\mathsf{cma}}(\mathcal{A})$, within running time t. Then there exist adversaries $\mathcal{B}_H, \mathcal{C}_H, \mathcal{D}_g$, operating within time bound t', and such that $\mathcal{B}_H$ is attempting to invert $H' = H \bmod q$ with success probability ε_H, $\mathcal{C}_H$ is attempting to find collisions for $H' = H \bmod q$ with success probability γ_H, and $\mathcal{D}_g$ is playing a distinguishing game for g, with advantage δ_g, where*

$$\varepsilon \leq 2\gamma_H + 2n(\delta_g + \varepsilon_H) + \frac{5(n+1)(n+q_s+1)}{q},$$

$$t' \leq t + n \times (5\tau_g \ln n + \tau_H),$$

with τ_g the running time of g and τ_H the running time for H.

Proof. Let $\mathcal{A}$ be a generic attacker able to forge a pair consisting of a message m and a valid signature (r, s). We assume that, once these outputs have been issued, $\mathcal{A}$ goes on checking the signature by requesting the encoding of $es^{-1} + xrs^{-1} \bmod$

q, where $e = H(m)$, and checking that its image under f is r. The request can be performed by mimicking the usual double-and-add algorithm, calling the generic encoding at each group operation. We assume furthermore, that, after each query m_j to the signing oracle, the adversary immediately performs a similar request to check the validity of the answer. To keep things simple, we do not perform any book-keeping of the additional requests and keep n to denote the overall number of queries to the group oracle. We now play games as before:

Game$_0$: An encoding σ is chosen and a key pair (pk, sk) is generated using $\mathcal{K}(1^k)$. Adversary $\mathcal{A}$ is fed with pk and, querying the generic encoding and the signing oracle, outputs a message m and a signature (r, s). We denote by S_0 the event $V_{pk}(m, (r, s)) = 1$ and use a similar notation S_i in any Game$_i$ below. By definition, we have $\Pr[S_0] = \varepsilon$.

Game$_1$: We slightly modify this game, by using the interpretation of the encoding proposed in Section A.1: this uses a sequence $\{z_1, \ldots, z_{n+2}, x, x_1, \ldots, x_{2n}\}$. As shown in Section A.1, in Lemma 2, the new game only differs from the old on unsafe sequences: $|\Pr[S_1] - \Pr[S_0]| \leq 5(n + 1)^2/q$.

Game$_2$: In this game, we perform additional random tests, without modifying the simulation of the generic oracle: a test is performed at each index ℓ, such that the corresponding affine polynomial appears for the first time (or is unmarked following the terminology of Section A.1). Let $F_\ell = b_\ell X + a_\ell$. We pick at random $\tilde{e}_\ell \in_R \mathbb{Z}_q$, and compute $c_\ell \leftarrow g(b_\ell a_\ell^{-1} \tilde{e}_\ell \bmod q)$ until the computation of g returns an answer different from Fail. However, we stop and abort the game after $5 \ln n$ trials. This game differs from the previous one if c_ℓ remains undefined after $5 \ln n$ attempts. Since $\tilde{e}_\ell$ is uniformly distributed, and since the successive trials are mutually independent, we may use Lemma 4 and bound the corresponding probability by $1/n^2$. This provides the overall bound $1/n$, when ℓ varies. Taking into account the fact that the experiments are independent from the execution of Game$_1$, we get $\Pr[S_2] \geq (1 - 1/n) \Pr[S_1]$.

Game$_3$: Here, we further modify the previous game by letting c_ℓ replace z_ℓ, for each index ℓ such that F_ℓ is unmarked. Note that we have $f(z_\ell) = b_\ell a_\ell^{-1} \tilde{e}_\ell \bmod q$. Since the $\tilde{e}_\ell$s are uniformly distributed, the inputs to g are uniformly distributed as well. Applying the so-called *hybrid* technique, which amounts to using n times the *almost-invertibility* of g, we bound the difference between the success probabilities of the two games by $n\delta_g$, and thus: $|\Pr[S_3] - \Pr[S_2]| \leq n\delta_g$.

Game$_4$: In this game, we simulate the signing oracle. For any query m_j to the signing oracle, one computes $e_j = H(m_j)$, and issues a random signature (r_j, s_j), using the simulation of Lemma 3. Recall that the simulation picks s_j at random and computes r_j as $f(u_j)$, where u_j is randomly drawn from S. By Lemma 3, this simulation is perfect. Observe that, while checking the signature, the adversary requests, at some later time, the encoding of $e_j s_j^{-1} + x r_j s_j^{-1} \bmod q$. We let ℓ the first index corresponding to such query, $F_\ell = b_\ell X + a_\ell$. We modify z_ℓ, replacing its earlier value by u_j and define $\tilde{e}_\ell$ as $e_j = H(m_j)$. Observe that we still have $f(z_\ell) = b_\ell a_\ell^{-1} \tilde{e}_\ell \bmod q$. This game only differs from the previous one if polynomial $e_j s_j^{-1} + X r_j s_j^{-1}$ collides with a previous one. Due to the randomness of s_j, we can bound $|\Pr[S_4] - \Pr[S_3]| \le n q_s / q$.

We note that the final simulation runs in time $t' \le t + n \times (5\tau_g \ln n + \tau_H)$ and we finally upper-bound $\Pr[S_4]$. We observe that, while checking the signature, the final request of the adversary, with index $n + 2$, is the encoding of $es^{-1} + xrs^{-1} \bmod q$, where $e = H(m)$. We let ℓ be the first occurrence of F_{n+2}. If the signature is valid, the following equalities hold:

$$es^{-1} = a_\ell \bmod q, \quad rs^{-1} = b_\ell \bmod q, \quad f(z_\ell) = b_\ell a_\ell^{-1} \tilde{e}_\ell \bmod q \text{ and } r = f(z_\ell).$$

From these equalities, it easily follows that $r = f(z_\ell) = r e^{-1} \tilde{e}_\ell \bmod q$, which in turn implies $e = \tilde{e}_\ell \bmod q$. We distinguish two cases:

- If z_ℓ has been created according to the rule of Game$_3$, then a pre-image m of some randomly chosen element $\tilde{e}_\ell$ among the n possible ones has been found.
- If z_ℓ has been created according to the rule of Game$_4$, then $\tilde{e}_\ell = e_j$ is the image under H of a message m_j queried from the signing oracle. Furthermore, we have: $e_j s_j^{-1} = a_\ell \bmod q$ and $r_j s_j^{-1} = b_\ell \bmod q$. Comparing to the above equalities, we get that $s = s_j \bmod q$ and $r = r_j \bmod q$. Note that m_j cannot be equal to m, since otherwise the output forged signature would coincide with an earlier signature (r_j, s_j) of the same message m. Thus, a collision has been found for H', where $H'(m) \stackrel{\text{def}}{=} H(m) \bmod q$.

The probability that an algorithm running in time t' finds a preimage under H' of an element among n is at most $n\varepsilon_H$. From this, we obtain that: $\Pr[S_4] \le n\varepsilon_H + \gamma_H$. Summing up inequalities, we get the announced result. $\square$

Separating Random Oracle Proofs from Complexity Theoretic Proofs: The Non-committing Encryption Case

Jesper Buus Nielsen

BRICS* Department of Computer Science
University of Aarhus
Ny Munkegade
DK-8000 Arhus C, Denmark
buus@brics.dk

Abstract. We show that there exists a natural protocol problem which has a simple solution in the random-oracle (RO) model and which has no solution in the complexity-theoretic (CT) model, namely the problem of constructing a non-interactive communication protocol secure against adaptive adversaries a.k.a. non-interactive non-committing encryption. This separation between the models is due to the so-called programability of the random oracle. We show this by providing a formulation of the RO model in which the oracle is not programmable, and showing that in this model, there does not exist non-interactive non-committing encryption.

1 Introduction

Before describing our separation result and the non-programmable random-oracle (NPRO) model, we introduce non-committing encryption (NCE) and the non-interactive NCE (NINCE) problem.

Non-committing Encryption. One way of constructing a secure protocol for the cryptographic model is to take a protocol which is secure in the information theoretical model, where secure channels are assumed, and then compile this protocol for the cryptographic model by adding encryption to the channels. A motivation for this approach has been, that only statically secure general multi-party computation (MPC) protocols have been constructed for the cryptographic model directly, whereas adaptively secure protocols for the information theoretical model were published already in [BGW88,CCD88]. The goal is therefore to replace the secure channels of the information theoretical model by open channels using an adaptively secure communication protocol a.k.a. NCE.

Before we can define NCE more formally we have to sketch our MPC model. We use the model of asynchronous MPC from [Can01]. The security of a protocol

---

* Basic Research in Computer Science,
  Centre of the Danish National Research Foundation.

M. Yung (Ed.): CRYPTO 2002, LNCS 2442, pp. 111–126, 2002.

is defined by requiring that the real-life execution of the protocol can be simulated efficiently given only access to an ideal-world abstraction of the protocol problem that the protocol is to solve. The real-life execution is controlled by an adversary $\mathcal{A}$ (a probabilistic polynomial time (PPT) interactive Turing machine (ITM)) which can see all communication between the parties and schedules message delivery. By PPT we mean PPT in the security parameter k, which is given to all entities in the system. The adversary can furthermore adaptively corrupt parties to learn their current state (or entire execution history if we do not model erasures) and start controlling the corrupted party. The execution takes place in context of an environment $\mathcal{Z}$ (also a PPT ITM) which provides inputs to and receives outputs from the parties. The environment and adversary can communicate during the execution. We denote the output of the environment after such an execution by $\mathrm{REAL}_{\pi,\mathcal{A},\mathcal{Z}}$, where π is the protocol. This execution is compared to an ideal-world execution where the parties have access to an ideal functionality with the desired input-output behavior of the protocol. Message delivery is controlled by an ideal-world adversary $\mathcal{S}$ which can again corrupt parties and learn their internal state (which is just the inputs from the environment), and the protocol is executed in context of an environment $\mathcal{Z}$ with the same role as in the real-life model. We denote the output of the environment after such an execution by $\mathrm{IDEAL}_{\mathcal{F},\mathcal{S},\mathcal{Z}}$, where $\mathcal{F}$ is a PPT ITM specifying the desired input-output behavior of the protocol problem to be solved. These executions are then compared by saying that for each real-life adversary $\mathcal{A}$ there should exist an ideal-world adversary $\mathcal{S}$ such that for all environments $\mathcal{Z}$ the executions $\mathrm{REAL}_{\pi,\mathcal{A},\mathcal{Z}}$ and $\mathrm{IDEAL}_{\mathcal{F},\mathcal{S},\mathcal{Z}}$ are computationally indistinguishable, i.e. the environment cannot tell whether its observing a real-life execution or the simulator $\mathcal{S}$ running in the ideal-world. The role of $\mathcal{S}$ is similar to the role of the simulator in the definition of zero-knowledge. The role of the environment is that of an distinguisher between the real-life execution $\mathrm{REAL}_{\pi,\mathcal{A},\mathcal{Z}}$ and the simulation $\mathrm{IDEAL}_{\mathcal{F},\mathcal{S},\mathcal{Z}}$. An important part of the model is that the environment receives the identity of all corrupted parties. This guarantees that the simulator does not accomplish its goal by corrupting other parties than the real-life adversary.

For the NCE problem the ideal functionality $\mathcal{F}_{\mathrm{nce}}$ works as follows: On input (mid, j, m) from P_i deliver (mid, i, m) to P_j, and reveal $(mid, i, j, |m|)$ to the adversary. Here mid is a message identifier, m is the message, and $|m|$ is the length of m. For the specific task of secure communication the above definition of security basically says that whatever a real-life adversary can obtain from attacking the protocol an ideal adversary $\mathcal{S}$ could obtain (simulate) given just the length of the messages sent.

If we let each party P_i have a private key for a semantically secure public-key encryption scheme, where the public key pk_i is known by all other parties, and if we encrypt all communication to P_i under pk_i (including in the messages the identity of the sender to protect against copying), then we will have a statically secure implementation. However, no encryption scheme exists for which this protocol is *adaptively* secure. This follows from a general result that no non-interactive communication protocol is adaptively secure; Throughout the paper we will let non-interactive communication protocol denote a communication protocol with the property that after a pre-processing phase, which might involve interaction (e.g. the receiver sending a public key to the sender), the sender can send an unbounded number of bits to the receiver without there being any communication from the receiver to the sender.

We show that no non-interactive communication protocol can be adaptively secure in the asynchronous model. Assume for this sake that we have an adaptively secure communication protocol for the asynchronous model. Consider two parties P_R and P_S acting as receiver resp. sender. Consider the environment $\mathcal{Z}$ which activates P_S with an arbitrary message m of length $l_m(k)$, where $l_m(k)$ is some polynomial. Consider the adversary $\mathcal{A}$ that corrupts no party but just waits for P_R and P_S to finish the preprocessing phase and for P_S to send a ciphertext c to P_R. The adversary outputs c to the environment and then corrupts P_R before c arrives, and the adversary outputs to the environment the value sk of the internal state of P_R. By the security of the encryption scheme we have that if the environment runs the code of P_R from internal state sk and with input c, then c will decrypt to m, except possibly with negligible probability. Now, by the definition of security there should exist a simulator S such that the simulator executed in the ideal-world execution with the same environment $\mathcal{Z}$ produces an output indistinguishable from that of the adversary $\mathcal{A}$. But in the ideal-world abstraction of secure communication given by $\mathcal{F}_{nce}$, the simulator does not see m during the execution as long as both parties are uncorrupted, and the simulator must therefore generate c given just $|m|$. Then on the corruption of P_R, the simulator sees m and computes sk to give to the environment. Since the definition of security requires that the environment cannot tell the difference between the real-life execution and the simulation it follows that running P_R from sk on input c will result in output m except with negligible probability, in particular with probability more than $\frac{1}{2}$. Since no internal state can make c decrypt to two different values, both with more than probability $\frac{1}{2}$, there exists an injective map from messages m to internal states $sk_{c,m}$ which make c decrypt to m. Intuitively this means that the length of sk must be at least l_m. If the protocol can send an unbounded number of bits, this holds for any polynomial l_m and thus the length of sk must be superpolynomial contradicting that P_R is a PPT ITM.

The NCE problem was first solved by Beaver and Haber in [BH92]. In their protocol P_R sends to P_S the public key pk. Then P_S generates a uniformly random message p and sends $c = E_{pk}(p)$ to P_R. Then P_R computes $p = D_{sk}(c)$ and erases everything except p. When m later becomes known to P_R he computes $c' = p \oplus m$, where $\oplus$ denotes bitwise xor, and sends c' to P_R who can then compute $m = p \oplus c'$. Since at no point P_R knows both sk and the encryption c' of m, the attacker cannot obtain both, which preempts the problem that for fixed c' there should be an injective map from sk to messages m. Since sk is deleted a new key-pair must be generated each time P_S has sent a total of $|p|$ bits. If further more synchronization between the parties are assumed, the protocol can be made non-interactive: Set aside a prefix of p to use as a seed s for a pseudo-random generator, and each time p is used up, expand s to obtain a new p as long as the original p and delete s. This method is then iterated each time p is used up. In this way no more than $|p|$ bits are communicated from P_S to P_R before P_R deletes its internal state and creates a new one.

The protocol from [BH92] depends essentially on the use of erasure. However, in many settings trusting the parties to be able to reliably erase parts of their state might be unrealistic, due to e.g. physical limitations on erasure and weak operating systems. The first solution to the NCE problem in the non-erasure model is presented in [CFGN96] by Canetti et al. The scheme is however ineffi-cient: It can encrypt 1 bit using a public key of $\Theta(k^2)$ bits. Later Beaver[Bea97] and Damgård and Nielsen[DN00] proposed more efficient schemes communicat-ing 1 plaintext bit using $\Theta(k)$ bits of communication. These protocols are all three-round protocols.

The Random-Oracle Model. The idea behind the random-oracle (RO) model is that by modeling primitives as DES, MD5 or SHA using the strong assumption that they (properly used/modified) behave like ROs, one can build efficient and secure protocols based on these primitives. The model has been used to argue the security of a number of constructions. Examples are the OAEP encryption mode for RSA[BR95,Sho01] and the Fiat-Shamir heuristic[FS86].

We define the RO model to be the real-life execution model described above where additionally the parties and the adversary has access to a uniformly ran-dom function $\{0,1\}^* \to \{0,1\}^k$. This can be modeled within the framework in [Can01] using a hybrid model. A hybrid model is the real-life model extended with an ideal functionality $\mathcal{F}$ (also called a trusted party) to which all par-ties have a secure channel. The calls to $\mathcal{F}$ works as in the ideal-world. We use $\text{HYB}^{\mathcal{F}}_{\pi,\mathcal{A},\mathcal{Z}}$ to denote an execution of protocol π in the hybrid model with trusted party $\mathcal{F}$, and say that π realizes $\mathcal{G}$ in the $\mathcal{F}$-hybrid model if for each hybrid ad-versary $\mathcal{A}$ there exist an ideal-world adversary $\mathcal{S}$ such that for all environments $\mathcal{Z}$ the executions $\text{HYB}^{\mathcal{F}}_{\pi,\mathcal{A},\mathcal{Z}}$ and $\text{IDEAL}_{\mathcal{G},\mathcal{S},\mathcal{Z}}$ are computationally indistinguish-able. We let the RO model by the hybrid model with the trusted party $\mathcal{F}_{\text{ro}}$ work-ing as follows: On input $x \in \{0,1\}^*$ from any of the parties or the adversary it outputs a uniformly random value $r \in \{0,1\}^k$ to the calling party; If queried on the same x twice, the same r is returned. Thus $\mathcal{F}_{\text{ro}}$ defines a uniformly random function $H : \{0,1\}^* \to \{0,1\}^k$.

Possibility of NINCE in the RO model. We prove that if trapdoor permutations exists, then NINCE exists in the RO model. Our protocol is reminiscent of a con-struction of chosen ciphertext secure encryption in [BR93]. In the pre-processing phase the receiver P_R sends a description f of a trapdoor permutation to the sender P_S. Each message from P_S to P_R is transmitted as $(f(x), H(x) \oplus m)$, where x is a uniformly random element in the domain of f and H is the uni-formly random function defined by $\mathcal{F}_{\text{ro}}$. To prove the scheme secure we construct a simulator $\mathcal{S}$. The simulator works by running internally a copy of the protocol and a copy of $\mathcal{A}$. It tries to make the internal protocol consistent with the values of m input to the ideal-world execution (knowing only $|m|$) and lets $\mathcal{A}$ attack the simulated execution and lets $\mathcal{A}$ do the interaction with the environment $\mathcal{Z}$. The simulator $\mathcal{S}$ simulates in such a way that $\mathcal{A}$ thinks that it observes a real-life execution, and such that in particular its interaction with the environment is distributed computationally indistinguishable from that observed by $\mathcal{Z}$ in the real-life execution, which in turn makes the output of $\mathcal{Z}$ computationally indistinguishable in the two worlds. The simulator $\mathcal{S}$ proceeds as follows: Distribute the public keys as in the real-life. Note that $\mathcal{A}$ and the parties of the protocol might request to evaluate the RO H on a value x, as they expect to run in the RO model. To simulate the RO, $\mathcal{S}$ returns a uniformly random element r; If queried on the same x twice, it returns the same r. To simulate the sending of m the simulator $\mathcal{S}$ generates random x and sends $(f(x), H(x) \oplus 0^l)$, where l is the length of m and 0^l is the all-zero string of length l. If the simulated oracle H was not defined on x the simulator sets it to be a uniformly random element r as above. Assume that after simulating the sending of an arbitrary number of messages $\mathcal{A}$ corrupts P_R. The simulator then corrupts P_R in the ideal evaluation, and for each $(f(x), H(x) \oplus 0^l)$ sent in the simulation it receives the real value m which should have been sent and must come up with an internal state of P_R consistent with m. Assume that the simulator defined $H(x) = r$, i.e. that $(f(x), r)$ was the message sent. The simulator then simulates by simply claiming that $H(x) = m \oplus r$. This is a perfect simulation as we get that $(f(x), r) = (f(x), (r \oplus m) \oplus m) = (f(x), H(x) \oplus m)$. However, there are two ways this simulation can fail. First of all, if the same x was used twice the simulator might be in the situation that it needs to define $H(x)$ to both $r \oplus m_1$ and $r \oplus m_2$ for $m_1 \neq m_2$. However this happens with negligible probability as the x's are chosen uniformly at random by the simulator. Second, it might be that $\mathcal{A}$ queried H on x and therefore knows that $H(x)$ was defined to r, which *commits* the simulator to this choice and makes the simulation fail. However, if the $\mathcal{A}$ queried H on x it intuitively had to invert the trapdoor function f on a uniformly random element: $\mathcal{A}$ returned x given only $f(x)$. This would contradict the hardness of inverting f on random elements, and thus except with negligible probability the simulation goes through.

Impossibility of NINCE in the CT and NPRO Model. The simulator sketched above uses essentially that it is possible to program the RO, by defining the value of $H(x)$ to be some value appropriately chosen by the simulator: It sets $H(x)$ to $r \oplus m$ after H "should" have been defined on x. We can prove that the use of the programability of the RO is necessary for the simulator. We start by formalizing the NPRO model.

The NPRO model is the real-life model, where all ITMs are extended to be ITMs with oracle access to a random oracle. An ITM M with oracle access is an ITM which in addition to the usual tapes and states has an oracle query tape, an oracle input tape, and a classification of some of its states as oracle query states. We write $M^{(\cdot)}$ to denote an ITM with oracle access. We write $M^{\mathcal{O}}$ to denote running M with oracle $\mathcal{O}$. If M enters an oracle query state, then the contents of the oracle query tape is given as input to $\mathcal{O}$, and the output of $\mathcal{O}$

is written on the oracle input tape of M. Now let $\mathcal{O}$ denote an ITM defining a uniformly random function $H : \{0,1\}^* \to \{0,1\}^k$. The NPRO model defines the two distribution ensembles $\text{REAL}_{\pi^{\mathcal{O}}, \mathcal{A}^{\mathcal{O}}, \mathcal{Z}^{\mathcal{O}}}$ and $\text{IDEAL}_{\mathcal{F}^{\mathcal{O}}, \mathcal{S}^{\mathcal{O}}, \mathcal{Z}^{\mathcal{O}}}$ and as above these are compared by requiring that for each real-life adversary $\mathcal{A}^{(\cdot)}$ there exists an ideal-world adversary $\mathcal{S}^{(\cdot)}$ such that for all environments $\mathcal{Z}^{(\cdot)}$ the executions $\text{REAL}_{\pi^{\mathcal{O}}, \mathcal{A}^{\mathcal{O}}, \mathcal{Z}^{\mathcal{O}}}$ and $\text{IDEAL}_{\mathcal{F}^{\mathcal{O}}, \mathcal{S}^{\mathcal{O}}, \mathcal{Z}^{\mathcal{O}}}$ are computationally indistinguishable.

The main difference between the RO model and the NPRO model is that in the NPRO model also the environment has access to the RO $\mathcal{O}$. Intuitively this allows the environment to verify whether the values that it is shown is consistent with the RO that it has access to, which basically makes it impossible for $\mathcal{S}$ to program the random oracle according to its desires.

The impossibility of NINCE in the NPRO model follows the proof for the CT model sketched in the introduction to NCE. Because $\mathcal{Z}$ has access to the same RO as $P_R^{(\cdot)}$ it can run $P_R^{\mathcal{O}}$ from internal state sk with input c and it follows that there exists an injective mapping from the possible messages to the fixed set of possible internal states of $P_R^{(\cdot)}$ after the pre-processing phase. This argument fails in the programmable RO model as the environment does not have access to $\mathcal{O}$ and thus cannot run $P_R^{\mathcal{O}}$.

To obtain our separation it would be enough to prove NINCE impossible in the asynchronous model without erasure. However, to strengthen the separation result we show that NINCE is impossible in a number of weaker models too. We show the result for the asynchronous model with erasure, the synchronous model without erasure, and for the synchronous model with erasure we show that no NCE protocol can communicate an unbounded number of bits *per round*; By the result from [BH92] mentioned above we cannot hope to prove a stronger result than this for the synchronous model with erasure.

Previous Separation Results. Other examples of constructions secure in the RO model and not secure in the CT model were known prior to our work. Most prominently, in [CGH98] Canetti, Goldreich and Halevi construct an encryption scheme which is secure in the RO model, but is not secure in the CT model no matter the instantiation of the RO. The scheme is constructed as to try to "detect" whether it is in the RO model or not, and then reveal the secret key if it is not in the RO model. A strength of the result from [CGH98] is that it is the semantic security of the encryption scheme that is violated in the CT model, whereas it in our example it is the less standard non-committing property that is violated. Their result thus establishes that even standard security properties do not carry over from the RO model to the CT model. Another strength of the result from [CGH98] is that their encryption scheme can be proven secure in the NPRO model, as they do not use the programability of the RO. This means that their result separates the CT model and the NPRO model[1].

Another well-known separation result is that the Fiat-Shamir[FS86] methodology can be proven secure in the RO model, and that not all non-interactive zero-knowledge proofs obtained by the methodology using a *fixed* function for

---

[1] Using CS proofs for the NPRO model.

implementing RO cannot be black-box zero-knowledge in the CT model unless $BPP \subset NP$.[GK90] This is however not a separation of the strength of the models: When the RO is implemented by a random function h drawn from some function family, say, by a trusted party, and handed to both the prover and the verifier, then f is a de facto common random string and the existence of one-round zero-knowledge proofs is no longer ruled out[BFM88]. It does in particular not follow that there does not exist in some preprocessing model some kind of non-interactive instantiation of the RO which makes the methodology secure.

Discussion and Future Work. We have shown that the programability of the RO in proofs in the RO model is a feature of the model which is so strong that there exist natural protocol problems which are trivially solvable in the RO model, but have no solution in a model without the programability.

We point out that the NPRO model is formulated in this paper primarily to pin-point a property of the RO model which allows for our separation result. It is not meant as a suggestion for 'the' formulation of a weaker RO model. Though it could be interesting to have a weaker formulation of the RO model, as to increase the trust that security in the model would imply a certain level of 'heuristic security' in the real world, our formulation has two shortcomings for this purpose: First of all, our formulation of the NPRO model only addresses security defined by simulation. Second, it is possibly to define even weaker versions of the RO model than the NPRO model and it is not clear which would be 'the appropriate' weak formulation.

As for the first shortcoming, the definition can to some extend be applied to different types of definitions of security as semantic security of encryption schemes and non-forgeability of signature schemes by giving an equivalent definition of the security notion in the MPC framework. As an example we describe how to define NPRO semantic security of public-key encryption: Let $\mathcal{E}$ be a public-key cryptosystem, and let $\pi_{\mathcal{E}}$ be the following protocol for two parties P_S and P_R: First P_R generates a random key pair (pk, sk) and sends pk to P_S. Each time P_S receives input m from the environment, it computes $c = E_{pk}(m)$ and sends c to P_R who computes and outputs to the environment the value $m' = D_{sk}(c)$. It can be proven that a public-key cryptosystem $\mathcal{E}$ is semantic secure in the CT model (resp. in the RO model of [BR93]) iff $\pi_{\mathcal{E}}$ is statically secure in the CT model (resp. in the RO model). Generalizing this, we can say that a public-key cryptosystem $\mathcal{E}$ is semantic secure in the NPRO model iff $\pi_{\mathcal{E}}$ is statically secure in the NPRO model.

As for existence of even weaker RO models, note that another strong property of the RO model which was used by our simulator was that the simulator learns on which points the simulated adversary evaluates the RO. This was what allowed us to make the reduction to the one-wayness of the trapdoor permutation f, as the simulator could obtain x from $f(x)$ if the adversary could evaluate H on x given $f(x)$. We call this property evaluation point knowledge (EPK). One interpretation of what is modeled by EPK is that it isn't possible to learn the value of $H(x)$ without knowing all of x. The fact that the simulator learns all points on which the adversary evaluates the oracle can then be viewed as a

knowledge extraction of the adversary's EPK. The NPRO model still has the
EPK property. We could formulate a RO model without EPK by requiring that
S must simulate given only oracle access to $A^{\mathcal{O}}$ and $\mathcal{O}$. However, we find that
this is far from a satisfactory formulation of the model, as it has the serious
restriction that as it only applies to black-box proofs.

We find giving a simple and general formulating of the NPRO model and a
RO model without EPK an interesting open problem.

The Rest of the Paper. The purpose of the rest of the paper is to give a formalization of the NPRO model and the separation between the RO and the NPRO
model.

2   Trapdoor Permutations

Definition 1 (Collection of trapdoor permutations). *We call a tuple*
$(\mathcal{K}, F, \mathcal{G}, \mathcal{X})$ *a collection of trapdoor permutations with security parameter k, if*
$\mathcal{K}$ *is an infinite index set, $F = \{f_{pk} : D_{pk} \to D_{pk}\}_{pk \in \mathcal{K}}$ is a set of permuta-*
tions, the key/trapdoor-generator $\mathcal{G}$ and the domain-generator $\mathcal{X}$ are PPT (in
k) algorithms, and the following hold:

Easy to generate and compute $\mathcal{G}$ *generates pairs of keys and trapdoors,*
$(pk, sk) \leftarrow \mathcal{G}(k)$, *where $pk \in \mathcal{K} \cap \{0,1\}^{p(k)}$ for some fixed polynomial $p(k)$.*
Furthermore, there is a polynomial time algorithm which on input pk and
$x \in D_{pk}$ *computes $f_{pk}(x)$.*
Easy to sample domain $\mathcal{X}$ *samples elements in the domains of the permuta-*
tions, we write $x \leftarrow \mathcal{X}(pk)$, where x is uniformly random in D_{pk}.
Hard to invert *For $(pk, sk) \leftarrow \mathcal{G}(k)$, $x \leftarrow \mathcal{X}(pk)$, and for any PPT algorithm*
A the probability that $A(pk, f_{pk}(x)) = x$ is negligible in k.
But easy with trapdoor *There is a polynomial time algorithm which on input*
$pk, sk, f_{pk}(x)$ computes x, for all $(pk, sk) \in \mathcal{G}(k)$ and $x \in D_{pk}$.

Let A be any PPT ITM and consider the following game, which we will
call the **trapdoor game**. The game is between A and the tuple $(\mathcal{K}, F, \mathcal{G}, \mathcal{X})$. The
algorithm A can ask for a number of public key generations and element gener-
ations, and the goal of A is to invert a permutation for which it does not know
the trapdoor information, on an element it did not generate itself.

- On a **key generation** request, A is given pk for a uniformly random key $(pk,$
  $sk) \leftarrow \mathcal{G}(k, r_{\mathcal{G}})$ (here $r_{\mathcal{G}}$ denotes the random bits used by $\mathcal{G}$).
- On a **give up** request on pk, where pk was generated in a key generation re-
  quest, A is given $r_{\mathcal{G}}$.
- On an **element generation** request for pk, A receives $y = f_{pk}(x)$, where x was
  generated as $x \leftarrow \mathcal{X}(pk, r_{\mathcal{X}})$.
- On a **give up** request on y, where y was generated in an element generation
  request, A is given $r_{\mathcal{X}}$.

- The ITM A wins the game, if it manages to return an element x such that $y = f_{pk}(x)$, where pk is a key from a key generation request on which it has not given up and where y is from an element generation request on which it has not given up.

It is straightforward to prove the following lemma.

Lemma 1. *The tuple $(\mathcal{K}, F, \mathcal{G}, \mathcal{X})$ is a collection of trapdoor permutations iff for all PPT algorithms A, the probability that A wins over $(\mathcal{K}, F, \mathcal{G}, \mathcal{X})$ in the trapdoor permutation game is negligible.*

3   The Multiparty Computation Model

We will use the framework for universally composable asynchronous MPC from [Can01].

The General Framework. A protocol $\pi = (P_1, \ldots, P_n)$ consists of n PPT ITMs. The most general computation model considered in [Can01] is the hybrid model with ideal functionality $\mathcal{F}$. The execution in the hybrid model involves the parties, the ideal functionality $\mathcal{F}$, the adversary $\mathcal{A}$, and the environment $\mathcal{Z}$. The ideal functionality, the adversary and the environment are PPT ITMs. All parties are connected by point-to-point channels. These channels are modeled as insecure authenticated asynchronous channels by letting the adversary $\mathcal{A}$ see all messages sent and schedule message delivery (without being able to introduce messages). Besides controlling message delivery the adversary can corrupt parties. When a party is corrupted the adversary learns the current internal state or the entire execution history of the party (depending on whether we allow erasures or not) and from the point of corruption the adversary sends messages on behalf of the corrupted party. Besides the communication channels all parties are connected to $\mathcal{F}$ with secure channels ($\mathcal{A}$ does not see the messages, but still schedules the delivery). When a party P_i or the ideal functionality $\mathcal{F}$ receives a message it runs it code and sends messages accordingly. The ideal functionality can also receive messages from $\mathcal{A}$ and send messages to $\mathcal{A}$. Finally, the role of the environment is to deliver input to the parties and receive outputs from the parties. The environment can also input to the adversary and the adversary can send messages to the environment. The environment is the driver of the execution. At the beginning of the protocol it receives an auxiliary input $z \in \{0, 1\}^*$, and it then activates the adversary and the parties of the protocol by giving them input. At some point the environment stops activating parties and halts by outputting some bit b. Let $\mathrm{HYB}_{\pi, \mathcal{A}, \mathcal{Z}}^{\mathcal{F}}(k, z)$ be a random variable describing the output of $\mathcal{Z}$.

We define the security of a protocol by *comparing* the input-output behavior (as seen be the environment) of its execution to an *ideally secure protocol* with the *desired input-output* behavior. We specify the desired input-output of the protocol by giving an ideal-functionality $\mathcal{F}$ defining the desired input-output behavior of the protocol. The ideally secure protocol implementing this desired input-output behavior is then defined to be $\mathrm{HYB}_{\tilde{\pi}, \mathcal{A}, \mathcal{Z}}^{\mathcal{F}}(k, z)$, where $\tilde{\pi}$ is the dummy protocol where the parties just send their input from the environment to $\mathcal{F}$ and send the response from $\mathcal{F}$ to the environment. Since the parties are connected to $\mathcal{F}$ via secure channels and $\mathcal{F}$ cannot be corrupted this protocol is trivially secure. We call $\text{IDEAL}_{\mathcal{F},\mathcal{S},\mathcal{Z}}(k,z) = \text{HYB}^{\mathcal{F}}_{\tilde{\pi},\mathcal{S},\mathcal{Z}}(k,z)$ the ideal-world execution. We then say that a protocol π securely realizes $\mathcal{G}$ in the $\mathcal{F}$-hybrid model if for all adversaries $\mathcal{A}$ there exists an adversary $\mathcal{S}$ such that for all environments $\mathcal{Z}$ and all $c \in N$ there exists $k_c \in N$ such that for all $z \in \{0,1\}^*$ it holds that that $|\Pr[\text{IDEAL}_{\mathcal{G},\mathcal{S},\mathcal{Z}}(k,z) = 1] - \Pr[\text{HYB}^{\mathcal{F}}_{\pi,\mathcal{A},\mathcal{Z}}(k,z) = 1]| < k^{-c}$.

Non-committing Encryption. We specify the desired input-output behavior of secure communication by the functionality $\mathcal{F}_{\text{nce}}$, which on input $(\texttt{send}, mid, j, m)$ from P_i delivers $(\texttt{receive}, mid, i, m)$ to P_j and delivers $(\texttt{receive}, mid, i, j, |m|)$ to $\mathcal{A}$. The value mid is a message identifier. We say that π is a an NCE protocol for some model if π securely realizes $\mathcal{F}_{\text{nce}}$ in that model.

Consider any communication protocol for two parties, sender P_S and receiver P_R, of the following form: First the parties execute a pre-processing phase. The protocol is executed independently of the messages to be send later, and in particular the length of the internal state of P_R after the pre-processing, which we denote by sk, is independent of the messages to be send. Then each time a message m becomes known to P_S he computes an encryption c of m and sends c to P_R who outputs a value m'. We allow access to ideal functionalities during the pre-processing phase. This means that in principle the keys could be distributed entirely by a trusted party. We only require that P_R receives no messages from P_S or ideal functionalities during decryption! We call such a protocol a **non-interactive communication protocol**. Since no messages are send from P_R to P_S between the encryptions sent to P_R from P_S and the computation is asynchronous we can assume that the protocol can handle arbitrary long messages, possibly by blockwise encryption using unique message identifiers.

The Random-Oracle Model. The random-oracle model is the hybrid model with access to an ideal functionality $\mathcal{O}$ specified as follows: On input $x \in \{0,1\}^*$ from any party (including the adversary) the functionality outputs to the calling party a uniformly random element $y \in \{0,1\}^k$ independent of all other evaluations (except that if queried on the same x twice the same value y will be returned). We say that a protocol π securely realizes $\mathcal{G}$ in the random-oracle model if π securely realizes $\mathcal{G}$ in the $\mathcal{O}$-hybrid model.

The Non-Programmable Random-Oracle Model. We say that a protocol $\pi^{(\cdot)} = (P_1^{(\cdot)}, \ldots, P_n^{(\cdot)})$ securely realizes $\mathcal{G}$ in the non-programmable random-oracle model if for all adversaries $\mathcal{A}^{(\cdot)}$ there exists an adversary $\mathcal{S}^{(\cdot)}$ such that for all environments $\mathcal{Z}^{(\cdot)}$ we have that $\text{IDEAL}_{\mathcal{G}^{\mathcal{O}},\mathcal{S}^{\mathcal{O}},\mathcal{Z}^{\mathcal{O}}}$ and $\text{REAL}_{\pi^{\mathcal{O}},\mathcal{A}^{\mathcal{O}},\mathcal{Z}^{\mathcal{O}}}$ are computationally indistinguishable, where $\mathcal{O}$ is the RO functionality.

4 Possibility of NINCE in the RO Model

Let F be a family of trapdoor permutations, where one can verify $y \in D_{pk}$ given just pk, and consider the following protocol $\pi_{F,\text{nce}}$: On initialization of the protocol each party P_i generates $(pk_i, sk_i) \leftarrow \mathcal{G}(k)$ and sends pk_i to all other parties. After the key distribution phase the protocol proceeds as follows:

Send On input $(\mathbf{send}, mid, j, m)$ party P_i generates a uniformly random element
$x \leftarrow \mathcal{X}(pk_j)$, computes $(mid, f_{pk_j}(x), H(mid\|i\|j\|x) \oplus m)$, and sends this
value to P_j [2].
Receive If P_j receives (mid, y, R) from P_i, where $y \in D_{pk_j}$, then P_j computes
$x = f_{sk_j}^{-1}(y)$ and $m = R \oplus H(mid\|i\|j\|x)$ and outputs $(\mathbf{receive}, mid, i, m)$.

Theorem 1. *If F is a family of trapdoor permutations, then $\pi_{F,\mathrm{nce}}$ is a NINCE protocol for the RO model.*

Proof. Let $\mathcal{A}$ be any PPT adversary. We construct an ideal process adversary $\mathcal{S}$, which running in the ideal process will simulate an execution of $\pi_{F,\mathrm{nce}}$ to $\mathcal{A}$ and let $\mathcal{A}$ do the communication with any $\mathcal{Z}$ to convince $\mathcal{Z}$ that it is viewing a real-life execution. Since the protocol runs in the RO model, $\mathcal{S}$ will also have to simulate a RO H. It does this by defining $H(h)$ to be some uniformly random value $r \in \{0, 1\}^k$, when $H(h)$ is needed. The simulator $\mathcal{S}$ will simulate the key-distribution phase by generating random keys as in the protocol. In fact, to make the proof of security easier we will assume that $\mathcal{S}$, besides running in the ideal process, participates in a trapdoor game. The public keys pk_i for the parties will then be obtained from the trapdoor game using n key generation requests. The trapdoors will therefore not be known to $\mathcal{S}$.

To be able to simulate without the trapdoors we represent H in a particular way using two dictionaries **raw** and **img**. At the beginning of the simulation both dictionaries are empty, and H is undefined on all values. We **record** a new definition $H(h) := r$ as follows.

- If h can be parsed as $mid\|i\|j\|x$, where i and j are indicies of parties and $x \in D_{pk_j}$, then the entry $(mid\|i\|j\|y, r)$, where $y = f_{pk_j}(x)$, is added to **img**.
- If h cannot be parsed as described above, then (h, r) is added to **raw**.

We say that $H(h)$ is **defined** and $H(h) = r$ iff $h = mid\|i\|j\|x$ (for $x \in D_{pk_j}$) and $(mid\|i\|j\|f_{pk_j}(x), r) \in \mathbf{img}$, or h cannot be parsed as specified and $(h, r) \in \mathbf{raw}$. Because f_{pk_j} is a permutation, this representation is consistent: $H(h) = r$ will become defined iff recorded. Equally important, this representation allows to define and evaluate H on $h = mid\|i\|j\|x$ given just (mid, i, j, y), where $y = f_{pk_j}(x)$. We call these manipulations **oblivious**.

Remember that $\mathcal{S}$ has access to the ideal-world execution. We name the parties in the ideal world $\tilde{P}_1, \ldots, \tilde{P}_n$ — remember that these dummy parties just pass messages between the environment and the ideal functionality $\mathcal{F}_{\mathrm{nce}}$. The parties of the simulated execution run by $\mathcal{S}$ we call $P_1, \ldots, P_n$. The simulation proceeds as follows:

RO Evaluation If $\mathcal{A}$ asks for an evaluation of the RO on some string h, then
if $H(h)$ is defined, return $H(h)$, otherwise generate uniformly random $r \in \{0, 1\}^k$, set $H(h) := r$, and return r.

---

[2] We let $\|$ denote an injective and easily parsable encoding $\{0, 1\}^* \times \{0, 1\}^* \to \{0, 1\}^*$.

Send On input $(\text{send}, mid, i, j, |m|)$ from the NCE functionality we know that $\tilde{P}_i$ has input (send, mid, j, m) for some $m \in \{0,1\}^{|m|}$ to the NCE functionality, which has then sent $(\text{receive}, mid, i, m)$ to $\tilde{P}_j$.

 - If $\tilde{P}_j$ is corrupted, then $\mathcal{S}$ will deliver the message to $\tilde{P}_j$ to learn m and will then simulate by following the protocol using m as the message.
 - If $\tilde{P}_j$ is honest, then $\mathcal{S}$ simulates the protocol to $\mathcal{A}$ by sending the message (mid, y, R), where y is obtained as a uniformly random element in the image of f_{pk_j} from the trapdoor game and $R \in \{0,1\}^k$ is chosen uniformly at random. If at a later point P_i or P_j is corrupted then:
    - If P_i was corrupted, then $\mathcal{S}$ corrupts $\tilde{P}_i$ in the ideal process and learns m. The simulator then gives up on y and learns x, r such that $x = \mathcal{X}(pk_j, r)$ and $y = f_{pk_j}(x)$. The simulator then gives up on pk_i and learns sk_i, r such that $(pk_i, sk_i) = \mathcal{G}(k, r)$. Then the simulator gives this internal view of P_i to $\mathcal{A}$ and records $H(mid\|i\|j\|x) := R \oplus m$.
    - If P_j was corrupted, then $\mathcal{S}$ corrupts $\tilde{P}_j$ in the ideal process and learns m. The simulator then gives up on pk_j and learns sk_j, r such that $(pk_j, sk_j) = \mathcal{G}(k, r)$. Then the simulator gives this internal view of P_j to $\mathcal{A}$ and records $H(mid\|i\|j\|x) := R \oplus m$.

 If $H(mid\|i\|j\|x)$ was already defined in either of the above cases (to a value different from $R \oplus m$), then the simulator gives up the simulation.

Receive On the message (mid, y, R) from P_i to P_j the simulator $\mathcal{S}$ needs to make the ideal functionality output $(\text{receive}, mid, i, m)$ to $\tilde{P}_j$, where $m = R \oplus H(mid\|i\|j\|f_{sk_j}^{-1}(y))$.

 - If P_i is honest, then (mid, y, R) was sent by $\mathcal{S}$ itself and in that case the message $(\text{receive}, mid, i, m)$ has already been sent to $\tilde{P}_j$ in the ideal process. The simulator then delivers this message to $\tilde{P}_j$.
 - If P_i is corrupted, then decrypt as follows: If $H(mid\|i\|j\|f_{sk_j}^{-1}(y))$ is not defined then obliviously define it to a uniformly random value. Obliviously look up $H(mid\|i\|j\|f_{sk_j}^{-1}(y))$, and let $m = R \oplus H(mid\|i\|j\|f_{sk_j}^{-1}(y))$. Then input (send, mid, j, m) to $\tilde{P}_i$ in the ideal process and make $\mathcal{F}_{nce}$ deliver the message $(\text{receive}, mid, i, m)$ to $\tilde{P}_j$.

If the simulation is not given up, then it is distributed exactly as a real-life execution. It is therefore enough to prove that the probability that the simulation is given up is negligible. Assume for the sake of contradiction that the simulation is given up with significant (i.e. not negligible) probability. This means that with significant probability the simulator obtained $y = f_{pk_j}(x)$ from the trapdoor game and send (mid, y, R) from honest P_i to honest P_j, and the dictionary was defined on the value $mid\|i\|j\|x$ before $\mathcal{S}$ needed to define it on that value. Since P_i is guaranteed to be honest up to the point in the simulation where $\mathcal{S}$ needs to define the dictionary on the value $mid\|i\|j\|x$, we can neglect the probability that the simulator has defined H on $mid\|i\|j\|x$ twice, as it would involve choosing the same value y in the image of f_{pk_j} twice under the uniform distribution, which happens with negligible probability. Therefore the other definition of H

on $mid\|i\|j\|x$ was made by the adversary in a RO Evaluation. The first definition of H on $mid\|i\|j\|x$ was therefore not oblivious, and thus $x = f_{sk_j}^{-1}(y)$ is known. Since both P_i and P_j are honest up to the point where the simulation is given up, the simulator has not given up on y or pk_j. This allows the simulator to win the trapdoor game with significant probability, a contradiction to Lemma 1.   □

5   Impossibility of NINCE in the NPRO Model

We start by proving a lemma. Let $S^{(\cdot)}$ be a probabilistic ITM with oracle access, let $D^{(\cdot)}$ be a probabilistic TM with oracle access, and let $l_m, l_{sk} : N \to N$. We say that $S^{(\cdot)}$ is a NPRO non-committing cryptosystem simulator with private key-length l_{sk}, message length l_m, and decryption algorithm $D^{(\cdot)}$ if the following holds: On input $k \in N$ and access to a RO $\mathcal{O}$ the ITM $S^{(\cdot)}$ outputs a string $c \in \{0,1\}^*$. Then on input $m \in \{0,1\}^{l_m(k)}$ the ITM $S^{(\cdot)}$ outputs a string $sk \in \{0,1\}^{\leq l_{sk}(k)}$, where $\{0,1\}^{\leq l_{sk}(k)}$ is the set of strings of length at most $l_{sk}(k)$. For all $m \in \{0,1\}^{l_m(k)}$, let $\Pr_{S,m}[c, sk, r]$ denote the joint probability distribution on the values (c, sk, r) when c, sk are generated by $S^{\mathcal{O}}$ on inputs (k, m) and r is a uniformly random string. Let $\Pr_S[c, r]$ be the distribution of the values (c, r), which are independent of m. We require that there exists k_0 such that for all $k > k_0$ and all $m \in \{0,1\}^{l_m(k)}$,

$$\Pr_{S,m}[D^{\mathcal{O}}(sk, c; r) = m] \geq \frac{3}{4} . \tag{1}$$

Lemma 2. *If $S^{(\cdot)}$ is a NPRO non-committing cryptosystem simulator with private key-length l_{sk}, message length l_m, and decryption algorithm $D^{(\cdot)}$, then for all $k > k_0$: $l_{sk}(k) + 2 \geq l_m(k)$.*

Proof. Assume for the sake of contradiction that there exists $k > k_0$ such that $l_{sk}(k) + 2 < l_m(k)$. For all $c \in \{0,1\}^*$ and all $m \in \{0,1\}^{l_m(k)}$ define

$$SK(c, m) = \{sk \in \{0,1\}^{\leq l_{sk}(k)} | \Pr_S[D^{\mathcal{O}}(sk, c; r) = m] > \frac{1}{2}\}$$

$$X_m = \sum_{c \in \{0,1\}^*} \Pr_S[c] |SK(c, m)|$$

$$X = \sum_{m \in \{0,1\}^{l_m(k)}} X_m .$$

If $sk \in SK(c, m_1) \cap SK(c, m_2)$ and $m_1 \neq m_2$, then $\Pr_S[D^{\mathcal{O}}(sk, c; r) \in \{m_1, m_2\}] > 1$. So, for fixed c each $sk \in \{0,1\}^{\leq l_{sk}(k)}$ belongs to only one of the sets $SK(c, m)$. Therefore

$$X = \sum_{c \in \{0,1\}^*} \Pr_S[c] \sum_{m \in \{0,1\}^{l_m(k)}} |SK(c, m)|$$
$$\leq \sum_{c \in \{0,1\}^*} \Pr_S[c](2^{l_{sk}(k)+1} - 1) = 2^{l_{sk}(k)+1} - 1 . \tag{2}$$

If $X_m \geq \frac{1}{4}$ for all $m \in \{0,1\}^{l_m(k)}$, then by (2) we have that $2^{l_{sk}(k)+1} - 1 \geq X \geq 2^{l_m(k)}\frac{1}{4}$ contradicting $l_{sk}(k) + 2 < l_m(k)$. So, there exists $m \in \{0,1\}^{l_m(k)}$ s.t. $X_m < \frac{1}{4}$. In particular, by the Markov inequality $\Pr_S[|SK(c,m)| \geq 1] < \frac{1}{4}$ and

$$
\Pr_{S,m}[D^{\mathcal{O}}(sk,c;r) = m]
$$
$$
= \Pr_{S,m}[|SK(c,m)| \geq 1] \cdot \Pr_{S,m}[D^{\mathcal{O}}(sk,c;r) = m \,||\, SK(c,m)| \geq 1]+
$$
$$
\Pr_{S,m}[|SK(c,m)| = 0] \cdot \Pr_{S,m}[D^{\mathcal{O}}(sk,c;r) = m \,||\, SK(c,m)| = 0]
$$
$$
< \frac{1}{4} \cdot 1 + 1 \cdot \frac{1}{2} = \frac{3}{4}
$$

contradicting (1).   □

Theorem 2. *There exists no NINCE protocol for the NPRO model with erasure.*

Proof. Consider the real-life execution of a NINCE protocol with two parties P_S and P_R (we drop the $(\cdot)$ superfix as all ITMs in the proof are ITMs with oracle access). Let $sk(k)$ denote the random variable describing the internal state of P_R after the pre-processing phase and before receiving the first encryption. Since this state is independent of the inputs to be sent, there exists a polynomial $l_{sk}(k)$ s.t. the expected value of $|sk(k)|$ is bounded by $\frac{l_{sk}(k)}{6}$ for large enough k, and by the Markov inequality $\Pr[|sk(k)| \geq l(k)] < \frac{1}{6}$ for large enough k.

Consider the following environment $\mathcal{Z}$: It activates P_S on a message $m \in \{0,1\}^{l_m(k)}$, where $l_m(k) = l_{sk}(k) + 3$ and m is a prefix of z. Let E_1 denote the event that after activating P_S with input m the adversary outputs a value $c \in \{0,1\}^*$. If E_1 does not occur, the environment terminates with output 0. If E_1 occurs, the environment activates the adversary with input "corrupt the receiver". Let E_2 be the event that $\mathcal{Z}$ as response to this activation observes that P_R is corrupted and that the adversary outputs a value $sk \in \{0,1\}^{\leq l_{sk}(k)}$. If E_2 does not occur, $\mathcal{Z}$ outputs 0. If E_2 does occur, $\mathcal{Z}$ generates uniformly random bits r and computes m' by running the code of P_R from internal state sk with input c and random bits r and using the RO $\mathcal{O}$; Since P_R does not access any ideal functionalities during decryption and $\mathcal{Z}$ has access to the same oracle as P_R, the environment can actually carry out this computation. If $m' = m$ then $\mathcal{Z}$ outputs 1, otherwise $\mathcal{Z}$ outputs 0.

Consider the following real-life adversary $\mathcal{A}$: It activates P_S and P_R until P_S sends c. Then $\mathcal{A}$ outputs c, where c is the value sent from P_S to P_R but not delivered. Then on the next activation, it corrupts P_R and outputs to the environment the internal state of P_R.

Let E be the event that E_1 and E_2 occurs. Note that in the real-life execution $\mathrm{REAL}_{\pi^{\mathcal{O}},\mathcal{A}^{\mathcal{O}},\mathcal{Z}^{\mathcal{O}}}(k,z)$ the event E occurs with probability at least $\frac{5}{6}$ for large enough k. Furthermore, by the security (which implies correctness) of the protocol, the probability that $m' \neq m$ for uniformly random r is bounded by a negligible function $\delta(k)$. Therefore $\Pr[\mathrm{REAL}_{\pi^{\mathcal{O}},\mathcal{A}^{\mathcal{O}},\mathcal{Z}^{\mathcal{O}}}(k,z) = 1] \geq \frac{5}{6} - \delta(k) > \frac{4}{5}$ for large enough k. Since the protocol is secure and $\frac{3}{4} < \frac{4}{5}$ there exists S and k_0

such that for all $k > k_0$ we have that $\Pr[\text{IDEAL}_{\mathcal{F}_{nce}, \mathcal{S}^{\mathcal{O}}, \mathcal{Z}^{\mathcal{O}}}(k, z) = 1] \geq \frac{3}{4}$. We use this to construct a NPRO non-committing cryptosystem simulator S as follows: On input k and access to oracle $\mathcal{O}$ run $\text{IDEAL}_{\mathcal{F}_{nce}, \mathcal{S}^{\mathcal{O}}, \mathcal{Z}^{\mathcal{O}}}(k, z)$ on a message m of length $l_m(k)$. Note that as long as no party is corrupted the execution does not require that we know the value of m. If during the execution the event E does not occur, S uses arbitrary values for c and sk. If E occurs, S proceeds as follows: Run $\text{IDEAL}_{\mathcal{F}_{nce}, \mathcal{S}^{\mathcal{O}}, \mathcal{Z}^{\mathcal{O}}}(k, z)$ until S outputs c and then output c. Then on input $m \in \{0, 1\}^{l_m(k)}$ run $\text{IDEAL}_{\mathcal{F}_{nce}, \mathcal{S}^{\mathcal{O}}, \mathcal{Z}^{\mathcal{O}}}(k, z)$ until S corrupts P_R, give S the value m, and run $\text{IDEAL}_{\mathcal{F}_{nce}, \mathcal{S}^{\mathcal{O}}, \mathcal{Z}^{\mathcal{O}}}(k, z)$ until S outputs sk. Then output sk. Let D be the TM which on input c, sk and access to oracle $\mathcal{O}$ runs P_R from internal state sk and input c using uniformly random bits and the RO $\mathcal{O}$. By $\Pr[\text{IDEAL}_{\mathcal{F}_{nce}, \mathcal{S}^{\mathcal{O}}, \mathcal{Z}^{\mathcal{O}}}(k, z) = 1] \geq \frac{3}{4}$ we have that S is a non-committing cryptosystem simulator with private key-length l_{sk}, message length l_m, and decryption algorithm D, contradicting Lemma 2. □

The proof of Theorem 2 was done in the model from [Can01]. Since Theorem 2 states a negative result, the result would be stronger if we could prove it in a weaker model. In [CFGN96] the NCE problem was formulated in the MPC model of [Can00], which is a considerably weaker model as it models synchronous computation and only guarantees security preservation under non-concurrent composition of protocols. The negative result however still holds in this model. We can prove that no NINCE protocol exists for the synchronous model without erasure, and we can prove that no NINCE protocol which can communicate an unbounded number of bits in one round (i.e. without synchronizing the sender and the receiver) exists for the synchronous model with erasure. The proof of the first claim follows the proof of Theorem 2, except that c is communicated between the execution and the so-called post-execution phase and P_R is corrupted in the post-execution phase. We sketch the proof of the second claim.

The main difference between the models in [CFGN96] and [Can00] is that the model in [Can00] does not have an explicit mechanism for the adversary to output values to the environment during the execution. Since it is essential to the proof with erasure that P_R is corrupted before c arrives and that it is essential that c is output before P_R is corrupted, this becomes an issue. However in [Can00] some information can indeed flow from the adversary to the environment as the environment learns the identity of the parties corrupted by $\mathcal{A}$. The NCE problem is cast as a multiparty problem, see [CFGN96], where a polynomial number of parties participate. So the adversary can use its corruption pattern to communicate c; To guarantee that the value output has a fixed polynomial length, a Markov inequality can be used as in the proof of Theorem 2.

Acknowledgments

I would like to thank the following persons for valuable discussions of the paper and for suggestions on how to improve the presentation: Ran Canetti, Ronald Cramer, Ivan Damgård, Yehuda Lindell, Moti Yung, and a number of anonymous referees.

126   Jesper Buus Nielsen

References

[ACM88]   *Proceedings of the Twentieth Annual ACM Symposium on Theory of Computing*, Chicago, Illinois, 2–4 May 1988.

[Bea97]   D. Beaver. Plug and play encryption. In *Crypto '97*, pages 75–89, Berlin, 1997. Springer. LNCS Vol. 1294.

[BFM88]   Manuel Blum, Paul Feldman, and Silvio Micali. Non-interactive zero-knowledge and its applications (extended abstract). In [ACM88], pages 103–112.

[BGW88]   Michael Ben-Or, Shafi Goldwasser, and Avi Wigderson. Completeness theorems for non-cryptographic fault-tolerant distributed computation (extended abstract). In [ACM88], pages 1–10.

[BH92]   D. Beaver and S. Haber. Cryptographic protocols provably secure against dynamic adversaries. In *EuroCrypt '92*, pages 307–323, Berlin, 1992. Springer. LNCS Vol. 658.

[BR93]   Mihir Bellare and Phillip Rogaway. Random oracles are practical: A paradigm for designing efficient protocols. In *First ACM Conference on Computing and Communications Security*, pages 62–73. ACM, 1993.

[BR95]   M. Bellare and P. Rogaway. Optimal asymmetric encryption. In *EuroCrypt '94*, pages 92–111, Berlin, 1995. Springer. LNCS Vol. 950.

[Can00]   Ran Canetti. Security and composition of multiparty cryptographic protocols. *Journal of Cryptology*, 13(1):143–202, winter 2000.

[Can01]   Ran Canetti. Universally composable security: A new paradigm for cryptographic protocols. In *42th Annual Symposium on Foundations of Computer Science*. IEEE, 2001.

[CCD88]   David Chaum, Claude Crépeau, and Ivan Damgård. Multiparty unconditionally secure protocols (extended abstract). In [ACM88], pages 11–19.

[CFGN96] Ran Canetti, Uri Feige, Oded Goldreich, and Moni Naor. Adaptively secure multi-party computation. In *Proceedings of the Twenty-Eighth Annual ACM Symposium on the Theory of Computing*, pages 639–648, Philadelphia, Pennsylvania, 22–24 May 1996.

[CGH98]   Ran Canetti, Oded Goldreich, and Shai Halevi. The random oracle methodology, revisited (preliminary version). In *Proceedings of the Thirtieth Annual ACM Symposium on the Theory of Computing*, pages 209–218, Dallas, TX, USA, 24–26 May 1998.

[DN00]   Ivan Damgård and Jesper B. Nielsen. Improved non-committing encryption schemes based on a general complexity assumption. In *Crypto 2000*, pages 432–450, Berlin, 2000. Springer. LNCS Vol. 1880.

[FS86]   A. Fiat and A. Shamir. How to prove yourself: practical solutions to identification and signature problems. In *Crypto '86*, pages 186–194, Berlin, 1986. Springer. LNCS Vol. 263.

[GK90]   O. Goldreich and H. Krawczyk. On the composition of zero knowledge proof systems. In *Proceedings of ICALP 90*, Berlin, 1990. Springer. LNCS Vol. 443.

[Sho01]   Victor Shoup. OAEP reconsidered. In *Crypto 2001*, pages 239–259, Berlin, 2001. Springer. LNCS Vol. 2139.

On the Security of RSA Encryption in TLS

Jakob Jonsson and Burton S. Kaliski Jr.

RSA Laboratories, 20 Crosby Drive, Bedford, MA 01730, USA
{jjonsson,bkaliski}@rsasecurity.com

Abstract. We show that the security of the TLS handshake protocol based on RSA can be related to the hardness of inverting RSA given a certain "partial-RSA" decision oracle. The reduction takes place in a security model with reasonable assumptions on the underlying TLS pseudo-random function, thereby addressing concerns about its construction in terms of two hash functions. The result is extended to a wide class of constructions that we denote *tagged key-encapsulation mechanisms*.

Keywords: key encapsulation, RSA encryption, TLS.

1 Introduction

One of the most popular methods for establishing secret information between two parties with no prior shared secret is the handshake protocol used in the Secure Sockets Layer (SSL) [15] and Transport Layer Security (TLS) [10] protocols (which we will refer to jointly as TLS). These protocols support a variety of algorithms (called "cipher suites"). In the suite of interest to this paper, the handshake protocol is based on the RSA-PKCS-1v1_5 (abbreviated RSA-P1) encryption scheme introduced in the PKCS #1 v1.5 [31] specification, which in turn is based on the RSA trapdoor permutation [30].

Due to their widespread use, both the TLS handshake protocol and the underlying encryption scheme RSA-P1 have been subject to a significant amount of cryptanalysis. A number of weaknesses in RSA-P1 for general message encryption have been found, including the results given in [3, 5, 6, 8]. These results suggest that RSA-P1 must be equipped with certain countermeasures to provide an adequate level of security.

Briefly, the common case of the TLS protocol we will analyze has the following form. A *server* has a public key / private key pair. A *client* establishes a secret s with the server through the following key agreement scheme (omitting certain details):

1. The client and server select a new *nonce* ρ, to which they both contribute.
2. The client generates a *pre-master secret* r, encrypts r with the server's public key under RSA-P1 to obtain a ciphertext y, and sends y to the server. The server decrypts y to recover r.
3. The client and the server both derive a *master secret* t from r.
4. The client and the server compute separate tags z and z' from t and ρ, exchange the tags, and verify them.
5. If the tags are correct, the client and the server both derive a shared secret s from the master secret t.

M. Yung (Ed.): CRYPTO 2002, LNCS 2442, pp. 127–142, 2002.
© Springer-Verlag Berlin Heidelberg 2002

Only the server has a public key, and hence only the server is authenticated in this scheme. The scheme follows reasonable design principles, such as including a nonce for freshness and a tag for assurance that the client knows the pre-master secret [24]. However, we are not aware of any formal security proof relating the difficulty of "breaking" this scheme to any underlying problem, e.g., RSA.

To facilitate a proof, we will model the interaction between the client and the server as a *tagged key-encapsulation mechanism* (TKEM), which may be viewed as an extension to key encapsulation mechanisms as defined in [28]. The client's steps are considered as an *encryption operation* that produces a ciphertext, a tag, and a secret from the nonce. A *decryption operation* corresponding to the server's steps produces the same secret from the ciphertext, the tag, and the nonce if the tag is correct. (We omit the tag computed by the server in this model as it is not needed for the proof, and in any case is no more helpful to an adversary than the one computed by the client.) The security of the key agreement scheme is thus transformed to the indistinguishability of the TKEM against a chosen-ciphertext attack.

Using this model, we show that the security of the TKEM underlying TLS can be related via a reasonably tight reduction to the hardness of inverting RSA with the assistance of a "partial-RSA decision oracle". This oracle takes as input an RSA ciphertext y and a bit string r of length 384 bits and checks whether r is equal to the last 384 bits of the RSA decryption of y. While based on a stronger assumption than the corresponding proofs for RSA-OAEP [2, 16], RSA-OAEP+ [27], RSA-KEM [32, 1, 28], RSA-REACT [23], and RSA-GEM [7], this is the first security proof relating an RSA-P1-based application to the RSA problem.

We consider two TKEMs in this paper. The first TKEM is based on a single pseudo-random function, which we model as a random oracle. However, TLS actually uses the xor of two pseudo-random functions, each based on a different hash function, to address the risk that one of the pseudo-random functions might turn out to be weak. The second TKEM we consider follows this design, and we model only one of the pseudo-random functions as a random oracle and assume no specific properties of the other. As a result, we are able to show that the TLS handshake is secure even if one of the hash functions turns out to be weak. This addresses a concern [18] about the TLS pseudo-random function construction.

2 Basic Concepts

In Section 2.1, we provide basic concepts and definitions related to trapdoor mappings; the special case RSA-P1, intended for use within the TKEM2 mechanism introduced in Section 3.2, is defined in Section 2.2. In Section 2.3 we define plaintext-checking oracles instrumental for the security reductions.

Notation

A *bit string* is an ordered sequence of elements from $B = \{0,1\}$. For $n \geq 0$, $B^n = \{0,1\}^n$ denotes the set of bit strings of length n. An *octet* is an element in B^8. Let B^* denote the set of all bit strings. Bit strings and octet strings are identified with the integers they represent in base 2 and 2^8, respectively. $\mathbb{Z}_n$ denotes the set $\{0, \ldots, n-1\}$; $\mathbb{Z}_n$ is the additive group of integers modulo n. To denote that an element a is chosen uniformly at random from a finite set A, we write $a \xleftarrow{R} A$.

2.1   Randomized Trapdoor Mappings

Here we give a brief introduction to randomized trapdoor mappings, generalizing the concept of trapdoor permutations; a trapdoor mapping is invertible but not necessarily deterministic. Let k be a security parameter. For each k, let $\mathcal{E}_k$ be a finite family of pairs (E, D) with the property that E is a randomized and reversible algorithm with inverse D. E takes as input an element r in a set $R = R_E$ and returns an element y in a set $Y = Y_E$, possibly via randomness generated within the algorithm; we will write $y \leftarrow E(r)$. D is a deterministic algorithm $Y \rightarrow R \cup \{\phi\}$ such that $D(y) = r$ if $y \leftarrow E(r)$ for some $r \in R$ and $D(y) = \phi$ otherwise. Each output y from E corresponds to at most one input r. y is *valid* if $D(y) \neq \phi$ and *invalid* otherwise. We assume that the running time of each of E and D is polynomial in k.

Let $\mathcal{G}$ be a probabilistic polynomial-time (PPT) algorithm that on input 1^k (i.e., k uniformly random bits) outputs a pair $(E, D) \in \mathcal{E}_k$. $\mathcal{G}$ is a *trapdoor mapping generator*. An *E-inverter* $\mathcal{I}$ is an algorithm that on input (E, y) tries to compute $D(y)$ for a random $y \in Y$. $\mathcal{I}$ has success probability $\epsilon = \epsilon(k)$ and running time $T = T(k)$ if

$$\Pr\left((E, D) \leftarrow \mathcal{G}(1^k), r \xleftarrow{R} R_E, y \leftarrow E(r) : \mathcal{I}(E, y) = r\right) \geq \epsilon$$

and the running time for $\mathcal{I}$ is at most T. In words, $\mathcal{I}$ should be able to compute $D(y)$ with probability ϵ within time T, where (E, D) is derived via the trapdoor mapping generator and y is random. $\mathcal{I}$ solves the E *problem*.

$\mathcal{E}_k$ is a *trapdoor mapping family* with respect to (ϵ, T) if the E-problem is (ϵ, T)-*hard*, meaning that there is no E-inverter with success probability ϵ within running time T. The individual mapping E is referred to as a *trapdoor mapping*. A *trapdoor permutation* is a deterministic trapdoor mapping E with $Y_E = R_E$.

An RSA permutation $f : \mathbb{Z}_N \rightarrow \mathbb{Z}_N$ is defined in terms of an *RSA public key* (N, e) as $f(x) = x^e \bmod N$. N is the *RSA modulus*, a product of two secret integer primes p and q, while e is an odd (typically small) integer such that $\gcd(e, (p-1)(q-1)) = 1$. RSA permutations are widely believed to be trapdoor permutations. This means that it is presumably hard to compute $f^{-1}(y)$ on a random input $y \in \mathbb{Z}_N$ provided N is large enough and generated at random. Yet, given secret information (e.g., the prime factors of N), the inverse $f^{-1}(y)$ is easy to compute.

2.2   RSA-PKCS-1v1_5

Here we describe the specific trapdoor mapping RSA-P1 (RSA-PKCS-1v1_5) introduced in PKCS #1 v1.5, which is based on the RSA permutation. For the purposes of this paper, the input to the RSA-P1 encryption operation has a fixed length; in PKCS #1 v1.5, the input may have a variable length.

Let l_N be the octet length of the RSA modulus N. The encryption operation takes as input an element $r \in B^{8l_r}$, where $l_r \leq l_N - 11$; put $k_r = 8l_r$. (In TLS, $l_r = 48$ and $k_r = 384$.) The trapdoor mapping RSA-P1 is defined as follows; 00 and 02 are octets given in hexadecimal notation. Note that P is an octet string of length $l_N - l_r - 3 \geq 8$ consisting of nonzero octets.

RSA-P1-ENCRYPT(r)

$- P \xleftarrow{R} \left(B^8 \setminus \{00\}\right)^{l_N - l_r - 3}$;

$- x \leftarrow 00\|02\|P\|00\|r$;

$- y \leftarrow x^e \bmod N$;

$-$ Return the integer y.

Aligning with the terminology in Section 2.1, an RSA-P1 inverter solves the *RSA-P1 problem*, whereas an RSA inverter solves the *RSA problem*.

2.3 Plaintext-Checking and Partial-RSA Decision Oracles

One property of many trapdoor mappings as defined in Section 2.1 is that it is presumably hard to tell for a given pair (r, y) whether $D(y) = r$ if D is secret. For example, RSA-P1 defined in Section 2.2 appears to have this property. The reason is that there might be many possibilities for $E(r)$ for each r as soon as E is randomized. While this may appear to be an attractive feature of a trapdoor mapping, in our setting it is in fact a drawback. Namely, to reduce an E-inverter to a chosen-ciphertext attack against the tagged key-encapsulation mechanisms introduced in Section 3, we must be able to simulate an oracle that on input (r, y) tells whether $D(y) = r$. There is no generic solution to this problem.

To address this concern, we make use of the *plaintext-checking oracle* concept introduced in [22, 23]. This oracle, which we denote PO_E, takes as input a pair $(r, y) \in R \times Y$ and checks whether $r = D(y)$. If this is true, the oracle outputs the bit 1. Otherwise it outputs the bit 0. The plaintext-checking oracle is correct on each input with probability 1.

For the specific RSA trapdoor permutation f, we introduce also a *partial-RSA decision oracle*. For any integer $k_0 < k$ (k is the bit length of the modulus), this oracle DO_{f,k_0} takes as input a string $r \in B^{k_0}$ and a ciphertext $y \in \mathbb{Z}_N$ and checks whether

$$f^{-1}(y) \bmod 2^{k_0} = r .$$

Thus this oracle compares a string r with a "partial" RSA inverse of y.

An interesting question is whether this oracle helps an adversary invert RSA. If k_0 is almost as large as k, the oracle clearly does not help since the adversary can simulate it efficiently, either by guessing or by applying the reduction technique of Coppersmith [5]. The latter case can accommodate k_0 down to the range of $k(1 - 1/e)$, where e is the RSA encryption exponent. If k_0 is small (say, smaller than 80), the oracle clearly does help: After a brute-force search using the oracle at most 2^{k_0} times, the adversary will be able to determine the last k_0 bits of the RSA inverse. Applying the method described in Appendix A, the adversary will then easily determine the whole of the inverse via another $k - k_0$ applications of the oracle.

For TLS, we have $k_0 = 192$ or 384 (depending on how we model the underlying pseudo-random function) and typically $k \geq 1024$, which implies that neither of the above extreme cases apply. Our conjecture is that k_0 is large enough to render the oracle useless.

An algorithm is $PO(q)$-*assisted* (resp. $DO(q)$-assisted) if it has access to a plaintext-checking oracle PO (resp. decision oracle DO) that accepts at most q queries. A PO_E-assisted E-inverter solves the *gap-E problem*. For example, a PO_{RSA-P1}-assisted RSA-P1 inverter solves the *gap-RSA-P1 problem*. A DO_{f,k_0}-assisted RSA inverter solves the *gap-partial-RSA problem* with parameter k_0.

3   Tagged Key-Encapsulation Mechanisms

In this section we discuss the concept of tagged key-encapsulation mechanisms (TKEM), which are useful for modeling key agreement schemes such as the one in the TLS handshake protocol. Section 3.1 is devoted to a TKEM defined in terms of a trapdoor mapping and a pseudo-random function approximating the mechanism underlying TLS, while Section 3.2 concentrates on identifying the specific RSA-P1-based TKEM within TLS. (The Diffie-Hellman [11] version of the TLS handshake, which is based on ElGamal [13] key agreement as defined in [21], can also be modeled with a TKEM; however, we do not address that version here.)

A TKEM consists of an *encryption operation* and a *decryption operation*. The encryption operation TKEM-ENCRYPT takes as input a public key K_{pub}, a nonce ρ, and possibly some other parameters and returns a triple (y, z, s), where y is the *ciphertext*, z is the *tag*, and s is the *secret*. The decryption operation TKEM-DECRYPT takes as input a private (secret) key K_{priv}, a ciphertext y, a tag z, a nonce ρ, and possibly some other parameters and returns the secret s or "Error" if the ciphertext is not valid or the tag is inconsistent with the nonce and the other information. Each new application of TKEM-DECRYPT requires a nonce that has not been used before.

As noted above, the security of the key agreement scheme in the TLS handshake can be transformed to the indistinguishability of the underlying TKEM. In particular, the key agreement scheme is secure against an adversary sending ciphertexts to the server if the underlying TKEM is secure against an adversary sending ciphertexts to a decryption oracle simulating TKEM-DECRYPT. It is clear that the latter adversary is at least as strong as the former adversary as she is allowed to select the entire nonce herself (as long as it is new).

3.1   TKEM1

We introduce the tagged key-encapsulation scheme TKEM1 as a first approximation to the TLS handshake. TKEM1 is defined in terms of a trapdoor mapping

E; write $R = R_E$ and $Y = Y_E$. Let $h : B^* \times B^* \times \mathbb{Z} \to B^*$ be a pseudo-random function; $h(r, \sigma, l)$ is a string of length l (there might be restrictions on the sizes of the inputs and outputs). Fix parameters k_s, k_t, k_z, k_δ, and k_ρ; these parameters denote bit lengths of different strings used within TKEM1. Let $\Delta : B^* \times B^* \times B^* \to b^{k_\delta}$ be an arbitrary function; we will refer to the output from Δ as a *digest* (this reflects the way Δ is used within TLS). Δ is part of the tag derivation construction. Let Q_s, Q_t, and Q_z be three distinct strings such that no string is a prefix of any of the others. This is to ensure that the inputs to the different applications of h are distinct.

The TKEM1 encryption operation takes as input a nonce $\rho \in B^{k_\rho}$ and a label $L \in B^*$. ρ provides *freshness*; each application of TKEM1 uses a new nonce. L contains public information that is intended to be integrity-protected by TKEM1. The TKEM1 encryption operation is defined as follows.

TKEM1-ENCRYPT(ρ, L)

$- r \xleftarrow{R} R;$
$- y \leftarrow E(r);$
$- t \leftarrow h(r, Q_t \| \rho, k_t);$
$- z \leftarrow h(t, Q_z \| \Delta(\rho, L, y), k_z);$
$- s \leftarrow h(t, Q_s \| \rho, k_s);$
$-$ Return the ciphertext y, the tag z, and the secret s.

The corresponding decryption operation TKEM1-DECRYPT is defined in the obvious way: First, $r = D(y)$ is recovered (if $D(y) = \phi$, then an error message is returned). Second, a tag z' is computed from r, ρ, L, and y. If $z' = z$, then the secret s is recovered and returned; otherwise, an error message is returned.

Since the output length of h is uniquely determined by the prefix of the second input (the prefix is either Q_s, Q_t, or Q_z) we will suppress the third argument of h below.

In practice, to thwart implementation attacks such as [3] and [20], the decryption operation should not output "Error" immediately if $D(y) = \phi$. Instead, the operation should proceed with a new r generated uniformly at random. With high probability, z will not match z', and "Error" will be output at the end, without revealing whether $D(y) = \phi$ or z' is incorrect.

3.2   TKEM2

TKEM2 is the specific TKEM used within TLS and may be viewed as a special case of TKEM1. However, while our security model for TKEM1 will assume that h is a strong pseudo-random function, the corresponding function in TLS, denoted Ω, has a seemingly more fragile structure, which will require special treatment.

Ω is defined as follows on input $(w, \sigma, l) \in B^{2*} \times B^* \times \mathbb{Z}$ (B^{2*} is the set of strings with an even bit length). Write $w = w^L \| w^R$, where the bit length of each of the strings w^L and w^R is half the bit length of w. Then

$$\Omega(w, \sigma, l) = g(w^{\mathrm{L}}, \sigma, l) \oplus h(w^{\mathrm{R}}, \sigma, l) \ ,$$

where g and h are pseudo-random functions defined in terms of HMAC [19] based on MD5 [26] and SHA-1 [29], respectively. We refer the interested reader to [10] for details. The rationale for the Ω construction is to achieve a reasonable amount of security even if one of the underlying pseudo-random functions turns out to be weak.

The function Δ outputs the concatenation of a SHA-1 hash value and an MD5 hash value of a string derived from the input nonce, label, and ciphertext; L corresponds to all messages in the handshake other than ρ and y. However, our results do not depend on the function Δ.

Let RSA-P1 denote the encryption scheme specified in Section 2.2. Let N be an RSA modulus and let e be an RSA public exponent; define $f(x) = x^e \bmod N$. With notation as in TKEM1, we fix $R = B^{384}$, $k_t = k_r = 384$ and $k_z = 96$, whereas k_s depends on the chosen cipher suite. An application of TKEM2 requires a version number v, a bit string of length 16. For simplicity, we assume that the version number is fixed. The nonce ρ is the concatenation of two strings ρ_1 and ρ_2 (one provided by the client and the other by the server). Each string is 32 octets long; the first 4 octets denote the number of seconds since 1/1/1970, and the rest are generated at random. Given the nonce ρ and a label L, the TKEM2 encryption operation proceeds as follows; the decryption operation is defined in alignment with TKEM1-DECRYPT.

TKEM2-ENCRYPT(ρ, L)

- $r_0 \leftarrow v \in \{0,1\}^{16}$; $r_1 \overset{R}{\leftarrow} \{0,1\}^{k_r-16}$; $r = r_0 \| r_1$;
- $y \leftarrow$ RSA-P1(r);
- $t \leftarrow \Omega(r, \text{"master secret"} \| \rho, k_t)$;
- $z \leftarrow \Omega(t, \text{"client finished"} \| \Delta(\rho, L, y), k_z)$;
- $s \leftarrow \Omega(t, \text{"key expansion"} \| \rho, k_s)$;
- Return the ciphertext y, the tag z, and the secret s.

4   Reduction from the Gap-Partial-RSA Problem to the Gap-RSA-P1 Problem

Before proceeding with security proofs for TKEM1 and TKEM2, we show how a PO_E-assisted RSA-P1-inverter ($E =$ RSA-P1 with parameter k_r) can be extended to a DO_{f,k_r}-assisted RSA inverter. Thus we reduce the gap-partial-RSA problem with parameter k_r to the gap-RSA-P1 problem[1].

Due to the shape of RSA-P1, the two oracles PO_E and DO_{f,k_r} act equivalently except on inputs (r, y) such that $D(y) = \phi$. In this case PO_E always outputs the bit 0, while DO_{f,k_r} outputs the bit 1 if $f^{-1}(y) \bmod 2^{k_r} = r$.

Let Y_0 be the set of valid ciphertexts y. Such ciphertexts have the property that $f^{-1}(y) = 00\|02\|P\|00\|r$ for some $r \in B^{k_r}$ and some string P of nonzero

---

[1] The related problem of finding a reduction from a PO_E-assisted RSA inverter to a PO_E-assisted RSA-P1-inverter is substantially more complex but can be solved via lattice reduction; consider the generalization in [9] of the approach in [16].

octets. Assuming that k_r and the bit length k of the RSA modulus are multiples of 8, the size of $Y_0/2^k$ is

$$2^{-24}\left(1-2^{-8}\right)^{(k-k_r-24)/8} > 2^{-24}\left(2^{-1/177}\right)^{(k-k_r-24)/8} = 2^{-24-(k-k_r-24)/1416},$$

which implies that $|Y_0|/|Y|$ is at least $2^{-24-(k-k_r-24)/1416}$, where $Y = \mathbb{Z}_N$. This is a lower bound on the probability that a uniformly random $y \in \mathbb{Z}_N$ is a valid RSA-P1 ciphertext. The reduction is as follows; the result is easily generalized to other RSA-based trapdoor mappings where the padding is prepended to the plaintext r. The proof is given in Appendix A.

Theorem 1. *Let E be an RSA-P1 trapdoor mapping, let k be the bit length of the RSA modulus N, and let q be a parameter. Assume that there is a $PO_E(q)$-assisted E-inverter with running time at most T' and success probability at least ϵ'. Then there is a $DO_{f,k_r}((q+1)(k-k_r+1))$-assisted f-inverter with running time at most T that is successful with probability at least ϵ, where*

$$\epsilon = \epsilon' \cdot \frac{|Y_0|}{|Y|};$$

$$T = T' + O((q+1)(k-k_r+1) \cdot T_{DO_{f,k_r}}) + O((q+1)k^3) ;$$

$T_{DO_{f,k_r}}$ is the running time for the partial-RSA decision oracle DO_{f,k_r}.

5   Security of TKEM1

5.1   TKEM1 Security Model

We define an attack model employing an adversary against TKEM1 who is given free access to a *decryption oracle*; hence we consider the family of adaptive chosen-ciphertext attacks (CCA2; see [17]). The task for the adversary is to distinguish a secret s_0^* corresponding to a certain ciphertext y^* with parameters (z^*, ρ^*, L^*) from a random string.

The decryption oracle responds to a query (y, z, ρ, L) with the corresponding secret $s = h(t, Q_s\|\rho)$ if y is a valid ciphertext and the tag z is correct ($t = h(D(y), Q_t\|\rho)$); otherwise, the oracle responds with a generic error message. The decryption oracle accepts any query, except that previous nonces must not be reused.

We will not make any specific assumptions about the digest function Δ. However, the function h will be instantiated as a random oracle. Thus the adversary has no information about $h(r, \sigma)$ unless she sends the *query* (r, σ) to an oracle instantiating h. The h-oracle responds to queries with strings chosen uniformly at random and independent from earlier queries and responses, except that a string that is repeatedly queried to the oracle should have the same response every time. To distinguish between different h-oracle queries, we let h_s-queries be queries prefixed by Q_s, while h_t-queries and h_z-queries are queries prefixed by Q_t and Q_z, respectively.

The attack experiment runs as follows. First, the adversary is given a trapdoor mapping E generated via a trapdoor mapping generator $\mathcal{G}$. The adversary is allowed to send queries to the h-oracle and the decryption oracle during the entire attack. At any time of the attack – but only once – the adversary sends a query (ρ^*, L^*) to a *challenge generator*[2]. Here, ρ^* must not be part of any previous or later decryption query. The challenge generator applies the TKEM1-ENCRYPT operation, producing a ciphertext y^*, a tag z^*, and a secret s_0^*. In addition, the generator selects a uniformly random string s_1^* and flips a fair coin b. The generator returns (y^*, z^*) and s_b^*; thus the response depends on b.

At the end, the adversary outputs a bit b'. The *distinguishing advantage* of the adversary is defined as $\Pr(b' = b) - \Pr(b' \neq b) = 2\Pr(b' = b) - 1$; the probability is computed over all possible trapdoor mappings for a given security parameter k. The adversary is an *IND-CCA2* adversary [17, 25] (IND = indistinguishability).

5.2 Reduction from a PO_E-Assisted E-Inverter to an IND-CCA2 Adversary Against TKEM1

In the full version of this paper, we show how to reduce a PO_E-assisted E-inverter to an IND-CCA2 adversary against TKEM1. If E is equal to RSA-P1, the result is easily translated into a reduction from the gap-partial-RSA problem to the security of TKEM1 via Theorem 1. Let $k_r = \lceil \log_2 |R| \rceil$. The result is as follows; the proof is suppressed in this extended abstract.

Theorem 2. *Let* q_s, q_t, q_z, q_D *be parameters. Assume that there is an IND-CCA2 adversary against TKEM1 with advantage at least* ϵ' *within running time* T' *making at most* q_s, q_t, q_z, *and* q_D *number of* h_s-*queries,* h_t-*queries,* h_z-*queries, and decryption queries, respectively. Then there is a* $PO_E(2q_t)$-*assisted* E-*inverter with running time at most* T *that is successful with probability at least* ϵ, *where*

$$\epsilon = \epsilon' - ((q_D + 2)(q_D + q_z) + q_s) \cdot 2^{-k_t} - q_D \cdot 2^{-k_z} \; ;$$
$$T = T' + O(2q_t \cdot T_{PO_E}) + \tau \; ;$$

T_{PO_E} *is the running time for the plaintext-checking oracle* PO_E, *while* τ *is the time needed for* $O(q_D + q_t + q_s + q_z)$ *elementary table operations.*

Remark. If E is deterministic, the E-inverter can implement PO_E directly by applying E; thus we may replace T_{PO_E} with the running time for an application of E. Note that when E is the RSA trapdoor permutation, we obtain a tight reduction from the ordinary RSA problem to the security of TKEM1. This result aligns with prior work on RSA-based key encapsulation mechanisms in [28, 23].

---

[2] Some authors denote this generator as an *encryption oracle*; since we find this notation somewhat confusing, we have chosen another term.

6 Security of TKEM2

In this section we restrict our attention to the TKEM2 mechanism defined in Section 3.2. Let f be an RSA permutation; $f(x) = x^e \bmod N$, where (N, e) is an RSA public key. Let E denote the corresponding RSA-P1 trapdoor mapping with $R_E = B^{k_r}$. We want to relate the security of TKEM2 to the gap-partial-RSA problem with parameter k_r.

Ideally, we would like to analyze the security of TKEM2 assuming that Ω is a random oracle. Indeed, with this assumption TKEM2 would be a special case of TKEM1. Unfortunately however, as the discussion in [18] indicates, this assumption does not seem appropriate for the specific PRF in TLS. First, the xor construction weakens the pseudo-randomness of the PRF output in terms of the input. Second, MD5 is known to have certain theoretical weaknesses [4, 12], which makes the random oracle assumption even less reasonable.

Instead, we will assume that only h based on SHA-1 is a random oracle; g based on MD5 will be treated as an ordinary function with no specific properties. It is possible to obtain a proof in case g is a random oracle and h is an ordinary function, but the reduction and the underlying decision oracle will be slightly different. Introduce six functions $g_t, h_t, g_s, h_s, g_z, h_z$ defined as

$$h_t(r^{\mathrm{R}}, \rho) = h(r^{\mathrm{R}}, \text{``master secret''} \| \rho, k_t) ;$$
$$h_s(t^{\mathrm{R}}, \rho) = h(t^{\mathrm{R}}, \text{``key expansion''} \| \rho, k_s) ;$$
$$h_z(t^{\mathrm{R}}, \delta) = h(t^{\mathrm{R}}, \text{``client finished''} \| \delta, k_z) ,$$

and analogously for g_t, g_s, and g_z. Let $h_t^{\mathrm{L}}(r^{\mathrm{R}}, \rho)$ denote the first $k_r/2$ bits of $h_t(r^{\mathrm{R}}, \rho)$ and let $h_t^{\mathrm{R}}(r^{\mathrm{R}}, \rho)$ denote the last $k_r/2$ bits. Use the corresponding notation for the two halves of $g_t(r^{\mathrm{L}}, \rho)$. With notations as in Section 3.2, note that

$$s = g_s\left(g_t^{\mathrm{L}}(r^{\mathrm{L}}, \rho) \oplus h_t^{\mathrm{L}}(r^{\mathrm{R}}, \rho), \rho\right) \oplus h_s\left(g_t^{\mathrm{R}}(r^{\mathrm{L}}, \rho) \oplus h_t^{\mathrm{R}}(r^{\mathrm{R}}, \rho), \rho\right) ;$$

the tag z satisfies a similar relation in terms of g_t, h_t, g_z and h_z.

Now, assume that h_t, h_s, and h_z are random oracles, while the other functions are just ordinary functions. Since we apply a random oracle only to the second half of the RSA-P1 input r, we will relate the security of TKEM2 to a gap-partial-RSA inverter with parameter $k_r/2$ rather than k_r. This is due to the PRF construction; with a stronger PRF, we would be able to give a security proof in terms of DO_{f,k_r} or PO_E. The result is as follows; a proof sketch is given in Appendix B (see the full version of this paper for complete details).

Theorem 3. *Let q_t, q_s, q_z, q_D be parameters. Assume that there is an IND-CCA2 adversary against TKEM2 with advantage at least ϵ' within running time T' making at most q_t, q_s, q_z, and q_D queries to the h_t-oracle, h_s-oracle, h_z-oracle, and decryption oracle, respectively. Let $q' = 2q_t + (k - k_r/2)(q_D + 1)$. Then there is a $DO_{f,k_r/2}(q')$-assisted RSA inverter with running time at most T that is successful with probability at least ϵ, where*

$$\epsilon = c \cdot \left(\epsilon' - ((q_D + 2)(q_D + q_z) + q_s) \cdot 2^{-k_t/2} - q_D \cdot 2^{-k_x} \right) ;$$

$$T = T' + O\left(q' \cdot T_{DO_{f,k_r/2}}\right) + O((q_D + 1)k^3) + \tau ;$$

$c = 2^{-40-(k-k_r-24)/1416}$. $T_{DO_{f,k_r/2}}$ *is the running time for the partial-RSA decision oracle* $DO_{f,k_r/2}$, *while* τ *is the time needed for* $O(q_D + q_t + q_s + q_z)$ *elementary table operations.*

Remark. The RSA problem is *randomly self-reducible* [14]: If an RSA inverter fails on an input y, one may run the inverter on new random inputs of the form $y' = (f(\alpha) \cdot y) \mod N$ until the inverter is successful; $f^{-1}(y)$ is easy to derive from $f^{-1}(y') = (\alpha \cdot f^{-1}(y)) \mod N$. The constant c in Theorem 3 is a lower bound on the probability that the inverter will not trivially fail just because the target ciphertext y is not a valid RSA-P1 ciphertext. Via $1/c$ applications of the random self-reducibility principle, the inverter in Theorem 3 can hence be translated into a new inverter with success probability and running time approximately $(1 - (1 - c)^{1/c}) \epsilon' > \epsilon'/2$ and T'/c, respectively.

7 Discussion and Conclusion

We have provided a security reduction from a variant of the RSA problem to the tagged key-encapsulation scheme based on RSA-P1 used within TLS. As a byproduct we have addressed the concern about the underlying function Ω. In particular, our proof holds even if MD5 is insecure.

An important aspect of any security reduction is what it implies in practice. Here, we might start with the typical assumption that the RSA problem, for 1024-bit keys, requires about 2^{80} steps to break. Assuming that the gap-partial-RSA problem is just as hard, and with typical parameters, Theorem 3 indicates that no IND-CCA2 adversary against TKEM2 can succeed in fewer than about 2^{40} steps. Of course, this does not mean that there is an attack that succeeds in so few steps, and perhaps there is a better proof than ours that results in a better bound[3].

The security of RSA-P1 used within TLS depends on the difficulty of the gap-partial-RSA problem. We conjecture that for typical parameters the gap-partial-RSA problem is as hard as the RSA problem. However, this problem needs further study. Indeed, an efficient solution to the problem might well lead to an effective chosen-ciphertext attack on TLS servers. The study of this problem is thus important in practice as well as in theory.

Though the security reduction for TKEM2 does not say very much for typical key sizes, the reduction does show, at least intuitively, that there is some strength in the way the RSA algorithm is employed in TLS. It also helps show how the algorithm might be employed better. First, we need to get a tighter bound. This

---

[3] In fact, we can get a better bound simply by changing assumptions: if we assume that the gap-partial-RSA problem *for a random, valid RSA-P1 ciphertext* is as hard as the RSA problem, then the bound will again be about 2^{80}.

can be done by reducing the number of fixed octets in the input to the RSA operation (there are currently up to five fixed octets). Second, we need to get to the ordinary RSA problem. This can be done by processing all of the input to the RSA operation with a secure function h. Essentially, TLS should use TKEM1 rather than TKEM2.

Security protocols have sometimes been designed with proofs of security in mind, and sometimes only according to "reasonable design principles." TLS was designed originally according to the latter philosophy, but we have shown that the former benefit is achieved as well, though this is somewhat accidental. In general we would argue for an approach that considers both philosophies at the same time.

Acknowledgements

We thank Håkan Andersson, Ari Juels, Mike Szydlo and the anonymous referees for valuable comments on preliminary versions of this paper. Phil Rogaway and Johan Håstad provided helpful feedback on aspects of this research. Don Johnson's observation about the hash function construction in TLS motivated our focus on TKEM2; Johnson also provided useful suggestions for the final version of this paper.

References

1. M. Bellare and P. Rogaway. Random Oracles are Practical: A Paradigm for Designing Efficient Protocols. *Proceedings of the First Annual Conference on Computer and Communications Security.* ACM, 1993.
2. M. Bellare and P. Rogaway. Optimal Asymmetric Encryption – How to Encrypt with RSA. *Advances in Cryptology – Eurocrypt '94*, pp. 92–111. Springer Verlag, 1994.
3. D. Bleichenbacher. Chosen Ciphertext Attacks against Protocols Based on the RSA Encryption Standard PKCS #1. *Advances in Cryptology – Crypto '98*, pp. 1–12. Springer Verlag, 1998.
4. B. den Boer and A. Bosselaers. Collisions for the Compression Function of MD5. *Advances in Cryptology – Eurocrypt '93*, pp. 293–304. Springer Verlag, 1994.
5. D. Coppersmith. Small Solutions to Polynomial Equations, and Low Exponent RSA Vulnerabilities. *Journal of Cryptology*, 10, pp. 233–260, 1997.
6. D. Coppersmith, M. Franklin, J. Patarin and M. Reiter. Low-Exponent RSA with Related Messages. *Advances in Cryptology – Eurocrypt '96*, pp. 1–9. Springer Verlag, 1996.
7. J.-S. Coron, H. Handschuh, M. Joye, P. Paillier, D. Pointcheval and C. Tymen. GEM: a Generic Chosen-Ciphertext Secure Encryption Method. *Topics in Cryptology – CT-RSA 2002*, pp. 263–276. Springer Verlag, 2002.
8. J.-S. Coron, M. Joye, D. Naccache and P. Paillier. New Attacks on PKCS #1 v1.5 Encryption. *Advances in Cryptology – Eurocrypt 2000*, pp. 369–379. Springer Verlag, 2000.
9. J.-S. Coron, M. Joye, D. Naccache and P. Paillier. Universal Padding Schemes for RSA. *Advances in Cryptology – Crypto 2002*, these proceedings.

10. T. Dierks and C. Allen. *IETF RFC 2246: The TLS Protocol Version 1.0*. January 1999.
11. W. Diffie and M. E. Hellman. New Directions in Cryptography. *IEEE Transactions on Information Theory*, IT–22(6), pp. 644–654. November 1976.
12. H. Dobbertin. *Cryptanalysis of MD5 Compress*. Presented at the rump session of Eurocrypt '96, May 14, 1996.
13. T. ElGamal. A Public Key Cryptosystem and a Signature Scheme Based on Discrete Logarithms. *IEEE Transactions on Information Theory*, IT–31(4), pp. 469–472. July 1985.
14. J. Feigenbaum. Locally Random Reductions in Interactive Complexity Theory. *Advances in Computational Complexity, DIMACS Series in Discrete Mathematics and Theoretical Computer Science*, vol. 13, pp. 73–98, 1993.
15. A. O. Freier, P. Karlton, and P. C. Kocher. *The SSL Protocol Version 3.0*. Netscape Communications Corp., November 1996.
16. E. Fujisaki, T. Okamoto, D. Pointcheval and J. Stern. RSA-OAEP Is Secure under the RSA Assumption. *Advances in Cryptology – Crypto 2001*, pp. 260–274. Springer Verlag, 2001.
17. S. Goldwasser and S. Micali. Probabilistic Encryption. *Journal of Computer and System Sciences*, 28 (2). April 1984.
18. D. B. Johnson. *Theoretical Security Concerns with TLS Use of MD5*. Contribution to ANSI X9F1 working group. June 21, 2001.
19. H. Krawczyk, M. Bellare and R. Canetti. *IETF RFC 2104: HMAC: Keyed-Hashing for Message Authentication*. February 1997.
20. J. Manger. A Chosen Ciphertext Attack on RSA Optimal Asymmetric Encryption Padding (OAEP) as Standardized in PKCS #1 v2.0. *Advances in Cryptology – Crypto 2001*, pp. 260–274. Springer Verlag, 2001.
21. A. J. Menezes, P. C. van Oorschot and S. A. Vanstone. *Handbook of Applied Cryptography*, CRC Press, 1996.
22. T. Okamoto and D. Pointcheval. The Gap-Problems: a New Class of Problems for the Security of Cryptographic Schemes. *Proceedings of the 2001 International Workshop on Practice and Theory in Public Key Cryptography (PKC'2001)*, pp. 104–118. Springer-Verlag, 2001.
23. T. Okamoto and D. Pointcheval. REACT: Rapid Enhanced-security Asymmetric Cryptosystem Transform. *Topics in Cryptology – CT-RSA 2001*, pp. 159–175. Springer Verlag, 2001.
24. L. C. Paulson. Inductive analysis of the Internet protocol TLS. *ACM Transactions on Information and System Security*, 2(3), pp. 332–351. August 1999.
25. C. Rackoff and D. R. Simon. Non-Interactive Zero-Knowledge Proof of Knowledge and Chosen Ciphertext Attack. *Advances in Cryptology – Crypto '91*, pp. 433–444. Springer-Verlag, 1992.
26. R. Rivest. *IETF RFC 1321: The MD5 Message-Digest Algorithm*. April 1992.
27. V. Shoup. OAEP Reconsidered. *Advances in Cryptology – Crypto 2001*, pp. 239–259. Springer Verlag, 2001.
28. V. Shoup. *A Proposal for an ISO Standard for Public Key Encryption*. Preprint, December 2001. Available from eprint.iacr.org/2001/112.
29. National Institute of Standards and Technology (NIST). *Draft FIPS 180-2: Secure Hash Standard*. Draft, May 2001.
30. R. Rivest, A. Shamir and L. Adleman. A Method for Obtaining Digital Signatures and Public-Key Cryptosystems. *Communications of the ACM*, 21(2), pp. 120 - 126. February 1978.

31. RSA Laboratories. *PKCS #1 v1.5: RSA Encryption Standard.* November 1993.
32. Y. Zheng and J. Seberry. Practical Approaches to Attaining Security Against Adaptively Chosen Ciphertext Attacks. *Advances in Cryptology – Crypto '92*, pp. 292–304. Springer Verlag, 1992.

A   Proof of Theorem 1

First we show that the partial-RSA decision oracle DO_{f,k_r} can simulate the E plaintext-checking oracle PO_E. As a consequence, DO_{f,k_r} is at least as strong as PO_E.

Lemma 1. *Let E be an RSA-P1 trapdoor mapping with parameter k_r. Given a pair (r, y), PO_E can be perfectly simulated via $k - k_r + 1$ queries to DO_{f,k_r}, where k is the bit length of the RSA modulus. The running time for the simulation is bounded by*

$$O((k - k_r + 1)T_{DO_{f,k_r}}) + O(k^3) .$$

Proof of Lemma 1. Suppose that we are given a pair (r, y). Send (r, y) to DO_{f,k_r}. If the response is 0, then either y is not a valid ciphertext (i.e., $D(y) = \phi$) or y is the encryption of a plaintext different from r. In this case, simulate PO_E by responding with 0.

If the response is 1, proceed as follows. We want to determine the whole of $x = f^{-1}(y)$. To achieve this, define $y_0 = y$ and $y_i = 2^e y_{i-1} \bmod N$ for $0 < i \le k - k_r$. Each y_i can be computed in time $O(k^2)$; $2^e \bmod N$ can be precomputed in time $O(\log_2 e \cdot k^2)$. Put $x_0 = x$ and

$$x_i = f^{-1}(y_i) = 2x_{i-1} \bmod N = 2^i x \bmod N ;$$

define $r_i = x_i \bmod 2^{k_r + i}$. Note that

$$r_{k-k_r} = x_{k-k_r} = 2^{k-k_r} x \bmod N ,$$

which implies that x can be determined from r_{k-k_r}. We use an induction argument to demonstrate how to determine r_{k-k_r} from r_0 in $k - k_r$ steps, each of complexity $O(k^2) + O(T_{DO_{f,k_r}})$; $O(k^2)$ bounds the time needed to compute y_i from y_{i-1}. Assume that we know r_{i-1}, where i is an integer between 1 and $k - k_r$. Since $x_i = 2x_{i-1} \bmod N$, r_i is either equal to $2r_{i-1}$ or $(2r_{i-1} - N) \bmod 2^{k_r + i}$. To distinguish between the cases, send $(2r_{i-1}, y_i)$ to the decision oracle. If the response is 1, then $r_i = 2r_{i-1}$; otherwise, $r_i = (2r_{i-1} - N) \bmod 2^{k_r + i}$. Thus r_{k-k_r} can be determined via $k - k_r$ applications of the decision oracle.

Once we know r_{k-k_r}, we can easily determine x via $k - k_r$ halving operations modulo N. Now check whether x is valid. Respond with 1 if this is true and 0 otherwise; it is clear that the simulation is perfect, which concludes the proof. $\square$

Proof of Theorem 1. Suppose that we are given a random integer $y^* = f(x^*)$, where x^* is unknown. Generate a random integer $\alpha < N$. With probability $|Y_0|/|Y|$, the integer $y' = \alpha^e \cdot y^* \bmod N$ is a valid encryption of a string r'.

Assume that there is an E-inverter that with probability ϵ' outputs the string r'. Thus with probability $\epsilon' \cdot |Y_0|/|Y|$, we have recovered the last k_r bits of $x' = \alpha x^*$ mod N. (To increase the probability of success, this procedure can be repeated with different values of α until a plaintext is found.) By Lemma 1, each application of PO_E can be replaced with $k - k_r + 1$ applications of DO_{f,k_r}.

Given that we know the last k_r bits r' of x', we want to determine the whole of x', and thereby the whole of x^*. This is done using the method in the proof of Lemma 1 via $k - k_r$ applications of DO_{f,k_r}. The time bound in the theorem includes the time needed to perform $q(k - k_r + 1) + k - k_r < (q + 1)(k - k_r + 1)$ DO_{f,k_r}-oracle queries plus a bound on the time needed for the arithmetic computations in the simulation of PO_E and the computations $y' = c^e \cdot y^*$ mod N and $x^* = c^{-1}x'$ mod N above. $\square$

B   Proof Sketch of Theorem 3

Let y^* be the target ciphertext. We deduced in Section 4 that the probability that y^* is a valid RSA-P1 ciphertext is lower-bounded by $2^{-24-(k-k_r-24)/1416}$. However, recall that there are 16 fixed bits in the RSA-P1 input r. This implies that the probability that y^* is a valid TKEM2 ciphertext is lower-bounded by $2^{-40-(k-k_r-24)/1416}$. This is equal to the factor c in the theorem. From now on, assume that y^* is valid.

At the beginning of the attack, the RSA inverter $\mathcal{I}$ generates random strings s_0^*, s_1^*, and z^*. There are additional strings related to the target ciphertext:

$$u^{*L} \| u^{*R} = g_t(r^{*L}, \rho^*) \; ;$$
$$v^{*L} \| v^{*R} = h_t(r^{*R}, \rho^*) \; ;$$

note that

$$s_0^* = g_s(u^{*L} \oplus v^{*L}, \rho^*) \oplus h_s(u^{*R} \oplus v^{*R}, \rho^*) \; ; \tag{1}$$
$$z^* = g_z(u^{*L} \oplus v^{*L}, \delta^*) \oplus h_z(u^{*R} \oplus v^{*R}, \delta^*) \; , \tag{2}$$

where $\delta^* = \Delta(\rho^*, L^*, y^*)$. $\mathcal{I}$ does not have to generate v^{*L} or v^{*R} in advance, but at the end of the simulation the identities (1) and (2) must be satisfied. Note that ρ^* and L^* are not known yet since they are not provided by the adversary until the challenge generator is queried.

We need to demonstrate how to simulate the oracles corresponding to h_t, h_s, and h_z. $\mathcal{I}$ responds to an h_t-query (r^R, ρ) as follows. First, $\mathcal{I}$ checks whether the query is old; in that case the output is already defined. Otherwise, $\mathcal{I}$ sends (r^R, y^*) to the partial-RSA decision oracle. If the decision oracle outputs 1, we can get the rest of the inverse $f^{-1}(y^*)$ via $k - k_r/2$ additional queries, following Lemma 1. In this case, $\mathcal{I}$ wins and exits. If the decision oracle outputs 0, $\mathcal{I}$ generates a random string v as the response to the query.

$\mathcal{I}$ responds to an h_s-query (t^R, ρ) in the straightforward manner, generating a random output (unless the query is old). $\mathcal{I}$ responds to an h_z-query (t^R, δ) in an analogous manner.

$\mathcal{I}$ proceeds as follows on a decryption query (y, z, ρ, L); recall that ρ is different for each decryption query and not equal to ρ^*. First, $\mathcal{I}$ sends (r^R, y) to the decision oracle for each string r^R such that (r^R, ρ) is a previous h_t-query. If the decision oracle outputs 0 on all queries, then $\mathcal{I}$ returns "Error". If the decision oracle outputs 1 for some r^R, then $\mathcal{I}$ proceeds as follows.

- Extract the entirety of $x = f^{-1}(y)$ via $k - k_r/2$ queries to the decision oracle using the approach in the proof of Lemma 1.
- If x is not a valid P1 encoding, then output "Error" and exit.
- With $r = x \bmod 2^{k_r}$, compute the corresponding tag z' via the appropriate query to the h_z-oracle combined with the relevant evaluation of g_z.
- If $z = z'$, then compute the corresponding secret s via the appropriate query to the h_s-oracle combined with the relevant evaluation of g_s, output s, and exit. Otherwise, output "Error" and exit.

In the following discussion, some technical details are omitted; see the full version of this paper for a rigorous treatment. It is easily seen that there are only two possible simulation failures:

First, there might be an inconsistency between the simulation of the oracles and the simulation of the challenge generator. This can be the case only if $t^{*R} = u^{*R} \oplus v^{*R}$ is part of an h_s- or h_z-query. Hence, let tBad be the event that there is an h_s- or h_z-query from the adversary or from the decryption oracle including t^{*R}. The probability of tBad is bounded as follows. Note that there are at most $2q_D + q_s + q_z$ different h_s- and h_z-queries. Moreover, note that the simulation is independent of t^{*R} (i.e., all responses to the adversary are independent of t^{*R}). This implies that the probability of tBad is at most

$$(2q_D + q_s + q_z) \cdot 2^{-k_t/2} . \tag{3}$$

Second, some valid ciphertext (y, z, ρ, L) might be erroneously rejected; let BadReject be the event that this is the case. This event occurs if there is a valid decryption query (y, z, ρ, L) such there is no matching previous h_t-query (r^R, ρ). Let (t^R, δ) be the input to h_z corresponding to this decryption query. If t^R is not part of a previous h_z-query, then the probability that the tag z is valid is 2^{-k_z}. The probability that t^R is part of a previous h_z-query is at most $(q_D + q_z) \cdot 2^{-k_t/2}$. This implies that the probability of BadReject is at most

$$q_D(q_D + q_z) \cdot 2^{-k_t/2} + q_D \cdot 2^{-k_z} \tag{4}$$

(this is a very rough bound but sufficient for our purposes).

It can be shown that the inverter will be successful if the adversary is successful and neither of tBad and BadReject occurs. Combining (3) and (4), rearranging, and multiplying with c, we obtain the desired probability.

Security Analysis of IKE's Signature-Based Key-Exchange Protocol*

Ran Canetti[1] and Hugo Krawczyk[2],**

[1] IBM T.J. Watson Research Center, NY, USA
canetti@watson.ibm.com
[2] EE Department, Technion, Haifa, Israel
hugo@ee.technion.ac.il

Abstract. We present a security analysis of the Diffie-Hellman key-exchange protocol authenticated with digital signatures used by the Internet Key Exchange (IKE) standard. The analysis is based on an adaptation of the key-exchange model from [Canetti and Krawczyk, Eurocrypt'01] to the setting where peers identities are not necessarily known or disclosed from the start of the protocol. This is a common practical setting, including the case of IKE and other protocols that provide confidentiality of identities over the network. The formal study of this "post-specified peer" model is a further contribution of this paper.

1 Introduction

The Internet Key-Exchange (IKE) protocol [11] specifies the key exchange mechanisms used to establish secret shared keys for use in the Internet Protocol Security (IPsec) standards [14]. IKE provides several key-exchange mechanisms, some based on public keys and others based on long-term shared keys. Its design emerged from the Photuris [13], SKEME [15] and Oakley [21] protocols. All the IKE key-exchange options support Diffie-Hellman exchanges but differ in the way authentication is provided. For authentication based on public-key techniques two modes are supported: one based on public-key encryption and the other based on digital signatures.

While the encryption-based modes of IKE are studied in [5], the security of IKE's signature-based mode has not been cryptographically analyzed so far. (But see [19] where the IKE protocol is scrutinized under an automated protocol analyzer.) This later mode originates with a variant of the STS protocol [8] adopted into Photuris. However, this STS variant, in which the DH key is signed, is actually insecure and was eventually replaced in IKE with the "sign-and-mac" mechanism proposed in [16, 18]. This mechanism forms the basis for a larger family of protocols referred to as SIGMA ("SIGn-and-MAc") [18] from which the IKE signature modes are particular cases.

The main goal of the current paper is to provide cryptographic analysis of IKE, and the underlying SIGMA protocols. The practical interest in this analysis work is natural given the wide deployment and use of IKE and the fact

---

* This proceedings version lacks most proof details essential for the results presented here; for a complete version see [4].
** Supported by Irwin and Bethea Green & Detroit Chapter Career Development Chair.

M. Yung (Ed.): CRYPTO 2002, LNCS 2442, pp. 143–161, 2002.
© Springer-Verlag Berlin Heidelberg 2002

that authentication via signatures is the most common mode of public-key authentication used in the context of IKE[1]. Yet, the more basic importance of this analytical work is in contributing to a further development of a theory that supports the analysis of complex and more functional protocols as required in real-world applications. Let us discuss two such issues, that are directly relevant to the design of IKE. One such issue (not dealt with in previous formal analysis work of key-exchange protocols) is the requirement for *identity concealment*. That is, the ability to protect the identities of the peers to a key-exchange session from eavesdroppers in the network (and, in some case, from active attackers as well). While this requirement may be perceived at first glance as having minor effects on the protocols, it actually poses significant challenges in terms of design and analysis. One piece of evidence pointing out to this difficulty is the fact that the STS protocol and its variants (see [8, 20]) that are considered as prime examples of key-exchange protocols offering identity protection, turned out to be insecure (they fail to ensure an authenticated binding between peers to the session and the exchanged secret key). The general reason behind this difficulty is the conflicting character of the authentication and identity-concealment requirements.

Another issue arising in the context of IKE is the possible unavailability of the peer identity at the onset of the protocol. In previous analytical work (such as [2, 22, 5]) the peer identities are assumed to be specified and given at the onset of a session activation, and the task of the protocol is to guarantee that it is this particular *pre-specified peer* the one which the key is agreed. In contrast, in IKE a party may be activated to exchange a key with an "address" of a peer but without a specified identity for that peer. This is a common case in practical situations. For example, the key-exchange session may take place with any one of a set of servers sitting behind a (url/ip) address specified in the session activation. Or, a party may respond to a request for a key exchange coming from a peer that is not willing to reveal its identity over the network and, sometimes, not even to the responder before the latter has authenticated itself (e.g., a roaming mobile user connecting from a temporary address, or a smartcard that authenticates the legitimacy of the card-reader before disclosing its own identity). So, how do the parties know who they are authenticating? The point is that each party learns the peer's identity *during* the protocol. A secure protocol in this setting will detect impersonation and will ensure that the learned identity is authentic (informally, if Alice completes a session with the view "I exchanged the session key k with Bob", then it is guaranteed that no other party than Bob learns k, and if Bob completes the session then it associates the key k with Alice)[2]. In this paper we refer to this general setting as the *"post-specified peer"* model.

---

[1] In particular, recent suggestions in the IPsec working group for variants of the key-exchange protocols in IKE fall also under the family of protocols analyzed here.

[2] The issue of whether a party may agree to establish a session with the particular peer whose identity is learned during the key-exchange process is an orthogonal issue taken care by a separate "policy engine" run by the party.

Remark. Note the crucial difference between this "post-specified peer" model and the "anonymous" model of protocols such as SSL where the server's identity is publicly known from the start of the protocol while the client's identity remains undisclosed even when the key exchange finishes. In the anonymous case, the client does not authenticate at all to the server; authentication happens only in the other direction: the server authenticates to the client. A formal treatment of this anonymous uni-directional model of authentication is proposed in [22].

The combination of the requirement for identity protection and the "post-specified peer" setting puts additional constraints on the design of protocols. For example, the natural and simple Diffie-Hellman protocol authenticated with digital signatures defined by ISO [12] and proven in [5], is not suitable for providing identity protection in the post-specified peer model. This is so since this protocol instructs each party to sign the peer identity, which in turn implies that the parties must know the peer identities before a session key is generated. In a setting where the peer identities are not known in advance, these identities must be sent over the network, in the clear, thus forfeiting identity concealment. As we will see in Section 3, the IKE and SIGMA protocols use a significantly different approach to authentication. In particular, parties never sign other parties identities; instead a MAC-based mechanism is added to "compensate" for the unsigned peer's identity. (See [18] for more information on the rationale behind the design of the SIGMA protocols.)

We present a notion of security for key exchange protocols that is appropriate for the post-specified peer setting. This notion is a simple relaxation of the key-exchange security model of [5] that suitably reflects the needs of the "post-specified" model as well as allows for a treatment of identity concealment. After presenting the adaptation of the security definition of [5] to our setting, we develop a detailed security proof for the basic protocol (denoted Σ_0) underlying the signature-based modes of IKE. This is a somewhat simplified variant that reflects the core cryptographic logic of the protocol and which already presents many of the technical issues and subtleties that need to be dealt with in the analysis. One prime example of such subtleties is the fact that the IKE protocols use the exchanged Diffie-Hellman key not only to derive a session key at the end of the session but also to derive keys used *inside* the key-exchange protocol itself to provide essential authentication functionality and for identity encryption. After analyzing and providing a detailed proof of this simplified protocol, we show how to extend the proof to deal with richer-functionality variants including the IKE protocols. The resultant analysis approach and techniques are applicable to other protocols, in particular other identity-concealing protocols and those that use the DH key during the session establishment protocol.

The security properties guaranteed by our analysis consider a strong realistic adversarial setting where the attacker has full control of the communication lines, and may corrupt session and parties at will (in an adaptive fashion). In particular, this security model and definition (even if relaxed with respect to [5]) guarantees that session keys derived in the protocol are secure for use in conjunction with symmetric encryption and authentication functions for implementing "secure channels" (as defined in [5]) that protect communications over realistic adversarially-controlled networks. Deriving such keys is the quintessential application of key-exchange protocols in general, and the fundamental requirement from the IKE protocols.

In the full version of this paper [4] we show how to embed the post-specified peer model in the framework of universally composable (UC) security [3]. Specifically, we formulate a UC notion of post-specified secure key exchange and show that protocol Σ_0 presented here satisfies this notion. The UC notion ensures strong composability guarantees with other protocols. In particular, we show that it suffices for implementing secure channels, both in the UC formalization of [6] and in the formalization of [5].

Paper's organization. In Section 2 we describe the adaptation of the security model of [5] to the post-specified peer setting, and establish the notion of security for key-exchange used throughout this paper. In Section 3 we describe Σ_0, the basic SIGMA protocol underlying all the other variants including the IKE signature-based protocols. In Section 4 we present the formal proof of the Σ_0 protocol in the model from Section 2 (due to space constraints we omit most of the lengthy technical details of this proof from this proceedings version – see [4] for a full version). In Section 5 we treat several variants of the basic protocol and extend the analysis from Section 4 to these cases. In particular, the two signature authentication variants of IKE are analyzed here (Section 5.2 and 5.4).

2   The Security Model

Here we present the adaptation of the security model for key-exchange protocols from [5] to the setting of post-specified peers as described above. We start by providing an overview of the model in [5] (refer to that paper for the full details). Then we describe the relaxation of the security definition required to support the post-specified setting.

2.1   The SK-Security Definition from [5]

Following the work of [2, 1], Canetti and Krawczyk [5] model key-exchange (KE) protocols as multi-party protocols where each party runs one or more copies of the protocol. Each activation of the protocol at a party results in a *local* procedure, called a *session*, that locally instantiates a run of the protocol and produces outgoing messages and processes incoming messages. In the case of key-exchange, a session is intended to agree on a "session key" with one other party (the "peer" to the session) and involves the exchange of messages with that party. Sessions can run concurrently and incoming messages are directed to its corresponding session via a session identifier. The activation of a KE session at a party has three input parameters (P, s, Q): the local party at which the session is activated, a unique session identifier, and the identity of the intended peer to the session. (There is also a fourth input field, specifying whether the party is the initiator or the responder in the exchange; however this field has no bearing on the security requirements and is thus ignored in this overview.) A party can be activated as initiator (e.g., by an application calling the KE procedure) or as a responder (upon an incoming key-exchange initiation message arriving from another party). The output of a KE session at a party P consists of a public triple (P, s, Q) that identifies the session, and of a secret value called the *session key*. Sessions can also be "aborted" without producing a session key value, in which case a special symbol is output instead of the session key. Sessions maintain a local state that is erased when the session *completes* (i.e., when the session produces output). Each party may have additional state, such as a long-term signature key, which is accessed by different sessions and which is not part of any particular session state.

The attacker model in [5] follows the *unauthenticated-links model* (UM) of [1] where the attacker is a (probabilistic) polynomial-time machine with full control of the communication lines between parties and free to intercept, delay, drop, inject or change all messages sent over these lines (i.e., a full-fledge "man-in-the-middle" attacker). The attacker can also schedule session activations at will and sees the output of sessions except for the values of session keys. In addition, the attacker can have access to secret information via *session exposure* attacks of three types: session-state reveal, session-key queries, and party corruption. The first type of attack is directed at a single session while still incomplete (i.e., before producing output) and its result is that the attacker learns the session state for that particular session (which does not include long-term secret information, such as private signature keys, shared by all sessions at the party). A session-key query can be performed against an individual session after completion and the result is that the attacker learns the corresponding session-key (this models leakage on the session key either via usage of the key by applications, cryptanalysis, break-ins, known-key attacks, etc.). Finally, party corruption means that the attacker learns *all* information in the memory of that party (including session states and session-key information and also long-term secrets); in addition, from the moment a party is corrupted all its actions are totally controlled by the attacker. (We stress that all attacker's actions can be decided by the attacker in a fully adaptive way, i.e., as a function of its current view).

In the model of [5] sessions can be *expired*. From the time a session is expired the attacker is not allowed to perform a session-key query or a state-reveal attack against the session, but is allowed to corrupt the party that holds the session. Protocols that ensure that expired sessions are protected even in case of party corruption are said to enjoy "perfect forward secrecy" [20] (this is a central property of the KE protocols analyzed here).

For defining the security of a KE protocol, [5] follows the indistinguishability style of definitions as used in [2] where the "success" of an attacker is measured via its ability to distinguish the real values of session keys from independent random values. In order to be considered successful the attacker should be able to distinguish session-key values for sessions that were *not exposed* by any of the above three types of attacks. (Indeed, the attacker could always succeed in its distinguishing task by exposing the corresponding session and learning the session key.) Moreover, [5] prohibits attackers from exposing the "matching session" either, where two sessions (P, s, Q) and (P', s', Q') are called *matching* if $s = s'$, $P = Q'$ and $Q = P'$ (this restriction of the attacker is needed since the matching session contains the session key as well).

As is customary, the ability of the attacker to distinguish between real and random values of the session key is formalized via the notion of a *test session* that the attacker is free to choose among all complete sessions in the protocol. When the attacker chooses the test session it is provided with a value v which is chosen as follows: a random bit b is tossed, if $b = 0$ then v is the real value of the output session key, otherwise v is a random value chosen under the same distribution of session-keys produced by the protocol but independent of the value of the real session key. After receiving v the attacker may continue with the regular actions against the protocol; at the end of its run the attacker outputs a bit b'. The attacker *succeeds* in its attack if (1) the test session is not exposed, and (2) the probability that $b = b'$ is significantly larger than $1/2$. We note that in the model of [5] the attacker is allowed to corrupt a peer to the test session once the test session expires at that peer (this captures perfect forward secrecy). The resultant security notion for KE protocols is called SK-security and is stated as follows:

Definition 1. (SK-security [5]) *An attacker with the above capabilities is called an* SK-attacker. *A key-exchange protocol π is called* SK-secure *if for all SK-attackers $\mathcal{A}$ running against π it holds:*

1. *If two uncorrupted parties complete matching sessions in a run of protocol π under attacker $\mathcal{A}$ then, except for a negligible probability, the session key output in these sessions is the same.*
2. *$\mathcal{A}$ succeeds (in its test-session distinguishing attack) with probability not more that $1/2$ plus a negligible fraction.*

(The term 'negligible' represents any function (in the security parameter) that diminishes asymptotically faster than any polynomial fraction, or a small specific quantity in a concrete security treatment).

Remark. In [5] there are two additional notions that play a central role in the analysis of KE protocols: the "authenticated-links model" (AM) and "authenticators" [1]. While these notions could have been used in our analysis too, they would have required their re-formulation to adapt to the post-specified peer setting treated here. We have chosen to save definitional complexity and develop our protocol analysis in the current paper directly in the UM model.

2.2   Adapting SK-Security to the Post-Specified Peer Setting

The model of [5] makes a significant assumption: *a party that is activated with a new session knows already at activation the identity of the intended peer to the session.* That is, the authentication process in [5] is directed to verify that the "intended peer" is the party we are actually talking to. In contrast, in the "post-specified setting" analyzed here (in particular in the setting of the IKE protocol) the information of who the other party is does not necessarily exist at the session initiation stage. It may be learned by the parties only after the protocol run evolves.

Adapting the security model from [5] to the post-specified peer setting requires: (A) generalizing the formalism of key-exchange protocols to allow for unspecified peers at the start of the protocol; and (B) relaxing the security definition to accept protocols where the peer of a session may be decided (or learned) only after a session evolves (possibly not earlier than the last protocol message as is the case of IKE). Fortunately this adaptation requires only small technical changes which we describe next; all the other definitional elements remain unchanged from [5]. In particular, we keep the UM model and most of the key-exchange formalism unchanged (including full adversarial control of the communication lines and the three types of session exposure: session-state reveal, session-key queries, and party corruption).

(A) Session activation and identification. Instead of activating sessions with input a triple (P, s, Q) as in [5] (where P is the identity of the local party, s a session identifier, and Q the identity of the intended peer for the session), in the post-specified case a session at a party P is activated with a triple (P, s, d) where d represents a "destination address" that may have no implications regarding the peer's identity sitting behind this address, and is used only as information for delivery of messages related to this session. This may be, for example, a temporary address used by arbitrary parties, or an address that may identify a set of parties, etc. Note that the above (P, s, d) formalism represents a generalization of the formalism from [5]; in the latter, d is uniquely associated with (and identifies) a specific party. We keep the convention from [5] that session id's are assumed to be unique among all the session id's used by party P (this is a simple abstraction of the practice where parties provide unique session id's for their own local sessions; we can see the identifier s as a concatenation of these local identifiers – see [5] for more discussion on this topic). We use the pair of entity identity and session-id (P, s) to *uniquely name sessions* for the purpose of attacker actions (as well as for identification of sessions for the purpose of protocol analysis). The output of a session (P, s) consists of a public triple (P, s, Q) where Q is the peer to the session, and the secret value of the session key. When the session produces such an output it is called *completed* and its state is erased (only the session output persists after the session completes and until the session expires). Sessions can abort without producing a session-key output in which case the session is referred to as *aborted* (and not completed).

(B) SK security and matching sessions. The formalism used in [2,5] to define the security of key-exchange protocols via a test session is preserved in our work. The significant (and necessary) change here is in the definition of "matching sessions" which in turn influences the limitations on the attacker's actions against the "test session" and its peers (recall, that the attacker is allowed to attack any session except for the test-session and its matching session). In [5] the matching session of a (complete) session (P, s, Q) within party P is defined as (Q, s, P) (running within Q). This is well-defined in the pre-specified setting where both peer identities are fixed from the start of the session. In our case, however, the peer of a session may only be decided (or learned) just before the completion of that session. In particular, a session (P, s) may complete with peer Q, while the session (Q, s) may not have completed and therefore its peer is not determined. In this case, corrupting Q or learning the state of (Q, s) could obviously provide the attacker with information about the session key output by (P, s, Q). We thus introduce the following modified definition of matching session.

Definition 2. *Let* (P, s) *be a* completed *session with public output* (P, s, Q). *The session* (Q, s) *is called the* matching session *of* (P, s) *if either*

1. (Q, s) *is not completed; or*
2. (Q, s) *is completed and its public output is* (Q, s, P).

Note that by this definition only completed sessions have a matching session; in particular the "matching" relation defined above is not symmetric (except if the matching session is completed too — in which case the above definition of matching session coincides with the definition in [5]). Also, note that if Q is uncorrupted then the matching session of (P, s) is unique.

Definition 3. (SK-security in the post-specified setting) *SK-security in the post-specified peer setting is defined identically as in Definition 1 but with the notion of matching sessions re-formulated via Definition 2.*

Notes on the definition: 1. We argue that the combination of the two matching conditions in Definition 2 above results in a sound definition of SK-security. In particular, it is sufficient to preserve the proof from [5] that SK-security guarantees secure channels (see below). On the other hand, none of the two matching conditions in isolation induces a satisfactory definition of security. In particular, defining the session (Q, s) to always be the matching session of (P, s) without requiring that the determined peer is correct (in condition (2)) would result in an over-restriction of the actions of the attacker against the test session to the point that such a definition would allow weak protocols to be called secure. An example of such an insecure protocol is obtained by modifying protocol Σ_0 from Section 3 by deleting from it the MAC applied to the parties identities. This modified protocol can be shown to succumb to a key-identity mis-binding (or "unknown key share") attack as in [8], yet it would be considered secure without the conditioning on the output of session (Q, s) as formulated in (2). On the other hand, condition (2) alone is too permissive for the attacker, thus resulting in a too strong definition that would exclude many natural protocols. Specifically, if we eliminate (1) then an attacker could perform a state-reveal query against (Q, s) and reveal the secret key (e.g., g^{xy}) when this information is still in the session's state memory. This would allow the attacker a strategy in which it chooses (P, s, Q) at the test session and forces (Q, s) to be incomplete, and then learn the test session key through a state-reveal attack against (Q, s). **2.** The above definition of secure key-exchange in the post-specified peer setting implies a strict relaxation of the SK-security definition in [5]. On the one hand, any SK-secure protocol according to [5] is also post-specified secure provided that we take care of the following formalities. First, we use the "address field" d in the input to the session to specify the identity of a party. Then, before completing a session, the protocol checks that the identity to be output is the same as the identity specified in the "address field" (if not, the session is aborted). On the other hand, there are protocols that are secure according to Definition 3 in the post-specified model but are not secure in the pre-specified setting of [5]. The IKE protocols studied here (in particular, protocols Σ_0 and Σ_1 presented in the following sections) constitute such examples (see the remark at the end of Section 3).

3. A natural question is whether the relaxation of SK-security adopted here is adequate. One strong evidence supporting the appropriateness of the definition is the fact that the proof in [5] that SK-security implies secure channels applies also for SK-security in the post-specified peer setting (Definition 3). One technical issue that arises when applying the notion of secure channels from [5] in our context is that this notion is formulated in the "pre-specified peer" model. Yet, one can use a post-specified SK-secure KE protocol also in this setting. All is needed is that each peer verifies, before completing a KE session, that the authenticated peer (i.e., the identity to be output as the session's peer) is the same as the identity specified in the activation of the secure channels protocol. If this verification fails, then the party aborts the KE session and the secure-channels session. Alternatively, one can easily adapt the model of secure channels in [5] to the post-specified peer setting. Also in this case an SK-secure KE protocol in the post-specified model suffices for constructing (post-specified) secure channels. In all we have:

Theorem 1. *SK-security in the post-specified peer setting implies secure channels in the formulation of [5] (either with pre-specified or post-specified secure-channel peers).*

3   The Basic SIGMA Protocol: Σ_0

Here we provide a description of a key-exchange protocol, denoted Σ_0, that represents a simplified version of the signature-mode of IKE. The protocol contains most of the core cryptographic elements and properties found in the full-fledge IKE and SIGMA protocols. In the next section we provide a proof of this basic protocol, and in the subsequent section we will treat some variants and the changes they require in the security analysis. These variants will include the actual IKE protocols (see Sections 5.2 and 5.4). The Σ_0 protocol is presented in Figure 1. Further notes and clarifications on the protocol follow.

Notes on the Description and Actions of the Protocol

- For simplicity we describe the protocol under a specific type of Diffie-Hellman groups, namely, a sub-group of Z_p^* of prime order. However, the protocol and subsequent analysis apply to any Diffie-Hellman group for which the DDH assumption holds (see Section 4).

Protocol Σ_0

Initial information: Primes p, q, $q/p-1$, and g of order q in Z_p^*. Each player has a private key for a signature algorithm SIG, and all have the public verification keys of the other players. The protocol also uses a message authentication family MAC, and a pseudorandom function family PRF.

The protocol messages

Start message $(I \to R)$: s, g^x
Response message $(R \to I)$: $s, g^y, ID_r, \text{SIG}_r(\text{"1"}, s, g^x, g^y), \text{MAC}_{k_1}(\text{"1"}, s, ID_r)$
Finish message $(I \to R)$: $s, ID_i, \text{SIG}_i(\text{"0"}, s, g^y, g^x), \text{MAC}_{k_1}(\text{"0"}, s, ID_i)$

The protocol actions

1. The start message is sent by the initiator ID_i upon activation with session-id s (after checking that no previous session at ID_i was initiated with identifier s); the DH exponent g^x is computed with $x \xleftarrow{R} Z_q$ and x is stored in the state of session (ID_i, s).

2. When a start message with session-id s is delivered to a party ID_r the (if session-id s did not exist before at ID_r) ID_r activates a local session s (as responder). It now generates the response message where the DH exponent g^y is computed with $y \xleftarrow{R} Z_q$, the signature SIG_r is computed under the signature key of ID_r, and the value g^x placed under the signature is the DH exponent received by ID_r in the incoming start message. The MAC_{k_1} value is produced with $k_1 = \text{PRF}_{g^{xy}}(1)$ where g^{xy} is computed by ID_r as $(g^x)^y$. Finally, the value $k_0 = \text{PRF}_{g^{xy}}(0)$ is computed and kept in memory, and the values y and g^{xy} are erased.

3. Upon receiving a (first) response message with session-id s, ID_i retrieves the public key of the party whose identity ID_r appears in this message and uses this key to verify the signature on the quadruple $(\text{"1"}, s, g^x, g^y)$ where g^x is the value sent by ID_r in the start message, and g^y the value received in this response message. ID_i also checks the received MAC under key $k_1 = \text{PRF}_{g^{xy}}(1)$ (where g^{xy} is computed as $(g^y)^x$) and on the identity ID_r as it appears in the response message. If any of these verification steps fails the session is aborted and a session output of "aborted (ID_i, s)" is generated; the session state is erased. If verification succeeds then ID_i completes the session with public output (ID_i, s, ID_r) and secret session key k_0 computed as $k_0 = \text{PRF}_{g^{xy}}(0)$. The finish message is sent and the session state erased.

4. Upon receiving the finish message of session s, ID_r verifies the signature (under the public key of party ID_i and with g^y being the DH value that ID_r sent in the response message), and verifies the MAC under key k_1 computed in step 2. If any of the verifications steps fails the session is aborted (with the "aborted (ID_r, s)" output), otherwise ID_r completes the session with public output (ID_r, s, ID_i) and secret session key k_0. The session state is erased.

Fig. 1. The basic SIGMA protocol

- The notation $I \to R$ and $R \to I$ is intended just to indicate the direction between initiator and responser of the messages. The protocol as described here does not specify where the messages are sent to. They can be sent to a pool of messages, to a local broadcast network, to a physical or logical address, etc. The protocol and its analysis accommodate any of these (or other) possibilities. What is important is that the protocol does not make any assumption on who will eventually get a message, how many times, and when (these are all actions decided by the attacker). Also, there is no assumption on the logical connection between the address where a message is delivered and the identity (either ID_i or ID_r) behind that address. This allows us to design the protocol (and prove its security) in the "post-specified peer" model introduced in Section 2.
- ID_i and ID_r represent the real identities of the parties to the exchange. In our model we assume that every party knows the other's party public key before hand. However, one can think of the above identities as full certificates signed by a trusted CA and verified by the recipient. (In this case, the full certificate may be included as the peer's identity under the MAC or just the identity in the certificate – e.g. the "distinguished name"). Our proofs work under this certification-based model as well.
- The strings "0" and "1" are intended to separate between authentication information created by the initiator and responder in the protocol. They serve as "symmetry breakers" in the protocol. However, in the case of Σ_0 these tags are not strictly needed for security; we will see later (Section 5.1) that the protocol is secure even without them. Yet, we include them here for two reasons. First, they simplify analysis; second, they make the protocol's security more robust to changes as we will also discuss later (e.g., they defeat reflection attacks in some of the protocol's variants).
- Recall the uniqueness of session-id's assumed by our model. We use this assumption in order to simplify the model and to accommodate different implementations of this assumption. A typical way to achieve this is to require each party in the exchange to choose a random number (say, s_i and s_r respectively) and then define s to be the concatenation of these values. In this case the values s_i and s_r can be exchanged before the protocol, or s_i can replace s in the start message, and (s_i, s_r) replace s in the response message.
- Parties use the session id's to bind incoming messages to existing (incomplete) sessions. However, only the first message of each type is processed. For example if a response message arrives with session id s at the initiator of session s, then the message is processed only if no previous response message under this session was received. Otherwise the message is discarded. Same for the other message types, or if a message arrives after the session is completed or aborted.
- We note that in this description of Σ_0 session identifiers serve a dual functionality: they serve to identify sessions and bind incoming messages to these sessions, but they also serve as "freshness guarantees" against replay attacks. We choose to "overload" session id's with the two functionalities in order to

simplify presentation. However, actual protocol implementations may use two different elements for these functions: a session identifier for the first functionality above, and random (or non-repeating) nonces for replay protection.

- In practice, it is recommended not to use the plain value g^{xy} of the DH key but a hashed value $H(g^{xy})$ where H is a hash function (e.g. a cryptographic hash function such as SHA or a universal hash function, etc.). This has the effect of producing a number of bits as required to key the PRF, and (depending on the properties of the hash function) may also help to "extracting the security entropy" from the g^{xy} output. If the plain g^{xy} is used, our security results hold under the DDH assumption. Using a hashed value of g^{xy} is secure under the (possibly weaker) HDH assumption [9].

Remark. As mentioned in Section 2 it is illustrative to note that protocol Σ_0 is not secure in the original (pre-specified) model of [5]. In that model an attacker could apply the following strategy: (1) initiate a session (P, s, Q) at P; (2) activate a session (Q, s, Eve) at Q as responder with the start message from (P, s, Q) where Eve is a corrupted party (let g^x be the DH exponent in this message); (3) deliver the response message produced by Q to P (let g^y be the DH exponent in this message). The result is that P completes (P, s, Q) with a session key derived from g^{xy}, while the session (Q, s, Eve) is still incomplete and its state contains the value g^{xy}. Therefore, in the [5] model, the attacker can choose (P, s, Q) as the test session and expose (Q, s, Eve) via a state-reveal attack to learn g^{xy}. This is allowed in [5] since (Q, s, Eve) is not a matching session to the test session (only (Q, s, P) is matching to the test session). In our post-specified model, however, the attacker is not allowed to expose (Q, s) which is incomplete and then by Definition 2 it is matching to the test session (P, s). This restriction of the adversary is needed in the post-specified setting since from the point of view of Q there is no information about who the peer is until the very end of the protocol and then its temporary internal state (before receiving the finish message) is identical whether its session is controlled by the adversary (via Eve as in the above example) or it is a regular run with a honest peer P. What is crucial to note is that protocol Σ_0 (and any SK-secure protocol in the post-specified model) guarantees that *if* Q completes the session (Q, s) then its view of the peer's identity is correct and consistent with the view in the matching session (e.g., in the above example it is guaranteed that if Q completes the session, it outputs P as the peer, and only P can compute the key g^{xy}).

4   Proof of Protocol Σ_0

4.1   The Statements

We start by formulating the Decisional Diffie-Hellman (DDH) assumption which is the standard assumption underlying the security of the DH key exchange against passive attackers. For simplicity, we formulate this assumption for a specific family of DH groups, but analogous assumptions can be formulated for other groups (e.g., based on elliptic curves).

Assumption 2 *Let κ be a security parameter. Let p, q be primes, where q is of length κ bits and $q/p-1$, and g be of order q in Z_p^*. Then the probability distributions of quintuples*

$$Q_0 = \{\langle p, g, g^x, g^y, g^{xy}\rangle : x, y \xleftarrow{R} Z_q\} \text{ and } Q_1 = \{\langle p, g, g^x, g^y, g^r\rangle : x, y, r \xleftarrow{R} Z_q\}$$

are computationally indistinguishable.

In addition to the DDH assumption we will assume the security of the other underlying cryptographic primitives in the protocol (digital signatures, message authentication codes, and pseudorandom functions) under the standard security notions in the cryptographic literature.

Theorem 3. (Main Theorem) *Assuming DDH and the security of the underlying cryptographic functions* SIG, MAC, PRF, *the Σ_0 protocol is SK-secure in the post-specified model, as defined in Section 2.*

Proving the theorem requires proving the two defining properties of SK-secure protocols (we use the term Σ_0-*attacker* to denote an SK-attacker working against the Σ_0 protocol):

P1. If two uncorrupted parties ID_i and ID_r complete matching sessions $((ID_i, s, ID_r)$ and (ID_r, s, ID_i), respectively) under protocol Σ_0 then, except for a negligible probability, the session key output in these sessions is the same. The proof of this property can be found in [4].

P2. No efficient Σ_0-attacker can distinguish a real response to the test-session query from a random response with non-negligible advantage. More precisely, if for a given Σ_0-attacker we define:

- $P_{\text{REAL}}(\mathcal{A}) = Prob(\mathcal{A}$ outputs 1 when given the real test session key)
- $P_{\text{RAND}}(\mathcal{A}) = Prob(\mathcal{A}$ outputs 1 when given a random test session key)

then we need to prove that for any Σ_0-attacker $\mathcal{A}$: $|P_{\text{REAL}}(\mathcal{A}) - P_{\text{RAND}}(\mathcal{A})|$ is negligible.

4.2 Proof Plan of Property P2

We prove property P2 by showing that if a Σ_0-attacker $\mathcal{A}$ can win the "real vs. random" game with significant advantage then we can build an attacker against one of the underlying cryptographic primitives used in the protocol: the Diffie-Hellman exchange (DDH assumption), the signature scheme SIG, the MAC scheme MAC, or the pseudorandom family PRF.

More specifically we will show that from any Σ_0-attacker $\mathcal{A}$ that succeeds in distinguishing between a real and a random response to the test-session query we can build a DDH distinguisher D that distinguishes triples g^x, g^y, g^{xy} from random triples g^x, g^y, g^r with the same success advantage as $\mathcal{A}$, or there is an algorithm (that we can construct explicitly) that breaks one of the other underlying cryptographic primitives. This distinguisher D gets as input a triple (g^x, g^y, z) where z is either g^{xy} or g^r for $r \xleftarrow{R} Z_q$. D starts by simulating a run of $\mathcal{A}$ on a virtual instantiation of protocol Σ_0 and uses the values g^x and g^y

from the input triple as the DH exponents in the start and response message of one randomly chosen session, say s_0, initiated by $\mathcal{A}$ in this run of protocol Σ_0. The idea is that if $\mathcal{A}$ happens to choose this session s_0 (or the corresponding responder's session) as its test session then D can provide $\mathcal{A}$ with z as the response to the test-session query. In this case, if $\mathcal{A}$ outputs that the response was real then D will decide that $z = g^{xy}$, otherwise D will decide that z is random. One difficulty here is that since D actually changes the regular behavior of the parties in session s_0 (e.g. it uses the value z to derive the key k_1 used in the MAC function) then we still have to show that D has a good probability to guess the right test session, and that the original ability of $\mathcal{A}$ to distinguish between "real" and "random" is not significantly reduced by the simulation changes. Proving this involves showing several properties of the protocol that relate to the authentication elements such as signatures and MAC.

In order to specify the distinguisher D we need to define the above simulation process and the exact rules on how to choose session s_0 and how to change the behavior of the parties to that session. In order to facilitate our analysis we will actually define a sequence of several simulators which differ from each other by the way they choose the keys (k_0 and k_1) used in the processing of the s_0 session. Each of these simulators will define a probability distribution on the runs of attacker $\mathcal{A}$. At one end of the sequence of simulators will be one that corresponds to a "real" run of $\mathcal{A}$ while at the other end the simulation corresponds to a "random" experiment where the session key in session s_0 provided to $\mathcal{A}$ is chosen as a random and independent value k_0. In between, there will be several "hybrid" simulators. We will show that either all the distributions generated by these simulators are computationally indistinguishable, or that a successful distinguisher against DDH or against the PRF family exists. From this we get a proof that the "real" and "random" simulators at the ends of the sequence are actually indistinguishable, and from this that the values P_{RAND} and P_{REAL} differ by at most a negligible quantity (this negligible difference will depend on the quantified security of DDH and of the cryptographic functions).

For a full proof of property P2 see [4].

Detailed proof. The detailed proof of properties P1 and P2 (and thus of the Main Theorem 3) is lengthy and therefore omitted from these proceedings. See [4] for a complete proof.

5   Variants and Discussions

We consider the security of several variants of the protocol and extensions to its functionality. In particular, we extend the analysis to the elements found in the IKE protocols and not included in the basic protocol Σ_0.

5.1   Eliminating the Initiator and Responder Tags in Σ_0

In protocol Σ_0 the initiator and responder include under their signatures and MAC a special tag "0" and "1", respectively. Here we show that protocol $\Sigma_0{}'$

defined identically to Σ_0 except for the lack of these tags is still secure. (We stress that the signature modes of IKE do not use these tags; this is one main reason to provide the analysis here without tags.)

For lack of space the rest of this subsection is omitted from these proceedings (see [4]).

5.2 Putting the MAC under the Signature

One seemingly significant difference between protocol Σ_0 and IKE signature-mode is that in the latter the MAC tag is not sent separately but rather it is computed *under* the signature operation. That is, in the response message of IKE the responder does not send $\mathrm{SIG}_r(\text{``1''}, s, g^x, g^y), \mathrm{MAC}_{k_1}(\text{``1''}, s, ID_r)$, as in Σ_0, but rather sends the value $\mathrm{SIG}_r(\mathrm{MAC}_{k_1}(s, g^x, g^y, ID_r))$. Similarly, the pair of signature-mac is replaced in the finish message by the value $\mathrm{SIG}_i(\mathrm{MAC}_{k_1}(s, g^y, g^x, ID_i))$. The reason for this inclusion of the MAC under the signature in IKE is twofold: to save the extra space taken by the MAC tag and to provide a message format consistent with other authentication modes of IKE[3].

Fortunately, the analysis of the protocol when the MAC goes under the signature is essentially the same as the simplified Σ_0 version analyzed before. The analysis adaptation is straightforward and is based on the following simple fact.

Lemma 1. *If* SIG *is a secure signature scheme and* MAC *a secure message authentication function then it is infeasible for an attacker to find different messages M and M' such that for a randomly chosen secret* MAC-key k_1 *the attacker can compute* $\mathrm{SIG}(\mathrm{MAC}_{k_1}(M'))$ *even after seeing* $\mathrm{SIG}(\mathrm{MAC}_{k_1}(M))$.

Indeed, if the attacker can do that then either $\mathrm{MAC}_{k_1}(M') \neq \mathrm{MAC}_{k_1}(M)$ with significant probability and this results in a signature forgery strategy, or $\mathrm{MAC}_{k_1}(M') = \mathrm{MAC}_{k_1}(M)$ with significant probability in which case the attacker has a strategy to break the MAC. (Note that the attacker cannot choose k_1; if it could, the lemma would not hold.)

This lemma implies that all the arguments in our proofs of Section 4 that use the unforgeability of signatures remain valid in this case. More precisely, they are extended through the above lemma to claim that if an attack is successful then either the signature scheme or the MAC are broken (the cases where the weakness comes from the insecurity of either the PRF family or the DDH assumption are treated identically as in the proof of Σ_0).

IKE's aggressive mode. With the above changes, in which the MAC is included under the signature and the "0"/"1" tags are not included, Σ_0 becomes basically the so called "aggressive mode of signature authentication" which is one of the two IKE's protocols based on authentication via digital signatures. One additional difference is that the IKE protocol uses the function PRF itself

---

[3] For example, the IKE mode where authentication is provided by a pre-shared key is obtained from the signature mode by using the same MAC expression but without applying the signature on it (in this case the MAC key is derived from the pre-shared key).

to implement the MAC function. Since a pseudorandom family is always a secure MAC then this implementation preserves security (in this case the key to the PRF is g^{xy} itself as in the other uses of this function in the protocol; the protocol also makes sure that the input to PRF when used as MAC is different that the inputs used for key derivation).

5.3   Encrypting the Identities

Here we consider the augmentation of Σ_0 for providing identity concealment over the network. We present the main ideas behind our treatment, and omit much of the formal and technical issues.

We start by considering the following variant of protocol Σ_0. Before transmitting the response message, the responder computes a key $k_2 = \text{PRF}_{g^{xy}}(2)$ and encrypts under key k_2 the response message excluding s and g^y. That is, the response message is changed to $s, g^y, \text{ENC}_{k_2}(ID_r, \text{SIG}_r("1", s, g^x, g^y), \text{MAC}_{k_1}("1", s, ID_r))$ where ENC is a symmetric-key encryption algorithm. Upon receiving the response message the initiator computes the key k_2 as above, decrypts the incoming message with this key, and then follows with the regular verification operations of Σ_0. If successful, it prepares the finish message as in Σ_0 but sends it encrypted under ENC_{k_2} (only s is sent in the clear). Upon reception of this message the responder decrypts it and follows with the regular operations of Σ_0.

The main goal of this use of encryption is to protect the identities of the peers from disclosure over the network (at least in cases that these identities are not uniquely derivable from the visible (say, IP) address from which communication takes place). We first argue that the addition of encryption preserves the SK-security of the protocol. Then we claim that the encryption provides semantic security of the encrypted information. For the response message semantic security is provided against passive attackers only (indeed, at the point that this encryption is applied by ID_r, the initiator has not yet authenticated to ID_r so this encryption can be decrypted by whoever chose the DH exponent g^x). For information encrypted in the finish message we can provide a stronger guarantee of security, namely, semantic security also against active attackers.

We start by claiming that the modified Σ_0 protocol with encryption as described above satisfies Theorem 3. The basic idea is that if we were encrypting under a random key independent from the Diffie-Hellman exchange then the security of the protocol would be preserved (in particular, since the attacker itself can simulate such an independent encryption on top of Σ_0). However, since we are using an encryption key that is derived from g^{xy} then we need to show that if the encryption helps the attacker in breaking the SK-security of (the encrypted) Σ_0 then we can use this attacker to distinguish g^{xy} from a random value. Technically, this requires an adaptation of the proof of Theorem 3 (see [4] for more details).

In order to show secrecy protection against a passive attacker (note that a passive attacker means an eavesdropper in the network that does not collaborate with the SK-attacker which is active by definition) we consider a run of the protocol where k_2 is chosen randomly. In this case semantic security against a passive attacker follows from the assumption that the encryption function (under a random secret key) is semantically secure against chosen plaintext attacks. Accounting for the fact that k_2 is actually derived from g^{xy} requires an argument similar to the "hybrid simulators" technique in the proof of Theorem 3 (see [4]).

In the case of the finish message, the security guarantee is stronger and the secrecy protection can stand active attackers too (assuming a suitable encryption function secure against active attacks [5, 17]). We can show that for any complete session (ID_i, s, ID_r) that is not exposed by the attacker (i.e., neither this session or its matching session are corrupted), breaking the semantic security of the information transmitted under ENC_{k_2} in the finish message of session (ID_i, s) implies a distinguishing test between k_2 and a random (encryption) key. This in turn can be used to build an attack against the SK-security of the protocol or against one of its underlying cryptographic primitives.

5.4   A Four Message Variant: IKE Main Mode

Here we study a four-message variant of the Σ_0 protocol. The interest in this protocol is two-fold: on one hand, if encryption is added to it (as discussed below) it allows concealing the responder's identity from active attackers and the initiator's identity from passive attacks. This is in contrast to Σ_0 where the strong active protection is provided to the initiator's identity (see Section 5.3). The other source of interest for this protocol is that it actually represents the core cryptographic skeleton of the so called "main mode with signature authentication" in IKE (which is one of the two signature-based protocols in IKE – see Section 5.2 for a discussion of the other IKE variant).

The four-message protocol, denoted Σ_1, is similar to Σ_0 except that the responder delays its authentication (via SIG_r) to a fourth message.

$$I \rightarrow R:\quad s, g^x$$
$$R \rightarrow I:\quad s, g^y$$
$$I \rightarrow R:\quad s, ID_i, \text{SIG}_i(\text{``0''}, s, g^y, g^x), \text{MAC}_{k_1}(\text{``0''}, s, ID_i)$$
$$R \rightarrow I:\quad s, ID_r, \text{SIG}_r(\text{``1''}, s, g^x, g^y), \text{MAC}_{k_1}(\text{``1''}, s, ID_r)$$

The security analysis of Σ_1 is similar to that of Σ_0 as presented in Section 4. It follows the same basic logic and structure of that proof but it requires some changes due to the addition of the fourth message and the fact that the responder authenticates after the initiator. The adaptation, however, of the previous proof to this new protocol is mostly straightforward. The details are omitted. One important point to note is that in this case (as opposed to Σ_0 – see Section 5.1) the use of the tags "0" and "1" is essential for security; at least if one regards reflection attacks (where the attacker impersonates the initiator of the exchange as responder by just replying to each of the initiator's messages with exactly the same message) as a real security threat (see discussion below).

Providing identity concealment in Σ_1 is possible via the encryption of the last two messages of the protocol (under a key $k_2 = \text{PRF}_{g^{xy}}(2)$ as in Section 5.3). In this case, the identity ID_r is protected against active attacks, while ID_i against passive attackers.

IKE's main mode. Protocol Σ_1 with the MAC included under the signature (as in Section 5.2), with encryption of the last two messages (not including the session-id s), and without the "0", "1" tags is essentially the "main mode signature authentication" in IKE. (There are some other secondary differences such as: (i) the session id s equals a pair s_1, s_2, where s_1, s_2 are "cookies" exchanged between the parties in two additional messages preceding the above four-message exchange, and (ii) the MAC function is implemented using $\mathrm{PRF}_{g^{xy}}$). Our analysis here applies to this IKE protocol except for the fact that IKE does not use the "0", "1" tags and thus it is open to reflection attacks. We note that without the use of these tags the protocol can be proven secure in our model if exchanges from a party with itself are considered invalid, or if the initiator verifies, for example, that the incoming DH exponent in the second message differs from the one sent in the initial message. From a practical point of view, these potential reflection attacks have been regarded as no real threats in the context of IKE; in particular based on other details of the IKE specification, such as the way encryption is specified, that make these attacks unrealistic. Yet, the addition of tags as in Σ_1 would have been advisable to close these "design holes" even if currently considered as theoretical threats only.

Note: In case that the MAC goes under the signature (as in IKE and in Section 5.2) then the "0", "1" tags can go under the MAC only. Moreover, in this case one can dispense of these tags and use instead different (and computationally independent) keys k_1 and k_1' to key the MAC going from ID_i to ID_r and from ID_r to ID_i, respectively.

5.5   Not Signing the Peer's DH Exponent

The protocols as presented before take care of signing each party's own DH exponent as well as the peer's DH exponent. While the former is strictly necessary for security (against "man in the middle" attacks), the later is not essential and is used mainly for simplifying the proofs. If the peer's exponent is not included under the signature then the proofs become more involved since the essential binding between g^x and g^y cannot be argued directly but via a binding of these exponents to the session id.

References

1. M. Bellare, R. Canetti and H. Krawczyk, "A modular approach to the design and analysis of authentication and key-exchange protocols", *30th STOC*, 1998.
2. M. Bellare and P. Rogaway, "Entity authentication and key distribution", *Advances in Cryptology, - CRYPTO'93*, Lecture Notes in Computer Science Vol. 773, D. Stinson ed, Springer-Verlag, 1994, pp. 232-249.
3. R. Canetti, "Universally Composable Security: A New paradigm for Cryptographic Protocols", *42nd FOCS*, 2001. Full version available at http://eprint.iacr.org/2000/067.

4. Canetti, R., and Krawczyk, H., "Security Analysis of IKE's Signature-based Key-Exchange Protocol", full version. *Cryptology ePrint Archive* (http://eprint.iacr.org/), 2002.
5. Canetti, R., and Krawczyk, H., "Analysis of Key-Exchange Protocols and Their Use for Building Secure Channels", *Advances in Cryptology – EUROCRYPT 2001*, Full version in: http://eprint.iacr.org/2001/040.
6. Canetti, R., and Krawczyk, H., "Universally Composable Notions of Key Exchange and Secure Channels", *Eurocrypt 02*, 2002. Full version available at http://eprint.iacr.org/2002/059.
7. R. Cramer and V. Shoup, "A Practical Public Key Cryptosystem Provable Secure Against Adaptive Chosen Ciphertext Attack", In *Crypto '98*, LNCS No. 1462, pages 13–25, 1998.
8. W. Diffie, P. van Oorschot and M. Wiener, "Authentication and authenticated key exchanges", *Designs, Codes and Cryptography*, 2, 1992, pp. 107–125.
9. Gennaro, R., Krawczyk H., and Rabin, T., "Hashed Diffie-Hellman: A Hierarchy of Diffie-Hellman Assumptions", manuscript, Feb 2002.
10. O. Goldreich, "Foundations of Cryptography: Basic Tools", Cambridge Press, 2001.
11. D. Harkins and D. Carrel, ed., "The Internet Key Exchange (IKE)", *RFC 2409*, Nov. 1998.
12. ISO/IEC IS 9798-3, "Entity authentication mechanisms — Part 3: Entity authentication using asymmetric techniques", 1993.
13. Karn, P., and Simpson W.A., "The Photuris Session Key Management Protocol", draft-ietf-ipsec-photuris-03.txt, Sept. 1995.
14. S. Kent and R. Atkinson, "Security Architecture for the Internet Protocol", *Request for Comments 2401*, Nov. 1998.
15. Krawczyk, H., "SKEME: A Versatile Secure Key Exchange Mechanism for Internet,", *Proceedings of the 1996 Internet Society Symposium on Network and Distributed System Security*, Feb. 1996, pp. 114-127.
16. Krawczyk, H., IPsec mailing list archives, http://www.vpnc.org/ietf-ipsec/, April-June 1995.
17. Krawczyk, H., "The order of encryption and authentication for protecting communications (Or: how secure is SSL?)", Crypto'2001. Full version in: *Cryptology ePrint Archive* (http://eprint.iacr.org/), Report 2001/045.
18. Krawczyk, H., "SIGMA: the 'SIGn-and-MAc' Approach to Authenticated Diffie-Hellman Protocols", http://www.ee.technion.ac.il/~hugo/sigma.html
19. Meadows, C., "Analysis of the Internet Key Exchange Protocol Using the NRL Protocol Analyzer", *Proceedings of the 1999 IEEE Symposium on Security and Privacy*, IEEE Computer Society Press, May 1999.
20. A. Menezes, P. Van Oorschot and S. Vanstone, "Handbook of Applied Cryptography," CRC Press, 1996.
21. Orman, H., "The OAKLEY Key Determination Protocol", *Request for Comments 2412*, Nov. 1998.
22. V. Shoup, "On Formal Models for Secure Key Exchange", Theory of Cryptography Library, 1999. Available at: http://philby.ucsd.edu/cryptolib/1999/99-12.html.

GQ and Schnorr Identification Schemes: Proofs of Security against Impersonation under Active and Concurrent Attacks

Mihir Bellare and Adriana Palacio

Department of Computer Science & Engineering, University of California, San Diego
9500 Gilman Drive, La Jolla, CA 92093, USA
{mihir,apalacio}@cs.ucsd.edu
http://www-cse.ucsd.edu/users/{mihir,apalacio}

Abstract. The Guillou-Quisquater (GQ) and Schnorr identification schemes are amongst the most efficient and best-known Fiat-Shamir follow-ons, but the question of whether they can be proven secure against impersonation under active attack has remained open. This paper provides such a proof for GQ based on the assumed security of RSA under one more inversion, an extension of the usual one-wayness assumption that was introduced in [5]. It also provides such a proof for the Schnorr scheme based on a corresponding discrete-log related assumption. These are the first security proofs for these schemes under assumptions related to the underlying one-way functions. Both results extend to establish security against impersonation under concurrent attack.

1 Introduction

The Guillou-Quisquater (GQ) [20] and Schnorr [26] identification schemes are amongst the most efficient and best known Fiat-Shamir [16] follow-ons, but the question of whether they can be proved secure against impersonation under active attack has remained open. This paper addresses this question, as well as its extension to even stronger attacks, namely concurrent ones. We begin with some background.

1.1 Identification Schemes and Their Security

An identification (ID) scheme enables a prover holding a secret key to identify itself to a verifier holding the corresponding public key. Fiat and Shamir (FS) [16] showed how the use of zero-knowledge techniques [19] in this area could lead to efficient schemes, paving the road for numerous successors including [20, 26], which are comparable to FS in computational cost but have much smaller key sizes.

The accepted framework for security notions for identification schemes is that of Feige, Fiat and Shamir [14]. As usual, one considers adversary goals as well as adversary capabilities, or attacks. The adversary goal is impersonation: playing

M. Yung (Ed.): CRYPTO 2002, LNCS 2442, pp. 162–177, 2002.

the role of prover but denied the secret key, it should have negligible probability of making the verifier accept. Towards this goal, one can allow it various attacks on the honest, secret-key equipped prover which, as per [14], take place and complete before the impersonation attempt. The weakest reasonable attack is a passive attack, in which the adversary obtains transcripts of interactions between the prover and verifier. However, the attack suggested by [14] as defining the main notion of security is an active attack in which the adversary plays the role of cheating verifier, interacting with the prover numerous times before the impersonation attempt.

Security against impersonation under active attack has been the classical goal of identification schemes. However, interest has been growing in stronger attacks, namely concurrent ones. Here, the adversary would still play the role of cheating verifier prior to impersonation, but could interact with many different prover "clones" concurrently. The clones all have the same secret key but are initialized with independent coins and maintain their own state. Security against impersonation under concurrent attack implies security against impersonation under active attack.

Analyses often approach the establishment of security against impersonation via consideration of whether or not the protocol is a proof of knowledge, honest-verifier zero knowledge, witness indistinguishable [15] and so on. These auxiliary properties are important and useful tools, but not the end goal, which remains establishing security against impersonation.

1.2   The GQ Scheme and Our Results about It

GQ is RSA based. The prover's public key is (N, e, X), where N is an RSA modulus, e is a prime RSA exponent, and $X \equiv x^e \pmod{N}$ where $x \in \mathbb{Z}_N^*$ is the prover's secret key. As typical for practical ID schemes, the protocol, depicted in Figure 2, has three moves: the prover sends a "commitment," the verifier sends a random challenge, the prover sends a "response," and the verifier then accepts or rejects. The protocol is honest-verifier zero knowledge and a proof of knowledge of x [20], and it follows easily that it is secure against impersonation under passive attack, assuming RSA is one-way.

The main question is whether the protocol is secure against impersonation under active attack. No attack has been found. However, no proof of security has been provided either. Furthermore, it is difficult to imagine such a proof being based solely on the assumption that RSA is one-way. (The prover response is the RSA inverse of a point that is a function of the verifier challenge, giving a cheating verifier some sort of limited chosen-ciphertext attack capability, something one-wayness does not consider.) In other words, the protocol seems to be secure against impersonation under active attack, but due to properties of RSA that go beyond mere one-wayness.

The research community is well aware that RSA has important strengths beyond one-wayness, and have captured some of them with novel assumptions. Examples include the strong RSA assumption, introduced in [17, 2] and exploited in [18, 13]; the dependent-RSA assumptions [24]; and the assumption of security under one more inversion [5]. The intent, or hope, of introducing such assumptions is that they underlie not one but numerous uses or protocols. Thus our approach is to attempt to build on this existing experience, and prove security based on one of these assumptions.

We prove that the GQ identification scheme is secure against impersonation, under both active and concurrent attacks, under the assumption that RSA is secure under one more inversion. The precise statement of the result is Corollary 1. Let us now explain the assumption.

Security of RSA under one more inversion, as introduced in [5], considers an adversary given input an RSA public key N, e, and access to two oracles. The *challenge oracle* takes no inputs and returns a random target point in $\mathbb{Z}_N^*$, chosen anew each time the oracle is invoked. The *inversion oracle* given $y \in \mathbb{Z}_N^*$ returns $y^d \bmod N$, where d is the decryption exponent corresponding to e. The assumption states that it is computationally infeasible for the adversary to output correct inverses of all the target points if the number of queries it makes to its inversion oracle is strictly less than the number of queries it makes to its challenge oracle. (When the adversary makes one challenge query and no inversion queries, this is the standard one-wayness assumption, which is why security under one more inversion is considered an extension of the standard one-wayness assumption.) This assumption was used in [5] to prove the security of Chaum's RSA-based blind-signature scheme [12] in the random oracle model. (Our results, however, do not involve random oracles.) It was also used in [6] to prove the security of an RSA-based transitive signature scheme due to [21].

Our result is based on a relatively novel and strong assumption that should be treated with caution. But the result still has value. It reduces the security of the GQ identification scheme to a question which is solely about the security of the RSA function. Cryptanalysts need no longer attempt to attack the identification scheme, but can instead concentrate on a simply stated assumption about RSA, freeing themselves from the details of the identification model. Furthermore, our result helps clarify and unify the global picture of protocol security by showing that the properties of RSA underlying the security of the GQ identification scheme and Chaum's RSA-based blind-signature scheme are the same. Thus our result brings the benefit we usually expect with a proof of security, namely reduction of the security of many cryptographic problems to a single number-theoretic problem. Finally, a proof under a stronger than standard assumption is better than no proof at all in the context of a problem whose provable security has remained an open question for more than ten years.

1.3   The Schnorr Scheme and Our Results about It

The Schnorr identification scheme is discrete logarithm based. The prover's public key is (g, X), where g is a generator of a suitable prime-order group and $X = g^x$ where x is the prover's secret key. The protocol, having the usual three-move format, is depicted in Figure 4. Again the protocol is honest-verifier zero knowledge and a proof of knowledge of x [26], and it follows easily that it is secure against impersonation under passive attack, assuming hardness of computation of discrete logarithms in the underlying group. (That is, one-wayness of the discrete exponentiation function.) As with GQ, the scheme appears to be secure against impersonation under active attack in the sense that no attacks are known, but proving security has remained open.

We prove that the Schnorr scheme is secure against impersonation, under both active and concurrent attacks, under the assumption that discrete exponentiation is secure under one more inversion in the underlying group. The precise statement of the result is Corollary 2. The assumption, also introduced in [5], is the natural analogue of the one we used for RSA. The adversary gets input the generator g. Its challenge oracle returns a random group element, and its inversion oracle computes discrete logarithms relative to g. The assumption states that it is computationally infeasible for the adversary to output correct discrete logarithms of all the target points if the number of queries it makes to its inversion oracle is strictly less than the number of queries it makes to its challenge oracle. (When the adversary makes one challenge query and no inversion queries, this is the standard discrete logarithm assumption, meaning the standard assumption of one-wayness of the discrete exponentiation function.)

The benefits of this result are analogous to those for GQ. Although the assumption is relatively novel and strong, our result reduces the security of the Schnorr identification scheme to a question about the hardness of a number-theoretic problem, thereby freeing a cryptanalyst from consideration of attacks related to the identification problem itself.

1.4 Discussion and Related Work

Within the large class of FS follow-on identification schemes, proven security properties vary. Some like GQ and Schnorr did not have proofs of security against active or concurrent attacks. However, the FS scheme itself can be proven secure against impersonation under active and concurrent attacks assuming factoring is hard by exploiting its witness-indistinguishability (WI) and proof-of-knowledge (POK) properties. Okamoto's discrete logarithm based scheme [22] is also WI and a POK, and can thus be proven secure against impersonation under active and concurrent attacks, assuming hardness of the discrete logarithm problem. Similar results hold for other schemes having the WI and POK properties. However, GQ and Schnorr are not WI, since there is only one secret key corresponding to a given public key, so these techniques do not work for them. On the other hand, they are preferable in terms of cost. Both have smaller key size than FS, and Schnorr is more efficient than Okamoto.

The so-called 2^m-th root identification scheme can be viewed as the analogue of the GQ scheme with the RSA encryption exponent e replaced by a power of two, or as a special case of the Ong-Schnorr scheme [23]. This has been proven secure against impersonation under active attack assuming factoring is hard [28, 27]. As far as we know, its security against impersonation under concurrent attack is an open question.

The signature schemes obtained from the GQ and Schnorr identification schemes via the Fiat-Shamir transform are already known to be provably-secure in the random oracle model assuming, respectively, the one-wayness of RSA and the hardness of the discrete logarithm problem [25], yet the security of the ID schemes against impersonation under active attack has remained open. This is not a contradiction, since the security of the signature scheme in the random oracle model relies on relatively weak security properties of the ID scheme, namely the security of the latter against impersonation under passive attack [1].

Reset attacks (where the cheating verifier can reset the internal state of prover clones with which it interacts [10, 3]) are not considered here since GQ and Schnorr, as with all proof-of-knowledge based schemes, are insecure against these attacks.

2   Definitions

The empty string is denoted ε. If x is a string then $|x|$ denotes its length, and if S is a set then $|S|$ denotes its size.

ID SCHEMES. An *identification* (ID) scheme $\mathcal{ID} = (\mathcal{K}, P, V)$ is a triple of randomized algorithms. On input security parameter $k \in \mathbb{N}$, the poly(k)-time key-generation algorithm $\mathcal{K}$ returns a pair consisting of a public key pk and a matching secret key sk. P and V are polynomial-time algorithms that implement the prover and verifier, respectively. We require the natural correctness condition, namely that the boolean decision produced by V, in the interaction in which P has input pk, sk and V has input pk, is one with probability one. This probability is over the coin tosses of both parties. We assume that the first and last moves in the interaction always belong to the prover.

The following security notion uses the basic two-phase framework of [14] in which, in a first phase, the adversary attacks the secret-key equipped P, and then, in a second phase, plays the role of cheating prover, trying to make V accept. We define and prove security only for impersonation under concurrent attack, since the usual (serial) active attack [14] is a special case of a concurrent attack.

IMPERSONATION UNDER CONCURRENT ATTACK. An *imp-ca adversary* $A = (\widehat{V}, \widehat{P})$ is a pair of randomized polynomial-time algorithms, the *cheating verifier* and *cheating prover*, respectively. We consider a game having two phases. In the first phase, $\mathcal{K}$ is run on input k to produce (pk, sk), a random tape is chosen for $\widehat{V}$ and it is given input pk. It then interacts concurrently with different *clones* of prover P, all clones having independent random tapes and being initialized with pk, sk. Specifically, we view P as a function that takes an incoming message and current state and returns an outgoing message and updated state. Cheating verifier $\widehat{V}$ can issue a request of the form (ε, i). As a result, a fresh random tape R_i is chosen, the initial state St_i of clone i is set to (pk, sk, R_i), the operation $(M_{\text{out}}, St_i) \leftarrow P(\varepsilon; St_i)$ is executed, M_{out} is returned to $\widehat{V}$, and the updated St_i is saved as the new state of clone i. Subsequently, $\widehat{V}$ can issue a request of the form (M, i), in which case message M is sent to clone i, who computes $(M_{\text{out}}, St_i) \leftarrow P(M; St_i)$, returns M_{out} to $\widehat{V}$, and saves the updated state

St_i. These requests of $\widehat{V}$ can be arbitrarily interleaved. Eventually, $\widehat{V}$ outputs some state information St and stops, ending the first phase. In the second phase of the game, the cheating prover $\widehat{P}$ is initialized with St, verifier V is initialized with pk and freshly chosen coins, and $\widehat{P}$ and V interact. We say that adversary A wins if V accepts in this interaction, and the *imp-ca advantage* of A, denoted

$$\mathbf{Adv}_{\mathcal{ID},A}^{\text{imp-ca}}(k)$$

is the probability that A wins, taken over the coins of $\mathcal{K}$, the coins of $\widehat{V}$, the coins of the prover clones, and the coins of V. (There is no need to give $\widehat{P}$ separate coins, or even pk, since it can get them from $\widehat{V}$ via St.) We say that $\mathcal{ID}$ is *secure against impersonation under concurrent attack* (IMP-CA secure) if the function

$$\mathbf{Adv}_{\mathcal{ID},A}^{\text{imp-ca}}(\cdot)$$

is negligible for all imp-ca adversaries A of time complexity polynomial in the security parameter k.

We adopt the convention that the *time complexity* of imp-ca adversary A does not include the time taken by the prover clones and the verifier to compute replies to the adversary's requests. Rather we view these as oracles, each returning replies in unit time. Barring this, the time complexity of A is the execution time of the entire two-phase game, including the time taken for key generation and initializations. This convention simplifies concrete security considerations.

An active attack [14] is captured by considering cheating verifiers that interact serially, one by one, with prover clones. (This means the cheating verifier initializes a clone and finishes interacting with it before starting up another one.)

COMMENTS. We clarify that we stay within the two-phase framework of [14] even while considering concurrent attacks, in the sense that the first phase (in which the adversary mounts a concurrent attack on the secret-key equipped P) is assumed to be completed before the start of the second phase (in which the adversary plays the role of cheating prover and tries to make V accept). This reflects applications such as smart card based identification for ATMs [14]. For identification over the Internet, it is more suitable to consider adversaries that can interact with the prover or prover clones even while they are interacting with the verifier in an attempt to make the latter accept. With this, one moves into the domain of authenticated key-exchange protocols which is definitionally more complex (see for example [9, 8, 29, 11]) and where identification without an associated exchange of a session-key is of little practical value.

3   Reset Lemma

We refer to a three-move protocol of the form depicted in Figure 1 as *canonical*. The prover's first message is called its *commitment*. The verifier selects a *challenge* uniformly at random from a set ChSet_v associated to its input v, and, upon receiving a *response* RsP from the prover, applies a deterministic *decision predicate* $\text{DEC}_v(\text{CMT}, \text{CH}, \text{RSP})$ to compute a boolean decision. The verifier is said to be *represented* by the pair $(\text{ChSet}, \text{DEC})$ which, given the verifier input v, defines the challenge set and decision predicate.

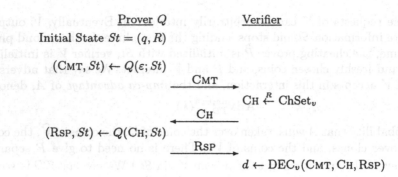

Fig. 1. A canonical protocol. Prover Q has input q and random tape R, and maintains state St. The verifier has input v and returns boolean decision d.

A prover is identified with a function Q that given an incoming message M_{in} (this is ε when the prover is initiating the protocol) and its current state St, returns an outgoing message M_{out} and an updated state. The initial state of the prover is (q, R), where q is an input for the prover and R is a random tape.

The following lemma, which we call the Reset Lemma, upper bounds the probability that a (cheating) prover Q can convince the verifier to accept as a function of the probability that a certain experiment based on resetting the prover yields two accepting conversation transcripts. We will use this lemma in our proofs of security of the GQ and the Schnorr schemes at the time of exploiting their proof-of-knowledge properties. In the past such analyses were based on the techniques of [14] who considered certain "execution trees" corresponding to the interaction, and their "heavy nodes." The Reset Lemma provides a slightly better bound, has a simple proof, and is general enough to be applicable in numerous settings, saving the need to apply the techniques of [14] from scratch in each analysis, and may thus be of independent interest. Note that the lemma makes no mention of proofs of knowledge; it is just about relating two probabilities. The formulation and proof of the lemma generalize some analyses in [4].

Lemma 1. (Reset Lemma) *Let Q be a prover in a canonical protocol with a verifier represented by $(\mathrm{ChSet}, \mathrm{DEC})$, and let q, v be inputs for the prover and verifier, respectively. Let $\mathrm{acc}(q, v)$ be the probability that the verifier accepts in its interaction with Q, namely the probability that the following experiment returns 1:*

> *Choose random tape R for Q ; $St \leftarrow (q, R)$; $(\mathrm{CMT}, St) \leftarrow Q(\varepsilon; St)$*
> *$\mathrm{CH} \xleftarrow{R} \mathrm{ChSet}_v$; $(\mathrm{RSP}, St) \leftarrow Q(\mathrm{CH}; St)$; $d \leftarrow \mathrm{DEC}_v(\mathrm{CMT}, \mathrm{CH}, \mathrm{RSP})$*
> *Return d*

Let $\mathrm{res}(q, v)$ be the probability that the following reset experiment returns 1:

> *Choose random tape R for Q ; $St \leftarrow (q, R)$; $(\mathrm{CMT}, St) \leftarrow Q(\varepsilon; St)$*
> *$\mathrm{CH}_1 \xleftarrow{R} \mathrm{ChSet}_v$; $(\mathrm{RSP}_1, St) \leftarrow Q(\mathrm{CH}_1; St)$; $d_1 \leftarrow \mathrm{DEC}_v(\mathrm{CMT}, \mathrm{CH}_1, \mathrm{RSP}_1)$*
> *$\mathrm{CH}_2 \xleftarrow{R} \mathrm{ChSet}_v$; $(\mathrm{RSP}_2, St) \leftarrow Q(\mathrm{CH}_2; St)$; $d_2 \leftarrow \mathrm{DEC}_v(\mathrm{CMT}, \mathrm{CH}_2, \mathrm{RSP}_2)$*
> *If $(d_1 = 1$ AND $d_2 = 1$ AND $\mathrm{CH}_1 \neq \mathrm{CH}_2)$ then return 1 else return 0*

Then

$$\mathrm{acc}(q, v) \leq \frac{1}{|\mathrm{ChSet}_v|} + \sqrt{\mathrm{res}(q, v)} . \quad \blacksquare$$

Proof (Lemma 1). With q, v fixed, let r denote the length of the prover's random tape. For $R \in \{0, 1\}^r$ let $\mathrm{CMT}(q, R)$ denote Q's commitment when it has input q and random tape R, and let $\mathrm{RSP}(q, R, \mathrm{CH})$ denote the response provided by Q to verifier challenge $\mathrm{CH} \in \mathrm{ChSet}_v$ when Q has input q and random tape R. We define functions $\mathsf{X}, \mathsf{Y} \colon \{0, 1\}^r \to [0, 1]$ as follows. For each $R \in \{0, 1\}^r$ we let

$$\mathsf{X}(R) = \Pr\left[\, \mathrm{DEC}_v(\mathrm{CMT}(q, R), \mathrm{CH}, \mathrm{RSP}(q, R, \mathrm{CH})) = 1 \,\right],$$

the probability being over a random choice of CH from ChSet_v. For each $R \in \{0, 1\}^r$ we let

$$\mathsf{Y}(R) = \Pr\left[\begin{array}{l} \mathrm{DEC}_v(\mathrm{CMT}(q, R), \mathrm{CH}_1, \mathrm{RSP}(q, R, \mathrm{CH}_1)) = 1 \text{ and} \\ \mathrm{DEC}_v(\mathrm{CMT}(q, R), \mathrm{CH}_2, \mathrm{RSP}(q, R, \mathrm{CH}_2)) = 1 \text{ and} \\ \mathrm{CH}_1 \neq \mathrm{CH}_2 \end{array} \right],$$

the probability being over random and independent choices of CH_1 and CH_2 from ChSet_v. A conditioning argument shows that for any $R \in \{0, 1\}^r$

$$\mathsf{Y}(R) \geq \mathsf{X}(R) \cdot \left[\mathsf{X}(R) - \frac{1}{|\mathrm{ChSet}_v|} \right] .$$

We view X, Y as random variables over the sample space $\{0, 1\}^r$ of coins of Q. Then letting $p = 1/|\mathrm{ChSet}_v|$ and using the above we have

$$\begin{aligned} \mathrm{res}(q, v) &= \mathbf{E}\left[\mathsf{Y}\right] \\ &\geq \mathbf{E}\left[\mathsf{X} \cdot (\mathsf{X} - p)\right] \\ &= \mathbf{E}\left[\mathsf{X}^2\right] - p \cdot \mathbf{E}\left[\mathsf{X}\right] \\ &\geq \mathbf{E}\left[\mathsf{X}\right]^2 - p \cdot \mathbf{E}\left[\mathsf{X}\right] \\ &= \mathrm{acc}(q, v)^2 - p \cdot \mathrm{acc}(q, v) . \end{aligned}$$

In the fourth line above, we used Jensen's inequality[1] applied to the convex function $f(x) = x^2$. Using the above we have

$$\left(\mathrm{acc}(q, v) - \frac{p}{2} \right)^2 = \mathrm{acc}(q, v)^2 - p \cdot \mathrm{acc}(q, v) + \frac{p^2}{4} \leq \mathrm{res}(q, v) + \frac{p^2}{4} .$$

Taking the square-root of both sides of the above, and using the fact that $\sqrt{a + b} \leq \sqrt{a} + \sqrt{b}$ for all real numbers $a, b \geq 0$, we get

$$\mathrm{acc}(q, v) - \frac{p}{2} \leq \sqrt{\mathrm{res}(q, v) + \frac{p^2}{4}} \leq \sqrt{\mathrm{res}(q, v)} + \sqrt{\frac{p^2}{4}} = \sqrt{\mathrm{res}(q, v)} + \frac{p}{2} .$$

Re-arranging terms gives us the desired conclusion. $\quad \blacksquare$

---

[1]   Jensen's inequality states that if f is a convex function and X is a random variable, then $\mathbf{E}\left[f(X)\right] \geq f(\mathbf{E}\left[X\right])$.

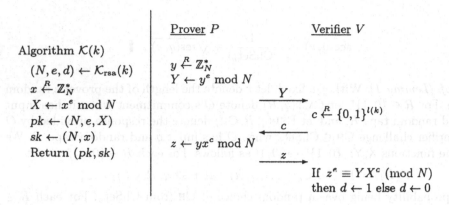

Fig. 2. GQ identification scheme. Prover P has input $pk = (N, e, X)$ and $sk = (N, x)$. Verifier V has input pk.

4   Security of GQ under Concurrent Attack

A randomized, poly(k)-time algorithm $\mathcal{K}_{\mathrm{rsa}}$ is said to be a *prime-exponent RSA key generator* if on input security parameter $k \in \mathbb{N}$, its output is a triple (N, e, d) where N is the product of two distinct primes, $2^{k-1} \leq N < 2^k$ (N is k bits long), $e < \varphi(N)$ is an odd prime, $\gcd(d, \varphi(N)) = 1$, and $ed \equiv 1 \pmod{\varphi(N)}$. We do not pin down any specific such generator. Rather it is a parameter of the GQ identification scheme, and security is proved based on an assumption about it.

GQ IDENTIFICATION SCHEME. Let $\mathcal{K}_{\mathrm{rsa}}$ be a prime-exponent RSA key generator and let $l \colon \mathbb{N} \to \mathbb{N}$ be a polynomial-time computable, polynomially-bounded function such that $2^{l(k)} < e$ for any e output by $\mathcal{K}_{\mathrm{rsa}}$ on input k. The *GQ identification scheme* associated to $\mathcal{K}_{\mathrm{rsa}}$ and challenge length l is the ID scheme whose constituent algorithms are depicted in Figure 2. The prover's commitment is a random element $Y \in \mathbb{Z}_N^*$. For any verifier input $pk = (N, e, X)$, $\mathrm{ChSet}_{pk} = \{0, 1\}^{l(k)}$. A challenge $c \in \mathrm{ChSet}_{pk}$ is interpreted in the natural way as an integer in the set $\{0, \ldots, 2^{l(k)} - 1\}$ in the ensuing computations. Due to the assumption that $2^{l(k)} < e$, the challenge is in $\mathbb{Z}_e$. The verifier's decision predicate $\mathrm{DEC}_{pk}(Y, c, z)$ evaluates to 1 if and only if z is the RSA-inverse of $Y X^c \bmod N$.

RSA ASSUMPTION. We recall the notion of security under one more inversion (omi) [5]. An *rsa-omi* adversary is a randomized, polynomial-time algorithm I that gets input N, e and has access to two oracles. The first is an RSA-inversion oracle $(\cdot)^d \bmod N$ that given $Y \in \mathbb{Z}_N^*$ returns $Y^d \bmod N$. The second is a challenge oracle that, each time it is invoked (it takes no inputs), returns a random challenge point $W \in \mathbb{Z}_N^*$. The game considered is to run $\mathcal{K}_{\mathrm{rsa}}(k)$ to get N, e, d and then run $I(N, e)$ with its oracles. Let $W_1, \ldots, W_n$ denote the challenges returned by I's challenge oracle. We say that I wins if its output is a sequence of points $w_1, \ldots, w_n \in \mathbb{Z}_N^*$ satisfying $w_i \equiv W_i^d \pmod{N}$ —meaning I inverts all the challenge points— and also the number of queries made by I to its RSA-inversion oracle is *strictly less than* n. The *rsa-omi advantage* of I, denoted $\mathbf{Adv}^{\text{rsa-omi}}_{\mathcal{K}_{\text{rsa}},I}(k)$, is the probability that I wins, taken over the coins of $\mathcal{K}_{\text{rsa}}$, the coins of I, and the coins used by the challenge oracle across its invocations. We say that $\mathcal{K}_{\text{rsa}}$ is OMI secure if the function $\mathbf{Adv}^{\text{rsa-omi}}_{\mathcal{K}_{\text{rsa}},I}(\cdot)$ is negligible for any rsa-omi adversary I of time complexity polynomial in k.

We adopt the convention that the *time complexity* of an rsa-omi adversary I is the execution time of the entire game, including the time taken for key generation and one time unit for each reply to an oracle query. (The time taken by the oracles to compute replies to the adversary's queries is not included.)

RESULT. The following theorem shows that the advantage of any imp-ca attacker against the GQ scheme can be upper bounded via the advantage of a related rsa-omi adversary and a function of the challenge length. The theorem shows the concrete security of the reduction.

Theorem 1. *Let* $\mathcal{ID} = (\mathcal{K}, P, V)$ *be the GQ identification scheme associated to prime-exponent RSA key generator* $\mathcal{K}_{\text{rsa}}$ *and challenge length* l. *Let* $A = (\widehat{V}, \widehat{P})$ *be an imp-ca adversary of time complexity* $t(\cdot)$ *attacking* $\mathcal{ID}$. *Then there exists an rsa-omi adversary* I *attacking* $\mathcal{K}_{\text{rsa}}$ *such that for every* k

$$\mathbf{Adv}^{\text{imp-ca}}_{\mathcal{ID},A}(k) \leq 2^{-l(k)} + \sqrt{\mathbf{Adv}^{\text{rsa-omi}}_{\mathcal{K}_{\text{rsa}},I}(k)} \, . \tag{1}$$

Furthermore, the time complexity of I *is* $2t(k) + O(k^4 + (n(k)+1) \cdot l(k) \cdot k^2)$, *where* $n(k)$ *is the number of prover clones with which* $\widehat{V}$ *interacts.*

Based on this theorem, which we will prove later, we can easily provide the following security result for the GQ scheme. In this result, we assume that the challenge length l is super-logarithmic in the security parameter, which means that $2^{-l(\cdot)}$ is negligible. This assumption is necessary, since otherwise the GQ scheme can be broken merely by guessing the verifier's challenge.

Corollary 1. *If prime-exponent RSA key generator* $\mathcal{K}_{\text{rsa}}$ *is OMI secure and challenge length* l *satisfies* $l(k) = \omega(\log(k))$, *then the GQ identification scheme associated to* $\mathcal{K}_{\text{rsa}}$ *and* l *is secure against impersonation under both active and concurrent attacks.*

We proceed to prove Theorem 1.

Proof (Theorem 1). We assume wlog that $\widehat{V}$ never repeats a request. Fix $k \in \mathbb{N}$ and let (N, e, d) be an output of $\mathcal{K}_{\text{rsa}}$ running on input k. Adversary I has access to an RSA-inversion oracle $(\cdot)^d \bmod N$ and a challenge oracle $\mathcal{O}_N$ that takes no inputs and returns a random challenge point $W \in \mathbb{Z}_N^*$ each time it is invoked. The adversary's goal is to invert all the challenges returned by $\mathcal{O}_N$, while making fewer queries to its RSA-inversion oracle then the number of such challenges.

A detailed description of the adversary is in Figure 3. It queries its challenge oracle to obtain a random element $W_0 \in \mathbb{Z}_N^*$ and uses it to create a public key pk for adversary A. It then uses A to achieve its goal by running $\widehat{V}$ and playing the role of the prover clones to answer its requests. In response to a request (ε, i),

Adversary $I^{(\cdot)^d \bmod N, \mathcal{O}_N}(N, e)$

　　Make a query to $\mathcal{O}_N$ and let W_0 be the response; $pk \leftarrow (N, e, W_0)$

　　Choose a random tape R for $\widehat{V}$; Initialize $\widehat{V}$ with (pk, R); $n \leftarrow 0$

　　Run $\widehat{V}$ answering its requests as follows:

　　　　When $\widehat{V}$ issues a request of the form (ε, i) do

　　　　　　$n \leftarrow n + 1$; Make a query to $\mathcal{O}_N$, let W_i be the response

　　　　　　and return W_i to $\widehat{V}$

　　　　When $\widehat{V}$ issues a request of the form (c, i), where $c \in \{0,1\}^{l(k)}$, do

　　　　　　$c_i \leftarrow c$; Make query $W_i W_0^{c_i}$ to $(\cdot)^d \bmod N$, let z_i be the response

　　　　　　and return z_i to $\widehat{V}$

　　Until $\widehat{V}$ outputs state information St and stops

　　$St \leftarrow (St, \varepsilon)$; $(Y, St) \leftarrow \widehat{P}(\varepsilon; St)$

　　$\text{CH}_1 \stackrel{R}{\leftarrow} \{0,1\}^{l(k)}$; $(\text{RSP}_1, \overline{St}) \leftarrow \widehat{P}(\text{CH}_1; St)$

　　If $\text{RSP}_1^e \equiv Y W_0^{\text{CH}_1} \pmod{N}$ then $d_1 \leftarrow 1$ else $d_1 \leftarrow 0$

　　$\text{CH}_2 \stackrel{R}{\leftarrow} \{0,1\}^{l(k)}$; $(\text{RSP}_2, \overline{St}) \leftarrow \widehat{P}(\text{CH}_2; St)$

　　If $\text{RSP}_2^e \equiv Y W_0^{\text{CH}_2} \pmod{N}$ then $d_2 \leftarrow 1$ else $d_2 \leftarrow 0$

　　If $(d_1 = 1 \text{ AND } d_2 = 1 \text{ AND } \text{CH}_1 \neq \text{CH}_2)$ then

　　　　$z \leftarrow \text{RSP}_1 \cdot \text{RSP}_2^{-1} \bmod N$; $(\overline{d}, a, b) \leftarrow \text{EGCD}(e, \text{CH}_1 - \text{CH}_2)$

　　　　$w_0 \leftarrow W_0^a z^b \bmod N$; For $i = 1$ to n do $w_i \leftarrow z_i w_0^{-c_i} \bmod N$

　　　　Return $w_0, w_1, \ldots, w_n$

　　else Return $\perp$

Fig. 3. rsa-omi adversary I for the proof of Theorem 1.

I queries its challenge oracle and returns the answer W_i to $\widehat{V}$. By the definition of prover P, from $\widehat{V}$'s perspective, this is equivalent to picking a random tape R_i for prover clone i, initializing clone i with state pk, R_i, computing clone i's commitment W_i, and returning the commitment to $\widehat{V}$. I is not in possession of the secret key $sk = (N, W_0^d \bmod N)$ corresponding to pk, which the prover clones would use to respond to $\widehat{V}$'s requests of the form (c, i), where $c \in \{0,1\}^{l(k)}$, but it compensates using its access to the RSA-inversion oracle to answer these requests. Specifically, in response to request (c, i), I makes the query $W_i W_0^c$ to its inversion oracle and returns the answer z_i to $\widehat{V}$. Since $z_i = (W_i W_0^c)^d \bmod N = W_i^d (W_0^d)^c \bmod N$, this is exactly the response that clone i would return to $\widehat{V}$. Hence I simulates the behavior of the prover clones perfectly.

　　If $n(k)$ is the number of prover clones with which $\widehat{V}$ interacts, when $\widehat{V}$ stops I has made $n(k)$ queries to its RSA-inversion oracle and it needs to invert $n(k) + 1$ points. I attempts to extract from $\widehat{P}$, initialized with the output of $\widehat{V}$, the RSA-inverse of challenge W_0. To do so, it runs $\widehat{P}$ obtaining its commitment, selects two independent random challenges from $\{0,1\}^{l(k)}$, runs $\widehat{P}$ to obtain its response to each of these challenges (with the same state), and evaluates the verifier's decision predicate on $\widehat{P}$'s commitment and each challenge-response pair. If the decision predicate evaluates to 1, meaning $\widehat{P}$ makes the verifier accept, on both accounts and the challenges are different, then I extracts the inverse of W_0 as follows. It computes the quotient mod N of the cheating prover's responses to the challenges and sets z to this value. We observe that $z^e \equiv W_0^{\mathrm{CH}_1 - \mathrm{CH}_2} \pmod{N}$. Then I uses the extended Euclid algorithm, EGCD, to compute $(\bar{d}, a, b)$, where $\bar{d} = \gcd(e, \mathrm{CH}_1 - \mathrm{CH}_2)$ and $a, b \in \mathbb{Z}$ are such that $ae + b(\mathrm{CH}_1 - \mathrm{CH}_2) = \bar{d}$. By the assumptions that e is prime and $2^{l(k)} < e$ (which implies $\mathrm{CH}_1, \mathrm{CH}_2 \in \mathbb{Z}_e$), $\bar{d} = 1$. Hence $ae + b(\mathrm{CH}_1 - \mathrm{CH}_2) = 1$. Therefore, modulo N we have

$$W_0 \equiv W_0^{ae} W_0^{b(\mathrm{CH}_1 - \mathrm{CH}_2)} \equiv W_0^{ae}(W_0^{\mathrm{CH}_1 - \mathrm{CH}_2})^b \equiv W_0^{ae}(z^e)^b \equiv (W_0^a z^b)^e .$$

This shows that $w_0 = W_0^a z^b \bmod N$ is the RSA-inverse of W_0. For $i = 1, \ldots, n(k)$, I computes the inverse of the i-th challenge point as $w_i = z_i w_0^{-c_i} \bmod N$. To prove that this computation yields the desired RSA-inverse, we show that $w_i^e \equiv W_i \pmod{N}$. Since z_i is the inverse of $W_i W_0^{c_i}$ and w_0 is the inverse of W_0,

$$w_i^e \equiv (z_i w_0^{-c_i})^e \equiv z_i^e (w_0^e)^{-c_i} \equiv W_i W_0^{c_i} W_0^{-c_i} \equiv W_i \pmod{N} .$$

If the decision predicate does not evaluate to 1 on both occasions or the challenges coincide, then I fails. Therefore, I wins if and only if $d_1 = 1$, $d_2 = 1$ and $\mathrm{CH}_1 \neq \mathrm{CH}_2$. We proceed to relate the probability of this event with the imp-ca advantage of adversary A.

We observe that pk has the same distribution as in the two-phase game that defines a concurrent attack. Since I simulates the environment provided to $\hat{V}$ in that game perfectly, $\hat{V}$ behaves as it does when performing a concurrent attack against $\mathcal{ID}$, and $\hat{P}$ is given state information with the same distribution as in that case. Therefore, the probability that $d_1 = 1$ is exactly $\mathbf{Adv}_{\mathcal{ID},A}^{\mathrm{imp\text{-}ca}}(k)$.

Let $\mathrm{acc}(St, pk)$ denote the probability that $d_1 = 1$ when the public key created by I is pk and the output of $\hat{V}$ is St. (This probability is over the choice of challenge CH_1.) Let $\mathrm{res}(St, pk)$ denote the probability that $d_1 = 1$, $d_2 = 1$ and $\mathrm{CH}_1 \neq \mathrm{CH}_2$ when the public key created by I is pk and the output of $\hat{V}$ is St. (The probability here is over the choice of challenges CH_1 and CH_2.) Then, if $\mathbf{E}[\cdot]$ denotes the expectation of random variable $\cdot$ over the choice of pk and St, the probability that $d_1 = 1$ is $\mathbf{E}[\mathrm{acc}(St, pk)]$, and the probability that I wins is $\mathbf{E}[\mathrm{res}(St, pk)]$. Applying the Reset Lemma to $\hat{P}$ with input St and verifier V with input pk, where the latter is implemented by I, we have

$$\mathrm{acc}(St, pk) \leq 2^{-l(k)} + \sqrt{\mathrm{res}(St, pk)} .$$

We obtain Equation (1) as follows.

$$
\begin{aligned}
\mathbf{Adv}_{\mathcal{ID},A}^{\mathrm{imp\text{-}ca}}(k) &= \mathbf{E}[\mathrm{acc}(St, pk)] \\
&\leq \mathbf{E}\left[2^{-l(k)} + \sqrt{\mathrm{res}(St, pk)}\right] \\
&= 2^{-l(k)} + \mathbf{E}\left[\sqrt{\mathrm{res}(St, pk)}\right] \\
&\leq 2^{-l(k)} + \sqrt{\mathbf{E}[\mathrm{res}(St, pk)]} \\
&= 2^{-l(k)} + \sqrt{\mathbf{Adv}_{\mathcal{K}_{\mathrm{rsa}},I}^{\mathrm{rsa\text{-}omi}}(k)} ,
\end{aligned}
$$

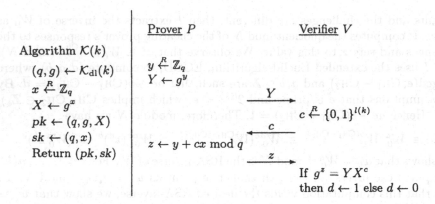

Fig. 4. Schnorr identification scheme. Prover P has input $pk = (q, g, X)$ and $sk = (q, x)$. Verifier V has input pk.

where the last inequality follows from Jensen's inequality[2] applied to the concave function $f(x) = \sqrt{x}$.

The proof of the claim about the time complexity of adversary I is in the full version of this paper [7]. ∎

5   Security of Schnorr under Concurrent Attack

A randomized, poly(k)-time algorithm $\mathcal{K}_{dl}$ is said to be a *discrete logarithm parameter generator* if given security parameter $k \in \mathbb{N}$, it outputs a pair (q, g) where q is a prime such that $q \mid p - 1$ for a prime p with $2^{k-1} \leq p < 2^k$ (p is k bits long), and g is a generator of G_q, a subgroup of $\mathbb{Z}_p^*$ of order q. As before, we do not pin down any specific such generator. The generator is a parameter of the Schnorr scheme, and security is proved based on an assumption about it.

SCHNORR IDENTIFICATION SCHEME. Let $\mathcal{K}_{dl}$ be a discrete logarithm parameter generator and let $l: \mathbb{N} \to \mathbb{N}$ be a polynomial-time computable, polynomially-bounded function such that $2^{l(k)} < q$ for any q output by $\mathcal{K}_{dl}$ on input k. The *Schnorr identification scheme* associated to $\mathcal{K}_{dl}$ and challenge length l is the ID scheme whose constituent algorithms are depicted in Figure 4. The prover's commitment is a random element $Y \in G_q$. For any verifier input $pk = (q, g, X)$, $\text{ChSet}_{pk} = \{0, 1\}^{l(k)}$. A challenge $c \in \text{ChSet}_{pk}$ is interpreted as an integer in the set $\{0, \ldots, 2^{l(k)} - 1\}$ in the ensuing computations. The assumption that $2^{l(k)} < q$ implies that the challenge is in $\mathbb{Z}_q$. The verifier's decision predicate $\text{DEC}_{pk}(Y, c, z)$ evaluates to 1 if and only if z is the discrete log of YX^c.

DL ASSUMPTION. We recall the notion of security under one more discrete logarithm (omdl) [5]. An *omdl* adversary is a randomized, polynomial-time algorithm I that gets input q, g and has access to two oracles. The first is a discrete log

---

[2] Jensen's inequality states that if f is a concave function and X is a random variable, then $\mathbf{E}[f(X)] \leq f(\mathbf{E}[X])$.

oracle $\text{DLog}_{G_q,g}(\cdot)$ that given $Y \in G_q$ returns $y \in \mathbb{Z}_q$ such that $g^y = Y$. The second is a challenge oracle that, each time it is invoked, returns a random challenge point $W \in G_q$. The game considered is to run $\mathcal{K}_{dl}(k)$ to get q, g and then run $I(q, g)$ with its oracles. Let $W_1, \ldots, W_n$ denote the challenges returned by I's challenge oracle. We say that I wins if its output is a sequence of points $w_1, \ldots, w_n \in \mathbb{Z}_q$ satisfying $g^{w_i} = W_i$, and also the number of queries made by I to its discrete log oracle is *strictly less than* n. The *omdl advantage* of I, denoted $\mathbf{Adv}^{omdl}_{\mathcal{K}_{dl}, I}(k)$, is the probability that I wins, taken over the coins of $\mathcal{K}_{dl}$, the coins of I, and the coins used by the challenge oracle across its invocations. We say that $\mathcal{K}_{dl}$ is OMDL secure if the function $\mathbf{Adv}^{omdl}_{\mathcal{K}_{dl}, I}(\cdot)$ is negligible for any omdl adversary I of time complexity polynomial in k.

We adopt the same convention regarding time complexity as in the case of an rsa-omi adversary.

RESULT. The following theorem guarantees that the advantage of any imp-ca adversary attacking the Schnorr scheme can be upper bounded via the advantage of a related omdl adversary and a function of the challenge length.

Theorem 2. *Let $\mathcal{ID} = (\mathcal{K}, P, V)$ be the Schnorr identification scheme associated to discrete logarithm parameter generator $\mathcal{K}_{dl}$ and challenge length l. Let $A = (\widehat{V}, \widehat{P})$ be an imp-ca adversary of time complexity $t(\cdot)$ attacking $\mathcal{ID}$. Then there exists an omdl adversary I attacking $\mathcal{K}_{dl}$ such that for every k*

$$\mathbf{Adv}^{imp\text{-}ca}_{\mathcal{ID},A}(k) \leq 2^{-l(k)} + \sqrt{\mathbf{Adv}^{omdl}_{\mathcal{K}_{dl},I}(k)} . \tag{2}$$

Furthermore, the time complexity of I is $2t(k) + O(k^3 + (l(k) + n(k)) \cdot k^2)$, where $n(k)$ is the number of prover clones with which $\widehat{V}$ interacts.

The proof of Theorem 2 is in the full version of this paper [7]. This theorem implies the following security result for the Schnorr scheme.

Corollary 2. *If discrete logarithm parameter generator $\mathcal{K}_{dl}$ is OMDL secure and challenge length l satisfies $l(k) = \omega(\log(k))$, then the Schnorr identification scheme associated to $\mathcal{K}_{dl}$ and l is secure against impersonation under both active and concurrent attacks.*

As in the case of the GQ scheme, the assumption that the challenge length l is super-logarithmic in the security parameter is necessary since otherwise the Schnorr scheme can be broken by guessing the verifier's challenge.

Acknowledgments

The authors were supported in part by NSF grant CCR-0098123, NSF grant ANR-0129617 and an IBM Faculty Partnership Development Award.

References

1. M. ABDALLA, J. AN, M. BELLARE AND C. NAMPREMPRE. From identification to signatures via the Fiat-Shamir Transform: Minimizing assumptions for security and forward-security. *Advances in Cryptology – EUROCRYPT '02*, Lecture Notes in Computer Science Vol. 2332 , L. Knudsen ed., Springer-Verlag, 2002.

2. N. BARIĆ AND B. PFITZMANN. Collision-free accumulators and fail-stop signature schemes without trees. *Advances in Cryptology – EUROCRYPT '97*, Lecture Notes in Computer Science Vol. 1233, W. Fumy ed., Springer-Verlag, 1997.

3. M. BELLARE, M. FISCHLIN, S. GOLDWASSER AND S. MICALI. Identification protocols secure against reset attacks. *Advances in Cryptology – EUROCRYPT '01*, Lecture Notes in Computer Science Vol. 2045, B. Pfitzmann ed., Springer-Verlag, 2001.

4. M. BELLARE AND S. MINER. A forward-secure digital signature scheme. *Advances in Cryptology – CRYPTO '99*, Lecture Notes in Computer Science Vol. 1666, M. Wiener ed., Springer-Verlag, 1999.

5. M. BELLARE, C. NAMPREMPRE, D. POINTCHEVAL AND M. SEMANKO. The one-more-RSA inversion problems and the security of Chaum's blind signature scheme. Available as *IACR eprint archive Report 2001/002*, http://eprint. iacr.org/2001/002/. Preliminary version, entitled "The power of RSA inversion oracles and the security of Chaum's RSA-based blind signature scheme," in *Financial Cryptography '01*, Lecture Notes in Computer Science Vol. 2339, P. Syverson ed., Springer-Verlag, 2001.

6. M. BELLARE AND G. NEVEN. Transitive signatures based on factoring and RSA. Manuscript, May 2002.

7. M. BELLARE AND A. PALACIO. GQ and Schnorr identification schemes: Proofs of security against impersonation under active and concurrent attacks. Full version of this paper, available via http://www-cse.ucsd.edu/uers/mihir.

8. M. BELLARE, D. POINTCHEVAL AND P. ROGAWAY. Authenticated key exchange secure against dictionary attacks. *Advances in Cryptology – EUROCRYPT '00*, Lecture Notes in Computer Science Vol. 1807, B. Preneel ed., Springer-Verlag, 2000.

9. M. BELLARE AND P. ROGAWAY. Entity authentication and key distribution. *Advances in Cryptology – CRYPTO '93*, Lecture Notes in Computer Science Vol. 773, D. Stinson ed., Springer-Verlag, 1993.

10. R. CANETTI, S. GOLDWASSER, O. GOLDREICH AND S. MICALI. Resettable zero-knowledge. *Proceedings of the 32nd Annual Symposium on the Theory of Computing*, ACM, 2000.

11. R. CANETTI AND H. KRAWCZYK. Universally composable notions of key-exchange and secure channels. *Advances in Cryptology – EUROCRYPT '02*, Lecture Notes in Computer Science Vol. 2332 , L. Knudsen ed., Springer-Verlag, 2002.

12. D. CHAUM. Blind signatures for untraceable payments. *Advances in Cryptology – CRYPTO '82*, Lecture Notes in Computer Science, Plenum Press, New York and London, 1983, Aug. 1982.

13. R. CRAMER AND V. SHOUP. Signature schemes based on the strong RSA assumption. In *5th ACM Conference on Computer and Communications Security*, pages 46–51, Singapore, Nov. 1999. ACM Press.

14. U. FEIGE, A. FIAT AND A. SHAMIR. Zero knowledge proofs of identity. *Journal of Cryptology*, 1(2):77–94, 1988.

15. U. FEIGE AND A. SHAMIR. Witness indistinguishable and witness hiding protocols. *Proceedings of the 22nd Annual Symposium on the Theory of Computing*, ACM, 1990.
16. A. FIAT AND A. SHAMIR. How to prove yourself: Practical solutions to identification and signature problems. *Advances in Cryptology – CRYPTO '86*, Lecture Notes in Computer Science Vol. 263, A. Odlyzko ed., Springer-Verlag, 1986.
17. E. FUJISAKI AND T. OKAMOTO. Statistical zero knowledge protocols to prove modular polynomial relations. *Advances in Cryptology – CRYPTO '97*, Lecture Notes in Computer Science Vol. 1294, B. Kaliski ed., Springer-Verlag, 1997.
18. R. GENNARO, S. HALEVI, AND T. RABIN. Secure hash-and-sign signatures without the random oracle. *Advances in Cryptology – EUROCRYPT '99*, Lecture Notes in Computer Science Vol. 1592, J. Stern ed., Springer-Verlag, 1999.
19. S. GOLDWASSER, S. MICALI AND C. RACKOFF. The knowledge complexity of interactive proof systems. *SIAM Journal of Computing*, 18(1):186–208, February 1989.
20. L. GUILLOU AND J. J. QUISQUATER. A "paradoxical" identity-based signature scheme resulting from zero-knowledge. *Advances in Cryptology – CRYPTO '88*, Lecture Notes in Computer Science Vol. 403, S. Goldwasser ed., Springer-Verlag, 1988.
21. S. MICALI AND R. RIVEST. Transitive signature schemes. *Topics in Cryptology – CT-RSA '02*, Lecture Notes in Computer Science Vol. 2271 , B. Preneel ed., Springer-Verlag, 2002.
22. T. OKAMOTO. Provably secure and practical identification schemes and corresponding signature schemes. *Advances in Cryptology – CRYPTO '92*, Lecture Notes in Computer Science Vol. 740, E. Brickell ed., Springer-Verlag, 1992.
23. H. ONG AND C. P. SCHNORR. Fast signature generation with a Fiat Shamir–like scheme. *Advances in Cryptology – EUROCRYPT '90*, Lecture Notes in Computer Science Vol. 473, I. Damgård ed., Springer-Verlag, 1990.
24. D. POINTCHEVAL. New public key cryptosystems based on the dependent-RSA problems. *Advances in Cryptology – EUROCRYPT '99*, Lecture Notes in Computer Science Vol. 1592, J. Stern ed., Springer-Verlag, 1999.
25. D. POINTCHEVAL AND J. STERN. Security arguments for digital signatures and blind signatures. *Journal of Cryptology*, 13(3):361–396, 2000.
26. C. P. SCHNORR. Efficient signature generation by smart cards. *Journal of Cryptology*, 4(3):161–174, 1991.
27. C. P. SCHNORR. Security of the 2^t-root identification and signatures. *Advances in Cryptology – CRYPTO '96*, Lecture Notes in Computer Science Vol. 1109, N. Koblitz ed., Springer-Verlag, 1996.
28. V. SHOUP. On the security of a practical identification scheme. *Journal of Cryptology*, 12:247–260, 1999.
29. V. SHOUP. On formal models for secure key exchange (version 4). *IACR eprint archive Report 1999/012*, http://eprint.iacr.org/1999/012/.

On 2-Round Secure Multiparty Computation

Rosario Gennaro[1], Yuval Ishai[2,*], Eyal Kushilevitz[3,**], and Tal Rabin[1]

[1] IBM T.J. Watson Research Center
{rosario,talr}@watson.ibm.com
[2] Princeton University
yishai@cs.princeton.edu
[3] Technion, Israel
eyalk@cs.technion.ac.il

Abstract. Substantial efforts have been spent on characterizing the round complexity of various cryptographic tasks. In this work we study the round complexity of *secure multiparty computation* in the presence of an *active* (Byzantine) adversary, assuming the availability of secure point-to-point channels and a broadcast primitive. It was recently shown that in this setting *three* rounds are sufficient for arbitrary secure computation tasks, with a linear security threshold, and two rounds are sufficient for certain nontrivial tasks. This leaves open the question whether *every* function can be securely computed in two rounds.

We show that the answer to this question is "no": even some very simple functions do not admit secure 2-round protocols (independently of their communication and time complexity) and thus 3 is the exact round complexity of general secure multiparty computation. Yet, we also present some positive results by identifying a useful class of functions which can be securely computed in two rounds. Our results apply both to the information-theoretic and to the computational notions of security.

Keywords: Secure multiparty computation, round complexity, lower bounds.

1   Introduction

The race for improving the *round complexity* of cryptographic protocols appears, quite miraculously, to never lose its steam[1]. In this work we study the round complexity of *secure multiparty computation*. Following the initial plausibility results in this area [32, 21, 5, 10], considerable efforts have been spent on obtaining round-efficient protocols [1, 4, 3, 13, 22, 31, 2, 8, 23, 17, 26, 11, 24]. In the multiparty setting, it was recently shown in [17] that every function can be securely computed in three rounds (tolerating a constant fraction of *mali-*

---

* Most of this work was done while the author was at AT&T Labs—Research.
** Most of this work was done while the author was at IBM T.J. Watson Research Center.
[1] It is somewhat reassuring to note in ·this context that the round complexity is restricted to be a *positive* integer.

M. Yung (Ed.): CRYPTO 2002, LNCS 2442, pp. 178–193, 2002.

cious players)[2] and that for certain nontrivial tasks two rounds suffice. This naturally raises the question whether *every* function can be securely computed in two rounds. In the current work we focus on this question, examining the capabilities and limitations of 2-round protocols.

The Model. We consider a system of n players, who interact in synchronous rounds via authenticated secure point-to-point channels and a broadcast medium[3]. Interfering with the interaction is an *active* (Byzantine) adversary, who may corrupt up to t players (where t is referred to as the *security threshold*), learn their internal data, and arbitrarily modify their behavior. By default, we make the standard assumption that the adversary has a *rushing* capability, namely in each round it may learn the messages sent at this round by uncorrupted players to the corrupted players before sending its own messages. This is the most commonly used model in the general secure multiparty computation literature (e.g., [21,5,10,30,28,12]), and in particular it is the standard model assumed in the context of *constant-round* secure multiparty computation (e.g., [1,4,3,2,23,17]). We will also address the situation in the *fully synchronous* setting, where the messages of each round are guaranteed to be simultaneous. As for other aspects of the model, such as perfect vs. computational security, and adaptive vs. non-adaptive adversary, they can be set appropriately so as to achieve the strongest statements for both our positive and negative results. Section 2 includes a more detailed description of the model and the standard definition of security in this model.

1.1 Our Results

We obtain both positive and negative results, resolving the main qualitative questions concerning the possibility of 2-round secure multiparty computation. Our main results are outlined below.

Positive results. We show that any function whose outputs are determined by the input of a *single* player can be securely computed in two rounds with a linear security threshold, efficiently in its circuit size. In contrast to its appearance, this function class is neither trivial nor useless. In particular, this result generalizes the 2-round protocols for verifiable secret-sharing and secure multicast from [17], implies 2-round distributed zero-knowledge protocols for NP [16], and also has other applications that we discuss. In addition, we observe that there are also functions outside this class which admit 2-round protocols.

---

[2] This is possible either with unconditional security, efficiently in the branching program size of the function being computed, or with computational security, efficiently in its circuit size.

[3] Broadcast allows each player to send an identical message to *all* players, without allowing it to violate the consistency requirement. This primitive can be simulated using secure point-to-point channels via a Byzantine Agreement protocol [29,25,7, 14]; however the cost of such a simulation would exceed the round complexity we consider here. From a more practical point of view, a broadcast medium may be implemented either physically or via semi-trusted external parties, such as Internet bulletin boards.

Negative results. Our main conclusion is that 3 rounds are necessary for general secure multiparty computation in any "standard" model (i.e., without public-key infrastructure or preprocessing and with full fairness requirement; see below). Specifically, if $t \geq 2$ (i.e., at least two players may be corrupted), then even some very simple functions cannot be securely computed in two rounds, regardless of the number of players, n, and the protocol's communication and time complexity. We consider the simple special cases of XOR_2^n and AND_2^n (exclusive-or and conjunction of two input bits, where all n players should learn the output) and show that both functions do not admit 2-round protocols. Interestingly, these two cases turn out to be quite different from each other, and each proof represents a distinct type of security concern. Indeed, the impossibility result in the former case is inherently linked to the adversary's rushing capability, as XOR_2^n (and more generally, any *linear* function) admits 2-round protocols in the fully synchronous model. In contrast, the second negative result (for the function AND_2^n) applies also to the fully synchronous model. Naturally, the above two special cases generalize to larger classes of functions with "similar" properties. Combining the above negative results with the 3-round upper bound of [17], we have that 3 is the *exact* round complexity of general secure multiparty computation with a linear security threshold.

Extension to more liberal models. We note that the above negative results can be extended to the common random string model or, more generally, to a setting where a public string from an arbitrary trusted distribution is given to *all* players. However, the results do not apply if we allow either a preprocessing stage or distribution of correlated random resources by a trusted party prior to the computation. In fact, in these models two rounds are sufficient for securely computing every function with a linear security threshold. For instance, [18] shows that two rounds are sufficient for achieving *independence* (i.e., a simultaneous broadcast among n players) given a public-key infrastructure. It follows from our results that some underlying infrastructure is indeed necessary, as otherwise the corresponding functionality is impossible to compute securely in two rounds. Our negative results also rely on the *fairness* requirement of secure computation, and do not apply if the adversary is allowed to "abort" the computation after learning its output.

1.2 Related Work

Most relevant to the current work are the works on the round complexity of secure multiparty computation, cited above. Among those, the only work to prove *lower bounds* on the round complexity is [17], where it is proved that *perfect* VSS and secure multicast with *optimal resilience* ($t < n/3$) require 3 rounds. However, settling for a slightly smaller security threshold ($t < n/4$), these tasks require only two rounds. In contrast, our negative results for 2 rounds apply even when the number of players n is arbitrarily larger than t, and even for the relaxed notions of statistical and computational security. We stress though that our negative results do not apply to more liberal settings of secure computation. For instance, if the adversary is *passive*, then two rounds are sufficient for computing any function with $t < n/3$; this can be achieved either with computational security, efficiently in the circuit size [4], or with perfect security, efficiently in the branching program size [24]. In the *two-party* case, 2-round secure computation is possible either against a passive adversary in the standard model [32, 31], or against an active adversary in the common reference string model, assuming that only one player has an output [8].

The round complexity of *zero-knowledge* protocols has been extensively studied in various settings (e.g., see [20]). However, this line of work is very different from ours both in the type of task being considered (zero-knowledge vs. general secure computation) and, more importantly, in the setting. Indeed, our multiparty setting is more liberal in the sense that it only requires security against limited collusions of players, and in particular allows to assume that a strict majority of the players remain uncorrupted.

Various papers deal with the round complexity of implementing *Byzantine Agreement* and broadcast using only point-to-point channels (e.g., see [29, 14, 27]). While these problems can be viewed as secure computation tasks, they are trivialized in our model since we assume broadcast as a primitive.

Finally, the round complexity of *collective coin-flipping* (which may also be viewed as a secure computation task) has been discussed in the *full information model* [6]. This model is dual to the Byzantine Agreement one: it allows *only* broadcast and no secure point-to-point communication. Similarly to our default model, the adversary is allowed rushing. We note that the availability of secure point-to-point channels in our model makes the coin-flipping task more feasible, and thus negative results from the relevant literature do not apply in our context.

Organization. In Section 2 we present the model and definitions. Section 3 includes our positive results, followed by the lower bounds in Section 4.

2   Model and Definitions

In this section we outline the definition of secure computation, following Canetti's definition approach [9], and highlight some details that are important for our purposes. The following version of the definition is somewhat simplified. In particular, this simplified version considers a protocol as a stand-alone application and does not support any kind of composition; however, our positive results (and obviously the negative results) hold for stronger versions of the definition as well. We refer the reader to [9, 19] for more complete definitions.

Communication model. We consider a network of n processors, denoted $P_1, \ldots, P_n$ and referred to as *players*. Each pair of players is connected via a private, authenticated point-to-point channel. In addition, all players share a common *broadcast* channel, which allows a player to send an identical message to *all* other players. In some sense, the broadcast channel can be viewed as a medium which "commits" the player to a specific value.

Function. A secure computation task is defined by some *n-party function* f : $(\{0, 1\}^*)^n \to (\{0, 1\}^*)^n$, specifying the desired mapping from the players' inputs

to their final outputs. While in certain interesting cases the players will have
to reach an agreement on a joint output, the definition allows for each player
to compute its own output. When referring to a single-output function, it is
assumed by default that all players output its value. One may also consider
randomized functions, which take an additional random input; however, in this
work we focus by default on the deterministic case.

Protocol. Initially, each player P_i holds an input x_i, a random input r_i, and
a common *security parameter* k. The players are restricted to (expected) polynomial time in k. The protocol proceeds in *rounds*, where in each round each
player P_i may send a "private" message to each player P_j (including itself) and
broadcast a "public" message, to be received by all players. The messages P_i
sends in each round may depend on all its inputs (x_i, r_i and k) and the messages
it received in previous rounds. From now on, we assume without loss of generality that each P_i sends x_i, r_i, k to itself in the first round, so that the messages it
sends in each subsequent round may be *determined* from the messages received
in previous rounds. We assume that the protocol terminates after a fixed number of rounds, and each player locally outputs some function of the messages it
received.

Adversary. We consider an *active t-adversary* $\mathcal{A}$, where the parameter t is
referred to as the *security threshold*. The adversary is an efficient interactive
algorithm[4], which is initially given the security parameter k and a random input r. Based on these, it may choose a set T of *at most* t players to corrupt[5].
The adversary then starts interacting with a protocol (either a "real" protocol
as above, or an *ideal-process* protocol to be defined below), where it takes control of all players in T. In particular, it can read their inputs, random inputs,
and received messages, and it can fully control the messages they send. (In the
weaker setting of *passive* security, the adversary cannot modify the corrupted
players' behavior, but only read their information.) We assume by default that
the adversary has a *rushing* capability: at any round it can first wait to hear all
messages sent by uncorrupted players to players in T, and use these to determine its own messages. However, we also consider the *fully synchronous* model,
in which the messages sent by the adversary in each round are independent of
the messages sent by uncorrupted players in the same round. Finally, upon the
protocol's termination, $\mathcal{A}$ outputs some function of its entire view.

Security. Informally, a protocol computing f is said to be t-secure if whatever a
t-adversary can "achieve" by attacking the protocol, it could have also achieved
(by corrupting the same set of players) in an ideal process in which f is evaluated using a trusted party. To formalize this definition, we have to define what
"achieve" means and what the ideal process is. The *ideal process* for evaluating

---

[4] It is usually assumed that the adversary is given an "advice" string a, or is alternatively modeled by a nonuniform algorithm. In fact, the proofs of our negative
results are formulated in this nonuniform setting, but can be modified to apply in
the uniform one as well.

[5] This corresponds to the *non-adaptive* security model; however, all our results apply
to the stronger adaptive model as well.

the function f is a protocol π_f involving the n players and an additional, incorruptible, trusted party TP. The protocol proceeds as follows: (1) each P_i sends its input x_i to TP; (2) TP computes f on the inputs (using its own random input in the randomized case), and sends to each player its corresponding output. Note that when an adversary corrupts the ideal process, it can pick the inputs sent by players in T to TP (possibly, based on their original inputs) and then output an arbitrary function of all its view (including the outputs it received from TP). To formally define security, we capture what the adversary "achieves" by a random variable concatenating the adversary's output together with the *outputs* and the *identities* of the uncorrupted players. For a protocol π, adversary $\mathcal{A}$, input vector x, and security parameter k, let $exec_{\pi,A}(k,x)$ denote the above random variable, where the randomness is over the random inputs of the uncorrupted players, the trusted party (if f is randomized), and the adversary. The security of a protocol Π (also referred to as a *real-life* protocol) is defined by comparing the $exec$ variable of the protocol Π to that of the ideal process π_f. Formally:

Definition 1. *We say that a protocol Π t-securely computes f if, for any (real-life) t-adversary $\mathcal{A}$, there exists (an ideal-process) t-adversary $\mathcal{A}'$ such that the distribution ensembles $exec_{\Pi,A}(k,x)$ and $exec_{\pi_f,A'}(k,x)$ are indistinguishable. The security is referred to as perfect, statistical, or computational according to the notion of indistinguishability being achieved. For instance, in the computational case it is required that for any family of polynomial-size circuits $\{C_k\}$ there exists some negligible function neg, such that for any x,*

$$|C_k(exec_{\Pi,A}(k,x)) - C_k(exec_{\pi_f,A'}(k,x))| \leq \mathsf{neg}(k).$$

An equivalent form of Definition 1 quantifies over all *input distributions* X rather than specific input vectors x, and gives X as an additional input to the distinguisher C_k. This equivalent form is convenient for proving our negative results.

Intuitive discussion. Definition 1 asserts that for any *real-life* t-adversary $\mathcal{A}$ attacking the real protocol there is an *ideal-process* t-adversary $\mathcal{A}'$ which can "achieve" in the ideal process as much as $\mathcal{A}$ does in the real life. The latter means that the output produced by $\mathcal{A}'$ together with the inputs and outputs of uncorrupted players in the ideal process is indistinguishable from the output (wlog, the entire view) of $\mathcal{A}$ concatenated with the inputs and outputs of uncorrupted players in the real protocol. This concatenation captures both *privacy* and *correctness* requirements. On the one hand, it guarantees that the view of $\mathcal{A}$ does not allow it to gain more information about inputs and outputs of uncorrupted players than is possible in the ideal process and, on the other hand, it ensures that the inputs and outputs of the uncorrupted players in the real protocol be consistent with some correct computation of f in the ideal process.

Additional intuition regarding the definition, including our general paradigm for proving *negative* results, is given in Section 4.

3   Positive Results

Which functions are the easiest to compute securely? In this section we obtain a 2-round protocol for every function whose outputs are determined by the input of a *single* player. We stress that this class of secure computation tasks is nontrivial, and in fact it can be used to implement some important tasks such as VSS and distributed zero-knowledge. To see that this class is nontrivial, note that in the multi-output case the protocol has to ensure that all local outputs be consistent with the same input, and at the same time must hide the players' outputs from each other. Moreover, even in the single-output case (i.e., where the same output is learned by all players) it is not enough to let the player holding the input compute and broadcast the global output; indeed, in this naive protocol the players may not be able to *efficiently* verify that the broadcasted value is consistent with some valid input.

We now show that every function in the above class can be securely computed in 2 rounds with perfect security and a linear security threshold, efficiently in its circuit size. To prove our claim, we reduce the task of securely computing such a function to that of securely computing a related vector of degree-2 polynomials, and in return show how to compute such a vector in 2 rounds. We start by describing the latter.

Lemma 1. *Let $p = (p_1, \ldots, p_s)$ be a vector of degree-2 multivariate polynomials in the inputs $x = (x_1, \ldots, x_m)$ over a finite field F, where $|F| > n$ [6]. Moreover, suppose that P_1 holds the entire input vector x and that each player gets some specified subset of the outputs. Then, $p(x)$ admits a perfectly secure 2-round protocol with a linear security threshold in which the communication and time complexity are linear in $s + m$ and polynomial in the number of players.*

Proof sketch: The protocol proceeds similarly to the 2-round VSS protocol from [17]. Here we outline a somewhat simplified version which does not achieve an optimal security threshold, but suffices for our purposes. For simplicity, assume that $s = 1$; the general case is handled by parallel repetition. In the first round, P_1 uses the bivariate polynomial secret-sharing of [5] to share each of its inputs; that is, it chooses a random bivariate polynomial $F^l(y, z)$ over F of degree at most t in each variable under the condition that $F^l(0, 0) = x_l$ for $l = 1, 2, \ldots, m$. It sends to player P_i the polynomials $f_i^l(y) = F^l(y, i)$ and $g_i^l(z) = F^l(i, z)$. In parallel, each pair of players privately exchange random pads. In the second round, each player P_i lets its primary share of x_l be $s_i^l = f_i^l(0)$ (note that if P_1 is honest then the points (i, s_i^l) lie on a degree-t polynomial) and sends, to each player who should receive the output, the value $p(s_i^1, \ldots, s_i^m)$, i.e. its share of the output $p(x)$ [7]. In parallel, each player P_j broadcasts the value of each *secondary* share $f_i^l(j)$ and $g_i^l(j)$ masked with the pad exchanged with

---

[6] This assumption on the size of F can be eliminated by the use of extension fields.

[7] To guarantee privacy, these values have to be randomized so that they lie on a *random* degree-2t polynomial with $p(x)$ as its free coefficient. We ignore this detail, since it complicates the presentation and is addressed in a standard way by letting P_1 share additional random values.

P_j in Round 1. These broadcasts induce an *inconsistency graph*, each edge of which represents a conflict between secondary shares of different players that were supposed to be equal (if P_1 and the two relevant players were honest).

We now describe how each player reconstructs the value of p from the n output shares it received and from the (public) inconsistency graph. Suppose that $n > 6t$. The players run a deterministic 2-approximation algorithm for vertex cover on the inconsistency graph [15]. If it returns a vertex cover of size$> 2t$ (implying that there is no vertex cover of size t), then it is clear that P_1 is dishonest, and the output is taken to be $p(0)$. Otherwise, let I be the complement of the vertex cover (which is an independent set in the graph). Note that $|I| > 4t$, and so the players in I contain at least $3t + 1$ uncorrupted players whose input shares were all consistent with some input vector $x' = (x'_1, \ldots, x'_m)$, and thus their output shares lie on a degree-2t polynomial with free coefficient $p(x')$. The output value is computed from the $|I|$ output shares of the players in I by applying a Reed-Solomon error correction procedure to find the "nearest" degree-2t polynomial, and taking its free coefficient. Note that if P_1 is uncorrupted then, since the distance of the relevant code is greater than $2t$, the correct output will be computed. Conversely, if I indeed contains more than $4t$ players, then the output will be consistent with the value of p on some input x' defined by the (consistent) shares of the uncorrupted players in I. □

Theorem 1. *Suppose that f is a deterministic function whose inputs are all held by a single player. Then, f admits a perfectly secure 2-round protocol with a linear security threshold, computing f efficiently in its circuit size.*

Proof. We prove the theorem for the case of a single-output function represented by a boolean circuit; a proof for the general case proceeds similarly. We reduce the secure computation of f to the secure computation of degree-2 polynomials. Suppose that C is a boolean circuit computing f, and let F be a finite field where $|F| > n$. We construct a vector p of degree-2 polynomials over F. The input variables of p are of three types: (1) variables x, such that x_i corresponds to the i-th input of f; (2) variables y, such that y_i corresponds to the i-th intermediate wire in C (i.e., excluding input and output wires); (3) variables z, such that z_i corresponds to the i-th output wire of C. The vector $p(x, y, z)$ will serve to verify that the input values x are valid (i.e., each of these variables is assigned either 0 or 1), and that the wire labels y, z are consistent with the gates of C and the inputs x. Specifically, it should hold that $p(x, y, z)$ is the zero vector if all the above consistency requirements are met, and otherwise it contains at least one nonzero entry. Note that each atomic consistency requirement can be verified by a single degree-2 polynomial. For instance, the validity of an input value x_i can be verified by the polynomial $x_i(1 - x_i)$, and the consistency of an internal NAND gate having input wires i, j and output wire k can be verified by $1 - y_i y_j - y_k$. Hence, the total length of p is proportional to the size of C.

Now, let $p'(x, y, z) = p(x, y, z) \circ z$ (where $\circ$ denotes concatenation). Given a secure protocol Π for computing $p'(x, y, z)$, a secure protocol for $C(x)$ proceeds as follows:

- On input $x \in \{0,1\}^m$, player P_1 computes the wire labels y, z, and invokes Π on inputs x, y, z. Let $v \circ z$ denote the output of Π (by its security, the same output must be obtained by all uncorrupted players).
- Each player computes its output as follows: If $v = 0$ output z, otherwise output $C(0)$.

The correctness of this reduction can be sketched as follows. If P_1 is honest, the correct output $z = C(x)$ will clearly be obtained (even in the presence of an active t-adversary), and no additional information about x will be revealed. Conversely, for either of the two possibilities for obtaining the output, it must be consistent with some input x. $\square$

As a corollary of Theorem 1, it is possible to obtain 2-round distributed zero-knowledge protocols for NP [16]. Indeed, if R is a polynomial-time predicate defining an NP-language, P_1 can prove that it knows a witness w such that $R(x, w) = 1$ by invoking a secure protocol for the function $f(x, y) = x \circ R(x, y)$ (substituting w for y). Theorem 1 can also be applied for obtaining a wide array of "certified secret-distribution" schemes, generalizing the VSS primitive. For instance, let $D(s, r)$ be a (t, n)-secret-sharing scheme (such that $D_i(s, r)$ is the share of the secret s held by P_i), and let $R(s)$ be an efficient predicate testing whether s satisfies some validity condition. Define a function f whose inputs s, r are held by P_1, and such that P_i's output is $(D_i(s, r), R(s))$. Then, the secure computation of f allows P_1 to securely distribute his secret s among n players ensuring consistency of the shares with some *valid* secret s, which can at a later stage be reconstructed even in the presence of faulty players. Note that if P_1 fails to pick the input r at random, then at most the *secrecy* of the secret s is compromised (which anyway cannot be avoided) but not its validity.

We end this section by noting that Theorem 1 does not cover *all* functions which admit 2-round protocols. We demonstrate this using the following "degenerate" example, which in fact requires only one round, but more interesting examples (requiring 2 rounds) can be obtained.

Example 1. Consider the function $f(x_1, x_2, \ldots, x_n) = (x_1 \oplus x_2, \perp, \ldots, \perp)$. The value x_i is the input of player P_i, only player P_1 should output the exclusive-or of the bits x_1, x_2 and other players have no output. Note that the output of f depends on inputs of *two* players. Yet, it can be verified that the trivial protocol, in which P_2 sends its input to P_1 and the latter computes the correct output, is t-secure for any threshold t.

In the next section we will show that the above function does *not* admit a 2-round protocol if all players should output $x_1 \oplus x_2$.

4   Negative Results

In this section we prove impossibility of 2-round secure computation for some simple specific functions. Since defining the notion of security is a delicate issue, it is not surprising that negative results may also involve some subtleties. In particular, one has to account for *all* possible strategies of the ideal-process adversary, who is not restricted to any particular behavior pattern. Our general paradigm for proving negative results is the following. For a given function f and a protocol Π, we define a specific real-life adversary $\mathcal{A}_0$ which "breaks" Π in the sense that it has some advantage over any ideal-process adversary $\mathcal{A}'$ attacking π_f. To demonstrate this, we typically define some distribution on the inputs of uncorrupted players, and then specify some concrete "challenge" which no $\mathcal{A}'$ can meet (in the ideal process) as successfully as $\mathcal{A}_0$ by corrupting the same players as $\mathcal{A}_0$ does. For specifying such a challenge, we may use any predicate on the inputs and outputs of uncorrupted players. For instance, $\mathcal{A}_0$ may challenge $\mathcal{A}'$ to guess the input of a specific uncorrupted player, or to fix the output of some uncorrupted player to 0. If we show that $\mathcal{A}_0$ can significantly outperform every $\mathcal{A}'$ in meeting such a challenge, then we have shown Π to be insecure.

4.1   The Functions SB and XOR, or: The Power of Rushing

In this section, we prove negative results for simultaneous broadcast (defined below) and XOR. A common characteristic of these tasks is that they become easier in the fully synchronous model; thus, in addition to proving the necessity of 3 rounds, these examples also serve to separate the fully synchronous model from the standard model.

We start by showing the impossibility of a 2-round simultaneous broadcast (SB) protocol. This natural task is formally defined by the function $SB(x_1, x_2, \ldots, x_n) = x_1 \circ x_2$, i.e., each player should output the concatenation of the first two inputs. (We refer to each of the two parts of the global output as an *output entry*.) Note that the main security requirement imposed by an SB protocol is *independence*: any ideal-process adversary attacking at most one of the first two players should be unable to induce a non-negligible correlation between the two output entries. We obtain our impossibility result by describing a strategy which allows the adversary to break this independence requirement. The high-level idea is the following. The adversary will corrupt one of the two input holders, say P_2, and some carefully chosen additional player P_j, where $j > 2$. It will pick its Round 1 messages in such a way that will allow its action in Round 2 to have a non-negligible effect on the second output entry (as seen by uncorrupted players). It will then use its rushing capabilities and the fact that P_j was "honest-looking" in Round 1 to first learn the first output entry, and then correlate the second entry with the first one. This will contradict the independence requirement.

We now formalize the above intuition and fill in some missing details. It will be convenient to use the following notation. By (B, M), where $M = (M_1, \ldots, M_n)$, we denote some joint distribution of messages and broadcast sent by P_2 in Round 1 (where M_i is the message sent to P_i). By (B^0, M^0) (resp., (B^1, M^1)) we denote the *honest* distributions corresponding to the input $x_2 = 0$ (resp., $x_2 = 1$). For a distribution (B, M), let $q_\sigma(B, M)$ denote the probability that the protocol's second output entry is equal to 1, given that: (1) $x_1 = \sigma$; (2) P_2's Round 1 messages and broadcast are distributed according to (B, M); (3) everyone else follows the protocol (including P_2 in Round 2). Finally, let

$q(B, M) = (q_0(B, M), q_1(B, M))$. All the above distributions and probabilities are parameterized by the security parameter k, which will usually be omitted. We start with the following lemma.

Lemma 2. *The distributions ensembles $(B^0, M_1^0)(k)$ and $(B^1, M_1^1)(k)$ are computationally indistinguishable.*

Proof. Assume towards a contradiction that there is a distinguisher D such that D always outputs 0 or 1, and $|\Pr[D(B^0, M_1^0, k) = 1] - \Pr[D(B^1, M_1^1, k) = 1]| > k^{-c}$, for some constant c and infinitely many values of k. We use D to show that an adversary corrupting P_1 can break the independence requirement. The adversary's strategy is simple: it waits to hear the broadcast b and message m received from P_2 in Round 1, evaluates $D(b, m, k)$, and uses the result as its input. Clearly, the correlation induced by the adversary cannot be emulated in the ideal process (even up to computational indistinguishability).   □

Theorem 2. *There is no 2-round (computationally) secure SB protocol, for any $t \geq 2$ and an arbitrarily large number of players $n \geq 3$.*

Proof. Assume towards a contradiction that a 2-round SB protocol is given. Consider the following four pairs of probabilities:

$$Q_1 = q(B^1, M_1^1, M_2^1, \ldots, M_n^1)$$
$$Q_2 = q(B^1, M_1^1, 0, 0, \ldots, 0)$$
$$Q_3 = q(B^0, M_1^0, 0, 0, \ldots, 0)$$
$$Q_4 = q(B^0, M_1^0, M_2^0, \ldots, M_n^0)$$

By the protocol's correctness, we must have $Q_1 \geq (1 - \mathsf{neg}, 1 - \mathsf{neg})$ and $Q_4 \leq (\mathsf{neg}, \mathsf{neg})$, where neg denotes some negligible function in k. Moreover, by Lemma 2, the difference between Q_2 and Q_3 is negligible. Hence, there is a substantial difference either between Q_1 and Q_2 or between Q_3 and Q_4. Assume wlog that the difference between the first entries of Q_1 and Q_2 (corresponding to the probability q_0) is large, say, more than $1/3$. By a hybrid argument, there exists $i \geq 2$ such that $q_0(B^1, M_1^1, \ldots, M_i^1, 0, \ldots, 0) - q_0(B^1, M_1^1, \ldots, M_{i-1}^1, 0, \ldots, 0) > 1/(3n)$. It follows that one of the two q_0 probabilities above must be different by at least $1/(6n)$ from one of the two corresponding q_1 probabilities. Assume, without loss of generality, that

$$|q_0(B^1, M_1^1, \ldots, M_i^1, 0, \ldots, 0) - q_1(B^1, M_1^1, \ldots, M_{i-1}^1, 0, \ldots, 0)| > \frac{1}{6n} \quad (1)$$

(the other cases are similar).

Now, we need to identify two players to complete two tasks in our attack. Yet, it might be the case that both these tasks can be embodied into a single player. We need an honest looking player P_j from whose local view we will compute the correct output, and a second player P_i who can toggle the output of the rest of the players in the protocol. The index of player P_i is given from Eq. (1). If $i > 2$ then this player has acted honestly until now and thus can also be "used" as the player P_j from which we extract the output. Otherwise, i.e. if $i = 2$, then we set $P_j = P_3$, as the player whose view we will examine. An adversary corrupting P_2, P_j can correlate the second output entry with the first as follows. Its Round 1 messages are distributed according to $(B^1, M_1^1, M_2^1, M_3^1, \ldots, M_i^1, 0, \ldots, 0)$. In Round 2, it waits to hear the messages from all uncorrupted players, and then computes the first output entry from the entire view of P_j. Since P_j was honest so far, the protocol's correctness guarantees that it learns the correct output with overwhelming probability[8]. Let α be the value of the first output entry computed by P_j. The adversary correlates its output with α by letting P_i, the toggling player, act as follows. If $\alpha = 0$, P_i behaves honestly (i.e., uses the original message M_i^1 received from P_2 in Round 1). Otherwise, it behaves as if this message was set to 0. It follows from Eq. (1) that the second output entry will be significantly correlated with the first, contradicting the independence requirement. $\square$

Note that the SB function admits a trivial 1-round protocol in the fully synchronous model; thus in this case coping with a rushing adversary costs *two* additional rounds. Another observation is that the requirement $t \geq 2$ is essential. Indeed, the following 5-player protocol computes SB with perfect 1-security in two rounds: (1) Each of P_1, P_2 privately sends its input to each of the remaining 3 players; (2) Each of P_3, P_4, P_5 passes the inputs it received to all other players; Each player outputs the majority of the 3 candidates for each input it received in Round 2. It is not hard to verify that the above protocol is a 1-secure SB protocol.

The function XOR. We now turn to the function $XOR(x_1, x_2, \ldots, x_n)$ defined as $x_1 \oplus x_2$. We show, by refining the previous arguments for the SB function, that this function as well cannot be securely computed in two rounds.

Theorem 3. *There is no 2-round (computationally) secure XOR protocol, for any $t \geq 2$ and an arbitrarily large number of players $n \geq 3$.*

Proof. Similarly to the proof of Theorem 2, the adversary corrupts the player P_2 which, together with an additional player (to be chosen carefully), is used to violate the properties of the alleged protocol. We also follow some of the notations used in the proof of Theorem 2. Specifically, by (B, M) we denote some joint distribution of the broadcast and the private messages (respectively) sent by P_2 in Round 1 of the protocol. By (B^b, M^b) ($b \in \{0, 1\}$) we denote the *honest* distribution corresponding to the input $x_2 = b$. Consider a scenario where P_1 chooses its input x_1, at random. In such a case, in the ideal process, the output is totally random (i.e., each value $\{0, 1\}$ is obtained with probability of exactly 0.5). On the other hand, we will show a strategy for the adversary to significantly bias the output (towards one of the output values). As in the proof of Theorem 2, let $q(B, M)$ be a pair (p_0, p_1) indicating the probability that the output of the function (as seen by the good players) is 1 provided that the input x_1 is 0 or 1 (respectively) and that the first round messages of P_2 are distributed as in (B, M). As before, consider the following four pairs:

---

[8] Note that there is no guarantee that the correct output can be inferred from the view of P_2, since P_2 has deviated from the protocol in Round 1.

$$Q_1 = q(B^0, M_1^0, M_2^0, \ldots, M_n^0)$$
$$Q_2 = q(B^0, M_1^0, 0, \ldots, 0)$$
$$Q_3 = q(B^1, M_1^1, 0, \ldots, 0)$$
$$Q_4 = q(B^1, M_1^1, M_2^1, \ldots, M_n^1)$$

By the protocol's correctness, it follows that $Q_1 = (\mathsf{neg}, 1 - \mathsf{neg})$ while $Q_4 = (1 - \mathsf{neg}, \mathsf{neg})$. In addition, similarly to Lemma 2, the distributions (B^0, M_1^0) and (B^1, M_1^1) must be indistinguishable (as otherwise P_1, by using its rushing capability in the first round is able to correlate its input x_1 with input x_2 and bias the output; this is impossible to achieve in the ideal process). Hence, the difference between q_2, q_3 is neg. Consider the L_1-distance between two pairs $(p_0, p_1), (p_0', p_1')$ (defined as $|p_0 - p_0'| + |p_1 - p_1'|$). It follows that either the distance between q_1 and q_2 is at least $1 - \mathsf{neg}$ or the distance between q_3 and q_4 is at least $1 - \mathsf{neg}$. Assume, without loss of generality, that the first is true. We now argue that, for some i ($2 \le i < n$), the two pairs

$$(p_0^i, p_1^i) \stackrel{\triangle}{=} q(B^0, M_1^0, \ldots, M_i^0, 0, \ldots, 0)$$

and

$$(p_0^{i-1}, p_1^{i-1}) \stackrel{\triangle}{=} q(B^0, M_1^0, \ldots, M_{i-1}^0, 0, \ldots, 0)$$

are such that either $\max\{p_0^i, p_0^{i-1}\} + \max\{p_1^i, p_1^{i-1}\} > 1 - \mathsf{neg} + 1/(5n)$ or $\min\{p_0^i, p_0^{i-1}\} + \min\{p_1^i, p_1^{i-1}\} < 1 - \mathsf{neg} - 1/(5n)$. Otherwise, this in particular implies that all the points (p_0^i, p_1^i) are such that $1 - \mathsf{neg} - 1/(5n) < p_0^i + p_1^i < 1 - \mathsf{neg} + 1/(rn)$. This in turn implies that the distance between two adjacent pairs is smaller than $1/(2n)$ and the total distance between q_1 and q_2 is less than $0.5 < 1 - \mathsf{neg}$, contradicting what we know about this distance. Suppose that for some i we have $\max\{p_0^i, p_0^{i-1}\} + \max\{p_1^i, p_1^{i-1}\} > 1 - \mathsf{neg} + 1/(5n)$; we describe a strategy for the adversary to bias the output towards 1 (in the other case there is a dual strategy to bias the output towards 0). Now, the adversary picks another corrupted player P_j: if $i > 2$ the adversary uses $P_j = P_i$; otherwise, if $i = 2$ the adversary uses, say, $P_j = P_3$. The adversary lets P_2 play in the first round as in $(B^0, M_1^0, \ldots, M_i^0, 0, \ldots, 0)$. In the second round, the adversary uses P_j as a Trojan horse; it lets P_j first get all the second round messages by other players (rushing) and checks which message from P_2 (either M_i^0 or 0) will cause P_j to output 1 (the idea being that a difference which is only in the first round message sent to P_i will only influence the second round message sent by P_i and no other message). If there is such a message then P_j proceeds as if it got this message (and since P_j is honest-looking its output is the same as the output of all good players); otherwise, the adversary picks one of the two messages arbitrarily. Since x_1 is randomly chosen and by the choice of i, the probability of getting the output 1 is now $0.5 \cdot (\max\{p_0^i, p_0^{i-1}\} + \max\{p_1^i, p_1^{i-1}\})$ which is significantly larger than 0.5, as needed.   □

4.2   The Function AND, or: The Advantage of Being Selfish

In this section we consider the function $\mathrm{AND}(x_1, x_2, \ldots, x_n)$ defined as $x_1 \wedge x_2$. We will show that this function cannot be securely computed in two rounds. This case differs from the previous ones in that the relevant impossibility result does *not* rely on the adversary's rushing capability, and thus holds also in the fully synchronous model. The intuition here is also different. In the previous examples, we showed that the adversary could violate the *correctness* of the protocol by inducing some invalid output distribution. In the current case, we will show that the real-life adversary can somehow gain an *information advantage* over its ideal-process counterpart. This will be achieved by practicing a typical *selfish* behavior: the adversary, corrupting P_2 and some other player P_j ($j > 2$), will manage to collect information about x_1 from all uncorrupted players and at the same time refuse to contribute its own share of information to the community. This will allow the real-life adversary to obtain a better prediction of the unknown input than any of the uncorrupted players, which (for the specific case of the AND function) is impossible to achieve in the ideal process.

Theorem 4. *There is no 2-round (computationally) secure AND protocol, for any $t \geq 2$ and an arbitrarily large number of players $n \geq 3$, even in the fully synchronous model.*

Proof. Similarly to the previous proofs, the adversary corrupts player P_2, which together with an additional player is used to violate the properties of the alleged protocol. We also follow some of the previous notation. Specifically, by (B, M) we denote some joint distribution of the broadcast and the private messages (respectively) sent by P_2 in Round 1 of the protocol. By (B^b, M^b) ($b \in \{0, 1\}$) we denote the *honest* distribution corresponding to the input $x_2 = b$. Consider a scenario where P_1 chooses its input x_1 at random. We now argue that in the ideal process the best prediction that the adversary has for the value of x_1 is OUT, the output of the good players, whereas we show a real-life adversary that can guess x_1 with significantly better probability than by using this output. For (B, M) as above, let $\mathrm{COR}(B, M)$ denote the correlation of OUT with x_1 provided that the first round messages by P_2 are distributed as in (B, M). Namely,

$$\mathrm{COR}(B, M) \triangleq |\Pr[\mathrm{OUT} = 1 | x_1 = 1, (B, M)] - \Pr[\mathrm{OUT} = 1 | x_1 = 0, (B, M)]|.$$

Consider the following four quantities:

$$q_1 = \mathrm{COR}(B^0, M_1^0, M_2^0, \ldots, M_n^0)$$
$$q_2 = \mathrm{COR}(B^0, M_1^0, 0, \ldots, 0)$$
$$q_3 = \mathrm{COR}(B^1, M_1^1, 0, \ldots, 0)$$
$$q_4 = \mathrm{COR}(B^1, M_1^1, M_2^1, \ldots, M_n^1)$$

By the protocol's correctness, it follows that $q_1 = \mathsf{neg}$ while $q_4 = 1 - \mathsf{neg}$. In addition, similarly to Lemma 2, the distributions (B^0, M_1^0) and (B^1, M_1^1) must be indistinguishable (otherwise an adversary corrupting, in a passive manner, the player P_1 gets a significantly better than 50% chance of guessing the input x_2 even when the honest players' output is 0; this is impossible in the ideal process). Hence, the difference between q_2, q_3 is neg. It follows that either the distance between q_1 and q_2 is at least $0.5 - \mathsf{neg}$ or the distance between q_3 and q_4 is at least $0.5 - \mathsf{neg}$. Assume, without loss of generality, that the first is true. Therefore, for some i ($2 \leq i < n$) the quantity $\mathrm{COR}(B^0, M_1^0, \ldots, M_i^0, 0, \ldots, 0)$

is significantly smaller than $\mathrm{COR}(B^0, M_1^0, \ldots, M_{i+1}^0, 0, \ldots, 0)$ (by at least $(1 - \mathrm{neg})/(2n)$). Now, the adversary picks another corrupted player P_j: if $i > 2$ the adversary uses $P_j = P_i$; otherwise, if $i = 2$ the adversary uses, say, $P_j = P_3$. The adversary plays as in the distribution $(B^0, M_1^0, \ldots, M_i^0, 0, \ldots, 0)$ [9] so as to guarantee a lower correlation between the output of good players, OUT, and x_1; however, it also uses P_j to compute the output OUT' of the good players (P_j behaves like such a player) in case the messages of P_2 come from the distribution $(B^0, M_1^0, \ldots, M_{i+1}^0, 0, \ldots, 0)$. This value OUT' is significantly better correlated with x_1 than the actual output OUT.

Finally, we argue that in the ideal process the adversary has no better predictor for the value of x_1 than OUT. For this, simply consider all the 4 potential views (x_2, OUT) that the adversary may see. The view $(0, 1)$ is impossible; for both views $(1, 0)$ and $(1, 1)$ we have $x_1 = \mathrm{OUT}$. If the view is $(0, 0)$ then the adversary has no information about x_1; in such a case, guessing the value $x_1 = \mathrm{OUT}$ is correct with probability $1/2$ and is as good as any other way of guessing. Hence, OUT is an optimal predictor for x_1.   □

5   Concluding Remarks

We have answered some of the main qualitative questions concerning the round complexity of secure multiparty computation in our standard model. In particular, we have shown that security against an active adversary requires strictly more interaction than security against a passive adversary, and that general secure computation tasks require more interaction than distributed zero-knowledge and similar tasks. As a future goal, it remains to find a characterization of secure computation tasks according to their exact round complexity. This question appears to be nontrivial, partly due to the difficulty of capturing the exact power of an adversary attacking an ideal-process implementation of complex functions.

References

1. J. Bar-Ilan and D. Beaver. Non-cryptographic fault-tolerant computing in a constant number of rounds. In *Proc. 8th ACM PODC*, pages 201–209. ACM, 1989.
2. D. Beaver. Minimal-Latency Secure Function Evaluation. In *Eurocrypt '00*, pages 335–350, 2000. LNCS No. 1807.
3. D. Beaver, J. Feigenbaum, J. Kilian, and P. Rogaway. Security with low communication overhead (extended abstract). In *Proc. of CRYPTO '90*.
4. D. Beaver, S. Micali, and P. Rogaway. The round complexity of secure protocols (extended abstract). In *Proc. 22nd STOC*, pages 503–513. ACM, 1990.
5. M. Ben-Or, S. Goldwasser, and A. Wigderson. Completeness Theorems for Non-cryptographic Fault-Tolerant Distributed Computations. *Proc. 20th STOC88*, pp. 1–10.
6. M. Ben-Or and N. Linial. Collective Coin-Flipping. In *Randomness and Computation*, pages 91–115, 1990.

---

[9] namely it samples from the distribution (B^0, M^0) and creates the hybrid distribution by replacing some of the messages by 0-messages.

7. P. Berman, J. Garay, and K. Perry. Bit Optimal Distributed Consensus. In R. Yaeza-Bates and U. Manber, editors, *Computer Science Research*, pages 313–322. Plenum Publishing Corporation, 1992.
8. C. Cachin, J. Camenisch, J. Kilian, and J. Muller. One-round secure computation and secure autonomous mobile agents. In *Proceedings of ICALP'00*, 2000.
9. R. Canetti. Security and composition of multiparty cryptographic protocols. *Journal of Cryptology*, 13(1):143–202, 2000.
10. D. Chaum, C. Crepeau, and I. Damgard. Multiparty Unconditionally Secure Protocols. In *Proc. 20th STOC88*, pages 11–19.
11. R. Cramer and I. Damgård. Secure distributed linear algebra in a constant number of rounds. In *Proc. Crypto 2001*.
12. R. Cramer, I. Damgård, S. Dziembowski, M. Hirt, and T. Rabin. Efficient multiparty computations with dishonest minority. In *Eurocrypt '99*, pages 311–326, 1999. LNCS No. 1592.
13. Uri Feige, Joe Kilian, and Moni Naor. A minimal model for secure computation (extended abstract). In *Proc. 26th STOC*, pages 554–563. ACM, 1994.
14. P. Feldman and S. Micali. An Optimal Algorithm for Synchronous Byzantine Agreement. *SIAM. J. Computing*, 26(2):873–933, 1997.
15. F. Gavril. Manuscript, 1974.
16. R. Gennaro, S. Halevi, and T. Rabin. Round-optimal zero knowledge with distributed verifiers. www.research.ibm.com/security, 2002.
17. R. Gennaro, Y. Ishai, E. Kushilevitz, and T. Rabin. The Round Complexity of Verifiable Secret Sharing and Secure Multicast. In *Proc. 33th STOC*. ACM, 2001.
18. Rosario Gennaro. Achieving Independence Efficiently and Securely. In *Proc. 14th ACM PODC*, pages 130–136. ACM, 1995.
19. O. Goldreich. Secure multi-party computation (manuscript). www.wisdom.weizmann.ac.il/~oded/pp.html, 1998.
20. O. Goldreich. *Foundation of Cryptography – Fragments of a Book*. ECCC 1995. Available online from *http://www.eccc.uni-trier.de/eccc/*.
21. O. Goldreich, S. Micali, and A. Wigderson. How to Play Any Mental Game. In *Proc. 19th STOC*, pages 218–229. ACM, 1987.
22. Y. Ishai and E. Kushilevitz. Private simultaneous messages protocols with applications. In *ISTCS97*, pages 174–184, 1997.
23. Y. Ishai and E. Kushilevitz. Randomizing polynomials: A new representation with applications to round-efficient secure computation. In *Proc. 41st FOCS*, 2000.
24. Y. Ishai and E. Kushilevitz. Perfect Constant-Round Secure Computation via Perfect Randomizing Polynomials. In *Proc. ICALP '02*.
25. L. Lamport, R.E. Shostack, and M. Pease. The Byzantine generals problem. *ACM Trans. Prog. Lang. and Systems*, 4(3):382–401, 1982.
26. Y. Lindell. Parallel Coin-Tossing and Constant-Round Secure Two-Party Computation. In *Crypto '01*, pages 171–189, 2001. LNCS No. 2139.
27. N. Lynch. *Distributed Algorithms*. Morgan Kaufman, 1996.
28. R. Ostrovsky and M. Yung. How to withstand mobile virus attacks. In *Proc. 10th ACM PODC*, pages 51–59. ACM, 1991.
29. M. Pease, R. Shostak, and L. Lamport. Reaching Agreement in the Presence of Faults. *Journal of the ACM*, 27(2):228–234, 1980.
30. T. Rabin and M. Ben-Or. Verifiable Secret Sharing and Multiparty Protocols with Honest Majority. In *Proc. 21st STOC*, pages 73–85. ACM, 1989.
31. T. Sander, A. Young, and M. Yung. Non-Interactive CryptoComputing For NC1. In *Proc. 40th FOCS*, pages 554–567. IEEE, 1999.
32. A. C-C. Yao. How to Generate and Exchange Secrets. In *Proc. 27th FOCS*, pages 162–167. IEEE, 1986.

Private Computation –
k-Connected versus 1-Connected Networks

Markus Bläser, Andreas Jakoby, Maciej Liśkiewicz*, and Bodo Siebert**

Institut für Theoretische Informatik
Universität zu Lübeck
Wallstraße 40, 23560 Lübeck, Germany
{blaeser,jakoby,liskiewi,siebert}@tcs.mu-luebeck.de

Abstract. We study the role of connectivity of communication networks in private computations under information theoretic settings. It will be shown that some functions can be computed by private protocols even if the underlying network is 1-connected but not 2-connected. Then we give a complete characterisation of non-degenerate functions that can be computed on non-2-connected networks.

Furthermore, a general technique for simulating private protocols on arbitrary networks will be presented. Using this technique every private protocol can be simulated on arbitrary k-connected networks using only a small number of additional random bits.

Finally, we give matching lower and upper bounds for the number of random bits needed to compute the parity function on k-connected networks.

1   Introduction

Consider a set of players, each knowing an individual secret. The goal is to compute a function depending on these secrets such that after the computation none of the players knows anything about the secrets that cannot be derived from the result of the function. An example for such a computation is the secret voting problem. The members of a committee wish to decide whether the majority votes for yes or for no. But the ballot should be proprietary, i.e. after the vote nobody should know anything about the opinion of the other committee members or about the exact number of yes- or no-votes. The only thing known after the computation is whether the majority votes for yes or for no. To exchange data we allow that the committee member can talk to each other in private.

More formally, the players exchange messages to compute the value of a function. But no player should learn anything about the concrete input values of the other players. Depending on the computational power of the players we distinguish between cryptographically secure privacy and privacy in an information theoretic sense. In the first case we assume that no player is able to recompute

---

* On leave from Instytut Informatyki, Uniwersytet Wrocławski, Poland.
** Supported by DFG research grant Re 672/3.

M. Yung (Ed.): CRYPTO 2002, LNCS 2442, pp. 194–209, 2002.

any information about the input within polynomial time (see e.g. [5, 15, 21, 22]). In the second case we do not restrict the computational power of the players (see e.g. [3, 6]). Hence, this notion of privacy is much stronger than in the cryptographic setting. In this paper we use the information theoretic approach.

Private computation has been the subject of a considerable amount of research. Traditionally, one investigates the number of rounds and random bits as complexity measures for private protocols. Chor and Kushilevitz [10] have studied the number of rounds necessary to compute the sum modulo an integer. This function has also been investigated by Blundo et al. [4] and Chor et al. [8]. The number of random bits needed to compute the parity function, i.e. the sum modulo 2, has been examined in [17, 19]. Gál and Rosén [14] have shown that the parity function cannot be computed by any private protocol in $o(\log n / \log d)$ rounds using d random bits. They have also given an almost tight randomness-round-tradeoff for private computations of arbitrary Boolean functions depending on their sensitivity. Bounds on the maximum number of rounds needed in the worst-case to compute a function by a private protocol are given by Bar-Ilan and Beaver [2] and by Kushilevitz [16].

The number of random bits necessary to compute a Boolean function by a private protocol is closely related to its circuit size. Kushilevitz et al. [18] have shown that every function that can be computed with linear circuit size can also be computed by a private protocol with only a constant number of random bits. Using this result one can show that the majority function can be computed by a private protocol using a constant number of random bits and simultaneously a linear number of exchanged bits between players (for the circuit complexity of majority see e.g. [20]).

So far we have assumed that players do not attempt to cheat. Depending on the way players attempt to acquire information about the input of the other players we distinguish between dishonest players and players that can work in teams (e.g. [3, 5, 6, 12]). The goal in this approach is to investigate the number of dishonest players or players in a team that are necessary to learn anything about the input of the remaining players. Chor and Kushilevitz [9] have shown that Boolean functions with one bit output can be computed with teams either of size at most $\lfloor \frac{n-1}{2} \rfloor$ or of any size up to n. For extensions, see [7, 8].

All papers mentioned above do not restrict the communication capabilities of the players. In other words, they use complete graphs as underlying communication networks. However, most realistic parallel architectures have a restricted connectivity and nodes of bounded degree. Franklin and Yung [13] have been the first who studied the role of connectivity in private computations. They have presented a protocol for k-connected bus networks. This protocol can simulate one communication step of a private protocol that was originally written for a complete graph. To simulate such a communication step, their protocol uses $O(n)$ additional random bits.

In this paper we investigate the number of random bits needed to compute functions by private protocols on k-connected networks. The consideration of k-connected networks instead of complete networks seems to be quite realistic for practical applications. We present a new simulation technique that allows us to reduce the number of random bits by taking the connectivity of the network into account. Furthermore, we show that the parity function can be computed by a private protocol on every k-connected network with $\lceil \frac{n-2}{k-1} \rceil - 1$ random bits. On the other hand, we will present k-connected networks where $\lceil \frac{n-2}{k-1} \rceil - 1$ random bits are necessary.

Furthermore, we investigate networks that are not 2-connected and present non-trivial functions that can still be computed by private protocols on such networks. We introduce the notion of a dominated function and prove that a function can be computed by a private protocol on non-2-connected networks if and only if the function is dominated. This result can be generalised to the case where the players can work in teams. Such a computation is not possible if some of the players are dishonest.

The paper is organised as follows. In the next section we define some notations and give a formal definition of private computation. In Section 3 we present a new technique to simulate private protocols on k-connected networks. Furthermore, we present a simple non-trivial function that can be computed by a private protocol on a non-2-connected network. In Section 4 we investigate the number of random bits needed to compute the parity function on arbitrary k-connected networks. Finally, in Section 5 we investigate non-2-connected networks and give a structural property that precisely determines whether a function can be computed on a non-2-connected network.

2   Preliminaries

2.1   Notations

For $i, j \in \mathbb{N}$ define $[i] := \{1, \ldots, i\}$ and $[i..j] := \{i, \ldots, j\}$. Throughout this paper, we will often use the following string operations. Let $x = x[1]x[2] \ldots x[n] \in \{0,1\}^n$ be a string of length n. Then, for $I \subseteq [n]$ and $\alpha \in \{0,1\}^{|I|}$, $x\lceil_{I \leftarrow \alpha}$ is defined as follows:

$$z = x\lceil_{I \leftarrow \alpha} \iff \forall i \in [n] : z[i] = \begin{cases} x[i] & \text{if } i \notin I \\ \alpha[j] & \text{if } i \in I \text{ and } i \text{ is the } j\text{th smallest} \\ & \text{element in } I . \end{cases}$$

For sets $I_1, I_2, \ldots, I_k \subseteq [n]$ and strings $\alpha_1, \alpha_2, \ldots, \alpha_k \in \{0,1\}^*$ with $|\alpha_i| = |I_i|$ we define

$$x\lceil_{I_1,I_2,\ldots,I_k \leftarrow \alpha_1,\alpha_2,\ldots,\alpha_k} := (x\lceil_{I_1 \leftarrow \alpha_1})\lceil_{I_2,\ldots,I_k \leftarrow \alpha_2,\ldots,\alpha_k} .$$

Let $\overline{x}$ denote the bitwise negation of x, i.e. $\forall i \in [n] : \overline{x}[i] = \overline{x[i]}$. For a function $f : \{0,1\}^n \to \{0,1\}$, a set of indices $I \subseteq [n]$, and a string $\alpha \in \{0,1\}^{|I|}$ define the partially restricted function $f\lceil_{I \leftarrow \alpha} : \{0,1\}^{n-|I|} \to \{0,1\}$ as the function obtained from f by assigning the values given by α to the positions in I, i.e.

$$\forall x \in \{0,1\}^{n-|I|} : \quad f\lceil_{I \leftarrow \alpha}(x) := f(0^n\lceil_{I,J \leftarrow \alpha,x}) ,$$

where $J = [n] \setminus I$. Finally, for a string $x \in \{0,1\}^n$ and a set $I \subseteq [n]$ define $x[I] \in \{0,1\}^{|I|}$ as follows:

$$\forall j \leq |I| \; : \; (x[I])[j] = x[i] \; :\Longleftrightarrow \; i \text{ is the } j\text{th smallest element in } I.$$

A graph is called k-connected if, after deleting an arbitrary subset of at most $k - 1$ nodes, the resulting node-induced graph remains connected.

2.2 Private Computation

We consider the computation of Boolean functions $f : \{0,1\}^n \to \{0,1\}$ on a network of n players. In the beginning each player knows a single bit of the input x. The players can send messages to other players via point-to-point communication using secure links where the link topology is given by an undirected graph $G = (V, E)$. When the computation stops, all players know the value $f(x)$. The goal is to compute $f(x)$ such that no player learns anything about the other input bits in an information theoretic sense except for the information it can deduce from its own bit and the result. Such a protocol is called private.

Definition 1. *Let C_i be a random variable of the communication string seen by player P_i, and let c_i be a particular string seen by P_i. A protocol $\mathcal{A}$ for computing a function f is **private with respect to player P_i** if for every pair of input vectors x and y with $f(x) = f(y)$ and $x[i] = y[i]$, for every c_i, and for every random string R_i provided to P_i,*

$$\Pr[C_i = c_i \mid R_i, x] = \Pr[C_i = c_i \mid R_i, y],$$

where the probability is taken over the random strings of all other players. A protocol $\mathcal{A}$ is private if it is private with respect to every player P_i.

We call a protocol *synchronous* if the communication takes place in rounds and each message consists of a single bit. We call a synchronous protocol *oblivious* if the number of bits that player P_i sends to P_j in round t depends only on i, j, and t but not on the input and the random strings. Furthermore, we do not bound the computational resources of the players. We assume that all of them are honest, i.e. the computation and the interactions between players are determined only by the protocol.

For a synchronous oblivious protocol $\mathcal{A}$ let $L(P_i, P_j, \mathcal{A})$ be the number of bits sent from P_i to P_j in $\mathcal{A}$ and

$$L(\mathcal{A}) := \sum_{i \in [n]} \sum_{j \in [n]} L(P_i, P_j, \mathcal{A}).$$

We distribute the given input bits among the nodes of the graph. For convenience, we call the node that gets the bit $x[i]$ player P_i. The players P_i and P_j can communicate directly if and only if they are connected by an edge in the graph.

3 Private Computation on k-Connected Networks

Most known private protocols are written for specific networks. A simulation of such a private protocol on a different network can be done in such way that each player of the new network simulates a player of the original network step-by-step. Hence, we have to find a way to realize the communication steps between all players that are not directly connected. Franklin and Yung [13] have presented a strategy to simulate a transmission of one single bit on a hypergraph by using $O(n)$ additional random bits. Thus, the whole simulation presented there requires $O(m + nL(\mathcal{A}))$ random bits where m is the number of random bits used by the original protocol. If we consider 2-connected graphs we can simulate each communication step between two players P_i and P_j by one additional random bit r as follows: Assume P_i has to send bit b to P_j. Then P_i chooses two disjoint paths to P_j and sends r to P_j along the one path and $r \oplus b$ along the other path. In this way, $O(m + L(\mathcal{A}))$ random bits are sufficient.

To reduce the number of random bits even more we consider the following problem:

Definition 2 (Max-Neighbour-Embedding). *Let* $G = (V, E)$ *be a graph with edge weights* $\sigma : E \to \mathbb{N}$ *and* $G' = (V', E')$ *a graph with* $|V| = |V'|$. *Let* $\pi : V \to V'$ *be a bijective mapping. Then the performance of* π *is defined as*

$$\rho(\pi) := \sum_{\substack{\{u, v\} \in E \text{ and} \\ \{\pi(u), \pi(v)\} \in E'}} \sigma(\{u, v\}).$$

The aim is to find a bijection $\pi : V \to V'$ *that maximizes* $\rho(\pi)$ *over all bijections.*

By a reduction from the 3-Dimensional-Matching-Problem, it can be shown that the decision problem corresponding to finding an optimal bijection is $\mathcal{NP}$-hard. The Max-Neighbour-Embedding-problem is $\mathcal{NP}$-hard even if both graphs have maximum degree 4.

In the following lemma we estimate the performance for the case that G' is k-connected.

Lemma 1. *Let* $G = (V, E)$ *be an undirected graph with* n *nodes and edge weights* σ. *Let* $G' = (V', E')$ *be a k-connected graph with* n *nodes. Then we have*

$$\max_{\substack{\pi : V \to V' \\ \pi \text{ is bijective}}} \rho(\pi) \geq \frac{k}{n - 1} \sum_{e \in E} \sigma(e).$$

Proof. By the definition above, there is no difference between edges with weight 0 and nonexistent edges. Therefore, we treat nonexistent edges like edges with weight 0 and restrict ourselves to the case that G is a complete graph.

The graph G' is k-connected. Thus, every node in V' has degree at least k.

Let Π be a random bijection from V to V'. Since every node in V' has degree at least k, the probability that two arbitrary nodes u and v are neighbours under

Π, i.e. $\{\Pi(u), \Pi(v)\} \in E'$, is at least $\frac{k}{n-1}$. Thus, the edge $e = \{u, v\} \in E$ yields weight $\sigma(e)$ with probability at least $\frac{k}{n-1}$ and its expected weight is at least $\frac{k}{n-1} \cdot \sigma(e)$. Hence, the expected performance $\rho(\Pi)$ fulfils

$$\mathbb{E}(\rho(\Pi)) \geq \sum_{e \in E} \frac{k}{n-1} \cdot \sigma(e) = \frac{k}{n-1} \cdot \sum_{e \in E} \sigma(e).$$

Therefore, there exists a bijection with performance at least $\frac{k}{n-1} \cdot \sum_{e \in E} \sigma(e)$. $\square$

A bijection that fulfils the requirements of the above lemma can be computed in polynomial time using the method of conditional expectation (see e.g. Alon et al. [1]).

Theorem 1. *Every oblivious private protocol $\mathcal{A}$ using m random bits can be simulated on every k-connected graph by using $m + (1 - \frac{k}{n-1}) \cdot \min\{L(\mathcal{A}), n^2 + \frac{L(\mathcal{A})}{k-1}\}$ random bits.*

Proof. Let $G = (V, E)$ be the network used in protocol $\mathcal{A}$ and $G' = (V', E')$ be the k-connected network for protocol $\mathcal{A}'$. To simulate $\mathcal{A}$ we first choose a bijection between the players in G and the players in G'. For every edge $\{P_i, P_j\} \in E$ define $\sigma(\{P_i, P_j\}) := L(P_i, P_j, \mathcal{A}) + L(P_j, P_i, \mathcal{A})$. In Lemma 1 we have seen that there exists a bijection $\pi : V \to V'$ with performance $\rho(\pi) \geq \frac{k}{n-1} L(\mathcal{A})$. Using this bijection, at least $\frac{k}{n-1} L(\mathcal{A})$ bits of the total communication in $\mathcal{A}$ are sent between players that are also neighbours in G'. Thus, this part of the communication can be simulated directly and without additional random bits.

For the remaining $(1 - \frac{k}{n-1}) L(\mathcal{A})$ bits we proceed as follows: Let P_i and P_j be two players that are not directly connected in G'. Then P_i partitions the bits it will send to P_j into blocks $B_1, \ldots, B_{\lceil L(P_i, P_j, \mathcal{A})/(k-1) \rceil}$ of size at most $k - 1$. Furthermore, P_i chooses k node-disjoint paths from P_i to P_j. P_i uses a separate random bit $r[\ell]$ for each block B_ℓ. It sends $r[\ell]$ along the first path and $b \oplus r[\ell]$ for each $b \in B_\ell$ along the remaining paths, each bit on a separate path. $\square$

We have seen that every function that can be computed by a private protocol on some network can also be computed by a private protocol on an arbitrary 2-connected network. On the other hand, there exist functions that cannot be computed by a private protocol, if the underlying network is not 2-connected.

Proposition 1. *The parity function over $n > 2$ bits cannot be computed by a private protocol on any network that is not 2-connected.*

The above theorem can be generalised to a large class of non-degenerate functions. This will be done in Section 5. There we give a characterisation for the class of non-degenerate functions that can be computed by private protocols on networks that are not 2-connected.

Definition 3. *A function $f : \{0,1\}^n \to \{0,1\}$ is called **non-degenerate** if for every $i \in [n]$ we have $f\lceil_{\{i\}\leftarrow 0} \neq f\lceil_{\{i\}\leftarrow 1}$.*

In other words, a non-degenerate function depends on all of its input bits. It turns out that there are functions that can be computed by a private protocol, even if the underlying network is not 2-connected.

Proposition 2. *There are non-degenerate functions that can be computed by a private protocol on networks that are not 2-connected.*

Consider the following non-degenerate function $f : \{0,1\}^{2n+1} \to \{0,1\}$:

$$f(z,x,y) := (z \wedge \bigwedge_{i=1}^{n} x[i]) \vee (\bar{z} \wedge \bigwedge_{i=1}^{n} y[i]) .$$

Here, z is a single bit and both x and y are bit strings of length n. We construct a communication network G for f as follows: Let G_x and G_y be complete networks with n players each. Then connect another player P_z with all players in both G_x and G_y. Obviously, the obtained network is not 2-connected. Using a slight modification of the protocol presented by Kushilevitz et al. [18] one can compute the subfunctions

$$f_x(z,x) := z \wedge \bigwedge_{i=1}^{n} x[i] \quad \text{and} \quad f_y(z,y) := \bar{z} \wedge \bigwedge_{i=1}^{n} y[i]$$

by a private protocol on the networks G_x with P_z and G_y with P_z, respectively. After the computation has been completed, P_z is the only player that knows the results of both subfunctions. Due to symmetry we consider the case that $z = 1$. Then $f_y(z,y) = 0$ and therefore, since f_y has been computed by a private protocol, P_z does not learn anything about y. Furthermore, P_z does not learn anything about x what he has not already known before the computation started.

4   Computing Parity on k-Connected Networks

It is well known that the parity function of n bits can be computed on a cycle by using only one random bit. On the other hand, using our simulation discussed in Section 3 one gets an upper bound of n random bits for general 2-connected networks. The aim of this section is to close this gap. We present a private protocol for parity that uses $\lceil \frac{n-2}{k-1} \rceil - 1$ random bits and show that there are k-connected networks on which parity cannot be computed with less than $\lceil \frac{n-2}{k-1} \rceil - 1$ random bits.

Lemma 2. *There exist k-connected networks with $n \geq 2k$ players on which the parity function cannot be computed by a private protocol with less than $\lceil \frac{n-2}{k-1} \rceil - 1$ random bits.*

Proof. We consider the bipartite graph $K_{k,n-k}$ (which is obviously k-connected) and show that every private protocol that computes the parity function on this network needs at least $\lceil \frac{n-2}{k-1} \rceil - 1$ random bits. Let $\{P_1, P_2, \ldots, P_k\}$ and $\{P_{k+1}, P_{k+2}, \ldots, P_n\}$ be the two sets of nodes of $K_{k,n-k}$. Recall that for each $i = 1, \ldots, k$ and $j = k+1, \ldots, n$ we have an edge $\{P_i, P_j\}$ in $K_{k,n-k}$ and that there are no other edges. Now assume to the contrary that there exists a private protocol $\mathcal{A}$ on $K_{k,n-k}$ using less than $\lceil \frac{n-2}{k-1} \rceil - 1$ random bits.

Let $R = (R_1, \ldots, R_n)$ be the contents $R_1, \ldots, R_n$ of all random tapes. For a string $x \in \{0, 1\}^n$ and $i \in [n]$, let $C_i(x, R)$ be a full description of the communication received by P_i during the computation of $\mathcal{A}$ with R on the input x. Moreover, let

$$C(x) = \{\langle c_1, c_2, \ldots, c_k \rangle \mid \exists R \, \forall i \in [k] \ \ c_i = C_i(x, R)\}.$$

We consider computations of $\mathcal{A}$ on inputs

$$X = \{x \mid x[1] = x[2] = \ldots = x[k] = 0 \ \text{ and } \ \bigoplus_{i=1}^n x[i] = 0\}.$$

Then for any $x \in X$ and any communication c_1 we define

$$C(c_1, x) = \{\langle c_2, \ldots, c_k \rangle \mid \langle c_1, c_2, \ldots, c_k \rangle \in C(x)\}.$$

From the fact that $\mathcal{A}$ is private it follows:

Claim. $\exists c_1 \, \forall x \in X \ \ C(c_1, x) \neq \emptyset$.

Indeed, because x is a valid input for the protocol $\mathcal{A}$, there exists at least one tuple $\langle c_1, \ldots, c_k \rangle$ in $C(x)$. Hence, there exists at least one c_1 with $C(c_1, x) \neq \emptyset$. On the other hand, if for some $y \in X$ the set $C(c_1, y)$ is empty then one can conclude that $\mathcal{A}$ is not private.

Note that $|X| = 2^{n-k-1}$ and that for every $x, y \in X$ and $i \in [k]$ we have $\bigcup_R C_i(x, R) = \bigcup_R C_i(y, R)$. Furthermore, using a bound on the number of different communication strings from Kushilevitz and Rosén [19] it follows that $|\bigcup_R C_i(x, R)| < 2^{\frac{n-k-1}{k-1}}$. Hence, we have $|\bigcup_{x \in X} C(c_1, x)| < 2^{n-k-1}$, because $\mathcal{A}$ uses less than $\frac{n-k-1}{k-1}$ random bits. Therefore, by the pigeon hole principle and the above claim we obtain

$$\exists c_1, c_2, \ldots, c_k \, \exists x, y \in X \ \ x \neq y \ \text{ and } \ \langle c_2, \ldots, c_k \rangle \in C(c_1, x) \cap C(c_1, y).$$

This means that there are two different input string $x, y \in X$ such that on both strings the players $P_1, \ldots, P_k$ receive $c_1, \ldots, c_k$, respectively. Let i, with $k + 1 \leq i \leq n$, be a position where x and y differ, i.e. $x[i] \neq y[i]$. Let $R = \langle R_1, \ldots, R_n \rangle$ and $R' = \langle R'_1, \ldots, R'_n \rangle$ be the contents of the random tapes such that $c_i = C_i(x, R) = C_i(y, R')$ for all $1 \leq i \leq k$.

It is easy to see that during a computation of $\mathcal{A}$ with random string $R'' = \langle R_1, \ldots, R_{i-1}, R'_i, R_{i+1}, \ldots R_n \rangle$ on the input $x \lceil_{\{i\} \leftarrow y[i]}$ the players $P_1, P_2, \ldots, P_k$ receive again communication strings $c_1, c_2, \ldots, c_k$, respectively. Hence, for this input they give the same result as for x – a contradiction. $\qquad \Box$

Now we show that this bound is best possible. To obtain a private protocol that computes the parity function with $\lceil \frac{n-2}{k-1} \rceil - 1$ random bits we use the result from Egawa, Glas, and Locke [11] that every k-connected graph G with minimum degree at least d and with at least $2d$ vertices has a cycle of length at least $2d$ through any specified set of k vertices. From this result we get the following observations:

Proposition 3. *Let $G = (V, E)$ be a k-connected graph with $k \leq |V| - 1$. Then for any subset $V' \subseteq V$ with $|V'| = k + 1$ there exists a simple path containing all nodes in V'.*

Proposition 4. *Let $G = (V, E)$ be a k-connected graph with $k \leq |V|$. Then for every subset $V' \subseteq V$ with $|V'| = k$ there exists a simple cycle containing all nodes in V'.*

Proposition 5. *Let $G = (V, E)$ be a k-connected graph. Then G has a simple path of length at least $\min\{2k + 1, |V|\}$.*

To compute the parity function by a private protocol on an arbitrary k-connected network G, we proceed as follows:

1. Mark all nodes in G red. Set $z[i] := x[i]$ for each player P_i.
2. Choose a path in G of length $2k + 1$. According to Proposition 5 such a path always exists. The first player P_i in the path generates a random bit r. Then P_i computes $r \oplus z[i]$ and sends the result to the next player in the path. Finally, P_i sets $z[i] := r$.

   Each internal player P_j on the path receives a bit b from its predecessor in the path, computes $b \oplus z[j]$, sends this bit to its successor, and changes its colour to black.

   The last player P_ℓ on the path receives a bit b from its predecessor and computes $z[\ell] := z[\ell] \oplus b$.

   After this step $2k - 1$ players have changed their colour.
3. We repeat the following step $\lceil \frac{n-3k+1}{k-1} \rceil$ times.

   Choose $k + 1$ red nodes and a path in G containing all these nodes. According to Proposition 3 such a path always exists. We can assume that the start and the end node of the path are among the $k + 1$ given players, hence both are red. Then the first player P_i on this path generates a random bit r, computes $r \oplus z[i]$, and sends the result to the next player in the path. Finally, P_i sets $z[i] := r$.

   Each internal player of the path P_j receives a bit b from its predecessor in the path. If P_j is a black player, it sends b to its successor. If P_j is a red player, it computes $b \oplus z[j]$, sends this bit to its successor, and changes its colour to black.

   The last player P_ℓ on the path receives a bit b from its predecessor and computes $z[\ell] := z[\ell] \oplus b$.

   After this step $k - 1$ players have changed their colour. Hence, after $\lceil \frac{n-3k+1}{k-1} \rceil$ iterations of this step we have at least

   $$\left\lceil \frac{n-3k+1}{k-1} \right\rceil \cdot (k - 1) + 2k - 1 \geq n - k$$

   black players. Thus, at most k are red.
4. Choose a cycle in G containing all red nodes. According to Proposition 4 such a cycle always exists. Let P_{i_0} be a red player. Then P_{i_0} generates a random bit r, computes $r \oplus z[i_0]$, and sends the result to the next player in the cycle.

Each other player P_j on the cycle receives a bit b from its predecessor. If P_j is a black player, it sends b to its successor. If P_j is a red player, it computes $b \oplus z[j]$, sends this bit to its successor, and changes its colour to black.
If P_{i_0} receives a bit b, it computes $b \oplus r$. The result of this step is the result of the parity function.

Let us now count the number of random bits used in the protocol above. In the second and in the last step we use one random bit. In the third step we need $\lceil \frac{n-3k+1}{k-1} \rceil$ random bits. Hence, the total number of random bits is

$$\left\lceil \tfrac{n-3k+1}{k-1} \right\rceil + 2 \;=\; \left\lceil \tfrac{n-2}{k-1} \right\rceil - 1 \,.$$

It remains to show that the protocol is private and computes the parity function. The correctness follows from the fact that each input bit $x[i]$ is stored by exactly one red player and each random bit is stored by either none or two players that are red after each step. By storing a bit b we mean that a player P_i knows a value $z[i]$ that depends on b. Since P_{i_0} is the last red player, it knows the result of the parity function.

Every bit a player receives in the second and third step is masked by a separate random bit. Hence, none of these players can learn anything from these bits. The same holds for all players except for player P_{i_0} in the last step. So we have to analyse the bits sent and received by P_{i_0} more carefully. In the last step $z[i_0]$ is either $x[i_0]$, a random bit, or the parity of a subset of input bits masked by a random bit. In neither case P_{i_0} can learn anything about the other input bits from the bit it receives and the value of $z[i_0]$ except for what can be derived from the result of the function and $x[i_0]$.

Theorem 2. *Let G be an arbitrary k-connected network. Then the parity function of n bits can be computed by a private protocol on G using at most $\lceil \frac{n-2}{k-1} \rceil - 1$ random bits. Moreover, there exist k-connected networks for which this bound is best possible.*

For 2-connected networks, we obtain the following corollary.

Corollary 1. *Let G be an arbitrary 2-connected network of n players ($n \geq 4$). Then the parity function over n bits can be computed by a private protocol on the network G using $n - 3$ random bits. Moreover, there exists 2-connected networks for which this bound is best possible.*

5 Private Computation on Non-2-connected Networks

In Section 3 we have claimed that the parity function cannot be computed by a private protocol on a network that is not 2-connected. On the other hand, we have presented a non-degenerate function that can be computed on a non-2-connected network. In this section, we study this phenomenon to a greater extend.

Throughout this section $f : \{0,1\}^n \to \{0,1\}$ denotes the function we want to compute. Furthermore, I_1, I_2, J_1, J_2 denote both subsets of input positions and indices of players.

We say that a pair (J_1, J_2) of two disjoint subsets $J_1, J_2 \subseteq [n]$ has the **flip-property** if there exist an input $x \in \{0,1\}^n$ and two strings $\alpha \in \{0,1\}^{|J_1|}$ and $\beta \in \{0,1\}^{|J_2|}$ with

$$f(x\lceil_{J_1,J_2 \leftarrow \alpha, \beta}) \neq f(x\lceil_{J_1,J_2 \leftarrow \bar{\alpha}, \beta}) = f(x\lceil_{J_1,J_2 \leftarrow \alpha, \bar{\beta}}).$$

We call the strings α and β **flip-witnesses** for (J_1, J_2).

Lemma 3. *If a function $f : \{0,1\}^n \to \{0,1\}$ is non-degenerate, then for every partition $I_1, I_2 \subseteq [n]$ and every $i \in I_1$ and $j \in I_2$ we have: There exist subsets $J_1 \subseteq I_1$ and $J_2 \subseteq I_2$ with $i \in J_1, j \in J_2$ such that (J_1, J_2) has the flip-property.*

Loosely speaking, this lemma says that each non-degenerate function behaves on subsets of input positions in some sense like the parity function.

Proof. By contradiction, assume that the lemma does not hold for a particular partition $I_1, I_2 \subseteq [n]$ and two indices $i \in I_1$ and $j \in I_2$. From Def. 3 it follows that for every $i \in I_1$ and $j \in I_2$ there exist input strings $y, z \in \{0,1\}^n$ such that

$$f(y\lceil_{\{i\} \leftarrow 0}) \neq f(y\lceil_{\{i\} \leftarrow 1}) \quad \text{and} \quad f(z\lceil_{\{j\} \leftarrow 0}) \neq f(z\lceil_{\{j\} \leftarrow 1}).$$

If the lemma does not hold, we can conclude that

$$f(y\lceil_{\{i\},\{j\} \leftarrow 0,0}) = f(y\lceil_{\{i\},\{j\} \leftarrow 0,1}) \neq f(y\lceil_{\{i\},\{j\} \leftarrow 1,0}) = f(y\lceil_{\{i\},\{j\} \leftarrow 1,1}).$$

Otherwise, at least one of the following cases holds:

- $f(y\lceil_{\{i\},\{j\} \leftarrow 0,1}) \neq f(y\lceil_{\{i\},\{j\} \leftarrow 1,1})$ and $f(y\lceil_{\{i\},\{j\} \leftarrow 0,1}) = f(y\lceil_{\{i\},\{j\} \leftarrow 1,0})$. Choosing $J_1 = \{i\}$, $J_2 = \{j\}$, and $\alpha = \beta = 1$ satisfies the claim of the lemma.
- $f(y\lceil_{\{i\},\{j\} \leftarrow 0,0}) \neq f(y\lceil_{\{i\},\{j\} \leftarrow 0,1})$ and $f(y\lceil_{\{i\},\{j\} \leftarrow 0,0}) = f(y\lceil_{\{i\},\{j\} \leftarrow 1,1})$ Choosing $J_1 = \{i\}$, $J_2 = \{j\}$, $\alpha = 0$ and $\beta = 1$ satisfies the claim of the lemma.

Analogously, one can show that

$$f(z\lceil_{\{i\},\{j\} \leftarrow 0,0}) = f(z\lceil_{\{i\},\{j\} \leftarrow 1,0}) \neq f(z\lceil_{\{i\},\{j\} \leftarrow 0,1}) = f(z\lceil_{\{i\},\{j\} \leftarrow 1,1}).$$

W.l.o.g. assume that $y[i] \neq z[i]$ and $y[j] \neq z[j]$. If $f(y) = f(z)$, we flip the bits $y[j]$ and $z[j]$. Since $f(y)$ does not depend on $y[j]$, we have $f(y) \neq f(z)$. We choose

$$Y_1 := \{ k \in I_1 \mid y[k] \neq z[k] \} \quad \text{and} \quad Y_2 := \{ k \in I_2 \mid y[k] \neq z[k] \}.$$

Let $Y_1 := \{i_1, \ldots, i_{|Y_1|}\}$ with $i_1 < i_1 < \cdots < i_{|Y_1|}$ and $Y_2 := \{j_1, \ldots, j_{|Y_2|}\}$ with $j_1 < j_1 < \cdots < j_{|Y_2|}$. Define $\rho \in \{0,1\}^{|Y_1|}$ and $\sigma \in \{0,1\}^{|Y_2|}$ such that

$$\forall \ell \in [1, |Y_1|] : \ \rho[\ell] := y[i_\ell] \quad \text{and} \quad \forall \ell \in [1, |Y_2|] : \ \sigma[\ell] := y[i_\ell].$$

Note that $y\lceil_{Y_1,Y_2\leftarrow\bar{\rho},\bar{\sigma}} = z$.

Recall that $f(y) \neq f(z)$. To prove the claim we have to distinguish between the following three cases: $f(y) \neq f(y\lceil_{Y_2\leftarrow\bar{\sigma}}) = f(z)$, $f(y) = f(y\lceil_{Y_1\leftarrow\bar{\rho}}) \neq f(z)$, and $f(y) \neq f(y\lceil_{Y_1\leftarrow\bar{\rho}}) = f(z)$. The last case can be reduced to the first case by exchanging y and z with each other.

1. If $f(y) \neq f(y\lceil_{Y_2\leftarrow\bar{\sigma}}) = f(z)$, we choose

$$\alpha := y[i], \quad \beta := \sigma, \quad J_1 := \{i\}, \quad J_2 := Y_2, \quad \text{and} \quad x := y.$$

From the definition of non-degenerate functions and the observation above we conclude that

$$y = x\lceil_{J_1,J_2\leftarrow\alpha,\beta} \text{ and } J_1 = \{i\} \implies f(x\lceil_{J_1,J_2\leftarrow\alpha,\beta}) \neq f(x\lceil_{J_1,J_2\leftarrow\bar{\alpha},\beta}),$$
$$y\lceil_{J_2\leftarrow\bar{\sigma}} = x\lceil_{J_1,J_2\leftarrow\alpha,\bar{\beta}} \implies f(x\lceil_{J_1,J_2\leftarrow\alpha,\beta}) \neq f(x\lceil_{J_1,J_2\leftarrow\alpha,\bar{\beta}}).$$

2. If $f(y) = f(y\lceil_{Y_2\leftarrow\bar{\sigma}}) = f(z\lceil_{Y_1\leftarrow\rho}) \neq f(z)$ then we choose

$$\alpha := \bar{\rho}, \quad \beta := z[j], \quad J_1 := Y_1, \quad J_2 := \{j\}, \quad \text{and} \quad x := z.$$

It follows

$$z = x\lceil_{J_1,J_2\leftarrow\alpha,\beta} \text{ and } J_2 = \{j\} \implies f(x\lceil_{J_1,J_2\leftarrow\alpha,\beta}) \neq f(x\lceil_{J_1,J_2\leftarrow\alpha,\bar{\beta}}),$$
$$z\lceil_{J_1\leftarrow\rho} = x\lceil_{J_1,J_2\leftarrow\bar{\alpha},\beta} \implies f(x\lceil_{J_1,J_2\leftarrow\alpha,\beta}) \neq f(x\lceil_{J_1,J_2\leftarrow\bar{\alpha},\beta}).$$

Hence, we can always find subsets $J_1 \subseteq I_1$ and $J_2 \subseteq I_2$ fulfilling the claim – a contradiction. $\qquad\square$

For a given subset I_1 of input positions define the **flip-witness-set** for I_1

$$\text{f-set}(I_1) := \{(\alpha, J_1) \mid J_1 \subseteq I_1, \alpha \in \{0,1\}^{|J_1|}$$
$$\text{and there exists } J_2 \subseteq [n] \setminus I_1, \beta \in \{0,1\}^{|J_2|}$$
$$\text{such that } \alpha, \beta \text{ are flip-witnesses for } J_1, J_2\}.$$

A set I_1 is **dominated** by an input position $k \in I_1$ if the following holds: For each pair of subsets $J_1 \subseteq I_1$ and $J_2 \subseteq [n] \setminus I_1$, such that (J_1, J_2) fulfils the flip-property, we have $k \in J_1$. A function is ℓ-**dominated** if there exists a set $I_1 \subseteq [n]$ of size ℓ that is dominated by some $k \in I_1$. A function f is called **dominated** if there exists $\ell > 1$ such that f is ℓ-dominated. Otherwise, f is called **non-dominated**.

Theorem 3. *Let f be a non-degenerate function and G be a network that can be separated into two networks G_1 and G_2 of size n_1 and n_2, respectively, by removing one bridge node from G. If f can be computed by a private protocol on G, then f is $(n_1 + 1)$- or $(n_2 + 1)$-dominated.*

Theorem 3 follows directly from Lemma 3 and the lemma below. Recall that for all $i \in [n]$ player P_i initially knows $x[i]$. Hence, we can obtain every possible allocation of players and input bits by permuting the enumeration of the players.

Lemma 4 (Fooling private protocols). *Let G be a network with n nodes. Assume that there exist $I_1, I_2 \subseteq [n]$ and $k \in [n]$, such that the following conditions hold:*

1. $I_1, I_2 \neq \emptyset$ and $k \notin I_1 \cup I_2$, $I_1 \cap I_2 = \emptyset$,
2. for every path $W_{i,j}$ from P_i to P_j, with $i \in I_1$ and $j \in I_2$, $P_k \in W_{i,j}$, and
3. (I_1, I_2) has the flip-property.

Then f cannot be computed on G by a private protocol.

Proof. Assume that there exists such a protocol. Let M_i^t be a message sent by player P_i in round t and $T(\mathcal{A})$ be the maximum number of rounds of $\mathcal{A}$ for all inputs of length n and all random tapes. Obviously M_i^t is a function of the input string z and the random tapes R. Player P_i receives in round $t \leq T(\mathcal{A})$ the messages

$$C_i^t(z, R) := M_{i_1}^t(z, R), \ldots, M_{i_s}^t(z, R),$$

where $P_{i_1}, \ldots, P_{i_s}$ are all the players incident to player P_i. We denote the sequence $C_i^1(z, R), C_i^2(z, R), \ldots, C_i^{T(\mathcal{A})}(z, R)$ by $C_i(z, R)$.

Now let k, I_1, I_2 fulfil conditions 1, 2, and 3 of the lemma and choose x, α, and β such that

$$f(x\lceil_{I_1, I_2 \leftarrow \alpha, \beta}) \neq f(x\lceil_{I_1, I_2 \leftarrow \alpha, \overline{\beta}}) = f(x\lceil_{I_1, I_2 \leftarrow \overline{\alpha}, \beta}).$$

Keep R fixed. Then consider $C_k(x\lceil_{I_1, I_2 \leftarrow \alpha, \overline{\beta}}, R)$, which is the sequence of messages received by the player k during the computation on $x\lceil_{I_1, I_2 \leftarrow \alpha, \overline{\beta}}$ with random bits R. Since the protocol is private and $k \notin I_1 \cup I_2$, there exists $R' = (R'_1, \ldots, R'_n)$, with $R_k = R'_k$, such that

$$C_k(x\lceil_{I_1, I_2 \leftarrow \overline{\alpha}, \beta}, R') = C_k(x\lceil_{I_1, I_2 \leftarrow \alpha, \overline{\beta}}, R). \tag{1}$$

Let $Y := \{\ell \mid$ there is a path $W_{\ell,i}$ from ℓ to a node $i \in I_1$ with $k \notin W_{\ell,i}\}$.

Obviously we have $I_1 \subseteq Y$ and $I_2 \cap Y = \emptyset$. Now let $R'' = (R''_1, \ldots, R''_n)$ be a content of random tapes defined as follows: for every $\ell \in Y$ let $R''_\ell := R_\ell$ and for every $j \in [n] \setminus Y$ let $R''_j := R'_j$. Note that $R''_k = R'_k = R_k$. From Equation (1) it follows that on input $x\lceil_{I_1, I_2 \leftarrow \alpha, \beta}$ and with random tapes R'' the protocol generates the following messages for every player $i \in [n]$ and every $t \geq 1$

$$M_i^t(x\lceil_{I_1, I_2 \leftarrow \alpha, \beta}, R'') = \begin{cases} M_i^t(x\lceil_{I_1, I_2 \leftarrow \alpha, \overline{\beta}}, R) & \text{if } i \in Y, \\ M_i^t(x\lceil_{I_1, I_2 \leftarrow \overline{\alpha}, \beta}, R') & \text{if } i \in [n] \setminus Y. \end{cases}$$

Hence, given the input string $x\lceil_{I_1, I_2 \leftarrow \alpha, \beta}$ the protocol computes the same value as on the input string $x\lceil_{I_1, I_2 \leftarrow \overline{\alpha}, \beta}$ and $x\lceil_{I_1, I_2 \leftarrow \alpha, \overline{\beta}}$ – a contradiction. $\square$

Corollary 2. *A non-dominated non-degenerate function cannot be computed by a private protocol on a network that is not 2-connected.*

Examples of non-dominated non-degenerate functions are the parity function, the or function, and the majority function. Hence, these functions cannot be computed by private protocols on networks that are not 2-connected.

In the remainder of this section, we show that for every dominated function f there is a non-2-connected network on which f can be computed by a private protocol.

The following three lemmas can be proved similar to Lemma 3.

Lemma 5. *Assume that a set I_1 with $|I_1| \geq 2$ is dominated by an input position $k \in I_1$. Then every pair $(a, J_1) \in f\text{-set}(I_1)$ assigns the same value to $x[k]$.*

For $c \in \{0, 1\}$, we call a set I_1 **(k, c)-dominated** if I_1 is dominated by k and for each pair $(\alpha, J_1) \in \text{f-set}(I_1)$, α assigns c to $x[k]$.

Lemma 6. *Assume that a set I_1 with $|I_1| \geq 2$ is (k, c)-dominated with $k \in I_1$ for some $c \in \{0, 1\}$. Then for every $\alpha \in \{0, 1\}^{|I_1|}$ with $\alpha[k] \neq c$, for every $w \in \{0, 1\}^n$, $J_2 \subseteq [n] \setminus I_1$, and $\beta \in \{0, 1\}^{|J_2|}$ we have*

$$f(w\lceil_{I_1, J_2 \leftarrow \alpha, \beta}) \; = \; f(w\lceil_{I_1, J_2 \leftarrow \alpha, \overline{\beta}}) \, .$$

By the previous lemma, we can conclude that for each set I_1 with $|I_1| \geq 2$ that is (k, c)-dominated with $k \in I_1$ there exists a function $f_1 : \{0, 1\}^{|I_1|} \to \{0, 1\}$ such that

$$f(x) \; = \; ((x[k] = c) \wedge f(x)) \vee ((x[k] \neq c) \wedge f_1(x[I_1])) \, .$$

This reduces the set of interesting variables to I_1 if $x[k] \neq c$. Let us now focus on input strings with $x[k] = c$.

Lemma 7. *Assume that a set I_1 with $|I_1| \geq 2$ is (k, c)-dominated with $k \in I_1$ for some $c \in \{0, 1\}$. Then for every pair $w_1, w_2 \in \{0, 1\}^n$ with $w_1[k] = w_2[k] = c$ and $w_1[i] = w_2[i]$ for all $i \in [n] \setminus I_1$ we have $f(w_1) = f(w_2)$.*

Thus, we can conclude that for each set I_1 with $|I_1| \geq 2$ that is (k, c)-dominated with $k \in I_1$, there exists a function $f_2 : \{0, 1\}^{|I_2|} \to \{0, 1\}$ such that $f(x) = ((x[k] \neq c) \wedge f_2(x[I_2])) \vee ((x[k] = c) \wedge f_1(x[I_1]))$. Summarising the above three lemmas we get the following result.

Theorem 4. *Assume that a set I_1 with $|I_1| \geq 2$ is (k, c)-dominated with $k \in I_1$ for some $c \in \{0, 1\}$. Let $I_2 = [n] \setminus I_1$. Then there are two functions $f_1 : \{0, 1\}^{|I_1|} \to \{0, 1\}$ and $f_2 : \{0, 1\}^{|I_2|} \to \{0, 1\}$ such that*

$$f(x) \; = \; ((x[k] = c) \wedge f_1(x[I_1])) \; \vee \; ((x[k] \neq c) \wedge f_2(x[I_2])) \, .$$

Note, that k, I_1, and I_2 are uniquely determined by the function f. Hence, every dominated function can be described by an if-then-else construction, i.e. it is of the form if $x[k] = c$ then $f_1(x[I_1])$ else $f_2(x[I_2])$.

Theorem 4 immediately implies that dominated functions can be computed on networks that are not 2-connected.

Theorem 5. *If f is ℓ-dominated with $\ell > 1$, then f can be computed by a private protocol on a network that consists of two 2-connected components with one node in common. One of the components has size ℓ and the other one size $n - \ell + 1$.*

Corollary 3. *Assume that f is a dominated function. Then there are non-2-connected networks on which f can be computed by a private protocol.*

Theorem 5 can be generalised to the case where we allow teams of players to work together. Assume that all members of a team belong to the component that computes, say, f_1. Then f is t-private if f_1 is t-private. If the members are distributed among both components, then this virtually decreases the team sizes for both components. f is t-private if both f_1 and f_2 are t-private.

6   Conclusions and Open Problems

We have investigated the relation between the connectivity of networks and the possibility of computing functions by private protocols on these networks. Special emphasis has been put on the amount of randomness needed.

We have presented a general simulation technique which allows us to transfer every oblivious private protocol on an arbitrary network into an oblivious private protocol on a given k-connected network of the same size, where $k \geq 2$. The new protocol needs $\left(1 - \frac{k}{n-1}\right) \cdot \min\{L, n^2 + \frac{L}{k-1}\}$ random bits more than the original protocol, where L is the total amount of bits sent in the original protocol. The obvious open question here is either to further reduce the number of extra random bits or to prove general lower bounds.

The parity function can be computed on a cycle using only one random bit and only one message per link. Thus, $1 + n - \frac{kn}{n-1}$ random bits are sufficient to compute the parity function on an arbitrary k-connected graph by a private protocol using our simulation. We have strengthened this bound by showing that on every k-connected graph, parity can be computed by an oblivious private protocol using at most $\lceil \frac{n-2}{k-1} \rceil - 1$ random bits. Furthermore, there exist k-connected networks for which this bound is sharp. The latter bound even holds for non-oblivious protocols.

While every Boolean function can be computed on a 2-connected network by a private protocol, this is no longer true for 1-connected networks. Starting from this observation, we have completely characterized the functions that can be computed by a private protocol on 1-connected networks.

Our simulation results focus on the extra amount of randomness needed. It would also be interesting to bound the number of rounds of the simulation in terms of the number of rounds of the original protocol and, say, the diameter of the new network.

Acknowledgement

We thank Adi Rosén for valuable discussions and hints to literature.

References

1. N. Alon, J. H. Spencer, and P. Erdös. *The Probabilistic Method*. John Wiley and Sons, 1992.
2. J. Bar-Ilan and D. Beaver. Non-cryptographic fault-tolerant computing in a constant number of rounds of interaction. In *Proc. 8th Ann. Symp. on Principles of Distributed Comput. (PODC)*, pages 201–209. ACM, 1989.
3. M. Ben-Or, S. Goldwasser, and A. Wigderson. Completeness theorems for non-cryptographic fault-tolerant distributed computation. In *Proc. 20th Ann. Symp. on Theory of Comput. (STOC)*, pages 1–10. ACM, 1988.
4. C. Blundo, A. de Santis, G. Persiano, and U. Vaccaro. Randomness complexity of private computation. *Comput. Complexity*, 8(2):145–168, 1999.
5. R. Canetti and R. Ostrovsky. Secure computation with honest-looking parties: What if nobody is truly honest? In *Proc. 31st Ann. Symp. on Theory of Comput. (STOC)*, pages 255–264. ACM, 1999.
6. D. Chaum, C. Crépeau, and I. Damgård. Multiparty unconditionally secure protocols. In *Proc. 20th Ann. Symp. on Theory of Comput. (STOC)*, pages 11–19. ACM, 1988.
7. B. Chor, M. Geréb-Graus, and E. Kushilevitz. On the structure of the privacy hierarchy. *J. Cryptology*, 7(1):53–60, 1994.
8. B. Chor, M. Geréb-Graus, and E. Kushilevitz. Private computations over the integers. *SIAM J. Comput.*, 24(2):376–386, 1995.
9. B. Chor and E. Kushilevitz. A zero-one law for boolean privacy. *SIAM J. Discrete Math.*, 4(1):36–47, 1991.
10. B. Chor and E. Kushilevitz. A communication-privacy tradeoff for modular addition. *Inform. Process. Lett.*, 45(4):205–210, 1993.
11. Y. Egawa, R. Glas, and S. C. Locke. Cycles and paths through specified vertices in k-connected graphs. *J. Combin. Theory Ser. B*, 52:20–29, 1991.
12. M. Franklin and R. N. Wright. Secure communication in minimal connectivity models. *J. Cryptology*, 13(1):9–30, 2000.
13. M. Franklin and M. Yung. Secure hypergraphs: Privacy from partial broadcast. In *Proc. 27th Ann. Symp. on Theory of Comput. (STOC)*, pages 36–44. ACM, 1995.
14. A. Gál and A. Rosén. A theorem on sensitivity and applications in private computation. In *Proc. 31st Ann. Symp. on Theory of Comput. (STOC)*, pages 348–357. ACM, 1999.
15. O. Goldreich, S. Micali, and A. Wigderson. How to play any mental game or a completeness theorem for protocols with honest majority. In *Proc. 19th Ann. Symp. on Theory of Comput. (STOC)*, pages 218–229. ACM, 1987.
16. E. Kushilevitz. Privacy and communication complexity. *SIAM J. Discrete Math.*, 5(2):273–284, 1992.
17. E. Kushilevitz and Y. Mansour. Randomness in private computations. *SIAM J. Discrete Math.*, 10(4):647–661, 1997.
18. E. Kushilevitz, R. Ostrovsky, and A. Rosén. Characterizing linear size circuits in terms of privacy. *J. Comput. System Sci.*, 58(1):129–136, 1999.
19. E. Kushilevitz and A. Rosén. A randomness-rounds tradeoff in private computation. *SIAM J. Discrete Math.*, 11(1):61–80, 1998.
20. I. Wegener. *The Complexity of Boolean Functions*. Wiley-Teubner, 1987.
21. A. C.-C. Yao. Protocols for secure computations. In *Proc. 23rd Ann. Symp. on Foundations of Comput. Sci. (FOCS)*, pages 160–164. IEEE, 1982.
22. A. C.-C. Yao. How to generate and exchange secrets. In *Proc. 27th Ann. Symp. on Foundations of Comput. Sci. (FOCS)*, pages 162–167. IEEE, 1986.

Analysis and Improvements
of NTRU Encryption Paddings

Phong Q. Nguyen and David Pointcheval

CNRS/Département d'informatique, École normale supérieure
45 rue d'Ulm, 75005 Paris, France
{phong.nguyen,david.pointcheval}@ens.fr
http://www.di.ens.fr/~{pnguyen,pointche}

Abstract. NTRU is an efficient patented public-key cryptosystem proposed in 1996 by Hoffstein, Pipher and Silverman. Although no devastating weakness of NTRU has been found, Jaulmes and Joux presented at Crypto '00 a simple chosen-ciphertext attack against NTRU as originally described. This led Hoffstein and Silverman to propose three encryption padding schemes more or less based on previous work by Fujisaki and Okamoto on strengthening encryption schemes. It was claimed that these three padding schemes made NTRU secure against adaptive chosen-ciphertext attacks (IND-CCA2) in the random oracle model. In this paper, we analyze and compare the three NTRU schemes obtained. It turns out that the first one is not even semantically secure (IND-CPA). The second and third ones can be proven IND-CCA2–secure in the random oracle model, under however rather unusual assumptions. They indeed require a partial-domain one-wayness of the NTRU one-way function which is likely to be a stronger assumption than the one-wayness of the NTRU one-way function. We propose several modifications to achieve IND-CCA2–security in the random oracle model under the original NTRU inversion assumption.

1 Introduction

The NTRU cryptosystem [13], patented by the company NTRU CRYPTOSYSTEMS (see http://www.ntru.com), is one of the fastest public-key encryption schemes known. Although this may not be a decisive advantage compared to hybrid encryption with say RSA, NTRU has attracted considerable interest and is being considered by the *Efficient embedded security standards* [6] and the *IEEE P1363 study group for future public-key cryptography standards* [18]. It is therefore important to know how to use the NTRU cryptosystem properly.

The security of NTRU is based on the hardness of some lattice problems, namely the shortest and closest vector problems (see for instance the survey [24]). More precisely, it was first noticed by Coppersmith and Shamir [3] that ideal lattice basis reduction algorithms could heuristically recover NTRU's private key from the public key. This does not necessarily imply that NTRU is insecure,

M. Yung (Ed.): CRYPTO 2002, LNCS 2442, pp. 210–225, 2002.

as currently known lattice basis reduction algorithms (such as LLL [20] or its improvements [28]) do not seem to perform sufficiently well in practice in very high dimension, while NTRU is so far the only lattice-based cryptosystem that can cope with high dimensions without sacrificing performances. Nor does it mean that the security of NTRU is strictly equivalent to the hardness of lattice problems, although the basic NTRU problem is equivalent to the lattice shortest vector problem in a very particular class of lattices called modular convolution lattices in [21].

The NTRU cryptosystem as originally described is easily seen to be semantically insecure. At Crypto '00 [19], Jaulmes and Joux further presented simple chosen-ciphertext attacks that can recover the private key. This shows that the NTRU cryptosystem as originally described should be viewed as a probabilistic trapdoor one-way function rather than a probabilistic cryptosystem (see also recent work by Micciancio on lattice-based cryptosystems [22]). NTRU CRYPTOSYSTEMS therefore proposed three padding schemes (two in [16] and a third one in [15]) to make NTRU secure against adaptive chosen-ciphertext attacks in the random oracle model. No security proof was provided by NTRU CRYPTOSYSTEMS (see [16, 15]). In this paper, we analyze the three NTRU schemes obtained. It turns out that the first scheme is not even semantically secure (IND-CPA). The second and third ones can be proven IND-CCA2–secure, in the random oracle model, but under rather unusual assumptions. Indeed, a partial-domain one-wayness of the NTRU one-way function is required, and that assumption is likely to be stronger than the one-wayness of the NTRU one-way function, as opposed to the situation of RSA (see [9]). Besides, the security proofs we obtain for such paddings are not efficient enough to be meaningful for the parameters recommended by NTRU CRYPTOSYSTEMS.

We therefore propose and compare new paddings to make NTRU IND-CCA2– secure in the random oracle model under the basic NTRU assumption, and not a stronger assumption: The new paddings give rise to better bounds for the security proof, and their computational overhead appears to be negligible. It should be stressed that no security proof in the standard model is known for NTRU, and that the search for an efficient and secure NTRU padding scheme is not a trivial matter. Although there now exist generic padding schemes (such as REACT [25]) that can enhance the security (in the random oracle model) of any cryptosystem, the case of NTRU differs from more usual cryptosystems such as RSA or El Gamal because the cost of hashing is no longer negligible compared to the cost of encryption and decryption, and because of special properties of the NTRU trapdoor function.

The rest of the paper is organized as follows. In Section 2, we recall security notions for public-key encryption schemes. In Section 3, we review the NTRU primitive and related computational assumptions. In Section 4, we present and analyze the various paddings proposed by NTRU. In Section 5, we consider new paddings and compare several constructions which make NTRU IND-CCA2– secure in the random oracle model.

2   Public-Key Encryption

The goal of encryption schemes is to achieve confidentiality of communications. In the public-key scenario, anyone knowing Alice's public key pk can send Alice a message that only she will be able to recover, thanks to her private key sk.

2.1   Definitions

A public-key encryption scheme Π over a message space S_M is formally defined by three algorithms:

- a *key generation algorithm* $\mathcal{K}(1^k)$ (k being the security parameter), which produces a pair (pk, sk) of public and private keys.
- an *encryption algorithm* $\mathcal{E}_{pk}(m; r)$ which outputs a ciphertext c corresponding to the plaintext $m \in S_M$, using random coins $r \in S_R$, according to the public key pk.
- a *decryption algorithm* $\mathcal{D}_{sk}(c)$ which outputs the plaintext m associated to the ciphertext c (or $\perp$, if c is an invalid ciphertext), given the private key sk.

2.2   Security Notions

The simplest security notion is *one-wayness*: with public data only, an attacker cannot recover the whole plaintext m of a given ciphertext c. More formally, the success of any adversary $\mathcal{A}$ in inverting $\mathcal{E}_{pk}$ without knowledge of the private key should be negligible over the probability space $S_M \times S_R$, and the internal random coins of the adversary and the algorithms $\mathcal{K}$ and $\mathcal{E}$:

$$\mathsf{Succ}_\Pi^{ow}(\mathcal{A}) = \Pr[(pk, sk) \leftarrow \mathcal{K}(1^k), c = \mathcal{E}_{pk}(m; r) : \mathcal{A}(pk, c) = m].$$

However, many applications require a higher security level, such as *semantic security* (*a.k.a. indistinguishability of encryptions* [11], denoted IND): if an attacker has some information about the plaintext, the view of the ciphertext should not leak any additional information. This security notion requires the computational intractability of winning with probability significantly better than $1/2$ the following game: the adversary chooses two messages; the challenger selects at random one of these two messages, encrypts it, and sends the ciphertext to the adversary; the adversary guesses which one of the two messages has been encrypted. In other words, an adversary is seen as a 2-stage Turing machine (A_1, A_2), and the advantage $\mathsf{Adv}_\Pi^{ind}(\mathcal{A})$ should be negligible for any adversary, where the advantage is formally defined as:

$$2 \times \Pr_{r \xleftarrow{R} S_R} \left[\begin{array}{l} (pk, sk) \leftarrow \mathcal{K}(1^k), (m_0, m_1, s) \leftarrow A_1(pk), \\ b \xleftarrow{R} \{0, 1\}, c = \mathcal{E}_{pk}(m_b; r) : A_2(m_0, m_1, s, c) = b \end{array} \right] - 1.$$

Another important security notion is *non-malleability* [5]. Here, we ask that an adversary, given a ciphertext, should not be able to create a new ciphertext such that the two plaintexts are meaningfully related. This notion is, in general, stronger than semantic security, but it was shown [1] to be equivalent in the strongest scenario (see below).

On the other hand, an attacker can use many kinds of attacks, depending on the information available to him. First, in the public-key setting, the adversary can encrypt any plaintext of its choice with the public key: this basic scenario is called *chosen-plaintext attack*, and denoted by CPA. Extended scenarios allow the adversary restricted or unrestricted access to various oracles:

- a *validity-checking oracle*, which answers whether or not its input c is a valid ciphertext. This leads to so-called *reaction attacks* [12].
- a *plaintext-checking oracle*, which given as input a pair (m, c) answers whether or not the ciphertext c is a ciphertext of the message m. This gives rise to *plaintext-checking attacks* [25], which we denote by PCA.
- a *decryption oracle*, which returns the decryption of any ciphertext, with the only restriction that it should be different from the challenge ciphertext. When the oracle is available only before knowledge of the challenge ciphertext, the attack is a *non-adaptive chosen-ciphertext attack* (*a.k.a. lunchtime attack* [23]), which we denote by CCA1. When the adversary still has access to the decryption oracle in the second stage, we talk about *adaptive chosen-ciphertext attacks* [27], denoted by CCA2.

The article [1] provides a general study of all these security notions and attacks. Its main result states that semantic security and non-malleability are equivalent in the CCA2–scenario. This security level is now widely accepted as the standard notion of security to be achieved by a public-key encryption scheme, and is sometimes called *chosen-ciphertext security*.

3 The NTRU Primitive

3.1 Description and Notation

In this section, we present the NTRU cryptosystem as originally described in [13] by Hoffstein, Pipher and Silverman. As mentioned in the introduction, this should be viewed as a primitive function rather than a cryptosystem. Several modifications of NTRU have recently been proposed in [15]: we will present those later.

Let k be the security parameter. The NTRU primitive works in the ring $\mathcal{P} = \mathbb{Z}[X]/(X^N - 1)$ where N is a safe prime typically around a few hundreds, whose value increases with k: NTRU CRYPTOSYSTEMS recommends specific values of N, such as $N = 251$ or $N = 503$. The ring $\mathcal{P}$ is identified with the set of integer polynomials of degree $< N$, and its multiplication is denoted by $*$. Polynomials will be denoted by letters in the Sans Serif font, such as f.

Note that the function that maps any polynomial $\mathsf{f} \in \mathcal{P}$ to the sum $\mathsf{f}(1) \in \mathbb{Z}$ of its coefficients is a ring homomorphism. NTRU uses two integer parameters: a small power of 2, denoted by q, such as $q = 128$ or $q = 256$, and a small integer

$p < q$ co-prime with q, such as $p = 3$. The restriction of N to safe primes was apparently made to guarantee that the multiplicative order of p modulo N is sufficiently large. NTRU performs operations in $\mathcal{P}$ modulo p or q.

The private key sk consists of two polynomials f, g $\in \mathcal{P}$ randomly chosen with very small coefficients such that f is invertible modulo both p and q: there exist f_p and f_q in $\mathcal{P}$ such that: $f * f_p \equiv 1 \pmod{p}$ and $f * f_q \equiv 1 \pmod{q}$. In [13], f and g only have coefficients in $\{0, \pm 1\}$ with a prescribed (publicly known) number of 1, -1 and 0 such that $f(1) = 1$ and $g(1) = 0$. All the integer constraints (N, p, q, the number of 1, -1, 0 in polynomials, *etc.*) are deduced from a security parameter k, and this process is described in [13, 15]. Then, f and g are uniformly distributed polynomials among the polynomials that satisfy the required constraints. The public key pk is $h = g * f_q \pmod{q}$. Therefore, $(h, (f, g)) \leftarrow \mathcal{K}(1^k)$.

The message space is $\mathcal{S}_M = \{-(p-1)/2 \ldots + (p-1)/2\}^N$, whose elements are viewed as elements of $\mathcal{P}$. To encrypt a message m, one selects at random a sparse polynomial r $\in \mathcal{P}$ with very small coefficients: In [13], r only has coefficients in $\{0, \pm 1\}$ with a prescribed (publicly known) number of 1, -1 and 0 such that $r(1) = 0$ (we denote the set of these specific polynomials by $\mathcal{S}_R$). The ciphertext is:

$$\mathcal{E}_{pk}(m; r) = e = m + pr * h \pmod{q}.$$

To decrypt, the following congruence is used:

$$e * f \equiv m * f + pr * g \pmod{q}.$$

In the right-hand part of this congruence, we have two convolution products of polynomials with very small coefficients and quite a few zeroes (except possibly m). Therefore, if the above reduction is centered (one takes the smallest residue in absolute value), the above congruence is likely[1] to be an equality over $\mathbb{Z}[X]$. By further reducing $e * f$ modulo p, one thus obtains $m * f \pmod{p}$, hence m thanks to f_p. Note that there is a potential probability of decryption failure, if the above equality (mod q) does not hold in $\mathbb{Z}$.

3.2 Efficiency

A multiplication in $\mathcal{P}$ requires $\mathcal{O}(N^2 \log q)$ elementary operations. It follows that the cost of encryption and decryption is $\mathcal{O}(N^2 \log q)$. Since the key generation process is such that $q = \mathcal{O}(N)$, the cost of both encryption and decryption is almost quadratic in the security parameter. Note that in most of the required convolution products, at least one of the polynomial is relatively sparse. And since q is a small power of two, the above complexity is rather pessimistic in practice.

---

[1] We stress that no provable precise estimate on the probability of such an event is known, and [13] only uses a heuristic estimate which seems to be validated by practice.

3.3 Optimizations

The authors of NTRU recently proposed several modifications in [15] to improve the efficiency of the scheme:

- Choosing p as an appropriate polynomial p instead of a small number co-prime with q (e.g. $p = X + 2$). The polynomial must be such that the ideal spanned by the polynomial must be co-prime with the ideal $\langle q \rangle$ spanned by q in $\mathcal{P}$. The aim of this modification is to reduce the probability of decryption failure. It also enables a simpler encoding of messages.
- Selecting f, g, and r with a special form instead of just a prescribed number of 0, -1 and 1. For instance, $r = r_1 r_2$ where r_1 and r_2 are sparse polynomials with a prescribed number of 0, -1 and 1. In all the proposals of [15], the values of f(1), g(1) and r(1) are always publicly known.

It is worth noting that such modifications may have an impact on the security of NTRU, but at the moment, no specific attack is known.

Our results apply to the original NTRU scheme, as well as to most of these optimizations of NTRU. However, to simplify the presentation, we will restrict to the case of the original NTRU primitive $\Pi = (\mathcal{K}, \mathcal{E}, \mathcal{D})$, where $p = 3$, f(1) = 1 and g(1) = r(1) = 0.

3.4 Computational Assumptions

To formally analyze the security of the NTRU cryptosystem, one needs clear and well-defined computational assumptions. First, we consider the one-wayness of the NTRU primitive:

Definition 1 (The NTRU Inversion Problem). *For a given security parameter k, which specifies N, p, q and several other constraints, as well as $(h, (f, g)) \leftarrow \mathcal{K}(1^k)$ and $e = m + pr * h$, where $m \in \mathcal{S}_M$ and $r \in \mathcal{S}_R$, find m.*

For any adversary $\mathcal{A}$, we denote by $\mathsf{Succ}^{\mathsf{ow}}_{\mathsf{ntru}}(\mathcal{A})$ its success for breaking the *NTRU Inversion Problem*, where

$$\mathsf{Succ}^{\mathsf{ow}}_{\mathsf{ntru}}(\mathcal{A}) = \Pr \left[\begin{array}{l} (h, (f, g)) \leftarrow \mathcal{K}(1^k), m \in \mathcal{S}_M, r \in \mathcal{S}_R, \\ e = m + pr * h : \mathcal{A}(e, h) = m \end{array} \right].$$

The *NTRU assumption* says that the NTRU inversion problem is hard to solve for any sufficiently large parameter.

Next, we consider the difficulty of only partially inverting this function, which will be useful to study NTRU paddings:

Definition 2 (The NTRU λ-Partial-Domain Inversion Problem). *For a given security parameter k, which specifies N, p, q and several other constraints, as well as $(h, (f, g)) \leftarrow \mathcal{K}(1^k)$ and $e = m + pr * h$, where $m \in \mathcal{S}_M$ and $r \in \mathcal{S}_R$, find $[m]_\lambda$, where $[m]_\lambda$ denotes the λ least significant coefficients of m.*

As above, for any adversary $\mathcal{A}$, we denote by $\mathsf{Succ}_{ntru}^{pd-ow_\lambda}(\mathcal{A})$ its success for breaking the *NTRU λ-Partial-Domain Inversion Problem*, where

$$\mathsf{Succ}_{ntru}^{pd-ow_\lambda}(\mathcal{A}) = \Pr \begin{bmatrix} (h,(f,g)) \leftarrow \mathcal{K}(1^k), m \in \mathcal{S}_M, r \in \mathcal{S}_R, \\ e = m + pr * h : \mathcal{A}(e,h) = [m]_\lambda \end{bmatrix}.$$

Note that the NTRU encryption primitive is malleable with respect to circular shifts: $\mathcal{E}_{pk}(X * m; X * r) = X * \mathcal{E}_{pk}(m; r)$. This implies that any λ-consecutive-coefficient search problem is equivalent to the above *NTRU λ-Partial-Domain Inversion Problem*. Because of the specific encodings used by NTRU, it will also be useful to consider the difficulty of obtaining some information bits of the pre-image only:

Definition 3 (The NTRU ℓ-Partial-Information Inversion Problem).
For a given security parameter k, which specifies N, p, q and several other constraints, as well as $(h,(f,g)) \leftarrow \mathcal{K}(1^k)$ and $e = m + pr * h$, where $m \in \mathcal{S}_M$ and $r \in \mathcal{S}_R$, find ℓ bits of information about m.

As above, we denote by $\mathsf{Succ}_{ntru}^{pi-ow_\ell}(\mathcal{A})$ its success for breaking the *NTRU ℓ-Partial-Information Inversion Problem*, where

$$\mathsf{Succ}_{ntru}^{pi-ow_\ell}(\mathcal{A}) = \Pr \begin{bmatrix} (h,(f,g)) \leftarrow \mathcal{K}(1^k), m \in \mathcal{S}_M, r \in \mathcal{S}_R, \\ e = m + pr * h, (f,y) \leftarrow \mathcal{A}(e,h) : y = f_\ell(m) \end{bmatrix}.$$

In the above definition, the adversary $\mathcal{A}$ outputs a (computable) bijective function $f : \mathcal{S}_M \rightarrow \{0,1\}^{mLen}$ (where mLen denotes the bit-length of an optimal encoding for polynomials in $\mathcal{S}_M$) and $y \in \{0,1\}^\ell$. Furthermore, f_ℓ denotes the truncation of f to its ℓ least significant bits.

More generally, we denote by $\mathsf{Succ}_{ntru}^{ow}(t)$, $\mathsf{Succ}_{ntru}^{pd-ow_\lambda}(t)$ and $\mathsf{Succ}_{ntru}^{pi-ow_\ell}(t)$, the maximal success among all the adversaries with a running time bounded by t.

Relations among Computational Assumptions. The definition of our assumptions implies that for all t and λ

$$\mathsf{Succ}_{ntru}^{ow}(t) \leq \mathsf{Succ}_{ntru}^{pd-ow_\lambda}(t) \leq \mathsf{Succ}_{ntru}^{pi-ow_\ell}(t),$$

where ℓ is the bit-length of an optimal encoding for λ coefficients. And in the specific case $\lambda = N$, all the inequalities become equalities:

$$\mathsf{Succ}_{ntru}^{ow}(t) = \mathsf{Succ}_{ntru}^{pd-ow_N}(t) = \mathsf{Adv}_{ntru}^{pi-ow_{mLen}}(t).$$

Remark that if $\mathsf{Succ}_{ntru}^{pd-ow_\lambda}(t) = 1$ for some (t, λ), then $\mathsf{Succ}_{ntru}^{ow}(\lceil N/\lambda \rceil t) = 1$, where $\lceil x \rceil$ denotes the smallest integer larger than x. This is because the malleability of the NTRU encryption primitive allows to reduce any instance of the NTRU inversion problem to $\lceil N/\lambda \rceil$ instances of the NTRU λ-partial-domain inversion problem, using appropriate shifts of e. Note however that the multiplication by X does not provide a random self-reduction: the previous reduction implies nothing on $\mathsf{Succ}_{ntru}^{ow}(\lceil N/\lambda \rceil t)$ if $\mathsf{Succ}_{ntru}^{pd-ow_\lambda}(t) < 1$. In fact, since no random self-reducibility property is known for NTRU, it is likely that if $\mathsf{Succ}_{\mathsf{ntru}}^{\mathsf{pd-ow}_\lambda}(t) < 1$, then the NTRU λ-partial-domain inversion problem is strictly harder than the NTRU inversion problem. Besides, the lack of random self-reducibility also suggests that the NTRU ℓ-partial-information inversion problem is strictly harder than the NTRU λ-partial-domain inversion problem.

In the following, we assume that all the above problems become intractable for a sufficiently large security parameter k, and for sufficiently large enough ℓ and λ (since random guesses lead to $\mathsf{Succ}_{\mathsf{ntru}}^{\mathsf{pd-ow}_\lambda}(1) = 1/p^\lambda$ and $\mathsf{Succ}_{\mathsf{ntru}}^{\mathsf{pi-ow}_\ell}(1) = 1/2^\ell$).

3.5 The Security of the NTRU Primitive

The best attack known against NTRU is based on lattice reduction, but this does not imply that lattice reduction is necessary to break NTRU. See [3, 13, 24] for further information. Based on numerous experiments, the authors of NTRU claimed in [13] that all known lattice-based attacks are exponential in N, and therefore suggested relatively small values of N. The parameter N must be prime, otherwise the lattice attacks can be improved due to non-trivial factors of $X^N - 1$ (see [10]). Because the key-size of NTRU is only $\mathcal{O}(N \log q)$, one can allow reasonably high lattice dimensions, while all other known knapsack-based or lattice-based cryptosystems have a key-size which is at least quadratic in the security parameter.

NTRU, like most public-key cryptosystems, should not be directly used as originally described. For instance, NTRU is easily seen to be semantically insecure, as $\mathsf{e}(1) \equiv \mathsf{m}(1) \pmod{q}$ because $\mathsf{r}(1) = 0$. This yields a significant bias for any adversary to distinguish between two possible plaintexts which one has been encrypted. In fact, although the NTRU cryptosystem is probabilistic, there is a public plaintext-checking oracle: one can easily check whether a given message m corresponds to a ciphertext e, which implies that any security level in the CPA scenario holds in the PCA one as well. This is because the shape of r is publicly verifiable and the public key h is "pseudo-invertible" modulo q with overwhelming probability. More precisely, one can compute from h a polynomial $\mathsf{H} \in \mathcal{P}$ such that for any polynomial $\mathsf{s} \in \mathcal{P}$ such that $\mathsf{s}(1) \equiv 0 \pmod{q}$:

$$\mathsf{h} * \mathsf{H} * \mathsf{s} \equiv \mathsf{s} \pmod{q}.$$

Then, if $\mathsf{e} = \mathsf{m} + p\mathsf{r} * \mathsf{h} \pmod{q}$ is a ciphertext of m, we have since $\mathsf{r}(1) \equiv 0 \pmod{q}$:

$$p\mathsf{r} \equiv (\mathsf{e} - \mathsf{m}) * \mathsf{H} \pmod{q}.$$

This allows to retrieve the random polynomial r modulo q, whose shape is publicly verifiable. By injectivity of encryption, we thus obtain a public plaintext-checking oracle.

We briefly explain this "pseudo-inversion" since we have not found any reference. Because N is an odd prime and q is a power of two, the ring $\mathcal{P}_q = \mathbb{Z}_q[X]/(X^N - 1)$ is isomorphic to $\mathcal{P}_1 \times \mathcal{P}_2$ where $\mathcal{P}_1 = \mathbb{Z}_q[X]/(X - 1)$ and

$\mathcal{P}_2 = \mathbb{Z}_q[X]/(X^{N-1} + X^{N-2} + \cdots + 1)$. Denote by ϕ_1 and ϕ_2 respectively the reduction modulo $X - 1$ and $X^{N-1} + X^{N-2} + \cdots + 1$. Of course, ϕ_1 is simply the evaluation at 1. Since $h(1) \equiv 0 \pmod{q}$, $\phi_1(h)$ is not invertible in $\mathcal{P}_1$, and therefore h is not invertible in $\mathcal{P}_q$. However, $\phi_2(h)$ is very likely to be invertible in $\mathcal{P}_2$ (the proportion of invertible elements can be computed as in [14]). And its inverse can easily be computed: for instance, by computing the inverse in $\mathbb{F}_2[X]/(X^{N-1} + X^{N-2} + \cdots + 1)$, and then lifting it modulo q. Eventually, one can derive an appropriate polynomial $H \in \mathcal{P}$ from this inverse.

As previously mentioned, NTRU is also easily malleable using multiplications by X: $\mathcal{E}_{pk}(X * m; X * r) = X * \mathcal{E}_{pk}(m; r)$. Jaulmes and Joux [19] further presented simple chosen-ciphertext attacks that can recover the private key. Curiously, one of these attacks could be applied to a specific padding proposed by NTRU CRYPTOSYSTEMS to avoid reaction attacks [17]. This stressed the need of an appropriate padding scheme to obtain high levels of security against a vast class of attacks, assuming the NTRU one-way function $\mathcal{E}_{pk}(m; r)$ is hard to invert (which, by definition, is equivalent to asking that the NTRU inversion problem is hard).

4   The NTRU Cryptosystems

For clarity, in the following, we consider the encryption scheme $\Pi' = (\mathcal{K}', \mathcal{E}', \mathcal{D}')$, which is the same as $\Pi = (\mathcal{K}, \mathcal{E}, \mathcal{D})$, up to the two public encodings:

$$\mathcal{M} : \{0,1\}^{\mathsf{mLen}} \longrightarrow \mathcal{S}_\mathsf{M} \text{ and } \mathcal{R} : \{0,1\}^{\mathsf{rLen}} \longrightarrow \mathcal{S}_\mathsf{R}.$$

$$\Pi' \begin{cases} \mathcal{K}'(1^k) = \mathcal{K}(1^k) & = (\mathsf{pk} = \mathsf{h}, \mathsf{sk} = (\mathsf{f}, \mathsf{g})), \\ \mathcal{E}'_{pk}(m; r) = \mathcal{E}_{pk}(\mathcal{M}(m); \mathcal{R}(r)) = \mathcal{M}(m) + p\mathcal{R}(r) * \mathsf{h} \pmod{q}, \\ \mathcal{D}'_{sk}(e) = \mathcal{M}^{-1}(\mathcal{D}_{sk}(e)) & = \mathcal{M}^{-1}((e * \mathsf{f} \pmod{p}) * \mathsf{f}_p). \end{cases}$$

Because of the encodings, without any assumption, recovering the bit-string m is as hard as recovering the polynomial $m = \mathcal{M}(m)$. However, recovering ℓ bits of m only provides ℓ bits of information about the polynomial $m = \mathcal{M}(m)$, which is why we introduced the NTRU ℓ-partial-information inversion problem. From these remarks:

$$\mathsf{Succ}_{\Pi'}^{\mathsf{ow-cpa}}(t) = \mathsf{Succ}_{\Pi'}^{\mathsf{ow-pca}}(t) = \mathsf{Succ}_{\mathsf{ntru}}^{\mathsf{ow}}(t).$$

4.1   NTRU Paddings

Following the publication of [19], NTRU proposed several padding schemes in [16, 15] to protect NTRU against adaptive chosen-ciphertext attacks. We note that at the time of the writing of [16], several generic transformations were known to make NTRU IND-CCA2–secure in the random oracle model [2, 7, 8, 26]. However, a few complications arise as the NTRU one-way function cannot be assumed IND-CPA.

All NTRU paddings require a hash function $H : \{0,1\}^{\mathsf{mLen}} \rightarrow \{0,1\}^{\mathsf{rLen}}$, and possibly F and G, whose output size will be made explicit later.

Let M be the original plaintext represented by a k_1-bit string. For each encryption, one generates a random string R, whose bit-length k_2 is between 40 and 80 according to [16, page 2]. However, $k_1 + k_2 \leq$ mLen. Let $\|$ denote bit-string concatenation. The paddings proposed by NTRU are as follows.

Padding I. The first padding is proposed in [16, page 3]. The ciphertext of M with random R is:

$$\mathcal{E}^1_{\mathsf{pk}}(M; R) = \mathcal{E}'_{\mathsf{pk}}(M \| R; H(M \| R)).$$

We denote by Π^1 the corresponding encryption scheme.

Padding II. The second padding is proposed in [16, page 3]. The ciphertext is:

$$\mathcal{E}^2_{\mathsf{pk}}(M; R) = \mathcal{E}'_{\mathsf{pk}}((F(R) \oplus M) \| R; H(M \| R)),$$

where F is a hash function that maps $\{0,1\}^{k_2} \rightarrow \{0,1\}^{k_1}$, and $\oplus$ denotes bitwise exclusive or. We denote by Π^2 the corresponding encryption scheme.

Padding III. The third padding is proposed in [15, page 3], and not in [16]. This is rather curious, as [15] actually suggests the reading of [16] for further details. It is this padding and not the previous ones which is being considered in the CEES standards [6].

This padding first applies an all-or-nothing transformation (OAEP [2]) on the concatenation $M \| R$. More precisely, it splits each of M and R into equal size pieces $M = \overline{M} \| \underline{M}$ and $R = \overline{R} \| \underline{R}$. It then uses two hash functions F and G that map $\{0,1\}^{k_1/2+k_2/2}$ into itself, to compute:

$$m_1 = (\overline{M} \| \overline{R}) \oplus F(\underline{M} \| \underline{R}) \text{ and } m_2 = (\underline{M} \| \underline{R}) \oplus G(m_1).$$

The ciphertext is then:

$$\mathcal{E}^3_{\mathsf{pk}}(M; R) = \mathcal{E}'_{\mathsf{pk}}(m_1 \| m_2; H(M \| R)).$$

We denote by Π^3 the corresponding encryption scheme.

4.2   Security Analyses

First of all, one may note that because of the random polynomial r that is generated from $H(M \| R)$, nobody can generate a valid ciphertext without knowing both the plaintext M and the random R, except with negligible probability. Indeed, for a given ciphertext e, at most one r is acceptable. Without having asked $H(M \| R)$, the probability for $\mathcal{R}(H(M \| R))$ to be equal to r is less than $1/2^N$. As a consequence, any security notion satisfied in the CPA scenario is satisfied in the CCA2-scenario. The latter scenario may increase the success probability of an adversary by at most $q_D/2^N$, where q_D is the number of queries asked to the decryption oracle. With a proper bookkeeping, the cost of the simulation increases by $q_H T_\mathcal{E}$, at most, where q_H is the number of queries asked to the random oracle H, and $T_\mathcal{E}$ the time required for one encryption: for $i = 1, 2, 3$,

$$\text{Adv}_{\Pi^i}^{\text{ind-cca2}}(t) \leq \text{Adv}_{\Pi^i}^{\text{ind-cpa}}(t + q_H T_\mathcal{E}) + \frac{2q_D}{2^N}.$$

Therefore, in the following, we focus on the *chosen-plaintext* attacks only.

Analysis of the First NTRU Padding. The first padding is exactly based on the first Fujisaki-Okamoto conversion [8], which requires the primitive to be IND-CPA, which is not the case of Π'! However, the one-wayness OW-CPA of Π^1 already relies on a stronger assumption than the hardness of the NTRU inversion problem: the hardness of the NTRU k_1-partial-information inversion problem.

More worryingly, contrarily to the claims of [16], the scheme Π^1 is not *semantically secure* (IND-CPA): let us consider the following adversary which chooses $M_0 = 0^{k_1}$ and $M_1 = 1^{k_1}$. R is unknown, but whatever it is, if the encoding $\mathcal{M}$ is such that R has an impact on at most k_2 coefficients: $\mathcal{M}(M_0 \| R)(1) \leq k_2$, $\mathcal{M}(M_1 \| R)(1) \geq k_1$. As already remarked, this value mod q is given by $e(1)$, which helps to distinguish which message has been encrypted, with advantage 1. Optimizations [15] (such as $r(1) \neq 0$ but still a public constant) may slightly worsen the advantage, but the advantage is still significant (more than $1/2$).

Analysis of the Second NTRU Padding. The second padding is more surprising. Even if it provides *one-wayness* under the sole NTRU assumption, *semantic security* still requires a stronger assumption: the hardness of the NTRU k_2-partial-information inversion problem.

First, to get any information about the bit b such that the message M_b is encrypted, any adversary has to ask either $F(R)$ or $H(M_i \| R)$. If one denotes by Ask such an event, one obtains: $\text{Adv}_{\Pi^2}^{\text{ind-cpa}}(t) \leq 2 \Pr[\text{Ask}]$. In the worst case, by randomly picking one candidate, one can extract R, and thus k_2 bits of information about the polynomial m:

$$\text{Adv}_{\Pi^2}^{\text{ind-cpa}}(t) \leq 2(q_F + q_H) \times \text{Succ}_{\text{ntru}}^{\text{pi-ow}_{k_2}}(t).$$

From a OW-adversary $\mathcal{A}$, which runs within a time bound t, one can get more, whereas the simulations of F and H may be inconsistent. Indeed, the challenge ciphertext e defines R uniquely, but M is a random variable later defined by $F(R)$, $M = F(R) \oplus \mathcal{M}^{-1}(\text{m})$, and then $H(M \| R) = \mathcal{R}^{-1}(r)$. The latter may not be correctly answered. If $H(M \| R)$ is not asked, the view of the adversary $\mathcal{A}$ is perfect, and the output M gives $\text{m} = \mathcal{M}(\rho \oplus M)$, where $F(R) = \rho$. But if $F(R)$ has not been asked, the success of the adversary is upper-bounded by $1/2^{k_1}$. If $H(M \| R)$ has been asked, one guesses one pair $(M \| R)$ among the queries asked to H, and performs the same as above. There are $q_F + q_H$ possibilities for R, in the first case, or q_H possibilities for $M \| R$ in the second one. The good candidate can easily be checked. As a result, one gets

$$\mathsf{Succ}_{\Pi^2}^{\mathsf{ow-cpa}}(t) \le \frac{1}{2^{k_1}} + \mathsf{Succ}_{\mathsf{ntru}}^{\mathsf{ow}}(t + (q_F + 2q_H)T_\varepsilon).$$

Analysis of the Third NTRU Padding. The third padding makes an incorrect use of the All-Or Nothing Transform, and therefore, the achieved security level is not as high as one might have expected.

Let us first consider an adversary $\mathcal{A}$ against semantic security, which tries to guess between M_0 and M_1 which one has been encrypted, within a time bound t. It is clear that without having asked $H(M_i \| R)$, $F(\underline{M_i} \| \underline{R})$, or $G(m_1)$ (which events are denoted AskH, AskF and AskG respectively), the plaintext M and the random R are totally impredictable. However, the probability to ask $F(\underline{M_i} \| \underline{R})$, without having asked $G(m_1)$, is less than $2q_F/2^{k_2/2}$. Similarly, the probability to ask $H(M_i \| R)$, without having asked either $G(m_1)$ or $F(\underline{M_i} \| \underline{R})$, is less than $2q_H/2^{k_2}$. Therefore, m_1 has been asked to G, except with a small probability:

$$\mathsf{Adv}_{\Pi^3}^{\mathsf{ind-cpa}}(\mathcal{A}) \le 2\Pr[\mathsf{AskF} \vee \mathsf{AskG} \vee \mathsf{AskH}]$$
$$\le 2\Pr[\mathsf{AskG}] + 2\Pr[\mathsf{AskF} \mid \neg\mathsf{AskG}] + 2\Pr[\mathsf{AskH} \mid \neg\mathsf{AskG} \wedge \neg\mathsf{AskF}]$$
$$\le 2\Pr[\mathsf{AskG}] + \frac{4q_F}{2^{k_2/2}} + \frac{4q_H}{2^{k_2}} = 2\Pr[\mathsf{AskG}] + 4 \times \frac{2^{k_2/2}q_F + q_H}{2^{k_2}}.$$

If the event AskG occurs, by correctly guessing the query asked to G, one gets m_1:

$$\frac{1}{q_G}\Pr[\mathsf{AskG}] \le \mathsf{Succ}_{\mathsf{ntru}}^{\mathsf{pi-ow_{mLen}/2}}(t).$$

As a consequence,

$$\mathsf{Adv}_{\Pi^3}^{\mathsf{ind-cpa}}(t) \le 2q_G \times \mathsf{Succ}_{\mathsf{ntru}}^{\mathsf{pi-ow_{mLen}/2}}(t) + \frac{2^{k_2/2}q_F + q_H}{2^{k_2-2}}.$$

From a OW-adversary $\mathcal{A}$, which outputs the whole plaintext M, one can get more: indeed, $\mathcal{A}$ has to ask $F(\underline{M} \| \underline{R})$ and $G(m_1)$, or $H(M \| R)$ to know M, otherwise $\overline{M}$ is totally impredictable: only m_1 and m_2 are determined by e. But $\underline{M} \| \underline{R}$ is a random variable defined later by $\underline{M} \| \underline{R} = m_2 \oplus G(m_1)$, and $\overline{M} \| \overline{R}$ is a random variable defined by $\overline{M} \| \overline{R} = m_1 \oplus F(\underline{M} \| \underline{R})$. Furthermore, the probability to ask $H(M \| R)$, without having asked $F(\underline{M} \| \underline{R})$ and $G(m_1)$ is very low, since half of the bits are still impredictable:

$$\mathsf{Succ}_{\Pi^3}^{\mathsf{ow-cpa}}(\mathcal{A}) = \Pr[M \leftarrow \mathcal{A}(\mathsf{e}) \wedge ((\mathsf{AskF} \wedge \mathsf{AskG}) \vee \mathsf{AskH})]$$
$$+ \Pr[M \leftarrow \mathcal{A}(\mathsf{e}) \wedge \neg((\mathsf{AskF} \wedge \mathsf{AskG}) \vee \mathsf{AskH})]$$
$$\le \Pr[M \leftarrow \mathcal{A}(\mathsf{e}) \wedge \mathsf{AskF} \wedge \mathsf{AskG}]$$
$$+ \Pr[\mathsf{AskH} \wedge \neg(\mathsf{AskF} \wedge \mathsf{AskG})] + \frac{1}{2^{k_1}}$$
$$\le \Pr[M \leftarrow \mathcal{A}(\mathsf{e}) \wedge \mathsf{AskF} \wedge \mathsf{AskG}] + \frac{q_H}{2^{k_1/2+k_2/2}} + \frac{1}{2^{k_1}}$$

With the solution $M = \overline{M} \| \underline{M}$, the query $\underline{M} \| \underline{R}$ and its answer by F, as well as the query m_1 and its answer by G, one gets m_2, which is possible to check.

However, the adversary may detect the simulation, since some inconsistency in the simulation of H may occur, if $M \| R$ is asked to H. But then one can fully invert the problem, by trying all the candidates in the list of queries to H. Finally, one first checks the q_H possibilities for $M \| R$, and then tries all the possible combinations between m_1 and $\underline{M} \| \underline{R}$, to get m_2:

$$\mathsf{Succ}_{\Pi^3}^{\mathsf{ow-cpa}}(t) \leq \mathsf{Succ}_{\mathsf{ntru}}^{\mathsf{ow}}(t + (q_H + q_F q_G)T_{\mathcal{E}}) + \frac{q_H 2^{(k_1 - k_2)/2} + 1}{2^{k_1}}.$$

Discussion. One should remark that k_2 is a crucial security parameter. With too small of a value, some security results become meaningless, namely the semantic security of the second and the third paddings. However, one can see that splitting the k_2-bit random value R in two parts $\overline{R}$ and $\underline{R}$, which are used independently, is a very bad idea: the provable security level of the third construction is less than $1/2^{k_2/2}$. The latter is thus at most 2^{-20}, or 2^{-40}, according to [16, page 2]. Thus, the security proofs we obtain are rather inefficient for the parameters suggested by [16]. Our proofs may however not be tight since we are unaware of any attack achieving the previous security bounds. Nevertheless, the lack of security proofs meaningful for the recommended parameters suggests to look at different paddings.

5   Suggestions and Comparisons

We showed that none of the three suggested paddings provides the maximal security level, that is, IND-CCA2 under the sole NTRU inversion problem. However, some constructions do exist: a better OAEP-based construction or REACT [25], which we will compare later.

5.1   Suggestions

An OAEP-Based Scheme. The first suggestion is a variant of the third padding, using two more hash functions

$$F : \{0,1\}^{k_1} \to \{0,1\}^{k_2} \text{ and } G : \{0,1\}^{k_2} \to \{0,1\}^{k_1}.$$

One first computes $s = M \oplus G(R)$ and $t = R \oplus F(s)$. The ciphertext consists of $\mathcal{E}_{\mathsf{pk}}'(s \| t; H(M \| R))$. The OAEP construction provides semantic security, while the H function strengthens it to chosen-ciphertext security. An usual argument (see the full version of the paper) shows that:

$$\mathsf{Adv}_{\mathsf{oaep}'}^{\mathsf{ind-cca2}}(t) \leq 2\mathsf{Succ}_{\mathsf{ntru}}^{\mathsf{ow}}(t + (q_H + q_F q_G)T_{\mathcal{E}}) + \frac{2q_D}{2^N} + \frac{4q_H}{2^{k_1}} + \frac{2q_G}{2^{k_2}}.$$

However, one can thus see that, as in the original OAEP construction with partial-domain one-way permutations [2,9], the reduction is quadratic in the number of queries to the hash functions F and G.

NTRU-REACT. Thanks to the OW-PCA–security level of the NTRU primitive, one can use the REACT construction. The straightforward application uses two hash functions:

$$G : \{0,1\}^N \to \{0,1\}^\ell \text{ and } H : \{0,1\}^* \to \{0,1\}^{k_2}.$$

On input a message $M \in \{0,1\}^\ell$, a random $R \in \{0,1\}^{\mathsf{mLen}}$ and another random $R' \in \{0,1\}^N$, one computes $a = \mathcal{E}'_{\mathsf{pk}}(R; R')$, $b = M \oplus G(R)$ and $c = H(a, b, M, R)$. The ciphertext is the triplet (a, b, c).

The semantic security is clear, since the adversary has no advantage without having asked $G(R)$ or $H(a, b, M_i, R)$. Therefore R, and thus m, can be recovered from the queries asked to G or H:

$$\mathsf{Adv}_{\mathsf{react}}^{\mathsf{ind-cpa}}(t) \leq 2\mathsf{Succ}_{\mathsf{ntru}}^{\mathsf{ow}}(t + (q_G + q_H)T_\mathcal{E}).$$

With chosen-ciphertext attacks, the adversary cannot produce a valid ciphertext without having asked for $H(a, b, M, R)$, except with probability $1/2^{k_2}$. With proper bookkeeping, this does not increase the cost:

$$\mathsf{Adv}_{\mathsf{react}}^{\mathsf{ind-cca2}}(t) \leq 2\mathsf{Succ}_{\mathsf{ntru}}^{\mathsf{ow}}(t + (q_G + q_H)T_\mathcal{E}) + \frac{2q_D}{2^{k_2}}.$$

Improved NTRU-REACT. Interestingly, the specific properties of NTRU can be used to improve the above construction. Namely, one can reduce the size and the number of random bits. It requires two hash functions:

$$G : \{0,1\}^{\mathsf{mLen}} \to \{0,1\}^\ell \text{ and } H : \{0,1\}^* \to \{0,1\}^N.$$

Like the original construction of REACT, it can use any symmetric encryption scheme (E, D), with an ℓ-bit key. On input a message M and a random element $R \in \{0,1\}^{\mathsf{mLen}}$, one computes $K = G(R)$, $b = \mathsf{E}_K(M)$ and $R' = H(R, b)$. Then $a = \mathcal{E}'_{\mathsf{pk}}(R; R')$, and the ciphertext consists of the pair $a \parallel b$. One can prove that with this improved scheme and the one-time pad:

$$\mathsf{Adv}_{\mathsf{react}'}^{\mathsf{ind-cca2}}(t) \leq 2\mathsf{Succ}_{\mathsf{ntru}}^{\mathsf{ow}}(t + (q_G + q_H)T_\mathcal{E}) + \frac{2q_D}{2^N}.$$

A high rate can be achieved thanks to the hybrid construction. However, one would need to compare the efficiency of the block cipher and the NTRU primitive.

5.2 Comparison of NTRU Cryptosystems

We now compare the efficiency and the security level of all the above constructions. Contrarily to previous complexity analyses, we need to consider the cost of the generation of random bits as well as the cost of hashings. Indeed, this is the only difference between each scheme, since all of them just need to apply once the encryption and decryption primitives in the encryption and decryption algorithms respectively. In the figure given below, mLen and rLen are the bit-length of the message and random inputs for the NTRU primitive, and cLen is the bit-length of the output; k is a security parameter. The columns #rand, Hin and Hout indicate the number of required random bits, the number of bits as input of hash functions and the number of output bits respectively.

With classical parameters, where N is the most crucial data, chosen among 167, 251, 347 and 503, $\mathsf{mLen} = N \log p = \lceil 1.585N \rceil$, $\mathsf{rLen} = N$, $\mathsf{cLen} = 7N$ and k is between 40 and 80.

Schemes	$|M|$	$|C|$	IND-CCA2	#rand	Hin	Hout
Π^1	mLen $- k$	cLen	NO	k	mLen	rLen
Π^2	mLen $- k$	cLen	PI-OW	k	mLen $+ k$	rLen $+$ mLen $- k$
Π^3	mLen	cLen	PI-OW	k	2mLen	rLen $+$ mLen
OAEP$'$	mLen $- k$	cLen	OW	k	2mLen	rLen $+$ mLen
REACT$'$	mLen	mLen $+$ cLen	OW	mLen	3mLen	rLen $+$ mLen

Acknowledgements

We would like to thank Mike Szydlo for helpful discussions.

References

1. M. Bellare, A. Desai, D. Pointcheval, and P. Rogaway. Relations among Notions of Security for Public-Key Encryption Schemes. In *Proc. of Crypto '98*, LNCS 1462, pages 26–45. Springer-Verlag, 1998.
2. M. Bellare and P. Rogaway. Optimal Asymmetric Encryption. In *Proc. of Eurocrypt '94*, LNCS 950, pages 92–111. Springer-Verlag, 1995.
3. D. Coppersmith and A. Shamir. Lattice Attacks on NTRU. In *Proc. of Eurocrypt '97*, LNCS 1233. Springer-Verlag, 1997.
4. NTRU Cryptosystems. Technical reports available at http://www.ntru.com. 2002.
5. D. Dolev, C. Dwork, and M. Naor. Non-Malleable Cryptography. *SIAM Journal on Computing*, 30(2):391–437, 2000.
6. Consortium for Efficient Embedded Security. Efficient embedded security standards #1: Implementation aspects of NTRU and NSS. Draft Version 3.0 available at http://www.ceesstandards.org, July 2001.
7. E. Fujisaki and T. Okamoto. Secure Integration of Asymmetric and Symmetric Encryption Schemes. In *Proc. of Crypto '99*, LNCS 1666, pages 537–554. Springer-Verlag, 1999.
8. E. Fujisaki and T. Okamoto. How to Enhance the Security of Public-Key Encryption at Minimum Cost. *IEICE Trans. Fundamentals of Electronics, Comunications and Computer Sciences*, E83-A(1), 2000. Special issue on cryptography and information security.
9. E. Fujisaki, T. Okamoto, D. Pointcheval, and J. Stern. RSA–OAEP is Secure under the RSA Assumption. In *Proc. of Crypto '01*, LNCS 2139, pages 260–274. Springer-Verlag, 2001.

10. C. Gentry. Key Recovery and Message Attacks on NTRU-Composite. In *Proc. of Eurocrypt '01*, LNCS 2045, pages 182–194. Springer-Verlag, 2001.
11. S. Goldwasser and S. Micali. Probabilistic Encryption. *Journal of Computer and System Sciences*, 28:270–299, 1984.
12. C. Hall, I. Goldberg, and B. Schneier. Reaction Attacks against Several Public-Key Cryptosystems. In *Proc. of ICICS '99*, LNCS, pages 2–12. Springer-Verlag, 1999.
13. J. Hoffstein, J. Pipher, and J.H. Silverman. NTRU: a Ring based Public Key Cryptosystem. In *Proc. of ANTS III*, LNCS 1423, pages 267–288. Springer-Verlag, 1998. First presented at the rump session of Crypto '96.
14. J. Hoffstein and J. H. Silverman. Invertibility in truncated polynomial rings. Technical report, NTRU Cryptosystems, October 1998. Report #009, version 1, available at [4].
15. J. Hoffstein and J. H. Silverman. Optimizations for NTRU. In *Public-key Cryptography and Computational Number Theory*. DeGruyter, 2000. To appear, available at [4].
16. J. Hoffstein and J. H. Silverman. Protecting NTRU against chosen ciphertext and reaction attacks. Technical report, NTRU Cryptosystems, June 2000. Report #016, version 1, available at [4].
17. J. Hoffstein and J. H. Silverman. Reaction attacks against the NTRU public key cryptosystem. Technical report, NTRU Cryptosystems, June 2000. Report #015, version 2, available at [4].
18. IEEE Standard 1363. Standard specifications for public key cryptography. IEEE. Available from http://grouper.ieee.org/groups/1363, August 2000.
19. E. Jaulmes and A. Joux. A Chosen Ciphertext Attack on NTRU. In *Proc. of Crypto '00*, LNCS 1880. Springer-Verlag, 2000.
20. A. K. Lenstra, H. W. Lenstra, Jr., and L. Lovász. Factoring Polynomials with Rational Coefficients. *Mathematische Ann.*, 261:513–534, 1982.
21. A. May and J. H. Silverman. Dimension Reduction Methods for Convolution Modular Lattices. In *Proc. of CALC '01*, LNCS 2146. Springer-Verlag, 2001.
22. D. Micciancio. Improving Lattice-based Cryptosystems using the Hermite Normal Form. In *Proc. of CALC '01*, LNCS 2146. Springer-Verlag, 2001.
23. M. Naor and M. Yung. Public Key Cryptosystems Provably Secure against Chosen Ciphertext Attacks. In *Proc. of the 22nd STOC*, pages 427–437. ACM Press, 1990.
24. P. Q. Nguyen and J. Stern. The Two Faces of Lattices in Cryptology. In *Proc. of CALC '01*, LNCS 2146. Springer-Verlag, 2001.
25. T. Okamoto and D. Pointcheval. REACT: Rapid Enhanced-security Asymmetric Cryptosystem Transform. In *Proc. of CT-RSA '01*, LNCS 2020, pages 159–175. Springer-Verlag, 2001.
26. D. Pointcheval. Chosen-Ciphertext Security for any One-Way Cryptosystem. In *Proc. of PKC '00*, LNCS 1751, pages 129–146. Springer-Verlag, 2000.
27. C. Rackoff and D. R. Simon. Non-Interactive Zero-Knowledge Proof of Knowledge and Chosen Ciphertext Attack. In *Proc. of Crypto '91*, LNCS 576, pages 433–444. Springer-Verlag, 1992.
28. C. P. Schnorr. A Hierarchy of Polynomial Lattice Basis Reduction Algorithms. *Theoretical Computer Science*, 53:201–224, 1987.

Universal Padding Schemes for RSA

Jean-Sébastien Coron, Marc Joye, David Naccache, and Pascal Paillier

Gemplus Card International, France
{jean-sebastien.coron,marc.joye,
david.naccache,pascal.paillier}@gemplus.com

Abstract. A common practice to encrypt with RSA is to first apply a padding scheme to the message and then to exponentiate the result with the public exponent; an example of this is OAEP. Similarly, the usual way of signing with RSA is to apply some padding scheme and then to exponentiate the result with the private exponent, as for example in PSS. Usually, the RSA modulus used for encrypting is different from the one used for signing. The goal of this paper is to simplify this common setting. First, we show that PSS can also be used for encryption, and gives an encryption scheme semantically secure against adaptive chosen-ciphertext attacks, in the random oracle model. As a result, PSS can be used indifferently for encryption or signature. Moreover, we show that PSS allows to safely use the same RSA key-pairs for both encryption and signature, in a concurrent manner. More generally, we show that using PSS the same set of keys can be used for both encryption and signature for any trapdoor partial-domain one-way permutation. The practical consequences of our result are important: PKIs and public-key implementations can be significantly simplified.

Keywords: Probabilistic Signature Scheme, Provable Security.

1 Introduction

A very common practice for encrypting a message m with RSA is to first apply a padding scheme μ, then raise $\mu(m)$ to the public exponent e. The ciphertext c is then:

$$c = \mu(m)^e \mod N$$

Similarly, for signing a message m, the common practice consists again in first applying a padding scheme μ' then raising $\mu'(m)$ to the private exponent d. The signature s is then:

$$s = \mu'(m)^d \mod N$$

For various reasons, it would be desirable to use the same padding scheme $\mu(m)$ for encryption and for signature: in this case, only one padding scheme needs to be implemented. Of course, the resulting padding scheme $\mu(m)$ should be provably secure for encryption and for signing. We say that a padding scheme is *universal* if it satisfies this property.

M. Yung (Ed.): CRYPTO 2002, LNCS 2442, pp. 226–241, 2002.
© Springer-Verlag Berlin Heidelberg 2002

The strongest public-key encryption security notion was defined in [15] as *indistinguishability under an adaptive chosen ciphertext attack*. An adversary should not be able to distinguish between the encryption of two plaintexts, even if he can obtain the decryption of ciphertexts of his choice. For digital signature schemes, the strongest security notion was defined by Goldwasser, Micali and Rivest in [10], as *existential unforgeability under an adaptive chosen message attack*. This notion captures the property that an adversary cannot produce a valid signature, even after obtaining the signature of (polynomially many) messages of his choice.

In this paper, we show that the padding scheme PSS [3], which is originally a provably secure padding scheme for producing signatures, can also be used as a provably secure encryption scheme. More precisely, we show that PSS offers indistinguishability under an adaptive chosen ciphertext attack, in the random oracle model, under the partial-domain one-wayness of the underlying permutation. Partial-domain one-wayness, introduced in [9], is a formally stronger assumption than one-wayness. However, for RSA, partial-domain one-wayness is equivalent to (full domain) one-wayness and therefore RSA-PSS encryption is provably secure under the sole assumption that RSA is one-way.

Generally, in a given application, the RSA modulus used for encrypting is different from the RSA modulus used for signing; our setting (and real-world PKIs) would be further simplified if one could use the same set of keys for both encryption and signature (see [11]). In this paper, we show that using PSS, the same keys can be safely used for encryption and for signature.

2   Public-Key Encryption

A public-key encryption scheme is a triple of algorithms $(\mathcal{K}, \mathcal{E}, \mathcal{D})$ where:

- $\mathcal{K}$ is a probabilistic key generation algorithm which returns random pairs of public and secret keys (pk, sk) depending on some security parameter k,
- $\mathcal{E}$ is a probabilistic encryption algorithm which takes as input a public key pk and a plaintext $M \in \mathcal{M}$, runs on a random tape $r \in \mathcal{R}$ and returns a ciphertext c. $\mathcal{M}$ and $\mathcal{R}$ stand for spaces in which messages and random strings are chosen respectively,
- $\mathcal{D}$ is a deterministic decryption algorithm which, given as input a secret key sk and a ciphertext c, returns the corresponding plaintext M, or Reject.

The strongest security notion for public-key encryption is the aforementioned notion of indistinguishability under an adaptive chosen ciphertext attack. An adversary should not be able to distinguish between the encryption of two plaintexts, even if he can obtain the decryption of ciphertexts of his choice. The attack scenario is the following:

1. The adversary $\mathcal{A}$ receives the public key pk with $(pk, sk) \leftarrow \mathcal{K}(1^\kappa)$.
2. $\mathcal{A}$ makes decryption queries for ciphertexts y of his choice.

3. $\mathcal{A}$ chooses two messages M_0 and M_1 of identical length, and receives the encryption c of M_b for a random unknown bit b.
4. $\mathcal{A}$ continues to make decryption queries. The only restriction is that the adversary cannot request the decryption of c.
5. $\mathcal{A}$ outputs a bit b', representing its "guess" on b.

The adversary's advantage is then defined as:

$$\mathsf{Adv}(\mathcal{A}) = |2 \cdot \Pr[b' = b] - 1|$$

An encryption scheme is said to be secure against adaptive chosen ciphertext attack (and denoted IND-CCA2) if the advantage of any polynomial-time bounded adversary is a negligible function of the security parameter. Usually, schemes are proven to be IND-CCA2 secure by exhibiting a polynomial reduction: if some adversary can break the IND-CCA2 security of the system, then the same adversary can be invoked (polynomially many times) to solve a related hard problem.

The random oracle model, introduced by Bellare and Rogaway in [1], is a theoretical framework in which any hash function is seen as an oracle which outputs a random value for each new query. Actually, a security proof in the random oracle model does not necessarily imply that a scheme is secure in the real world (see [6]). Nevertheless, it seems to be a good engineering principle to design a scheme so that it is provably secure in the random oracle model. Many encryption and signature schemes were proven to be secure in the random oracle model.

3   Encrypting with PSS-R

In this section we prove that given any trapdoor partial-domain one-way permutation f, the encryption scheme defined by first applying PSS with message recovery (denoted PSS-R) and then encrypting the result with f achieves the strongest security level for an encryption scheme, in the random oracle model.

3.1   The PSS-R Padding Scheme

PSS-R, defined in [3], is parameterized by the integers k, k_0 and k_1 and uses two hash functions:

$$H : \{0,1\}^{k-k_1} \to \{0,1\}^{k_1} \quad \text{and} \quad G : \{0,1\}^{k_1} \to \{0,1\}^{k-k_1}$$

PSS-R takes as input a $(k - k_0 - k_1)$-bit message M and a k_0-bit random integer r. As illustrated in Figure 1, PSS-R outputs:

$$\mu(M, r) = \omega \| s$$

where $\|$ stands for concatenation, $\omega = H(M\|r)$ and $s = G(\omega) \oplus (M\|r)$. Actually, in [3], $M\|r$ is used as the argument to H and $r\|M$ is used as the mask to xor with $G(\omega)$. Here for simplicity we use $M\|r$ in both places, but the same results apply either way.

$$M \| r$$

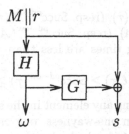

Fig. 1. The PSS-R padding scheme

3.2 The PSS-E Encryption Scheme

The new encryption scheme $(\mathcal{K}, \mathcal{E}, \mathcal{D})$, that we denote PSS-E, is based on μ and a k-bit trapdoor permutation f:

- $\mathcal{K}$ generates the public key f and the secret key f^{-1}.
- $\mathcal{E}(M,r)$: given a message $M \in \{0,1\}^{k-k_0-k_1}$ and a random $r \in \{0,1\}^{k_0}$, the encryption algorithm outputs the ciphertext:

$$c = f(\mu(M,r))$$

- $\mathcal{D}(c)$: the decryption algorithm recovers $(\omega, s) = f^{-1}(c)$ and then $M \| r = G(\omega) \oplus s$. If $\omega = H(M\|r)$, the algorithm returns M, otherwise it returns Reject.

3.3 The Underlying Problem

The security of PSS-E is based on the difficulty of inverting f without knowing f^{-1}. As in [9], we use two additional related problems: the partial-domain one-wayness and the set partial-domain one-wayness of f:

- (τ, ε)-**one-wayness of f**, means that for any adversary $\mathcal{A}$ who wishes to recover the full pre-image (ω, s) of $f(\omega, s)$ in time less than τ, $\mathcal{A}$'s success probability $\mathsf{Succ}^{\mathsf{ow}}(\mathcal{A})$ is upper-bounded by ε:

$$\mathsf{Succ}^{\mathsf{ow}}(\mathcal{A}) = \Pr_{\omega,s}[\mathcal{A}(f(\omega, s)) = (\omega, s)] < \varepsilon$$

- (τ, ε)-**partial-domain one-wayness of f**, means that for any adversary $\mathcal{A}$ who wishes to recover the partial pre-image ω of $f(\omega, s)$ in time less than τ, $\mathcal{A}$'s success probability $\mathsf{Succ}^{\mathsf{pd-ow}}(\mathcal{A})$ is upper-bounded by ε:

$$\mathsf{Succ}^{\mathsf{pd-ow}}(\mathcal{A}) = \Pr_{\omega,s}[\mathcal{A}(f(\omega, s)) = \omega] < \varepsilon$$

- $(\ell, \tau, \varepsilon)$-**set partial-domain one-wayness of f**, means that for any adversary $\mathcal{A}$ who wishes to output a set of ℓ elements which contains the partial pre-image ω of $f(\omega, s)$, in time less than τ, $\mathcal{A}$'s success probability $\mathsf{Succ}^{\mathsf{s-pd-ow}}(\mathcal{A})$ is upper-bounded by ε:

$$\mathsf{Succ}^{\mathsf{s-pd-ow}}(\mathcal{A}) = \Pr_{\omega,s}[\omega \in \mathcal{A}(f(\omega, s))] < \varepsilon$$

As in [9], we denote by $\mathsf{Succ}^{\mathsf{ow}}(\tau)$, (resp. $\mathsf{Succ}^{\mathsf{pd-ow}}(\tau)$ and $\mathsf{Succ}^{\mathsf{s-pd-ow}}(\ell,\tau)$) the maximal probability $\mathsf{Succ}^{\mathsf{ow}}(\mathcal{A})$, (resp. $\mathsf{Succ}^{\mathsf{pd-ow}}(\mathcal{A})$ and $\mathsf{Succ}^{\mathsf{s-pd-ow}}(\mathcal{A})$), over all adversaries whose running times are less than τ. For any τ and $\ell \geq 1$, we have:

$$\mathsf{Succ}^{\mathsf{s-pd-ow}}(\ell,\tau) \geq \mathsf{Succ}^{\mathsf{pd-ow}}(\tau) \geq \mathsf{Succ}^{\mathsf{ow}}(\tau)$$

Moreover, by randomly selecting any element in the set returned by the adversary against the set partial-domain one-wayness, one can break the partial-domain one-wayness with probability $1/\ell$, which gives:

$$\mathsf{Succ}^{\mathsf{pd-ow}}(\tau) \geq \mathsf{Succ}^{\mathsf{s-pd-ow}}(\ell,\tau)/\ell \tag{1}$$

We will see in Section 5 that for RSA, the three problems are polynomially equivalent.

3.4   Security of PSS-E

The following theorem shows that PSS-E is semantically secure under adaptive chosen ciphertext attacks (IND-CCA2), in the random oracle model, assuming that the underlying permutation is partial-domain one-way.

Theorem 1. *Let $\mathcal{A}$ be a CCA2-adversary against the semantic security of PSS-E, with advantage ε and running time t, making q_D, q_H and q_G queries to the decryption oracle and the hash functions H and G, respectively. Then:*

$$\mathsf{Succ}^{\mathsf{pd-ow}}(t') \geq \frac{1}{q_H + q_G} \cdot \left(\varepsilon - q_H 2^{-k_0} - q_D 2^{-k_1}\right)$$

where $t' \leq t + q_H \cdot T_f$, and T_f denotes the time complexity of f.

The theorem follows from inequality (1) and the next lemma:

Lemma 1. *Using the notations introduced in Theorem 1, we have:*

$$\mathsf{Succ}^{\mathsf{s-pd-ow}}(q_H + q_G, t') \geq \varepsilon - q_H \cdot 2^{-k_0} - q_D \cdot 2^{-k_1} \tag{2}$$

Proof. We describe a reduction $\mathcal{B}$ which using $\mathcal{A}$, constructs an adversary against the set partial-domain one-wayness of f. We start with a top-level description of the reduction and then show how to simulate the random oracles G, H and the decryption oracle D. Eventually we compute the success probability of $\mathcal{B}$.

Top-level description of the reduction $\mathcal{B}$:

1. $\mathcal{B}$ is given a function f and $c^* = f(\omega^*, s^*)$, for random integers ω^* and s^*. $\mathcal{B}$'s goal is to output a list which contains the partial pre-image ω^* of c^*.
2. $\mathcal{B}$ runs $\mathcal{A}$ with f and gets $\{M_0, M_1\}$. It chooses a random bit b and gives c^* as a ciphertext for M_b. $\mathcal{B}$ simulates the oracles G, H and D as described below.
3. $\mathcal{B}$ receives from $\mathcal{A}$ the answer b' and outputs the list of queries asked to G.

Simulation of the random oracles G, H and D:
The simulation of G and H is very simple: a random answer is returned for each new query of G and H. Moreover, when ω is the answer of a query to H, we simulate a query for ω to G, so that $G(\omega)$ is defined.

On query c to the decryption oracle, the reduction $\mathcal{B}$ looks at each query $M'\|r'$ to H and computes:

$$\omega' = H(M'\|r') \text{ and } s' = G(\omega') \oplus (M'\|r')$$

Then if $c = f(\omega', s')$ the reduction $\mathcal{B}$ returns M'. Otherwise, the reduction outputs Reject.

Analysis:
Since $c^* = f(\omega^*, s^*)$ is the ciphertext corresponding to M_b, we have the following constraint for random oracles G and H:

$$H(M_b\|r^*) = \omega^* \text{ and } G(\omega^*) = s^* \oplus (M_b\|r^*) \tag{3}$$

We denote by AskG the event: "ω^* has been asked to G" and by AskH the event: "there exists M' such that $M'\|r^*$ has been queried to H".

If ω^* was never queried to G, then $G(\omega^*)$ is undefined and r^* is then a uniformly distributed random variable. Therefore the probability that there exists M' such that (M', r^*) has been asked to H is at most $q_H \cdot 2^{-k_0}$. This gives:

$$\Pr[\mathsf{AskH}|\neg\mathsf{AskG}] \leq q_H \cdot 2^{-k_0} \tag{4}$$

Our simulation of D can only fail by rejecting a valid ciphertext. We denote by DBad this event. Letting $c = f(\omega, s)$ be the ciphertext queried to D and

$$M\|r = G(\omega) \oplus s$$

we reject a valid ciphertext if $H(M\|r) = \omega$ while $M\|r$ was never queried to H. However, if $M\|r$ was never queried to H, then $H(M\|r)$ is randomly defined. Namely if the decryption query occured before c^* was sent to the adversary, then constraint (3) does not apply and $H(M\|r)$ is randomly defined. Otherwise, if the decryption query occured after c^* was sent to the adversary, then $c \neq c^*$ implies $(M, r) \neq (M_b, r^*)$ and $H(M\|r)$ is still randomly defined. In both cases the probability that $H(M, r) = \omega$ is then 2^{-k_1}, which gives:

$$\Pr[\mathsf{DBad}] \leq q_D \cdot 2^{-k_1} \tag{5}$$

Let us denote by Bad the event: "ω^* has been queried to G or (M', r^*) has been queried to H for some M' or the simulation of D has failed". Formally:

$$\mathsf{Bad} = \mathsf{AskG} \vee \mathsf{AskH} \vee \mathsf{DBad} \tag{6}$$

Let us denote by S the event: "the adversary outputs the correct value for b, i.e., $b = b'$". Conditioned on $\neg\mathsf{Bad}$, our simulations of G, H and D are independent of b, and therefore $\mathcal{A}$'s view is independent of b as well. This gives:

$$\Pr[\mathsf{S}|\neg\mathsf{Bad}] = \frac{1}{2} \tag{7}$$

Moreover, conditioned on ¬Bad, the adversary's view is the same as when interacting with (perfect) random and decryption oracles, which gives:

$$\Pr[S \wedge \neg Bad] \geq \frac{1}{2} + \frac{\varepsilon}{2} - \Pr[Bad] \qquad (8)$$

From (7) we obtain

$$\Pr[S \wedge \neg Bad] = \Pr[S|\neg Bad] \cdot \Pr[\neg Bad] = \frac{1}{2}(1 - \Pr[Bad])$$

which gives using (8):

$$\Pr[Bad] \geq \varepsilon \qquad (9)$$

From (6) we have:

$$\Pr[Bad] \leq \Pr[AskG \vee AskH] + \Pr[DBad]$$
$$\leq \Pr[AskG] + \Pr[AskH \wedge \neg AskG] + \Pr[DBad]$$
$$\leq \Pr[AskG] + \Pr[AskH|\neg AskG] + \Pr[DBad]$$

which yields using (4), (5) and (9):

$$\Pr[AskG] \geq \varepsilon - q_H \cdot 2^{-k_0} - q_D \cdot 2^{-k_1}$$

and hence (2) holds. This completes the proof of lemma 1.   □

4   Signing and Encrypting with the Same Set of Keys

In this section we show that when using PSS, the *same* public key can be used for encryption and signature in a concurrent manner. For RSA, this means that the same set (N, e, d) can be used for both operations. In other words, when Alice sends a message to Bob, she encrypts it using Bob's public key (N, e); Bob decrypts it using the corresponding private key (N, d). To sign a message M, Bob will use the *same* private key (N, d). As usual, anybody can verify Bob's signatures using his public pair (N, e).

Although provably secure (as we will see hereafter), this is contrary to the folklore recommendation that signature and encryption keys should be distinct. This recommendation may prove useful is some cases; this is particularly true when a flaw has been found in the encryption scheme or in the signature scheme. In our case, we will prove that when using PSS-R, a decryption oracle does not help the attacker in forging signatures, and a signing oracle does not help the attacker in gaining information about the plaintext corresponding to a ciphertext.

Nevertheless, we advise to be very careful when implementing systems using the same keys for encrypting and signing. For example, if there are some implementation errors in a decryption server (see for example [13]), then an attacker could use this server to create forgeries.

4.1   The PSS-ES Encryption and Signature Scheme

The PSS-ES encryption and signature scheme $(\mathcal{K}, \mathcal{E}, \mathcal{D}, \mathcal{S}, \mathcal{V})$ is based on PSS-R and a k-bit trapdoor permutation f. As for the PSS-R signature scheme, the signature scheme in PSS-ES is with message recovery: this means that the message is recovered when verifying the signature. In this case, only messages of fixed length $k - k_0 - k_1$ can be signed. To sign messages M of arbitrary length, it suffices to apply a collision-free hash function to M prior to signing.

- $\mathcal{K}$ generates the public key f and the secret key f^{-1}.
- $\mathcal{E}(M, r)$: given a message $M \in \{0,1\}^{k-k_0-k_1}$ and a random value $r \in \{0,1\}^{k_0}$, the encryption algorithm computes the ciphertext:

$$c = f(\mu(M, r))$$

- $\mathcal{D}(c)$: the encryption algorithm recovers $(\omega, s) = f^{-1}(c)$ and computes

$$M\|r = G(\omega) \oplus s$$

If $\omega = H(M\|r)$, the algorithm returns M, otherwise it returns Reject.
- $\mathcal{S}(M, r)$: given a message $M \in \{0,1\}^{k-k_0-k_1}$ and a random value $r \in \{0,1\}^{k_0}$, the signing algorithm computes the signature:

$$\sigma = f^{-1}(\mu(M, r))$$

- $\mathcal{V}(\sigma)$: given the signature σ, the verification algorithm recovers $(\omega, s) = f(\sigma)$ and computes:

$$M\|r = G(\omega) \oplus s$$

If $\omega = H(M\|r)$, the algorithm accepts the signature and returns M. Otherwise, the algorithm returns Reject.

4.2   Semantic Security

We must ensure that an adversary is still unable to distinguish between the encryption of two messages, even if he can obtain the decryption of ciphertexts of his choice, and the signature of messages of his choice. The attack scenario is consequently the same as previously, except that the adversary can also obtain the signature of messages he wants.

The following theorem, whose proof is given in Appendix A, shows that PSS-ES is semantically secure under adaptive chosen ciphertext attacks, in the random oracle model, assuming that the underlying permutation is partial domain one-way.

Theorem 2. *Let $\mathcal{A}$ be an adversary against the semantic security of PSS-ES, with success probability ε and running time t, making q_D, q_{sig}, q_H and q_G queries to the decryption oracle, the signing oracle, and the hash functions H and G, respectively. Then, $\mathrm{Succ}^{pd-ow}(t')$ is greater than:*

$$\frac{1}{q_H + q_G + q_{sig}} \left(\varepsilon - (q_H + q_{sig}) \cdot 2^{-k_0} - q_D 2^{-k_1} - (q_H + q_{sig})^2 \cdot 2^{-k_1} \right)$$

where $t' \leq t + (q_H + q_{sig}) \cdot T_f$, and T_f denotes the time complexity of f.

4.3   Unforgeability

For signature schemes, the strongest security notion is the previously introduced existential unforgeability under an adaptive chosen message attack. An attacker cannot produce a valid signature, even after obtaining the signature of (polynomially many) messages of his choice. Here the adversary can also also obtain the decryption of ciphertexts of his choice under the same public-key. Consequently, the attack scenario is the following:

1. The adversary $\mathcal{A}$ receives the public key pk with $(pk, sk) \leftarrow \mathcal{K}(1^{\kappa})$.
2. $\mathcal{A}$ makes signature queries for messages M of his choice. Additionally, he makes decryption queries for ciphertexts y of his choice.
3. $\mathcal{A}$ outputs the signature of a message M' which was not queried for signature before.

An encryption-signature scheme is said to be secure against chosen-message attacks if for any polynomial-time bounded adversary, the probability to output a forgery is negligible.

The following theorem shows that PSS-ES is secure against an adaptive chosen message attack. The proof is similar to the security proof of PSS [3] and is given in Appendix B.

Theorem 3. *Let $\mathcal{A}$ be an adversary against the unforgeability of PSS-ES, with success probability ε and running time t, making q_D, q_{sig}, q_H and q_G queries to the decryption oracle, the signing oracle, and the hash oracles H and G, respectively. Then $\mathsf{Succ}^{\mathrm{ow}}(t')$ is greater than:*

$$\frac{1}{q_H} \left(\varepsilon - ((q_H + q_{sig})^2 + q_D + 1) \cdot 2^{-k_1} \right) \qquad (10)$$

where $t' \le t + (q_H + q_{sig}) \cdot T_f$, and T_f denotes the time complexity of f.

5   Application to RSA

5.1   The RSA Cryptosystem

The RSA cryptosystem [16] is the most widely used cryptosystem today. In this section, we show that by virtue of RSA's homomorphic properties, the partial-domain one-wayness of RSA is equivalent to the one-wayness of RSA. This enables to prove that the encryption scheme RSA-PSS-E and the encryption and signature scheme RSA-PSS-ES are semantically secure against chosen ciphertext attacks, in the random oracle model, assuming that inverting RSA is hard.

Definition 1 (The RSA Primitive). *The RSA primitive is a family of trapdoor permutations, specified by:*

– The RSA generator $\mathcal{RSA}$, which on input 1^k, randomly selects two distinct $k/2$-bit primes p and q and computes the modulus $N = p \cdot q$. It randomly picks an

encryption exponent $e \in \mathbb{Z}_{\phi(N)}^*$, *computes the corresponding decryption exponent* $d = e^{-1} \mod \phi(N)$ *and returns* (N, e, d).
- *The encryption function* $f : \mathbb{Z}_N^* \rightarrow \mathbb{Z}_N^*$ *defined by* $f(x) = x^e \mod N$.
- *The decryption function* $f^{-1} : \mathbb{Z}_N^* \rightarrow \mathbb{Z}_N^*$ *defined by* $f^{-1}(y) = y^d \mod N$.

In the following, we state our result in terms of the RSA primitive with a randomly chosen public exponent. The same results apply to the common practice of choosing a small public exponent. Actually, using Coppersmith's algorithm [7] as in [17] for OAEP [2], it would be possible to obtain tigther bounds for a small public exponent.

5.2   Partial-Domain One-Wayness of RSA

The following lemma shows that the partial-domain one-wayness of RSA is equivalent to the one-wayness of RSA. This is a generalization of the result that appears in [9] for OAEP and in [4] for SAEP$^+$, wherein the size of the partial pre-image is greater than half the size of the modulus. The extension was announced in [9] and [4], even if the proper estimates were not worked out.

The technique goes as follows. Given $y = x^e \mod N$, we must find x. We obtain the least significant bits of $x \cdot \alpha_i \mod N$ for random integers $\alpha_i \in \mathbb{Z}_N$, by querying for the partial pre-image of $y_i = y \cdot (\alpha_i)^e \mod N$. Finding x from the least significant bits of the $x \cdot \alpha_i \mod N$ is a Hidden Number Problem modulo N. We use an algorithm similar to [5] to efficiently recover x.

Lemma 2. *Let $\mathcal{A}$ be an algorithm that on input y, outputs a q-set containing the k_1 most significant bits of $y^d \mod N$, within time bound t, with probability ε, where $2^{k-1} \leq N < 2^k$, $k_1 \geq 64$ and $k/(k_1)^2 \leq 2^{-6}$. Then there exists an algorithm $\mathcal{B}$ that solves the RSA problem with success probability ε' within time bound t', where:*

$$\varepsilon' \geq \varepsilon \cdot (\varepsilon^{n-1} - 2^{-k/8}) \tag{11}$$

$$t' \leq n \cdot t + q^n \cdot \mathrm{poly}(k)$$

$$n = \left\lceil \frac{5k}{4k_1} \right\rceil$$

Proof. See the full paper [8].

5.3   RSA-PSS-E and RSA-PSS-ES

The RSA-PSS-E encryption scheme $(\mathcal{K}, \mathcal{E}, \mathcal{D})$ based on the PSS-R padding μ with parameters k, k_0, and k_1 is defined as follows:
- $\mathcal{K}$ generates a $(k + 1)$-bit RSA modulus and exponents e and d. The public key is (N, e) and the private key is (N, d).
- $\mathcal{E}(M, r)$: given a message $M \in \{0, 1\}^{k-k_0-k_1}$ and a random $r \in \{0, 1\}^{k_0}$, the encryption algorithm outputs the ciphertext:

$$c = (\mu(M, r))^e \mod N$$

- $\mathcal{D}(c)$: the decryption algorithm recovers $x = c^d \mod N$. It returns Reject if the most significant bit of x is not zero. It writes x as $0\|\omega\|s$ where ω is a k_1-bit string and s is a $k - k_1$ bit string. It writes $M\|r = G(\omega) \oplus s$. If $\omega = H(M\|r)$, the algorithm returns M, otherwise it returns Reject.

The RSA-PSS-ES encryption and signature scheme $(\mathcal{K}, \mathcal{E}, \mathcal{D}, \mathcal{S}, \mathcal{V})$ is defined as follows:

- $\mathcal{K}$, $\mathcal{E}(M, r)$ and $\mathcal{D}(c)$ are identical to RSA-PSS-E.
- $\mathcal{S}(M, r)$: given a message $M \in \{0,1\}^{k - k_0 - k_1}$ and a random value $r \in \{0,1\}^{k_0}$, the signing algorithm computes the signature:

$$\sigma = \mu(M, r)^d \mod N$$

- $\mathcal{V}(\sigma)$: given the signature σ, the verification algorithm recovers $x = \sigma^e \mod N$. It returns Reject if the most significant bit of x is not zero. It writes x as $0\|\omega\|s$ where ω is a k_1-bit string and s is a $(k - k_1)$-bit string. It writes $M\|r = G(\omega) \oplus s$. If $\omega = H(M\|r)$, the algorithm accepts the signature and returns M, otherwise it returns Reject.

5.4   Security of RSA-PSS-E and RSA-PSS-ES

Combining Lemma 1 and Lemma 2, we obtain the following theorem which shows that the encryption scheme RSA-PSS-E is provably secure in the random oracle model, assuming that inverting RSA is hard.

Theorem 4. *Let $\mathcal{A}$ be a CCA2-adversary against the semantic security of the RSA-PSS-E scheme $(\mathcal{K}, \mathcal{E}, \mathcal{D})$, with advantage ε and running time t, making q_D, q_H and q_G queries to the decryption oracle and the hash function H and G, respectively. Provided that $k_1 \geq 64$ and $k/(k_1)^2 \leq 2^{-6}$, RSA can be inverted with probability ε' greater than:*

$$\varepsilon' \geq \left(\varepsilon - q_H \cdot 2^{-k_0} - q_D 2^{-k_1}\right)^n - 2^{-k/8}$$

within time bound $t' \leq n \cdot t + (q_H + q_G)^n \cdot \mathrm{poly}(k)$, where $n = \lceil 5k/(4k_1) \rceil$.

We obtain a similar theorem for the semantic security of the RSA-PSS-ES encryption and signature scheme (from Lemma 2 and Lemma 3 in appendix A).

Theorem 5. *Let $\mathcal{A}$ be a CCA2-adversary against the semantic security of the RSA-PSS-ES scheme $(\mathcal{K}, \mathcal{E}, \mathcal{D}, \mathcal{S}, \mathcal{V})$, with advantage ε and running time t, making q_D, q_{sig}, q_H and q_G queries to the decryption oracle, the signing oracle and the hash function H and G, respectively. Provided that $k_1 \geq 64$ and $k/(k_1)^2 \leq 2^{-6}$, RSA can be inverted with probability ε' greater than:*

$$\varepsilon' \geq \left(\varepsilon - (q_H + q_{sig}) \cdot 2^{-k_0} - (q_D + (q_H + q_{sig})^2) \cdot 2^{-k_1}\right)^n - 2^{-k/8}$$

within time bound $t' \leq n \cdot t + (q_H + q_G + q_{sig})^n \cdot \mathrm{poly}(k)$, where $n = \lceil 5k/(4k_1) \rceil$.

For the unforgeability of the RSA-PSS-ES encryption and signature scheme, we obtain a better security bound than the general result of Theorem 3, by relying upon the homomorphic properties of RSA. The proof of the following theorem is similar to the security proof of PSS in [3] and is given in the full version of this paper [8]

Theorem 6. *Let $\mathcal{A}$ be an adversary against the unforgeability of the PSS-ES scheme $(\mathcal{K}, \mathcal{E}, \mathcal{D}, \mathcal{S}, \mathcal{V})$, with success probability ε and running time t, making q_D, q_{sig}, q_H and q_G queries to the decryption oracle, the signing oracle, and the hash functions H and G, respectively. Then RSA can be inverted with probability ε' greater than:*

$$\varepsilon' \geq \varepsilon - \left((q_H + q_{sig})^2 + q_D + 1 \right) \cdot \left(2^{-k_0} + 2^{-k_1} \right) \tag{12}$$

within time bound $t' \leq t + (q_H + q_{sig}) \cdot \mathcal{O}(k^3)$.

Note that as for OAEP [9], the security proof for encrypting with PSS is far from being tight. This means that it does not provide a meaningful security result for a moderate size modulus (*e.g.*, 1024 bits). For the security proof to be meaningful in practice, we recommend to take $k_1 \geq k/2$ and to use a larger modulus (*e.g.*, 2048 bits).

6   Conclusion

In all existing PKIs different padding formats are used for encrypting and signing; moreover, it is recommended to use different keys for encrypting and signing. In this paper we have proved that the PSS padding scheme used in PKCS#1 v.2.1 [14] and IEEE P1363 [12] can be safely used for encryption as well. We have also proved that the same key pair can be safely used for both signature and encryption. The practical consequences of this are significant: besides halving the number of keys in security systems and simplifying their architecture, our observation allows resource-constrained devices such as smart cards to use the same code for implementing both signature and encryption.

Acknowledgements

We wish to thank Jacques Stern for pointing out an error in an earlier version of this paper, and the anonymous referees for their useful comments.

References

1. M. Bellare and P. Rogaway, *Random oracles are practical: a paradigm for designing efficient protocols*. Proceedings of the First Annual Conference on Computer and Commmunications Security, ACM, 1993.
2. M. Bellare and P. Rogaway, *Optimal Asymmetric Encryption*, Proceedings of Eurocrypt'94, LNCS vol. 950, Springer-Verlag, 1994, pp. 92–111.

3. M. Bellare and P. Rogaway, *The exact security of digital signatures - How to sign with RSA and Rabin.* Proceedings of Eurocrypt'96, LNCS vol. 1070, Springer-Verlag, 1996, pp. 399-416.

4. D. Boneh, *Simplified OAEP for the RSA and Rabin functions,* Prooceedings of Crypto 2001, LNCS vol 2139, pp. 275-291, 2001.

5. D. Boneh and R. Venkatesan, *Hardness of computing the most significant bits of secret keys in Diffie-Hellman and related schemes.* Proceedings of Crypto '96, pp. 129-142, 1996.

6. R. Canetti, O. Goldreich and S. Halevi, *The random oracle methodology, revisited,* STOC' 98, ACM, 1998.

7. D. Coppersmith, *Finding a small root of a univariate modular equation,* in Eurocrypt '96, LNCS 1070.

8. J.S. Coron, M. Joye, D. Naccache and P. Paillier, *Universal padding schemes for RSA.* Full version of this paper. Cryptology ePrint Archive, http://eprint.iacr.org.

9. E. Fujisaki, T. Okamoto, D. Pointcheval and J. Stern, *RSA-OAEP is secure under the RSA assumption,* Proceedings of Crypto' 2001, LNCS vol. 2139, Springer-Verlag, 2001, pp. 260-274.

10. S. Goldwasser, S. Micali and R. Rivest, *A digital signature scheme secure against adaptive chosen-message attacks,* SIAM Journal of computing, 17(2), pp. 281-308, April 1988.

11. S. Haber and B. Pinkas, *Combining Public Key Cryptosystems,* Proceedings of the ACM Computer and Security Conference , November 2001.

12. IEEE P1363a, *Standard Specifications For Public Key Cryptography: Additional Techniques,* available at http://www.manta.ieee.org/groups/1363

13. J. Manger, *A chosen ciphertext attack on RSA Optimal Asymmetric Encryption Padding (OAEP) as Standardized in PKCS #1 v2.0.* Proceedings of Crypto 2001, LNCS 2139, pp. 230-238, 2001.

14. PKCS #1 v2.1, *RSA Cryptography Standard (draft),* available at http:www.rsasecurity.com /rsalabs/pkcs.

15. C.Rackoff and D. Simon, *Noninteractive zero-knowledge proof of knowledge and chosen ciphertext attack.* Advances in Cryptology, Crypto '91, pages 433-444, 1991.

16. R. Rivest, A. Shamir and L. Adleman, *A method for obtaining digital signatures and public key cryptosystems,* CACM 21, 1978.

17. V. Shoup, *OAEP reconsidered,* Proceedings of Crypto 2001, LNCS vol. 2139, pp 239-259, 2001.

A Proof of Theorem 2

The theorem follows from inequality (1) and the following lemma.

Lemma 3. *Let A be an adversary against the semantic security of PSS-ES, with success probability ε and running time t, making q_D, q_{sig}, q_H and q_G queries to the decryption oracle, the signing oracle, and the hash functions H and G, respectively. Then, the success probability $\text{Succ}^{\text{s-pd-ow}}(q_H + q_G + q_{sig}, t')$ is greater than:*

$$\varepsilon - (q_H + q_{sig}) \cdot 2^{-k_0} - q_D 2^{-k_1} - (q_H + q_{sig})^2 \cdot 2^{-k_1}$$

where $t' \leq t + (q_H + q_{sig}) \cdot T_f$, and T_f denotes the time complexity of f.

Proof. The proof is very similar to the proof of lemma 1. The top-level description of the reduction $\mathcal{B}$ is the same and the simulation of the decryption oracle is the same. However, oracles H and G are simulated differently. Instead of simulating H and G so that $\mu(M, r) = y$ is a random integer, we simulate H and G so that $\mu(M, r) = f(x)$ for a known random x, which allows to answer the signature query for M.

Simulation of oracles G and H and signing oracle:
When receiving the query $M\|r$ to H, we generate a random $x \in \{0, 1\}^k$ and compute $y = f(x)$. We denote $y = \omega\|s$. If ω never appeared before, we let $G(\omega) = s \oplus (M\|r)$ and return ω, otherwise we abort.

When receiving a query ω for G, if $G(\omega)$ has already been defined, we return $G(\omega)$, otherwise we return a random $(k - k_1)$-bit integer.

When we receive a signature query for M, we generate a random k_0-bit integer r. If $M\|r$ was queried to H before, we know ω, s, y and x such that:

$$H(M\|r) = \omega \quad \text{and} \quad G(\omega) = s \oplus (M\|r) \quad \text{and} \quad y = f(x) = \omega\|s$$

so we return the corresponding signature x. If $M\|r$ was never queried before, we simulate an H-query for $M\|r$ as previously: we pick a random $x \in \{0, 1\}^k$ and compute $y = f(x)$. We denote $y = \omega\|s$. If ω never appeared before, we let $H(M\|r) = \omega$, $G(\omega) = s \oplus (M\|r)$ and return the signature x, otherwise we abort.

Analysis:
As in lemma 1, we denote by AskG the event: "ω^* has been asked to G" and by AskH the event: "there exists M' such that $M'\|r^*$ has been queried to H"; we denote by DBad the event: "a valid ciphertext has been rejected by our simulation of the decryption oracle D". Moreover, we denote by SBad the event: "the reduction aborts when answering an H-oracle query or a signature query". As previously, we have:

$$\Pr[\mathsf{AskH}|\neg\mathsf{AskG}] \leq (q_H + q_{sig}) \cdot 2^{-k_0}$$

and

$$\Pr[\mathsf{DBad}] \leq q_D \cdot 2^{-k_1}$$

When answering an H-oracle query or a signature query, the integer ω which is generated is uniformly distributed because f is a permutation. Moreover, at most $q_H + q_{sig}$ values of ω can appear during the reduction. Therefore the probability that the reduction aborts when answering an H-oracle query or a signature query is at most $(q_H + q_{sig}) \cdot 2^{-k_1}$, which gives:

$$\Pr[\mathsf{SBad}] \leq (q_H + q_{sig})^2 \cdot 2^{-k_1}$$

We denote by Bad the event:

$$\mathsf{Bad} = \mathsf{AskG} \vee \mathsf{AskH} \vee \mathsf{DBad} \vee \mathsf{SBad}$$

Let S denote the event: "the adversary outputs the correct value for b, i.e. $b = b'$". Conditioned on $\neg$Bad, our simulation of oracles G, H, D and of the signing oracle are independent of b, and therefore the adversary's view is independent of b. This gives:

$$\Pr[S|\neg Bad] = \frac{1}{2} \tag{13}$$

Moreover, conditioned on $\neg Bad$, the adversary's view is the same as when interacting with (perfect) random oracles, decryption oracle and signing oracle, which gives:

$$\Pr[S \wedge \neg Bad] \geq \frac{1}{2} + \frac{\varepsilon}{2} - \Pr[Bad] \tag{14}$$

which yields as in Lemma 1:

$$\Pr[Bad] \geq \varepsilon \tag{15}$$

and eventually:

$$\Pr[AskG] \geq \varepsilon - (q_H + q_{sig}) \cdot 2^{-k_0} - q_D \cdot 2^{-k_1} - (q_H + q_{sig})^2 \cdot 2^{-k_1}$$

B   Proof of Theorem 3

From $\mathcal{A}$ we construct an algorithm $\mathcal{B}$, which receives as input c and outputs η such that $c = f(\eta)$.

Top-level description of the reduction $\mathcal{B}$:

1. $\mathcal{B}$ is given a function f and $c = f(\eta)$, for a random integer η.
2. $\mathcal{B}$ selects uniformly at random an integer $j \in [1, q_H]$.
3. $\mathcal{B}$ runs $\mathcal{A}$ with f. It simulates the decryption oracle, the signing oracle and random oracles H and G as described below. $\mathcal{B}$ maintains a counter i for the i-th query $M_i \| r_i$ to H. The oracles H and G are simulated in such a way that if $i = j$ then $\mu(M_i \| r_i) = c$.
4. $\mathcal{B}$ receives from $\mathcal{A}$ a forgery σ. Letting M and r be the corresponding message and random, if $(M, r) = (M_j, r_j)$ then $f(\sigma) = \mu(M_j \| r_j) = c$ and $\mathcal{B}$ outputs σ.

Simulation of the oracles G, H, D and signing oracle:
When receiving the i-th query $M_i \| r_i$ to H, we distinguish two cases: if $i \neq j$, we generate a random $x_i \in \{0,1\}^k$ and compute $y_i = f(x_i)$. If $i = j$, we let $y_i = c$. In both cases we denote $y_i = \omega_i \| s_i$. If ω_i never appeared before, we let $G(\omega_i) = s_i \oplus (M_i \| r_i)$ and return ω_i, otherwise we abort.

When receiving a query ω for G, if $G(\omega)$ has already been defined, we return $G(\omega)$, otherwise we return a random $(k - k_1)$-bit integer.

When we receive a signature query for M, we generate a random k_0-bit integer r. If $M \| r$ was queried to H before, we have $M \| r = M_i \| r_i$ for some i. If $i \neq j$, we have:

$$H(M_i \| r_i) = \omega_i, \quad G(\omega_i) = s_i \oplus (M_i \| r_i) \quad \text{and} \quad y_i = \omega_i \| s_i = f(x_i)$$

so we return the corresponding signature x_i, otherwise we abort. If $M\|r$ was never queried before, we simulate an H-query for $M\|r$ as previously: we generate a random $x \in \{0,1\}^k$ and compute $y = f(x)$. We denote $y = \omega\|s$. If ω never appeared before, we let $H(M\|r) = \omega$ and $G(\omega) = s \oplus (M\|r)$ and return the signature x, otherwise we abort.

The simulation of the decryption oracle is identical to that of Lemma 1.

Analysis:
Let σ be the forgery sent by the adversary. If ω was not queried to G, we simulate a query to G as previously. Let $\omega\|s = f(\sigma)$ and $M\|r = G(\omega) \oplus s$. If $M\|r$ was never queried to H, then $H(M\|r)$ is undefined because there was no signature query for M; the probability that $H(M\|r) = \omega$ is then 2^{-k_1}. Otherwise, let $(M,r) = (M_i, r_i)$ be the corresponding query to H. If $i = j$, then $\mu(M_j, r_j) = c = f(\sigma)$ and $\mathcal{B}$ succeeds in inverting f.

Conditioned on $i = j$, our simulation of H and the signing oracle are perfect, unless some ω appears twice, which happens with probability less than $(q_H + q_{sig})^2 \cdot 2^{-k_1}$. As in lemma 1, our simulation of D fails with probability less than $q_D \cdot 2^{-k_1}$. Consequently, the reduction $\mathcal{B}$ succeeds with probability greater than:

$$\frac{1}{q_H} \cdot \left(\varepsilon - 2^{-k_1} - (q_H + q_{sig})^2 \cdot 2^{-k_1} - q_D \cdot 2^{-k_1} \right)$$

which gives (10).

Cryptanalysis of Unbalanced RSA
with Small CRT-Exponent

Alexander May

Department of Mathematics and Computer Science
University of Paderborn
33102 Paderborn, Germany
alexx@uni-paderborn.de

Abstract. We present lattice-based attacks on RSA with prime factors p and q of unbalanced size. In our scenario, the factor q is smaller than N^β and the decryption exponent d is small modulo $p - 1$. We introduce two approaches that both use a modular bivariate polynomial equation with a small root. Extracting this root is in both methods equivalent to the factorization of the modulus $N = pq$. Applying a method of Coppersmith, one can construct from a bivariate modular equation a bivariate polynomial $f(x, y)$ over $\mathbb{Z}$ that has the same small root. In our first method, we prove that one can extract the desired root of $f(x, y)$ in polynomial time. This method works up to $\beta < \frac{3 - \sqrt{5}}{2} \approx 0.382$. Our second method uses a heuristic to find the root. This method improves upon the first one by allowing larger values of d modulo $p - 1$ provided that $\beta \leq 0.23$.

Keywords: RSA, lattice reduction, Coppersmith's method, small secret exponent

1 Introduction

An RSA key is a tuple (N, e) where $N = pq$ is the product of two primes and e is the public key. The corresponding secret key d satisfies the equation $ed = 1 \bmod \frac{(p-1)(q-1)}{2}$ with $\gcd(p - 1, \frac{q-1}{2}) = 1$. The Chinese Remainder Theorem (CRT) gives us the equations $ed = 1 \bmod p - 1$ and $ed = 1 \bmod \frac{q-1}{2}$.

To speed up the RSA decryption and signature generation process, one is tempted to use small secret decryption exponents d. Unfortunately, Wiener [17] showed that $d < \frac{1}{3} N^{\frac{1}{4}}$ leads to a polynomial time attack on the RSA cryptosystem. This result was generalized by Verheul and Tilborg [16] to the case where one guesses high-order bits of the prime factors. They showed that in order to improve Wiener's bound for r bits one has to guess approximately $2r$ bits.

Recently, Boneh and Durfee [3] showed how to improve the bound of Wiener up to $d < N^{0.292}$. Their attack works in polynomial time and builds upon Coppersmith's method for finding small roots of modular polynomial equations. This method in turn is based on the famous L^3-lattice reduction algorithm of Lenstra, Lenstra and Lovász [9]. Coppersmith's method is rigorous for the univariate case

M. Yung (Ed.): CRYPTO 2002, LNCS 2442, pp. 242–256, 2002.

but the proposed generalization in the modular multivariate case is a heuristic. Since Boneh and Durfee use Coppersmith's method in the bivariate modular case, their attack is a heuristic. In contrast, the approach of Wiener is a provable method. However, the Boneh-Durfee attack works very well in practice. In fact, many other works (e.g. [1, 5, 8]) are based on this useful heuristical multivariate approach.

The results above show that one cannot use a small decryption exponent d. But there is another way to speed up the decryption and signature generation process. One can use a decryption exponent d such that $d_p = d \bmod p - 1$ and $d_q = d \bmod \frac{q-1}{2}$ are small. Such an exponent d is called a small CRT-exponent. In order to sign a message m, one computes $m^{d_p} \bmod p$ and $m^{d_q} \bmod q$. Both terms are combined using the Chinese Remainder Theorem to yield the desired term $m^d \bmod N$. The attacks described before do not work in this case, since d is likely to be large.

It is an open problem if there is a polynomial time algorithm that breaks RSA if d_p and d_q are small. This problem is mentioned several times in the literature, see e.g. [17, 2, 3]. The best algorithm that is known runs in time $O(\min(\sqrt{d_p}, \sqrt{d_q}))$ which is exponentially in the bit-size.

In this work, we give the first polynomial time attack on RSA with small CRT-exponent. Unfortunately, our results are restricted to the case of unbalanced prime numbers p and q. The use of unbalanced primes was first proposed by Shamir [13] to guard the modulus N against different kinds of factorization algorithms and to speed up the computation. There are also other systems that use unbalanced primes [10, 15]. Interestingly, sometimes the use of unbalanced primes decreases the security. For instance, Durfee and Nguyen [5] showed that the Boneh-Durfee attack works for larger exponents d if the prime factors are unbalanced. This breaks the RSA-type scheme of Sun, Yang and Laih [15].

We show in the following work that there is also a decrease in security for unbalanced primes when using small CRT-exponents. The more unbalanced the prime factors are, the larger are the CRT-exponents that can be attacked by our methods.

Let $q < N^\beta$ and $d_p \leq N^\delta$. We show in Section 3 that an RSA public key tuple (N, e) satisfying the condition $3\beta + 2\delta \leq 1 - \log_N(4)$ yields the factorization of N in time $O(\log^2(N))$. Thus, this method does only work provided that $\beta < \frac{1}{3}$.

Like the methods in [1, 3, 5, 8], our approach is based on Coppersmith's technique [4] in the modular multivariate case. More precisely, we use a modular bivariate polynomial equation with a small root. This root gives us the factorization of N. Using a Theorem of Howgrave-Graham [7], we can turn the modular bivariate polynomial into a polynomial $f(x, y)$ over $\mathbb{Z}$ such that the desired small root must be among the roots of $f(x, y)$. Interestingly, for the polynomial $f(x, y)$ we are able to prove that this small root can be extracted easily. This shows that our method provably factors the modulus N. Note, that this is in contrast to other works using the multivariate approach [1, 3, 5, 8] which rely on a heuristic assumption. To our knowledge, this is the first rigorous method using a modular bivariate approach. We think that this method will be useful in other settings as well. As an example, we show that our technique yields an elegant and simple proof of the results of Wiener[17] and Verheul, Tilborg [16].

The attack in Section 3 uses a two-dimensional lattice. In Section 4, we generalize our method to lattices of arbitrary dimension. This improves the condition above to $3\beta - \beta^2 + 2\delta \leq 1 - \epsilon$ for some small error term ϵ. Therefore, this approach works as long as $\beta < \frac{3-\sqrt{5}}{2} = \hat{\phi}^2$, where $\hat{\phi} = \frac{1-\sqrt{5}}{2}$ is the conjugate of the golden ratio. Again, we can show that the desired root can be extracted in polynomial time. This yields a rigorous method for factoring N.

In Section 5, we use a different modular bivariate polynomial. This approach works for larger CRT-exponents than our first attack provided that $\beta \leq 0.23$. Unfortunately, we cannot give a rigorous proof for this method. It relies on Coppersmith's heuristic for modular multivariate polynomials.

Finally, we compare our approaches in Section 6.

2   Preliminaries

Let $\mathbb{Z}_N$ denote the ring of integers modulo N. Let $\mathbb{Z}_N^*$ denote the multiplicative group of invertible integers modulo N. The order of $\mathbb{Z}_N^*$ is given by the Euler phi-function $\phi(N)$. Using RSA, we have $N = pq$ and $\phi(N) = (p-1)(q-1)$. If x is a random element in $\mathbb{Z}_N^*$, we use the notation $x \in_R \mathbb{Z}_N^*$.

Let $f(x, y) = \sum_{i,j} a_{i,j} x^i y^j \in \mathbb{Z}[x, y]$ be a bivariate polynomial with coefficients $a_{i,j}$ in the ring of integers. We will often use the short-hand notation f when the parameters follow from the context. The degree of f is the maximal sum $i + j$ taken over all monomials $a_{i,j} x^i y^j$ with non-zero coefficients. The coefficient vector of f is the vector of the coefficients $a_{i,j}$. The Euclidean norm of f is defined as the norm of the coefficient vector: $\|f\|^2 = \sum_{i,j} a_{i,j}^2$.

In the following, we state a few basic facts about lattices and lattice basis reduction and refer to the textbooks [6, 14] for an introduction into the theory of lattices.

Let $v_1, \ldots, v_n \in \mathbb{R}^m$, $m \geq n$ be linearly independent vectors. A lattice L spanned by $\{v_1, \ldots, v_n\}$ is the set of all integer linear combinations of $v_1, \ldots, v_n$. If $m = n$, the lattice is called a full rank lattice. The set of vectors $B = \{v_1, \ldots, v_n\}$ is called a basis for L.

We denote by $v_1^*, \ldots, v_n^*$ the vectors obtained by applying Gram-Schmidt orthogonalization to the basis vectors. The determinant of L is defined as

$$\det(L) = \prod_{i=1}^{n} \|v_i^*\|,$$

where $\|v\|$ denotes the Euclidean norm of v. Any lattice L has infinitely many bases but all bases have the same determinant. If a lattice is full rank, $\det(L)$ is the absolute value of the determinant of the $(n \times n)$-matrix whose rows are the basis vectors $v_1, \ldots, v_n$. Hence if the basis matrix is triangular, the determinant is very easy to compute.

A well-known result by Minkowski relates the determinant of a lattice L to the length of a shortest vector in L. Minkowski's Theorem shows that every

n-dimensional lattice L contains a non-zero vector v with $\|v\| \leq \sqrt{n} \det(L)^{\frac{1}{n}}$. Unfortunately, the proof of this theorem is non-constructive.

In dimension 2, the Gauss reduction algorithm yields a shortest vector of a lattice. In arbitrary dimension, we can use the famous L^3-reduction algorithm of Lenstra, Lenstra and Lovász [9] to approximate a shortest vector.

Fact 1 (Lenstra, Lenstra and Lovász) *Let L be a lattice spanned by $\{v_1, \ldots, v_n\}$. The L^3-reduction algorithm will output in polynomial time a lattice basis $\{v_1', \ldots, v_n'\}$ with*

$$\|v_1'\| \leq 2^{\frac{n-1}{4}} \det(L)^{\frac{1}{n}} \quad and \quad \|v_2'\| \leq 2^{\frac{n}{2}} \det(L)^{\frac{1}{n-1}}.$$

2.1 Key Generation Using the Chinese Remainder Theorem (CRT)

We briefly describe the key generation process. In our scenario, the RSA modulus N is composed of a large prime factor p and a small prime factor q. The secret decryption exponent d is chosen to be small modulo $p - 1$ and of arbitrary size modulo $q - 1$.

CRT Key Generation Process

Fix a bit-size n for the public key modulus N. Additionally, fix two positive parameters β, δ with $\beta \leq \frac{1}{2}$ and $\delta \leq 1$.

Modulus: Choose randomly prime numbers p and q with bit-sizes approximately $(1 - \beta)n$ and βn. Additionally, $p - 1$ and $\frac{q-1}{2}$ must be coprime. Compute the modulus $N = pq$. If the smaller prime factor q does not satisfy $q < N^\beta$, repeat the prime generation.

Secret exponent: Choose a small secret $d_p \in_R Z_{p-1}^*$ such that $d_p \leq N^\delta$. Choose another secret $d_q \in_R Z_{\frac{q-1}{2}}^*$ arbitrarily.

Chinese remaindering: Compute the unique $d \mod \frac{\phi(N)}{2}$ that satisfies $d = d_p \mod p - 1$ and $d = d_q \mod \frac{q-1}{2}$.

Public exponent: Compute the inverse e of d in $\mathbb{Z}_{\frac{\phi(N)}{2}}^*$.

Public parameters: Publish the tuple (N, e).

In this work, we will study the following question:
Up to which parameter choices for β and δ does the public key tuple (N, e) yield the factorization of N ?

Note, that the decryption and the signature generation process of a message m are very efficient for small β and δ. Since d_p is small, the computation of $m^{d_p} \mod p - 1$ requires only a small amount of multiplications. On the other hand, the computation of $m^{d_q} \mod \frac{q-1}{2}$ is cheap because q is small. Both terms can easily be combined to yield the desired term $m^d \mod \frac{\phi(N)}{2}$ using the Chinese Remainder Theorem(CRT).

In the next section, we will show that given the public key (N, e) there is a provable polynomial time algorithm that factors N if the condition $3\beta + 2\delta \leq 1 - \epsilon$

holds, where ϵ is a small error term. This implies that our method works as long as $\beta < \frac{1}{3}$. The smaller β is chosen, the larger δ can be in the attack. For $\beta = 0$, we obtain $\delta < \frac{1}{2}$. Later, we will improve the bound for β up to $\frac{3-\sqrt{5}}{2} \approx 0.382$ and for δ up to 1.

3   An Approach Modulo p

Given a public key (N, e) that is constructed according to the CRT Key Generation process. We know that

$$ed_p = 1 \bmod p - 1.$$

Thus, there is an integer k such that

$$ed_p + k(p - 1) = 1 \quad \text{over } \mathbb{Z}. \tag{1}$$

We can rewrite this equation as

$$ed_p - (k + 1) = -kp \tag{2}$$

In the following, we assume that q does not divide k. Otherwise, the right hand side of the equation is a multiple of N and we can obtain much stronger results. This case will be analyzed later.

Equation (2) gives us the polynomial

$$f_p(x, y) = ex - y$$

with a root $(x_0, y_0) = (d_p, k + 1)$ modulo p.

By construction, we have $d_p \leq N^\delta$. Since $e < \frac{(p-1)(q-1)}{2}$, we obtain

$$|k + 1| = \left| \frac{ed_p - 2}{p - 1} \right| < \frac{ed_p}{p - 1} < \frac{q - 1}{2} d_p < N^{\beta+\delta}.$$

Let as define two upper bounds $X = N^\delta$ and $Y = N^{\beta+\delta}$. Then, we have a modular bivariate polynomial equation f_p with a small root (x_0, y_0) that satisfies $|x_0| \leq X$ and $|y_0| \leq Y$. This modular equation can be turned into an equation over the integers using a theorem of Howgrave-Graham.

Fact 2 (Howgrave-Graham) *Let $f(x, y)$ be a polynomial that is a sum of at most ω monomial. Suppose $f(x_0, y_0) = 0 \bmod p^m$ for some positive integer m, where $|x_0| \leq X$ and $|y_0| \leq Y$. If $\|f(xX, yY)\| < \frac{p^m}{\sqrt{\omega}}$, then $f(x_0, y_0) = 0$ holds over the integers.*

Using our polynomial $f_p(x, y)$, we want to construct a polynomial $f(x, y)$ that satisfies the conditions of Howgrave-Graham's theorem. Since we have to find a small Euclidean norm polynomial $f(xX, yY)$, we use lattice reduction methods. Our first approach uses a lattice of dimension 2. In that dimension, the Gauss reduction algorithm finds a shortest vector.

Let m be the integer defined in Fact 2. We choose $m = 1$. Next, we use the helper polynomial $f_0(x) = Nx$ that also has the root x_0 modulo p, since N is a multiple of p. Therefore, every integer linear combination of f_0 and f_p has the root (x_0, y_0) modulo p. We construct a lattice L_p that is spanned by the coefficient vectors of the polynomials $f_0(xX)$ and $f_p(xX, yY)$. These coefficient vectors are the row vectors of the following (2×2)-lattice basis B_p:

$$
B_p = \begin{bmatrix} NX \\ eX & -Y \end{bmatrix}
$$

We will now give a condition under which the lattice L_p has a vector v with norm smaller than $\frac{p}{\sqrt{2}}$. This vector v can then be transformed into a polynomial $f(x, y)$ satisfying Fact 2.

Lemma 3 *Let* $X = N^\delta$ *and* $Y = N^{\beta+\delta}$ *with*

$$
3\beta + 2\delta \leq 1 - \log_N(4).
$$

Then L_p *has a smallest vector* v *with* $\|v\| < \frac{p}{\sqrt{2}}$.

Proof: By Minkowski's Theorem, L_p must contain a vector v with $\|v\| \leq \sqrt{2 \det(L_p)}$. Thus, v has norm smaller than $\frac{p}{\sqrt{2}}$ if the condition

$$
\sqrt{2 \det(L_p)} < \frac{p}{\sqrt{2}}
$$

holds.

We have $\det(L_p) = NXY$. This implies $NXY < \frac{p^2}{4}$.

By the CRT Key Generation Process, we know $p > N^{1-\beta}$. On the other hand, we have $X = N^\delta$ and $Y = N^{\beta+\delta}$.

Hence, we obtain

$$
N^{1+\beta+2\delta} \leq \frac{1}{4} N^{2-2\beta} < \frac{p^2}{4}.
$$

This implies the condition $3\beta + 2\delta \leq 1 - \log_N(4)$ and the claim follows. □

Using Lemma 3, we obtain for every fixed $\epsilon > 0$ the condition $3\beta + 2\delta \leq 1 - \epsilon$ for suitably large moduli N.

Assume we have found a vector v in L_p with norm smaller than $\frac{p}{\sqrt{2}}$ by lattice reduction. Let v be the coefficient vector of the polynomial $f(xX, yY)$. Applying Fact 2, we know that $f(x, y)$ has a root $(x_0, y_0) = (d_p, k + 1)$ over the integers. The next theorem shows that the root (x_0, y_0) can easily be determined.

Lemma 4 *Let* $v = (c_0, c_1) \cdot B_p$ *be a shortest vector in* L_p *with* $\|v\| < \frac{p}{\sqrt{2}}$. *Then* $|c_0| = k$ *and* $|c_1| = qd_p$.

Proof: We have $v = c_0(NX, 0) + c_1(eX, -Y)$. Define the polynomial $f(xX, yY)$ that has the coefficient vector v. By construction, $\|f(xX, yY)\| < \frac{p}{\sqrt{2}}$ and we can apply Fact 2.

Therefore, the polynomial

$$f(x, y) = c_0 Nx + c_1(ex - y)$$

has the root (x_0, y_0) over $\mathbb{Z}$. Plugging (x_0, y_0) into the equation yields

$$c_0 Nx_0 = -c_1(ex_0 - y_0).$$

We know that $(x_0, y_0) = (d_p, k+1)$. That leads to

$$c_0 Nd_p = -c_1(ed_p - (k+1)).$$

Using equation (2) and dividing by p gives us

$$c_0 qd_p = c_1 k.$$

Since we assumed that q does not divide k, we have $\gcd(qd_p, k) = \gcd(d_p, k)$. Now, let us look at equation (1). Every integer that divides both d_p and k must also divide 1. Hence, $\gcd(d_p, k) = 1$.

Thus, we obtain

$$c_0 = ak \quad \text{and} \quad c_1 = aqd_p$$

for some integer a. But v is a shortest vector in L_p. Therefore, we must have $|a| = 1$ and the claim follows. $\quad\square$

Summing up the results gives us the following theorem.

Theorem 5 *Given an RSA public key tuple (N, e) with $N = pq$ and secret exponent d. Let $q < N^\beta$, $d_p \leq N^\delta$ and*

$$3\beta + 2\delta \leq 1 - \log_N(4).$$

Then N can be factored in time $O(\log^2(N))$.

Proof: Construct the lattice basis B_p and find a shortest vector $v = (c_0, c_1) \cdot B_p$ using Gauss reduction. Compute $\gcd(N, c_1) = q$. The total running time for Gauss reduction and greatest common divisor computation is $O(\log^2(N))$. $\quad\square$

In the previous analysis, we made the assumption that q does not divide k. If we are in the very unlikely case that $k = qr$ for some $r \in \mathbb{Z}$, then we obtain analogous to the reasoning before the following stronger result.

Theorem 6 *Given an RSA public key tuple (N, e) with $N = pq$ and secret exponent d. Let $q < N^\beta$, $d_p \leq N^\delta$,*

$$k = qr \quad and \quad \beta + 2\delta \leq 1 - \log_N(4).$$

Then N can be factored in time $O(\log^2(N))$.

Proof: The polynomial $f_p(x, y) = ex - y$ has the root $(x_0, y_0) = (d_p, k+1)$ not just modulo p but also modulo N. Thus, we can use the modulus N in Fact 2. Analogous to Lemma 3, we conclude that L_p has a shortest vector v with norm smaller than $\frac{N}{\sqrt{2}}$ as long as the condition $\beta + 2\delta \leq 1 - \log_4(N)$ holds. Following the proof of Lemma 4, we see that $v = (c_0, c_1) \cdot B_p$ with $|c_0| = r$ and $|c_1| = d_p$. Since $\frac{1-ed_p}{r} = q(p - 1)$ by equation (1), the computation $\gcd(\frac{1-ed_p}{r}, N) = q$ reveals the factorization. □

Interestingly , choosing $\beta = \frac{1}{2}$ in Theorem 6 gives us the bound $\delta \leq \frac{1}{4} - \log_N(4)$. This is similar to Wiener's bound in the attack on low secret exponent RSA [17]. In fact, one can prove the results of Wiener and Verheul, Tilborg [16] in terms of lattice theory in the same manner. We briefly sketch how to obtain their results in a simpler fashion.

Verheul and Tilborg studied the case where they guess high order bits of p. Assume we know $\tilde{p}$ with $|p - \tilde{p}| \leq N^{\frac{1}{2}-\gamma}$ and by calculating $\tilde{q} = \frac{N}{\tilde{p}}$ we know an approximation of q with accuracy $N^{\frac{1}{2}-\gamma}$ as well. The RSA-equation $ed + k(N + 1 - p - q) - 1 = 0$ gives us a polynomial $f_{N'}(x, y) = ex - y$ with root $(x'_0, y'_0) = (d, k(p - \tilde{p} + q - \tilde{q}) + 1)$ modulo $N + 1 - \tilde{p} - \tilde{q}$. We have $|x'_0| \leq N^\delta$ and $|y'_0| \leq N^{\delta + \frac{1}{2} - \gamma}$. Working through the arithmetic, this gives us the condition $\delta \leq \frac{1}{4} + \frac{\gamma}{2} - \epsilon$, where ϵ is a small error term. Wiener's result follows as the special case where $\gamma = 0$.

4   Improving the Bound to $\beta < N^{0.382}$

Using Theorem 5, our approach with the two-dimensional lattice L_p only works provided that $\beta < \frac{1}{3}$. In this section, we use lattices of larger dimension to make our method work for less unbalanced moduli. We are able to improve the bound up to $\beta < \frac{3-\sqrt{5}}{2} \approx 0.382$.

In section 3, we used Fact 2 with the parameter choice $m = 1$. Now, we generalize the method for arbitrary m.
We define the x-shifted polynomials

$$g_{m,i,j}(x, y) = N^{\max(0, m-j)} x^i f_p^j(x, y),$$

where f_p is defined as in section 3. Note, that every integer linear combination of polynomials $g_{m,i,j}$ has the zero $(x_0, y_0) = (d_p, k+1)$ modulo p^m.

We fix a lattice dimension n. Next, we build a lattice $L_p(n)$ of dimension n using as basis vectors the coefficient vectors of $g_{m,i,j}(xX, yY)$ for $j = 0 \ldots n-1$ and $i = n - j - 1$. The parameter m is a function of n and must be optimized.

For example, take $n = 4$ and $m = 2$. The lattice $L_p(n)$ is spanned by the row vectors of the following (4×4)-matrix

$$B_p(4) = \begin{bmatrix} N^2 X^3 & & & \\ eNX^3 & -NX^2Y & & \\ e^2X^3 & -2eX^2Y & XY^2 & \\ e^3X^3 & -3e^2X^2Y & 3eXY^2 & -Y^3 \end{bmatrix}$$

Note, that the lattice L_p of section 3 is equal to $L_p(2)$.

To apply Fact 2, we need a coefficient vector v with norm smaller than $\frac{p^m}{\sqrt{n}}$. The following Lemma gives us a condition for finding such a vector.

Lemma 7 *For every fixed $\epsilon > 0$, there are parameters n and N_0 such that for every $N \geq N_0$ the following holds: Let $X = \frac{n+1}{2}N^\delta$ and $Y = \frac{n+1}{2}N^{\beta+\delta}$ with*

$$3\beta - \beta^2 + 2\delta \leq 1 - \epsilon.$$

Then using the L^3-reduction algorithm, we can find a vector v in $L_p(n)$ with norm smaller than $\frac{p^m}{\sqrt{n}}$, where m is a function of n.

Proof: An easy computation shows that

$$\det(L_p(n)) = N^{\frac{m(m+1)}{2}}(XY)^{\frac{n(n-1)}{2}} = \left(\frac{n+1}{2}\right)^{n(n-1)} N^{\frac{m(m+1)}{2}+(2\delta+\beta)\frac{n(n-1)}{2}}$$

for $m < n$. By Fact 1, the L^3-algorithm will find a vector v in $L_p(n)$ with

$$\|v\| \leq 2^{\frac{n-1}{4}}\det(L_p(n))^{\frac{1}{n}}.$$

Using $p > N^{1-\beta}$, we must have

$$2^{\frac{n-1}{4}}\det(L_p(n))^{\frac{1}{n}} \leq \frac{N^{(1-\beta)m}}{\sqrt{n}}.$$

We plug in the value for $\det(L_p(n))$ and obtain the inequality

$$N^{\frac{m(m+1)}{2}+(2\delta+\beta)\frac{n(n-1)}{2}} \leq cN^{(1-\beta)mn},$$

where the factor $c = \left((2^{-\frac{3}{4}}(n+1))^{n-1}\sqrt{n}\right)^{-n}$ does not depend on N. Thus, c contributes to the error term ϵ and will be neglected in the following.

We obtain the condition

$$\frac{m(m+1)}{2} + (2\delta+\beta)\frac{n(n-1)}{2} - (1-\beta)mn \leq 0.$$

Using straightforward arithmetic to minimize the left hand side, one obtains that $m = (1-\beta)n$ is asymptotically optimal for $n \to \infty$. Again doing some calculations, we finally end up with the desired condition $3\beta - \beta^2 + 2\delta \leq 1$. $\quad\boxdot$

Now, we can use the above Lemma 7 in combination with Fact 2 to construct a bivariate polynomial $f(x,y)$ of degree n with at most n monomials and root (x_0, y_0). The problem is how to extract the root (x_0, y_0).

Analogous to Lemma 4, one can show for a vector $v = (c_1, c_2, \ldots, c_n) \cdot B_p(n)$ with norm smaller than $\frac{p^m}{\sqrt{n}}$ that k divides c_1 and d_p divides c_n. But we may not be able to find these factors k and d_p easily.

Therefore, we use another method to obtain the root. This is described in the following Lemma.

Lemma 8 *Let $X = \frac{n+1}{2}N^\delta$ and $Y = \frac{n+1}{2}N^{\beta+\delta}$. Let $f_p(x,y) = ex - y$ be a polynomial with root (x_0, y_0) modulo p that satisfies $|x_0| \leq N^\delta$, $|y_0| \leq N^{\beta+\delta}$. Let v be a vector in $L_p(n)$ with norm smaller than $\frac{p^m}{\sqrt{n}}$, where v is the coefficient vector of a polynomial $f(xX, yY)$. Then, the polynomial $p(x,y) = y_0 x - x_0 y \in \mathbb{Z}[x,y]$ must divide $f(x,y)$. We can find p by factoring f over $\mathbb{Z}[x,y]$.*

Proof: The point (x_0, y_0) is a root of f_p. For every integer a, the point (ax_0, ay_0) is also a root of f_p. Every root (ax_0, ay_0) with $|a| \leq \frac{n+1}{2}$ satisfies the conditions $|ax_0| \leq X$ and $|ay_0| \leq Y$ of Fact 2. These are at least $n+1$ roots. According to Fact 2, f must contain these roots over $\mathbb{Z}$.

But these roots lie on the line $y = \frac{y_0}{x_0}x$ through the origin. Hence, they are also roots of the polynomial $p(x,y) = y_0 x - x_0 y \in \mathbb{Z}[x,y]$. Note, that p is an irreducible polynomial of degree 1 and f is a polynomial of degree n. Using the Theorem of Bézout (see [12], page 20), either p and f share at most n points or p must divide f. But we know $n+1$ common points of p and f. Thus, the polynomial p must divide f. Since p is irreducible, we can find an integer multiple $p' = (by_0)x - (bx_0)y$ of p by factoring f over $\mathbb{Z}[x,y]$. Note that $\gcd(x_0, y_0) = 1$ since by equation (2) we know that $\gcd(d_p, k+1)$ must divide kp, but $\gcd(d_p, kp) = \gcd(d_p, k) = 1$. Hence, we obtain p by computing $p = \frac{p'}{\gcd(by_0, bx_0)}$. $\qquad\qquad\square$

Summarizing the results in this section, we obtain the following theorem.

Theorem 9 *Given an RSA public key tuple (N, e) with $N = pq$ and secret exponent d. Let $q < N^\beta$, $\delta \leq N^\delta$ and*

$$3\beta - \beta^2 + 2\delta \leq 1 - \epsilon,$$

where $\epsilon > 0$ is arbitrary small for N suitably large. Then in deterministic time polynomial in $\log(N)$, we can find the factorization of N.

Proof: Construct the lattice basis $B_p(n)$ according to Lemma 7 and find a short vector v with norm smaller than $\|v\| < \frac{p^m}{\sqrt{n}}$ using the L^3-reduction algorithm. Find the polynomial $p(x,y) = y_0 x - x_0 y$ using Lemma 8 which gives us $(x_0, y_0) = (d_p, k+1)$.

It is known that the factorization of the polynomial $f(x,y) \in \mathbb{Z}[x,y]$ in Lemma 8 can be done in deterministic time polynomial in $\log(N)$. Note that the coefficients of $f(x,y)$ must be of bit-size polynomial in $\log(p)$ since the coefficient vector of $f(xX, yY)$ has norm smaller than $\frac{p^m}{\sqrt{n}}$.

We may assume that we are in the case that k does not divide q in equation (2). Otherwise Theorem 6 proves the claim. Hence $f(x_0, y_0) = -kp$ and $\gcd(f(x_0, y_0), N) = p$ yields the factorization of N. $\qquad\qquad\square$

In practice, the factorization of polynomials over $\mathbb{Z}[x,y]$ is very fast. Thus, our method is practical even for large n.

5   An Approach Modulo e

Throughout this section, we assume that e is of the same order of magnitude as N. The results in this section as well as the results in section 3 and 4 can be easily generalized to arbitrary exponents e.

Analogous to the works [3,17] dealing with small secret exponent RSA, the smaller the exponent e is the better our methods work. On the other hand, one can completely counteract the attacks by adding to e a suitably large multiple of $\phi(N)$. We will give a detailed analysis of this in the full version of the paper.

Let us look again at equation (1) and rewrite it as

$$(k+1)(p-1) - p = -ed_p.$$

Multiplying with q yields

$$(k+1)(N-q) - N = -ed_pq$$

This gives as the polynomial

$$f_e(y,z) = y(N-z) - N$$

with a root $(y_0, z_0) = (k+1, q)$ modulo e.

Let us define the upper bounds $Y = N^{\beta+\delta}$ and $Z = N^\beta$. Note, that $|y_0| \leq Y$ and $|z_0| \leq Z$. Analogous to section 3, we can define a three-dimensional lattice L_e that is spanned by the row vectors of the (3×3)-matrix

$$B_e = \begin{bmatrix} e & & \\ & eY & \\ -N & NY & -YZ \end{bmatrix}.$$

Using a similar argumentation as in section 3, one can find a vector $v \in L_e$ with norm smaller than the bound $\frac{e}{\sqrt{3}}$ of Fact 2 provided that $3\beta + 2\delta \leq 1 - \epsilon$. Hence as before, this approach does not work if $\beta \geq \frac{1}{3}$ or $\delta \geq \frac{1}{2}$. In section 4, we used x-shifted polynomials to improve the bound for β. Now, z-shifted polynomials will help us to improve the bound for δ up to $\delta < 1$.

Fix the parameter m. Let us define the y-shifted polynomials

$$g_{i,j}(y,z) = e^{m-i}y^j f_e^i(y,z)$$

and the z-shifted polynomials

$$h_{i,j}(y,z) = e^{m-i}z^j f_e^i(y,z).$$

All these polynomials have the common root (y_0, z_0) modulo e^m. Thus, every integer linear combination of these polynomials also has the root (y_0, z_0).

We build a lattice $L_e(m)$ that is defined by the span of the coefficient vectors of the y-shifted polynomials $g_{i,j}(yY, zZ)$ and $h_{i,j}(yY, zZ)$ for certain parameters i, j. We take the coefficient vectors of $g_{i,j}$ for all non-negative i, j with $i + j \leq m$

and the coefficient vectors $h_{i,j}$ for $i = 0 \ldots m$ and $j = 1 \ldots t$ for some t. The parameter t has to be optimized as a function of m.

For example, choose $m = 2$ and $t = 1$. We take the coefficient vectors of $g_{0,0}$, $g_{0,1}$, $g_{1,0}$, $g_{0,2}$, $g_{1,1}$, $g_{2,0}$ and the coefficient vectors of $h_{0,1}$, $h_{1,1}$, $h_{2,1}$ to build the lattice basis $B_e(2)$:

$$
\left[
\begin{array}{ccccccccc}
e^2 & & & & & & & & \\
 & e^2Y & & & & & & & \\
-eN & eNY & -eYZ & & & & & & \\
 & & & e^2Y^2 & & & & & \\
 & -eNY & & eN^2Y^2 & -eY^2Z & & & & \\
N^2 & -2N^2Y & 2NYZ & N^2Y^2 & -2NY^2Z & Y^2Z^2 & & & \\
\hline
 & & & & & & e^2Z & & \\
 & eNYZ & & & & & -eNZ & -eYZ^2 & \\
 & -2N^2YZ & & N^2Y^2Z & -2NY^2Z^2 & N^2Z & 2NYZ^2 & Y^2Z^3 &
\end{array}
\right]
$$

The row vectors of $B_e(2)$ span the lattice $L_e(2)$.

In order to apply Fact 2, we need a vector in $L_e(m)$ with norm smaller than $\frac{e^m}{\sqrt{\dim L_e(m)}}$. The following lemma gives us a condition under which we can find such a vector.

Lemma 10 *For every constant $\epsilon > 0$ there exist m, N_0 such that for every $N \geq N_0$ the following holds: Let $Y = N^{\beta+\delta}$, $Z = N^\beta$ with*

$$
\frac{5}{3}\beta + \frac{2}{3}\sqrt{3\beta - 5\beta^2} + \delta \leq 1 - \epsilon,
$$

where ϵ is arbitrary small for N suitably large. Then we can find a vector v in $L_e(m)$ with norm smaller than $\frac{e^m}{\sqrt{\dim L_e(m)}}$ using the L^3-algorithm.

Proof: A straightforward computation shows that

$$
\det L_e(m) = (eY)^{\frac{1}{6}(2m^3+(6+3t)m^2+(4+3t)m)} Z^{\frac{1}{6}(m^3+(3+6t)m^2+(2+9t+3t^2)m+3t+3t^2)}.
$$

Let $t = \tau m$ and $e = N^{1-o(1)}$. Using $Y = N^{\beta+\delta}$ and $Z = N^\beta$, we obtain

$$
\det L_e(m) = N^{\frac{1}{6}m^3((1+\beta+\delta)(2+3\tau)+\beta(1+6\tau+3\tau^2)+o(1))}.
$$

Analogous to the reasoning in Lemma 7, we obtain the condition

$$
\det L_e(m) < cN^{(1-o(1))m \dim L_e(m)},
$$

where c does not depend on N and contributes to the error term ϵ. An easy calculation shows that $\dim(L) = \frac{(m+1)(m+2)}{2} + t(m+1)$. We plug in the value for $\det L_e(m)$ and $\dim L_e(m)$. Neglecting all low order terms yields the condition

$$
3\beta(\tau^2 + 3\tau + 1) + \delta(3\tau + 2) - 3\tau - 1 < 0
$$

for $m \to \infty$. Using elementary calculus to minimize the left hand side, we obtain an optimal choice for the value $\tau = \frac{1-3\beta-\delta}{2\beta}$. Plugging this value in, we finally end up with the condition $\frac{5}{3}\beta + \frac{2}{3}\sqrt{3\beta - 5\beta^2} + \delta \leq 1$.   □

Using Lemma 10, we can again apply Fact 2 and obtain a polynomial $f(y, z)$ with root (y_0, z_0) over $\mathbb{Z}$. But in contrast to the previous sections, we are not able to give a rigorous method to extract this root. Instead, we follow a heuristic approach due to Coppersmith [4]. Using the bounds of Fact 1 and a slightly different error term ϵ in Lemma 10, the L^3-algorithm must find a second vector that is short enough. This gives us another polynomial $g(y, z)$ with the same root (y_0, z_0) over $\mathbb{Z}$.

We take the resultant of f and g with respect to y. The resultant is a polynomial in z that can be solved by standard root finding algorithms. This gives as the unknown $z_0 = q$ and with it the factorization of N. Unfortunately, we cannot prove that the resultant is not the zero polynomial. It may happen that f and g share a non-trivial greatest common divisor. In this case, the resultant vanishes.

We carried out several experiments. If both y-shifted and z-shifted polynomials were used, we did not find any example where the resultant vanished. Thus although we cannot state the result as a theorem due to the gap in theory, the method works very well in practice. In fact, there are many results in cryptanalysis that rely on this heuristic, this includes among others [1, 3, 5, 8].

One can improve the shape of the curve for the approach modulo e slightly by using only a certain subset of the z-shifted polynomials. This approach leads to non-triangular lattice bases. We will analyze this in the full version of the paper.

We do not know if our lattice based approach yields the optimal bound. But there is a heuristic argument that gives us an upper bound for our method when using the polynomial $f_e(y, z)$.

Assume that the function $h(y, z) = y(N - z) \bmod e$ takes on random values in $\mathbb{Z}_e$ for $|y| \leq Y$ and $|z| \leq Z$. Every tuple (y, z) with $h(y, z) = N \bmod e$ is a root of f_e. The expected number of those tuples is $\Omega(\frac{YZ}{e}) = \Omega(N^{2\beta+\delta-1})$. As soon as $2\beta + \delta - 1$ is larger than some positive fixed constant, the number of small roots satisfying f_e is exponentially in $\log(N)$. All of these roots fulfill the criterion in Fact 2. But we require that $f(y, z)$ has a unique root over the integers in order to extract this root by resultant computation.

Thus heuristically, we cannot expect to obtain a bound better than $2\beta + \delta \leq 1$ using the polynomial f_e.

It is an open problem if one can really reach this bound.

6   Comparison of the Methods

We compare the methods introduced in section 4 and section 5. In the figure below, we plotted the maximal δ as a function of β for which our two approaches succeed. The method modulo p is represented by the dotted line $\delta = \frac{1}{2} - \frac{3}{2}\beta + \frac{1}{2}\beta^2$

resulting from Theorem 9. The approach modulo e gives as the curve $\delta = 1 - \frac{5}{3}\beta - \frac{2}{3}\sqrt{3\beta - 5\beta^2}$ by Lemma 10. The points below the curves are the feasible region of parameter choices for our attacks. We see that our method modulo e yields better results for small β. The breaking point is approximately $\beta = 0.23$.

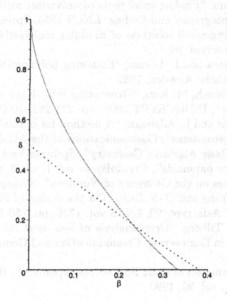

Fig. 1. Comparison of the two methods

One might we tempted to combine the two approaches and use the polynomials $ez \cdot f_p(x, y)$ and $N \cdot f_e(y, z)$ in a single lattice basis (i.e. working modulo eN). However, such a lattice will always contain an extremely short coefficient vector corresponding to the polynomial $f(x, y, z) = exz + y(N - z) - z$ over $\mathbb{Z}$. But this polynomial can be obtained by multiplying equation (1) with q and does not help us any further. It is an open problem if there is a successful way how to combine the methods.

Acknowledgement

I want to thank Johannes Blömer for many helpful discussions.

References

1. D. Bleichenbacher, "On the Security of the KMOV public key cryptosystem", Proc. of Crypto '97
2. D. Boneh, "Twenty years of attacks on the RSA cryptosystem", Notices of the AMS, 1999
3. D. Boneh, G. Durfee, "Cryptanalysis of RSA with private key d less than $N^{0.292}$", IEEE Trans. on Information Theory, vol. 46(4), 2000

4. D. Coppersmith, "Small Solutions to Polynomial Equations, and Low Exponent RSA Vulnerabilities", Journal of Cryptology 10(4), 1997
5. G. Durfee, P. Nguyen, "Cryptanalysis of the RSA Schemes with Short Secret Exponent from Asiacrypt '99", Proc. of Asiacrypt '2000
6. M. Gruber, C.G. Lekkerkerker, "Geometry of Numbers", North-Holland, 1987
7. N. Howgrave-Graham, "Finding small roots of univariate modular equations revisited", Proc. of Cryptography and Coding, LNCS 1355, Springer-Verlag, 1997
8. C. Jutla, "On finding small solutions of modular multivariate polynomial equations", Proc. of Eurocrypt '98
9. A. Lenstra, H. Lenstra and L. Lovasz, "Factoring polynomials with rational coefficients", Mathematische Annalen, 1982
10. N. Modadugu, D. Boneh, M. Kim, "Generating RSA Keys on a Handheld Using an Untrusted Server", INDOCRYPT 2000, pp. 271-282, 2000
11. R. Rivest, A. Shamir and L. Adleman, "A method for obtaining digital signatures and public key cryptosystems", Communications of the ACM, volume 21, 1978
12. I.R. Shafarevich, "Basic Algebraic Geometry", Springer-Verlag, 1994
13. A. Shamir, "RSA for paranoids", CryptoBytes vol. 1, no. 3, pp. 1-4, 1995
14. C.L. Siegel, "Lectures on the Geometry of Numbers", Springer Verlag, 1989
15. H.-M. Sun, W.-C. Yang and C.-S. Laih, "On the design of RSA with short secret exponent", Proc. of Asiacrypt '99, LNCS vol. 1716, pp. 150-164, 1999
16. E. Verheul, H. van Tilborg, "Cryptanalysis of less short RSA secret exponents", Applicable Algebra in Engineering, Communication and Computing, Springer Verlag, vol. 8, 1997
17. M. Wiener, "Cryptanalysis of short RSA secret exponents", IEEE Transactions on Information Theory, vol. 36, 1990

Hyper-encryption against Space-Bounded Adversaries from On-Line Strong Extractors

Chi-Jen Lu

Institute of Information Science, Academia Sinica, Taipei, Taiwan
cjlu@iis.sinica.edu.tw

Abstract. We study the problem of information-theoretically secure encryption in the bounded-storage model introduced by Maurer [10]. The sole assumption of this model is a limited storage bound on an eavesdropper Eve, who is even allowed to be computationally unbounded. Suppose a sender Alice and a receiver Bob agreed on a short private key beforehand, and there is a long public random string accessible by all parties, say broadcast from a satellite or sent by Alice. Eve can only store some partial information of this long random string due to her limited storage. Alice and Bob read the public random string using the shared private key, and produce a one-time pad for encryption or decryption. In this setting, Aumann, Ding, and Rabin [2] proposed protocols with a nice property called everlasting security, which says that the security holds even if Eve later manages to obtain that private key. Ding and Rabin [5] gave a better analysis showing that the same private key can be securely reused for an exponential number of times, against some adaptive attacks.
We study this problem from the approach of constructing randomness extractors ([13, 11, 16, 15] and more), which seems to provide a more intuitive understanding together with some powerful tools. A strong extractor is a function which purifies randomness from a slightly random source using a short random seed as a catalyst, so that its output and its seed together look almost random. We show that any strong extractor immediately yields an encryption scheme with the nice security properties of [2, 5]. To have an efficient encryption scheme, we need strong extractors which can be evaluated in an on-line and efficient way. We give one such construction. This yields an encryption scheme, which has the same nice security properties as before but now can encrypt longer messages using a shorter private key. In addition, our scheme works even when the long public random string is not perfectly random, as long as it contains enough amount of randomness.

1 Introduction

Almost all cryptographic protocols in use today are based on some intractability assumptions. That is, adversaries are assumed to be computationally bounded, and some problems are assumed to be computationally hard. However, no such complexity lower bound has been proved, and in fact it seems to remain far beyond the reach of current techniques in complexity theory. So it is possible that future advances in cryptanalysis or computer technology may jeopardize the security of today's cryptographic systems. One extreme then is to look

M. Yung (Ed.): CRYPTO 2002, LNCS 2442, pp. 257–271, 2002.
© Springer-Verlag Berlin Heidelberg 2002

for protocols with provable information-theoretical security, which is against an computationally-unbounded adversary. However, a well-known pessimistic result of Shannon says that no interesting protocol can be expected. Taking a step back, it would still be nice if one could prove information-theoretical security basing only on some minimal and reasonable assumption. Maurer proposed one such model, called the bounded-storage model [10], where the only assumption is that an adversary has a bounded amount of storage. To exploit such weakness of an adversary, a very long public random string, accessible by all parties, is usually employed. With respect to this model, several interesting cryptographic protocols have been proposed, with provable information-theoretical security.

One important task in cryptography is secure transmission against eavesdropping, where a sender Alice wants to send a message to a receiver Bob in a way to keep an eavesdropper Eve from learning the content. Any public-key encryption scheme unavoidably must be based on some computational lower bound, as an adversary with unbounded computation power certainly can invert the publicly-known encryption procedure. To achieve information-theoretical security, we have to use private-key encryption. In this paper, we study private-key encryption using one-time pads in the bounded-storage model. Assume Alice and Bob share a private key beforehand, say sent from Alice to Bob via today's public-key encryption. Then a long public random string X of length n is generated, say broadcast from a satellite or sent by Alice, which is accessible by all parties. Eve only has a storage of νn bits, for some constant $\nu < 1$, so she can only store some partial information about X. To be reasonable, the same storage bound is also imposed on Alice and Bob. Alice and Bob read the string X on the fly, and compute a one-time pad Z. Then Alice encrypts her message as $C = M \oplus Z$ and sends C to Bob. When X is sent, Eve computes and stores some νn bits of information about X, hoping later to recover M after eavesdropping the cipher-text C. In this setting, Aumann, Ding, and Rabin [2] gave protocols, improving those of Maurer [10] and Cachin and Maurer [4], which enjoy a nice provable property called *everlasting security*. This is an information-theoretical security that will last even if Eve later after the transmission manages to obtain that private key, say by breaking that public-key encryption. Such a feature seems quite attractive, as the security is guaranteed by the limitation of current storage technology, and will not be affected by future advances of any kind. It is possible as some crucial information has been lost forever. Shortly after, Ding and Rabin [5] gave a better analysis showing that the same private key in [2] can be securely reused for an exponential number of times in a way that each encryption remains secure even after revealing all previous plain-texts.

Observe that as Eve can only store νn bits of information Q from the n-bit random string X, very likely a substantial amount of randomness still remains in X relative to Q. The remaining randomness may be crude and not directly applicable, so one would like to have it purified. This is exactly the issue addressed in the research on constructing the so-called *extractors*, first explicitly defined by Nisan and Zuckerman [13]. Extractors turn out to have many important applications (see [11] for a survey) and have become a subject of intense study, with a break-through by Trevisan [16]. An extractor is a function that given any source with enough randomness uses a short perfect random seed to extract many bits which are close to random. We can use the output from an extractor as a one-time pad for encryption as it looks random to Eve. A strong extractor is an extractor with a stronger guarantee that its output and its seed together are close to random, which gives the following nice properties. On one hand, with high probability, the output of the extractor still looks random even given the value of the seed, which guarantees the property of everlasting security of [2] discussed above. On the other hand, with high probability, the seed still looks random even given the output value of the extractor, which implies that the same private key can be reused again for the next encryption even if Eve knows the pads of previous encryptions. Formally, we show that any strong extractor immediately yields an encryption scheme with everlasting security, against an exponential number of adaptive attacks, the same property enjoyed by protocols of Aumann, Ding, and Rabin [2, 5].

However, for practical consideration, not every strong extractor is suitable for an encryption scheme in the setting we consider here. As the public random string is broadcast at a very high rate and each party does not have enough memory to store the whole string, we need a strong extractor that can be evaluated in an on-line and very efficient way. Existing extractor constructions fail to meet this criterion. One way of constructing extractors, introduced by Trevisan [16], is to encode the input string using some list-decodable code and then project that codeword onto some dimensions determined by the random seed as output. A list-decodable code, roughly speaking, is a code guaranteeing some upper bound on the number of codewords in any Hamming ball of certain radius. The work of Aumann, Ding, and Rabin [2, 5] can also be understood within this framework. Suppose the security parameter is k, so that Eve can only distinguish different messages with an advantage 2^{-k}. One of the main technical work in [2, 5] can be seen as constructing a list-decodable code mapping from an n-bit string to an $n^{O(k)}$-bit codeword, where each bit of a codeword is the parity of $O(k)$ input bits. For an extractor with one output bit, they encode the input, pick a random dimension of that codeword, and output the bit there, needing a seed of length $O(k \log n)$ for sampling. To output m bits, they independently choose m random dimensions, and project that codeword onto those m dimensions, needing a seed of length $O(mk \log n)$. So in their encryption scheme, a private key of length $O(mk \log n)$ is required to encode a message of length m. That is, in order to achieve the everlasting security, they need a private key much longer than the message, which may limit the applicability of their protocol. Our improvement comes from two directions. First, we construct a list-decodable code mapping an n-bit string onto a codeword of length $n2^{O(k)}$, where each bit of a codeword is again the parity of $O(k)$ input bits. To have an m-bit output, if we simply project the codeword onto m random dimensions, needing a seed of length $O(m(k+\log n))$, we already have improvement over that of [2, 5]. For large m, we follow the approach of [16, 15] by picking some pseudo-random collection of m dimensions, instead of m random dimensions, for projection. As a result, for any $m \leq n^\gamma$ with $\gamma \in (0,1)$, we only need a seed of length $O((\log n + k)^2 / \log n)$. Note that n is typically much larger than k. If we assume or choose $n = 2^{\Omega(k)}$, we get an encryption scheme using a private key of length $O(\log n)$, which is a dramatic improvement over [2, 5].

To encrypt an m-bit message, Alice and Bob need to prepare m families of $O(k)$ indices and remember them. We use random walk on expander graphs and some set system to determine those m families, which seems to cost some extra computation compared to [2, 5]. However, this small extra effort is only required during the preprocessing phase, when computation time may not be a concern, so hopefully would not be an issue from a practical point of view. During the broadcast of the long public random string X, what we do is exactly the same as that of [2, 5]. That is, Alice and Bob compute the m-bit one-time pad, where each of the m bits is a parity of $O(k)$ bits from X, which can be done in an extremely fast way.

Our protocol enjoys the following key expansion property, which is addressed by a new result of Dziembowski and Maurer [6]. In [2], to generate a one-time pad of length m, Alice and Bob need to first agree on a private key of length $O(mk \log n)$, much longer than the pad they want to generate. In [5], a key doubling technique is used to generate a one-time pad of length m using a private key of length $O(k \log n)$, but it requires a public random string of length $\Theta(nk \log m \log n)$, much longer than Eve's storage bound νn. So an open question was whether one could generate a pad longer than the private key while using a public random string only slightly longer than Eve's storage bound. The main contribution of Dziembowski and Maurer [6] is to settle this question, and they need a private key of length $O(k \log n)$. Independently[1], we also settles this question but via a very different method, needing a private key of length $O((k + \log n)^2 / \log n)$, which is smaller than theirs for $k = O(\log^2 n)$. Both [6] and our result allow the ratio between the length of public random string and Eve's storage bound to be arbitrarily close to one. On the other hand, we do not require the long public random string X to be perfectly random, while this is not clear in [6]. In reality, perfect random sources may not be available and one may have to rely on sources of lower quality. Our scheme works as long as X contains enough amount of randomness, which only needs to be slightly larger than Eve's storage bound.

In addition to obtaining a better construction, we try to place this line of research in an appropriate framework, where ideas could be understood in a more intuitive and perhaps deeper way and various powerful tools have been developed. Most of the techniques we use are standard in the research area on pseudo-randomness. The idea of using extractors in such a setting actually appeared before in the work of Cachin and Maurer [4]. Since then, the theory of pseudo-randomness has made some advancement, which enables us to have a better understanding and derive stronger results.

Notation, definitions, and some simple facts are given in Section 2. In Section 3, we prove that any strong extractor gives an encryption scheme with ev-

---

[1] We were not aware of their result when we obtained ours.

erlasting security. In Section 4, we construct an efficient on-line strong extractor which yields an efficient encryption scheme.

2   Preliminaries

For a positive integer n, let $[n]$ denote the set $\{1, \ldots, n\}$. For an n-dimensional vector m, let $m(i)$, for $i \in [n]$, denote the component in the i-th dimension of m, and let $m(I)$, for $I \subseteq [n]$, denote the vector consisting of those $m(i)$'s for $i \in I$. For our convenience, we use $\{-1, 1\}$, instead of the usual $\{0, 1\}$, for the binary values unless noted otherwise. For a positive integer n, let U^n denote the uniform distribution over $\{-1, 1\}^n$, and we omit the superscript n when no confusion is caused. All the logarithms in this paper have base 2.

The distance between two vectors x, y is defined as $d_H(x, y) \equiv |\{i : x(i) \neq y(i)\}|$, which is also known as their *Hamming* distance.

Definition 1. *A mapping $C : \{-1, 1\}^n \to \{-1, 1\}^{\bar{n}}$ defines a code, with $\{C(x) : x \in \{-1, 1\}^n\}$ as the set of codewords. It is called a list-decodable code with parameter $(\frac{1}{2} - \varepsilon, L)$, or a $(\frac{1}{2} - \varepsilon, L)$-list code, if for any $z \in \{-1, 1\}^{\bar{n}}$, there are at most L codewords within distance $(\frac{1}{2} - \varepsilon)\bar{n}$ from z.*

We also need a way to measure how close two distributions are.

Definition 2. *The distance between two distributions A, B is defined as $\|A - B\| \equiv \frac{1}{2} \sum_x |\Pr[A = x] - \Pr[B = x]|$.*

This is usually called the variational distance, and it is easy to verify that $\|A - B\| \leq 1$. We say that a distribution is ε-random if its distance to the uniform one is at most ε. Here is a simple but useful lemma, which is proved in Appendix A.

Lemma 1. *Suppose the distribution A is independent of distributions B and B'. Then for any function g, $\|\langle g(A, B), B \rangle - \langle g(A, B'), B' \rangle\| = \|B - B'\|$.*

We need a measure to quantify the randomness contained in a slightly random source. Shannon's entropy function captures the average randomness of a source, while we need some worst-case measure of randomness instead.

Definition 3. *For a distribution X, its min-entropy is defined as $H_\infty(X) \equiv \min_x \log \frac{1}{\Pr[X=x]}$.*

So $H_\infty(X) \geq r$ iff $\Pr[X = x] \leq 2^{-r}$ for any x. For a distribution with enough randomness, guaranteed by its min-entropy, we look for a procedure that can purify the randomness by using a short random seed as a catalyst.

Definition 4. *A function $\mathrm{EXT} : \{-1, 1\}^n \times \{-1, 1\}^s \to \{-1, 1\}^m$ is called a strong (r, ε)-extractor if for any distribution X over $\{-1, 1\}^n$ with $H_\infty(X) \geq r$ and for $Y = U^s$, the distribution of $\langle \mathrm{EXT}(X, Y), Y \rangle$ is ε-random.*

The usual definition of extractors only requires the distribution of $\text{EXT}(X,Y)$ being ε-random. The stronger requirement of a strong extractor EXT gives the following nice property. With high probability, the distribution of $\text{EXT}(X,Y)$ still looks random even given the value of Y, and also with high probability, the distribution of Y still looks random even given the value of $\text{EXT}(X,Y)$. This is guaranteed by the following lemma, which is proved in Appendix B.

Lemma 2. *Suppose the distribution $\langle A, B \rangle$ is ε-random. Then the probability over $b \in B$ that the distribution $\langle A \mid B = b \rangle$ is not $\sqrt{\varepsilon}$-random is at most $2\sqrt{\varepsilon}$.*

We consider private-key encryption schemes using one-time pads. In such a scheme, a sender Alice and a receiver Bob first agreed on a private key Y and then later generate a one-time pad Z_Y so that the message M is encrypted as $C = M \oplus Z_Y$, where $\oplus$ denote the bit-wise exclusive or operation. Note that an eavesdropper Eve with C knows M iff she knows Z_Y. So for the security of such a scheme, it suffices to show that Z_Y looks random to Eve. An even stronger notion of security is the following, introduced by Aumann, Ding, and Rabin [2].

Definition 5. *A private-key encryption scheme is said to have everlasting security of degree k, if with probability at least $1 - 2^{\Omega(k)}$, the distribution of Z_Y remains 2^{-k}-random conditioned on Eve's information and Y.*

This guarantees that even if Eve is given that private key Y afterwards, she is still unlikely to know anything about the message M.

3   Everlasting Security from Strong Extractors

In this section, we show that using a strong extractor for encryption guarantees the everlasting security of [2, 5]. Recall the following scenario. Before the transmission of an m-bit message M, a sender Alice and a receiver Bob agree on a random private key Y. Then they read the n-bit public random string X, compute the one-time pad Z, and the encrypted message $C = M \oplus Z$ is sent from Alice to Bob. An adversary Eve uses a function f to store νn bits of information $Q = f(X)$ from the public string X, and then eavesdrops the encrypted message C. If later Eve is given that private key Y, can she obtain any information about M?

For the security parameter k, which is usually much smaller than n, choose $\varepsilon = 2^{-2k}$. Pick $\mu < 1 - \nu$, say $\mu = 0.99 - \nu$, so that $2^{-(1-\mu-\nu)n} \leq \frac{\varepsilon}{2}$, and let $r = \mu n$. We will use a strong $(r, \frac{\varepsilon}{2})$-extractor $\text{EXT} : \{-1, 1\}^n \times \{-1, 1\}^s \to \{-1, 1\}^m$. The private key Y is selected randomly according to the distribution U^s and the one-time pad is generated as $Z = \text{EXT}(X, Y)$. For $M = C \oplus Z$, Eve knows M iff she knows Z, so for the security, it suffices to show that Z looks random to Eve even given Y. The idea is that X conditioned on Q is likely to contain enough randomness, so for a strong extractor EXT, very likely the distribution of $Z = \text{EXT}(X, Y)$ would still look random even given the value of Y. Here is the key lemma.

Lemma 3. $\|\langle f(X), Y, \text{EXT}(X,Y) \rangle - \langle f(X), U^s, U^m \rangle\| \leq \varepsilon.$

Proof. First, note that $\|\langle f(X), Y, \mathrm{Ext}(X,Y)\rangle - \langle f(X), U^m, U^s\rangle\|$ equals

$$\sum_q \Pr[f(X) = q] \cdot \|\langle Y, \mathrm{Ext}(X,Y) \mid f(X) = q\rangle - \langle U^m, U^s \mid f(X) = q\rangle\|.$$

We call a value $q \in \{-1,1\}^{\nu n}$ *bad* if $\Pr[f(X) = q] \leq 2^{-(1-\mu)n}$, and let B denote this set of bad q's. For $q \notin B$, $\Pr[X = x \mid f(X) = q] \leq 2^{-n}/2^{-(1-\mu)n} = 2^{-\mu n}$ for any x, so $H_\infty(X \mid f(X) = q) \geq \mu n$. Then for $q \notin B$,

$$\|\langle Y, \mathrm{Ext}(X,Y) \mid f(X) = q\rangle - \langle U^s, U^m \mid f(X) = q\rangle\|$$
$$= \|\langle Y, \mathrm{Ext}(\langle X \mid f(X) = q\rangle, Y)\rangle - \langle U^s, U^m\rangle\|$$
$$\leq \frac{\varepsilon}{2}.$$

For $q \in B$, we only have $\|\langle \mathrm{Ext}(X,Y), Y \mid f(X) = q\rangle - \langle U^m, U^s \mid f(X) = q\rangle\| \leq 1$, but fortunately it is unlikely to have a bad q as

$$\sum_{q \in B} \Pr[f(X) = q] \leq \sum_{q \in B} 2^{-(1-\mu)n}$$
$$\leq 2^{(\nu - 1 + \mu)n}$$
$$\leq \frac{\varepsilon}{2}.$$

Thus,

$$\|\langle f(X), Y, \mathrm{Ext}(X,Y)\rangle - \langle f(X), U^s, U^m\rangle\|$$
$$= \sum_q \Pr[f(X) = q] \cdot \|\langle Y, \mathrm{Ext}(X,Y) \mid f(X) = q\rangle - \langle U^s, U^m \mid f(X) = q\rangle\|$$
$$\leq \sum_{q \in B} \Pr[f(X) = q] \cdot 1 + \sum_{q \notin B} \Pr[f(X) = q] \cdot \frac{\varepsilon}{2}$$
$$\leq \frac{\varepsilon}{2} + \frac{\varepsilon}{2}$$
$$= \varepsilon. \qquad \square$$

Now according to Lemma 2, even if Eve saved the information $f(X)$ and later obtains the private key Y, the distribution of $\mathrm{Ext}(X,Y)$ is still 2^{-k}-random with probability at least $1 - 2^{-k+1}$. That is, it gives an encryption scheme with everlasting security of degree k.

Just like previous work, we assume above that the public string X is perfectly random. However, such a perfect random source may not be available in reality, and one may need to rely on sources of lower quality. Note that our proof above actually works for any source X of length $n' > n$ with $H_\infty(X) \geq n$. That is, our result holds as long as Eve's storage bound is at most a fraction ν of the source's min-entropy.

3.1   Reusing the Same Private Key

Next, we show that the same key can be reused an exponential number of times, even given all previous one-time pads. Now the idea is that for a strong extractor

EXT, very likely the distribution of Y would still look random even given the value of $\text{EXT}(X, Y)$, so the same Y could be used again to extract randomness in the next round.

Consider the following scenario, with K messages $M_1, M_2, \ldots, M_K$ to be transmitted. For $i \in [K]$, Alice reads the i-th block of n-bit public random string X_i, computes the i-th pad $Z_i = \text{EXT}(X_i, Y)$, and sends the encrypted message as $C_i = M_i \oplus Z_i$ to Bob. For $i \in [K]$, suppose Eve is also given $Z_{[i-1]} = (Z_1, \ldots, Z_{i-1})$ and uses a function f_i to store νn bits of information $Q_i = f_i(X_i, Q_{i-1}, Z_{[i-1]})$ from the public string X_i. Finally, Eve eavesdrops the encrypted message C_K and is given Y, can she learn anything about M_K? For the security, again it suffices to show that Z_K looks random even given $Z_{[K-1]}$, Q_K, and Y. We choose $\varepsilon = 2^{-3k}$, $r = \mu n$, and use a strong $(r, \frac{\varepsilon}{2})$-extractor EXT. Here is the key lemma.

Lemma 4. *For any $i \in \mathbb{N}$, $\|\langle Z_{[i-1]}, Q_i, Y, Z_i \rangle - \langle Z_{[i-1]}, Q_i, U^s, U^m \rangle\| \leq i\varepsilon$.*

Proof. We use induction on i. Lemma 3 handles precisely the case $i = 1$. So assume it holds for some $i \geq 1$ and let's consider the case $i + 1$. By triangle inequality, we have

$$\|\langle Z_{[i]}, Q_{i+1}, Y, Z_{i+1} \rangle - \langle Z_{[i]}, Q_{i+1}, U^s, U^m \rangle\|$$
$$\leq \|\langle Z_{[i]}, Q_{i+1}, Y, \text{EXT}(X_{i+1}, Y) \rangle - \langle Z_{[i]}, Q_{i+1}, Y', \text{EXT}(X_{i+1}, Y') \rangle\| +$$
$$\|\langle Z_{[i]}, Q_{i+1}, Y', \text{EXT}(X_{i+1}, Y') \rangle - \langle Z_{[i]}, Q_{i+1}, U^s, U^m \rangle\|,$$

where Y' is the distribution U^s which is independent of $Z_{[i]}$. The first term is about how random Y remains after i iterations, while the second term is about how good the $(i + 1)$-th extraction is using a new random key Y'. In the first term, $Q_{i+1} = f_{i+1}(X_{i+1}, Q_i, Z_{[i]})$, and X_{i+1} is independent of Q_i, $Z_{[i]}$, Y, and Y'. Then according to Lemma 1, the first term equals

$$\|\langle Z_{[i]}, Q_i, Y \rangle - \langle Z_{[i]}, Q_i, Y' \rangle\| = \|\langle Z_{[i]}, Q_i, Y \rangle - \langle Z_{[i]}, Q_i, U^s \rangle\|$$
$$= \|\langle Z_{[i-1]}, Q_i, Y, Z_i \rangle - \langle Z_{[i-1]}, Q_i, U^s, Z_i \rangle\|$$
$$\leq i\varepsilon,$$

where the inequality is from the inductive hypothesis. In the second term, X_{i+1} is independent of $Z_{[i]}$, so only Q_{i+1} affects the distribution of X_{i+1}. Then using an argument similar to the proof of Lemma 3, the second term can be bounded above by ε. Therefore, we have

$$\|\langle Z_{[i]}, Q_{i+1}, Y, Z_{i+1} \rangle - \langle Z_{[i]}, Q_{i+1}, U^s, U^m \rangle\| \leq (i + 1)\varepsilon,$$

which proves the inductive step and thus the lemma. $\square$

For $K \leq 2^k$, the distance is at most 2^{-2k}. Then according to Lemma 2, even given the previous pads $Z_1, \ldots, Z_{K-1}$ and the private key Y, the K-th pad Z_K is still 2^{-k}-random with probability at least $1 - 2^{-k+1}$. So we have the following theorem.

Theorem 1. *Any strong $(\mu n, 2^{-3k})$-extractor yields an encryption scheme with everlasting security of degree k, whose private key can be securely reused for 2^k times, against the adaptive attacks discussed above.*

4 An On-Line Strong Extractor

Although using any strong extractor for encryption does provide a good security guarantee, not any arbitrary strong extractor is suitable in the setting we consider here. Recall that the public random string X is very long so it needs to be broadcast at a very high rate. Alice and Bob do not have enough memory to store the whole string, so they would like to be able to apply the extractor in an on-line and efficient way. One way of constructing extractors, introduced by Trevisan [16], is based on list-decodable codes. In order to have an efficient on-line strong extractor, we need a list-decodable code with an efficient on-line encoding procedure. The main work of [2, 5] can be seen as constructing one such code, where each output bit of a codeword is a parity of a small number of input bits. Here we give another code with a better parameter.

4.1 An On-Line List-Decoding Code

We will use some type of graphs called expander graphs which have their second largest eigenvalue bounded away from their largest one[2].

Lemma 5. *[7, 8] There exists an explicit family of expander graphs $(G_n)_{n \in \mathbb{N}}$ with the following property. There is a constant d and a constant $\lambda < d$ such that for every $n \in \mathbb{N}$, G_n is a d-regular graph of n vertices with the second largest eigenvalue λ.*

Expander graphs enjoys some pseudo-random properties, and we will use the following one.

Lemma 6. *[1] Suppose G is a d-regular graph with the second largest eigenvalue λ, and B is a set containing at least α fraction of G's vertices. Then a t-step random walk on G misses B with probability at most $\beta_\alpha \equiv (1 - \alpha + (\frac{\lambda}{d})^2)^{t/2}$.*

In fact, this probability is not far off away from the probability $(1 - \alpha)^t$ achieved by sampling t vertices randomly and independently, but now we only need $\log n + t \log d = \log n + O(t)$ random bits for sampling, instead of $t \log n$ random bits. Observe that smaller $\frac{\lambda}{d}$ gives smaller β_α. This ratio could be reduced to $(\frac{\lambda}{d})^i$ by considering the graph G^i, but then each step of a walk on G^i would need $i \log d$ bits, instead of $\log d$ bits, to sample.

Let G denote one such d-regular expander graph on n vertices. For a vertex v and for $w \in [d]^t$, let $W_G(v, w)$ denote the sequence of t vertices visited by a t-step walk on G starting from v and then following the directions $w(1), \ldots, w(t)$. Consider the encoding $\text{ECC}_t : \{-1, 1\}^n \to \{-1, 1\}^{\bar{n}}$, for $\bar{n} = nd^t 2^t$, defined in Figure 1.

---

[2] The eigenvalues of a graph's adjacency matrix are called the eigenvalues of that graph.

- **Input:** $x \in \{-1, 1\}^n$.
- **Output index:** $(v, w, b) \in [n] \times [d]^t \times \{0, 1\}^t$.
- **Algorithm:**
  - Do a t-step walk on G to get $W_G(v, w) = (v_1, \ldots, v_t)$.
  - Output the bit indexed by (v, w, b) as

  $$\text{Ecc}_t(x)(v, w, b) \equiv \prod_{i \in [t]} x(v_i)^{b(i)}.$$

Fig. 1. The code Ecc_t

Note that the bit $\text{Ecc}_t(x)(v, w, b) = \prod_{i \in [t]} x(v_i)^{b(i)}$ is just the parity of at most t input bits, those bits $x(j)$'s with $j = v_i$ and $b(i) = 1$. Suppose one had (v, w, b) and derived those indices j beforehand. Then when x is given, the bit $\text{Ecc}_t(x)(v, r, b)$ can be computed in an on-line and extremely fast way.

To show that this is a good list-decodable code, we would like to bound the number of codewords within any Hamming ball of some radius. The Johnson bound in coding theory (e.g. Chapter 17 of [9]) provides one such bound for codes with some minimum distance guarantee between any two codewords. It cannot be applied directly to our code Ecc_t as some codewords are in fact close to each other. However, we do have some distance guarantee between some codewords, as shown by the following lemma.

Lemma 7. *For any $x, y \in \{-1, 1\}^n$ with $d_H(x, y) \geq \alpha n$, $d_H(\text{Ecc}_t(x), \text{Ecc}_t(y)) \geq (\frac{1}{2} - \beta_\alpha)\bar{n}$.*

Proof. Let $B = \{v \in [n] : x(v) \neq y(v)\}$, which has at least αn elements. Consider any walk $(v_1, \ldots, v_t) \in [n]^t$ that hits B, say at v_{i_0} with $x(v_{i_0}) \neq y(v_{i_0})$. For any assignment to b, flipping the bit $b(i_0)$ flips exactly one value of $x(v_{i_0})^{b(i_0)}$ and $y(v_{i_0})^{b(i_0)}$, and thus exactly one value of $\prod_{i \in [t]} x(v_i)^{b(i)}$ and $\prod_{i \in [t]} y(v_i)^{b(i)}$. Then for such a walk that hits B,

$$\Pr_{b \in \{0,1\}^t} \left[\prod_{i \in [t]} x(v_i)^{b(i)} = \prod_{i \in [t]} y(v_i)^{b(i)} \right] = \frac{1}{2}.$$

According to Lemma 6, a t-step random walk on G misses B with probability at most β_α. So,

$$\Pr_{v,w,b} [\text{Ecc}_t(x)(v, r, b) = \text{Ecc}_t(y)(v, r, b)]$$

$$\leq \Pr_{v,w,b} [W_G(v, w) \text{ misses } B] +$$

$$\Pr_{v,w,b} \left[\prod_{i \in [t]} x(v_i)^{b(i)} = \prod_{i \in [t]} y(v_i)^{b(i)} \mid W_G(v, w) \text{ hits } B \right]$$

$$\leq \beta_\alpha + \frac{1}{2},$$

where $W_G(v, w) = (v_1, \ldots, v_t)$ in the first inequality. Then,

$$d_H(\mathrm{ECC}_t(x), \mathrm{ECC}_t(y)) = \left(1 - \Pr_{v, w, b}[\mathrm{ECC}_t(x)(v, r, b) = \mathrm{ECC}_t(y)(v, r, b)]\right) \bar{n}$$

$$\geq \left(\frac{1}{2} - \beta_\alpha\right) \bar{n}.$$

$\square$

Now for each codeword $\mathrm{ECC}_t(x)$, the only possible codewords with distance less than $(\frac{1}{2} - \beta_\alpha)\bar{n}$ from $\mathrm{ECC}_t(x)$ are those $\mathrm{ECC}_t(x')$'s with $d_H(x, x') \leq \alpha n$, and there are at most $2^{h(\alpha)n}$ such codewords, where $h(\alpha) = \alpha \log \frac{1}{\alpha} + (1 - \alpha) \log \frac{1}{1-\alpha}$ is the binary entropy function. Choose $\alpha = 1 + (\frac{\lambda}{d})^2 - \delta^{4/t}$, so $\beta_\alpha \leq \delta^2$. Then the following lemma implies that our code ECC_t is a $(\frac{1}{2} - \delta, \frac{2^{h(\alpha)n}}{2\delta^2})$-list code. It can be seen as a generalization to the Johnson bound, and our proof generalizes the version given in Appendix A of [3].

Lemma 8. *Suppose* $\mathrm{ECC} \subseteq \{-1, 1\}^{\bar{n}}$ *is a code such that for any codeword* $c \in \mathrm{ECC}$*, there are at most* M *codewords in* ECC *within distance* $(\frac{1}{2} - \beta)\bar{n}$ *from* c*. Then for any* $z \in \{-1, 1\}^{\bar{n}}$ *and any* $\delta > \sqrt{\beta/2}$*, the number of codewords within distance* $(\frac{1}{2} - \delta)\bar{n}$ *from* z *is at most* $M/(4\delta^2 - 2\beta)$*.*

Proof. For any two vectors $u, v \in \{-1, 1\}^{\bar{n}}$, let $u \odot v$ denote their inner product, and note that

$$u \odot v \equiv \sum_{j \in [\bar{n}]} u(j)v(j) = |\{j : u(j) = v(j)\}| - |\{j : u(j) \neq v(j)\}|.$$

Suppose $c_1 = \mathrm{ECC}_t(x_1), \ldots, c_L = \mathrm{ECC}_t(x_L) \in \{-1, 1\}^{\bar{n}}$ are those codewords within distance $(1/2 - \delta)\bar{n}$ from z. For $i \in [L]$, let $u_i \in \{-1, 1\}^{\bar{n}}$ represent the discrepancy between c_i and z, with $u_i(j) = c_i(j)z(j)$ for $j \in [\bar{n}]$ [3]. Define

$$T \equiv \left(\sum_{i \in [L]} u_i\right) \odot \left(\sum_{i \in [L]} u_i\right).$$

On one hand,

$$T = \sum_{j \in [\bar{n}]} \left(\sum_{i \in [L]} u_i(j)\right)^2$$

$$\geq \frac{1}{\bar{n}} \left(\sum_{j \in [\bar{n}]} \sum_{i \in [L]} u_i(j)\right)^2,$$

---

[3] In [3], $u_i(j)$ is defined as 1 if $c_i(j) = z(j)$ and 0 otherwise. Our definition makes the proof slightly cleaner.

from Cauchy-Schwartz inequality. As $\sum_{j \in [\bar{n}]} u_i(j) = c_i \odot z \geq 2\delta\bar{n}$ for any $i \in [L]$, we have

$$T \geq \frac{1}{\bar{n}} \left(\sum_{i \in [L]} \sum_{j \in [\bar{n}]} u_i(j) \right)^2$$

$$\geq \frac{1}{\bar{n}} (2\delta\bar{n}L)^2$$

$$= 4\delta^2\bar{n}L^2.$$

On the other hand, we can write

$$T = \sum_{i \in [L]} \sum_{i' \in [L]} u_i \odot u_{i'}.$$

For $i \in [L]$, let $N(i) = \{i' \in [L] : d_H(x_i, x_{i'}) \leq \alpha n\}$. For $i' \notin N(i)$, we have $u_i \odot u_{i'} \leq 2\beta\bar{n}$. For $i' \in N(i)$, we only have $u_i \odot u_{i'} \leq \bar{n}$, but $|N(i)| \leq M$. So

$$T = \sum_{i \in [L]} \sum_{i' \in N(i)} u_i \odot u_{i'} + \sum_{i \in [L]} \sum_{i' \notin N(i)} u_i \odot u_{i'}$$

$$\leq \bar{n}ML + 2\beta\bar{n}L^2,$$

Combining the two inequalities, we have $4\delta^2\bar{n}L^2 \leq T \leq \bar{n}ML + 2\beta\bar{n}L^2$, which implies $L \leq M/(4\delta^2 - 2\beta)$.   □

4.2 Expanding the Output

Given any list-decodable code $\text{Ecc}' : \{-1,1\}^n \to \{-1,1\}^{\bar{n}}$, it is known according to [16, 15] that one immediately gets a strong extractor $\text{Ext}' : \{-1,1\}^n \times [\bar{n}] \to \{-1,1\}$ defined as

$$\text{Ext}'(x, y) = \text{Ecc}'(x)(y).$$

That is, the extractor encodes the input string x as $\text{Ecc}'(x)$ and projects it onto a random dimension y. To obtain m output bits, the approach of [2, 5] is to project $\text{Ecc}'(x)$ onto m independent and random dimensions, needing $m \log \bar{n}$ random bits for sampling. That is, they use the extractor $\text{Ext}'' : \{-1,1\}^n \times [\bar{n}]^m \to \{-1,1\}^m$ defined as

$$\text{Ext}''(x, (y_1, \ldots, y_m)) = (\text{Ecc}'(x)(y_1), \ldots, \text{Ecc}'(x)(y_m)).$$

They use a code with $\bar{n} = n^{O(k)}$, so their extractor needs a seed of length $O(mk \log n)$. Using our code Ecc_t for some $t = O(k)$, we immediately have a extractor of the same quality but needing a seed of length only $O(m(k + \log n))$.

For large m, we can get a dramatic improvement by picking some pseudo-random collection of m dimensions instead of m random ones for projection. This is the idea behind Trevisan's extractor construction [16] and the improvement by Raz, Reingold, and Vadhan [15]. The pseudo-random projection is determined by some set system defined next.

Definition 6. *[15] A family of sets $S_1, \ldots, S_m \subseteq [s]$ is called a weak (ℓ, ρ)-design if*

- $\forall i, |S_i| = \ell$, *and*
- $\forall i, \sum_{j<i} 2^{|S_i \cap S_j|} \leq \rho(m - 1)$.

Lemma 9. *[15] For every ℓ, m and $\rho > 1$, there exists a weak (ℓ, ρ)-design $S_1, \ldots, S_m \subseteq [s]$ with $s = \left\lceil \frac{\ell}{\ln \rho} \right\rceil \ell$. Such a family can be found in time $\text{poly}(m, s)$.*

For a random $y \in \{-1, 1\}^s$, such a weak $(\log \bar{n}, \rho)$-design gives a pseudo-random collection of m dimensions $y(S_1), \ldots, y(S_m)$, where each $y(S_i)$ is the integer in $[\bar{n}]$ represented by the $\log \bar{n}$ bits of y indexed by S_i.

Lemma 10. *[16, 15] Suppose $S_1, \ldots, S_m \subseteq [s]$ is a weak $(\log \bar{n}, \rho)$-design, and $\text{ECC}' : \{-1, 1\}^n \rightarrow \{-1, 1\}^{\bar{n}}$ is a $(\frac{1}{2} - \frac{\varepsilon}{2m}, L)$-list code. Then the function $\text{EXT} : \{-1, 1\}^n \times \{-1, 1\}^s \rightarrow \{-1, 1\}^m$ defined as*

$$\text{EXT}(x, y) = \left(\text{ECC}'(x)(y(S_1)), \ldots, \text{ECC}'(x)(y(S_m)) \right)$$

is a strong (r, ε)-extractor, for any $r \geq \rho(m - 1) + \log \frac{2L}{\varepsilon}$.

To have a strong $(\mu n, 2^{-3k})$-extractor for our encryption scheme, we use our $(\frac{1}{2} - \delta, L)$-list code ECC_t, with

- $\delta = \frac{1}{2m2^{3k}}$,
- $t = O(k)$ large enough and $\frac{\lambda}{d} = O(1)$ small enough so that $h(\alpha) < \mu$ for $\alpha = 1 + (\frac{\lambda}{d})^2 - \delta^{4/t}$, and
- $L = \frac{2^{h(\alpha)n}}{2\delta^2} = 2^{h(\alpha)n + 6k + 1}m^2$.

Note that $\rho(m - 1) + \log \frac{2L}{\varepsilon} = \rho(m - 1) + h(\alpha)n + 9k + 2 + 2 \log m$. This can be made smaller than μn for any $m \leq n^\gamma$ with constant $\gamma \in (0, 1)$, by choosing a proper $\rho = n^{O(1)}$. Using such a weak $(\log \bar{n}, \rho)$-design and our $(\frac{1}{2} - \delta, L)$ list-code ECC_t, we have the following theorem.

Theorem 2. *For any constant $\gamma \in (0, 1)$, there is a strong $(\mu n, 2^{-3k})$-extractor with a seed of length $O((k + \log n)^2 / \log n)$ and a output of length n^γ, where each output bit is the parity of $O(k)$ input bits.*

This, together with Theorem 1, gives the encryption scheme we claim. Recall that n is typically much larger than k. A larger n provides a higher security but only costs a negligible slow-down during encryption time. If we assume or choose $n = 2^{\Omega(k)}$, then we have an encryption scheme using a private key of length only $O(\log n)$.

References

1. N. Alon and J. H. Spencer. *The Probabilistic Method.* Wiley Interscience, New York, 1992.
2. Y. Aumann, Y. Z. Ding, and M. O. Rabin. Everlasting security in the bounded storage model. To appear in *IEEE Transactions on Information Theory*, 2002.
3. M. Bellare, O. Goldreich and M. Sudan. Free bits, PCPs and non-approximability — towards tight results. *SIAM Journal on Computing*, 27(3), pages 804–915, 1998.
4. C. Cachin and U. Maurer. Unconditional security against memory-bounded adversaries. In *Advances in Cryptology - CRYPTO'97*, Lecture Notes in Computer Science, Springer-Verlag, vol. 1294, pages 292–306, 1997.
5. Y. Z. Ding and M. O. Rabin. Hyper-encryption and everlasting security. In *Proceedings of the 19th Annual Symposium on Theoretical Aspects of Computer Science*, pages 1–26, 2002.
6. S. Dziembowski and U. Maurer. Tight security proofs for the bounded-storage model. To appear in *Proceedings of the 34th Annual ACM Symposium on Theory of Computing*, 2002.
7. O. Gabber and Z. Galil. Explicit constructions of linear-sized superconcentrators. *Journal of Computer and System Sciences*, 22(3), pages 407–420, 1981.
8. A. Lubotzky, R. Philips, and P. Sarnak. Explicit expanders and the Ramanujan conjecture. In *Proceedings of the 18th Annual ACM Symposium on Theory of Computing*, pages 240–246, 1986.
9. F. J. MacWilliams and N. J. A. Sloan. *The Theory of Error-Correcting Codes.* Noth-Holland, 1981.
10. U. Maurer. Conditionally-perfect secrecy and a provably-secure randomized cipher. *Journal of Cryptology*, 5(1), pages 53–66, 1992.
11. N. Nisan. Extracting randomness: how and why — a survey. In *Proceedings of the 11th Annual IEEE Conference on Computational Complexity*, pages 44–58, 1996.
12. N. Nisan and A. Wigderson. Hardness vs. randomness. *Journal of Computer and System Sciences*, 49(2), pages 149–167, 1994.
13. N. Nisan and D. Zuckerman. Randomness is linear in space. *Journal of Computer and System Sciences*, 52(1), pages 43–52, 1996.
14. R. Impagliazzo, R. Shaltiel, and A. Wigderson. Extractors and pseudo-random generators with optimal seed length. In *Proceedings of the 32nd Annual ACM Symposium on Theory of Computing*, pages 1–10, 2000.
15. R. Raz, O. Reingold, and S. P. Vadhan. Extracting all the randomness and reducing the error in Trevisan's extractors. In *Proceedings of the 31st Annual ACM Symposium on Theory of Computing*, pages 149–158, 1999.
16. L. Trevisan. Extractors and pseudorandom generators. *Journal of ACM*, 48(4), pages 860–879, 2001.
17. A. Yao. Theory and applications of trapdoor functions. In *Proceedings of the 23rd Annual IEEE Symposium on the Foundations of Computer Science*, pages 80–91, 1982.

A   Proof of Lemma 1

$$\|\langle g(A, B), B\rangle - \langle g(A, B'), B'\rangle\|$$

$$= \frac{1}{2}\sum_b \sum_c |\Pr[g(A, B) = c \wedge B = b] - \Pr[g(A, B') = c \wedge B' = b]|$$

$$= \frac{1}{2}\sum_b \sum_c |\Pr[g(A, b) = c]\Pr[B = b] - \Pr[g(A, b) = c]\Pr[B' = b]|$$

$$= \frac{1}{2}\sum_b \sum_c \Pr[g(A, b) = c]\,|\Pr[B = b] - \Pr[B' = b]|$$

$$= \frac{1}{2}\sum_b |\Pr[B = b] - \Pr[B' = b]|$$

$$= \|B - B'\|,$$

where the second equality holds because A is independent of both B and B'.

B   Proof of Lemma 2

The expectation of $\|\langle A \mid B = b\rangle - U\|$, with b sampled from the distribution B, is

$$\sum_b \Pr[B = b]\frac{1}{2}\sum_a \left|\Pr[A = a \mid B = b] - \frac{1}{|A|}\right|$$

$$= \frac{1}{2}\sum_{a,b} \left|\Pr[(A, B) = (a, b)] - \Pr[B = b]\frac{1}{|A|}\right|$$

$$\leq \frac{1}{2}\sum_{a,b} \left|\Pr[(A, B) = (a, b)] - \frac{1}{|A||B|}\right| + \frac{1}{2}\sum_{a,b} \left|\Pr[B = b]\frac{1}{|A|} - \frac{1}{|A||B|}\right|$$

$$= \|\langle A, B\rangle - U\| + \|B - U\|$$

$$\leq 2\varepsilon.$$

Then the lemma follows from the Markov inequality.

Optimal Black-Box Secret Sharing
over Arbitrary Abelian Groups

Ronald Cramer and Serge Fehr

BRICS*, Department of Computer Science, Aarhus University, Denmark
{cramer,fehr}@brics.dk

Abstract. A *black-box* secret sharing scheme for the threshold access
structure $T_{t,n}$ is one which works over any finite Abelian group G. Briefly,
such a scheme differs from an ordinary linear secret sharing scheme (over,
say, a given finite field) in that distribution matrix and reconstruction
vectors are defined over $\mathbb{Z}$ and are designed *independently* of the group
G from which the secret and the shares are sampled. This means that
perfect completeness and perfect privacy are guaranteed *regardless* of
which group G is chosen. We define the black-box secret sharing problem
as the problem of devising, for an arbitrary given $T_{t,n}$, a scheme with
minimal expansion factor, i.e., where the length of the full vector of
shares divided by the number of players n is minimal.

Such schemes are relevant for instance in the context of distributed cryptosystems based on groups with secret or hard to compute group order. A
recent example is secure general multi-party computation over black-box
rings.

In 1994 Desmedt and Frankel have proposed an elegant approach to
the black-box secret sharing problem based in part on polynomial interpolation over cyclotomic number fields. For arbitrary given $T_{t,n}$ with
$0 < t < n - 1$, the expansion factor of their scheme is $O(n)$. This is the
best previous general approach to the problem.

Using certain low degree integral extensions of $\mathbb{Z}$ over which there exist
pairs of sufficiently large Vandermonde matrices with co-prime determinants, we construct, for arbitrary given $T_{t,n}$ with $0 < t < n - 1$, a
black-box secret sharing scheme with expansion factor $O(\log n)$, which
we show is minimal.

1 Introduction

A *black-box* secret sharing scheme for the threshold access structure $T_{t,n}$ is one
which works over any finite Abelian group G. Briefly, such a scheme differs from
an ordinary linear secret sharing scheme (over, say, a given finite field; see e.g.
[5, 24, 6, 3, 2, 20, 19, 1, 16, 8]) in that distribution matrix and reconstruction vectors are defined over $\mathbb{Z}$ and are designed *independently* of the group G from which
the secret and the shares may be sampled. In other words, the dealer computes
the shares for the n players as $\mathbb{Z}$-linear combinations of the secret group element of his interest and secret randomizing group elements, and reconstruction

---

* Basic Research in Computer Science (www.brics.dk), funded by the Danish National
  Research Foundation.

M. Yung (Ed.): CRYPTO 2002, LNCS 2442, pp. 272–287, 2002.
© Springer-Verlag Berlin Heidelberg 2002

of the secret from the shares held by a large enough set of players is by taking $\mathbb{Z}$-linear combinations over those shares. Note that each player may receive one or more group elements as his share in the secret. Perfect completeness and perfect privacy are guaranteed *regardless* of which group G is chosen. Here, perfect completeness means that the secret is uniquely determined by the joint shares of at least $t+1$ players, and perfect privacy means that the joint shares of at most t players contain no Shannon information at all about the secret of interest. Note that these schemes are homomorphic in the sense that the sum of share vectors is a share vector for the sum of the corresponding secrets.

We define the black-box secret sharing problem as the problem of devising, for an arbitrary given $T_{t,n}$, a scheme with minimal expansion factor, i.e., where the length of the full vector of shares divided by the number of players n is minimized[1]. Note the case $t = n - 1$ is easily solved by "additive n-out-of-n sharing," which has expansion factor 1. The cases $t = 0, n$ have no meaning for secret sharing. For the rest of this discussion we assume $0 < t < n - 1$.

The idea of black-box secret sharing was first considered by Desmedt and Frankel [11] in the context of distributed cryptosystems based on groups with secret order. Shamir's polynomial based secret sharing scheme over finite fields [24] cannot immediately be adapted to the setting of black-box secret sharing. Later, Desmedt and Frankel [12] showed a black-box secret sharing scheme that elegantly circumvents problems with polynomial interpolation over the integers by passing to an integral extension ring of $\mathbb{Z}$ over which a sufficiently large *invertible* Vandermonde matrix exists. Their scheme is then constructed using the fact that (sufficiently many copies of) an arbitrary Abelian group can be viewed as a module over such an extension ring.

For a given commutative ring R with 1, the largest integer l such that there exists an invertible $l \times l$ Vandermonde matrix with entries in R is called the *Lenstra constant* $l(R)$ of the ring R. Equivalently, $l(R)$ is the maximal size of a subset E of R that is "exceptional" in that for all $\alpha, \alpha' \in E$, $\alpha \neq \alpha'$, it holds that $\alpha - \alpha'$ is a unit of R.

Given an integral extension ring R of degree m over $\mathbb{Z}$, they construct a black-box secret sharing scheme with expansion factor m for a threshold access structure on *at most* $l(R) - 1$ players. For any prime p, Lenstra's constant for the ring of integers of the pth cyclotomic number field is p [2]. Given an arbitrary $T_{t,n}$ and choosing R as the ring of integers of the pth cyclotomic number field,

---

[1] That minimal expansion is at most polynomial in n, even when appropriately generalizing the concept to encompass non-Abelian groups as well, is verified by combination of the technique of Benaloh-Leichter [2] with the classical result of complexity theory that all monotone threshold functions are representable by poly-size monotone Boolean formulas. See also [10].

[2] It is not hard to find an exceptional set of size p in this ring. To see that the maximal size of such a set is p, let K be a number field of degree m, and let $\mathbb{Z}_K$ denote its ring of algebraic integers. For an arbitrary non-trivial ideal I of $\mathbb{Z}_K$, it is easy to see that $l(\mathbb{Z}_K) \leq |\mathbb{Z}_K/I| \ (\leq 2^m)$. In the case where K is the pth cyclotomic number field, the integer prime p totally ramifies. Hence $l(\mathbb{Z}_K) \leq |\mathbb{Z}_K/P| = p$, where P is the unique prime ideal of $\mathbb{Z}_K$ lying above p.

where p is the smallest prime greater than n, they construct a black-box secret sharing scheme for $T_{t,n}$ with expansion factor between n and $2n$. This is the best previous general approach to the problem. Further progress on the black-box secret sharing problem via the approach of [12] depends on the problem of finding for each n an extension whose degree is *substantially* smaller than n and whose Lenstra constant is greater than n. To the best of our knowledge, this is an open problem of algebraic number theory (see also [12] and the references therein).

Except for some quite special cases, namely when t is constant or when t (resp. $n - t$) is small compared to n [14, 4] or the constant factor gain from [15], no substantial improvement on the general black-box secret sharing problem has been reported since.

The crucial difference with our approach to the black-box secret sharing problem is that we avoid dependence on Lenstra's constant altogether. Namely, first, we observe that a sufficient condition for black-box secret sharing is the existence (over an extension of $\mathbb{Z}$) of *a pair* of sufficiently large Vandermonde matrices with *co-prime determinants*. And, second, we show how to construct *low degree* integral extensions of $\mathbb{Z}$ satisfying this condition. For arbitrary given $T_{t,n}$, this leads to a black-box secret sharing scheme with expansion factor $O(\log n)$. Using a result of Karchmer and Wigderson [20], we show that this is minimal.

There are several applications of black-box secret sharing. For instance, the result of [12] is exploited in [13] to obtain an efficient and secure solution for sharing any function out of a certain abstract class of functions, including RSA. The interest in application of the result of [12] to practical distributed RSA-based protocols seems to have decreased somewhat due to recent developments, see for instance [25] and the references therein. However, apart from the fact that optimal black-box secret sharing is perhaps interesting in its own right, we note that in [9] our black-box secret sharing scheme is applied in protocols for secure general multi-party computation over black-box rings. Also, optimal black-box secret sharing may very well be relevant to new distributed cryptographic schemes for instance based on class groups.

This paper is organized as follows. In Section 2 we give a formalization of the notion of black-box secret sharing, and show a natural correspondence between such schemes and our notion of *integer span programs* (ISPs). This generalizes the well-known correspondence between monotone span programs over finite fields [20] and linear secret sharing schemes over finite fields. In Section 3 we show lower bounds on the size of ISPs computing threshold access structures. Our main result is presented in Section 4, where we construct an ISP with minimal size for an arbitrary given threshold access structure. This leads to an optimal black-box secret sharing scheme for an arbitrary given threshold access structure. At the end, we point out further combinatorial properties of our scheme that are useful when exhibiting *efficient simulators* as required in the security proofs of threshold crypto-systems such as threshold RSA.

2 Black-Box Secret Sharing

2.1 Definitions

Definition 1. *A* monotone access structure *on* $\{1,\ldots,n\}$ *is a non-empty collection* Γ *of sets* $A \subset \{1,\ldots,n\}$ *such that* $\emptyset \notin \Gamma$ *and such that for all* $A \in \Gamma$ *and for all sets* B *with* $A \subset B \subset \{1,\ldots,n\}$ *it holds that* $B \in \Gamma$.

Definition 2. *Let* t *and* n *be integers with* $0 < t < n$. *The* threshold access structure $T_{t,n}$ *is the collection of sets* $A \subset \{1,\ldots,n\}$ *with* $|A| > t$ [3].

Let Γ be a monotone access structure on $\{1,\ldots,n\}$. Let $M \in \mathbb{Z}^{d,e}$ be an integer matrix, and let $\psi : \{1,\ldots,d\} \to \{1,\ldots,n\}$ be a surjective function. We say that the jth row $(j = 1\ldots d)$ of M is *labelled* by $\psi(j)$ or that "$\psi(j)$ owns the jth row." For $A \subset \{1,\ldots,n\}$, M_A denotes the restriction of M to the rows jointly owned by A. Write d_A for the number of rows in M_A. Similarly, for $\mathbf{x} \in \mathbb{Z}^d$, $\mathbf{x}_A \in \mathbb{Z}^{d_A}$ denotes the restriction of $\mathbf{x}$ to the coordinates jointly owned by A. For each $A \in \Gamma$, let $\lambda(A) \in \mathbb{Z}^{d_A}$ be an integer (column-) vector. We call this the *reconstruction vector for* A. Collect all these vectors in a set $\mathcal{R}$.

Definition 3. *Let* Γ *be a monotone access structure on* $\{1,\ldots,n\}$, *and let* $\mathcal{B} = (M, \psi, \mathcal{R})$ *be as defined above.* $\mathcal{B}$ *is called an* integer Γ-scheme. *Its* expansion rate *is defined as* d/n, *where* d *is the number of rows of* M.

Let G be a finite Abelian group. We use additive notation for its group operation, and use 0_G to denote its neutral element. The group G is of course a $\mathbb{Z}$-module (see e.g. [23]), by defining the map $\mathbb{Z} \times G \to G$, $(\mu, g) \mapsto \mu \cdot g$, where $0 \cdot g = 0_G$, $\mu \cdot g = g + \ldots + g$ (μ times) for $\mu > 0$ and $\mu \cdot g = -((-\mu) \cdot g)$ for $\mu < 0$ [4]. We also write μg or $g\mu$ instead of $\mu \cdot g$. Note that it is well-defined how an integer matrix acts on a vector of group elements.

Definition 4. *Let* Γ *be a monotone access structure on* $\{1,\ldots,n\}$ *and let* $\mathcal{B} = (M, \psi, \mathcal{R})$ *be an integer* Γ-scheme. *Then* $\mathcal{B}$ *is a* black-box secret sharing scheme *for* Γ *if the following holds. Let* G *be an arbitrary finite Abelian group* G, *and let* $A \subset \{1,\ldots,n\}$ *be an arbitrary non-empty set. For arbitrarily distributed* $s \in G$, *let* $\mathbf{g} = (g_1,\ldots,g_e)^T \in G^e$ *be drawn uniformly at random, subject to* $g_1 = s$. *Define* $\mathbf{s} = M\mathbf{g}$. *Then:*

- *(Completeness) If* $A \in \Gamma$, *then* $\mathbf{s}_A^T \cdot \lambda(A) = s$ *with probability 1, where* $\lambda(A) \in \mathcal{R}$ *is the reconstruction vector for* A.
- *(Privacy) If* $A \notin \Gamma$, *then* $\mathbf{s}_A$ *contains no Shannon information on* s.

---

[3] Note that some authors define $T_{t,n}$ as consisting of all sets of size *at least* t. Our definition is consistent with a convention in the multi-party computation literature.

[4] If the group operation in G is efficient, multiplication by an integer can also be efficiently implemented using standard "double-and-add."

Note that these schemes[5] are homomorphic in the sense that the sum $\mathbf{s} + \mathbf{s}'$ of two share vectors $\mathbf{s}$ and $\mathbf{s}'$, is a share vector for the sum $s + s'$ of their corresponding secrets s and s'.

2.2 Monotone Span Programs over Rings

In this section we provide quite natural necessary and sufficient conditions under which an integer Γ-scheme is a black-box secret sharing scheme for Γ. To this end, we introduce the notion of *monotone span programs over rings*. This is a certain variation of monotone span programs over finite fields, introduced by Karchmer and Wigderson [20]. These are well-known to have a natural one-to-one correspondence with linear secret sharing schemes over *finite fields* (see e.g. [19, 1]). Monotone span programs over $\mathbb{Z}$ (*ISPs*) will turn out to have a similar correspondence with black-box secret sharing schemes. We also show an efficient conversion of a monotone span program over an integral extension ring of $\mathbb{Z}$ to an ISP.

As an aside, monotone span programs over rings are the basis for multi-party computation over black-box rings, as studied in [9]. In particular, the techniques of [8] for secure multiplication and VSS apply to this flavor of monotone span program as well.

Throughout this paper, R denotes a (not necessarily finite) commutative ring with 1. Let Γ be a monotone access structure on $\{1, \ldots, n\}$, and let $M \in R^{d,e}$ be a matrix whose d rows are labelled by a surjective function $\psi : \{1, \ldots, d\} \to \{1, \ldots, n\}$.

Definition 5. $\varepsilon = (1, 0, \ldots, 0)^T \in R^e$ *is called the* target vector. *Furthermore,* $\mathcal{M} = (R, M, \psi, \varepsilon)$ *is called a* monotone span program *(over the ring R). If $R = \mathbb{Z}$, it is called an* integer span program, *or ISP, for short. We define* $\text{size}(\mathcal{M}) = d$, *where d is the number of rows of M.*

For $N \in R^{a,b}$, $\text{im}N$ denotes its column space, i.e., the space of all vectors $N\mathbf{x} \in R^a$, where $\mathbf{x}$ ranges over R^b, and $\text{ker}N$ denotes its null-space, i.e., the space of all vectors $\mathbf{x} \in R^b$ with $N\mathbf{x} = \mathbf{0} \in R^a$.

Definition 6. *As above, let Γ be a monotone access structure and let $\mathcal{M} = (R, M, \psi, \varepsilon)$ be a monotone span program over R. Then $\mathcal{M}$ is a monotone span program for Γ, if for all $A \subset \{1, \ldots, n\}$ the following holds.*

- *If $A \in \Gamma$, then $\varepsilon \in \text{im}M_A^T$.*
- *If $A \notin \Gamma$, then there exists $\kappa = (\kappa_1, \ldots, \kappa_e)^T \in \text{ker}M_A$ with $\kappa_1 = 1$.*

We also say that $\mathcal{M}$ computes Γ.

---

[5] See [21] for an equivalent definition. We also note that only requiring reconstruction to be linear, as some authors do, results in an equivalent definition of black-box secret sharing. This is an easily proved consequence of Lemma 2, but we omit the details here.

If R is a *field*, our definition is equivalent to the computational model of monotone span programs over fields [20]. Indeed, this model is characterized by the condition that $A \in \Gamma$ if and only if $\boldsymbol{\varepsilon} \in \mathrm{im}M_A^T$. The equivalence follows from the remark below.

Remark 1. By basic linear algebra, if R is a *field*, then $\boldsymbol{\varepsilon} \notin \mathrm{im}M_A^T$ implies that there exists $\boldsymbol{\kappa} \in \mathrm{ker}M_A$ with $\kappa_1 = 1$. If R is *not a field* this does not necessarily hold[6]. The implication in the other direction trivially holds regardless of R.

Using (generally inefficient) representations of monotone access structures as monotone Boolean formulas and using induction in a similar style as in e.g. [2], it is straightforward to verify that for all Γ and for all R, there is a monotone span program over R that computes Γ.

Definition 7. *For any Γ and for any R, $\mathrm{msp}_R(\Gamma)$ denotes the minimal size of a monotone span program over R computing Γ. If $R = \mathbb{Z}$, we write $\mathrm{isp}(\Gamma)$.*

Define a *non-degenerate monotone span program* as one for which the rows of M span the target-vector. As opposed to the case of fields, a non-degenerate monotone span program over a ring need not compute any monotone access structure. This is of no concern here, though.

The following proposition characterizes black-box secret sharing schemes in terms of ISPs.

Proposition 1. *Let Γ be a monotone access structure on $\{1, \ldots, n\}$, and let $\mathcal{B} = (M, \psi, \mathcal{R})$ be an integer Γ-scheme. Then $\mathcal{B}$ is a black-box secret sharing scheme for Γ if and only if $\mathcal{M} = (\mathbb{Z}, M, \psi, \boldsymbol{\varepsilon})$ is an ISP for Γ and for all $A \in \Gamma$, its reconstruction vector $\boldsymbol{\lambda}(A) \in \mathcal{R}$ satisfies $M_A^T \boldsymbol{\lambda}(A) = \boldsymbol{\varepsilon}$.*

Proof. The argument that the stated ISP is sufficient for black-box secret sharing is quite similar to the well-known case of linear secret sharing over finite fields. The other direction of the implication follows in essence from Lemma 1 below. We include full details for convenience.

Consider the ISP from the statement of the proposition, together with the assumption on the reconstruction vectors. Consider an arbitrary set $A \subset \{1, \ldots, n\}$ and an arbitrary finite Abelian group G. Define $\mathbf{s} = M\mathbf{g}$ for arbitrary $\mathbf{g} = (s, g_2, \ldots, g_e)^T \in G^e$. Suppose $A \in \Gamma$, and let $\boldsymbol{\lambda}(A) \in \mathcal{R}$ be its reconstruction vector. It follows that $\mathbf{s}_A^T \boldsymbol{\lambda}(A) = (M_A\mathbf{g})^T \boldsymbol{\lambda}(A) = \mathbf{g}^T(M_A^T\boldsymbol{\lambda}(A)) = \mathbf{g}^T\boldsymbol{\varepsilon} = s$. Thus the completeness condition from Definition 4 is satisfied. If $A \notin \Gamma$, then there exists $\boldsymbol{\kappa} \in \mathbb{Z}^e$ with $M_A\boldsymbol{\kappa} = \mathbf{0} \in \mathbb{Z}^{d_A}$ and $\kappa_1 = 1$, by Definition 6. For arbitrary $s' \in G$, define $\mathbf{s}' = M(\mathbf{g} + (s' - s)\boldsymbol{\kappa}) \in G^{d_A}$. The secret defined by $\mathbf{s}'$ equals s', while on the other hand $\mathbf{s}'_A = \mathbf{s}_A$. This implies perfect privacy: the assignment $\mathbf{g}' = \mathbf{g} + (s' - s)\boldsymbol{\kappa}$ provides a bijection between the set of possible vectors of "random coins" consistent with $\mathbf{s}_A$ and s, and the set of those consistent with $\mathbf{s}_A$ and s'. Therefore, the privacy condition from Definition 4 is also satisfied.

---

[6] Consider for example the integer matrix $M = (2\ 0)$.

In the other direction of the proposition, we start with a black-box secret sharing scheme for Γ according to Definition 4. Consider an arbitrary set $A \subset \{1, \ldots, n\}$. Suppose $A \in \Gamma$, and let $\lambda(A) \in \mathcal{R}$ be its reconstruction vector. For an arbitrary prime p, set $G = \mathbb{Z}_p$. By the completeness condition from Definition 4, it follows that $(1, 0, \ldots, 0)^T \equiv (M_A I_e)^T \lambda(A) \equiv M_A^T \lambda(A) \bmod p$, where $I_e \in \mathbb{Z}_p^{e,e}$ is the identity matrix. This holds for all primes p. Hence, $M_A^T \lambda(A) = (1, 0, \ldots, 0)^T = \varepsilon$. Therefore, the condition on the sets $A \in \Gamma$ in Definition 6 and the condition on the reconstruction vectors $\mathcal{R}$ from the statement of the proposition are satisfied.

To conclude the proof we show that the privacy condition from Definition 4 implies the condition on the sets $A \notin \Gamma$ from Definition 6. The following formulation is equivalent. Let $\mathbf{y} \in \mathbb{Z}^{d_A}$ denote the left-most column of M_A, and let $N_A \in \mathbb{Z}^{d_A, e-1}$ denote the remaining $e - 1$ columns. Then it is to be shown that the linear system of equations $N_A \mathbf{x} = \mathbf{y}$ is solvable over $\mathbb{Z}$.

By Lemma 1 below, it is sufficient to show that this holds modulo m, for all $m \in \mathbb{Z}$, $m \neq 0$. With notation as in Definition 4 and considering $G = \mathbb{Z}_m$, it follows from the privacy condition that there exists $\mathbf{g}' \in \mathbb{Z}_m^e$ such that $g_1' \equiv s - 1$ and $\mathbf{s}_A \equiv M_A \mathbf{g}'$. Setting $\kappa \equiv \mathbf{g} - \mathbf{g}' \in \mathbb{Z}_m^e$, we have $M_A \kappa \equiv 0$ with $\kappa_1 \equiv 1$. In other words, $N_A \mathbf{x} = \mathbf{y}$ is solvable over $\mathbb{Z}_m$ for all integers $m \neq 0$. $\square$

We note that [21] also gives a characterization. Although there are some similarities in the technical analysis, the conditions stated there are still in terms of the black-box secret sharing scheme, rather than by providing simple algebraic conditions on the matrix M as we do. Therefore, we feel that our approach based on integer span programs is perhaps more useful and insightful, especially since monotone span programs over finite fields have since long been known to be equivalent to linear secret sharing schemes over finite fields.

Lemma 1. *Let $N \in \mathbb{Z}^{a,b}$ and $\mathbf{y} \in \mathbb{Z}^a$. Then the linear system of equations $N\mathbf{x} = \mathbf{y}$ is solvable over $\mathbb{Z}$ if and only if it is solvable over $\mathbb{Z}_m$ for all integers $m \neq 0$.*

Proof. The forward direction of the proposition is trivial. In the other direction, consider the $\mathbb{Z}$-module H generated by the columns of N. By basic theory of $\mathbb{Z}$-modules (see e.g. [23]), there exists a $\mathbb{Z}$-basis $\mathcal{B} = (\mathbf{b}_1, \ldots, \mathbf{b}_a)$ of $\mathbb{Z}^a$, and non-zero integers $a_1, \ldots, a_l$ such that $\mathcal{B}_H = (a_1 \mathbf{b}_1, \ldots, a_l \mathbf{b}_l)$ is a $\mathbb{Z}$-basis of H. Let L denote the $\mathbb{Z}$-module with basis $\mathcal{B}_L = (\mathbf{b}_1, \ldots, \mathbf{b}_l)$. Note that $H \subset L$. Let p be an arbitrary prime, and let $\overline{(\cdot)}$ denote reduction modulo p. Since the determinant of $\mathcal{B}$ is ± 1, $\overline{\mathcal{B}}$ (resp. $\overline{\mathcal{B}}_L$) provides a basis for the vector-space $\mathbb{F}_p^a$ (resp. the vector-space $\overline{L}$). Note that $\overline{\mathcal{B}}_L \subset \overline{\mathcal{B}}$.

It follows from the assumptions that $\overline{\mathbf{y}} \in \overline{H} \subset \overline{L}$. Let $(y_1, \ldots, y_a) \in \mathbb{Z}^a$ denote the coordinates of $\mathbf{y}$ wrt. $\mathcal{B}$. Since the latter observation holds for all primes p, it follows that $y_{l+1} = \ldots = y_a = 0$. Hence, $\mathbf{y} \in L$. Now set $\hat{m} = \prod_{i=1}^l a_i$. By the assumptions, there exists $\mathbf{c}_{\hat{m}} \in \mathbb{Z}^a$ such that $\mathbf{y} + \hat{m} \cdot \mathbf{c}_{\hat{m}} \in H$. Therefore, $\hat{m} \cdot \mathbf{c}_{\hat{m}} \in L$, and by the definition of L, $\mathbf{c}_{\hat{m}} \in L$. By the choice of $\hat{m}$, it follows that $\hat{m} \cdot \mathbf{c}_{\hat{m}} \in H$. We conclude that $\mathbf{y} \in H$, as desired. $\square$

Remark 2. Let $\mathcal{M} = (R, M, \psi, \varepsilon)$ compute Γ. If R is a field or a principal ideal domain (such as $\mathbb{Z}$), then we may assume without loss of generality that $e \leq d$, i.e., there are at most as many columns in M as there are rows.

This is easily shown using elementary linear algebra, and using the basic properties of modules over principal ideal domains (see e.g. [23] and the proof of Lemma 1). Briefly, since $\mathcal{M}$ is non-degenerate, the last statement in Remark 1 implies that the space generated by the 2nd up to the eth column of M does not contain even a non-zero multiple of the first column. Without changing the access structure that is computed, we can always replace the 2nd up to the eth column of M by any set of vectors that generates the same space. If R is a field or a principal ideal domain, this space has a basis of cardinality at most $d - 1$.

Remark 3. We may now identify a black-box secret sharing scheme for Γ with an ISP $\mathcal{M} = (\mathbb{Z}, M, \psi, \varepsilon)$ for Γ. A reconstruction vector for $A \in \Gamma$ is simply any vector $\lambda(A) \in \mathbb{Z}^{d_A}$ such that $M_A^T \lambda(A) = \varepsilon$. Note that the expansion rate of the corresponding black-box secret sharing scheme is equal to size$(\mathcal{M})/n$. By Remark 2 it uses at most size$(\mathcal{M})$ random group elements.

We now state some lemmas that are useful in the sequel.

Definition 8. *The dual Γ^* of a monotone access structure Γ on $\{1, \dots, n\}$ is the collection of sets $A \subset \{1, \dots, n\}$ such that $A^c \notin \Gamma$.*

Note that Γ^* is a monotone access structure on $\{1, \dots, n\}$, that $(\Gamma^*)^* = \Gamma$, and that $(T_{t,n})^* = T_{n-t-1,n}$. The lemma below generalizes a similar property shown in [20] for the case of fields.

Lemma 2. $\mathrm{msp}_R(\Gamma) = \mathrm{msp}_R(\Gamma^*)$, *for all R and Γ.*

Proof. Let $\mathcal{M} = (R, M, \psi, \varepsilon)$ be a monotone span program for Γ. Select an arbitrary generating set of vectors $\mathbf{b}_1, \dots, \mathbf{b}_l$ for kerM^T, and choose λ with $M^T \lambda = \varepsilon$. Let M^* be the matrix defined by the $l+1$ columns $(\lambda, \mathbf{b}_1, \mathbf{b}_2, \dots, \mathbf{b}_l)$, and use ψ to label M^* as well. Define $\mathcal{M}^* = (R, M^*, \psi, \varepsilon^*)$, where $\varepsilon^* = (1, 0, \dots, 0)^T \in R^{l+1}$. Note that size$(\mathcal{M}^*) = $ size$(\mathcal{M})$. We claim that $\mathcal{M}^*$ computes Γ^*. This is easy to verify.

If $A^c \notin \Gamma$, then by Definition 6, there exists $\kappa \in R^{l+1}$ such that $M_{A^c}\kappa = 0$ and $\kappa_1 = 1$. Define $\lambda^* = M_A \kappa$. Then $(M^*)_A^T \lambda^* = ((M^*)^T \cdot M)\kappa = \varepsilon^*$. On the other hand, if $A^c \in \Gamma$, then there exists $\hat{\lambda} \in R^d$ such that $M^T \hat{\lambda} = \varepsilon$ and $\hat{\lambda}_A = 0$. By definition of M^*, there exists $\kappa \in R^{l+1}$ such that $M^* \kappa = \hat{\lambda}$ and $\kappa_1 = 1$. Hence, $M_A^* \kappa = \hat{\lambda}_A = 0$ and $\kappa_1 = 1$. This concludes the proof. $\square$

The lemma below holds in a more general setting, but we tailor it to ours.

Lemma 3. *Let $f(X) \in \mathbb{Z}[X]$ be a monic, irreducible polynomial. Write $m = \deg(f)$. Consider the ring $R = \mathbb{Z}[X]/(f(X))$. Suppose $\mathcal{M} = (R, M, \psi, \varepsilon)$ is a monotone span program over R for a monotone access structure Γ. Then there exists an ISP $\hat{\mathcal{M}} = (\mathbb{Z}, \hat{M}, \hat{\psi}, \hat{\varepsilon})$ for Γ with size$(\hat{\mathcal{M}}) = m \cdot $ size$(\mathcal{M})$.*

Proof. The proof is based on a standard algebraic technique for representing a linear map defined over an extension ring in terms of a linear map defined over the ground ring. This technique is also used in [20] for monotone span programs over extension fields. Since our definition of monotone span programs over rings differs slightly from the definitions in [20], we explain it in detail.

Note that R is a commutative ring with 1 and that it has no zero divisors, but that it is not a field. Fix $w \in R$ such that $f(w) = 0$ (such as $w = \overline{X}$, the class of X modulo $f(X)$). Then for each $x \in R$, there exists a unique coordinate-vector $\vec{x} = (x_0, \ldots, x_{m-1})^T \in \mathbb{Z}^m$ such that $x = x_0 \cdot 1 + x_1 \cdot w + \cdots + x_{m-1} \cdot w^{m-1}$. In other words, $\mathcal{W} = \{1, w, \ldots, w^{m-1}\}$ is a basis for R when viewed as a $\mathbb{Z}$-module.

For each $x \in R$ there exists a matrix in $\mathbb{Z}^{m,m}$, denoted as $[x]$, such that, for all $y \in R$, $[x]\vec{y} = \overrightarrow{xy}$ (the coordinate vector of $xy \in R$). The columns of $[x]$ are simply the coordinate vectors of $x, x \cdot w, \ldots, x \cdot w^{m-1}$. If $x \in \mathbb{Z}$, then $[x]$ is a diagonal matrix with x's on its main diagonal. Furthermore, for all $x, y \in R$, we have the identities $[x + y] = [x] + [y]$ and $[xy] = [x][y]$.

Consider the monotone span program $\mathcal{M} = (R, M, \psi, \varepsilon)$ from the statement of the lemma. As before, write d (resp. e) for the number of rows (resp. columns) of M. We define the ISP $\hat{\mathcal{M}} = (\mathbb{Z}, \hat{M}, \hat{\psi}, \hat{\varepsilon})$ as follows. Construct $\hat{M} \in \mathbb{Z}^{md,me}$ from M by replacing each entry $x \in R$ in M by the matrix $[x]$. The labeling ψ is extended to $\hat{\psi}$ in the obvious way, i.e., if a player owns a certain row in M, then that same player owns the m rows that it is substituted with in $\hat{M}$. The target vector $\hat{\varepsilon}$ is defined by $\hat{\varepsilon} = (1, 0 \ldots, 0)^T \in \mathbb{Z}^{me}$.

We verify that $\hat{\mathcal{M}}$ is an ISP for Γ. First, consider a set $A \in \Gamma$. By definition, there exists a vector $\boldsymbol{\lambda} = (\lambda_1, \ldots, \lambda_{d_A})^T \in R^{d_A}$ such that $M_A^T \boldsymbol{\lambda} = \varepsilon$. Using the identities stated above and carrying out matrix multiplication "block-wise," it follows that $\hat{M}_A^T([\lambda_1], \ldots, [\lambda_{d_A}])^T = ([1], [0], \ldots, [0])^T$. Define $\hat{\boldsymbol{\lambda}} \in \mathbb{Z}^{md_A}$ as the first column of the matrix $([\lambda_1], \ldots, [\lambda_{d_A}])^T$. Then $\hat{M}_A^T \hat{\boldsymbol{\lambda}} = \hat{\varepsilon}$. Now consider a set $A \notin \Gamma$. By definition, there exists $\boldsymbol{\kappa} = (\kappa_1, \kappa_2, \ldots, \kappa_e)^T \in R^e$ such that $\kappa_1 = 1$ and $M_A \boldsymbol{\kappa} = \mathbf{0} \in R^{d_A}$. Using similar reasoning as above, it follows that $\hat{M}_A([\kappa_1]^T, \ldots, [\kappa_e]^T)^T = ([0]^T, \ldots, [0]^T)^T$. Define $\hat{\boldsymbol{\kappa}} \in \mathbb{Z}^{me}$ as the first column of the matrix derived from $\boldsymbol{\kappa}$ in the above equation. Then, the first m entries of $\hat{\boldsymbol{\kappa}}$ are $1, 0, \ldots, 0$ (since $\kappa_1 = 1$) and $\hat{M}_A \hat{\boldsymbol{\kappa}} = \mathbf{0} \in \mathbb{Z}^{d_A}$.

This proves the lemma. As an aside, it follows directly from the analysis above that we may delete the 2nd up to mth leftmost colums of $\hat{M}$ and the corresponding coordinates of $\hat{\varepsilon}$ without changing the access structure computed. Hence, $1 + m(e - 1)$ columns suffice, rather than me. □

3　Lower Bounds for the Threshold Case

Proposition 2. *For all integers t, n with $0 < t < n - 1$, $\mathrm{isp}(T_{t,n}) = \Omega(n \cdot \log n)$. Hence, the expansion factor of a black-box secret sharing scheme for $T_{t,n}$ with $0 < t < n - 1$ is $\Omega(\log n)$.*

Proposition 2 follows quite directly from the bound shown in Theorem 1 for binary monotone span programs, as proved in [20][7]. Before we give the details of the proof of Proposition 2, we include a proof of their bound for convenience. Note that we have made constants for their asymptotic bound explicit.

Throughout this section, K denotes a field. Let $\mathcal{M} = (K, M, \psi, \varepsilon)$ be a non-degenerate monotone span program. The access structure of $\mathcal{M}$, denoted $\Gamma(\mathcal{M})$, is the collection of sets A such that $\varepsilon \in \mathrm{im} M_A^T$. Note that by Remark 1 this is consistent with our Definition 6. We write $\mathrm{msp}_2(\Gamma)$ instead of $\mathrm{msp}_{\mathbb{F}_2}(\Gamma)$.

Proposition 3. *[20]* $\mathrm{msp}_2(T_{1,n}) \geq n \cdot \log n$.

Proof. Consider a monotone span program $\mathcal{M} = (\mathbb{F}_2, M, \psi, \varepsilon)$ such that $\Gamma(\mathcal{M}) = T_{1,n}$. Define e as the number of columns of M, d as its number of rows, and d_i as the number of rows of M_i for $i = 1 \ldots n$, where we write M_i instead of $M_{\{i\}}$ and d_i instead of $d_{\{i\}}$. Without loss of generality, assume that the rows of each M_i are linearly independent over $\mathbb{F}_2$. Let H_1 collect the vectors in $\mathbb{F}_2^e$ with first coordinate equal to 1. Since $\{i\} \notin T_{1,n}$, Remark 1 implies that $|\ker M_i \cap H_1| \neq \emptyset$. By assumption on M_i, $|\ker M_i \cap H_1| = 2^{e-1-d_i}$ for $i = 1 \ldots n$. On the other hand, $\{i, j\} \in T_{1,n}$. Hence, by Remark 1, we have $\ker M_i \cap \ker M_j \cap H_1 = \emptyset$, for all i, j with $1 \leq i < j \leq n$. By counting and normalizing, $2^{-d_1} + \cdots + 2^{-d_n} \leq 1$. By the Log Sum Inequality (see e.g. [7]), $d = d_1 + \cdots + d_n \geq n \log n$. $\qquad\square$

Theorem 1. *[20]* $n \cdot (\lfloor \log n \rfloor + 1) \geq \mathrm{msp}_2(T_{t,n}) \geq n \cdot \log \frac{n+3}{2}$, *for all t, n with* $0 < t < n - 1$.

Proof. The upper bound, which is not needed for our purposes, follows by considering an appropriate Vandermonde matrix over the field $\mathbb{F}_{2^u}$, where $u = (\lfloor \log n \rfloor + 1)$. This is turned into a binary monotone span program for $T_{t,n}$ using a similar conversion technique as in Lemma 3.

For the lower bound, note that we may assume that $t \geq (n-1)/2$, since $\mathrm{msp}_2(T_{t,n}) = \mathrm{msp}_2(T_{n-t-1,n})$ by Lemma 2. We have the following estimates.

$$\mathrm{msp}_2(T_{t,n}) \geq \frac{n}{t+2} \cdot \mathrm{msp}_2(T_{t,t+2}) = \frac{n}{t+2} \cdot \mathrm{msp}_2(T_{1,t+2})$$

$$\geq \frac{n}{t+2} \cdot (t+2) \cdot \log(t+2) \geq n \cdot \log \frac{n+3}{2}.$$

The first inequality is argued as follows. Consider an arbitrary monotone span program $\mathcal{M} = (\mathbb{F}_2, M, \psi, \varepsilon)$ for $T_{t,n}$. Assume without loss of generality that the number of rows in M_i is at most the number of rows in M_{i+1}, $i = 1, \ldots, n-1$. The first $t+2$ blocks $M_1, \ldots, M_{t+2}$ clearly form a monotone span program for $T_{t,t+2}$. Hence, the total number of rows in these blocks is at least $\mathrm{msp}_2(T_{t,t+2})$. Each other block M_j with $j > t+2$ has at least as many rows as any of the first $t+2$ blocks. Therefore, M_j has at least $\mathrm{msp}_2(T_{t,t+2})/(t+2)$ rows. Summing up over all i according to the observations above gives the first inequality.

The equality is implied by Lemma 2, the second to last inequality follows from Proposition 3, and the last one from $t \geq (n-1)/2$. $\qquad\square$

---

[7] Note that $\mathrm{isp}(T_{n-1,n}) = n$: the case $t = n-1$ is solved by simple additive "n-out-of-n" secret sharing."

For the proof of Proposition 2, let an ISP for $T_{t,n}$ be given, and consider the ISP matrix, but with all entries reduced modulo 2. By our ISP definition and by arguing the cases $A \notin T_{t,n}$ using Remark 1, it follows that a binary monotone span program for $T_{t,n}$ is obtained in this way. The argument is concluded by applying Theorem 1[8]. The statement about black-box secret sharing follows from Proposition 1.

Note that our lower bound on black-box secret sharing can also be appreciated without reference to Proposition 1, by essentially the same argument as above. Namely, setting $G = \mathbb{Z}_2$ in Definition 4, we clearly obtain a (binary) linear secret sharing scheme. This is well-known to be equivalent to a binary monotone span program, as mentioned before. Hence, we can directly apply the bound from Theorem 1.

4   Optimal Black-Box Threshold Secret Sharing

Theorem 2. *For all integers t, n with $0 < t < n - 1$, $\mathrm{isp}(T_{t,n}) = \Theta(n \cdot \log n)$. Hence, there exists a black-box secret sharing scheme for $T_{t,n}$ with expansion factor $O(\log n)$, which is minimal.*

Proof. By Proposition 1 it is sufficient to focus on the claim about the ISPs. The lower bound follows from Proposition 2. For the upper bound, we consider rings of the form $R = \mathbb{Z}[X]/(f(X))$, where $f(X) \in \mathbb{Z}[X]$ is a monic, irreducible polynomial. Write $m = \deg(f)$, the degree of R over $\mathbb{Z}$.

On account of Lemma 3, it is sufficient to exhibit a ring R together with a monotone span program $\mathcal{M}$ over R for $T_{t,n}$ such that $m = O(\log n)$ and $\mathrm{size}(\mathcal{M}) = O(n)$.

The proof is organized as follows. We first identify a certain technical property of a ring R that facilitates the construction of a monotone span program over R for $T_{t,n}$, with size $O(n)$. We finalize the proof by constructing a ring R that enjoys this technical property, and that has degree $O(\log n)$ over $\mathbb{Z}$.

For $x_1, \ldots, x_n \in R$, define

$$\Delta(x_1, \ldots, x_n) = \prod_{i=1}^{n} x_i \cdot \prod_{1 \leq j < i \leq n} (x_i - x_j).$$

Assume, for the moment, that there exist $\alpha_1, \ldots, \alpha_n \in R$ and $r_0, r_1 \in R$ such that

$$r_0 \cdot \Delta(1, \ldots, n)^2 + r_1 \cdot \Delta(\alpha_1, \ldots, \alpha_n)^2 = 1.$$

This assumption implies the existence of a monotone span program over R for $T_{t,n}$ with size $2n$, as we now show. Define

$$\Delta_0 = \Delta(1, \ldots, n) \in \mathbb{Z}, \quad \text{and} \quad \Delta_1 = \Delta(\alpha_1, \ldots, \alpha_n) \in R.$$

---

[8] See [21, 22] for lower bounds on the randomness required in black-box secret sharing schemes.

Let $N_0 \in R^{n,t+1}$ (resp. $N_1 \in R^{n,t+1}$) be the matrix in which the i-th row is equal to $(\Delta_0, i, i^2, \ldots, i^t)$ (resp. $(\Delta_1, \alpha_i, \alpha_i^2, \ldots, \alpha_i^t)$), $i = 1 \ldots n$. In both cases, the ith row is labelled by i. When studied as possible monotone span programs over R for $T_{t,n}$, N_0 (resp. N_1) satisfies Definition 6 for the sets $A \notin T_{t,n}$. On the other hand, in both cases, the rows owned by a set $A \in T_{t,n}$ do not necessarily span the target vector $(1, 0, \ldots, 0) \in R^{t+1}$. However, these rows do span[9] the vector $(\Delta_0^2, 0, \ldots, 0) \in R^{t+1}$ (resp. $(\Delta_1^2, 0, \ldots, 0) \in R^{t+1}$). Both properties stated can be verified immediately, for instance using the well-known expression for a Vandermonde determinant in combination with Cramér's rule (see e.g. [23]); passing to the fraction field K of R (note that R has no zero-divisors), this rule implies that a $c \times c$ linear system of equations $N\mathbf{x} = \mathbf{y}$ over the ring R, has a solution at least in case where $\mathbf{y} \in \det(N) \cdot R^c$. Another way is by using Lagrange Interpolation over K, and clearing denominators.

Define a new monotone span program matrix $M \in R^{2n,2t+1}$ consisting of all pairs of rows

$$(\Delta_0, i, i^2, \ldots, i^t, 0, \ldots, 0), \quad \text{and} \quad (\Delta_1, 0, \ldots, 0, \alpha_i, \alpha_i^2, \ldots, \alpha_i^t),$$

for $i = 1 \ldots n$. The shown padding consists of t zeroes in both cases, and each of the rows in a pair is labelled by i. Define $\varepsilon = (1, 0, \ldots, 0)^T \in R^{2t+1}$. The sets $A \notin T_{t,n}$ clearly satisfy Definition 6, and this time the rows owned by sets $A \in T_{t,n}$ span the target vector: they span in particular all vectors of the form $(r \cdot \Delta_0^2 + s \cdot \Delta_1^2, 0, \ldots, 0)$, with $r, s \in R$. By setting $r = r_0$ and $s = r_1$, these include the target vector ε.

To conclude, we exhibit a ring R with degree $O(\log n)$ over the integers and $\alpha_1, \ldots, \alpha_n, r_0, r_1 \in R$ with $r_0 \cdot \Delta_0^2 + r_1 \cdot \Delta_1^2 = 1$, where $\Delta_0 = \Delta(1, \ldots, n)$ and $\Delta_1 = \Delta(\alpha_1, \ldots, \alpha_n)$.

These conditions are reformulated as follows. Let Π_n denote the set of integer primes p with $2 \leq p \leq n$ and define

$$Q_n = \prod_{p \in \Pi_n} p \in \mathbb{Z}.$$

Then we are looking for a ring R with degree $O(\log n)$ over the integers and $\alpha_1, \ldots, \alpha_n \in R$ such that

$$\overline{\Delta}_1 \in (R/(Q_n))^*,$$

i.e., the residue-class of Δ_1 in the ring $R/(Q_n)$ is a unit.

Indeed, if $\overline{\Delta}_1 \in (R/(Q_n))^*$, then $\overline{\Delta}_1 \in (R/(Q_n^k))^*$ as well, for any positive integer k. To verify this by induction, suppose that $\Delta_1 \cdot v = 1 + w \cdot Q_n^i$ for some $v, w \in R$ and $i \geq 1$: then $\Delta_1 \cdot (v - vw \cdot Q_n^i) = 1 - w^2 \cdot Q_n^{2i}$ and $2i \geq i + 1$. As a consequence, $\overline{\Delta}_1 \in (R/(\Delta_0^2))^*$. Namely, as an integer, Δ_0^2 factors completely over the primes $p \in \Pi_n$. Then choose k_* large enough such that Δ_0^2 divides $Q_n^{k_*}$, and apply the previous observation. It follows that $\overline{\Delta}_1^2 \in (R/(\Delta_0^2))^*$ as well, or equivalently, there exist $r_0, r_1 \in R$ such that $r_0 \cdot \Delta_0^2 + r_1 \cdot \Delta_1^2 = 1$.

---

[9] A similar property was first noticed and exploited in [17, 18] and later in [25].

Set $m = \lfloor \log n \rfloor + 1$. Let $\hat{f}(X) \in \mathbb{Z}[X]$ be any monic, irreducible polynomial of degree m such that for all $p \in \Pi_n$, $\hat{f}_p(X)$ (the polynomial $\hat{f}(X)$ with its coefficients reduced modulo p) is irreducible in $\mathbb{F}_p[X]$.

One way of constructing such a polynomial is as follows. For all $p \in \Pi_n$, select a monic, irreducible polynomial $\hat{f}_p(X) \in \mathbb{F}_p[X]$ of degree m. By the theory of finite fields, this is always possible. Applying the Chinese Remainder Theorem to each of the coefficients separately, select an arbitrary lift to a monic polynomial $\hat{f}(X) \in \mathbb{Z}[X]$ of degree m such that $\hat{f}(X) \equiv \hat{f}_p(X) \bmod p$. Note that the monic polynomial $\hat{f}(X)$ is irreducible in $\mathbb{Z}[X]$: if not, reduction modulo p with $p \in \Pi_n$, gives a non-trivial factorization of $\hat{f}_p(X)$ in $\mathbb{F}_p[X]$.

Set $R = \mathbb{Z}[X]/(\hat{f}(X))$. By definition of $\hat{f}(X)$, it follows that $R/(p)$ is a finite field, for all $p \in \Pi_n$. Indeed, for all $p \in \Pi_n$,

$$R/(p) \simeq \mathbb{Z}[X]/(p, \hat{f}(X)) \simeq \mathbb{F}_p[X]/(\hat{f}_p(X)) \simeq \mathbb{F}_{p^m}.$$

Note that all ideals (p) of R with $p \in \Pi_n$ are distinct and maximal. It follows, using the Chinese Remainder Theorem for general rings, that

$$R/(Q_n) \simeq \prod_{p \in \Pi_n} \mathbb{F}_{p^m}.$$

For all $p \in \Pi_n$ we have $|\mathbb{F}_{p^m}^*| = p^m - 1 \geq 2^m - 1 \geq n$. Therefore, for each $p \in \Pi_n$, distinct non-zero

$$\beta_1^{(p)}, \ldots, \beta_n^{(p)} \in \mathbb{F}_{p^m}$$

can be selected. Finally, select arbitrary $\alpha_1, \ldots, \alpha_n \in R$ such that, for $i = 1 \ldots n$,

$$R/(Q_n) \ni \overline{\alpha}_i \longleftrightarrow (\beta_i^{(p)})_{p \in \Pi_n} \in \prod_{p \in \Pi_n} \mathbb{F}_{p^m},$$

where the correspondence is via the (implicit) isomorphism. By construction, for all i, j with $1 \leq i, j \leq n$ and $i \neq j$, it holds that $\overline{\alpha}_i \in (R/(Q_n))^*$ and $\overline{\alpha}_i - \overline{\alpha}_j \in (R/(Q_n))^*$. Hence, $\overline{\Delta}_1 \in (R/(Q_n))^*$, as desired. $\square$

Corollary 1. *For all integers t, n with $0 < t < n - 1$, there exists an ISP of size $n \cdot (\lfloor \log n \rfloor + 2)$ for $T_{t,n}$.*

Proof. Let $R, \alpha_1, \ldots, \alpha_n, r_0, r_1, N_0, N_1$ be as constructed in the proof of Theorem 2. Apply the construction from the proof of Lemma 3 to N_1, and take into account the final remark of that proof. This gives an ISP matrix $\hat{N}_1$ with $n \cdot (\lfloor \log n \rfloor + 1)$ rows and $1 + t(\lfloor \log n \rfloor + 1)$ columns. Clearly, the sets $A \notin T_{t,n}$ satisfy Definition 6. For the sets $A \in T_{t,n}$, the rows owned by A span $\delta_1 \cdot \hat{\varepsilon}$, where $\delta_1 \in \mathbb{Z}$ is the first coordinate of $r_1 \cdot \Delta_1^2$.

The ISP matrix N_0 has the properties stated in the proof of Theorem 2 also over $\mathbb{Z}$. Hence, the sets $A \notin T_{t,n}$ satisfy Definition 6 over $\mathbb{Z}$. For the sets $A \in T_{t,n}$, the rows owned by them clearly span $(\delta_0, 0, \ldots, 0) \in \mathbb{Z}^{t+1}$, where $\delta_0 \in \mathbb{Z}$ is the first coordinate of $r_0 \cdot \Delta_0^2$. Since $\delta_0 + \delta_1 = 1$, this leads directly to an ISP for $T_{t,n}$, where the ISP matrix has $n \cdot (\lfloor \log n \rfloor + 2)$ rows and $t(\lfloor \log n \rfloor + 2) + 1$ columns. $\square$

5   Concluding Remarks

5.1   A Note on Simulateability

The ISPs $\hat{\mathcal{M}} = (\mathbb{Z}, \hat{M}, \hat{\psi}, \hat{\varepsilon})$ constructed in the proofs of Theorem 2 and Corollary 1 satisfy the following additional properties, which are helpful when proving the security of certain threshold cryptosystems.

Let the share vector $\mathbf{s} = \hat{M}\mathbf{g}$ be computed according to the corresponding black-box secret sharing scheme, then the following holds.

1. The entries of $\mathbf{s}_A$ are *independent* random group elements for any subset A of $\{1, \ldots, n\}$ with $|A| \leq t$.
2. Every player i can compute a *reconstruction share* $\mathbf{s}'_i$ by taking $\mathbb{Z}$-linear combinations (of course independent of the group) of the entries of his original share $\mathbf{s}_i$, such that any t reconstruction shares $\mathbf{s}'_i$ still allow to reconstruct the secret s, and such that any t original shares $\mathbf{s}_i$ together with s allow to compute the complete reconstruction share vector $\mathbf{s}'$ (by taking $\mathbb{Z}$-linear combinations).

The former property is inherited from the two Vandermonde matrices upon which the construction of $\hat{\mathcal{M}}$ is based on, and the latter holds for $\mathbf{s}'$ defined as $\mathbf{s}' = \hat{M}'\mathbf{g}$, where the ISP $\hat{\mathcal{M}}' = (\mathbb{Z}, \hat{M}', \hat{\psi}, \hat{\varepsilon})$ is constructed from the matrices $\Delta_0 N_0$ and $\Delta_1 N_1$ in a way similar to which $\hat{\mathcal{M}}$ is constructed from N_0 and N_1 in the proof of Theorem 2.

Assuming that the group operation is efficiently computatble and that (almost) random group elements can be sampled efficiently, these properties allow the players of a set A with $|A| \leq t$ to *efficiently* simulate their joint view $\mathbf{s}_A$ of the distribution phase, by sampling (almost) random elements from the group and to *efficiently* simulate their view of the corresponding reconstruction phase by computing $\mathbf{s}'$ from $\mathbf{s}_A$ and the secret s.

When proving the security of a direct application of our black-box secret sharing scheme to distributed RSA for instance, these properties enable an efficient simulator for the adversary's view of the distributed decryption or signing process (see also [12, 25]).

5.2   Implementation

We stress that in this paper we are primarily interested in the asymptotically optimal result from Theorem 2. Several choices in its proof have been made to simplify the mathematical exposition, while suppressing computational aspects.

There are a number of possible practical implementations of black-box secret sharing based on our result. We do not optimize its performance here, but merely indicate below that straightforward implementations run in time polynomial in n.

Note that the scheme consumes $O(n \log n)$ random coins (group elements) and that the expansion factor is $O(\log n)$ in any case, i.e., each player receives

$O(\log n)$ groups elements as his share in a secret group element. For an implementation, it is important to limit the necessary *computational resources* for dealer and players.

One implementation is based on the well-known fact that for any finite Abelian group G, G^m can be viewed as a module over the ring R (see also [12]). The multiplication of an element of R by an element of G^m can be performed having only black-box access to the group operation of G. This way, the monotone span program over R acts directly on vectors of elements of G^m. This leads in a straightforward fashion to an attractive implementation of black-box secret sharing where the actual ISP it is based upon can be left implicit. See for instance [12] for the computational details of this general procedure, taking into account the remarks below.

By the constructive method from the proof of Theorem 2, we may assume without loss of generality that the coefficients of the polynomial $f(X)$ have bit length smaller than $\log Q_n \leq \log(n!) = O(n \log n)$ bits. Recall that its degree m is $\lfloor \log n \rfloor + 1$. For given threshold parameters t, n, it can be fixed once and for all. One simple possible choice for the α_i's is to identify them with distinct, non-zero integer polynomials of degree at most $\lfloor \log n \rfloor$, such that each of the coefficients is either 0 or 1. For instance, α_i can point to i by basing it on the bit representation of i. Δ_0^2 is simply represented by an integer with bit length $O(n^2 \cdot \log n)$. The value Δ_1^2 is the product of $O(n^2)$ elements of R, each of which has integer coordinates -1, 0 or 1. The values r_0 and r_1 can be obtained by computing the inverse $\overline{u}$ of $\overline{\Delta}_1^2 \in R/(\Delta_0^2)$, for instance by solving a linear system of equations over $\mathbb{Z}_{\Delta_0^2}$, and by computing $u \cdot \Delta_1^2 \in R$. The reconstruction vectors are computed from r_0, r_1 and obvious "interpolation coefficients" obtained from the α_i's.

Acknowledgments

We thank Ivan Damgaard for many helpful suggestions and discussions. Also thanks to Yvo Desmedt, Yair Frankel, Anna Gál, Yuval Ishai, Brian King and the anonymous referees of CRYPTO '02 for comments.

References

1. A. Beimel. Secure schemes for secret sharing and key distribution. Ph.D.-thesis, Technion, Haifa, June 1996.
2. J. Benaloh and J. Leichter. Generalized secret sharing and monotone functions. In: Proc. CRYPTO '88, Springer LNCS, vol. 765, pp. 274–285, 1988.
3. M. Bertilsson, I. Ingemarsson. A construction of practical secret sharing schemes using linear block codes. In Proc. AUSCRYPT '92, Springer LNCS, vol. 718, pp. 67-79, 1993.
4. S. Blackburn, M. Burmester, Y. Desmedt, and P. Wild. Efficient multiplicative sharing scheme. In: Proc. EUROCRYPT '96, Springer LNCS, vol. 1070, pp. 107–118, 1996.

5. G. R. Blakley. Safeguarding cryptographic keys. In: Proc. National Computer Conference '79, AFIPS Proceedings, vol. 48, pp. 313-317, 1979.
6. E. F. Brickell. Some ideal secret sharing schemes. In: J. Combin. Maths. & Combin. Comp. vol. 9, pp. 105–113, 1989.
7. T. Cover and J. Thomas. Elements of information theory. Wiley Series in Telecommunications, 1991.
8. R. Cramer, I. Damgaard, and U. Maurer. Efficient general secure multi-party computation from any linear secret-sharing scheme. In: Proc. EUROCRYPT '00, Springer LNCS, vol. 1807, pp. 316–334, 2000.
9. R. Cramer, S. Fehr, Y. Ishai, and E. Kushilevitz. Efficient multi-party computation over rings. Manuscript, February 2002.
10. Y. Di Crescenzo, and Y. Frankel. Existence of Multiplicative Secret Sharing Schemes with Polynomial Share Expansion. In: Proc. SODA '99, ACM Press, pp. 895–896, 1999.
11. Y. Desmedt and Y. Frankel. Theshold cryptosystem. In: Proc. CRYPTO '89, Springer LNCS, vol. 435, pp. 307–315, 1990.
12. Y. Desmedt and Y. Frankel. Homomorphic zero-knowledge threshold schemes over any finite Abelian group. In: SIAM Journal on Discrete Mathematics, 7(4), pp. 667–679, 1994.
13. Y. Desmedt, A. De Santis, Y. Frankel, and M. Yung. How to share a function securely. In: Proc. STOC '94, ACM Press, pp. 22–33, 1994.
14. Y. Desmedt, G. Di Crescenzo, and M. Burmester. Multiplicative non-abelian sharing schemes and their application to threshold cryptography. In: Proc. ASIACRYPT '94, Springer LNCS, vol. 917, pp. 21–31, 1995.
15. Y. Desmedt, B. King, W. Kishimoto, and K. Kurosawa. A comment on the efficiency of secret sharing scheme over any finite Abelian group. In: Proc. ACISP '98, Springer LNCS, vol. 1438, pp. 391–402, 1998.
16. M. van Dijk. Secret key sharing and secret key generation. Ph. D. Thesis, Eindhoven University of Technology, 1997.
17. Y. Frankel, P. Gemmell, P. MacKenzie, and M. Yung. Optimal resilience proactive public-key cryptosystems. In: Proc. FOCS '97, IEEE Press, pp. 384–393, 1997.
18. Y. Frankel, P. Gemmell, P. MacKenzie, and M. Yung. Proactive RSA. In: Proc. CRYPTO '97, Springer LNCS, vol. 1294, pp. 440–454, 1997.
19. A. Gál. Combinatorial methods in boolean function complexity. Ph.D.-thesis, University of Chicago, 1995.
20. M. Karchmer and A. Wigderson. On span programs. In: Proc. Structures in Complexity Theory '93, IEEE Computer Society Press, pp. 102–111, 1993.
21. B. King. Some results in linear secret sharing. Ph.D.-thesis, University of Wisconsin-Milwaukee, 2001.
22. B. King. Randomness required for linear threshold sharing schemes defined over any finite abelian group. In: Proc. ACISP '01, Springer LNCS, vol. 2119, pp. 376–391, 2001.
23. S. Lang. Algebra. Addison-Wesley Publishing Co., 2nd edition, 1984.
24. A. Shamir. How to share a secret. In: Communications of the ACM, (22) pp. 612–613, 1979.
25. V. Shoup. Practical threshold signatures. In: Proc. EUROCRYPT '00, Springer LNCS, vol. 1807, pp. 207–220, 2000.

A Generalized Birthday Problem

(Extended Abstract)

David Wagner

University of California at Berkeley

Abstract. We study a k-dimensional generalization of the birthday problem: given k lists of n-bit values, find some way to choose one element from each list so that the resulting k values XOR to zero. For $k = 2$, this is just the extremely well-known birthday problem, which has a square-root time algorithm with many applications in cryptography. In this paper, we show new algorithms for the case $k > 2$: we show a cube-root time algorithm for the case of $k = 4$ lists, and we give an algorithm with subexponential running time when k is unrestricted.

We also give several applications to cryptanalysis, describing new subexponential algorithms for constructing one-more forgeries for certain blind signature schemes, for breaking certain incremental hash functions, and for finding low-weight parity check equations for fast correlation attacks on stream ciphers. In these applications, our algorithm runs in $O(2^{2\sqrt{n}})$ time for an n-bit modulus, demonstrating that moduli may need to be at least 1600 bits long for security against these new attacks. As an example, we describe the first-known attack with subexponential complexity on Schnorr and Okamoto-Schnorr blind signatures over elliptic curve groups.

1 Introduction

One of the best-known combinatorial tools in cryptology is the birthday problem:

Problem 1. Given two lists L_1, L_2 of elements drawn uniformly and independently at random from $\{0, 1\}^n$, find $x_1 \in L_1$ and $x_2 \in L_2$ such that $x_1 \oplus x_2 = 0$.

(Here the $\oplus$ symbol represents the bitwise exclusive-or operation.) The birthday problem is well understood: A solution x_1, x_2 exists with good probability once $|L_1| \times |L_2| \gg 2^n$ holds, and if the list sizes are favorably chosen, the complexity of the optimal algorithm is $\Theta(2^{n/2})$. The birthday problem has numerous applications throughout cryptography and cryptanalysis.

In this paper, we explore a generalization of the birthday problem. The above presentation suggests studying a variant of the birthday problem with an arbitrary number of lists. In this way, we obtain the following k-dimensional analogue, which we call the k-sum problem:

Problem 2. Given k lists $L_1, \ldots, L_k$ of elements drawn uniformly and independently at random from $\{0, 1\}^n$, find $x_1 \in L_1, \ldots, x_k \in L_k$ such that $x_1 \oplus x_2 \oplus \cdots \oplus x_k = 0$.

M. Yung (Ed.): CRYPTO 2002, LNCS 2442, pp. 288–303, 2002.
© Springer-Verlag Berlin Heidelberg 2002

We allow the lists to be extended to any desired length, and so it may aid the intuition to think of each element of each list as being generated by a random (or pseudorandom) oracle R_i, so that the j-th element of L_i is $R_i(j)$. It is easy to see that a solution to the k-sum problem exists with good probability so long as $|L_1| \times \cdots \times |L_k| \gg 2^n$. However, the challenge is to find a solution $x_1, \ldots, x_k$ explicitly and efficiently.

This first half of this paper is devoted to a theoretical study of this problem. First, Section 2 describes a new algorithm, called the k-tree algorithm, that allows us to solve the k-sum problem with lower complexity than previously known to be possible. Our algorithm works only when one can extend the size of the lists freely, i.e., in the special case where there are sufficiently many solutions to the k-sum problem. For example, we show that, for $k = 4$, the k-sum problem can be solved in $O(2^{n/3})$ time using lists of size $O(2^{n/3})$. We also discuss a number of generalizations of the problem, e.g., to operations other than XOR. Next, in Section 3, we study the complexity of the k-sum problem and give several lower bounds. This theoretical study provides a tool for cryptanalysis which we will put to use in the second half of the paper.

The k-sum problem may not appear very natural at first sight, and so it may come as no surprise that, to our knowledge, this problem has not previously been stated or studied in full generality. Nonetheless, we show in the second half of this paper a number of cases where the k-sum problem has applications to cryptanalysis of various systems: in Section 4, we show how to break various blind signature schemes, how to attack several incremental hash functions, and how to find low-weight parity checks. Other examples may be found in the full version of this paper [36]. We do not claim that this is an exhaustive list of possible applications; rather, it is intended to motivate the relevance of this problem to cryptography.

Finally, we conclude in Sections 5 and 6 with several open problems and final remarks.

2   Algorithms

The classic birthday problem. We recall the standard technique for finding solutions to the birthday problem (with 2 lists). We define a join operation $\bowtie$ on lists so that $S \bowtie T$ represents the list of elements common to both S and T. Note that $x_1 \oplus x_2 = 0$ if and only if $x_1 = x_2$. Consequently, a solution to the classic (2-list) birthday problem may be found by simply computing the join $L_1 \bowtie L_2$ of the two input lists L_1, L_2. We represent this algorithm schematically in Figure 1.

The join operation has been well-studied in the literature on database query evaluation, and several efficient methods for computing joins are known. A merge-join sorts the two lists, L_1, L_2, and scans the two sorted lists, returning any matching pairs detected. A hash-join stores one list, say L_1, in a hash table, and then scans through each element of L_2 and checks whether it is present in the hash table. If memory is plentiful, the hash-join is very efficient:

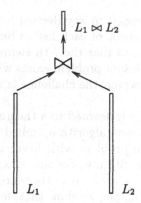

Fig. 1. An abstract representation of the standard algorithm for the (2-list) birthday problem: given two lists L_1, L_2, we use a join operation to find all pairs (x_1, x_2) such that $x_1 = x_2$ and $x_1 \in L_1$ and $x_2 \in L_2$. The thin vertical boxes represent lists, the arrows represent flow of values, and the $\bowtie$ symbol represents a join operator.

it requires $|L_1| + |L_2|$ simple steps of computation and $\min(|L_1|, |L_2|)$ units of storage. A merge-join is slower in principle, running in $O(n \log n)$ time where $n = \max(|L_1|, |L_2|)$, but external sorting methods allow computation of merge-joins even when memory is scarce.

The consequence of these observations is that the birthday problem may be solved with square-root complexity. In particular, if we operate on n-bit values, then the above algorithms will require $O(2^{n/2})$ time and space, if we are free to choose the size of the lists however we like. Techniques for reducing the space complexity of this algorithm are known for some important special cases [25].

The birthday problem has many applications. For example, if we want to find a collision for a hash function $h : \{0,1\}^* \to \{0,1\}^n$, we may define the j-th element of list L_i as $h(i, j)$. Assuming that h behaves like a random function, the lists will contain an inexhaustible supply of values distributed uniformly and independently at random, so the premises of the problem statement will be met. Consequently, we can expect to find a solution to the corresponding birthday problem with $O(2^{n/2})$ work, and any such solution immediately yields a collision for the hash function [38].

The 4-list birthday problem. To extend the above well-known observations, consider next the 4-sum problem. We are given lists $L_1, \ldots, L_4$, and our task is to find values $x_1, \ldots, x_4$ that XOR to zero. (Hereafter $x_i \in L_i$ holds implicitly.) It is easy to see that a solution should exist with good probability if each list is of length at least $2^{n/4}$. Nonetheless, no good algorithm for explicitly finding such a solution was previously known: The most obvious approaches all seem to require $2^{n/2}$ steps of computation.

We develop here a more efficient algorithm for the 4-sum problem. Let $\text{low}_\ell(x)$ denote the least significant ℓ bits of x, and define the generalized join operator

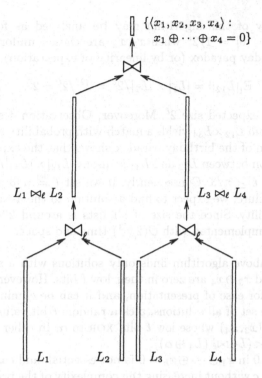

$$\{\langle x_1, x_2, x_3, x_4 \rangle : \\ x_1 \oplus \cdots \oplus x_4 = 0\}$$

$L_1 \bowtie_\ell L_2$ $L_3 \bowtie_\ell L_4$

L_1 L_2 L_3 L_4

Fig. 2. A pictorial representation of our algorithm for the 4-sum problem.

$\bowtie_\ell$ so that $L_1 \bowtie_\ell L_2$ contains all pairs from $L_1 \times L_2$ that agree in their ℓ least significant bits. We will use the following basic properties of the problem:

Observation 1 *We have* $low_\ell(x_i \oplus x_j) = 0$ *if and only if* $low_\ell(x_i) = low_\ell(x_j)$.

Observation 2 *Given lists* L_i, L_j, *we can easily generate all pairs* $\langle x_i, x_j \rangle$ *satisfying* $x_i \in L_i$, $x_j \in L_j$, *and* $low_\ell(x_i \oplus x_j) = 0$ *by using the join operator* $\bowtie_\ell$.

Observation 3 *If* $x_1 \oplus x_2 = x_3 \oplus x_4$, *then* $x_1 \oplus x_2 \oplus x_3 \oplus x_4 = 0$.

Observation 4 *If* $low_\ell(x_1 \oplus x_2) = 0$ *and* $low_\ell(x_3 \oplus x_4) = 0$, *then we necessarily have* $low_\ell(x_1 \oplus x_2 \oplus x_3 \oplus x_4) = 0$, *and in this case* $\Pr[x_1 \oplus x_2 \oplus x_3 \oplus x_4 = 0] = 2^\ell / 2^n$.

These properties suggest a new algorithm for the 4-sum problem. First, we extend the lists $L_1, \ldots, L_4$ until they each contain about 2^ℓ elements, where ℓ is a parameter to be determined below. Then, we apply Observation 2 to generate a large list L_{12} of values $x_1 \oplus x_2$ such that $low_\ell(x_1 \oplus x_2) = 0$. Similarly, we generate a large list L_{34} of values $x_3 \oplus x_4$ where $low_\ell(x_3 \oplus x_4) = 0$. Finally, we search for matches between L_{12} and L_{34}. By Observation 3, any such match will satisfy $x_1 \oplus x_2 \oplus x_3 \oplus x_4 = 0$ and hence will yield a solution to the 4-sum problem. See Figure 2 for a visual depiction of this algorithm.

The complexity of this algorithm may be analyzed as follows. We have $\Pr[\text{low}_\ell(x_1 \oplus x_2) = 0] = 1/2^\ell$ when x_1, x_2 are chosen uniformly at random. Thus, by the birthday paradox (or by linearity of expectation),

$$\mathbb{E}[|L_{12}|] = |L_1| \times |L_2|/2^\ell = 2^{2\ell}/2^\ell = 2^\ell.$$

Similarly, L_{34} has expected size 2^ℓ. Moreover, Observation 4 ensures that any pair of elements from $L_{12} \times L_{34}$ yields a match with probability $2^\ell/2^n$. Therefore, a second invocation of the birthday paradox shows that the expected number of elements in common between L_{12} and L_{34} is about $|L_{12}| \times |L_{34}|/2^{n-\ell}$. The latter is at least 1 when $\ell \geq n/3$. Consequently, if we set $\ell \stackrel{\text{def}}{=} n/3$ as the beginning of the above procedure, we expect to find a solution to the 4-sum problem with non-trivial probability. Since the size of all lists is around $2^{n/3}$, the resulting algorithm can be implemented with $O(2^{n/3})$ time and space.

Extensions. The above algorithm finds only solutions with a special property, namely, $x_1 \oplus x_2$ and $x_3 \oplus x_4$ are zero in their low ℓ bits. However, this restriction was made merely for ease of presentation, and it can be eliminated. To sample randomly from the set of all solutions, pick a random ℓ-bit value α, and look for pairs (x_1, x_2) and (x_3, x_4) whose low ℓ bits XOR to α. In other words, compute $(L_1 \bowtie_\ell (L_2 \oplus \alpha)) \bowtie (L_3 \bowtie_\ell (L_4 \oplus \alpha))$.

Also, the value 0 in $x_1 \oplus \cdots \oplus x_k = 0$ is not essential, and can be replaced by any other constant c without increasing the complexity of the problem. This may be easily seen as follows: if we replace L_k with $L'_k = L_k \oplus c \stackrel{\text{def}}{=} \{x_k \oplus c : x_k \in L_k\}$, then any solution to $x_1 \oplus \cdots \oplus x_{k-1} \oplus x'_k = 0$ will be a solution to $x_1 \oplus \cdots \oplus x_k = c$ and vice versa. Consequently, we may assume (without loss of generality) that $c = 0$.

As a corollary, when $k > k'$ the complexity of the k-sum problem can be no larger than the complexity of the k'-sum problem. This can be proven using a trivial list-elimination trick. We pick arbitrary values $x_{k'+1}, \ldots, x_k$ from $L_{k'+1}, \ldots, L_k$ and fix this choice. Then, we set $c \stackrel{\text{def}}{=} x_{k'+1} \oplus \cdots \oplus x_k$ and use a k'-sum algorithm to find a solution to the equation $x_1 \oplus \cdots \oplus x_{k'} = c$. For instance, this shows that we can solve the k-sum problem with complexity at most $O(2^{n/3})$ for all $k \geq 4$.

More interestingly, we can use the above ideas to solve the k-sum problem even faster than cube-root time for larger values of k. We extend the 4-list tree algorithm above as follows. When k is a power of two, we replace the complete binary tree of depth 2 in Figure 2 with a complete binary tree of depth $\lg k$. At internal nodes of height h, we use the join operator $\bowtie_{\ell_h}$ (where $\ell_h = hn/(1 + \lg k)$), except that at the root we use the full join operator $\bowtie$. Each element x of an internal list $L_{i \ldots j}$ contains back-pointers to elements x' and x'' of the two child lists used to form $L_{i \ldots j}$, such that $x = x' \oplus x''$. In this way we will obtain an algorithm for the k-sum problem that requires $O(k \cdot 2^{n/(1+\lg k)})$ time and space and uses lists of size $O(2^{n/(1+\lg k)})$. The complexity of this algorithm improves only slowly as k increases, though, so this does not seem to yield large improvements unless k becomes quite large.

We can also obtain an algorithm for the general k-sum problem when k is not a power of two. We take $k' \stackrel{\text{def}}{=} 2^{\lfloor \lg k \rfloor}$ to be the largest power of two less than k, and we use the list-elimination trick above. However, the results are less satisfying: The algorithm obtained in this way runs essentially no faster for $k = 2^i + j$ than for $k = 2^i$.

Operations other than XOR. The k-sum problem has so far been described over the group $(GF(2)^n, \oplus)$, but it is natural to wonder whether these techniques will apply over other groups as well.

We note first that the tree algorithm above transfers immediately to the additive group $(\mathbb{Z}/2^n\mathbb{Z}, +)$. In particular, we compute $L_{12} \stackrel{\text{def}}{=} L_1 \bowtie_\ell -L_2$, $L_{34} \stackrel{\text{def}}{=} L_3 \bowtie_\ell -L_4$, and finally $L_{12} \bowtie -L_{34}$, where $-L \stackrel{\text{def}}{=} \{-x \bmod 2^n : x \in L\}$. The result will be a set of solutions to the equation $x_1 + \cdots + x_k \equiv 0 \pmod{2^n}$. The reason this works is that $a \equiv b \pmod{2^\ell}$ implies $(a+c \bmod 2^n) \equiv (b+c \bmod 2^n)$ $\pmod{2^\ell}$: the carry bit propagates in only one direction.

We can also apply the tree algorithm to the group $(\mathbb{Z}/m\mathbb{Z}, +)$ where m is arbitrary. Let $[a, b] \stackrel{\text{def}}{=} \{x \in \mathbb{Z}/m\mathbb{Z} : a \leq x \leq b\}$ denote the interval of elements between a and b (wrapping modulo m), and define the join operation $L_1 \bowtie_{[a,b]} L_2$ to represent the solutions to $x_1 + x_2 \in [a, b]$ with $x_i \in L_i$. Then we may solve a 4-sum problem over $\mathbb{Z}/m\mathbb{Z}$ by computing $(L_1 \bowtie_{[a,b]} L_2) \bowtie (L_3 \bowtie_{[a,b]} L_4)$ where $[a, b] = [-m/2^{\ell+1}, m/2^{\ell+1} - 1]$ and $\ell = \frac{1}{3} \lg m$. In general, one can adapt the k-tree algorithm to work in $(\mathbb{Z}/m\mathbb{Z}, +)$ by replacing each $\bowtie_\ell$ operator with $\bowtie_{[-m/2^{\ell+1}, m/2^{\ell+1}-1]}$, and this will let us solve k-sum problems modulo m about as quickly as we can for XOR.

Finding many solutions. In some applications, it may be useful to find many solutions to the k-sum problem. It is not too hard to see[1] that we can find α^3 solutions to the 4-sum problem with about α times the work of finding a single solution, so long as $\alpha \leq 2^{n/6}$. Similarly, we can find $\alpha^{1+\lfloor \lg k \rfloor}$ solutions to the k-sum problem with α times as much work as finding a single solution, as long as $\alpha \leq 2^{n/(\lfloor \lg k \rfloor \cdot (1 + \lfloor \lg k \rfloor))}$.

Reducing the space complexity. As we have described it so far, these algorithms require a lot of memory. Since memory is often more expensive than computing time, this may be a barrier in practice. While we have not extensively studied the memory complexity of the k-sum problem, we note that in some cases a trivial technique can greatly reduce the space complexity of our k-tree algorithm. In particular, when $k \gg 2$, we can evaluate the tree in postfix order, discarding lists when they are no longer needed. In this way, we will need storage for only about $\lg k$ lists. For example, if we take $k = 2^{\sqrt{n}-1}$, then the k-tree algorithm will run in approximately $2^{2\sqrt{n}}$ time and $\sqrt{n}2^{\sqrt{n}}$ space using this optimization, a significant improvement over naive implementations.

---

[1] Simply use lists $L_1, \ldots, L_4$ of size $\alpha \cdot 2^{n/3}$, and filter on $\ell' = n/3 + \lg \alpha$ bits at the lower level of the tree.

Related work. The idea of using a priority queue to generate pairwise sums $x_1 + x_2$ in sorted order (for $x_1 \in L_1$, $x_2 \in L_2$ with lists L_1, L_2 given as input) first appeared in Knuth, exercise 5.2.3-29, and was credited to W.S. Brown [20, p.158].

Later, Schroeppel and Shamir showed how to generate 4-wise sums $x_1 + \cdots + x_4$ in sorted order using a tree of priority queues [30, 31]. In particular, given 4 lists of integers and a n-bit integer c, they considered how to find all solutions to $x_1 + \cdots + x_4 = c$, and they gave an algorithm running in $\Theta(2^{n/2})$ time and $\Theta(2^{n/4})$ space when the lists are of size $\Theta(2^{n/4})$. In contrast, the problem we consider differs in four ways: we relax the problem to ask only for a single solution rather than all solutions; we allow an arbitrary number of lists; we consider other group operations; and, most importantly, our main goal in this paper is to break the $\Theta(2^{n/2})$ running time barrier. When looking for only a single solution, it is possible to beat Schroeppel and Shamir's algorithm—using Floyd's cycle-finding algorithm, distinguished points cycling algorithms [24], or parallel collision search [25], one can often achieve $\Theta(2^{n/2})$ time and $\Theta(1)$ space— but there was previously no known algorithm with running time substantially better than $2^{n/2}$. Consequently, Schroeppel and Shamir's result is not directly applicable to our problem, but their idea of using tree-based algorithms can be seen as a direct precursor of our k-tree algorithm.

Bernstein has used similar techniques in the context of enumerating solutions in the integers to equations such as $a^3 + 2b^3 + 3c^3 - 4d^3 = 0$ [3].

Boneh, Joux and Nguyen have used Schroeppel and Shamir's algorithm for solving integer knapsacks to reduce the space complexity of their birthday attacks on plain RSA and El Gamal [6]. They also used (a version of) our Theorem 3 to transform a 4-sum problem over $((\mathbb{Z}/p\mathbb{Z})^*, \times)$ to a knapsack (i.e., 4-sum) problem over $(\mathbb{Z}/q\mathbb{Z}, +)$, which allowed them to apply Schroeppel and Shamir's techniques.

Bleichenbacher used similar techniques in his attack on DSA [4].

Chose, Joux, and Mitton have independently discovered a space-efficient algorithm for finding all solutions to $x_1 \oplus \cdots \oplus x_k = 0$ and shown how to use it to speed up search for parity checks for stream cipher cryptanalysis [11]. For $k = 4$, their approach runs in $O(2^{n/2})$ time and $O(2^{n/4})$ space if $|L_1| = \cdots = |L_4| = 2^{n/4}$ and all values are n bits long, and so their scheme is in a similar class as Schroeppel and Shamir's result. Interestingly, the algorithm of Chose, et al., is essentially equivalent to repeatedly running our 4-list tree algorithm once for each possible predicted value of $\alpha = \mathrm{low}_l(x_1 \oplus x_2)$, taking $\ell = n/4$. Thus, their work is complementary to ours: their algorithm does not beat the square-root barrier, but it takes a different point in the tradeoff space, thereby reinforcing the importance of the k-sum problem to cryptography.

Joux and Lercier have used related ideas to reduce the space complexity of a birthday step in point-counting algorithms for elliptic curves [19].

Blum, Kalai, and Wasserman previously have independently discovered something closely related to the k-tree algorithm for XOR in the context of their work on learning theory [5]. In particular, they use the existence of a subexponential algorithm for the k-sum problem when k is unrestricted to find the first known subexponential algorithm for the "learning parity with noise" problem. We note that any improvement in the k-tree algorithm would immediately lead to improved algorithms for learning parity with noise, a problem that has resisted algorithmic progress for many years. Others in learning theory have since used similar ideas [37], and the hardness of this problem has even been proposed as the basis for a human-computer authentication scheme [17].

Ajtai, Kumar, and Sivakumar have used Blum, Kalai, and Wasserman's algorithm as a subroutine to speed up the shortest lattice vector problem from $2^{O(n \log n)}$ to $2^{O(n)}$ time [1].

Bellare, et al., showed that the k-sum problem over $(GF(2)^n, \oplus)$ can be solved in $O(n^3 + kn)$ time using Gaussian elimination when $k \geq n$ [2, Appendix A].

Wagner and Goldberg have shown how to efficiently find solutions to $x_1 = x_2 = \cdots = x_k$ (where $x_i \in L_i$) using parallel collision search [35]. This is an alternative way to generalize the birthday problem to higher dimensions, but the techniques do not seem to carry over to the k-sum problem.

There is also a natural connection between the k-sum problem over $(\mathbb{Z}/m\mathbb{Z}, +)$ and the subset sum problem over $\mathbb{Z}/m\mathbb{Z}$. This suggests that techniques known for the subset sum problem, such as LLL lattice reduction, may be relevant to the k-sum problem. We have not explored this direction, and we leave it to future work.

Summary. We have shown how to solve the k-sum problem (for the XOR operation) in $O(k \cdot 2^{n/(1+\lfloor \lg k \rfloor)})$ time and space, using lists of size $O(2^{n/(1+\lfloor \lg k \rfloor)})$. In particular, for $k = 4$, we can solve the 4-sum problem with complexity $O(2^{n/3})$. If k is unrestricted, we obtain a subexponential algorithm running in $O(2^{2\sqrt{n}})$ time by setting $k = 2^{\sqrt{n}-1}$.

3   Lower Bounds

In this section we study how close to optimal the k-tree algorithm is. This section may be safely skipped on first reading.

Information-theoretic bounds. We can easily use information-theoretic arguments to bound the complexity of the k-sum problem as follows.

Theorem 1. *The computational complexity of the k-sum problem is $\Omega(2^{n/k})$.*

Proof. For the k-sum problem to have a solution with constant probability, we need $|L_1| \times \cdots \times |L_k| = \Omega(2^n)$, i.e., $\max_i |L_i| = \Omega(2^{n/k})$. The bound follows easily.

This bound applies to the k-sum problem over all groups.

However, this gives a rather weak bound. There is a considerable gap between the information-theoretic lower bound $\Omega(2^{n/k})$ and the constructive upper bound $O(k \cdot 2^{n/(1+\lfloor \lg k \rfloor)})$ established in the previous section. Therefore, it is natural to wonder whether this gap can be narrowed. In the general case, this seems to be a difficult question, but we show next that the lower bound can be improved in some special cases.

Relation to discrete logs. There are close connections to the discrete log problem, as shown by the following observation from Wei Dai [12].

Theorem 2 (W. Dai). *If the k-sum problem over a cyclic group $G = \langle g \rangle$ can be solved in time t, then the discrete logarithm with respect to g can be found in $O(t)$ time as well.*

Proof. We describe an algorithm for finding the discrete logarithm $\log_g y$ of a group element $y \in G$ using any algorithm for the k-sum problem in G. Each list will contain elements of the form g^w for w chosen uniformly at random. Then any solution to $x_1 \times \cdots \times x_k = y$ with $x_i \in L_i$ will yield a relation of the form $g^w = y$, where $w = w_1 + \cdots + w_k$, and this reveals the discrete log of y with respect to g, as claimed.

This immediately allows us to rule out the possibility of an efficient generic algorithm for the k-sum problem over any group G with order divisible by any large prime. Recall that a generic algorithm is one that uses only the basic group operations (multiplication, inversion, testing for equality) and ignores the representation of elements of G.

Corollary 1. *Every generic algorithm for the k-sum problem in a group G has running time $\Omega(\sqrt{p})$, where p denotes the largest prime factor of the order of G.*

Proof. Any generic algorithm for the discrete log problem in a group of prime order p has complexity $\Omega(\sqrt{p})$ [23, 33]. Now see Theorem 2.

Moreover, Theorem 2 shows that we cannot hope to find a polynomial-time algorithm for the k-sum problem over any group where the discrete log problem is hard. For example, finding a solution to $x_1 \times \cdots \times x_k \equiv 1 \pmod{p}$ is as hard as taking discrete logarithms in $(\mathbb{Z}/p\mathbb{Z})^*$, and thus we cannot expect any especially good algorithm for this problem.

The relationship to the discrete log problem goes both ways:

Theorem 3. *Suppose the discrete log problem in a multiplicative group $G = \langle g \rangle$ of order m can be solved in time t. Suppose moreover that the k-sum problem over $(\mathbb{Z}/m\mathbb{Z}, +)$ with lists of size ℓ can be solved in time t'. Then the k-sum problem over G with lists of size ℓ can be solved in time $t' + k\ell t$.*

Proof. Let $L_i' = \{\log_g x : x \in L_i\}$; then any solution to the k-sum problem over $(\mathbb{Z}/m\mathbb{Z}, +)$ with lists $L_1', \ldots, L_k'$ yields a solution to the k-sum problem over G with lists $L_1, \ldots, L_k$.

As we have seen earlier, there exists an algorithm for solving the k-sum problem over $(\mathbb{Z}/m\mathbb{Z}, +)$ in time $t' = O(k \cdot m^{1/(1+\lfloor \lg k \rfloor)})$ so long as each list has size at least $\ell \geq t'/k$. As a consequence, there are non-trivial algorithms for solving the k-sum problem in any group where the discrete log problem is easy.

4   Attacks and Applications

Blind signatures. Schnorr has recently observed that the security of several discrete-log-based blind signature schemes depends not only on the hardness of the discrete log but also on the hardness of a novel algorithmic problem, called *the ROS problem* [28]. This observation applies to Schnorr blind signatures and Okamoto-Schnorr blind signatures, particular when working over elliptic curve groups and other groups with no known subexponential algorithm for the discrete log.

We recall the ROS problem. Suppose we are working in a group of prime order q. Let $F : \{0,1\}^* \to GF(q)$ represent a cryptographic hash function, e.g., a random oracle. The goal is to find a singular $k \times k$ matrix M over $GF(q)$ satisfying two special conditions. First, the entries of the matrix should satisfy

$$M_{i,k} = F(M_{i,1}, M_{i,2}, \ldots, M_{i,k-1}) \qquad \text{for } i = 1, \ldots, k.$$

Second, there should be a vector in the kernel of M whose last component is non-zero: in other words, there should exist $v = (v_1, \ldots, v_k)^T \in GF(q)^k$ with $Mv = 0$ and $v_k = -1$.

Any algorithm to solve the ROS problem immediately leads to a one-more forgery attack using $k - 1$ parallel interactions with the signer. Previously, the best algorithm known for the ROS problem required $\Theta(q^{1/2})$ time.

We show that the ROS problem can be solved in subexponential time using our k-tree algorithm. To illustrate the idea, we first show how to solve the ROS problem in cube-root time for the case $k = 4$. Consider matrices of the following form:

$$M = \begin{bmatrix} w_1 & 0 & 0 & F(w_1,0,0) \\ 0 & w_2 & 0 & F(0,w_2,0) \\ 0 & 0 & w_3 & F(0,0,w_3) \\ w_4 & w_4 & w_4 & F(w_4,w_4,w_4) \end{bmatrix},$$

where $w_1, \ldots, w_4$ vary over $GF(q)^*$. We note that M is of the desired form if the unknowns $w_1, \ldots, w_4$ satisfy the equation

$$F(w_1,0,0)/w_1 + F(0,w_2,0)/w_2 +$$

$$+F(0,0,w_3)/w_3 - F(w_4,w_4,w_4)/w_4 \equiv 0 \pmod{q}.$$

Thus, this can be viewed as an instance of a 4-sum problem over $GF(q)$: we fill list L_1 with candidates for the first term of the equation above, i.e., with values of the form $F(w_1,0,0)/w_1$, and similarly for L_2, L_3, L_4; then we search for a solution to $x_1 + \cdots + x_4 \equiv 0 \pmod{q}$ with $x_i \in L_i$. Applying our 4-list tree algorithm lets us break Schnorr and Okamoto-Schnorr blind signatures over a group of prime order q in $\Theta(q^{1/3})$ time and using 3 parallel interactions with the signer.

Of course, the above attack can be generalized to any number $k > 4$ of lists. As a concrete example, if we consider an elliptic curve group of order $q \approx 2^{160}$, then there is a one-more forgery attack using $k - 1 = 2^9 - 1$ parallel interactions,

2^{25} work, and 2^{12} space. Compare this to the conjectured 2^{80} security level that seems to be usually expected if one assumes that the best attack is to compute the discrete log using a generic algorithm. We see that the k-tree algorithm yields unexpectedly devastating attacks on these blind signature schemes.

In the general case, we obtain a signature forgery attack with subexponential complexity. If we take $k = 2^{\sqrt{\lg q}-1}$, the k-tree algorithm runs in roughly $2^{2\sqrt{\lg q}}$ time, requires $2^{\sqrt{\lg q}-1}\sqrt{\lg q}$ space, and uses $2^{\sqrt{\lg q}-1}-1$ parallel interactions with the signer. Consequently, it seems that we need a group of order $q \gg 2^{1600}$ if we wish to enjoy 80-bit security. In other words, the size of the group order in bits must be an order of magnitude larger than one might otherwise expect from the best currently-known algorithms for discrete logs in elliptic curve groups.

NASD incremental hashing. One proposal for network-attached secure disks (NASD) uses the following hash function for integrity purposes [13, 14]:

$$H(x) \stackrel{\text{def}}{=} \sum_{i=1}^{k} h(i, x_i) \bmod 2^{256}.$$

Here x denotes a padded k-block message, $x = \langle x_1, \ldots, x_k \rangle$. We reduce inverting this hash to a k-sum problem over the additive group $(\mathbb{Z}/2^{256}\mathbb{Z}, +)$.

The inversion attack proceeds as follows. Generate k lists $L_1, \ldots, L_k$, where L_i consists of $y_i = h(i, x_i)$ with x_i ranging over many values chosen at random. Then any solution to $y_1 + \cdots + y_k \equiv c \pmod{2^{256}}$ with $y_i \in L_i$ reveals a pre-image of the digest c. If we take $k = 128$, for example, we can find a 128-block message that hashes to a desired digest using the k-tree algorithm and about 2^{40} work.

The attack can be further improved by exploiting the structure of h, which divides its one-block input x_i into two halves y_i, z_i and then computes

$$h(i, \langle y_i, z_i \rangle) \stackrel{\text{def}}{=} (\text{SHA}(2i, y_i) \ll 96) \oplus \text{SHA}(2i + 1, z_i).$$

Our improved attack proceeds as follows. First, we find values $z_1, \ldots, z_k$ satisfying $\text{SHA}(3, z_1) + \text{SHA}(5, z_2) + \cdots + \text{SHA}(2k + 1, z_k) \equiv 0 \pmod{2^{96}}$. If we set $k = 128$, this can be done with about 2^{20} work using the k-tree algorithm. We fix the values $z_1, \ldots, z_k$ obtained this way, and then we search for values $y_1, \ldots, y_k$ such that $h(1, \langle y_1, z_1 \rangle) + \cdots + h(k, \langle y_k, z_k \rangle) \equiv 0 \pmod{2^{256}}$. Due to the structure of h, the left-hand side is guaranteed to be zero modulo 2^{96}, so 96 bits come for free and we have a k-sum problem over only 160 bits. The latter problem can be solved with about 2^{28} work using a second invocation of the k-tree algorithm.

This allows an adversary to find a pre-image with 2^{28} work. Similar techniques can be used to find collisions in about the same complexity. We conclude, therefore, that the NASD hash should be considered thoroughly broken.

AdHash. The NASD hash may be viewed as a special case of a general incremental hashing construction proposed by Bellare, et al., and named AdHash [2]:

$$H(x) \stackrel{\text{def}}{=} \sum_{i=1}^{k} h(i, x_i) \bmod m,$$

where the modulus m is public and chosen randomly. However, Bellare, et al., give no concrete suggestions for the size of m, and so it is no surprise that some implementors have used inadequate parameters: for instance, NASD used a 256-bit modulus [13, 14], and several implementations have used a 128-bit modulus [8, 9, 32]. Our first attack on the NASD hash applies to AdHash as well, so we find that AdHash's modulus m must be very large indeed: the asymptotic complexity of the k-sum problem is as low as $O(2^{2\sqrt{\lg m}})$ if we take $k = 2^{\sqrt{\lg m}-1}$, so we obtain an attack on AdHash with complexity $O(2^{2\sqrt{\lg m}})$.

To our knowledge, this appears to be the first subexponential attack on AdHash. As a consequence of this attack, we will need to ensure that $m \gg 2^{1600}$ if we want 80-bit security. The need for such a large modulus may reduce or negate the performance advantages of AdHash.

The PCIHF hash. We next cryptanalyze the PCIHF hash construction, proposed recently for incremental hashing [15]. PCIHF hashes a padded n-block message x as follows:

$$H(x) \stackrel{\text{def}}{=} \sum_{i=1}^{n-1} \text{SHA}(x_i, x_{i+1}) \bmod 2^{160} + 1.$$

Our attack on AdHash does not apply directly to this scheme, because the blocks cannot be varied independently: changing x_i affects two terms in the above sum. However, it is not too difficult to extend our attack on AdHash to apply to PCIHF as well.

We first show how to compute pre-images. Let us fix every other block of x, say $x_2 = x_4 = x_6 = \cdots = 0$, and vary the remaining blocks of x. Then the hash computation takes the form

$$H(x) = \sum_{j=1}^{\lfloor (n+1)/2 \rfloor} h(x_{2j-1}) \bmod 2^{160} + 1 \quad \text{where } h(w) \stackrel{\text{def}}{=} \text{SHA}(0, w) + \text{SHA}(w, 0).$$

Now we may apply the AdHash attack to this equation, and if we take $n = 255$ and apply the 128-list tree algorithm, we can find a 255-block preimage of H with about 2^{28} work.

Similarly, it is straightforward to adapt the attack to find collisions for PCIHF. The above ideas can be used to find a pair of 127-block messages that hash to the same digest, after about 2^{28} work. These results demonstrate that PCIHF is highly insecure as proposed. Though the basic idea underlying PCIHF may be sound, it seems that one must choose a much larger modulus, or some other combining operation with better resistance to subset sum attacks.

Low-weight parity checks. Let $p(x)$ be an irreducible polynomial of degree n over $GF(2)$. A number of attacks on stream cipher begin by solving an instance of the following problem:

The parity-check problem. Given an irreducible polynomial $p(x)$, find a multiple $m(x)$ of $p(x)$ so that $m(x)$ has very low weight (say, 3 or 4 or 5) and so that the degree of $m(x)$ is not too large (say, 2^{32} or so).

Here, the weight of a polynomial is defined as the number of non-zero coefficients it has. Recently, there has been increased interest in finding parity-check equations of weight 4 or 5 [7, 18, 10, 22], and the cost of the precomputation for finding parity checks has been identified as a significant barrier in some cases [10]. Efficient solutions for finding low-weight multiples of $p(x)$ provide low-weight parity checks and thereby enable fast correlation attacks on stream ciphers, so stream cipher designers are understandably interested in the complexity of this problem.

We show a new algorithm for the parity-check problem that is faster on some problem instances than any previously known technique. Let $\mathbb{F} \stackrel{\text{def}}{=} GF(2)[t]/(p(t))$ be the finite field of size 2^n induced by $p(t)$, and let $\oplus$ denote addition in $\mathbb{F}$. We generate k lists $L_1, \ldots, L_k$, each containing values from $\mathbb{F}$ of the form $t^a \bmod p(t) \in \mathbb{F}$ where a ranges over many small integer values. Then any solution of the form $u_1 \oplus \cdots \oplus u_k = 1$ with $u_i \in L_i$, i.e., $u_i = t^{a_i} \bmod p(t) \in \mathbb{F}$, yields a non-trivial low-weight multiple $m(x) \stackrel{\text{def}}{=} x^{a_1} + \cdots + x^{a_k} + 1$ of $p(x)$: it is a multiple of $p(x)$ since $m(t) = t^{a_1} \oplus \cdots \oplus t^{a_k} \oplus 1 = 0$ in $\mathbb{F}$ and hence $m(x) \equiv 0$ (mod $p(x)$), it is non-trivial since it is very unlikely to find fully repeated u_i's, and it has weight at most $k + 1$. If we ensure that $a \in \{1, 2, \ldots, A\}$ for every a used in any list L_i, then $m(x)$ will also be guaranteed to have degree at most A, so we have a parity check with our desired properties. With our k-tree algorithm, we will typically need to take $A \approx 2^{n/(1+\lg k)}$ to find the first parity check.

Consequently, we obtain an algorithm to find a parity check of weight $k + 1$ and degree about $2^{n/(1+\lfloor \lg k \rfloor)}$ after about $k \cdot 2^{n/(1+\lfloor \lg k \rfloor)}$ work. If we wish to obtain many parity checks, about $d^{1/(1+\lfloor \lg k \rfloor)}$ times as much work will suffice to find d parity checks, as long as $d \leq 2^{n/\lfloor \lg k \rfloor}$. This algorithm is an extension of previous techniques which used the (2-list) birthday problem [16, 21, 27, 18].

As a concrete example, if $p(x)$ represents a polynomial of degree 120, we can find a multiple $m(x)$ with degree 2^{40} and weight 5 after about 2^{42} work by using the 4-tree algorithm. Compare this to previous birthday-based techniques, which can find a multiple with degree 2^{30} and weight 5, or a multiple with degree 2^{60} and weight 3, in both cases using 2^{61} work. Thus, our k-tree algorithm runs faster than previous algorithms, but the multiples it finds have higher degrees or larger weights, so where previous techniques for finding parity-checks are computationally feasible, they are likely to be preferable. However, our algorithm may make it feasible to find non-trivial parity checks in some cases that are intractible for the previously known birthday-based methods.

Interestingly, Penzhorn and Kühn also gave a totally different cubic-time algorithm [26], using discrete logarithms in $GF(2^n)$. Their method finds a parity check with weight 4 and degree $2^{n/3}$ in $O((1+\alpha) \cdot 2^{n/3})$ time, where α represents the time to compute a discrete log in $GF(2^n)$. Using batching, they predict α will be a small constant. Also, they can obtain d times as many parity checks with about $d^{1/2}$ times as much work. Hence, when finding only a single parity check, their method improves on our algorithm: it reduces the weight from 5 to 4, while all other parameters remain comparable. However, when finding multiple parity checks, our method may be competitive. Further implementation work may be required to determine which of these algorithms performs better in practice.

5   Open Problems

Other values of k. We have shown improved algorithms only for the case where k is a power of two. An open question is whether this restriction can be removed. A case of particular interest might be where $k = 3$: is there any group operation $+$ where we can find solutions to $x_1 + x_2 + x_3 = 0$ more efficiently than a naive square-root birthday search? It would also be nice to have more efficient algorithms for the case where k is large: our techniques provide only very modest improvements as k increases, yet the existence of other approaches (such as the Gaussian elimination trick of Bellare, et al. [2]) inspires hope for improvements.

Other combining operations. We can ask for other operations $+$ where the k-sum problem $x_1 + x_2 + \cdots + x_k = c$ has efficient solutions. For example, for modular addition modulo n, can we find better algorithms using lattice reduction or other methods?

Golden solutions. Suppose there is a single "golden" solution to $x_1 + \cdots + x_k = 0$ that we wish to find, hidden amongst many other useless solutions. How efficiently can we find the golden solution for various group operations? Similarly, how efficiently can we find all solutions to $x_1 + \cdots + x_k = 0$? Better answers would have implications for some attacks [6, 18].

Memory and communication complexity. We have not put much thought into optimizing the memory consumption of our algorithms. However, in practice, N bytes of memory often cost much more than N steps of computation, and so it would be nice to know whether the memory requirements of our k-tree algorithm can be reduced. Another natural question to ask is whether the algorithm can be parallelized effectively without enormous communication complexity. Over the past two decades, researchers have found clever ways (e.g., Pollard's rho, distinguished points [24], van Oorschot & Wiener's parallel collision search [25]) to dramatically reduce the memory and parallel complexity of standard (2-list) birthday algorithms, so improvements are not out of the question. In the meantime, we believe it would be prudent for cryptosystem designers to assume that such algorithmic improvements may be forthcoming: for example, in the absence of other evidence, it appears unwise to rely on the large memory consumption of our algorithms as the primary defense against k-list birthday attacks.

Lower bounds. Finally, since the security of a number of cryptosystems seems to rest on the hardness of the k-sum problem, it would be very helpful to have better lower bounds on the complexity of this problem. As it stands, the existing lower bounds are very weak when $k \gg 2$. Lacking provable lower bounds, we hope the importance of this problem will motivate researchers to search for credible conjectures regarding the true complexity of this problem.

6   Conclusions

We have introduced the k-sum problem, shown new algorithms to solve it more efficiently than previously known to be possible, and discussed several applications to cryptanalysis of various cryptosystems. We hope this will motivate further work on this topic.

Acknowledgements

I thank Wei Dai for Theorem 2. Also, I would like to gratefully acknowledge helpful comments from Dan Bernstein, Avrim Blum, Wei Dai, Shai Halevi, Nick Hopper, Claus Schnorr, Luca Trevisan, and the anonymous reviewers.

References

1. M. Ajtai, R. Kumar, D. Sivakumar, "A Sieve Algorithm for the Shortest Lattice Vector Problem," *STOC 2001*, pp.601–610, ACM Press, 2001.
2. M. Bellare, D. Micciancio, "A New Paradigm for Collision-free Hashing: Incrementality at Reduced Cost," *EUROCRYPT'97*, LNCS 1233, Springer-Verlag, 1997.
3. D. Bernstein, "Enumerating solutions to $p(a) + q(b) = r(c) + s(d)$," *Math. Comp.*, 70(233):389–394, AMS, 2001.
4. D. Bleichenbacher, "On the generation of DSA one-time keys," unpublished manuscript, Feb. 7, 2002.
5. A. Blum, A. Kalai, H. Wasserman, "Noise-Tolerant Learning, the Parity Problem, and the Statistical Query Model," *STOC 2000*, ACM Press, 2000.
6. D. Boneh, A. Joux, P.Q. Nguyen, "Why Textbook ElGamal and RSA Encryption are Insecure," *ASIACRYPT 2000*, LNCS 1976, Springer-Verlag, pp.30–44, 2000.
7. A. Canteaut, M. Trabbia, "Improved Fast Correlation Attacks Using Parity-Check Equations of Weight 4 and 5," *EUROCRYPT 2000*, LNCS 1807, Springer-Verlag, pp.573–588, 2000.
8. M. Casto, B. Liskov, "Practical Byzantine Fault Tolerance," *Proc. 3rd OSDI* (Operating Systems Design & Implementation), Usenix, Feb. 1999.
9. M. Casto, B. Liskov, "Proactive Recovery in a Byzantine-Fault-Tolerant System," *Proc. 4th OSDI* (Operating Systems Design & Implementation), Usenix, Oct. 2000.
10. V.V. Chepyzhov, T. Johansson, B. Smeets, "A Simple Algorithm for Fast Correlation Attacks on Stream Ciphers," *FSE 2000*, LNCS 1978, Springer-Verlag, 2001.
11. P. Chose, A. Joux, M. Mitton, "Fast Correlation Attacks: an Algorithmic Point of View," *EUROCRYPT 2002*, LNCS 2332, Springer-Verlag, 2002.
12. W. Dai, personal communication, Aug. 1999.
13. H. Gobioff, "Security for a High Performance Commodity Storage Subsystem," Ph.D. thesis, CS Dept., Carnegie Mellon Univ., July 1999.
14. H. Gobioff, D. Nagle, G. Gibson, "Embedded Security for Network-Attached Storage," Tech. report CMU-CS-99-154, CS Dept., Carnegie Mellon Univ., June 1999.
15. B.-M. Goi, M.U. Siddiqi, H.-T. Chuah, "Incremental Hash Function Based on Pair Chaining & Modular Arithmetic Combining," *INDOCRYPT 2001*, LNCS 2247, Springer-Verlag, pp.50–61, 2001.
16. J. Golić, "Computation of low-weight parity-check polynomials," *Electronics Letters*, 32(21):1981–1982, 1996.
17. N.J. Hopper, M. Blum, "Secure Human Identification Protocols," *ASIACRYPT 2001*, LNCS 2248, Springer-Verlag, pp.52–66, 2001.
18. T. Johansson, F. Jönsson, "Fast Correlation Attacks Through Reconstruction of Linear Polynomials," *CRYPTO 2000*, LNCS 1880, Springer-Verlag, 2000.
19. A. Joux, R. Lercier, "'Chinese & Match', an alternative to Atkin's 'Match and Sort' method used in the SEA algorithm," *Math. Comp.*, 70(234):827–836, AMS, 2001.
20. D.E. Knuth, *The Art of Computer Programming*, vol 3, Addison-Wesley, 1973.

21. W. Meier, O. Staffelbach. "Fast correlation attacks on certain stream ciphers," *J. Cryptology*, 1(3):159–167, 1989.
22. M.J. Mihalević, M.P.C. Fossorier, H. Imai, "A Low-Complexity and High-Performance Algorithm for the Fast Correlation Attack," *FSE 2000*, LNCS 1978, Springer-Verlag, pp.196–212, 2001.
23. V.I. Nechaev, "Complexity of a determinate algorithm for the discrete logarithm," *Math. Notes*, 55(2):165–172, 1994.
24. J.-J. Quisquater, J.-P. Delescaille, "How easy is collision search? Application to DES (Extended summary)," *EUROCRYPT'89*, LNCS 434, Springer-Verlag, pp.429–434, 1990.
25. P.C. van Oorschot, M.J. Wiener, "Parallel Collision Search with Cryptanalytic Applications," *Journal of Cryptology*, 12(1):1–28, 1999.
26. W.T. Penzhorn, G.J. Kühn, "Computation of Low-Weight Parity Checks for Correlation Attacks on Stream Ciphers," *Cryptography and Coding*, LNCS 1024, Springer, pp.74–83, 1995.
27. M. Salmasizadeh, J. Golic, E. Dawson, L. Simpson. "A Systematic Procedure for Applying Fast Correlation Attacks to Combiners with Memory," *SAC'97* (Selected Areas in Cryptography).
28. C.P. Schnorr, "Security of Blind Discrete Log Signatures against Interactive Attacks," *ICICS 2001*, LNCS 2229, Springer-Verlag, pp.1–12, 2001.
29. C.P. Schnorr, S. Vaudenay, "Black box cryptanalysis of hash networks based on multipermutations," *EUROCRYPT'94*, LNCS 950, Springer-Verlag, 1994.
30. R. Schroeppel, A. Shamir, "A $TS^2 = O(2^n)$ Time/Space Tradeoff for Certain NP-Complete Problems," *FOCS '79*, pp. 328–336, 1979.
31. R. Schroeppel, A. Shamir, "A $T = O(2^{n/2}), S = O(2^{n/4})$ Algorithm for Certain NP-Complete Problems," *SIAM J. Comput.*, 10(3):456–464, 1981.
32. L. Shrira, B. Yoder, "Trust but Check: Mutable Objects in Untrusted Cooperative Caches," *Proc. POS8* (Persistent Object Systems), Morgan Kaufmann, pp.29–36, Sept. 1998.
33. V. Shoup, "Lower Bounds for Discrete Logarithms and Related Problems," *EUROCRYPT'97*, LNCS 1233, Springer-Verlag, pp.256–266, 1997.
34. S. Vaudenay, "On the need for multipermutations: Cryptanalysis of MD4 and SAFER." *FSE'94*, LNCS 1008, Springer-Verlag, pp.286–297, 1994.
35. D. Wagner, I. Goldberg, "Parallel Collision Search: Making money the old-fashioned way—the NOW as a cash cow," unpublished report, 1997. http://www.cs.berkeley.edu/~daw/papers/kcoll97.ps
36. D. Wagner, "A Generalized Birthday Problem," Full version at http://www.cs.berkeley.edu/~daw/papers/genbday.html.
37. K. Yang, "On Learning Correlated Functions Using Statistical Query," *ALT'01* (12th Intl. Conf. Algorithmic Learning Theory), LNAI 2225, Springer-Verlag, 2001.
38. G. Yuval, "How to Swindle Rabin," *Cryptologia*, 3(3):187–189, 1979.

(Not So) Random Shuffles of RC4

Ilya Mironov[*]

Computer Science Department, Stanford University
mironov@cs.stanford.edu

Abstract. Most guidelines for implementation of the RC4 stream cipher recommend discarding the first 256 bytes of its output. This recommendation is based on the empirical fact that known attacks can either cryptanalyze RC4 starting at any point, or become harmless after these initial bytes are dumped. The motivation for this paper is to find a conservative estimate for the number of bytes that should be discarded in order to be safe. To this end we propose an idealized model of RC4 and analyze it applying the theory of random shuffles. Based on our analysis of the model we recommend dumping at least 512 bytes.

1 Introduction

RC4 is a stream cipher designed by Ron Rivest in 1987. This cipher is extremely fast and exceptionally simple, which makes it ideal for protecting network traffic. In particular, RC4 is part of SSL and WEP implementations, which probably makes it, in the words of its author, "the most widely-used stream cipher in the world" [Riv01]. The design of the cipher was kept as a trade secret until 1994, when it was leaked to the cypherpunks mailing list.

Several vulnerabilities and possible attacks, surveyed in Section 3, have appeared in the open literature. Most of these attacks revolve around the concept of a *distinguisher*, which is an algorithm that can reliably distinguish a pseudo-random number generator from a truly random source. We separate *weak* and *strong* distinguishers. Weak distinguishers may be applied to any continuous segment of the RC4 output stream. A strong distinguisher can only detect a bias at the beginning of the output stream and usually requires access to several streams generated for different keys. The most staggering discovery that falls in the second category was a statistical anomaly in the second byte of the RC4 output, which is zero with probability twice as much as it should [FMS01]. To thwart this and other strong attacks, researchers recommend discarding the first 256 bytes of the output. This proposal can be traced back to 1995, when a Usenet post suggested it as a hedge against a weak-key attack [Roo95]. RSA Security, Inc. maintains that it has been its routine recommendation.

We propose a novel idealized model for studying RC4. We use this abstract model to estimate the number of initial bytes that ought to be dumped from the output stream. In other words, we want to know where the beginning of RC4 ends. We conclude that this number must be more than previously thought. As

[*] Supported by NSF contract #CCR-9732754.

M. Yung (Ed.): CRYPTO 2002, LNCS 2442, pp. 304–319, 2002.
© Springer-Verlag Berlin Heidelberg 2002

a practical corollary we describe and explain a bias in the *first* byte of RC4 output. This phenomenon is almost as persistent and reliable as the bias in the second byte, even though they are completely unrelated.

2   The RC4 Cipher

RC4 is a stream cipher in which the keystream is independent of the plaintext. The internal state consists of a permutation on numbers $0 \ldots 255$ represented as an array of length 256 and two indices in this array. We parameterize the length of the permutation by a variable n that may take any integer values (not necessarily powers of 2). The high-level view of the encryption algorithm is as follows:

> Input: L-byte message $m_1, \ldots, m_L$, key K
> Output: ciphertext $c_1, \ldots, c_L$
> $\mathsf{state}_0 \leftarrow \mathsf{KeySched}(K)$
> for $i := 1$ to L do
> $\quad \langle \mathsf{state}_i, z_i \rangle \leftarrow \mathsf{PseudoRand}(\mathsf{state}_{i-1})$
> $\quad c_i \leftarrow z_i \oplus m_i$

The size of the key K is variable and typically ranges from 40 to 256 bits.

The internals of the algorithm (functions $\mathsf{KeySched}$ and $\mathsf{PseudoRand}$) are shown in Fig. 1. Throughout the paper we denote the permutation as S, the two indices of the internal state as i and j and assume that all arrays are indexed starting at 0. All arithmetic is done modulo n.

$\mathsf{KeySched}(K)$	$\mathsf{PseudoRand}(i,j,S)$
for $i := 0$ to $n-1$ **do**	
$\quad S[i] := i$	
$j := 0$	
for $i := 0$ to $n-1$ **do**	$i := i+1$
$\quad j := j + S[i] + K[i \bmod \ell]$	$j := j + S[i]$
$\quad \mathtt{swap}(S[i], S[j])$	$\mathtt{swap}(S[i], S[j])$
$i, j := 0$	output $z := S[S[i] + S[j]]$

Fig. 1. The RC4 stream cipher.

3   Existing Attacks

Since the encryption algorithm, namely XORing the plaintext with the output stream, is so simple, any algorithm that predicts a bit in the RC4 output can be used to launch an attack. Consequently, any statistical anomaly in the output stream is a potential vulnerability.

The first weak distinguisher was Golić's [Goli97] that exploited a correlation between z_i and z_{i+2}. Later, twelve much stronger correlations between consecutive bytes of the output stream were discovered by Fluhrer and McGrew [FM00].

Two backtracking algorithms were independently proposed in [MT98] and [K+98]. In this approach the attacker tries to guess the internal state of RC4 by emulating its execution and checking it for consistency with the known output. Both attacks are not practical because of their enormous computation complexity.

Another approach has been taken by Roos and Wagner [Roo95,Wag95]. They described several classes of weak keys that either generate predictable output or, when these keys are used, several bytes of the key can be extracted from the output stream. Ironically, according to our definitions, these classes of weak keys lead to strong distinguishers since they can only be applied to the beginning of the output stream.

The results due to Mantin and Shamir [MS01] and Fluhrer, Mantin, and Shamir [FMS01] are of most practical importance. The first work describes the bias in the second byte of RC4, which is zero with probability twice what you would expect. The second work presents an analysis of a broad class of weak keys based on some parity-preserving properties in the key scheduling.

The fullest study of RC4 to date with comprehensive description of attacks of [MS01] and [FMS01] can be found in [Man01]. [Dur01] proves exact bounds on the effectiveness of the distinguisher of [FMS01].

If cryptanalisys of [FMS01] favors short keys, [GW00] demonstrates a related-key attack that works better on very long keys. It is generally bad practice to use related keys in cryptographic applications. In the case of RC4 it has turned out to be disastrous: the second part of the [FMS01] paper cryptanalyzed the Wired Equivalent Privacy protocol (WEP) that did just that; the attack was implemented and shown to be feasible by [SIR02].

To contrast our work with known results, we stress that we do *not* exploit weak keys. We depart from the concept of modelling RC4 as a finite automaton (as in [Fin94]) and consider the cipher as a random walk on a symmetric group. We do not describe any practical distinguishers other than the one that detects a bias in the first byte. Instead we prove a necessary condition for existence of strong distinguishers in the idealized model.

4   Random Shuffles in RC4

In this section we describe our idealized model for RC4. We model KeySched and PseudoRand as a random shuffle and study the resulting distribution of the permutation S. After one application of KeySched this distribution is not uniform, which results in a detectable bias in the first byte of the output. This raises the question of convergence, i.e., how fast the distribution becomes indistinguishable from the uniform. In Section 5 we tackle this problem.

4.1   Motivational Observation

It is often convenient to assume that the internal state is a random permutation. Sometimes it is a valid assumption, but it is hardly so when we look at the behavior of the cipher just after the key scheduling algorithm. The following observation explains why.

Claim. After execution of KeySched we may correctly guess the *sign* of the permutation S with probability 56%.

Before proceeding with a heuristic argument we recall the definition of the sign of a permutation. Here we use one of many equivalent definitions. For any representation of a permutation π as a product of non-trivial transpositions

$$\pi = (a_1 b_1)(a_2 b_2) \ldots (a_m b_m),$$

the sign is defined as

$$\text{sign}(\pi) = (-1)^m = \begin{cases} +1, & \text{if } 2 \mid m \\ -1, & \text{otherwise} \end{cases}.$$

The RC4 cipher initializes the permutation S with the identity permutation. Therefore $\text{sign}(S) = +1$ before the main loop of KeySched. Each iteration of the loop transposes $S[i]$ and $S[j]$. Unless $i = j$, the transposition changes the sign of the permutation. Heuristically, $i \neq j$ with probability $1 - \frac{1}{n}$ and these events are independent for different i. Therefore, the probability that the sign changes every time, for a total of n times, is $(1 - \frac{1}{n})^n \approx e^{-1}$. Similarly, we may compute the probability of the sign changing $1, 3, 5, \ldots$ times taking on -1 at the end and $2, 4, 6, \ldots$ resulting in the $+1$ value. The limiting distribution for the two possible values of the sign of the permutation S after KeySched is

$$\Pr[\text{sign}(S) = (-1)^n] = \left(1 - \frac{1}{n}\right)^n + \binom{n}{2}\left(1 - \frac{1}{n}\right)^{n-2}\frac{1}{n^2} + \cdots + \frac{1}{n^n} \xrightarrow{}$$

$$e^{-1}\left(1 + \frac{1}{2!} + \frac{1}{4!} + \ldots\right) \xrightarrow{n \to \infty} \frac{1}{2}(1 + e^{-2})$$

$$\Pr[\text{sign}(S) = (-1)^{n-1}] = 1 - \Pr[\text{sign}(S_n) = (-1)^n] \xrightarrow{n \to \infty} \frac{1}{2}(1 - e^{-2}).$$

Therefore we may predict the sign of the permutation S after execution of the key scheduling algorithm with advantage $\frac{1}{2}e^{-2} \approx 6.7\%$ over a random guess.

Significance of this anomaly. First, this computation is asymptotic and only valid when n goes to infinity. However, $n = 256$ is large enough to make the exact probabilities sufficiently close to their limit values.

Second, and more importantly, the argument hinges on the heuristic assumption that $i = j$ with probability $\frac{1}{n}$ and that these events for different $i = 0 \ldots n-1$ are independent of one another. When the key length is maximal, i.e., n bytes, and the key material is drawn from the uniform distribution, this argument is rigorously true. For the actual RC4 parameters it is rarely the case, but the table in Appendix B supports our heuristic.

Third, we do not suggest that this predictor can be useful for an attack on RC4. What we want to stress, however, is that **RC4's state is by no means**

random, at least just after executing the key generation algorithm. We hope that the above argument proves this point convincingly. In the next sections we describe a model that captures some aspects of the permutation's behavior and explain other irregularities in its distribution.

4.2   Idealized Model

In idealizing (in other words, simplifying) the inner workings of RC4 we take an approach which is seemingly consistent with the initial intent of the cipher's designer. Indeed, all arithmetic performed on the index j, is supposed to randomize it, rendering j unpredictable. The only modification we make to RC4 is to explicitly assign to j a random value whenever j gets changed (see Fig. 2).

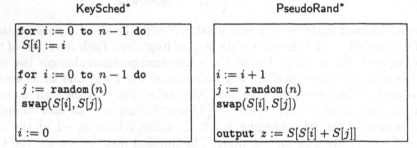

Fig. 2. Idealized RC4 stream cipher.

A closer look at the two modified algorithms KeySched* and PseudoRand* comprising the idealized RC4 reveals that their actions on S are identical and can be unified in a single procedure (Fig. 3). This procedure reflects transpositions that occur in the permutation over the running time of the cipher, both in the key scheduling and the pseudo-random number generator[1]. We call this procedure the P_t **shuffle**, where t is the length of the output including the key scheduling phase or the time, for short. The focus of the rest of the paper is on understanding this shuffle and its connection with real RC4.

$$S \leftarrow \mathrm{id}$$
$$\text{for } i := 0 \text{ to } t - 1 \text{ do}$$
$$\text{swap}(S[i \bmod n], S[\text{random}(n)])$$

Fig. 3. P_t shuffle.

---

[1] There is a slight inaccuracy in this approximation. Index i points to $S[1]$ rather than $S[0]$ at the beginning of PseudoRand. We ignore this difference in our theoretical analysis but will take it into account in computations.

5 Exchange Shuffle

The seemingly innocuous problem of shuffling cards has become a subject in its own right, with books and Ph.D. theses devoted entirely to it. The richness and difficulty of the problem lies in the fact that all shuffles are not alike, and no general method exists for dealing with some of the most interesting cases.

Card shuffling is often disguised as a problem in the theory of random walks on groups. We refer the reader to a monograph by Diaconis [Dia88] for both introductory and advanced material and to a recently published survey [Sal01] of an even broader range of topics.

It is even more surprising that the P_t shuffle (Fig. 3) has attracted little attention compared to other shuffles. Examples of better studied shuffling schemes include the random transposition shuffle, when pairs of randomly chosen card are transposed, the riffle shuffle that is modelled after the way professional players shuffle cards, analyzed by Shannon, and several others.

Keeping up with the tradition of a large body of literature, we refer to the permuted elements as *cards* and to the permutation itself as a *deck*. For consistency with our discussion of RC4 we enumerate cards starting at 0.

5.1 Combinatorial Results

To the best of our knowledge, the P_t shuffle first appeared in print as a problem in the American Mathematical Monthly [Tho65]. The problem asked for which n the P_n shuffle is truly random, i.e. induces the uniform distribution on permutations[2]. Later on, a series of papers [RB81,SS92,GM01] explored some combinatorial properties of the shuffle, whose summary is given below. The first paper due to Robbins and Bolker christened P_n the *exchange shuffle*. We expand usage of this term, calling P_t the exchange shuffle even when $t \neq n$. The shuffle P_n, as well as any other full sweep through the deck starting with the first card, is called a *pass*.

The simplest argument that demonstrates that the exchange shuffle cannot be truly random is the following. Any single execution of P_t is chosen randomly among n^t equally likely possibilities. When $n > 2$, it is impossible for this scheme to generate the uniform distribution on $n!$ possible decks of n cards since $n!$ does not divide n^n. It is then only natural to ask what are the most and the least likely permutations after one application of the exchange shuffle.

Let us first explain how to fix the exchange shuffle to make it truly random, resulting in a uniform distribution after n steps. Instead of swapping $S[i]$ with a random card, it must be transposed with a card randomly chosen from $S[i] \ldots S[n-1]$. This algorithm is discussed and a detailed proof is given in [Knu75, section 3.4.2].

We list several facts of varied difficulty known about the exchange shuffle P_n.

---

[2] Several years later another problem in the same venue [Gro77] asked to compute the probability that i ended up in position j after the P_n shuffle. Apparently, that problem prompted combinatorialists Robbins and Bolker to study the shuffle deeper [RB81].

Fact 1 [RB81] *The probability of obtaining the right cycle (i.e. the final permutation being* $(1\ 2\ldots n-1\ 0)$ *in cycle notation) at the end of* P_n *is* $2^{n-1}/n^n$.

Fact 2 [RB81] *The right cycle* $(1\ 2\ldots n-1\ 0)$ *is the least likely permutation after one execution of* P_n.

Fact 3 [RB81] *The probability of obtaining the identity permutation is* Q_n/n^n, *where* Q_n *is the number of involutions on n elements (an involution is a permutation which is its own inverse).*

Fact 4 [Knu75, section 5.1.4] *The number of involutions satisfies*

$$Q_n = \frac{1}{\sqrt{2}} n^{n/2} e^{-n/2+\sqrt{n}-1/4}(1 + O(n^{-1/2})).$$

Fact 5 [SS92] *The probability of obtaining a derangement (a permutation with no fixed points) using* P_n *is asymptotically* $\exp(3e^{-1} + e^{-2}/2 - 2) = 0.436\ldots$ *The expected number of fixed points is asymptotically* $2 - 3/e = 0.896\ldots$

For comparison, if a permutation is drawn from the uniform distribution, its probability of being a derangement is $1/e = 0.367\ldots$ and the expected number of fixed points is 1.

Fact 6 [GM01] *If* $n \geq 18$, *the identity permutation is the most likely result of the* P_n *shuffle.*

From the facts above we conclude that the probability of any deck π is contained within the following bounds:

$$\frac{1}{2}\left(\frac{2}{n}\right)^n \leq \Pr[\pi \text{ is generated by } P_n] \leq \frac{1}{\sqrt{2}} n^{-n/2} e^{-n/2+\sqrt{n}-1/4}.$$

The expected value for this probability is approximated using the Sterling formula

$$E[\Pr[\pi \text{ is generated by } P_n]] = \frac{1}{n!} \approx \frac{1}{\sqrt{2\pi n}}\left(\frac{e}{n}\right)^n.$$

There are asymptotically exponential gaps between the probabilities of the random, the most and the least likely permutations.

However instructive and interesting these combinatorial results can be, they are not very useful in analyzing RC4. Indeed, the chances of observing one of the extreme permutations are negligibly small. The only notable exception is Fact 5 that gives us a powerful distinguisher between random permutations and the ones generated by P_n. We show more examples of this approach in the next section.

5.2   Distinguishers

If there is a statistical anomaly in distributions of permutations that can be used to attack RC4, the most direct way to demonstrate existence of this anomaly is to design an algorithm that can reliably distinguish a deck shuffled by P_t from a truly random one. We already know one such distinguisher, namely computing the sign of the permutation (Section 4.1). A practical attack based on this distinguisher is quite unlikely, but due to its theoretical importance (unveiled in Section 5.4) we formally define it and complete the analysis.

Sign distinguisher. First, we describe the algorithm in pseudo-code (Fig. 4).

> Input: n-card deck S, time t
> Output: guessed bit b: true if S is random and false if S is output by P_t
> if $\text{sign}(S) = (-1)^t$
>   then return *false*
>   else return *true*

Fig. 4. Sign distinguisher.

A computation analogous to our first analysis of this algorithm in Section 4.1 proves that the *advantage* of guessing bit b correctly is approximately $\frac{1}{2}e^{-2t/n}$. Therefore, after two passes of the shuffle, which models discarding the first 256 bytes of the output of RC4, the sign is 1.83% more likely to be $+1$ than -1. With more passes the advantage of guessing the sign vanishes exponentially fast.

To correctly compute the sign, we must know exactly the entire permutation. More useful distinguishers should be more robust to partial or inaccurate information. Our next algorithm has precisely this property.

Position distinguisher. We observe that the likelihood of the ith card ending up in the jth slot after the P_t shuffle is not constant, as it would be in the case of a random deck, but instead depends on i, j, and t. Let

$$p_{i,j}^{(t)} = \Pr[S[j] = i \text{ after } P_t].$$

The following recurrence equation can be used to efficiently compute p for all t:

$$p_{i,j}^{(0)} = \begin{cases} 1, & \text{if } i = j \\ 0, & \text{otherwise} \end{cases} \tag{1}$$

and for $t > 0$

$$p_{i,j}^{(t)} = \begin{cases} p_{i,j}^{(t-1)}(1 - \frac{1}{n}) + \frac{1}{n}p_{t_0,j}^{(t-1)}, & j \neq t_0 \\ \frac{1}{n}, & \text{otherwise} \end{cases}, \tag{2}$$

where $t_0 = t \pmod{n}$. In Section A the recurrence is solved for the important case $t = n$ (also independently found by [Man01]).

Given the probabilities $p_{i,j}^{(t)}$, we assign to any permutation π a measure

$$p_t(\pi) = \prod_{i=0}^{n-1} p_{i,\pi(i)}^{(t)}$$

that approximates the likelihood of π after P_t. The approximation is rather crude, since the events $\pi(i) = a$ and $\pi(j) = b$ are not independent (in particular, $a \neq b$). In practice, instead of computing $p_t(\pi)$ directly, we evaluate $\log p_t(\pi) = \sum_{i=0}^{n-1} \log p_{i,\pi(i)}^{(t)}$. We also multiply the probabilities by a scaling factor n, which shifts the distribution toward zero.

The expected value of $p_t(\pi)$ for π drawn from the uniform distribution can be written in the following closed form. It is $\frac{1}{n!} perm((p_{i,j}^{(t)}))$, where $(p_{i,j}^{(t)})$ is the $n \times n$ matrix of $p^{(t)}$-values and $perm$ is the permanent of a matrix. However, in addition to the expected values we also need the distribution of this measure on random π, as well as its distribution on decks shuffled by the exchange shuffle. We resort to a numerical computation and experimentally find the threshold value A used in the distinguisher in Fig. 5.

Input: n-card deck S, time t, table $(p_{i,j}^{(t)})$, and the threshold value A
Output: guessed bit b: true if S is random and false if S is output by P_t

```
p := 0
for i := 0 to n - 1 do
    p := p + log np_{i,S[i]}^{(t)}
    if p < A
        then return false
        else return true
```

Fig. 5. Position distinguisher. The table $(p_{i,j}^{(t)})$ and the threshold value A are precomputed.

The effectiveness of this measure for different t is summarized in Section B. The table demonstrates that for $t \leq 3n$ the measure p_t is a good distinguisher. Distributions of measure p_n on P_n outputs and on truly random decks are plotted in Fig. 7.

The irregularities in permutations' distribution exploited by this distinguisher are directly observable and show up in the first byte of the RC4 output stream. We analyze its bias in Section 6.

5.3   Variation Distance

We have demonstrated two statistical tests that the decks shuffled by P_t fail to pass when $t \leq 3n$. Once it is shown that two distributions are sufficiently distant, existence of more potent distinguishers cannot be excluded. Our next problem is to determine what number of passes of the exchange shuffle is enough to rule out existence of *any* effective statistical tests. To put it another way, we are concerned about computational indistinguishability of two distributions: the uniform distribution and the one induced by P_t.

One technique to prove that two distributions can not be told apart is to show that they are *statistically* close. We define the distance between two distributions.

Definition 1 (variation distance). *Let P and Q be probability distributions on set F. The variation distance between P and Q is*

$$\|P - Q\| = \max_{G \subset F} |P(G) - Q(G)| = \max_{G \subset F} \left| \sum_{g \in G} P(g) - \sum_{g \in G} Q(g) \right|.$$

The following simple fact clarifies the relation between the variation distance and statistical tests:

Fact 7 *For two probability distributions P and Q on F and any probabilistic distinguisher $\mathcal{A} \colon F \mapsto \{0, 1\}$, its probability of success is no more than the variation distance between P and Q. Formally,*

$$\left| \Pr_{f \leftarrow P(F)} [\mathcal{A}(f) = 1] - \Pr_{f \leftarrow Q(F)} [\mathcal{A}(f) = 1] \right| \leq \|P - Q\|,$$

where $f \leftarrow P(F)$ means that f is sampled from the set F according to the probability distribution P.

The converse is not true, i.e. computational indistinguishability does not imply statistical closeness [Gold01].

Let S_n be the symmetric group on n cards. Let $U(S_n)$ be the uniform distribution on this group and $P_t(S_n)$ be the probability distribution of decks shuffled by P_t. Then the main problem of this section can be formulated as follows:

> What is the minimal t, as a function of n and ε, such that $\|P_t(S_n) - U(S_n)\| < \varepsilon$?

The exchange shuffle can be modelled as a Markov chain (we organize the shuffle in passes, which have identical transition matrices on S_n). This chain is finite, aperiodic (a permutation may stay unchanged after one pass), irreducible (each permutation can be reached from any other within one pass) and therefore converges to the unique stationary distribution. Since the uniform distribution can be shown to be stationary, $P_t(S_n) \xrightarrow[t \to \infty]{} U(S_n)$ and the question posed above is correct when $\varepsilon > 0$. One technical step omitted from this argument is a proof that the variation distance between P_t and U monotonically decreases in t (see full version of the paper [Mir02]).

5.4   Upper and Lower Bounds

In this section we prove explicit bounds on the convergence rate of $P_t(S_n)$, i.e. its variation distance from $U(S_n)$.

Lower bound. For the lower bound we use the sign distinguisher (Fig. 4). Its probability of success in distinguishing $P_t(S_n)$ and the uniform distribution is known and asymptotically equal to $\frac{1}{2}e^{-2t/n}$. Combining it with Fact 7 we prove that

Theorem 8. *For $n \to \infty$ and $t/n \to \lambda$:*

$$\|P_t(S_n) - U(S_n)\| > \frac{1}{2}e^{-2\lambda}.$$

Corollary 1. *If the variation distance between $P_t(S_n)$ and $U(S_n)$ is to be made smaller than ε, it must hold that $t/n > \frac{1}{2}\ln\frac{1}{2\varepsilon}$.*

The conclusion is that the number of discarded shuffles must grow *at least* linearly in n. For example, taking $\varepsilon = 2^{-20}$ leads to $t \geq 6.5n$.

Upper bound. To prove an upper bound on the convergence rate we use a variant of an ingenious argument credited to Andrei Broder by [Dia88,Mat88]. We begin by introducing strong uniform times instrumental in the proof.

Definition 2 (strong uniform time [Dia88]). *Suppose we have a randomized shuffling process Q_t on S_n, each execution of which can be unambiguously described by a sequence $q_0, \ldots, q_{t-1}$ in alphabet Q. A strong uniform time is a function $T \colon Q^* \mapsto \{0, 1, \ldots, \infty\}$ with the following properties:*

1. *If $T(q_0, \ldots, q_t) = r < \infty$, then $T(q_0, \ldots, q_s) = r$ for all $r \leq s \leq t$.*
2. *If $T(q_0, \ldots, q_t) = r < \infty$, then T takes the same value r on all sequences which are prefixed with $q_0, \ldots, q_t$.*
3. *Conditionally on T being finite, the shuffle is uniform:*

$$\Pr[\pi = P_t(\mathrm{id}) \mid T(q_0, \ldots, q_t) < \infty] = \frac{1}{n!}.$$

We say that T *happens* when it becomes finite (and equal to the current time). Intuitively, the first two properties mean that T cannot change its value after this has happened. The last property implies that if we look at the decks where T has happened, their distribution is uniform.

The following construction adapted to the exchange shuffle (Fig. 3) from Broder's original algorithm is crucial for this section:

> At the beginning all cards numbered $0 \ldots n-2$ are *unchecked*, the $(n-1)^{\text{th}}$ card is *checked*. Whenever the shuffling algorithm exchanges two cards, $S[i]$ and $S[j]$, one of the two rules may apply before the swap takes place:
>
> **a.** If $S[i]$ is unchecked and $i = j$, check $S[i]$.
> **b.** If $S[i]$ is unchecked and $S[j]$ is checked, check $S[i]$.
>
> The event T happens when all cards become checked.

Theorem 9. *The event T defined above is the strong uniform time for the P_t shuffle.*

Proof. At any given moment preceding t, let k be the number of checked cards, $A = \{a_1, \ldots, a_k\}$ be the set of their positions, $B = \{b_1, \ldots, b_k\}$ be the set of their labels, and $\Pi_k \colon A \mapsto B$ be the correspondence between the two.

Lemma 1. *For all t, the permutations Π_k are uniformly distributed conditional on $k, A,$ and B.*

Proof. [Lemma] Intuitively, when a card is checked, it joins the permutation on checked card in a random position, keeping the set of these permutations uniformly distributed. All other transpositions preserve the conditional distribution of checked cards.

More formally, the proof is by double induction on k and t. When $k = 1$ or $k > 1$ but $t = 0$, the claim is vacuous. Suppose the claim is true for all k and $t < t_0$. Any permutation on k checked cards at time t_0 was either a permutation on the same cards at time $t_0 - 1$, or a new card was checked. If no new card is checked, then all possible transpositions act identically on the permutations with the same A and B sets. It is easy to see that when a new card gets checked, all k positions for this card in the new permutation Π_k are equally likely. Therefore, the distribution on Π_k conditional on $k, A,$ and B is uniform for t_0. $\square$[Lemma]

If the distribution of $\Pi_k|_{t,k,A,B}$ is uniform for all t, we can drop the dependency on t from the condition. When all cards are checked, $k = n$ and $A = B = \{0, \ldots, n-1\}$, the permutation on checked cards is the same as the permutation on all cards. Since the number of checked cards is monotone, the first two requirements of Definition 2 are automatically satisfied; the third requirement has just been proved. Therefore, the event T is the strong uniform time for the shuffle P_t. $\square$

We want to point out that many simpler or similar checking rules would not define a strong uniform time. For instance, always checking $S[i]$ or $S[j]$, or checking $S[j]$ if $S[i]$ is unchecked do not result in the uniform distribution on permutations.

The following lemma demonstrates how a strong uniform time can be used to bound the convergence rate of a shuffling process.

Lemma 2. [Dia88] *Let Q_t be a shuffling process with a strong uniform time T_Q. Then for all t*

$$\|Q_t(S_n) - U(S_n)\| \leq \Pr[T_Q > t].$$

The last step in proving an upper bound on the variation distance between P_t and the uniform distribution is the following fact, whose proof is given in the full version of the paper [Mir02]:

Theorem 10. *There exists some constant c such that*

$$\Pr[T > cn \log n] \to 0 \text{ when } n \to \infty.$$

An upper bound on the variation distance follows from Lemma 2 and Theorem 10:

For $n \to \infty$ and $t > cn \log n$:

$$\|P_t(S_n) - U(S_n)\| \to 0.$$

5.5 Experimental Results

Since our estimate of the rate of growth of the strong uniform time T is quite loose and "too" asymptotic, we compute the distribution of T for $n = 256$ experimentally. The expected strong uniform time turns out to be

$$E[T] \approx 11.16n \approx 1.4n \log n, \qquad \text{where } n = 256.$$

The threshold $\varepsilon = 1/n$ was chosen as a reasonable goal. Experimentally,

$$\Pr[T > 2n \log n] < 1/n, \qquad \text{for } n = 256.$$

These results[3] are a far cry both from the provable upper and lower bounds on the convergence rate of P_t. There is an apparent gap between theoretical bounds and empirical evidence.

6   First Byte of RC4 Output

As was shown in Section 5.2, the permutation S is not random after the key scheduling algorithm, and quite noticeably so (explicit formula for individual probabilities is in [Mir02,Man01]). The first byte of RC4 output is $z_0 = S[S[1] + S[S[1]]]$ (in some cases a swap would affect the result). If the permutation is not uniform, there is no reason to believe that z_0 will be uniformly distributed. It is not, and in Fig. 6 we may see the fluctuations in distribution of z_0 around its mean value (the picture is zoomed in, the actual difference in probabilities is 0.6%).

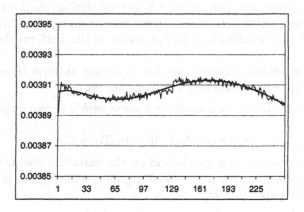

Fig. 6. Bias in the first byte.

Since the distribution of z_0 is different from the uniform distribution across the map, the difference between the two distributions is easily observable. Applying the information-theoretic bound from [FM00] on the number of necessary samples required for a reliable distinguisher, we have that about 1,700 first bytes are sufficient to recognize the distribution with 10% double-sided error.

We note that the exact probability distributions following from (1) and (2) are not sufficient to compute the bias in the first byte. The positions occupied by individual cards are not independent, and this must be taken into account.

---

[3] Tabulated for different n, the mean and the tail probability support the hypothesis that their asymptotic is $\Theta(n \log n)$.

7   Conclusion

We identified a weakness in RC4 stemming from an imperfect shuffling algorithm used in the key scheduling phase and the pseudo-random number generator. The weakness is noticeable in the first byte but does not disappear until *at least* the third or the fourth pass (512 or 768 bytes away from the beginning of the output). To find out when the nonuniformity vanishes completely we analyze an idealization of RC4 in the form of the exchange shuffle. There is an asymptotic gap between the upper and lower bounds on the convergence rate of the exchange shuffle, and there is no doubt that the constant factor in the upper bound can be improved.

Our most conservative recommendation is based on the experimental data on the tail probability of the strong uniform time T (Section 5.5). This means that discarding the initial $12 \cdot 256$ bytes most likely eliminates the possibility of a strong attack.

Dumping several times more than 256 bytes from the output stream (twice or three times this number) appears to be just as reasonable a precaution. We recommend doing so in most applications.

As a final remark we want to stress that the analysis of the idealized model of RC4 should on no account be accepted as a proof of its security. Many known vulnerabilities, such as weak attacks [Goli97,FM00] as well as results due to Fluhrer, Mantin, and Shamir are not captured by our model.

Acknowledgement

This work would have never started without excellent talks given at Stanford by Glenn Durfee and Scott Fluhrer. I am extremely grateful to Dan Boneh and Persi Diaconis for numerous valuable discussions and suggestions. It would like to thank Itsik Mantin for showing me [Man01]. It is my pleasure to thank David Aldous, Philippe Golle, Ronald Rivest, Adi Shamir, David Wagner, and anonymous referees for their helpful comments.

References

[Dia88]   P. Diaconis. *Group Representations in Probability and Statistics*. Lecture Notes-Monograph Series, vol. 11, IMS, Hayward, CA, 1988.

[Dur01]   G. Durfee. Distinguishers for the RC4 stream cipher. Manuscript, 2001.

[Fin94]   H. Finney. An RC4 cycle that can't happen. Post in `sci.crypt`, message-id `35hq1u$c72@news1.shell`, 18 September, 1994.

[FM00]   S. Fluhrer and D. McGrew. Statistical analysis of the alleged RC4 keystream generator. In proceedings *Fast Software Encryption 2000*, pp. 19–30, Lecture Notes in Computer Science, vol. 1978, Springer-Verlag, 2000.

[FMS01]   S. Fluhrer, I. Mantin, and A. Shamir. Weaknesses in the key scheduling algorithm of RC4. In proceedings *SAC 2001*, pp. 1–24, Eighth Annual Workshop on Selected Areas in Cryptography, August 2001.

[Gold01] O. Goldreich. *The Foundations of Cryptography. Basic tools.* Cambridge University Press, Cambridge, England, 2001.

[GM01] D. Goldstein and D. Moews. The identity is the most likely exchange shuffle for large n. arXiv:math.co/0010066 available from arXiv.org.

[Goli97] J. Golić. Linear statistical weakness of alleged RC4 keystream generator. In proceedings *Eurocrypt '97*, LNCS 1233, Springer-Verlag, 1997.

[Gro77] J. Grossman. Problem E 2645. Amer. Math. Month., vol. 84(3), p. 217, 1977.

[GW00] A. Grosul and D. Wallach. A related-key analysis of RC4. TR00-358, Rice University, 2000.

[K+98] L. Knudsen, W. Meier, B. Preneel, V. Rijmen, and S. Verdoolaege. Analysis methods for (alleged) RC4. In proceedings *Asiacrypt '98*, Lecture Notes in Computer Science, vol. 1514, Springer-Verlag, 1998.

[Knu75] D. Knuth. *The Art of Computer Programming. Second Edition.* Addison-Wesley, Reading, MA, 1975.

[Man01] I. Mantin. Analysis of the stream cipher RC4. Master's Thesis, Weizmann Insitute, Israel, 2001.

[MS01] I. Mantin and A. Shamir. A practical attack on broadcast RC4. In proceedings *Fast Software Encryption 2001*, Springer-Verlag, 2001.

[Mat88] P. Matthews. A strong uniform time for random transpositions. *Journal of Theoretical Probability,* vol. 1(4), 1988.

[Mir02] I. Mironov. (Not So) Random Shuffles of RC4. Full version of this paper. Cryptology ePrint Archive, Report 2002/106, available from http://eprint.iacr.org, 2002.

[Mis98] S. Mister. Cryptanalysis of RC4-like ciphers. Master's Thesis, Queen's University, Kingston, Ontario, Canada. May 1998.

[MT98] S. Mister and S. Tavares. Cryptanalysis of RC4-like ciphers. In proceedings *SAC '98*, Fifth Annual Workshop on Selected Areas in Cryptography, 1998.

[Riv01] R. Rivest. RSA Security response to weaknesses in key scheduling algorithm of RC4. Technical note available from RSA Security, Inc. site. http://www.rsasecurity.com/rsalabs/technotes/wep.html, 2001.

[RB81] D. Robbins and E. Bolker. The bias of three pseudo-random shuffles. *Acquationes Mathematicae,* vol. 22, pp. 268–292, 1981.

[Roo95] A. Roos. Class of weak keys in the RC4 stream cipher. Two posts in sci.crypt, message-id 43u1eh$1j3@hermes.is.co.za and 44ebge$llf@hermes.is.co.za, 1995.

[Rue86] R. Rueppel. *Analysis and Design of Stream Ciphers.* Springer-Verlag, 1986.

[Sal01] L. Saloff-Coste, Probability on groups: random walks and invariant diffusions. *Notices of the American Mathemtatical Society,* vol. 48(9), pp. 968–977. 2001.

[SS92] F. Schmidt and R. Simion, Card shuffling and a transformation on S_n. *Acquationes Mathematicae,* vol. 44, pp. 11–34, 1992.

[SIR02] A. Stubblefield, J. Ioannidis, and A. Rubin. Using the Fluhrer, Mantin, and Shamir attack to break WEP. In proceedings *NDSS '02.* 2002.

[Tho65] E. Thorp. Problem E 1763. Amer. Math. Month., vol. 72(2), p. 183, 1965.

[Wag95] D. Wagner. My RC4 weak keys. Post in sci.crypt, message-id 447o11$cbj@cnn.princeton.EDU, 26 September, 1995.

A   A Solution to the Recurrence

Claim. Solution to the equations (1) and (2) at time $t = n$ is given by

$$p_{a,b}^{(n)} = \begin{cases} \frac{1}{n}\left[(1-\frac{1}{n})^{n-b-1} + (1-\frac{1}{n})^a\right], & \text{if } b < a \\ \frac{1}{n}\left[(1-\frac{1}{n})^a + \left(1-(1-\frac{1}{n})^a\right)(1-\frac{1}{n})^{n-b-1}\right], & \text{if } b \geq a. \end{cases}$$

Proof. Notice that the card that has been indexed by i at least once has equal chances of ending up in any slot at time n. With this in mind consider the case of $b < a$ (the card moves left). There are two possibilities for this outcome. First, it may happen if at time $i = b$ the ath card is indexed by j and after the swap, when $S[b] = a$, the bth slot is never visited again. The probability of this event is $\frac{1}{n}(1-\frac{1}{n})^{n-b-1}$. Second, when $i = a$ and $a = S[a]$, the ath card become "randomized" and has equal chances of being in any position at the end of the pass. In this case, the probability that the ath card ends up in the bth slot is $\frac{1}{n}(1-\frac{1}{n})^a$.

The case $a \leq b$ is treated analogously.   □

B   Experimental Data

The following table summarizes success probability of the position distinguisher for different t (Section 5.2) with $n = 256$.

t	Cut-off A	Advantage
n	0.0	60%
$2n$	0.0	3.4%
$3n$	0.0	0.1%
$4n$	—	unreliable

Figure 7 plots two distributions of the measure p_t (Fig. 5), one on the random decks and another on $P_n(S_n)$ (decks shuffled with one pass of the exchange shuffle).

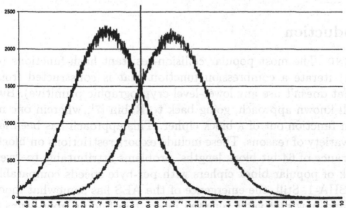

Fig. 7. Two distributions of measure p_t for $U(S_n)$ and $P_n(S_n)$, where $n = 256$.

Black-Box Analysis of the Block-Cipher-Based Hash-Function Constructions from PGV

John Black[1], Phillip Rogaway[2], and Thomas Shrimpton[3]

[1] Dept. of Computer Science, University of Colorado, Boulder CO 80309, USA
jrblack@cs.colorado.edu, www.cs.colorado.edu/~jrblack
[2] Dept. of Computer Science, University of California, Davis, CA 95616, USA, and
Dept. of Computer Science, Fac of Science, Chiang Mai University, 50200 Thailand
rogaway@cs.ucdavis.edu, www.cs.ucdavis.edu/~rogaway
[3] Dept. of Electrical and Computer Engineering, University of California, Davis,
CA 95616, USA
teshrim@ucdavis.edu, www.ece.ucdavis.edu/~teshrim

Abstract. Preneel, Govaerts, and Vandewalle [6] considered the 64 most basic ways to construct a hash function $H: \{0,1\}^* \to \{0,1\}^n$ from a block cipher $E: \{0,1\}^n \times \{0,1\}^n \to \{0,1\}^n$. They regarded 12 of these 64 schemes as secure, though no proofs or formal claims were given. The remaining 52 schemes were shown to be subject to various attacks. Here we provide a formal and quantitative treatment of the 64 constructions considered by PGV. We prove that, in a black-box model, the 12 schemes that PGV singled out as secure really *are* secure: we give tight upper and lower bounds on their collision resistance. Furthermore, by stepping outside of the Merkle-Damgård approach to analysis, we show that an additional 8 of the 64 schemes are just as collision resistant (up to a small constant) as the first group of schemes. Nonetheless, we are able to differentiate among the 20 collision-resistant schemes by bounding their security as one-way functions. We suggest that proving black-box bounds, of the style given here, is a feasible and useful step for understanding the security of any block-cipher-based hash-function construction.

1 Introduction

BACKGROUND. The most popular collision-resistant hash-functions (eg., MD5 and SHA-1) iterate a compression function that is constructed from scratch (i.e., one that doesn't use any lower-level cryptographic primitive). But there is another well-known approach, going back to Rabin [7], wherein one makes the compression function out of a block cipher. This approach has been less widely used, for a variety of reasons. These include export restrictions on block ciphers, a preponderance of 64-bit block lengths, problems attributable to "weak keys", and the lack of popular block ciphers with per-byte speeds comparable to that of MD5 or SHA-1. Still, the emergence of the AES has somewhat modified this landscape, and now motivates renewed interest in finding good ways to turn a block cipher into a cryptographic hash function. This paper casts some fresh light on the topic.

M. Yung (Ed.): CRYPTO 2002, LNCS 2442, pp. 320–335, 2002.
© Springer-Verlag Berlin Heidelberg 2002

THE PGV PAPER. We return to some old work by Preneel, Govaerts, and Vandewalle [6] that considered turning a block cipher E: $\{0,1\}^n \times \{0,1\}^n \to \{0,1\}^n$ into a hash function H: $(\{0,1\}^n)^* \to \{0,1\}^n$ using a compression function f: $\{0,1\}^n \times \{0,1\}^n \to \{0,1\}^n$ derived from E. For v a fixed n-bit constant, PGV considers all 64 compression functions f of the form $f(h_{i-1}, m_i) = E_a(b) \oplus c$ where $a, b, c \in \{h_{i-1}, m_i, h_{i-1} \oplus m_i, v\}$. Then define the *iterated hash of* f as:

> **function** $H(m_1 \cdots m_\ell)$
> **for** $i \leftarrow 1$ **to** ℓ **do** $h_i \leftarrow f(h_{i-1}, m_i)$
> **return** h_ℓ

Here h_0 is a fixed constant, say 0^n, and $|m_i| = n$ for each $i \in [1..\ell]$. Of the 64 such schemes, the authors of [6] regard 12 as secure. Another 13 schemes they classify as *backward-attackable*, which means they are subject to an identified (but not very severe) potential attack. The remaining 39 schemes are subject to damaging attacks identified by [6] and others.

SOME MISSING RESULTS. The authors of [6] focused on attacks, not proofs. All the same, it seems to be a commonly held belief that it should be possible to produce proofs for the schemes they regarded as secure. Indeed [6] goes so far as to say that "For each of these schemes it is possible to write a 'security proof' based on a black box model of the encryption algorithm, as was done for the Davies-Meyer scheme [by Winternitz [10]]". This latter paper uses a black-box model of a block cipher—a model dating back to Shannon [8]—to show that the scheme we will later call H_5 is secure in the sense of preimage-resistance. Specifically, [10] shows that any algorithm (with E and E^{-1} oracles) that always finds a preimage under H_5 for a fixed value $y \in \{0,1\}^n$ will necessarily make at least 2^{n-1} expected oracle queries.

The model introduced by Winternitz for analyzing block-cipher-based hash functions was subsequently used by Merkle [5]. He gives black-box model arguments for H_1, and other functions, and considers questions of efficiency and concrete security. The black-box model of a block cipher has also found use in other contexts, such as [3,4]. But, prior to the current work, we are unaware of any careful analysis in the literature, under any formalized model, for the collision-resistance of any block-cipher-based hash-function.

SUMMARY OF OUR RESULTS. This paper takes a more proof-centric look at the schemes from PGV [6], providing both upper and lower bounds for each. Some of our results are as expected, while others are not.

First we prove collision-resistance for the 12 schemes singled out by PGV as secure (meaning those marked "✓" or "FP" in [6]). We analyze these *group-1* schemes, $\{H_1, \ldots, H_{12}\}$, within the Merkle-Damgård paradigm. That is, we show that for each group-1 scheme H_i its compression function f_i is already collision resistant, and so H_i must be collision resistant as well.

PGV's backward-attackable schemes (marked "B" in [6]) held more surprises. We find that eight of these 13 schemes are secure, in the sense of collision resistance. In fact, these eight *group-2* schemes, $\{H_{13}, \ldots, H_{20}\}$, are just as collision-resistant as the group-1 schemes.

Despite having essentially the same collision-resistance, the group-1 and group-2 schemes can be distinguished based on their security as one-way functions: we get a better bound on inversion-resistance for the group-1 schemes than we get for the group-2 schemes. Matching attacks (up to a constant) demonstrate that this difference is genuine and not an artifact of the security proof.

The remaining $44 = 64 - 20$ hash functions considered by PGV are completely insecure: for these *group-3* schemes one can find a (guaranteed) collision with two or fewer queries. This includes five of PGV's backward-attackable schemes, where [6] had suggested a (less effective) meet-in-the-middle attack (see Appendix A).

Other surprises emerged in the mechanics of carrying out our analyses. Unlike the group-1 schemes, we found that the group-2 schemes could not be analyzed within the Merkle-Damgård paradigm; in particular, these schemes are collision resistant even though their compression functions are not. We also found that, for one set of schemes, the "obvious attack" on collision resistance needed some subtle probabilistic reasoning to rigorously analyze.

The security of the 64 PGV schemes is summarized in Fig. 1 and Fig. 2, which also serve to define the different hash functions H_i and their compression functions f_i. Fig. 3 gives a more readable description of $f_1, \ldots, f_{20}$. A high-level summary of our findings is given by the following chart. The model (and the meaning of q) will be described momentarily.

PGV Category	Our Category	Collision Bound	OWF Bound
✓or FP (12 schemes)	group-1: $H_{1..12}$ (12 schemes)	$\Theta(q^2/2^n)$	$\Theta(q/2^n)$
B (13 schemes)	group-2: $H_{13..20}$ (8 schemes)	$\Theta(q^2/2^n)$	$\Theta(q^2/2^n)$
F, P, or D (39 schemes)	group-3 (44 schemes)	$\Theta(1)$	$\Theta(1)$

BLACK-BOX MODEL. Our model is the one dating to Shannon [8] and used for works like [3, 4, 10]. Fix a key-length κ and a block length n. An adversary A is given access to oracles E and E^{-1} where E is a random block cipher $E: \{0,1\}^\kappa \times \{0,1\}^n \rightarrow \{0,1\}^n$ and E^{-1} is its inverse. That is, each key $k \in \{0,1\}^\kappa$ names a randomly-selected permutation $E_k = E(k, \cdot)$ on $\{0,1\}^n$, and the adversary is given oracles E and E^{-1}. The latter, on input (k, y), returns the point x such that $E_k(x) = y$.

For a hash function H that depends on E, the adversary's job in attacking the collision resistance of H is to find distinct M, M' such that $H(M) = H(M')$. One measures the optimal adversary's chance of doing this as a function of the number of E or E^{-1} queries it makes. Similarly, the adversary's job in inverting H is to find an inverse under H for a random range point $Y \in \{0,1\}^n$. (See Section 2 for a justification of this definition.) One measures the optimal adversary's chance of doing this as a function of the total number of E or E^{-1} queries it makes.

i	j	$h_i =$	CR low-bnd	CR up-bnd	IR low-bnd	IR up-bnd	
	1	$E_{m_i}(m_i) \oplus v$	1	1			a
	2	$E_{h_{i-1}}(m_i) \oplus v$	1	1			b
13	3	$E_{w_i}(m_i) \oplus v$	$.3q(q-1)/2^n$	$3q(q+1)/2^n$	$0.15q^2/2^n$	$9(q+3)^2/2^n$	c
	4	$E_v(m_i) \oplus v$	1	1			a
	5	$E_{m_i}(m_i) \oplus m_i$	1	1			a
1	6	$E_{h_{i-1}}(m_i) \oplus m_i$	$.039(q-1)(q-3)/2^n$	$q(q+1)/2^n$	$0.4q/2^n$	$2q/2^n$	d
9	7	$E_{w_i}(m_i) \oplus mi$	$.3q(q-1)/2^n$	$q(q+1)/2^n$	$0.6q/2^n$	$2q/2^n$	e
	8	$E_v(m_i) \oplus m_i$	1	1			a
	9	$E_{m_i}(m_i) \oplus h_{i-1}$	1	1			f
	10	$E_{h_{i-1}}(m_i) \oplus h_{i-1}$	1	1			b
11	11	$E_{w_i}(m_i) \oplus h_{i-1}$	$.3q(q-1)/2^n$	$q(q+1)/2^n$	$0.6q/2^n$	$2q/2^n$	e
	12	$E_v(m_i) \oplus h_{i-1}$	1	1			b
	13	$E_{m_i}(m_i) \oplus w_i$	1	1			f
3	14	$E_{h_{i-1}}(m_i) \oplus w_i$	$.039(q-1)(q-3)/2^n$	$q(q+1)/2^n$	$0.4q/2^n$	$2q/2^n$	d
14	15	$E_{w_i}(m_i) \oplus w_i$	$.3q(q-1)/2^n$	$3q(q+1)/2^n$	$0.15q^2/2^n$	$9(q+3)^2/2^n$	c
	16	$E_v(m_i) \oplus w_i$	1	1			f
15	17	$E_{m_i}(h_{i-1}) \oplus v$	$.3q(q-1)/2^n$	$3q(q+1)/2^n$	$0.15q^2/2^n$	$9(q+3)^2/2^n$	c
	18	$E_{h_{i-1}}(h_{i-1}) \oplus v$	$.3q(q-1)/2^n$	1			a
16	19	$E_{w_i}(h_{i-1}) \oplus v$	$.3q(q-1)/2^n$	$3q(q+1)/2^n$	$0.15q^2/2^n$	$9(q+3)^2/2^n$	c
	20	$E_v(h_{i-1}) \oplus v$	1	1			a
17	21	$E_{m_i}(h_{i-1}) \oplus m_i$	$.3q(q-1)/2^n$	$3q(q+1)/2^n$	$0.15q^2/2^n$	$9(q+3)^2/2^n$	c
	22	$E_{h_{i-1}}(h_{i-1}) \oplus m_i$	1	1			b
12	23	$E_{w_i}(h_{i-1}) \oplus m_i$	$.3q(q-1)/2^n$	$q(q+1)/2^n$	$0.6q/2^n$	$2q/2^n$	e
	24	$E_v(h_{i-1}) \oplus m_i$	1	1			b
5	25	$E_{m_i}(h_{i-1}) \oplus h_{i-1}$	$.3q(q-1)/2^n$	$q(q+1)/2^n$	$0.6q/2^n$	$2q/2^n$	e
	26	$E_{h_{i-1}}(h_{i-1}) \oplus h_{i-1}$	1	1			a
10	27	$E_{w_i}(h_{i-1}) \oplus h_{i-1}$	$.3q(q-1)/2^n$	$q(q+1)/2^n$	$0.6q/2^n$	$2q/2^n$	e
	28	$E_v(h_{i-1}) \oplus h_{i-1}$	1	1			a
7	29	$E_{m_i}(h_{i-1}) \oplus w_i$	$.3q(q-1)/2^n$	$q(q+1)/2^n$	$0.6q/2^n$	$2q/2^n$	e
	30	$E_{h_{i-1}}(h_{i-1}) \oplus w_i$	1	1			b
18	31	$E_{w_i}(h_{i-1}) \oplus w_i$	$.3q(q-1)/2^n$	$3q(q+1)/2^n$	$0.15q^2/2^n$	$9(q+3)^2/2^n$	c
	32	$E_v(h_{i-1}) \oplus w_i$	1	1			b

Fig. 1. Summary of results. Column 1 is our number i for the function (we write f_i for the compression function and H_i for its induced hash function). Column 2 is the number from [6] (we write $\hat{f}_j$ and $\hat{H}_j$). Column 3 defines $f_i(h_{i-1}, m_i)$ and $\hat{f}_j(h_{i-1}, m_i)$. We write w_i for $m_i \oplus h_{i-1}$. Columns 4–7 give our collision-resistance and inversion-resistance bounds. Column 8 comments on collision-finding attacks: (a) $H(M)$ is determined by the last block only; two E queries; (b) Attack uses two E queries and one E^{-1} query; (c) Attack uses $q/2$ E queries and $q/2$ E^{-1} queries; (d) Attack given by Theorem 3; (e) Attack given by Theorem 4; (f) $H(M)$ independent of block order; two E queries; (g) Attack uses (at most) two E queries. We do not explore inversion resistance for schemes that are trivially breakable in the sense of collision resistance.

i	j	$h_i =$	CR low-bnd	CR up-bnd	IR low-bnd	IR up-bnd	
19	33	$E_{m_i}(w_i) \oplus v$	$.3q(q-1)/2^n$	$3q(q+1)/2^n$	$0.15q^2/2^n$	$9(q+3)^2/2^n$	c
	34	$E_{h_{i-1}}(w_i) \oplus v$	1	1			b
	35	$E_{w_i}(w_i) \oplus v$	1	1			g
	36	$E_v(w_i) \oplus v$	1	1			b
20	37	$E_{m_i}(w_i) \oplus m_i$	$.3q(q-1)/2^n$	$3q(q+1)/2^n$	$0.15q^2/2^n$	$9(q+3)^2/2^n$	c
4	38	$E_{h_{i-1}}(w_i) \oplus m_i$	$.039(q-1)(q-3)/2^n$	$q(q+1)/2^n$	$0.4q/2^n$	$2q/2^n$	d
	39	$E_{w_i}(w_i) \oplus m_i$	1	1			g
	40	$E_v(w_i) \oplus m_i$	1	1			g
8	41	$E_{m_i}(w_i) \oplus h_{i-1}$	$.3q(q-1)/2^n$	$q(q+1)/2^n$	$0.6q/2^n$	$2q/2^n$	e
	42	$E_{h_{i-1}}(w_i) \oplus h_{i-1}$	1	1			b
	43	$E_{w_i}(w_i) \oplus h_{i-1}$	1	1			g
	44	$E_v(w_i) \oplus h_{i-1}$	1	1			b
6	45	$E_{m_i}(w_i) \oplus w_i$	$.3q(q-1)/2^n$	$q(q+1)/2^n$	$0.6q/2^n$	$2q/2^n$	e
2	46	$E_{h_{i-1}}(w_i) \oplus w_i$	$.039(q-1)(q-3)/2^n$	$q(q+1)/2^n$	$0.4q/2^n$	$2q/2^n$	d
	47	$E_{w_i}(w_i) \oplus w_i$	1	1			g
	48	$E_v(w_i) \oplus w_i$	1	1			g
	49	$E_{m_i}(v) \oplus v$	1	1			a
	50	$E_{h_{i-1}}(v) \oplus v$	1	1			a
	51	$E_{w_i}(v) \oplus v$	1	1			g
	52	$E_v(v) \oplus v$	1	1			a
	53	$E_{m_i}(v) \oplus m_i$	1	1			a
	54	$E_{h_{i-1}}(v) \oplus m_i$	1	1			b
	55	$E_{w_i}(v) \oplus m_i$	1	1			g
	56	$E_v(v) \oplus m_i$	1	1			a
	57	$E_{m_i}(v) \oplus h_{i-1}$	1	1			f
	58	$E_{h_{i-1}}(v) \oplus h_{i-1}$	1	1			a
	59	$E_{w_i}(v) \oplus h_{i-1}$	1	1			g
	60	$E_v(v) \oplus h_{i-1}$	1	1			a
	61	$E_{m_i}(v) \oplus w_i$	1	1			f
	62	$E_{h_{i-1}}(v) \oplus w_i$	1	1			b
	63	$E_{w_i}(v) \oplus w_i$	1	1			g
	64	$E_v(v) \oplus w_i$	1	1			b

Fig. 2. Summary of results, continued. See the caption of Fig. 1 for an explanation of the entries in this table.

DISCUSSION. As with [6], we do not concern ourselves with MD-strengthening [2, 5], wherein strings are appropriately padded so that any $M \in \{0,1\}^*$ may be hashed. Simple results establish the security of the MD-strengthened hash function H^* one gets from a secure multiple-of-block-length hash-function H. All of our attacks work just as well in the presence of MD-strengthening.

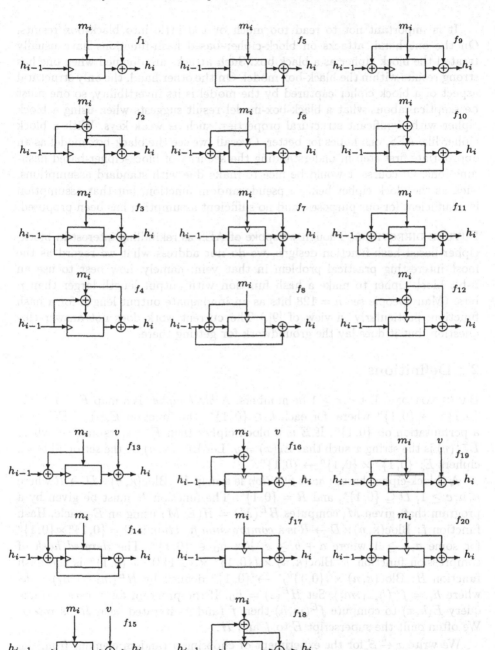

Fig. 3. The compression functions $f_1, \ldots, f_{20}$ for the 20 collision-resistant hash functions $H_1, \ldots, H_{20}$. A hatch marks the location for the key.

It is important not to read too much or too little into black-box results. On the one hand, attacks on block-cipher-based hash-functions have usually treated the block cipher as a black box. Such attacks are doomed when one has strong results within the black-box model. On the other hand, the only structural aspect of a block cipher captured by the model is its invertibility, so one must be skeptical about what a black-box-model result suggests when using a block cipher with significant structural properties, such as weak keys. With a block cipher like AES, one hopes for better. Overall, we see the black-box model as an appropriate first step in understanding the security of block-cipher-based hash-functions. Of course it would be nice to make due with standard assumptions, such as the block cipher being a pseudorandom function, but that assumption is insufficient for our purposes, and no sufficient assumption has been proposed.

FUTURE DIRECTIONS. Though we spoke of AES as rekindling interest in block-cipher-based hash-function designs, we do not address what we regard as the most interesting practical problem in that vein: namely, how best to use an n-bit block cipher to make a hash function with output length larger than n bits. (Many people see $n = 128$ bits as an inadequate output length for a hash function, particularly in view of [9].) The current work does not answer this question, but it does lay the groundwork for getting there.

2   Definitions

BASIC NOTIONS. Let $\kappa, n \geq 1$ be numbers. A *block cipher* is a map $E\colon \{0,1\}^\kappa \times \{0,1\}^n \to \{0,1\}^n$ where, for each $k \in \{0,1\}^\kappa$, the function $E_k(\cdot) = E(k, \cdot)$ is a permutation on $\{0,1\}^n$. If E is a block cipher then E^{-1} is its inverse, where $E_k^{-1}(y)$ is the string x such that $E_k(x) = y$. Let $\mathrm{Bloc}(\kappa, n)$ be the set of all block ciphers $E\colon \{0,1\}^\kappa \times \{0,1\}^n \to \{0,1\}^n$.

A (block-cipher-based) *hash function* is a map $H\colon \mathrm{Bloc}(\kappa, n) \times D \to R$ where $\kappa, n, c \geq 1$, $D \subseteq \{0,1\}^*$, and $R = \{0,1\}^c$. The function H must be given by a program that, given M, computes $H^E(M) = H(E, M)$ using an E-oracle. Hash function $f\colon \mathrm{Bloc}(\kappa, n) \times D \to R$ is a *compression function* if $D = \{0,1\}^a \times \{0,1\}^b$ for some $a, b \geq 1$ where $a + b \geq c$. Fix $h_0 \in \{0,1\}^a$. The *iterated hash* of compression function $f\colon \mathrm{Bloc}(\kappa, n) \times (\{0,1\}^a \times \{0,1\}^b) \to \{0,1\}^a$ is the hash function $H\colon \mathrm{Bloc}(\kappa, n) \times (\{0,1\}^b)^* \to \{0,1\}^a$ defined by $H^E(m_1 \cdots m_\ell) = h_\ell$ where $h_i = f^E(h_{i-1}, m_i)$. Set $H^E(\varepsilon) = h_0$. If the program for f uses a single query $E(k, x)$ to compute $f^E(m, h)$ then f (and its iterated hash H) is *rate-1*. We often omit the superscript E to f and H.

We write $x \overset{\$}{\leftarrow} S$ for the experiment of choosing a random element from the finite set S and calling it x. An *adversary* is an algorithm with access to one or more oracles. We write these as superscripts.

COLLISION RESISTANCE. To quantify the collision resistance of a block-cipher-based hash function H we instantiate the block cipher by a randomly chosen $E \in \mathrm{Bloc}(\kappa, n)$. An adversary A is given oracles for $E(\cdot, \cdot)$ and $E^{-1}(\cdot, \cdot)$ and wants to find a *collision* for H^E—that is, M, M' where $M \neq M'$ but $H^E(M) = H^E(M')$.

We look at the number of queries that the adversary makes and compare this with the probability of finding a collision.

Definition 1 (Collision resistance of a hash function). Let H be a block-cipher-based hash function, $H: \mathrm{Bloc}(\kappa, n) \times D \to R$, and let A be an adversary. Then the advantage of A in finding collisions in H is the real number

$$\mathbf{Adv}_H^{\mathrm{coll}}(A) = \Pr \left[E \xleftarrow{\$} \mathrm{Bloc}(\kappa, n); (M, M') \xleftarrow{\$} A^{E, E^{-1}} : \right.$$

$$\left. M \neq M' \wedge H^E(M) = H^E(M') \right] \qquad \diamondsuit$$

For $q \geq 1$ we write $\mathbf{Adv}_H^{\mathrm{coll}}(q) = \max_A \{ \mathbf{Adv}_H^{\mathrm{coll}}(A) \}$ where the maximum is taken over all adversaries that ask at most q oracle queries (ie, E-queries + E^{-1} queries). Other advantage functions are silently extended in the same way.

We also define the advantage of an adversary in finding collisions in a compression function $f: \mathrm{Bloc}(\kappa, n) \times \{0,1\}^a \times \{0,1\}^b \to \{0,1\}^c$. Naturally (h, m) and (h', m') collide under f if they are distinct and $f^E(h, m) = f^E(h', m')$, but we also give credit for finding an (h, m) such that $f^E(h, m) = h_0$, for a fixed $h_0 \in \{0,1\}^c$. If one treats the hash of the empty string as the constant h_0 then $f^E(h, m) = h_0$ amounts to having found a collision between (h, m) and the empty string.

Definition 2 (Collision resistance of a compression function). Let f be a block-cipher-based compression function, $f: \mathrm{Bloc}(\kappa, n) \times \{0,1\}^a \times \{0,1\}^b \to \{0,1\}^c$. Fix a constant $h_0 \in \{0,1\}^c$ and an adversary A. Then the advantage of A in finding collisions in f is the real number

$$\mathbf{Adv}_f^{\mathrm{comp}}(A) = \Pr \left[E \xleftarrow{\$} \mathrm{Bloc}(\kappa, n); ((h, m), (h', m')) \xleftarrow{\$} A^{E, E^{-1}} : \right.$$

$$\left. ((h, m) \neq (h', m') \wedge f^E(h, m) = f^E(h', m')) \vee f^E(h, m) = h_0 \right] \qquad \diamondsuit$$

INVERSION RESISTANCE. Though we focus on collision resistance, we are also interested in the difficulty of inverting hash functions. We use the following measure for the difficulty of inverting a hash function at a random point.

Definition 3 (Inverting random points). Let H be a block-cipher-based hash function, $H: \mathrm{Bloc}(\kappa, n) \times D \to R$, and let A be an adversary. Then the advantage of A in inverting H is the real number

$$\mathbf{Adv}_H^{\mathrm{inv}}(A) = \Pr \left[E \xleftarrow{\$} \mathrm{Bloc}(\kappa, n); \sigma \xleftarrow{\$} R; M \leftarrow A^{E, E^{-1}}(\sigma) : H^E(M) = \sigma \right] \qquad \diamondsuit$$

THE PGV HASH FUNCTIONS. Fig. 1 and Fig. 2 serve to define $f_i[n]: \mathrm{Bloc}(\kappa, n) \times \{0,1\}^n \times \{0,1\}^n \to \{0,1\}^n$ and $\hat{f}_j[n]: \mathrm{Bloc}(\kappa, n) \times (\{0,1\}^n \times \{0,1\}^n) \to \{0,1\}^n$ for $i \in [1..20]$ and $j \in [1..64]$. These compression functions induce hash functions $H_i[n]$ and $\hat{H}_j[n]$. Usually we omit writing the $[n]$.

DISCUSSION. The more customary formalization for a one-way function speaks to the difficulty of finding a preimage for the image of a random domain point (as opposed to finding a preimage of a random range point). But a random-domain-point definition becomes awkward when considering a function H with an infinite domain: in such a case one would normally have to partition the domain into finite sets and insist that H be one-way on each of them. For each of the functions $H_1, \ldots, H_{20}$, the value $H_i^E(M)$ is uniform or almost uniform in $\{0,1\}^n$ when M is selected uniformly in $(\{0,1\}^n)^m$ and E is selected uniformly in $\mathrm{Bloc}(n, n)$. Thus there is no essential difference between the two notions in these cases. This observation justifies defining inversion resistance in the manner that we have. See Appendix B.

Definition 3 might be understood as giving the technical meaning of *preimage resistance*. However, a stronger notion of preimage resistance also makes sense, where the range value σ is a fixed point, not a random one, and one maximizes over all such points. Similarly, the usual, random-domain-point notion for a one-way function (from the prior paragraph) might be understood as a technical meaning of *2nd preimage resistance*, but a stronger notion makes sense, where the domain point M is a fixed string, not a random one, and one must maximize over all domain points of a given length. A systematic exploration of different notions of inversion resistance is beyond the scope of this paper.

CONVENTIONS. For the remainder of this paper we assume the following significant conventions. First, an adversary does not ask any oracle query in which the response is already known; namely, if A asks a query $E_k(x)$ and this returns y, then A does not ask a subsequent query of $E_k(x)$ or $E_k^{-1}(y)$; and if A asks $E_k^{-1}(y)$ and this returns x, then A does not ask a subsequent query of $E_k^{-1}(y)$ or $E_k(x)$. Second, when a (collision-finding) adversary A for H outputs M and M', adversary A has already computed $H^E(M)$ and $H^E(M')$, in the sense that A has made the necessary E or E^{-1} queries to compute $H^E(M)$ and $H^E(M')$. Similarly, we assume that a (collision-finding) adversary A for the compression function f computes $f^E(h, m)$ and $f^E(h', m')$ prior to outputting (h, m) and (h', m'). Similarly, when an (inverting adversary) A for H outputs a message M, we assume that A has already computed $H^E(M)$, in the sense that A has made the necessary E or E^{-1} queries to compute this value. These assumption are all without loss of generality, in that an adversary A not obeying these conventions can easily be modified to given an adversary A' having similar computational complexity that obeys these conventions and has the same advantage as A.

3   Collision Resistance of the Group-1 Schemes

The group-1 hash-functions $H_1, \ldots, H_{12}$ can all be analyzed using the Merkle-Damgård paradigm. Our security bound is identical for all of these schemes.

Theorem 1 (Collision resistance of the group-1 hash functions). Fix $n \geq 1$ and $i \in [1..12]$. Then $\mathbf{Adv}_{H_i[n]}^{\mathrm{coll}}(q) \leq q(q + 1)/2^n$ for any $q \geq 1$.   ◇

The proof combines a lemma showing the collision-resistance of $f_1, \ldots, f_{12}$ with the classical result, stated for the black-box model, showing that a hash function is collision resistant if its compression function is.

Lemma 1 (Merkle-Damgård [2, 5] in the black-box model). Let f be a compression function f: $\text{Bloc}(n, n) \times \{0,1\}^n \times \{0,1\}^n \to \{0,1\}^n$ and let H be the iterated hash of f. Then $\text{Adv}_H^{\text{coll}}(q) \leq \text{Adv}_f^{\text{comp}}(q)$ for all $q \geq 1$. $\diamond$

Lemma 2 (Collision resistance of the group-1 compression functions). Fix $n \geq 1$ and $\imath \in [1..12]$. Then $\text{Adv}_{f_\imath[n]}^{\text{comp}}(q) \leq q(q+1)/2^n$ for any $q \geq 1$. $\diamond$

Proof of Lemma 2: Fix a constant $h_0 \in \{0,1\}^n$. We focus on $f = f_1$; assume that case. Let $A^{?,?}$ be an adversary attacking the compression function f. Assume that A asks its oracles a total of q queries. We are interested in A's behavior when its left oracle is instantiated by $E \xleftarrow{\$} \text{Bloc}(n, n)$ and its right oracle is instantiated by E^{-1}. That experiment is identical, from A's perspective, to the one defined in Fig. 4. Define $((x_1, k_1, y_1), \ldots, (x_q, k_q, y_q), out)$ by running $SimulateOracles(A, n)$. If A is successful it means that A outputs $(k, m), (k', m')$ such that one of the following holds: $(k, m) \neq (k', m')$ and $f(k, m) = f(k', m')$, or else $f(k, m) = h_0$. By our definition of f this means that $E_k(m) \oplus m = E_{k'}(m') \oplus m'$ for the first case, or $E_k(m) \oplus m = h_0$ for the second. By our conventions at the end of Section 2, either there are distinct $r, s \in [1..q]$ such that $(x_r, k_r, y_r) = (m, k, E_k(m))$ and $(x_s, k_s, y_s) = (m', k', E_{k'}(m'))$ and $E_{k_r}(m_r) \oplus m_r = E_{k_s}(m_s) \oplus m_s$ or else there is an $r \in [1..q]$ such that $(x_r, k_r, y_r) = (m, k, h_0)$ and $E_{k_r}(x_r) = h_0$. We show that this event is unlikely.

In the execution of $SimulateOracles(A, n)$, for any $i \in [1..q]$, let C_i be the event that $y_i \oplus x_i = h_0$ or that there exists $j \in [1..i - 1]$ such that either $y_i \oplus x_i = y_j \oplus x_j$. In carrying out the simulation of A's oracles, either y_i or x_i was randomly selected from a set of at least size $2^n - (i-1)$, so $\Pr[\mathsf{C}_i] \leq i/(2^n - i)$. By the contents of the previous paragraph, we thus have that $\text{Adv}_{f[n]}^{\text{comp}}(A) \leq \Pr[\mathsf{C}_1 \vee \ldots \vee \mathsf{C}_q] \leq \sum_{i=1}^{q} \Pr[\mathsf{C}_i] \leq \sum_{i=1}^{q} \frac{i}{2^n-(i-1)} \leq \frac{1}{2^n-2^{n-1}} \sum_{i=1}^{q} i$ if $q \leq 2^{n-1}$. Continuing, our expression is at most $\frac{1}{2^{n-1}} \frac{q(q+1)}{2} = \frac{q(q+1)}{2^n}$. Since the above inequality is vacuous when $q > 2^{n-1}$, we may now drop the assumption that $q \leq 2^{n-1}$. We conclude that $\text{Adv}_{f[n]}^{\text{comp}}(q) \leq q(q+1)/2^n$.

The above concludes the proof for the case of f_1. Compression functions $f_{2..12}$ are similar. ∎

Algorithm $SimulateOracles(A, n)$
Initially, $i \leftarrow 0$ and $E_k(x) = $ undefined for all $(k, x) \in \{0,1\}^n \times \{0,1\}^n$
Run $A^{?,?}$, answering oracle queries as follows:
    When A asks a query (k, x) to its left oracle:
        $i \leftarrow i + 1$; $k_i \leftarrow k$; $x_i \leftarrow x$; $y_i \xleftarrow{\$} \overline{\text{Range}}(E_k)$; $E_k(x) \leftarrow y_i$; return y_i to A
    When A asks a query (k, y) to its right oracle:
        $i \leftarrow i + 1$; $k_i \leftarrow k$; $y_i \leftarrow y$; $x_i \xleftarrow{\$} \overline{\text{Domain}}(E_k)$; $E_k(x_i) \leftarrow y$; return x_i to A
    When A halts, outputting a string out:
        return $((x_1, k_1, y_1), ..., (x_i, k_i, y_i), out)$

Fig. 4. Simulating a block-cipher oracle. $\text{Domain}(E_k)$ is the set of points x where $E_k(x)$ is no longer undefined and $\overline{\text{Domain}}(E_k) = \{0,1\}^n - \text{Range}(E_k)$. $\text{Range}(E_k)$ is the set of points where $E_k(x)$ is no longer undefined and $\overline{\text{Range}}(E_k) = \{0,1\}^n - \text{Range}(E_k)$.

4 Collision Resistance of the Group-2 Schemes

We cannot use the Merkle-Damgård paradigm for proving the security of $H_{13..20}$ because their compression functions are *not* collision-resistant. Attacks for each compression function are easy to find. For example, one can break $f_{17}(h, m) = E_m(h) \oplus m$ as a compression function by choosing any two distinct $m, m' \in \{0,1\}^n$, computing $h = E_m^{-1}(m)$ and $h' = E_{m'}^{-1}(m')$, and outputting (h, m) and (h', m'). All the same, hash functions $H_{13..20}$ enjoy almost the same collision-resistance upper bound as $H_{1..12}$.

Theorem 2 (Collision resistance of the group-2 hash functions). Fix $n \geq 1$ and $\imath \in [13..20]$. Then $\mathbf{Adv}_{H_\imath[n]}^{\mathrm{coll}}(q) \leq 3q(q+1)/2^n$ for all $q \geq 1$.   ◇

Proof of Theorem 2: Fix constants $h_0, v \in \{0,1\}^n$. We prove the theorem for the case of H_{13}, where $f(h, m) = f_{13}(h, m) = E_{h \oplus m}(m) \oplus v$.

We define a directed graph $G = (V_G, E_G)$ with vertex set $V_G = \{0,1\}^n \times \{0,1\}^n \times \{0,1\}^n$ and an arc $(x, k, y) \to (x', k', y')$ in E_G if and only if $k' \oplus x' = y \oplus v$.

Let $A^{?,?}$ be an adversary attacking H_{13}. We analyze the behavior of A when its left oracle is instantiated by $E \xleftarrow{\$} \mathrm{Bloc}(n, n)$ and its right oracle is instantiated by E^{-1}. Assume that A asks its oracles at most q total queries. We must show that $\mathbf{Adv}_{H_{13}[n]}^{\mathrm{coll}}(A) \leq 3q(q+1)/2^n$. Run the algorithm *SimulateOracles*(A, n). As A executes with its (simulated) oracle, color the vertices of G as follows:

- Initially, each vertex of G is *uncolored*.
- When A asks an E-query (k, x) and this returns a value y, or when A asks an E^{-1}-query of (k, y) and this returns x, then: if $x \oplus k = h_0$ then vertex (x, k, y) gets colored *red*; otherwise vertex (x, k, y) gets colored *black*.

According to the conventions at the end of Section 2, every query the adversary asks results in exactly one vertex getting colored red or black, that vertex formerly being uncolored.

We give a few additional definitions. A vertex of G is colored when it gets colored red or black. A path P in G is colored if all of its vertices are colored. Vertices (x, k, y) and (x', k', y') are said to collide if $y = y'$. Distinct paths P and P' are said to collide if all of their vertices are colored and they begin with red vertices and they end with colliding vertices. Let C be the event that, as a result of the adversary's queries, there are formed in G some two colliding paths.

Claim 1. $\mathbf{Adv}_{H_{13}[n]}^{\mathrm{coll}}(A) \leq \Pr[\mathsf{C}]$.

Claim 2. $\Pr[\mathsf{C}] \leq 3q(q+1)/2^n$.

The theorem follows immediately from these two claims, whose proofs can be found in the full paper [1]. Proofs for for $H_{14..20}$ can be obtained by adapting the proof for H_{13} using the rules from Fig. 5.   ∎

ı	$h_i =$	$(x,k,y) \to (x',k',y')$ if	(x,k,y) red if	$(x,k,y), (x',k',y')$ collide if
13	$E_{w_i}(m_i) \oplus v$	$y \oplus v = x' \oplus k'$	$x \oplus k = h_0$	$y = y'$
14	$E_{w_i}(m_i) \oplus w_i$	$k \oplus y = x' \oplus k'$	$x \oplus k = h_0$	$k \oplus y = k' \oplus y'$
15	$E_{m_i}(h_{i-1}) \oplus v$	$y \oplus v = x'$	$x = h_0$	$y \oplus v = x'$
16	$E_{w_i}(h_{i-1}) \oplus v$	$y \oplus v = x'$	$x = h_0$	$y \oplus v = x'$
17	$E_{m_i}(h_{i-1}) \oplus m_i$	$k \oplus y = x'$	$x = h_0$	$k \oplus y = k' \oplus y'$
18	$E_{w_i}(h_{i-1}) \oplus w_i$	$k \oplus y = x'$	$x = h_0$	$k \oplus y = k' \oplus y'$
19	$E_{m_i}(w_i) \oplus v$	$y \oplus v = x' \oplus k'$	$x \oplus k = h_0$	$y = y'$
20	$E_{m_i}(w_i) \oplus m_i$	$k \oplus y = x' \oplus k'$	$x \oplus k = h_0$	$k \oplus y = k' \oplus y'$

Fig. 5. Rules for the existence of arcs, the coloring of a vertex red, and when vertices are said to collide. These notions are used in the proof of Theorem 2.

5 Matching Attacks on Collision Resistance

In this section we show that the security bounds given in Sections 3 and 4 are tight: we devise and analyze attacks that achieve advantage close to the earlier upper bounds. Our results are as follows.

Theorem 3 (Finding collisions in $H_{1..4}$). Let $\imath \in [1..4]$ and $n \geq 1$. Then $\mathbf{Adv}^{\mathrm{coll}}_{H_\imath[n]}(q) \geq 0.039(q-1)(q-3)/2^n$ for any even $q \in [1..2^{(n-1)/2}]$.   ◇

Let $\mathrm{Perm}(n)$ be the set of all permutations on $\{0,1\}^n$. Let $\mathcal{P}_q(\{0,1\}^n)$ denote the set of all q-element subsets of $\{0,1\}^n$. The proof of Theorem 3 uses the following technical lemma whose proof appears in the full paper [1].

Lemma 3. Fix $n \geq 1$. Then

$$\Pr\left[\pi \xleftarrow{\$} \mathrm{Perm}(n); Q \xleftarrow{\$} \mathcal{P}_q(\{0,1\}^n) \; : \; \exists x, x' \in Q \text{ such that } x \neq x' \text{ and} \right.$$
$$\left. \pi(x) \oplus x = \pi(x') \oplus x' \right] \geq .039(q-1)(q-3)/2^n$$

for any even $q \in [1..2^{(n-1)/2}]$.   ◇

Proof of Theorem 3: Consider the case $H^E = H_1^E$ and fix $h_0 \in \{0,1\}^n$. Let A be an adversary with the oracles E, E^{-1}. Let A select $m_1, \ldots, m_q \xleftarrow{\$} \{0,1\}^n$ and compute $y_j = E_{h_0}(m_j) \oplus m_j$, $j \in [1..q]$. If A finds $r, s \in [1..q]$ such that $r < s$ and $y_r = y_s$ then it returns (m_r, m_s); otherwise it returns (m_1, m_1) (failure). Let $\pi = E_{h_0}$. By definition π is a uniform element in $\mathrm{Perm}(n)$, so we can invoke Lemma 3 to see that the probability that A succeeds to find a collision among $m_1, \ldots, m_q$ under H is at least $.039(q-1)(q-3)/2^n$.

This attack and analysis extends to $H_{2..4}$ by recognizing that for each scheme and distinct one-block messages m and m' we have $H^E(m) = H^E(m')$ if and only if $\pi(x) \oplus x = \pi(x') \oplus x'$ where $\pi = E_{h_0}$ and x, x' are properly defined. For example, for H_2^E define $x = h_0 \oplus m$ and $x' = h_0 \oplus m'$.   ∎

Analysis of collision-finding attacks on $H_{5..20}$ is considerably less technical than for $H_{1..4}$. The crucial difference is that in each of $H_{5..20}$ the block cipher is keyed in the first round by either the message m, or $m \oplus h_0$, where h_0 is a fixed constant. Hence when A hashes q distinct one-block messages it always observes q random values. See the full paper [1] for a proof of the following.

Theorem 4 (Finding collisions in $H_{5..20}$). Let $\imath \in [5..20]$ and $n \geq 1$. Then $\mathbf{Adv}_{H_\imath[n]}^{\mathrm{coll}}(q) \geq 0.3q(q-1)/2^n$ for any $q \in [1..2^{n/2}]$.   ◇

6   Security of the Schemes as OWFs

From the perspective of collision resistance there is no reason to favor any particular scheme from $H_{1..20}$. However, in this section we show that these schemes can be separated based on their strength as one-way functions. In particular, for an n-bit block cipher, an adversary attacking a group-1 hash function requires nearly 2^n oracle queries to do well at inverting a random range point, while an adversary attacking a group-2 hash function needs roughly $2^{n/2}$ oracle queries to do the same job.

We begin with the theorem establishing good inversion-resistance for the group-1 schemes. The theorem is immediate from the two lemmas that follow it. The first result is analogous to Lemma 1. The second result shows that $f_{1..12}$ have good inversion-resistance. All omitted proofs can be found in the full version of the paper [1].

Theorem 5 (OWF security of the group-1 hash functions). Fix $n \geq 1$ and $\imath \in [1..12]$. Then $\mathbf{Adv}_{H_\imath[n]}^{\mathrm{inv}}(q) \leq q/2^{n-1}$ for any $q \geq 1$.   ◇

Lemma 4 (Merkle-Damgård for inversion resistance). Let f be a compression function $f\colon \mathrm{Bloc}(n,n) \times \{0,1\}^n \times \{0,1\}^n \to \{0,1\}^n$ and let H be the iterated hash of f. Then $\mathbf{Adv}_H^{\mathrm{inv}}(q) \leq \mathbf{Adv}_f^{\mathrm{inv}}(q)$ for all $q \geq 1$.   ◇

Lemma 5 (Inversion resistance of the group-1 compression functions). Fix $n \geq 1$ and $\imath \in [1..12]$. Then $\mathbf{Adv}_{f_\imath[n]}^{\mathrm{inv}}(q) \leq q/2^{n-1}$ for any $q \geq 1$.   ◇

Proof of Lemma 5: Fix a constant $h_0 \in \{0,1\}^n$. We focus on compression function $f^E = f_1^E$; assume that case. Let A be an adversary with oracles E, E^{-1} and input σ. Assume that A asks its oracles q total queries.

Define $((x_1, k_1, y_1), \ldots, (x_q, k_q, y_q), out)$ by running $SimulateOracles(A, n)$. By our conventions at the end of Section 2, if A outputs (h, m) such that $E(h, m) \oplus m = \sigma$ then $(m, h, E(h, m)) = (x_i, k_i, y_i)$ for some $i \in [1..q]$. Let C_i be the event that (x_i, k_i, y_i) is such that $x_i \oplus y_i = \sigma$. In carrying out the simulation of A's oracles, either x_i or y_i was randomly assigned from a set of at least size $2^n - (i-1)$, so $\Pr[C_i] \leq 1/(2^n - (i-1))$. Thus $\Pr[(h, m) \leftarrow A^{E, E^{-1}}(z) : E(h, m) \oplus m = \sigma] \leq \Pr[C_1 \vee \ldots \vee C_q] \leq \sum_{i=1}^{q} \Pr[C_i] \leq \sum_{i=1}^{q} \frac{1}{2^n - (i-1)} \leq \frac{q}{2^n - 2^{n-1}}$ if $q \leq 2^{n-1}$. Continuing, our expression is at most $\frac{q}{2^{n-1}}$. Since the above inequality is vacuous when $q > 2^{n-1}$, we may now drop the assumption that $q \leq 2^{n-1}$.

The above concludes the proof for the case of f_1. Compression functions $f_{2..12}$ are similar. ∎

We cannot use Lemma 4 to prove the security of the group-2 schemes because the associated compression functions are *not* inversion-resistant. An attack for each is easy to find. For example, consider $f_{13}(h, m) = E(h \oplus m, m) \oplus v$. For any point σ, the adversary fixes $k = 0$, computes $m = E_0^{-1}(\sigma \oplus v)$, and returns (m, m), which is always a correct inverse to σ. Still, despite these compression functions being invertible with a single oracle query, there is a reasonable security bound for the group-2 schemes.

Theorem 6 (OWF security of the group-2 hash functions). Fix $n \geq 1$ and $\imath \in [13..20]$. Then $\mathbf{Adv}_{H_\imath[n]}^{\mathrm{inv}}(q) \leq 9(q+3)^2/2^n$ for any $q > 1$.   ◇

The proof of Theorem 6 makes use of the following lemma, which guarantees that, up to a constant, for messages of length greater than n-bits, the bounds we have computed for collision resistance hold for inversion resistance as well.

Lemma 6 (Collision resistance ⇒ inversion resistance). Fix $\imath \in [1..20]$ and $n \geq 1$. Let $\tilde{H} = H_\imath[n]$ restricted to domain $\mathrm{Bloc}(n, n) \times \bigcup_{i \geq 2}\{0, 1\}^{in}$. Then $\mathbf{Adv}_{\tilde{H}}^{\mathrm{inv}}(q) \leq 3\mathbf{Adv}_H^{\mathrm{coll}}(q+2) + q/2^{n-1}$ for any $q \geq 1$.   ◇

Finally, we prove that the security bounds given in Theorems 5 and 6 are tight, by describing adversaries that achieve advantage very close to the upper bounds. The analysis falls into three groupings.

Theorem 7 (Attacking $H_{1..4}$ as OWFs). Fix $n \geq 1$ and $\imath \in [1..4]$. Then $\mathbf{Adv}_{H_\imath[n]}^{\mathrm{inv}}(q) \geq 0.4q/2^n$ for any $q \in [1..2^{n-2}]$.   ◇

Theorem 8 (Attacking $H_{5..12}$ as OWFs). Fix $n \geq 1$ and $\imath \in [5..12]$. Then $\mathbf{Adv}_{H_\imath[n]}^{\mathrm{inv}}(q) \geq 0.6q/2^n$ for any $q \in [1..2^n - 1]$.   ◇

Theorem 9 (Attacking $H_{13..20}$ as OWFs). Fix $n \geq 1$ and $\imath \in [13..20]$. Then $\mathbf{Adv}_{H_\imath[n]}^{\mathrm{inv}}(q) \geq 0.15q^2/2^n$ for any even $q \in [2..2^{n/2}]$.   ◇

The proofs for the above three theorems appear in the full paper.

Acknowledgments

Thanks to the anonymous reviewers for helpful comments and references. John Black received support from NSF CAREER award CCR-0133985. This work was carried out while John was at the University of Nevada, Reno. Phil Rogaway and his student Tom Shrimpton received support from NSF grant CCR-0085961 and a gift from CISCO Systems. Many thanks for their kind support.

References

1. J. Black, P. Rogaway, and T. Shrimpton. Black-box analysis of the block-cipher-based hash-function constructions from PGV. Full version of this paper, www.cs.ucdavis.edu/~rogaway, 2002.
2. I. Damgård. A design principle for hash functions. In G. Brassard, editor, *Advances in Cryptology - CRYPTO '89*, volume 435 of *Lecture Notes in Computer Science*. Springer-Verlag, 1990.
3. S. Even and Y. Mansour. A construction of a cipher from a single pseudorandom permutation. In *Advances in Cryptology - ASIACRYPT '91*, volume 739 of *Lecture Notes in Computer Science*, pages 210–224. Springer-Verlag, 1992.
4. J. Kilian and P. Rogaway. How to protect DES against exhaustive key search. *Journal of Cryptology*, 14(1):17–35, 2001. Earlier version in CRYPTO '96.
5. R. Merkle. One way hash functions and DES. In G. Brassard, editor, *Advances in Cryptology - CRYPTO '89*, volume 435 of *Lecture Notes in Computer Science*. Springer-Verlag, 1990.
6. B. Preneel, R. Govaerts, and J. Vandewalle. Hash functions based on block ciphers: A synthetic approach. In *Advances in Cryptology - CRYPTO '93*, Lecture Notes in Computer Science, pages 368–378. Springer-Verlag, 1994.
7. M. Rabin. Digitalized signatures. In R. DeMillo, D. Dobkin, A. Jones, and R. Lipton, editors, *Foundations of Secure Computation*, pages 155–168. Academic Press, 1978.
8. C. Shannon. Communication theory of secrecy systems. *Bell Systems Technical Journal*, 28(4):656–715, 1949.
9. P. van Oorschot and M. Wiener. Parallel collision search with cryptanalytic applications. *Journal of Cryptology*, 12(1):1–28, 1999. Earlier version in ACM CCS '94.
10. R. Winternitz. A secure one-way hash function built from DES. In *Proceedings of the IEEE Symposium on Information Security and Privacy*, pages 88–90. IEEE Press, 1984.

A   Fatal Attacks on Five of PGV's B-Labeled Schemes

In [6] there are a total of 13 schemes labeled as "backward attackable." We have already shown that eight of these, $H_{13..20}$, are collision resistant. But the remaining five schemes are completely insecure; each can be broken with two queries. Consider, for example, $H = \hat{H}_{39}$, constructed by iterating the compression function $f = \hat{f}_{39}$ defined by $f^E(h_{i-1}, m_i) = E_{m_i \oplus h_{i-1}}(m_i \oplus h_{i-1}) \oplus m_i$. For any $c \in \{0,1\}^n$ the strings $(h_0 \oplus c) \parallel (E_c(c) \oplus h_0)$ hashes to h_0, and so it so it takes only two queries to produce a collision. Variants of this attack, break the schemes $\hat{H}_{40}$, $\hat{H}_{43}$, $\hat{H}_{55}$ and $\hat{H}_{59}$ defined in Fig. 1. Namely, for $\hat{H}_{40}$, messages $(h_0 \oplus c) \parallel (E_v(c) \oplus h_0)$ collide; for $\hat{H}_{43}$, $(h_0 \oplus c) \parallel (E_c(c) \oplus h_0 \oplus c)$; for $\hat{H}_{55}$, $(h_0 \oplus c) \parallel (E_c(v) \oplus h_0)$; for $\hat{H}_{59}$, $(h_0 \oplus c) \parallel (E_c(v) \oplus h_0 \oplus c)$.

B   Two Notions of Inversion Resistance

We defined $\mathbf{Adv}_H^{inv}$ by giving the adversary a random range point $\sigma \in \{0,1\}^n$ and asking the adversary to find an H-preimage for σ. The usual definition for a one-way function has one choose a random domain point M, apply H, and ask then ask the adversary to invert the result.

Definition 4 (Conventional definition of a OWF). Let H be a block-cipher-based hash function, $H\colon \mathrm{Bloc}(\kappa, n) \times D \to R$, and let ℓ be a number such that $\{0,1\}^\ell \subseteq D$. Let A be an adversary. Then the advantage of A in inverting H on the distribution induced by applying H to a random ℓ-bit string is the real number

$$\mathbf{Adv}_H^{\mathrm{owf}}(A, \ell) = \Pr\Big[E \xleftarrow{\$} \mathrm{Bloc}(\kappa, n);\ M \xleftarrow{\$} (\{0,1\}^n)^\ell;\ \sigma \leftarrow H^E(M);$$

$$M' \leftarrow A^{E, E^{-1}}(\sigma) \colon H^E(M') = \sigma\Big] \qquad \Diamond$$

For $q \geq 0$ a number, $\mathbf{Adv}_H^{\mathrm{owf}}(q, \ell)$ is defined in the usual way, as the maximum value of $\mathbf{Adv}_H^{\mathrm{owf}}(A, \ell)$ over all adversaries A that ask at most q queries.

Though the $\mathbf{Adv}^{\mathrm{owf}}$ and $\mathbf{Adv}^{\mathrm{inv}}$ measures can, in general, be far apart, it is natural to guess that they coincide for "reasonable" hash-functions like $H_{1..20}$. In particular, one might think that the random variable $H_i^E(M)$ is uniformly distributed in $\{0,1\}^n$ if $M \xleftarrow{\$} \{0,1\}^{n\ell}$ and $E \xleftarrow{\$} \mathrm{Bloc}(n, n)$. Interestingly, this is not true. For example, experiments show that when $E \xleftarrow{\$} \mathrm{Bloc}(2, 2)$ and $M \xleftarrow{\$} \{0,1\}^4$ the string $H_1^E(M)$ takes on the value 00 more than a quarter of the time (in fact, 31.25% of the time) while each of the remaining three possible outputs (01, 10, 11) occur less than a quarter of the time (each occurs $22.91\overline{6}$% of the time). Still, for $H_{1..20}$, the two notions are close enough that we have used Definition 3 as a surrogate for Definition 4. The result is as follows.

Lemma 7. Fix $n \geq 1$ and $i \in [1..20]$. Then for any $q, \ell \geq 1$,

$$\left| \mathbf{Adv}_{H_i[n]}^{\mathrm{inv}}(q) - \mathbf{Adv}_{H_i[n]}^{\mathrm{owf}}(q, \ell) \right| \leq \ell/2^{n-1} \qquad \Diamond$$

The proof of Lemma 7 is found in [1].

Supersingular Abelian Varieties in Cryptology

Karl Rubin[1],[*] and Alice Silverberg[2],[**]

[1] Department of Mathematics
Stanford University
Stanford CA, USA
rubin@math.stanford.edu

[2] Department of Mathematics
Ohio State University
Columbus, OH, USA
silver@math.ohio-state.edu

Abstract. For certain security applications, including identity based encryption and short signature schemes, it is useful to have abelian varieties with security parameters that are neither too small nor too large. Supersingular abelian varieties are natural candidates for these applications. This paper determines exactly which values can occur as the security parameters of supersingular abelian varieties (in terms of the dimension of the abelian variety and the size of the finite field), and gives constructions of supersingular abelian varieties that are optimal for use in cryptography.

1 Introduction

The results of this paper show that it is the best of times and the worst of times for supersingular abelian varieties in cryptology. The results in Part 1 give the bad news. They state exactly how much security is possible using supersingular abelian varieties. Part 2 gives the good news, producing the optimal supersingular abelian varieties for use in cryptographic applications, and showing that it is sometimes possible to accomplish this with all computations taking place on an elliptic curve.

One-round tripartite Diffie-Hellman, identity based encryption, and short digital signatures are some problems for which good solutions have recently been found. These solutions make critical use of supersingular elliptic curves and Weil (or Tate) pairings. It was an open question whether or not these new schemes could be improved (more security for the same signature size or efficiency) using abelian varieties in place of elliptic curves. This paper answers the question in the affirmative. We construct families of examples of the "best" supersingular abelian varieties to use in these cryptographic applications (§§5–6), and determine exactly how much security can be achieved using supersingular abelian varieties (§§3–4).

[*] Rubin was partially supported by NSF grant DMS-9800881.
[**] Silverberg was partially supported by Xerox PARC and by NSF grant DMS-9988869.
Some of this work was conducted while she was a visiting researcher at Xerox PARC.

M. Yung (Ed.): CRYPTO 2002, LNCS 2442, pp. 336–353, 2002.
© Springer-Verlag Berlin Heidelberg 2002

Abelian varieties are higher dimensional generalizations of elliptic curves (elliptic curves are the one-dimensional abelian varieties). Weil and Tate pairings exist and have similar properties for abelian varieties that they have for elliptic curves. Supersingular abelian varieties are a special class of abelian varieties. For standard elliptic curve cryptography, supersingular elliptic curves are known to be weak. However, for some recent interesting cryptographic applications [18, 15, 2, 3, 22, 9], supersingular elliptic curves turn out to be very good. New schemes using supersingular elliptic curves and Weil or Tate pairings are being produced rapidly. The abelian varieties in this paper can be utilized in all these applications, to give better results (e.g., shorter signatures, or shorter ciphertexts) for the same security.

The group of points on an abelian variety over a finite field can be used in cryptography in the same way one uses the multiplicative group of a finite field. The security of the system relies on the difficulty of the discrete logarithm (DL) problem in the group of points. One of the advantages of using the group $A(\mathbf{F}_q)$ of an abelian variety in place of the multiplicative group $\mathbf{F}_q^*$ of a finite field $\mathbf{F}_q$ is that there is no known subexponential algorithm for computing discrete logarithms on general abelian varieties.

One of the attacks on the DL problem in $A(\mathbf{F}_q)$ is to map $A(\mathbf{F}_q)$ (or the relevant large cyclic subgroup of $A(\mathbf{F}_q)$) into a multiplicative group $\mathbf{F}_{q^k}^*$, using the Weil or Tate pairing [17, 8, 7]. If this can be done for some small k, then the subexponential algorithm for the DL problem in $\mathbf{F}_{q^k}^*$ can be used to solve the DL problem in $A(\mathbf{F}_q)$. Thus, to have high security, $\#A(\mathbf{F}_q)$ should be divisible by a large prime that does not divide $\#\mathbf{F}_{q^k}^* = q^k - 1$ for any very small values of k.

On the other hand, for cryptographic applications that make use of the Weil or Tate pairing, it is important that $A(\mathbf{F}_q)$ (or the relevant large cyclic subgroup of $A(\mathbf{F}_q)$) *can* be mapped into $\mathbf{F}_{q^k}^*$ with k not too large, in order to be able to compute the pairing efficiently. Thus for these applications it is of interest to produce families of abelian varieties for which the security parameter $\frac{k}{g}$ is not too large, but not too small, where g is the dimension of the abelian variety. (In defining the security parameter, one takes the minimal k.) Taking supersingular elliptic curves (so $g = 1$), one can attain security parameter up to 6. However, it seems to be difficult to systematically produce elliptic curves with security parameter larger than 6 but not enormous. To obtain security parameters that are not too large but not too small, it is natural to consider supersingular abelian varieties.

In [9], Galbraith defined a certain function $k(g)$ and showed that if A is a supersingular abelian variety of dimension g over a finite field $\mathbf{F}_q$, then there exists an integer $k \leq k(g)$ such that the exponent of $A(\mathbf{F}_q)$ divides $q^k - 1$. For example, $k(1) = 6$, $k(2) = 12$, $k(3) = 30$, $k(4) = 60$, $k(5) = 120$, and $k(6) = 210$.

Note that, since cryptographic security is based on the cyclic subgroups of $A(\mathbf{F}_q)$, for purposes of cryptology it is only necessary to consider simple abelian varieties, i.e., abelian varieties that do not decompose as products of lower dimensional abelian varieties.

Table 1. Upper bounds on the cryptographic exponents

g	1	2	3	4	5	6
q a square	3	6	9	15	11	21
q not a square, $p > 11$	2	6	*	12	*	18
q not a square, $p = 2$	4	12	*	20	*	36
q not a square, $p = 3$	6	4	18	30	*	42
q not a square, $p = 5$	2	6	*	15	*	18
q not a square, $p = 7$	2	6	14	12	*	42
q not a square, $p = 11$	2	6	*	12	22	18

In §4, we determine exactly which security parameters can occur, for simple supersingular abelian varieties. For example, we show that if A is a simple supersingular abelian variety over $\mathbf{F}_q$ of dimension g, then the exponent of $A(\mathbf{F}_q)$ divides $q^k - 1$ for some positive integer k less than or equal to the corresponding entry in Table 1 (where $p = \mathrm{char}(\mathbf{F}_q)$), and each entry can be attained. The maximum of each column shows how these bounds compare with the bounds of Galbraith stated above, and how they improve on his bounds when $g \geq 3$. For these bounds, see Theorems 11, 12, and 6 below. A '*' means that there are no simple supersingular abelian varieties of dimension g over $\mathbf{F}_q$.

In particular, we show that the highest security parameter for simple supersingular 4-dimensional abelian varieties is $7.5 = 30/4$, and this can be attained if and only if $p = 3$ and q is not a square. In particular, this answers in the affirmative an open question from [3] on whether one can use higher dimensional abelian varieties to obtain short signatures with higher security. When the dimension is 6 the highest security parameter is 7, and this can be attained if and only if $p = 3$ or 7 and q is not a square. In dimension 2 the highest security parameter is 6, which ties the elliptic curve case. However, these abelian surfaces are in characteristic 2, while the best supersingular elliptic curves occur only in characteristic 3. Therefore, there may be efficiency advantages in using abelian surfaces over binary fields.

In §§5–6 we find the best supersingular abelian varieties for use in cryptography. Theorem 17 gives an algorithm whose input is an elliptic curve and whose output is an abelian variety with higher security. The abelian variety is constructed as a subvariety of a Weil restriction of scalars of the elliptic curve (in the same way that the "XTR supergroup" [16] turns out to be the Weil restriction of scalars from $\mathbf{F}_{p^6}$ to $\mathbf{F}_p$ of the multiplicative group). The group of points of the abelian variety lies inside the group of points of the elliptic curve over a larger field, and thus all computations on the abelian variety can be done directly on the curve. We construct 4-dimensional abelian varieties with security parameter 7.5, thereby beating the security of supersingular elliptic curves, and construct abelian surfaces over binary fields with security parameter 6. We obtain efficient implementations of a variant of the BLS short signature scheme [3] using these abelian varieties (embedded in elliptic curves over larger fields). This gives the first practical application to cryptography of abelian varieties that are not known to be Jacobians of curves.

Theorem 20 gives a method for generating supersingular curves whose Jacobian varieties are good for use in cryptography. This result produces varieties in infinitely many characteristics. Example 21 gives families of examples of Jacobian varieties that are "best possible" in the sense that they achieve the upper bounds listed in the top row of Table 1.

Since $\frac{k}{\varphi(k)} \to \infty$ as $k \to \infty$ (where φ is Euler's φ-function), Theorems 11 and 12 imply that security parameters for simple supersingular abelian varieties are unbounded (as the dimension of the varieties grows). However, $\frac{k}{\varphi(k)}$ grows very slowly, and computational issues and security considerations preclude using high dimensional abelian varieties with high security parameters, at least at this time. We therefore restrict the examples in this paper to small dimensional cases.

The results in §4 rely on the theory of cyclotomic fields, Honda-Tate theory, and work of Zhu. The proof of Theorem 17 uses the theory of Weil restriction of scalars. The proof of Theorem 20 uses the theory of complex multiplication of abelian varieties, applied to Fermat curves.

Part 1: Bounds on the Security

We begin with some preliminaries on abelian varieties.

Suppose A is an abelian variety over a finite field $\mathbf{F}_q$, where q is a power of a prime p. Then A is **simple** if it is not isogenous over $\mathbf{F}_q$ to a product of lower dimensional abelian varieties, and A is **supersingular** if A is isogenous over $\overline{\mathbf{F}}_q$ to a power of a supersingular elliptic curve. (An elliptic curve E is supersingular if $E(\overline{\mathbf{F}}_q)$ has no points of order p.) A **supersingular q-Weil number** is a complex number of the form $\sqrt{q}\zeta$ where ζ is a root of unity. (Throughout the paper, $\sqrt{q}$ denotes the positive square root.)

Theorem 1 ([13, 21, 24]) *Suppose A is a simple supersingular abelian variety over $\mathbf{F}_q$, where q is a power of a prime p, and $P(x)$ is the characteristic polynomial of the Frobenius endomorphism of A. Then:*

(i) $P(x) = G(x)^e$, where $G(x) \in \mathbf{Z}[x]$ is a monic irreducible polynomial and $e = 1$ or 2;

(ii) the roots of G are supersingular q-Weil numbers;

(iii) $A(\mathbf{F}_q) \cong (\mathbf{Z}/G(1)\mathbf{Z})^e$ unless q is not a square and either

    (i) $p \equiv 3 \pmod 4$, $\dim(A) = 1$, and $G(x) = x^2 + q$, or

    (ii) $p \equiv 1 \pmod 4$, $\dim(A) = 2$, and $G(x) = x^2 - q$;

in these exceptional cases, $A(\mathbf{F}_q) \cong (\mathbf{Z}/G(1)\mathbf{Z})^a \times (\mathbf{Z}/\frac{G(1)}{2}\mathbf{Z} \times \mathbf{Z}/2\mathbf{Z})^b$ with non-negative integers a and b such that $a + b = e$;

(iv) $\#A(\mathbf{F}_q) = P(1)$.

The roots of G are called the q-**Weil numbers** for A. For a given abelian variety, its q-Weil numbers are the Galois conjugates of a given one (under the action of the Galois group of $\overline{\mathbf{Q}}$ over $\mathbf{Q}$). We retain the notation of this section, including P, G, and e, throughout the paper. Note that

$$\dim(A) = \frac{\deg(P)}{2} = \frac{e\deg(G)}{2}.$$

Theorem 2 ([13, 21]) *The map that associates to a simple supersingular abelian variety over $\mathbf{F}_q$ one of its q-Weil numbers gives a one-to-one correspondence between the $\mathbf{F}_q$-isogeny classes of simple supersingular abelian varieties over $\mathbf{F}_q$ and Galois conjugacy classes of supersingular q-Weil numbers.*

2 Definition of the Cryptographic Exponent c_A

We introduce a useful new invariant, c_A, which we will call the cryptographic exponent. In the next section we show that c_A captures the MOV security [17] of the abelian variety.

Suppose A is a simple supersingular abelian variety over $\mathbf{F}_q$ and $\sqrt{q}\zeta$ is a q-Weil number for A. Let m denote the order of the root of unity ζ. Note that if $\sqrt{q}\zeta'$ is another q-Weil number for A, and m' is the order of ζ', then ζ^2 and $(\zeta')^2$ are Galois conjugate, and therefore have the same order, namely $\frac{m}{\gcd(2,m)} = \frac{m'}{\gcd(2,m')}$. If q is a square, then ζ and ζ' are Galois conjugate, and thus $m = m'$. Therefore when q is a square, m depends only on A.

Definition 3
$$c_A = \begin{cases} \dfrac{m}{2} & \text{if } q \text{ is a square,} \\ \dfrac{m}{\gcd(2,m)} & \text{if } q \text{ is not a square.} \end{cases}$$

We will call c_A the **cryptographic exponent** *of A. Let $\alpha_A = c_A/g$ and call it the* **security parameter** *of A.*

Roughly speaking, for a group G to have security parameter α means that the DL problem in G can be reduced to the DL problem in the multiplicative group of a field of size approximately $|G|^\alpha$. The group $G = A(\mathbf{F}_q)$ has order approximately q^g, and we will see in §3 below that q^{c_A} is the size of the smallest field F such that every cyclic subgroup of $A(\mathbf{F}_q)$ can be embedded in F^*.

When q is not a square, c_A is a natural number. When q is a square, c_A is either a natural number or half of a natural number.

If $\gcd(t, 2c_A) = 1$, then the cryptographic exponent for A over $\mathbf{F}_{q^t}$ is the same as the cryptographic exponent for A over $\mathbf{F}_q$.

Let $\mathbf{N}$ denote the set of natural numbers. If $k \in \mathbf{N}$, write $\Phi_k(x)$ for the k-th cyclotomic polynomial $\prod_\zeta (x - \zeta)$, where the product is over the primitive k-th roots of unity ζ. Note that $\deg(\Phi_k) = \varphi(k)$, where φ is Euler's φ-function.

Lemma 4 *Suppose that $\Phi_m(d)$ is divisible by a prime number ℓ, and $\ell \nmid m$. Then m is the smallest natural number k such that $d^k - 1$ is divisible by ℓ.*

Proof. The roots of Φ_m in $\overline{\mathbf{F}}_\ell$ are exactly the primitive m-th roots of unity, since $\ell \nmid m$. By assumption, d is a root of Φ_m in $\mathbf{F}_\ell$, and so m is the order of d in $\mathbf{F}_\ell^*$.

We include a useful closely related result.

Proposition 5 *If $m, d \in \mathbf{N}$, $d > 1$, and $(m,d) \neq (6,2)$, then m is the smallest natural number k such that $d^k - 1$ is divisible by $\Phi_m(d)$.*

Proof. Since $x^m - 1 = \prod_{r|m} \Phi_r(x)$, we have that $\Phi_m(d)$ divides $d^m - 1$. The proposition is true if $m = 1$ or 2. If $m > 2$ and $(m, d) \neq (6, 2)$, it follows from an 1892 result of Zsigmondy (see Theorem 8.3, §IX of [14]) that $\Phi_m(d)$ has a prime divisor that does not divide m. The proposition now follows from Lemma 4.

In the exceptional case $(m, d) = (6, 2)$, we have $\Phi_m(d) = 3 = d^2 - 1$.

Theorem 6 *Suppose A is a simple supersingular abelian variety over $\mathbf{F}_q$.*

(i) *If q is a square then the exponent of $A(\mathbf{F}_q)$ divides $\Phi_{2c_A}(\sqrt{q})$, which divides $\sqrt{q}^{2c_A} - 1$.*

(ii) *If q is not a square then the exponent of $A(\mathbf{F}_q)$ divides $\Phi_{c_A}(q)$, which divides $q^{c_A} - 1$.*

Proof. By Theorem 1(iii), the exponent of $A(\mathbf{F}_q)$ divides $G(1)$. Let π be a q-Weil number for A. If q is a square, then $\Phi_{2c_A}(\frac{\pi}{\sqrt{q}}) = 0$. Thus, $G(x) = \sqrt{q}^{\varphi(2c_A)}\Phi_{2c_A}(\frac{x}{\sqrt{q}})$ and $G(1) = \sqrt{q}^{\varphi(2c_A)}\Phi_{2c_A}(\frac{1}{\sqrt{q}}) = \pm\Phi_{2c_A}(\sqrt{q})$. If q is not a square, then $\Phi_{c_A}(\frac{\pi^2}{q}) = 0$, so $G(x)$ divides $q^{\varphi(c_A)}\Phi_{c_A}(\frac{x^2}{q})$. Therefore $G(1)$ divides $q^{\varphi(c_A)}\Phi_{c_A}(\frac{1}{q}) = \pm\Phi_{c_A}(q)$. As in Proposition 5, $\Phi_m(d)$ divides $d^m - 1$.

3   The Cryptographic Exponent and MOV Security

The next result shows that the cryptographic exponent c_A captures the MOV security of the abelian variety. In other words, if $A(\mathbf{F}_q)$ has a subgroup of large prime order ℓ, then q^{c_A} is the size of the smallest field of characteristic p containing a multiplicative subgroup of order ℓ. Recall $e \in \{1, 2\}$ from Theorem 1.

Theorem 7 *Suppose A is a simple supersingular abelian variety of dimension g over $\mathbf{F}_q$, $q = p^n$, $\ell > 5$ is a prime number, $\ell \mid \#A(\mathbf{F}_q)$, and $\ell > (1 + \sqrt{p})^{ng/e}$. Let r denote the smallest natural number k such that $\ell \mid p^k - 1$. Then $p^r = q^{c_A}$.*

Since the proof is rather technical, we do not give it here, but instead prove the following slightly weaker result.

Theorem 8 *Suppose A is a simple supersingular abelian variety over $\mathbf{F}_q$, ℓ is a prime number, $\ell \mid \#A(\mathbf{F}_q)$, and $\ell \nmid 2c_A$. Then c_A is the smallest half-integer k such that $q^k - 1$ is an integer divisible by ℓ.*

Proof. By Theorem 6, we have $\ell \mid \Phi_{2c_A}(\sqrt{q})$ if q is a square, and $\ell \mid \Phi_{c_A}(q)$ otherwise. The theorem now follows from Lemma 4.

Remark 9 *For purposes of cryptography we are only interested in the case where ℓ is large. If $\ell > 2g + 1$, then $\ell \nmid 2c_A$, so the condition $\ell \nmid 2c_A$ is not a problem. This follows since $2g = \deg(P) = e\deg(G)$, $\deg(G) = \varphi(2c_A)$ if q is a square, $\deg(G) = \varphi(c_A)$ or $2\varphi(c_A)$ if q is not a square, and $\varphi(M) \geq \ell - 1$ if $\ell \mid M$.*

4 Bounding the Cryptographic Exponent

Next we determine exactly which values can occur as cryptographic exponents for simple supersingular abelian varieties. Let

$$W_n = \{k \in \mathbf{N} : \varphi(k) = n\}.$$

For example, $W_1 = \{1, 2\}$, $W_n = \emptyset$ if n is odd and $n > 1$,

$$W_2 = \{3, 4, 6\}, \quad W_4 = \{5, 8, 10, 12\}, \quad W_6 = \{7, 9, 14, 18\}.$$

Let k' denote the odd part of a natural number k. If p is a prime, define

$$X_p = \begin{cases} \{k \in \mathbf{N} : 4 \nmid k \text{ and } 2 \text{ has odd order in } (\mathbf{Z}/k'\mathbf{Z})^*\} & \text{if } p = 2, \\ \{k \in \mathbf{N} : p \nmid k \text{ and } p \text{ has odd order in } (\mathbf{Z}/k\mathbf{Z})^*\} & \text{if } p \text{ is odd}; \end{cases}$$

$$V_p = \begin{cases} \{k \in \mathbf{N} : k \equiv 4 \pmod{8}\} & \text{if } p = 2, \\ \{k \in \mathbf{N} : p \mid k \text{ and } k \equiv 2 \pmod 4\} & \text{if } p \equiv 3 \pmod 4, \\ \{k \in \mathbf{N} : p \mid k \text{ and } k \text{ is odd}\} & \text{if } p \equiv 1 \pmod 4; \end{cases}$$

$$K_g(p) = \begin{cases} (W_{2g} \cap V_p) \cup (W_g - V_p) & \text{if } g > 2, \\ (W_4 \cap V_p) \cup (W_2 - V_p) \cup \{1\} & \text{if } g = 2, \\ (W_2 \cap V_p) \cup (W_1 - V_p - \{1\}) & \text{if } g = 1. \end{cases}$$

The next result can be shown to follow from Proposition 3.3 of [24].

Proposition 10 ([24]) *Suppose A is a simple supersingular abelian variety of dimension g over $\mathbf{F}_q$.*

(i) If q is a square, then $e = 2$ if and only if $2c_A \in X_p$.

(ii) If q is not a square, then $e = 2$ if and only if $c_A = 1$ and $g = 2$.

Theorem 11 *Suppose g and n are natural numbers and n is even. Then $c = \frac{m}{2}$ occurs as the cryptographic exponent of a simple supersingular abelian variety of dimension g over $\mathbf{F}_{p^n}$ if and only if $m \in (W_g \cap X_p) \cup (W_{2g} - X_p)$.*

Proof. If ζ is a primitive m-th root of unity, then $\sqrt{p^n}\zeta$ corresponds by Theorem 2 to a simple supersingular abelian variety over $\mathbf{F}_{p^n}$ of dimension $d = e \deg(G)/2 = e\varphi(m)/2$. By Proposition 10(i), $d = g$ if and only if $m \in (W_g \cap X_p) \cup (W_{2g} - X_p)$.

Theorem 12 *Suppose g and n are natural numbers and n is odd. Then c occurs as the cryptographic exponent of a simple supersingular abelian variety of dimension g over $\mathbf{F}_{p^n}$ if and only if $c \in K_g(p)$.*

Proof. Suppose A is a simple supersingular abelian variety of dimension g over $\mathbf{F}_q = \mathbf{F}_{p^n}$ with a q-Weil number $\pi = \sqrt{q}\zeta$ with ζ a primitive m-th root of unity. Then $\varphi(c_A) = [\mathbf{Q}(\pi^2) : \mathbf{Q}]$. We have $2g = e[\mathbf{Q}(\pi) : \mathbf{Q}] = e[\mathbf{Q}(\pi) : \mathbf{Q}(\pi^2)][\mathbf{Q}(\pi^2) : \mathbf{Q}]$. It follows from Lemma 2.6 of [24] that $\mathbf{Q}(\pi) = \mathbf{Q}(\pi^2)$ if and only if $c_A \in V_p$. It follows from Proposition 10(ii) that $c_A \in K_g(p)$. The converse follows by the same reasoning.

For any given g and q, it is easy to work out from Theorems 11 and 12 exactly which values can occur as cryptographic exponents c_A for g-dimensional simple supersingular abelian varieties A over $\mathbf{F}_q$, as is done in the following two corollaries.

Corollary 13 *If n is even, then the only possible cryptographic exponents c_A for simple supersingular abelian surfaces A over $\mathbf{F}_{p^n}$ are the numbers of the form $\frac{m}{2}$ with $m \in \{3,4,5,6,8,10,12\}$. For $m \in \{3,4,6\}$, $\frac{m}{2}$ occurs as a c_A if and only if $p \equiv 1 \pmod{m}$, and for $m \in \{5,8,10,12\}$, $\frac{m}{2}$ occurs as a c_A if and only if $p \not\equiv 1 \pmod{m}$. An analogous statement holds for 4-dimensional varieties, with $\{3,4,6\}$ and $\{5,8,10,12\}$ replaced by $\{5,8,10,12\}$ and $\{15,16,20,24,30\}$, respectively.*

Corollary 14 *If n is odd, then the exact sets of cryptographic exponents c_A that occur for simple supersingular abelian varieties A of dimension g over $\mathbf{F}_{p^n}$ with $2 \le g \le 5$ are given below.*

(i) *Suppose $g = 2$.*

    (a) $c_A \in \{1,3,4,6\}$ *if $p \ge 7$;*
    (b) $c_A \in \{1,3,4,5,6\}$ *if $p = 5$;*
    (c) $c_A \in \{1,3,4\}$ *if $p = 3$;*
    (d) $c_A \in \{1,3,6,12\}$ *if $p = 2$.*

(ii) *Suppose $g = 3$.*

    (a) *There does not exist such an A if $p \ne 3,7$;*
    (b) $c_A = 14$ *if $p = 7$;*
    (c) $c_A = 18$ *if $p = 3$.*

(iii) *Suppose $g = 4$.*

    (a) $c_A \in \{5,8,10,12\}$ *if $p \ge 7$;*
    (b) $c_A \in \{8,10,12,15\}$ *if $p = 5$;*
    (c) $c_A \in \{5,8,10,12,30\}$ *if $p = 3$;*
    (d) $c_A \in \{5,10,20\}$ *if $p = 2$.*

(iv) *Suppose $g = 5$.*

    (a) *There does not exist such an A if $p \ne 11$;*
    (b) $c_A = 22$ *if $p = 11$.*

Corollary 15 *Suppose p is prime, n and g are odd natural numbers, and $g > 1$.*

(i) *If $p \not\equiv 3 \pmod 4$, then there does not exist a simple supersingular abelian variety of dimension g over $\mathbf{F}_{p^n}$.*

(ii) *If $p \equiv 3 \pmod 4$, and there exists a simple supersingular abelian variety of dimension g over $\mathbf{F}_{p^n}$, then $g = p^{b-1}(p-1)/2$ for some natural number b.*

Proof. Suppose there is a simple supersingular abelian variety A of dimension g over $\mathbf{F}_{p^n}$. Since $g > 1$ is odd, we conclude from Theorem 12 that $\varphi(c_A) = 2g \equiv 2 \pmod 4$ and $p \mid c_A$. This is only possible if $c_A = p^b$ or $2p^b$, and $p \equiv 3 \pmod 4$.

Part 2: Optimal Supersingular Abelian Varieties

Definition 16 *Suppose A is a supersingular abelian variety of dimension g over $\mathbf{F}_q$. We say that A is **optimal** if A is simple, and $c_A \geq c_B$ for every simple supersingular abelian variety B of dimension g over $\mathbf{F}_q$.*

Optimal supersingular elliptic curves are well-known. The Jacobian of the genus 2 curve $y^2 + y = x^5 + x^3$ over $\mathbf{F}_2$ is optimal ($c_A = 12$), and was given in [9]. Recall that the genus of a curve is the same of the dimension of the Jacobian variety of the curve.

The next two sections give two different constructions of families of examples of optimal supersingular abelian varieties. The first comes from taking a piece of the Weil restriction of scalars of an elliptic curve. This construction has the advantage of producing abelian varieties of dimensions 2, 3, 4, and 6 with the largest security parameter possible for abelian varieties of that dimension, namely 6, 6, 7.5, and 7, respectively. The best such examples occur in characteristics 2 and 3, which gives a computational advantage. The second construction comes from Jacobian varieties of superelliptic curves, and has the advantage of giving a choice of infinitely many abelian varieties and characteristics.

5   A Subvariety of the Weil Restriction of Scalars

If $\mathbb{k} \subset \mathbb{k}'$ are finite fields, E is an elliptic curve over $\mathbb{k}$, and $Q \in E(\mathbb{k}')$, write $\mathrm{Tr}_{\mathbb{k}'/\mathbb{k}} Q = \sum_{\sigma \in \mathrm{Gal}(\mathbb{k}'/\mathbb{k})} \sigma(Q)$. See the appendix for a proof of a generalization of the following result.

Theorem 17 *Suppose E is a supersingular elliptic curve over $\mathbf{F}_q$, π is a q-Weil number for E, and π is not a rational number. Fix $r \in \mathbf{N}$ with $\gcd(r, 2pc_E) = 1$. Then there is a simple supersingular abelian variety A over $\mathbf{F}_q$ such that:*

  (i) $\dim(A) = \varphi(r)$;
  (ii) for every primitive r-th root of unity ζ, $\pi\zeta$ is a q-Weil number for A;
  (iii) $c_A = rc_E$;
  (iv) $\alpha_A = \frac{r}{\varphi(r)}\alpha_E$;
  (v) there is a natural identification of $A(\mathbf{F}_q)$ with the subgroup of $E(\mathbf{F}_{q^r})$

$$\{Q \in E(\mathbf{F}_{q^r}) : \mathrm{Tr}_{\mathbf{F}_{q^r}/\mathbf{F}_{q^r/\ell}} Q = O \quad \text{for every prime } \ell \mid r\}.$$

Abelian varieties of this form were considered by Frey in §3.2 of [6].

Remark 18 *By Theorem 17(iii), $A(\mathbf{F}_q)$ has the same MOV security as $E(\mathbf{F}_{q^r})$. By Theorem 17(v), computation in $A(\mathbf{F}_q)$ is as efficient as computation in $E(\mathbf{F}_{q^r})$. The advantage of using $A(\mathbf{F}_q)$ is that (by Theorem 17(iv)) its security parameter α_A is higher than that of $E(\mathbf{F}_{q^r})$ by a factor $r/\varphi(r)$, so (for example) it provides shorter signatures for the same security in the BLS short signature scheme [3].*

Using $E(\mathbf{F}_{q^r})$, a signature in the BLS scheme is the x-coordinate of a point on the elliptic curve, which is an element of $\mathbf{F}_{q^r}$ and therefore is $r\log_2(q)$ bits.

Fixing a basis for $\mathbf{F}_{q^r}$ over $\mathbf{F}_q$, an element of $\mathbf{F}_{q^r}$ can be viewed as a vector with r coordinates in $\mathbf{F}_q$. Using $A(\mathbf{F}_q)$ in the short signature scheme and identifying it with a subgroup of $E(\mathbf{F}_{q^r})$ as in Theorem 17(v), a signature will now be only the first $\varphi(r)$ coordinates of the x-coordinate of a point in $E(\mathbf{F}_{q^r})$ (along with a few extra bits to resolve an ambiguity that may arise), so the signature is about $\varphi(r)\log_2(q)$ bits. Thus, for signature generation there is no additional computation required: just follow the algorithm in [3] to produce the x-coordinate of a point in $E(\mathbf{F}_{q^r})$, and drop the extra coordinates. However, for signature verification there is now an extra step: given a signature one must reconstruct the missing coordinates to get the x-coordinate of a point in our subgroup of $E(\mathbf{F}_{q^r})$, and then follow the verification algorithm in [3]. For more information on this extra verification step, see the examples below.

Theorem 17 can be applied in particular to the low dimensional cases where the tuple $(\dim(A), p, r, c_A)$ is $(2, 2, 3, 12)$, $(2, p > 3, 3, 6)$, $(4, 2, 5, 20)$, $(4, 3, 5, 30)$, $(6, 2, 9, 36)$, or $(6, 3, 7, 42)$. Next we use Theorem 17 to give implementations in the cases $(4, 3, 5, 30)$ and $(2, 2, 3, 12)$.

5.1 $\dim(A) = 4$, $p = 3$

The largest security parameter for a 4-dimensional abelian variety is 7.5, and this occurs only in characteristic 3.

When $\gcd(n, 6) = 1$ there are exactly 2 isogeny classes of elliptic curves over $\mathbf{F}_{3^n}$ with security parameter 6. Equations for a curve from each isogeny class, along with one of its Weil numbers and its characteristic polynomial of Frobenius, are given below, where $(\frac{3}{n})$ denotes the Jacobi symbol, which is $+1$ if $n \equiv \pm 1 \pmod{12}$, and -1 if $n \equiv \pm 5 \pmod{12}$.

curve	equation	Weil number	characteristic polynomial
E_n^+	$y^2 = x^3 - x + (\frac{3}{n})$	$\sqrt{3^n}e^{7\pi i/6}$	$G_n(x) = x^2 + 3^{\frac{n+1}{2}}x + 3^n$
E_n^-	$y^2 = x^3 - x - (\frac{3}{n})$	$\sqrt{3^n}e^{\pi i/6}$	$H_n(x) = x^2 - 3^{\frac{n+1}{2}}x + 3^n$

By Theorem 11 there is no elliptic curve over $\mathbf{F}_{3^n}$ with security parameter 6 when n is even. If n is an odd multiple of 3 then there are again two isogeny classes of curves with the same Weil numbers and characteristic polynomials as in the above table, but with different curves E_n^+ and E_n^-.

Applying Theorem 17 to the elliptic curves E_n^+ and E_n^- over $\mathbf{F}_{3^n}$ with $r = 5$ produces 4-dimensional abelian varieties A_n^+ and A_n^- over $\mathbf{F}_{3^n}$, described in the following table.

	Weil number	characteristic polynomial
A_n^+	$\sqrt{3^n}e^{\pi i/30}$	$H_{5n}(x^5)/G_n(x)$
A_n^-	$\sqrt{3^n}e^{17\pi i/30}$	$G_{5n}(x^5)/H_n(x)$

Write E for $E_n^\pm$ and A for $A_n^\pm$. By Theorem 17(iv), $\alpha_A = \frac{5}{4}\alpha_E = 7.5$. Using the characteristic polynomials to compute $\#A(\mathbf{F}_{3^n})$ for various n, we find the following sample values of n for which $\#A(\mathbf{F}_{3^n})$ is of a size suitable for cryptographic applications, and has a large prime factor. Here the signature length is $4\log_2(3^n)$ (see Remark 18), the DL security column contains $\log_2(\ell)$ where ℓ is the largest prime dividing $\#A(\mathbf{F}_{3^n})$, and the MOV security column contains $\log_2(q^{c_A}) = \log_2(3^{30n})$.

variety	n	signature length	DL security	MOV security
A_n^+	15	95	95	713
A_n^+	17	108	100	808
A_n^+	19	120	112	903
A_n^+	33	209	191	1569
A_n^+	43	273	265	2045

Let $\mathbb{k} = \mathbf{F}_{3^n}$ and $\mathbb{k}_1 = \mathbf{F}_{3^{5n}}$. As discussed in Remark 18, the extra computation required for signature verification amounts to solving the problem: given 4 of the 5 $\mathbb{k}$-coordinates of x, where $(x, y) \in E(\mathbb{k}_1)$ and $\mathrm{Tr}_{\mathbb{k}_1/\mathbb{k}}(x, y) = O$, compute the fifth.

We next give an algorithm to do this. Suppose $Q = (x, y) \in E(\mathbb{k}_1)$ and $\sum_{i=0}^4 \sigma^i(Q) = O$ where σ generates $\mathrm{Gal}(\mathbb{k}_1/\mathbb{k})$. Then there is a function $\mathcal{F}$ on E with zeros at the points $\sigma^i(Q)$ for $0 \le i \le 4$, a pole of order 5 at O, and no other zeros or poles. Let $g(z) = \prod_{i=0}^4 (z - \sigma^i(x)) \in \mathbb{k}[z]$, and let X and Y denote the coordinate functions on E. Then $g(X)$ is a function on E with zeros at $\pm\sigma^i(Q)$ for $0 \le i \le 4$, a pole of order 10 at O, and no other zeros or poles. Thus $g(X) = \mathcal{F}\tilde{\mathcal{F}}$, where $\tilde{\mathcal{F}}$ is $\mathcal{F}$ composed with multiplication by -1 on E. Write $\mathcal{F} = f_1(X) + f_2(X)Y$ with $f_1(X), f_2(X) \in \mathbb{k}[X]$. Since X has a double pole at O and Y a triple pole, we have $\deg(f_1) \le 2$ and $\deg(f_2) = 1$. Setting $g(X) = f_1(X)^2 - Y^2 f_2(X)^2 = f_1(X)^2 - (X^3 - X \pm 1)f_2(X)^2$ gives equations relating the coefficients of g, f_1, and f_2.

Suppose we know 4 of the 5 coordinates of x with respect to some fixed basis of $\mathbb{k}_1$ over $\mathbb{k}$, and let $b \in \mathbb{k}$ denote the missing coordinate. The coefficients of g are polynomials in b with coefficients in $\mathbb{k}$. Solving the above system of equations for b reduces to computing the resultant of 2 polynomials in 2 variables, and then finding the roots of a degree 9 polynomial in $\mathbb{k}[z]$. (The extra bits in the signature are used here in case the polynomial has more than one root.) This extra verification step takes a few seconds on a desktop computer, using the number theory software package KASH to compute the resultant and find its roots, but this could be optimized by writing a dedicated program.

Remark 19 $\mathbb{k}$-*coordinates of* x, *An alternative way to generate a signature from the point Q above is to take 4 of the 5 symmetric functions of x and its conjugates (i.e., 4 of the 5 coefficients of the polynomial g), instead of taking 4 of the 5 $\mathbb{k}$-coordinates of x. It is computationally very fast to recover the missing coefficient of g using the algorithm above. Then x can be computed by factoring g over $\mathbb{k}_1$. In our experiments the method above, which works over $\mathbb{k}$ rather than $\mathbb{k}_1$, seems to be more efficient.*

One could alternatively apply Theorem 17 with $q = 3$ and $r = 5n$ and gain an additional factor of $n/\varphi(n)$ in the signature length. However, the verification problem becomes harder.

5.2 $\dim(A) = 2$, $p = 2$

The largest security parameter for an abelian surface is 6, and this occurs only in characteristic 2.

When n is odd there are exactly 2 isogeny classes of elliptic curves over $\mathbf{F}_{2^n}$ with $\alpha_E = 4$, namely those of $y^2 + y = x^3 + x + 1$ and $y^2 + y = x^3 + x$. Applying Theorem 17 with these curves and $r = 3$ produces two abelian surfaces $A_n^\pm$ over $\mathbf{F}_{2^n}$ with Weil number $\pm\sqrt{2^n}e^{\pi i/12}$ and characteristic polynomial of Frobenius

$$x^4 \mp 2^{\frac{n+1}{2}}x^3 + 2^n x^2 \mp 2^{\frac{3n+1}{2}}x + 2^{2n}.$$

(One of these abelian varieties was given in [9] as the Jacobian of a hyperelliptic curve.)

By Theorem 17(iv) (or directly from the definition), $\alpha_{A_n^\pm} = 6$. Using the characteristic polynomials to compute $\#A_n^\pm(\mathbf{F}_{2^n})$ for various n, we find the following sample values of n that are suitable for cryptographic applications. Here the signature length is $2n$.

variety	n	signature length	DL security	MOV security
A_+	43	86	82	516
A_-	53	106	93	636
A_+	79	158	141	948
A_+	87	174	167	1044
A_-	87	174	156	1044
A_-	103	206	192	1236
A_-	121	242	220	1452

As discussed in Remark 18, there is no extra computation required to generate short signatures using $A_n^\pm$, and the extra computation required for signature verification amounts to solving the following problem: given two of the three $\mathbf{F}_{2^n}$-coordinates of a point in the subgroup of $E_n^\pm(\mathbf{F}_{2^{3n}})$ corresponding to $A_n^\pm(\mathbf{F}_{2^n})$ under Theorem 17(v), find the third coordinate. Using the method described above in the case of $p = 3$, $g = 4$, and $r = 5$, in the present case the computation reduces to taking one square root in $\mathbf{F}_{2^n}$ and solving one quadratic polynomial over $\mathbf{F}_{2^n}$. Taking square roots in a field of characteristic 2 is just a single exponentiation, and solving a quadratic equation is not much harder. Neither of these operations took measurable time on a desktop computer with the field $\mathbf{F}_{2^{103}}$.

6   Jacobian Varieties That Are Optimal When q Is a Square

The next result gives families of examples of Jacobian varieties that are optimal. They have the advantage of giving a choice of infinitely many field characteristics.

Theorem 20 *Suppose that $a, b, n \in \mathbf{N}$ have no common divisor greater than 1, n is odd, and $n + 2 - ((n, a) + (n, b) + (n, a + b)) = \varphi(n)$. Let q be a prime power congruent to $-1 \pmod{n}$, and let $F = \mathbf{F}_{q^2}$. For $\gamma \in F^*$, let C_γ be the curve*

$$y^n = \gamma x^a (1 - x)^b$$

over F and write A_γ for its Jacobian variety. Then the dimension of A_γ is $\varphi(n)/2$ and A_γ is supersingular. If in addition γ generates F^* modulo n-th powers, then A_γ is simple, $c_{A_\gamma} = n$, and $A_\gamma(F)$ is cyclic.

Proof. The dimension of A_γ is the genus of C_γ. The genus g of C_γ being $\varphi(n)/2$ follows from the fact that g is independent of γ, and the formula for the genus of $C_{\pm 1}$ given on p. 55 of [4].

Since $q \equiv -1 \pmod{n}$, Theorem 20.15 of [19] shows that the Frobenius endomorphism of A_1 is multiplication by $-q$. In particular, the characteristic polynomial of Frobenius is $(x + q)^{2g}$, and A_1 is supersingular. Since every A_γ is isomorphic to A_1 over the algebraic closure $\bar{F}$, every A_γ is supersingular.

The endomorphism ring $\mathrm{End}(A_\gamma)$ contains the group of n-th roots of unity μ_n, where $\xi \in \mu_n$ acts on C_γ by sending (x, y) to $(x, \xi y)$. Fix an n-th root δ of γ. Then δ^{q^2} is also an n-th root of γ. Let $\zeta = \gamma^{(q^2-1)/n} = \delta^{q^2-1}$. Then $\zeta^n = 1$, so we can view $\zeta \in \mu_n \subset \mathrm{End}(A_\gamma)$. We have a commutative diagram

$$
\begin{array}{ccc}
C_1 & \xrightarrow{\phi_1} & C_1 \\
\lambda \downarrow & & \downarrow \lambda' \\
C_\gamma & \xrightarrow{\phi_\gamma} & C_\gamma
\end{array}
$$

where ϕ_1, ϕ_γ are the q^2-power maps $(x, y) \mapsto (x^{q^2}, y^{q^2})$ of C_1 and C_γ, respectively, and $\lambda, \lambda' : C_1 \to C_\gamma$ are the isomorphisms $(x, y) \mapsto (x, \delta y)$, $(x, y) \mapsto (x, \delta^{q^2} y)$. Writing $[\phi_\gamma]$, $[\lambda']$, etc. for the induced maps on A_1 and A_γ, we noted above that $[\phi_1] = -q$, and so the Frobenius endomorphism of A_γ is

$$[\phi_\gamma] = [\lambda' \circ \phi_1 \circ \lambda^{-1}] = [\lambda'^{-1}] \circ [\phi_1] \circ [\lambda'] = [\lambda^{-1}] \circ (-q) \circ [\lambda'] = -q \circ [\lambda' \circ \lambda^{-1}] = -\zeta q.$$

Suppose now that γ generates F^* modulo n-th powers. Then ζ is a primitive n-th root of unity, and since n is odd, $-\zeta$ is a primitive $2n$-th root of unity. The characteristic polynomial $P(x)$ of Frobenius on A_γ has degree $2g = \varphi(n) = \varphi(2n)$, and has $-\zeta q$ as a root, so $P(x) = \prod_\xi (x - \xi q)$, product over primitive $2n$-th roots of unity ξ. Thus $P(x) = q^{\varphi(2n)} \Phi_{2n}(x/q)$. Since $\Phi_{2n}(x)$ is irreducible, so is $P(x)$. Therefore A_γ is simple and $c_A = n$. By Theorem 1, $A_\gamma(F)$ is cyclic.

Example 21 *Suppose (g, n, a, b) is one of the following 4-tuples:*

g	n	a	b
3	9	3	1
4	15	5	3
6	21	7	3
9	27	9	1
10	33	11	3
$\frac{\ell-1}{2}$	ℓ	α	β

where in the last row ℓ is a prime, $1 \leq \alpha, \beta \leq \ell-1$, and $\alpha+\beta \neq \ell$. Let q be a prime power congruent to -1 (mod n), $F = \mathbf{F}_{q^2}$, and γ a generator of F^* modulo n-th powers. Let C be the curve $y^n = \gamma x^a (1 - x)^b$ and A its Jacobian variety. Then by Theorem 20, A is simple and supersingular, $\text{genus}(C) = \dim(A) = g$, $c_A = n$, $A(F)$ is cyclic, and $2n$ is the smallest integer k such that $\#A(F)$ divides $q^k - 1$. In the table, if $g = 3, 4, 6, 9, 10$, or if $g > 3$ and g is a prime of the form $(\ell-1)/2$, then $2n$ is the largest element of W_{2g}, so A is optimal. Optimal examples with $g = 1$ and 5 are obtained by taking $\ell = 3$ and 11 in the last row, and non-optimal examples with $g = 2$ and 3 by taking $\ell = 5$ and 7 in the last row.

7 Security

Proofs of security for cryptosystems based on elliptic curves rely on the difficulty of some problem (EC Diffie-Hellman and/or Weil Diffie-Hellman, for the systems in [18, 15, 2, 3, 22]). These hard problems generalize to abelian varieties, where they are also believed to be hard. However, we note some additional security considerations.

Allowing the cryptographic exponent c_A to take half-integer values when q is a square means that c_A correctly captures the MOV security of the variety. For example, for every prime p there is a supersingular elliptic curve E over $\mathbf{F}_{p^2}$ such that $c_A = \frac{1}{2}$, by Theorem 11. By Theorem 1, $E(\mathbf{F}_{p^2}) \cong (\mathbf{Z}/(p-1)\mathbf{Z})^2$, and the smallest field in which the Weil and Tate pairings take their values is $\mathbf{F}_p$. Therefore, solving the DL problem in $\mathbf{F}_p^*$ will break cryptographic schemes that base their security on the difficulty of solving the DL problem in a subgroup of $E(\mathbf{F}_{p^2})$. In other words, the MOV security here really comes from $\mathbf{F}_p$, and not $\mathbf{F}_q$. Theorem 7 says that in general the MOV security comes from a field of size q^{c_A}.

It follows from Theorem 8 that in the special case where A is an elliptic curve, q is not a square, and Q is a point in $A(\mathbf{F}_q)$ of large order, the cryptographic exponent c_A coincides with the "security multiplier" for Q that was defined in [3].

Abelian varieties that are Jacobians of hyperelliptic curves over a finite field whose size is small compared to the curve's genus are considered to be weak for use in cryptography, due to attacks in [1, 11]. The examples coming from §5 do not appear in general to be Jacobians of curves. The examples in §6 are Jacobians, but outside of the cases equivalent to the $a = b = 1$ case they do not appear to be Jacobians of hyperelliptic curves. In any case, these attacks do not apply to abelian varieties of small dimension.

Weil descent attacks [12, 10] have been carried out for certain elliptic curves over binary fields. In these attacks one starts with an elliptic curve over $\mathbf{F}_{q^r}$ and takes its Weil restriction of scalars down to $\mathbf{F}_q$. This is an abelian variety B of dimension r over $\mathbf{F}_q$. The attack proceeds by looking for a hyperelliptic curve whose Jacobian variety is related to B, solving the DL problem for this Jacobian variety, and using it to solve the DL problem for the original elliptic curve. For an abelian variety A produced by Theorem 17 from an elliptic curve E, we have $A(\mathbf{F}_q) \subseteq E(\mathbf{F}_{q^r})$. It is tempting to try to break the associated cryptosystems by solving the DL problem on $E(\mathbf{F}_{q^r})$ using Weil descent. However, the Weil descent attack replaces (the subfield curve) E by its Weil restriction of scalars from $\mathbf{F}_{q^r}$ to $\mathbf{F}_q$, which has A as a large simple factor, so we are back where we started. In addition, it is not known how to carry out Weil descent attacks except when $p = 2$ and $\dim(A) \geq 4$, and the most important applications of Theorem 17 (the examples in §5) have either $p = 3$ and $\dim(A) = 4$, or $p = 2$ and $\dim(A) = 2$. For these examples, one could ask whether there is an efficient way to find hyperelliptic curves, if they exist, whose Jacobians are related to the given abelian variety in a helpful way. This is likely to be a hard problem in general. Its analogue in characteristic zero would solve a long-standing problem by producing a sequence of elliptic curves of unbounded rank.

Acknowledgments

The authors thank Steven Galbraith for his observations and Dan Boneh for helpful conversations.

References

1. L. Adleman, J. DeMarrais and M-D. Huang. A subexponential algorithm for discrete logarithms over the rational subgroup of the Jacobians of large genus hyperelliptic curves over finite fields, in Algorithmic number theory. Lecture Notes in Computer Science, Vol. 877. Springer-Verlag (1994) 28–40.
2. D. Boneh and M. Franklin. Identity based encryption from the Weil pairing, in Advances in Cryptology – Crypto 2001. Lecture Notes in Computer Science, Vol. 2139. Springer-Verlag (2001) 213–229.
3. D. Boneh, B. Lynn and H. Shacham. Short signatures from the Weil pairing, in Advances in Cryptology – Asiacrypt 2001. *Lect. Notes in Comp. Sci.* **2248** (2001), Springer-Verlag, 514–532.
4. R. Coleman and W. McCallum, Stable reduction of Fermat curves and Jacobi sum Hecke characters. *J. Reine Angew. Math.* **385** (1988) 41–101.
5. D. Cox, J. Little and D. O'Shea. Ideals, varieties, and algorithms: an introduction to computational algebraic geometry and commutative algebra. Springer-Verlag (1997).
6. G. Frey. Applications of arithmetical geometry to cryptographic constructions, in Finite fields and applications (Augsburg, 1999). Springer-Verlag (2001) 128–161.

7. G. Frey, M. Müller and H-G. Rück. The Tate pairing and the discrete logarithm applied to elliptic curve cryptosystems. *IEEE Trans. Inform. Theory* **45** (1999) 1717–1719.

8. G. Frey and H-G. Rück. A remark concerning m-divisibility and the discrete logarithm in the divisor class group of curves. *Math. Comp.* **62** (1994) 865–874.

9. S. Galbraith. Supersingular curves in cryptography, in Advances in Cryptology – Asiacrypt 2001. Lecture Notes in Computer Science, Vol. 2248. Springer-Verlag (2001) 495–513.

10. S. Galbraith, F. Hess and N. P. Smart. Extending the GHS Weil descent attack, in Advances in Cryptology – Eurocrypt 2002. Lecture Notes in Computer Science, Vol. 2332. Springer-Verlag (2002) 29–44.

11. P. Gaudry. A variant of the Adleman–DeMarrais–Huang algorithm and its application to small genera, in Advances in Cryptology – Eurocrypt 2000. Lecture Notes in Computer Science, Vol. 1807. Springer-Verlag (2000) 19–34.

12. P. Gaudry, F. Hess and N. P. Smart. Constructive and destructive facets of Weil descent on elliptic curves. *J. Cryptology* **15** (2002) 19–46.

13. T. Honda. Isogeny classes of abelian varieties over finite fields. *J. Math. Soc. Japan* **20** (1968) 83–95.

14. B. Huppert and N. Blackburn. Finite groups II. Springer-Verlag (1982).

15. A. Joux. A one round protocol for tripartite Diffie-Hellman, in Algorithmic Number Theory (ANTS-IV), Leiden, The Netherlands, July 2–7, 2000, Lecture Notes in Computer Science, Vol. 1838. Springer-Verlag (2000) 385–394.

16. A. K. Lenstra and E. R. Verheul. *The XTR public key system*, in Advances in Cryptology – Crypto 2000. Lecture Notes in Computer Science, Vol. 1880. Springer-Verlag (2000) 1–19.

17. A. J. Menezes, T. Okamoto and S. A. Vanstone. Reducing elliptic curve logarithms to logarithms in a finite field. *IEEE Trans. Inform. Theory* **39** (1993) 1639–1646.

18. R. Sakai, K. Ohgishi and M. Kasahara, Cryptosystems based on pairing. SCIS2000 (The 2000 Symposium on Cryptography and Information Security), Okinawa, Japan, January 26–28, 2000, C20.

19. G. Shimura. Abelian varieties with complex multiplication and modular functions. Princeton Univ. Press, Princeton, NJ (1998).

20. J. Silverman. The arithmetic of elliptic curves. Springer-Verlag (1986).

21. J. Tate. Classes d'isogénie des variétés abéliennes sur un corps fini (d'après T. Honda), in Séminaire Bourbaki, 1968/69, Soc. Math. France, Paris (1968) 95–110.

22. E. R. Verheul. Self-blindable credential certificates from the Weil pairing, in Advances in Cryptology – Asiacrypt 2001, Lecture Notes in Computer Science, Vol. 2248. Springer-Verlag (2001) 533–551.

23. A. Weil. Adeles and algebraic groups. Progress in Math. **23**, Birkhäuser, Boston (1982).

24. H. J. Zhu. Group structures of elementary supersingular abelian varieties over finite fields. *J. Number Theory* **81** (2000) 292–309.

Appendix

In this appendix we will state and prove a more general version (Theorem 24 below) of Theorem 17.

Write $\mathrm{Res}(f, g)$ for the resultant of two polynomials f and g.

Lemma 22 *Suppose a, b, c are pairwise relatively prime integers. Then there are $g_1(x), g_2(x) \in \mathbf{Z}[x]$ such that*

$$g_1(x) \prod_{a|d|abc} \Phi_d(x) + g_2(x) \prod_{b|d|abc} \Phi_d(x) = \prod_{ab|d|abc} \Phi_d(x).$$

Proof. Let $f_1(x) = \prod_{a|d|abc, b\nmid d} \Phi_d(x)$ and $f_2(x) = \prod_{b|d|abc, a\nmid d} \Phi_d(x)$. If η_i is a root of f_i for $i = 1$ and 2, then η_1/η_2 is a root of unity of order divisible by both a prime divisor of a and a prime divisor of b. Hence $\eta_1/\eta_2 - 1$ is a (cyclotomic) unit in the ring of algebraic integers. Therefore $\mathrm{Res}(f_1, f_2)$, an integer which is the product of the differences of the roots of f_1 and the roots of f_2, is ± 1. By Proposition 9 in §3.5 of [5], there are $g_1, g_2 \in \mathbf{Z}[x]$ such that $g_1(x)f_1(x) + g_2(x)f_2(x) = \mathrm{Res}(f_1, f_2)$.

Lemma 23 *Suppose M is a square matrix over a field F with characteristic polynomial f_M, and $g(x) \in F[x]$. Then $\det(g(M)) = \mathrm{Res}(g, f_M)$.*

Proof. This is clear if M is upper-triangular. To obtain the general case, replace F by its algebraic closure and upper-triangularize M.

Recall the notation e from Theorem 1.

Theorem 24 *Suppose $\mathcal{E}$ is a supersingular abelian variety over $\mathbf{F}_q$ with $e = 1$. Fix $r \in \mathbf{N}$ such that $\gcd(r, 2pc_{\mathcal{E}}) = 1$. Then there is a simple supersingular abelian variety A over $\mathbf{F}_q$ such that:*

(i) $\dim(A) = \varphi(r)\dim(\mathcal{E})$;
(ii) if π is a q-Weil number for $\mathcal{E}$, then $\pi\zeta$ is a q-Weil number for A for every primitive r-th root of unity ζ;
(iii) $c_A = rc_{\mathcal{E}}$;
(iv) $\alpha_A = \frac{r}{\varphi(r)}\alpha_{\mathcal{E}}$;
(v) there is a natural identification of $A(\mathbf{F}_q)$ with the subgroup of $\mathcal{E}(\mathbf{F}_{q^r})$

$$\{Q \in \mathcal{E}(\mathbf{F}_{q^r}) : \mathrm{Tr}_{\mathbf{F}_{q^r}/\mathbf{F}_{q^{r/\ell}}} Q = O \quad \text{for every prime } \ell \mid r\}.$$

Proof. Let Ω be the set of q-Weil numbers for $\mathcal{E}$, and $d = \dim(\mathcal{E})$. Since $e = 1$, the characteristic polynomial of the Frobenius endomorphism $\phi_{\mathcal{E}}$ on $\mathcal{E}$ is $P_{\mathcal{E}}(x) = \prod_{\pi \in \Omega}(x - \pi)$.

Let $\Bbbk = \mathbf{F}_q$ and $\Bbbk_1 = \mathbf{F}_{q^r}$, and let B denote the Weil restriction of scalars (§1.3 of [23]) of $\mathcal{E}$ from $\Bbbk_1$ to $\Bbbk$. Then B is an rd-dimensional abelian variety defined over $\Bbbk$, there is a natural isomorphism

$$B(\Bbbk) \cong \mathcal{E}(\Bbbk_1), \tag{1}$$

and $P_B(x) = \prod_{\pi \in \Omega}(x^r - \pi^r)$ is the characteristic polynomial of the Frobenius endomorphism on B over $\Bbbk$. Fix a $\pi \in \Omega$ and a primitive r-th root of unity ζ. Then $P_B(\pi\zeta) = 0$, so B has a simple supersingular abelian subvariety A with $\pi\zeta$ as a q-Weil number. We will show that the conclusions of the theorem hold for A.

Assertion (iii) holds by Definition 3. By Proposition 10 and the fact that $p \nmid r$, $e = 1$ for A. Thus, $2\dim(A) = [\mathbf{Q}(\pi\zeta) : \mathbf{Q}]$. Since $\gcd(r, 2pc_\mathcal{E}) = 1$ we have $\mathbf{Q}(\zeta) \cap \mathbf{Q}(\pi) = \mathbf{Q}$, so $[\mathbf{Q}(\pi\zeta) : \mathbf{Q}] = [\mathbf{Q}(\pi) : \mathbf{Q}][\mathbf{Q}(\zeta) : \mathbf{Q}] = 2\dim(\mathcal{E})\varphi(r)$. This proves (i), and (ii) and (iv) follow. The isomorphism (1) identifies $A(\Bbbk)$ with a subgroup of $\mathcal{E}(\Bbbk_1)$, and it remains only to determine this subgroup.

If ℓ is a prime divisor of r, write $r = \ell^i m$ with $\ell \nmid m$, let $\Bbbk_\ell = \mathbf{F}_{q^{r/\ell}}$, and let $h_\ell(x) = \prod_{d|m} \Phi_{\ell^i d}(x) = (x^r - 1)/(x^{r/\ell} - 1)$. Let

$$T = \{Q \in \mathcal{E}(\Bbbk_1) : \mathrm{Tr}_{\Bbbk_1/\Bbbk_\ell} Q = O \text{ for every prime } \ell \mid r\} = \cap_{\ell|r} \ker(h_\ell(\phi_\mathcal{E})).$$

Applying Lemma 22 inductively one can show that there are $\gamma_\ell(x) \in \mathbf{Z}[x]$ such that $\sum_{\ell|r} \gamma_\ell(x)h_\ell(x) = \Phi_r(x)$. It follows that $T = \ker(\Phi_r(\phi_\mathcal{E}))$. Since $\phi_\mathcal{E}$ is (purely) inseparable (see Proposition II.2.11 of [20]), it follows that $\Phi_r(\phi_\mathcal{E})$ is separable (its action on the space of differential forms on $\mathcal{E}$ is $\Phi_r(0) \not\equiv 0 \pmod{p}$; see Proposition II.4.2(c) of [20]). By Theorem III.4.10(c) of [20], $\#T$ is the degree of the endomorphism $\Phi_r(\phi_\mathcal{E})$.

Applying Lemma 23 to the matrix $M_\mathcal{E}$ giving the action of $\phi_\mathcal{E}$ on the ℓ-adic Tate module of $\mathcal{E}$ for some prime $\ell \neq p$ shows that

$$\#T = \deg(\Phi_r(\phi_\mathcal{E})) = \det(\Phi_r(M_\mathcal{E}))$$

$$= \mathrm{Res}(\Phi_r, P_\mathcal{E}) = \prod_{\Phi_r(\eta)=0} P_\mathcal{E}(\eta) = P_A(1) = \#A(\Bbbk).$$

If $\mathcal{E}(\Bbbk_1)$ is cyclic, it follows that the isomorphism (1) identifies $A(\Bbbk)$ with T. In the special cases where $\mathcal{E}(\Bbbk_1)$ is not cyclic (Theorem 1(iii)) one can show that $\#A(\Bbbk) = P_A(1)$ is odd, and since the odd part of $\mathcal{E}(\Bbbk_1)$ is always cyclic, (1) identifies $A(\Bbbk)$ with T in this case also.

Efficient Algorithms
for Pairing-Based Cryptosystems

Paulo S.L.M. Barreto[1], Hae Y. Kim[1], Ben Lynn[2], and Michael Scott[3]

[1] Universidade de São Paulo, Escola Politécnica
Av. Prof. Luciano Gualberto, tr. 3, 158
BR 05508-900, São Paulo(SP), Brazil
pbarreto@larc.usp.br, hae@lps.usp.br
[2] Computer Science Department, Stanford University, USA
blynn@cs.stanford.edu
[3] School of Computer Applications
Dublin City University
Ballymun, Dublin 9, Ireland
mscott@indigo.ie

Abstract. We describe fast new algorithms to implement recent crypto-systems based on the Tate pairing. In particular, our techniques improve pairing evaluation speed by a factor of about 55 compared to previously known methods in characteristic 3, and attain performance comparable to that of RSA in larger characteristics. We also propose faster algorithms for scalar multiplication in characteristic 3 and square root extraction over $\mathbb{F}_{p^m}$, the latter technique being also useful in contexts other than that of pairing-based cryptography.

1 Introduction

The recent discovery [11] of groups where the Decision Diffie-Hellman (DDH) problem is easy while the Computational Diffie-Hellman (CDH) problem is hard, and the subsequent definition of a new class of problems variously called the Gap Diffie-Hellman [11], Bilinear Diffie-Hellman [2], or Tate-Diffie-Hellman [6] class, has given rise to the development of a new, ever expanding family of cryptosystems based on pairings, such as:

- Short signatures [3].
- Identity-based encryption and escrow ElGamal encryption [2].
- Identity-based authenticated key agreement [29].
- Identity-based signature schemes [8, 22, 24].
- Tripartite Diffie-Hellman [10].
- Self-blindable credentials [33].

The growing interest and active research in this branch of cryptography has led to new analyses of the associated security properties and to extensions to more general (e.g. hyperelliptic and superelliptic) algebraic curves [6, 23].

However, a central operation in these systems is computing a bilinear pairing (e.g. the Weil or the Tate pairing), which are computationally expensive. More-over, it is often the case that curves over fields of characteristic 3 are used to

M. Yung (Ed.): CRYPTO 2002, LNCS 2442, pp. 354–368, 2002.

achieve the best possible ratio between security level and space requirements for supersingular curves, but such curves have received considerably less attention than their even or (large) prime characteristic counterparts. Our goal is to make such systems entirely practical and contribute to fill the theoretical gap in the study of the underlying family of curves, and to this end we propose several efficient algorithms for the arithmetic operations involved.

The contributions of this paper are:

- The definition of *point tripling* for supersingular elliptic curves over $\mathbb{F}_{3^m}$, that is, over fields of characteristic 3. A point tripling operation can be done in $O(m)$ steps (or essentially for free in hardware), as opposed to conventional point doubling that takes $O(m^2)$ steps. Furthermore, a faster point addition algorithm is proposed for normal basis representation. These operations lead to a noticeably faster scalar multiplication algorithm in characteristic 3.
- An algorithm to compute square roots over $\mathbb{F}_{p^m}$ in $O(m^2 \log m)$ steps, where m is odd and $p \equiv 3 \pmod 4$ or $p \equiv 5 \pmod 8$. The best previously known algorithms for square root extraction under these conditions take $O(m^3)$ steps. This operation is important for the point compression technique, whereby a curve point $P = (x, y)$ is represented by its x coordinate and one bit of its y coordinate, and its usefulness transcends pairing-based cryptography.
- A deterministic variant of Miller's algorithm to compute the Tate pairing that avoids many *irrelevant operations* present in the conventional algorithm whenever one of the pairing's arguments is restricted to a base field (as opposed to having both in an extension field). Besides, in characteristics 2 and 3 both the underlying scalar multiplication and the final powering in the Tate pairing experience a complexity reduction from $O(m^3)$ to $O(m^2)$ steps.

All of these improvements are very practical and result in surprisingly faster implementations. Independent results on this topic have been obtained by Galbraith, Harrison and Soldera, and are reported in [7]; in particular, they provide a very clear and nice description of the Tate pairing.

This paper is organized as follows. Section 2 summarizes the mathematical concepts we will use in the remainder of the paper. Section 3 describes point tripling and derives a fast scalar multiplication algorithm for characteristic 3. Section 4 introduces a fast method to compute square roots that works for half of all finite fields, and an extension to half of the remaining cases. Section 5 presents our improvements for Tate pairing computation. Section 6 discusses experimental results. We conclude in section 7.

2   Mathematical Preliminaries

Let p be a prime number, m a positive integer and $\mathbb{F}_{p^m}$ the finite field with p^m elements; p is said to be the *characteristic* of $\mathbb{F}_{p^m}$, and m is its *extension degree*. We simply write $\mathbb{F}_q$ with $q = p^m$ when the characteristic or the extension degree are known from the context or irrelevant for the discussion. We also write $\mathbb{F}_q^* \equiv \mathbb{F}_q - \{0\}$.

Table 1. Some cryptographically interesting supersingular elliptic curves.

curve equation	underlying field	curve order	k
$E_{1,b} : y^2 = x^3 + (1-b)x + b,\ b \in \{0,1\}$	$\mathbb{F}_p$	$p+1$	2
$E_{2,b} : y^2 + y = x^3 + x + b,\ b \in \{0,1\}$	$\mathbb{F}_{2^m}$	$2^m + 1 \pm 2^{(m+1)/2}$	4
$E_{3,b} : y^2 = x^3 - x + b,\ b \in \{-1,1\}$	$\mathbb{F}_{3^m}$	$3^m + 1 \pm 3^{(m+1)/2}$	6

An *elliptic curve* $E(\mathbb{F}_q)$ is the set of solutions (x, y) over $\mathbb{F}_q$ to an equation of form $E : y^2 + a_1xy + a_3y = x^3 + a_2x^2 + a_4x + a_6$, where $a_i \in \mathbb{F}_q$, together with an additional *point at infinity*, denoted O. The same equation defines curves over $\mathbb{F}_{q^k}$ for $k > 0$.

There exists an abelian group law on E. Explicit formulas for computing the coordinates of a point $P_3 = P_1 + P_2$ from the coordinates of P_1 and P_2 are given in [27, algorithm 2.3]; we shall present in section 3 a subset of those formulas.

The number of points of an elliptic curve $E(\mathbb{F}_q)$, denoted $\#E(\mathbb{F}_q)$, is called the *order* of the curve over the field $\mathbb{F}_q$. The *Hasse bound* states that $\#E(\mathbb{F}_q) = q + 1 - t$, where $|t| \leqslant 2\sqrt{q}$. The quantity t is called the *trace of Frobenius* (for brevity, we will call it simply 'trace'). Of particular interest to us are *supersingular* curves, which are curves whose trace t is a multiple of the characteristic p.

Let $n = \#E(\mathbb{F}_q)$. The order of a point $P \in E$ is the least nonzero integer r such that $rP = O$. The set of all points of order r in E is denoted $E[r]$, or $E(K)[r]$ to stress the particular subgroup $E(K)$ for a field K. The order of a point always divides the curve order. It follows that $\langle P \rangle$ is a subgroup of $E[r]$, which in turn is a subgroup of $E[n]$.

Let P be a point on E of prime order r where $r^2 \nmid n$. The subgroup $\langle P \rangle$ is said to have *security multiplier* k for some $k > 0$ if $r \mid q^k - 1$ and $r \nmid q^s - 1$ for any $0 < s < k$. If E is supersingular, the value of k is bounded by $k \leqslant 6$ [16]. This bound is attained in characteristic 3 but not in characteristic 2, where the maximum achievable value is $k = 4$ [15, section 5.2.2].

The group $E(\mathbb{F}_q)$ is (isomorphic to) a subgroup of $E(\mathbb{F}_{q^k})$. Let $P \in E(\mathbb{F}_q)$ be a point of order r such that $\langle P \rangle$ has security multiplier k. Then $E(\mathbb{F}_{q^k})$ contains a point Q of the same order r but linearly independent of P.

We will consider in detail the curves listed in table 1, where k is the security multiplier, both m and p are prime numbers, and either $p \equiv 2 \pmod 3$ or $p \equiv 3 \pmod 4$. The curve orders are explicitly computed in [15, section 5.2.2].

For our purposes, a *divisor* is a formal sum of points on the curve $E(\mathbb{F}_{q^m})$, $m > 0$. The *degree* of a divisor $\mathcal{A} = \sum_P a_P(P)$ is the sum $\sum_P a_P$. An abelian group structure is imposed on the set of divisors by the addition of corresponding coefficients in their formal sums; in particular, $n\mathcal{A} = \sum_P (na_P)(P)$.

Let $f : E(\mathbb{F}_{q^k}) \to \mathbb{F}_{q^k}$ be a function on the curve and let $\mathcal{A} = \sum_P a_P(P)$ be a divisor of degree 0. We define $f(\mathcal{A}) \equiv \prod_P f(P)^{a_P}$. Note that, since $\sum_P a_P = 0$, $f(\mathcal{A}) = (cf)(\mathcal{A})$ for any factor $c \in \mathbb{F}_{q^k}^*$. The divisor of a function f is $(f) \equiv \sum_P \text{ord}_P(f)(P)$ where $\text{ord}_P(f)$ is the order of the zero or pole of f at P (if f has no zero or pole at P, then $\text{ord}_P(f) = 0$). A divisor $\mathcal{A}$ is called *principal* if $\mathcal{A} = (f)$ for some function (f). It is known [15, theorem 2.25] that a divisor $\mathcal{A} = \sum_P a_P(P)$ is principal if and only if the degree of $\mathcal{A}$ is zero and $\sum_P a_P P = O$.

Two divisors $\mathcal{A}$ and $\mathcal{B}$ are equivalent, and we write $\mathcal{A} \sim \mathcal{B}$, if their difference $\mathcal{A} - \mathcal{B}$ is a principal divisor. Let $P \in E[n]$ where n is coprime to q, and let $\mathcal{A}_P$ be a divisor equivalent to $(P) - (O)$; under these circumstances the divisor $n\mathcal{A}_P$ is principal, and hence there is a function f_P such that $(f_P) = n\mathcal{A}_P = n(P) - n(O)$.

Let ℓ be a natural number coprime to q. The *Tate pairing* of order ℓ is the map $e_\ell : E(\mathbb{F}_q)[\ell] \times E(\mathbb{F}_{q^k})[\ell] \to \mathbb{F}_{q^k}^*$ defined[1] as $e_\ell(P, Q) = f_P(\mathcal{A}_Q)^{(q^k - 1)/\ell}$. It satisfies the following properties:

- (Bilinearity) $e_\ell(P_1 + P_2, Q) = e_\ell(P_1, Q) \cdot e_\ell(P_2, Q)$ and $e_\ell(P, Q_1 + Q_2) = e_\ell(P, Q_1) \cdot e_\ell(P, Q_2)$ for all $P, P_1, P_2 \in E(\mathbb{F}_q)[\ell]$ and all $Q, Q_1, Q_2 \in E(\mathbb{F}_{q^k})[\ell]$. It follows that $e_\ell(aP, Q) = e_\ell(P, aQ) = e_\ell(P, Q)^a$ for all $a \in \mathbb{Z}$.
- (Non-degeneracy) If $e_\ell(P, Q) = 1$ for all $Q \in E(\mathbb{F}_{q^k})[\ell]$, then $P = O$. Alternatively, for each $P \neq O$ there exists $Q \in E(\mathbb{F}_{q^k})[\ell]$ such that $e_\ell(P, Q) \neq 1$.
- (Compatibility) Let $\ell = h\ell'$. If $P \in E(\mathbb{F}_q)[\ell]$ and $Q \in E(\mathbb{F}_{q^k})[\ell']$, then $e_{\ell'}(hP, Q) = e_\ell(P, Q)^h$.

Notice that, because $P \in E(\mathbb{F}_q)$, f_P is a rational function with coefficients in $\mathbb{F}_q$.

3   Scalar Multiplication in Characteristic 3

Arithmetic on the curve $E_{3,b}$ is performed according to the following rules. Let $P_1 = (x_1, y_1)$, $P_2 = (x_2, y_2)$, $P_3 = P_1 + P_2 = (x_3, y_3)$. By definition, $-O = O$, $-P_1 = (x_1, -y_1)$, $P_1 + O = O + P_1 = P_1$. Furthermore,

$$P_1 = -P_2 \quad \Rightarrow P_3 = O.$$
$$P_1 = P_2 \quad \Rightarrow \lambda \equiv 1/y_1, \; x_3 = x_1 + \lambda^2, \; y_3 = -(y_1 + \lambda^3).$$
$$P_1 \neq -P_2, P_2 \Rightarrow \lambda \equiv \frac{y_2 - y_1}{x_2 - x_1}, \; x_3 = \lambda^2 - (x_1 + x_2), \; y_3 = y_1 + y_2 - \lambda^3.$$

These rules in turn give rise to the *double-and-add* method to compute scalar multiples $V = kP$, $k \in \mathbb{Z}$. Let the binary representation of $k > 0$ be $k = (k_t \ldots k_1 k_0)_2$ where $k_i \in \{0, 1\}$ and $k_t \neq 0$. Computation of $V = kP \equiv P + P + \cdots + P$ (with k terms) proceeds as follows.

Double-and-add scalar multiplication:

```
set V ← P
for i ← t - 1, t - 2, ..., 1, 0 do {
    set V ← 2V
    if k_i = 1 then set V ← V + P
}
return V
```

---

[1] This definition differs from those given in [5,6] in that we restrict the first argument of e_ℓ to $E(\mathbb{F}_q)[\ell]$ and the second argument to $E(\mathbb{F}_{q^k})[\ell]$ instead of $E(\mathbb{F}_{q^k})[\ell]$ and $E(\mathbb{F}_{q^k})/\ell E(\mathbb{F}_{q^k})$ respectively, and we raise $f_P(\mathcal{A}_Q)$ to the power $(q^k - 1)/\ell$, so that e_ℓ maps to certain uniquely determined coset representatives. However, our definition keeps the properties listed above unchanged, and captures the essential properties needed in practice for cryptographical purposes.

By extension, one defines $0P = O$ and $(-k)P = k(-P) = -(kP)$.

Several improvements to this basic algorithm are well known [1, 17]. However, one can do much better than this, as we will now see.

3.1 Point Tripling

In characteristic 3, *point tripling* for the supersingular curve $E_{3,b}$ can be done in time $O(m)$ in polynomial basis, or simply $O(1)$ in hardware using normal basis. Indeed, since the cubing operation is linear in characteristic 3, given $P = (x, y)$ one computes $3P = (x_3, y_3)$ with the formulas:

$$x_3 = (x^3)^3 - b$$
$$y_3 = -(y^3)^3$$

These formulas are derived from the basic arithmetic formulas above in a straightforward way.

The linearity of point tripling corresponds to that of point doubling for supersingular curves in characteristic 2, as discovered by Menezes and Vanstone [18], and it leads to a triple-and-add scalar multiplication algorithm much faster than the double-and-add method. Let the signed ternary representation of k be $k = (k_t \ldots k_1 k_0)_2$ where $k_i \in \{-1, 0, 1\}$ and $k_t \neq 0$. Computation of $V = kP$ proceeds as follows.

Triple-and-add scalar multiplication:

> set $V \leftarrow P$ if $k_t = 1$, or $V \leftarrow -P$ if $k_t = -1$
> for $i \leftarrow t - 1, t - 2, \ldots, 1, 0$ do {
>   set $V \leftarrow 3V$
>   if $k_i = 1$ then set $V \leftarrow V + P$
>   if $k_i = -1$ then set $V \leftarrow V - P$
> }
> return V

Obviously, the same advanced techniques used for the double-and-add method can be easily applied to triple-and-add.

3.2 Projective Coordinates

Koblitz [12] describes a method to add curve points in characteristic 3 in projective coordinates with 10 multiplications. Actually, point addition can be done with only 9 multiplications. Let $P_1 = (x_1, y_1, z_1)$, $P_2 = (x_2, y_2, 1)$; one computes $P_3 = P_1 + P_2 = (x_3, y_3, z_3)$ as:

$$A \leftarrow x_2 z_1 - x_1, \ B \leftarrow y_2 z_1 - y_1, \ C \leftarrow A^3, \ D \leftarrow C - z_1 B^2,$$

$$x_3 \leftarrow x_1 C - AD, \ y_3 \leftarrow BD - y_1 C, \ z_3 \leftarrow z_1 C.$$

To recover P_3 in affine coordinates one just sets $P_3 = (x_3/z_3, y_3/z_3)$. This involves one single inversion, which is usually only performed at the end of a scalar multiplication.

4   Square Root Extraction

One can use the elliptic curve equation $E : y^2 = f(x)$ over $\mathbb{F}_q$, where $f(x)$ is a cubic polynomial, to obtain a compact representation of curve points. The idea is to use a single bit from the ordinate y as a selector[2] between the two solutions of the equation $y^2 = f(x)$ for a given x.

In a finite field $\mathbb{F}_{p^m}$ where $p \equiv 3 \pmod 4$ and odd m, the best algorithm known [4, 17] to compute a square root executes $O(m^3)$, or more precisely $O(m^3 \log p)$, $\mathbb{F}_p$ operations. By that method, a solution of $x^2 = a$ is given by $x = a^{(p^m+1)/4}$, assuming a is a quadratic residue.

We first notice that, if $m = 2k + 1$ for some k:

$$\frac{p^m + 1}{4} = \frac{p+1}{4} \left[p(p-1) \sum_{i=0}^{k-1} (p^2)^i + 1 \right],$$

so that

$$a^{(p^m+1)/4} = [(a^{\sum_{i=0}^{k-1} (p^2)^i})^{p(p-1)} \cdot a]^{(p+1)/4}.$$

These relations can be verified by straightforward induction. The quantity $a^{\sum_{i=0}^{k-1} u^i}$ where $u = p^2$ can be efficiently computed in an analogous fashion to Itoh-Teechai-Tsujii inversion [9], based on the Frobenius map in characteristic p:

$$a^{1+u+\cdots+u^{k-1}} = \begin{cases} (a^{1+u+\cdots+u^{\lfloor k/2 \rfloor - 1}}) \cdot (a^{1+u+\cdots+u^{\lfloor k/2 \rfloor - 1}})^{u^{\lfloor k/2 \rfloor}}, & k \text{ even,} \\ ((a^{1+u+\cdots+u^{\lfloor k/2 \rfloor - 1}}) \cdot (a^{1+u+\cdots+u^{\lfloor k/2 \rfloor - 1}})^{u^{\lfloor k/2 \rfloor}})^u \cdot a, & k \text{ odd.} \end{cases}$$

Notice that raising to a power of p is a linear operation in characteristic p (and almost for free in normal basis representation). It can be easily verified by induction that this method requires $\lfloor \lg k \rfloor + \omega(k) - 1$ field multiplications, where $\omega(k)$ is the Hamming weight of the binary representation of k. Additional $O(\log p)$ multiplications are needed to complete the square root evaluation due to the extra multiplication by a and to the raisings to $p - 1$ and $(p + 1)/4$, which can be done with a conventional exponentiation algorithm[3]. The overall cost is $O(m^2(\log m + \log p))$ $\mathbb{F}_p$ operations to compute a square root. If the characteristic p is fixed and small compared to m, the complexity is simply $O(m^2 \log m)$ $\mathbb{F}_p$ operations.

Similar recurrence relations hold for a variant of Atkin's algorithm [21, section A.2.5] for computing square roots in $\mathbb{F}_{p^m}$ when $p \equiv 5 \pmod 8$ and odd m, with the same $O(m^2(\log m + \log p))$ complexity. The details are left to the reader.

---

[2] In certain cryptographic applications one can simply discard y. This happens, for instance, in BLS signatures [3], where one only keeps the abscissa x as signature representative. Notice that one could discard the ordinates of public keys as well without affecting the security level.

[3] If p is large, it may be advantageous to compute z^{p-1} as z^p/z, trading $O(\log p)$ multiplications by one inversion.

The general case is unfortunately not so easy. Neither the Tonelli-Shanks algorithm [4] nor Lehmer's algorithm [21, section A.2.5] can benefit entirely from the above technique, although partial improvements that don't change the overall complexity are possible.

The above improvements are useful not only for pairing-based cryptosystems, but for more conventional schemes as well (see e.g. [12, section 6]).

5   Computing the Tate Pairing

In this section we propose several improvements to Miller's algorithm [19] for computing the Tate pairing in the cases of cryptographical interest. Let $P \in E(\mathbb{F}_q)[\ell]$ and $Q \in E(\mathbb{F}_{q^k})[\ell]$ be linearly independent points, and let $n \equiv \#E(\mathbb{F}_q)$. As we saw in section 2, the Tate pairing is defined as $e_\ell(P, Q) = f_P(\mathcal{A}_Q)^{(q^k-1)/\ell}$, where $\mathcal{A}_Q \sim (Q) - (O)$ and $(f_P) = \ell(P) - \ell(O)$. Computation of the Tate pairing is helped by the following observations.

Lemma 1. *The value $q - 1$ is a factor of $(q^k - 1)/r$ for any factor r of n for a supersingular elliptic curve over $\mathbb{F}_q$ with security multiplier $k > 1$.*

Proof. Since $\mathbb{F}_q^*$ is a multiplicative subgroup of $\mathbb{F}_{q^k}^*$, it follows that $\#\mathbb{F}_q^* \mid \#\mathbb{F}_{q^k}^*$, i.e. $q - 1 \mid q^k - 1$. On the other hand, it is known [15, section 5.2.2] that the order n of a supersingular curve with security multiplier $k > 1$ does not divide $q - 1$, and hence no factor r of n does. Therefore $(q^k - 1)/r$ contains a factor $q - 1$. $\square$

Theorem 1. *Let r be a factor of n. As long as $k > 1$, $e_r(P, Q) = f_P(Q)^{(q^k-1)/r}$ for $Q \neq O$.*

Proof. Suppose $R \notin \{O, -P\}$ is some point on the curve. Let f_P' be a function with divisor $(f_P') = r(P + R) - r(R) \sim (f_P)$, so that $e_r(P, Q) = f_P'((Q) - (O))^{(q^k-1)/r}$. Since P has coordinates in $\mathbb{F}_p$, and because f_P' does not have a zero or pole at O, we know that $f_P'(O) \in \mathbb{F}_q^*$. Thus $f_P'((Q) - (O)) = f_P'(Q)/f_P'(O)$. By Fermat's Little Theorem for finite fields [13, lemma 2.3], $f_P'(O)^{q-1} = 1$. Lemma 1 then ensures that $f_P'(O)^{(q^k-1)/r} = 1$. Hence, $f_P'(O)$ is an irrelevant factor and can be omitted from the Tate pairing computation, i.e. $e_r(P, Q) = f_P'(Q)^{(q^k-1)/r}$. Now consider P, Q to be fixed and R to be variable. Since the above statement holds for all $R \notin \{O, -P\}$ we have that $f_P'(Q)$ is a constant when viewed as a function of R, coinciding with the value of $f_P(Q)$. Therefore, $e_r(P, Q) = f_P(Q)^{(q^k-1)/r}$. $\square$

Corollary 1. *One can freely multiply or divide $f_P(Q)$ by any nonzero $\mathbb{F}_q$ factor without affecting the pairing value.*

The above corollary is not the same property that enables one to replace (f) by (cf); in particular, it does not hold for the Weil pairing. Notice that the special case $Q = O$ where the theorem does not apply is trivially handled, since then $e_r(P, Q) = 1$.

In the next theorem, for each pair $U, V \in E(\mathbb{F}_q)$ we define $g_{U,V} : E(\mathbb{F}_{q^k}) \to \mathbb{F}_{q^k}$ to be (the equation of) the line through points U and V (if $U = V$, then $g_{U,V}$ is the tangent to the curve at U, and if either one of U, V is the point at infinity O, then $g_{U,V}$ is the vertical line at the other point). The shorthand g_U stands for $g_{U,-U}$: if $U = (u, v)$ and $Q = (x, y)$, then $g_U(Q) = x - u$.

Theorem 2 (Miller's formula). *Let P be a point on $E(\mathbb{F}_q)$ and f_c be a function with divisor $(f_c) = c(P) - (cP) - (c - 1)(O)$, $c \in \mathbb{Z}$. For all $a, b \in \mathbb{Z}$, $f_{a+b}(Q) = f_a(Q) \cdot f_b(Q) \cdot g_{aP,bP}(Q)/g_{(a+b)P}(Q)$.*

Proof. The divisors of the line functions satisfy:

$$(g_{aP,bP}) = (aP) + (bP) - (-(a + b)P) - 3(O),$$
$$(g_{(a+b)P}) = ((a + b)P) + (-(a + b)P) - 2(O).$$

Hence, $(g_{aP,bP}) - (g_{(a+b)P}) = (aP) + (bP) - ((a+b)P) - (O)$. From the definition of f_c we see that:

$$
\begin{aligned}
(f_{a+b}) &= (a + b)(P) - ((a + b)P) - (a + b - 1)(O) \\
&= a(P) - (aP) - (a - 1)(O) \\
&\quad + b(P) - (bP) - (b - 1)(O) \\
&\quad + (aP) + (bP) - ((a + b)P) - (O) \\
&= (f_a) + (f_b) + (g_{aP,bP}) - (g_{(a+b)P}).
\end{aligned}
$$

Therefore $f_{a+b}(Q) = f_a(Q) \cdot f_b(Q) \cdot g_{aP,bP}(Q) / g_{(a+b)P}(Q)$. □

Notice that $(f_0) = (f_1) = 0$, so that $f_0(Q) = f_1(Q) = 1$. Furthermore, $f_{a+1}(Q) = f_a(Q) \cdot g_{aP,P}(Q)/g_{(a+1)P}(Q)$ and $f_{2a}(Q) = f_a(Q)^2 \cdot g_{aP,aP}(Q)/g_{2aP}(Q)$.

Let the binary representation of $\ell \geqslant 0$ be $\ell = (\ell_t, \ldots, \ell_1, \ell_0)$ where $\ell_i \in \{0, 1\}$ and $\ell_t \neq 0$. Miller's algorithm computes $f_P(Q) = f_\ell(Q), Q \neq O$ by coupling the above formulas with the double-and-add method to calculate ℓP:

Miller's algorithm:

> set $f \leftarrow 1$ and $V \leftarrow P$
> for $i \leftarrow t - 1, t - 2, \ldots, 1, 0$ do {
> > set $f \leftarrow f^2 \cdot g_{V,V}(Q)/g_{2V}(Q)$ and $V \leftarrow 2V$
> > if $\ell_i = 1$ then set $f \leftarrow f \cdot g_{V,P}(Q)/g_{V+P}(Q)$ and $V \leftarrow V + P$
> }
> return f

5.1   Irrelevant Denominators

We will now show that, when computing $e_n(P, \phi(Q))$ where $Q \in E(\mathbb{F}_q)$ and where ϕ is a distortion map [32], the g_{2V} and g_{V+P} denominators in Miller's algorithm can be discarded. The choice of parameters is important, and is summarized in table 2. Notice that there is no entry for $E_{1,1}$.

Table 2. Choice of distortion maps.

curve (see table 1)	underlying field	distortion map	conditions
$E_{1,0}$	$\mathbb{F}_p, p > 3$	$\phi_1(x,y) = (-x, iy)$	$i \in \mathbb{F}_{p^2},$ $i^2 = -1$
$E_{2,b}, b \in \{0,1\}$	$\mathbb{F}_{2^m}$	$\phi_2(x,y) = (x+s^2, y+sx+t)$	$s, t \in \mathbb{F}_{2^{4m}},$ $s^4 + s = 0,$ $t^2 + t + s^6 + s^2 = 0$
$E_{3,b}, b \in \{-1,1\}$	$\mathbb{F}_{3^m}$	$\phi_3(x,y) = (-x+r_b, iy)$	$r_b, i \in \mathbb{F}_{3^{6m}}$ $r_b^3 - r_b - b = 0,$ $i^2 = -1$

Theorem 3. *With the settings listed in table 2, the denominators in Miller's formula can be discarded altogether without changing the value of $e_n(P,Q)$.*

Proof. We will show that the denominators become unity at the final powering in the Tate pairing.

- (Characteristic 2) Let $q \equiv 2^m$. From the defining condition $s^4 = s$ it follows by induction that $s^{4^t} = s$ for all $t \geqslant 0$; in particular, $s^{q^2} = s^{2^{2m}} = s$, and hence $(s^2)^{q^2} = s^2$. The denominators in Miller's formula have the form $g_U(\phi(Q)) \equiv x + s^2 + c$, where $x \in \mathbb{F}_q$ is the abscissa of Q and $c \in \mathbb{F}_q$, so that $x^{q^2} = x$ and $c^{q^2} = c$. Hence, $g_U(\phi(Q))^{q^2} = x^{q^2} + (s^2)^{q^2} + c^{q^2} = x + s^2 + c = g_U(\phi(Q))$, using the linearity of raising to powers of q in $\mathbb{F}_q$. It follows that $g_U(\phi(Q))^{q^2-1} = 1$. Now the exponent of the final powering in the Tate pairing has the form $z = (q^4 - 1)/n = (q + 1 \pm \sqrt{2q})(q^2 - 1)$, i.e. $q^2 - 1 \mid z$. Therefore, $g_U(\phi(Q))^z = 1$.
- (Characteristic 3) Let $q \equiv 3^m$. From the defining condition $r_b^3 - r_b - b = 0$ it follows by induction that $r_b^{3^t} = r_b + b(t \bmod 3)$ for all $t \geqslant 0$; in particular, $r_b^{q^3} = r_b^{3^{3m}} = r_b$. The denominators in Miller's formula have the form $g_U(\phi(Q)) \equiv r_b - x - c$, where $x \in \mathbb{F}_q$ is the abscissa of Q and $c \in \mathbb{F}_q$, so that $x^{q^t} = x$ and $c^{q^t} = c$ for all $t \geqslant 0$. Hence, $g_U(\phi(Q))^{q^3} = r_b^{q^3} - x^{q^3} - c^{q^3} = r_b - x - c = g_U(\phi(Q))$, using the linearity of raising to powers of q in $\mathbb{F}_q$. It follows that $g_U(\phi(Q))^{q^3-1} = 1$. Now the exponent of the final powering in the Tate pairing has the form $z = (q^6 - 1)/n = (q + 1 \pm \sqrt{3q})(q^3 - 1)(q + 1)$, i.e. $q^3 - 1 \mid z$. Therefore, $g_U(\phi(Q))^z = 1$.
- (Characteristic $p > 3$) The denominators in Miller's formula have the form $g_U(\phi(Q)) \equiv -x - c$, where $x \in \mathbb{F}_p$ is the abscissa of Q and $c \in \mathbb{F}_p$. Hence, $g_U(\phi(Q))^p = -x^p - c^p = -x - c = g_U(\phi(Q))$, using the linearity of raising to p in $\mathbb{F}_p$. It follows that $g_U(\phi(Q))^{p-1} = 1$. Now the exponent of final powering in the Tate pairing is precisely $z = (p^2 - 1)/n = p - 1$. Therefore, $g_U(\phi(Q))^z = 1$. $\square$

One can alternatively couple the evaluation of f_n with the more efficient triple-and-add method in characteristic 3. To this end one needs a recursive formula for $f_{3a}(Q)$, which is easy to obtain from Miller's formula: the divisor of

f_{3a} is $(f_{3a}) = 3(f_a) + (g_{aP,aP}) + (g_{2aP,aP}) - (g_{2aP}) - (g_{3aP})$, hence discarding the irrelevant denominators one obtains:

$$f_{3b}(Q) = f_b^3(Q) \cdot g_{aP,aP}(Q) \cdot g_{2aP,aP}(Q).$$

Notice that it is not necessary to actually compute $2aP$, because the coefficients of $g_{2aP,aP}$ can be obtained from aP and $3aP$.

In characteristic 3, the tripling formula is by itself more efficient than the doubling formula, since the squaring operation, which takes $O(m^2)$ time, is replaced by cubing, which has only linear complexity at most; besides, it is invoked only a fraction $\log_3 2$ times compared to the doubling case. Furthermore, for the Tate pairing of order $n = (3^{(m-1)/2} \pm 1)3^{(m+1)/2} + 1$ the contribution of the underlying scalar multiplication to the complexity of Miller's algorithm is only $O(m^2)$ instead of $O(m^3)$, as it involves only two additions or one addition and one subtraction. An analogous observation holds for supersingular elliptic curves in characteristic 2.

An interesting observation is that, even if Miller's algorithm computes $f_r(Q)$ for $r \mid n$, it is often the case that a technique similar to that used for square root extraction can be applied, reducing the number of point additions or subtractions from $O(m)$ down to $O(\log m)$. However, we won't elaborate on this possibility, as the above choice is clearly faster.

5.2   Choice of the Subgroup Order

Pairing evaluation over fields $\mathbb{F}_{p^2}$ of general characteristic (as used, for instance, in the Boneh-Franklin identity-based cryptosystem [2]) with Miller's algorithm can benefit from the above observations with a careful choice of parameters, particularly the size q of the subfield where calculations are performed. Instead of choosing a random subfield prime, use a Solinas prime [30] of form $q = 2^\alpha \pm 2^\beta \pm 1$ (it is always possible to find such primes for practical subgroup sizes), since $qP = (2^\beta(2^{\alpha-\beta} \pm 1) \pm 1)P$ involves only two additions or subtractions plus α doublings.

5.3   Speeding up the Final Powering

Evaluation of the Tate pairing $e_n(P,Q)$, where $n \equiv \#E(\mathbb{F}_{p^m})$, includes a final raising to the power of $(p^{km} - 1)/n$. The powering is usually computed in $O(m^3)$ steps. However, this exponent shows a rather periodical structure in base p. One can exploit this property in a fashion similar to the square root algorithm of section 4, reducing the computational effort to $O(m^2 \log m)$ steps. As it turns out, it is actually possible to compute the power in only $O(m^2)$ steps, by carefully exploiting the structure of the exponent. Details of this process are given in appendix A.2.

5.4   Fixed-Base Pairing Precomputation

Actual pairing-based cryptosystems often need to compute pairings $e_n(P,Q)$ where P is either fixed (e.g. the base point on the curve) or used repeatedly (e.g.

a public key). In these cases, the underlying scalar multiplication in Miller's algorithm can be executed only once to precompute the coefficients of the line functions $g_U(Q)$. The speedup resulting from this technique is more prominent for characteristic $p > 3$.

5.5 MNT Curves

Until recently, the only elliptic curves known to have embedding degree $k \leqslant 6$ were supersingular like $E_{2,b}$ and $E_{3,b}$. As it turns out, it is possible to construct ordinary (non-supersingular) curves with $k \in \{3, 4, 6\}$. Such curves were first described by Miyaji, Nakabayashi and Takano in [20]; we call them MNT curves.

Briefly, MNT curves are built with the complex multiplication (CM) method [1, chapter VIII]. The idea is to impose certain constraints on the form of the underlying finite field $\mathbb{F}_q$, the curve order n, and the trace of Frobenius t, which are linked to each other by the relation $n = q + 1 - t$. These in turn lead to further constraints on the form of the CM equation $DV^2 = 4q - t^2$, which for $k \in \{3, 4, 6\}$ reduces to a Pell equation[4], whose solution is well known [28].

MNT curves address concerns that supersingular curves may not be as secure as ordinary curves. They are suitable for variants of pairing-based cryptosystems that do not involve distortion maps, like the BLS variant of [3, section 3.5] or the general IBE variants of [2, section 4] and [6, section 3]. In such systems, the pairings have the form $e_\ell(P, Q)$ where $P \in E(\mathbb{F}_q)$ and $Q \in E(\mathbb{F}_{q^k})$, and both are chosen so that $e_\ell(P, Q) \neq 1$.

An important property of the MNT criteria is that $n \mid \Phi_k(q)$ but $n \nmid (q^k - 1)/\Phi_k(q)$, where Φ_k is the k-th cyclotomic polynomial. Due to this property, lemma 1 holds for MNT curves as well, and consequently, so do theorem 1 and corollary 1. Therefore, the deterministic version of Miller's algorithm presented in section 5 is equally valid for the MNT case. Furthermore, for even k it often happens that the point $Q = (x, y)$ in the variant cryptosystems can be chosen so that $x \in \mathbb{F}_{q^{k/2}}$ but $y \notin \mathbb{F}_{q^{k/2}}$; with this setting[5], denominator elimination as suggested in section 5.1 is also applicable.

6 Experimental Results

The heaviest operation in any pairing-based cryptosystem is the pairing computation. We give our timings for these operations in table 3.

---

[4] There is reason to believe that one can effectively construct MNT-like curves with $k \in \{5, 8, 10, 12\}$, for which the CM equation reduces to a quartic elliptic Diophantine equation [31]. However, we refrain from further investigating this possibility here.

[5] Representing $\mathbb{F}_{q^k}$ in polynomial basis as $\mathbb{F}_q[t]/R_k(t)$ and carefully choosing $R_k(t)$, it is quite easy to find a point Q satisfying these constraints. For instance, if $R_k(t) = t^k + t^2 + c$ for some $c \in \mathbb{F}_q$, one can show that a suitable Q can be found by restricting the coordinates to the form $x = a_{k-2}t^{k-2} + a_{k-4}t^{k-4} + \cdots + a_2t^2 + a_0$ and $y = b_{k-1}t^{k-1} + b_{k-3}t^{k-3} + \cdots + b_3t^3 + b_1t$.

Table 3. Tate pairing computation times (in ms) on a PIII 1 GHz.

underlying base field	timing
$\mathbb{F}_{3^{97}}$	26.2
$\mathbb{F}_{2^{271}}$	23.0
$\mathbb{F}_p$, $|p| = 512$ bits	20.0
$\mathbb{F}_p$ with preprocessing	8.6

Table 4. Comparison of signing and verification times (in ms) on a PIII 1 GHz.

algorithm	signing	verification
RSA, $|n| = 1024$ bits, $|d| = 1007$ bits	7.90	0.40
DSA, $|p| = 1024$ bits, $|q| = 160$ bits	4.09	4.87
$\mathbb{F}_p$ ECDSA, $|p| = 160$ bits	4.00	5.17
$\mathbb{F}_{2^{160}}$ ECDSA	5.77	7.15
$\mathbb{F}_{3^{97}}$ BLS (supersingular)	3.57	53.0
$\mathbb{F}_p$ BLS (MNT), $|p| = 157$ bits	2.75	81.0

Table 5. BLS and IBE times (in ms) on a PIII 1 GHz.

operation	original [3, 14]	ours
BLS verification	2900	53
IBE encryption	170	48 (preprocessed: 36)
IBE decryption	140	30 (preprocessed: 19)

Boneh-Lynn-Shacham (BLS) signature generation is comparable to RSA or DSA signing at the same security level. Table 4 compares the signing times for the RSA, DSA (without precomputation), ECDSA (without precomputation), and BLS signature schemes. We consider two BLS implementations, namely, one using the curve $E_{3,b}$ and one using an MNT curve. Timings for BLS verification and Boneh-Franklin identity-based encryption (IBE) are listed in table 5. BLS signature verification speed for $\mathbb{F}_{3^{97}}$ shows an improvement by a factor of about 55 over published timings. The performance of IBE is also comparable to other cryptosystems; the data refers to a curve over $\mathbb{F}_p$ where $|p| = 512$ bits, using a subgroup of order q where q is a Solinas prime and $|q| = 160$ bits.

The implementations in this section were written in C/C++ and based on the MIRACL [26] library.

7   Conclusions and Acknowledgements

We have proposed several new algorithms to implement pairing-based cryptosystems. Our algorithms are all practical and lead to significant improvements, not only for the pairing evaluation process but to other operations as well, such as elliptic curve scalar multiplication and square root extraction.

An interesting line of further research is the application of these techniques to more general algebraic curves; for instance, a fast n-th root algorithm in the lines of the square root algorithm presented here would be useful for super-elliptic curves. Investigating the conditions leading to composition operations computable in linear time in abelian varieties would also be of great interest.

We are very grateful to Dan Boneh, Steven Galbraith, Antoine Joux, Frederik Vercauteren, and the anonymous referees for their valuable comments and/or feedback regarding this work.

References

1. I. Blake, G. Seroussi and N. Smart, "Elliptic Curves in Cryptography," Cambridge University Press, 1999.
2. D. Boneh and M. Franklin, "Identity-based encryption from the Weil pairing," Advances in Cryptology – Crypto'2001, Lecture Notes in Computer Science **2139**, pp. 213–229, Springer-Verlag, 2001.
3. D. Boneh, B. Lynn, and H. Shacham, "Short signatures from the Weil pairing," Asiacrypt'2001, Lecture Notes in Computer Science **2248**, pp. 514–532, Springer-Verlag, 2002.
4. H. Cohen, "A Course in Computational Algebraic Number Theory," Springer-Verlag, 1993.
5. G. Frey, M. Müller, and H. Rück, "The Tate Pairing and the Discrete Logarithm Applied to Elliptic Curve Cryptosystems," IEEE Transactions on Information Theory **45(5)**, pp. 1717–1719, 1999.
6. S. Galbraith, "Supersingular curves in cryptography," Asiacrypt'2001, Lecture Notes in Computer Science **2248**, pp. 495–513, Springer-Verlag, 2002.
7. S. Galbraith, K. Harrison and D. Soldera, "Implementing the Tate pairing," Algorithm Number Theory Symposium – ANTS V, Lecture Notes in Computer Science **2369**, Springer-Verlag, to appear.
8. F. Hess, "Exponent Group Signature Schemes and Efficient Identity Based Signature Schemes Based on Pairings," Cryptology ePrint Archive, Report 2002/012, available at http://eprint.iacr.org/2002/012/.
9. T. Itoh, O. Teechai and S. Tsujii, "A fast algorithm for computing multiplicative inverses in GF(2^m) using normal bases," *Information and Computation* **78**, pp. 171–177, 1988.
10. A. Joux, "A one-round protocol for tripartite Diffie-Hellman," Algorithm Number Theory Symposium – ANTS IV, Lecture Notes in Computer Science **1838**, pp. 385–394, Springer-Verlag, 2000.
11. A. Joux and K. Nguyen, "Separating Decision Diffie-Hellman from Diffie-Hellman in Cryptographic Groups," Cryptology ePrint Archive, Report 2001/003, available at http://eprint.iacr.org/2001/003/.
12. N. Koblitz, "An Elliptic Curve Implementation of the Finite Field Digital Signature Algorithm," Advances in Cryptology – Crypto'98, Lecture Notes in Computer Science **1462**, pp. 327–337, Springer-Verlag, 1998.
13. R. Lidl and H. Niederreiter, "Finite Fields," Encyclopedia of Mathematics and its Applications **20**, 2nd Ed. Cambridge University Press, 1997.
14. B. Lynn, "Stanford IBE library," available at http://crypto.stanford.edu/ibe/.
15. A.J. Menezes, "Elliptic Curve Public Key Cryptosystems," Kluwer International Series in Engineering and Computer Science, 1993.
16. A.J. Menezes, T. Okamoto and S.A. Vanstone, "Reducing elliptic curve logarithms to logarithms in a finite field," IEEE Transactions on Information Theory **39**, pp. 1639–1646, 1993.

17. A.J. Menezes, P.C. van Oorschot and S.A. Vanstone, "Handbook of Applied Cryptography," CRC Press, 1997.
18. A.J. Menezes and S.A. Vanstone, "The implementation of elliptic curve cryptosystems," Advances in Cryptology – Auscrypt'90, Lecture Notes in Computer Science **453**, pp. 2–13, Springer-Verlag, 1990.
19. V. Miller, "Short Programs for Functions on Curves," unpublished manuscript, 1986.
20. A. Miyaji, M. Nakabayashi, and S. Takano, "New explicit conditions of elliptic curve traces for FR-reduction," IEICE Trans. Fundamentals, Vol. E84 A, no. 5, May 2001.
21. IEEE Std 2000–1363, "Standard Specifications for Public Key Cryptography," 2000.
22. K.G. Paterson, "ID-based signatures from pairings on elliptic curves," Cryptology ePrint Archive, Report 2002/004, available at http://eprint.iacr.org/2002/004/.
23. K. Rubin and A. Silverberg, "Supersingular abelian varieties in cryptology," Advances in Cryptology – Crypto'2002, these proceedings.
24. R. Sakai, K. Ohgishi and M. Kasahara, "Cryptosystems based on pairing," 2000 Symposium on Cryptography and Information Security (SCIS2000), Okinawa, Japan, Jan. 26–28, 2000.
25. R. Schroeppel, H. Orman, S. O'Malley, O. Spatscheck, "Fast Key Exchange with Elliptic Curve Systems," Advances in Cryptology – Crypto '95, Lecture Notes in Computer Science **963**, pp. 43–56, Springer-Verlag, 1995.
26. M. Scott, "Multiprecision Integer and Rational Arithmetic C/C++ Library (MIRACL)," available at http://indigo.ie/~mscott/.
27. J.H. Silverman, "The Arithmetic of Elliptic Curves," Graduate Texts in Mathematics, vol. 106, Springer-Verlag, 1986.
28. N.P. Smart, "The Algorithmic Resolution of Diophantine Equations," London Mathematical Society Student Text **41**, Cambridge University Press, 1998.
29. N.P. Smart, "An Identity Based Authenticated Key Agreement Protocol Based on the Weil Pairing," Electronics Letters, to appear.
30. J. Solinas, "Generalized Mersenne numbers," technical report CORR-39, Department of C&O, University of Waterloo, 1999, available at http://www.cacr.math.uwaterloo.ca/.
31. N. Tzanakis, "Solving elliptic diophantine equations by estimating linear forms in elliptic logarithms. The case of quartic equations," Acta Arithmetica **75** (1996), pp. 165–190.
32. E. Verheul, "Evidence that XTR is more secure than supersingular elliptic curve cryptosystems," Advances in Cryptology – Eurocrypt'2001, Lecture Notes in Computer Science **2045** (2001), pp. 195–210.
33. E. Verheul, "Self-blindable Credential Certificates from the Weil Pairing," Asiacrypt'2001, Lecture Notes in Computer Science **2248**, pp. 533–551, Springer-Verlag, 2002.

A   Implementation Issues

A.1   Field Representation

The authors of the BLS scheme suggest representing $\mathbb{F}_{3^{6m}}$ as $\mathbb{F}_{3^6}[x]/\tau_m(x)$ for a suitable irreducible polynomial $\tau_m(x)$ [3, section 5.1]. It is our experience that the alternative representation as $\mathbb{F}_{3^m}[x]/\tau_6(x)$ using an irreducible trinomial $\tau_6(x)$ (for instance, $\tau_6(x) = x^6 + x - 1$) leads to better performance for practical values of m; moreover, both signing and verification benefit at once from any improvement made to the implementation of $\mathbb{F}_{3^m}$. Karatsuba multiplication can also be used to great effect, as one $\mathbb{F}_{3^{6m}}$ multiplication can be implemented with only 18 $\mathbb{F}_{3^m}$ multiplications. Similar observations apply to characteristic 2, where one $\mathbb{F}_{2^{4m}}$ multiplication takes 9 $\mathbb{F}_{2^m}$ multiplications.

As it turns out, however, Karatsuba is *not* the fastest multiplication technique in all circumstances. As seen in section 5.1, it is often the case that the actual pairing to be computed is $e_n(P, \phi(Q))$ where both P and Q are on the curve over $\mathbb{F}_q$ (rather than the curve over the extension field $\mathbb{F}_{q^k}$), and the pairing algorithm can explicitly use the form of the ϕ distortion map to reduce the number of $\mathbb{F}_q$ products involved in Miller's formula down to only two per line equation evaluation.

A.2   Speeding up the Final Powering in the Tate Pairing

The exponentiation needed by the Tate pairing $e_n(P, Q) = f_P(Q)^z$ where $z = (q^k - 1)/n$ can be efficiently computed with the following observations:

1. (Characteristic $p > 3$) Assume that $p \equiv 2 \pmod 3$ and $p \equiv 3 \pmod 4$. The order of a curve $E_{1,b}$ is $n = p + 1$. Let the order of the curve subgroup of interest be r, and notice that $r \mid p + 1$. Consider the scenario where the representation of a point $t \in \mathbb{F}_{p^2}$ is $t = u + iv$ where $u, v \in \mathbb{F}_p$ and i satisfies $i^2 + 1 = 0$. The Tate exponent is $z = (p^2 - 1)/r = ((p+1)/r) \cdot (p-1)$. To calculate $s = w^z \bmod p$, compute $t = w^{(p+1)/r} \equiv u + iv$ and set $s = (u + iv)^{p-1} = (u + v)^p/(u + iv) = (u - v)/(u + iv)$, using the linearity of raising to p and the fact that $i^p = -i$ for $p \equiv 3 \pmod 4$. We can further simplify to obtain $s = (u^2 - v^2)/(u^2 + v^2) - 2uvi/(u^2 + v^2)$.

2. (Characteristic 2) Let $q = 2^m$. As we saw in the proof of theorem 3, the Tate exponent is of form $z = (q + 1 \pm \sqrt{2q})(q^2 - 1)$. Therefore, to calculate $s = w^z$ one computes $t = w^q \cdot w \cdot w^{\pm\sqrt{2q}}$ and $s = t^{q^2}/t$. Raising to the exponents q, $\sqrt{2q}$ and q^2 can be done in $O(m)$ steps using normal basis, or in $O(m^2)$ steps using polynomial basis with a careful choice of the reduction polynomial (see [25], for instance), while the small (and constant) number of multiplications and inversions can be done in $O(m^2)$ steps. Therefore, the complete operation takes time $O(m^2)$.

3. (Characteristic 3) Let $q = 3^m$. As we saw in the proof of theorem 3, the Tate exponent is of form $z = (q + 1 \pm \sqrt{3q})(q^3 - 1)(q + 1)$. Therefore, to calculate $s = w^z$ one computes $u = w^q \cdot w \cdot w^{\pm\sqrt{3q}}$, $t = u^{q^3}/u$, and $s = t^q \cdot t$. Raising to the exponents q, $\sqrt{3q}$ and q^3 can be done in $O(m)$ steps using normal basis, or in $O(m^2)$ steps using polynomial basis with a careful choice of the reduction polynomial (see [25], for instance), while the small (and constant) number of multiplications and inversions can be done in $O(m^2)$ steps. Therefore, the complete operation takes time $O(m^2)$.

Computing Zeta Functions of Hyperelliptic Curves over Finite Fields of Characteristic 2

Frederik Vercauteren[1,2,*]

[1] Department of Electrical Engineering
University of Leuven
Kasteelpark Arenberg 10, B-3001 Leuven-Heverlee, Belgium
frederik.vercauteren@esat.kuleuven.ac.be
[2] Computer Science Department
University of Bristol
Woodland Road, Bristol BS8 1UB, United Kingdom
frederik@cs.bris.ac.uk

Abstract. We present an algorithm for computing the zeta function of an arbitrary hyperelliptic curve over a finite field $\mathbb{F}_q$ of characteristic 2, thereby extending the algorithm of Kedlaya for small odd characteristic. For a genus g hyperelliptic curve over $\mathbb{F}_{2^n}$, the asymptotic running time of the algorithm is $O(g^{5+\varepsilon}n^{3+\varepsilon})$ and the space complexity is $O(g^3n^3)$.

Keywords: Hyperelliptic curves, Kedlaya's algorithm, Monsky-Washnitzer cohomology

1 Introduction

Since elliptic curve cryptosystems were introduced by Koblitz [16] and Miller [23], various other systems based on the discrete logarithm problem in the Jacobian of curves have been proposed, e.g. hyperelliptic curves [17], superelliptic curves [10] and $\mathcal{C}_{ab}$ curves [1]. One of the main initialization steps of these cryptosystems is to generate a suitable curve defined over a given finite field. To ensure the security of the system, the curve must be chosen such that the group order of the Jacobian is divisible by a large prime.

The problem of counting the number of points on elliptic curves over finite fields of any characteristic can be solved in polynomial time using Schoof's algorithm [32] and its improvements due to Atkin [2] and Elkies [6]. An excellent account of the resulting SEA-algorithm can be found in [3] and [20]. For finite fields of small characteristic, Satoh [29] described an algorithm based on p-adic methods which is asymptotically faster than the SEA-algorithm. Skjernaa [33] and Fouquet, Gaudry and Harley [7] extended the algorithm to characteristic 2 and Vercauteren [35] presented a memory efficient version. Mestre proposed a variant of Satoh's algorithm based on the Arithmetic-Geometric Mean, which has the same asymptotic behavior as [35], but is faster by some constant. Satoh, Skjernaa and Taguchi [30] described an algorithm which has a better complexity than all previous algorithms, but requires some precomputations.

* F.W.O. research assistant, sponsored by the Fund for Scientific Research - Flanders (Belgium).

The equivalent problem for higher genus curves seems to be much more difficult. Pila [28] described a theoretical generalization of Schoof's approach, but the algorithm is not practical, not even for genus 2 as shown by Gaudry and Harley [12]. An extension of Satoh's method to higher genus curves needs the Serre-Tate canonical lift of the Jacobian of the curve, which need not be a Jacobian itself and thus is difficult to compute with. The AGM method does generalize to hyperelliptic curves, but currently only the genus 2 case is practical.

Recently Kedlaya [14] described a p-adic algorithm to compute the zeta function of hyperelliptic curves over finite fields of small *odd* characteristic, using the theory of Monsky-Washnitzer cohomology. The running time of the algorithm is $O(g^{4+\varepsilon} \log^{3+\varepsilon} q)$ for a hyperelliptic curve of genus g over $\mathbb{F}_q$. The algorithm readily generalizes to superelliptic curves as shown by Gaudry and Gurel [11].

A related approach by Lauder and Wan [18] is based on Dwork's proof of the rationality of the zeta function and leads to a polynomial time algorithm for computing the zeta function of an arbitrary variety over a finite field. Note that Wan [36] suggested the use of p-adic methods, including the method of Dwork and Monsky, already several years ago. Despite the polynomial complexity of the Lauder and Wan algorithm, it is not practical for cryptographical sizes. Using Dwork cohomology, Lauder and Wan [19] adapted their original algorithm for the special case of Artin-Schreier curves, resulting in an $O(g^{5+\varepsilon} \log^{3+\varepsilon} q)$ time algorithm. Denef and Vercauteren [4] described an extension of Kedlaya's algorithm for Artin-Schreier curves in characteristic 2 which has the same running time $O(g^{5+\varepsilon} \log^{3+\varepsilon} q)$.

In this paper we describe an extension of Kedlaya's algorithm to compute the zeta function of an *arbitrary* hyperelliptic curve C defined over a finite field $\mathbb{F}_q$ of characteristic 2. The resulting algorithm has running time $O(g^{5+\varepsilon} \log^{3+\varepsilon} q)$ and needs $O(g^3 \log^3 q)$ storage space for a genus g curve. Furthermore, a first implementation of this algorithm in the C programming language shows that cryptographical sizes are now feasible for any genus g. For instance, computing the order of a 160-bit Jacobian of a hyperelliptic curve of genus 2, 3 or 4 takes less than 100 seconds. The theoretical version of this paper, co-authored by Jan Denef [5], provides a detailed description of the underlying mathematics of the algorithm and contains several proofs which we have omitted from the current article.

The remainder of the paper is organized as follows: after recalling some basics about curves and zeta functions in Section 2, we give a brief overview of the formalism of Monsky-Washnitzer cohomology in Section 3. In Section 4 we study the cohomology of hyperelliptic curves over finite fields of characteristic 2 and in Section 5 we present a ready to implement description of the resulting algorithm. We conclude in Section 6 with some numerical examples obtained by an implementation of our algorithm in the C programming language.

2   Hyperelliptic Curves, Zeta Functions and p-Adics

In this section we briefly recall the definition of a hyperelliptic curve, the main properties of its zeta function and some basic facts about p-adic numbers. More details can be found in the elementary books by Fulton [9], Lorenzini [21] and Koblitz [15] or in the standard reference by Hartshorne [13].

2.1 Hyperelliptic Curves

Let $\mathbb{F}_q$ be a finite field with $q = p^n$ elements and fix an algebraic closure $\overline{\mathbb{F}}_q$ of $\mathbb{F}_q$. For $k \in \mathbb{N}_0$, let $\mathbb{F}_{q^k}$ be the unique subfield of $\overline{\mathbb{F}}_q$ of order q^k. An affine hyperelliptic curve C of genus g is a plane curve defined by an equation of the form

$$C : y^2 + \overline{h}(x)y = \overline{f}(x), \tag{1}$$

where $\overline{f}(x) \in \mathbb{F}_q[x]$ is a monic polynomial of degree $2g + 1$ and $\overline{h}(x) \in \mathbb{F}_q[x]$ is a polynomial of degree at most g. Furthermore, the curve should be non-singular, i.e. there is no point in $C(\overline{\mathbb{F}}_q)$ such that both partial derivatives

$$2y + \overline{h}(x) \quad \text{and} \quad \overline{h}'(x)y - \overline{f}'(x),$$

simultaneously vanish. Note that for $g = 1$ we recover the definition of an elliptic curve and that for $g > 1$ the hyperelliptic curve C is singular at the point at infinity. However, there always exists a unique smooth projective curve $\widetilde{C}$ birational to C. Since the degree of $\overline{f}(x)$ is odd, $\widetilde{C}$ has a unique place of degree 1 (i.e. a point) at infinity. Note that there exists an involution $\imath$ on $\widetilde{C}$ which sends the point (x, y) to the point $(x, -y - \overline{h}(x))$.

Let $\widetilde{C}(\mathbb{F}_{q^k})$ denote the set of points on $\widetilde{C}$ with coordinates in $\mathbb{F}_{q^k}$. If $\widetilde{C}$ is an elliptic curve, one can define an additive abelian group law on the set $\widetilde{C}(\mathbb{F}_{q^k})$ by the chord-tangent method. For a hyperelliptic curve with $g > 1$ this is no longer possible; instead one computes in the group of points on the Jacobian $J_{\widetilde{C}}(\mathbb{F}_{q^k})$ of the curve.

A divisor D on a curve $\widetilde{C}$ is a finite formal sum of points

$$D = \sum_{P \in \widetilde{C}(\overline{\mathbb{F}}_q)} n_P P,$$

where $n_P \in \mathbb{Z}$. The degree of D is defined as $\sum n_P$. A divisor is called $\mathbb{F}_{q^k}$-rational if it is stable under the action of the q^k-th power Frobenius endomorphism $F_k : \overline{\mathbb{F}}_q \to \overline{\mathbb{F}}_q : x \mapsto x^{q^k}$. Every function on the curve gives rise to a so called principal divisor, i.e. the degree zero divisor consisting of the formal sum of the poles and zeros of the function. The Jacobian $J_{\widetilde{C}}(\mathbb{F}_{q^k})$ is then defined as the group of $\mathbb{F}_{q^k}$-rational divisors of degree zero modulo principal divisors. This is a finite abelian group and forms the basis of the cryptographic schemes based on hyperelliptic curves. In this article we give an efficient algorithm for computing the group order of $J_{\widetilde{C}}(\mathbb{F}_{q^k})$ for $k \in \mathbb{N}_0$ and $\mathbb{F}_q$ a finite field of characteristic 2.

2.2 Zeta-Functions

Let N_k denote the number of $\mathbb{F}_{q^k}$-rational points on $\widetilde{C}$, i.e. $N_k = \#\widetilde{C}(\mathbb{F}_{q^k})$. The zeta function $Z(\widetilde{C}/\mathbb{F}_q; T)$ of $\widetilde{C}$ is then defined as

$$Z(\widetilde{C}/\mathbb{F}_q; T) := \exp\left(\sum_{k=1}^{\infty} \frac{N_k T^k}{k}\right). \tag{2}$$

Weil [37] conjectured and proved that $Z(\widetilde{C}/\mathbb{F}_q; T)$ has many remarkable properties, which we summarize in the next theorem.

Theorem 1 (Weil) *Let $\widetilde{C}$ be a smooth projective curve of genus g defined over a finite field $\mathbb{F}_q$, then*

$$Z(\widetilde{C}/\mathbb{F}_q; T) = \frac{\Psi(T)}{(1 - qT)(1 - T)}, \tag{3}$$

where $\Psi(T) \in \mathbb{Z}[T]$ is a degree $2g$ polynomial with integer coefficients. Since $Z(\widetilde{C}/\mathbb{F}_q; 0) = 1$, we have $\Psi(0) = 1$. Write $\Psi(T) = \prod_{i=1}^{2g}(1 - \omega_i T)$, then $|\omega_i| = \sqrt{q}$ for $i = 1, \ldots, 2g$, and we can label the ω_i such that $\omega_i \cdot \omega_{g+i} = q$ for $i = 1, \ldots, g$. Substituting the expression for $\Psi(T)$ in equation (3) and taking the logarithm of equations (2) and (3), it follows that

$$N_k = \#\widetilde{C}(\mathbb{F}_{q^k}) = q^k + 1 - \sum_{i=1}^{2g} \omega_i^k. \tag{4}$$

Furthermore, $\Psi(1) = \#J_{\widetilde{C}}(\mathbb{F}_q)$ is the group order of the Jacobian of $\widetilde{C}$ over $\mathbb{F}_q$.

The above theorem shows that it is sufficient to compute the zeta function of a hyperelliptic curve $\widetilde{C}/\mathbb{F}_q$ to recover the group order of its Jacobian $J_{\widetilde{C}}(\mathbb{F}_q)$ as $\Psi(1)$. Let $F : \overline{\mathbb{F}}_q \to \overline{\mathbb{F}}_q : x \mapsto x^q$ be the q-th power Frobenius automorphism, then F extends naturally to the Jacobian $J_{\widetilde{C}}(\mathbb{F}_q)$. Denote with $\chi(T)$ the characteristic polynomial of F on $J_{\widetilde{C}}(\mathbb{F}_q)$, then one can prove that $\Psi(T) = T^{2g}\chi(1/T)$.

2.3 p-Adic Numbers

Let $\mathcal{K}$ be the degree n unramified extension of $\mathbb{Q}_p$ with valuation ring $\mathcal{R}$ and residue field $\mathcal{R}/p\mathcal{R} = \mathbb{F}_q$. The field $\mathcal{K}$ can be constructed as follows: let $\overline{P}(t)$ be a monic, irreducible polynomial of degree n over $\mathbb{F}_p$, such that $\mathbb{F}_q \simeq \mathbb{F}_p[t]/(\overline{P}(t))$. Take any lift $P(t) \in \mathbb{Z}_p[t]$ of $\overline{P}(t)$ of degree n, then $\mathcal{K}$ is isomorphic with $\mathbb{Q}_p[t]/(P(t))$. In practice, we represent an element α of $\mathcal{R}$ as a polynomial $\sum_{i=0}^{n-1} \alpha_i t^i$, with $\alpha_i \in \mathbb{Z}/(p^N\mathbb{Z})$, where N is called the precision of the representation. The Galois group of $\mathcal{K}$ over $\mathbb{Q}_p$ is cyclic of order n and there exists a unique generator Σ which reduces to the p-th power Frobenius $\sigma : \mathbb{F}_q \to \mathbb{F}_q : x \mapsto x^p$. This generator Σ is called the Frobenius substitution on $\mathcal{K}$. By definition Σ is a $\mathbb{Q}_p$-linear automorphism, so we can compute $\Sigma(\alpha)$ as $\sum_{i=0}^{n-1} \alpha_i \Sigma(t)^i$. Since $P(t)$ is defined over $\mathbb{Z}_p$, it follows that $P(\Sigma(t)) = 0$, which implies that $\Sigma(t)$ can be computed by the Newton iteration $Y \to Y - P(Y)/P'(Y)$ initialized with t^p. Note that Σ is not a simple powering like σ.

3 Monsky-Washnitzer Cohomology

We specialize the formalism of Monsky-Washnitzer cohomology to smooth affine plane curves. The more general case of a smooth affine variety can be found in the seminal papers by Monsky and Washnitzer [27, 24, 26], the lectures by Monsky [25] and the survey by van der Put [34].

Let C be a smooth affine plane curve over a finite field $\mathbb{F}_q$ with $q = p^n$ and let $\mathcal{K}$ be a degree n unramified extension of $\mathbb{Q}_p$ with valuation ring $\mathcal{R}$, such that $\mathcal{R}/p\mathcal{R} = \mathbb{F}_q$. The aim of Monsky-Washnitzer cohomology is to express the zeta function of the curve C in terms of a Frobenius operator $\mathcal{F}$ acting on p-adic cohomology groups $H^i(C/\mathcal{K})$ associated to C. Although "p-adic cohomology groups" sounds very complicated, these groups are simply finite dimensional $\mathcal{K}$-vectorspaces. Furthermore, the Frobenius operator $\mathcal{F}$ acts linearly on these vectorspaces which implies that $\mathcal{F}$ can be represented as a matrix over $\mathcal{K}$. For smooth affine plane curves the only non-trivial cohomology groups are $H^0(C/\mathcal{K})$ and $H^1(C/\mathcal{K})$. Let $M_{\mathcal{F}}$ be the matrix through which the Frobenius operator $\mathcal{F}$ acts on $H^1(C/\mathcal{K})$. The crux of the whole construction is that the characteristic polynomial of $M_{\mathcal{F}}$ is equal to $\chi(T)$, i.e. the characteristic polynomial of Frobenius on C.

In the remainder of this section we will give a middlebrow overview of the construction of the cohomology group $H^1(C/\mathcal{K})$. Suppose that the smooth affine plane curve C is given by an equation $\bar{g}(x,y) = 0$ and let $A := \mathbb{F}_q[x,y]/(\bar{g}(x,y))$ be its coordinate ring. Take any lift $g(x,y) \in \mathcal{R}[x,y]$ of $\bar{g}(x,y)$ and let C be the curve defined by $g(x,y) = 0$ with coordinate ring $\mathcal{A} := \mathcal{R}[x,y]/(g(x,y))$. To compute the zeta function of C in terms of a Frobenius operator, one needs to lift the Frobenius endomorphism F on A to the $\mathcal{R}$-algebra $\mathcal{A}$, but in general this is not possible. Note that in the special case of elliptic curves, Satoh [29] solves this problem by using the Serre-Tate canonical lift, which does admit a lift of the Frobenius endomorphism.

A first attempt to remedy this difficulty is to work with the p-adic completion $\mathcal{A}^\infty$ of $\mathcal{A}$, since we can lift F to $\mathcal{A}^\infty$. But then a new problem arises since the de Rham cohomology of $\mathcal{A}^\infty$, which provides the vectorspaces we are looking for, is too big. For example, consider the affine line over $\mathbb{F}_p$, then $\mathcal{A} = \mathcal{R}[x]$ and $\mathcal{A}^\infty$ is the ring of power series $\sum_{k=0}^{\infty} r_k x^k$ with $r_i \in \mathcal{R}$ and $\lim_{k \to \infty} r_k = 0$. We would like to define $H^1(A/\mathcal{K})$ as $(\mathcal{A}^\infty \otimes \mathcal{K})\, dx / \frac{d}{dx}(\mathcal{A}^\infty \otimes \mathcal{K})$, but this turns out to be infinite dimensional. For instance, it is clear that each term in the differential form $\sum_{n=0}^{\infty} p^n x^{p^n-1} dx$ is exact, but its sum is not since $\sum_{n=0}^{\infty} x^{p^n}$ is not in $\mathcal{A}^\infty$. The fundamental problem is that $\sum_{n=0}^{\infty} p^n x^{p^n-1}$ does not converge fast enough for its integral to converge as well.

Monsky and Washnitzer therefore work with a subalgebra $\mathcal{A}^\dagger$ of $\mathcal{A}^\infty$, whose elements satisfy growth conditions. This *dagger ring* or *weak completion* $\mathcal{A}^\dagger$ is defined as follows: let $\mathcal{A} = \mathcal{R}[x,y]/(g(x,y))$, then $\mathcal{A}^\dagger := \mathcal{R}\langle x, y\rangle^\dagger/(g(x,y))$, where $\mathcal{R}\langle x, y\rangle^\dagger$ is the ring of overconvergent power series

$$\left\{ \sum r_{i,j} x^i y^j \in \mathcal{R}[[x,y]] \mid \exists\, \delta, \varepsilon \in \mathbb{R}, \varepsilon > 0, \forall (i,j) : \operatorname{ord}_p r_{i,j} \geq \varepsilon(i+j) + \delta \right\}.$$

The ring $\mathcal{A}^\dagger$ satisfies $\mathcal{A}^\dagger/p\mathcal{A}^\dagger = A$ and depends up to $\mathcal{R}$-isomorphism only on A. Furthermore, Monsky and Washnitzer show that if E is an $\mathbb{F}_q$-endomorphism of A, then there exists an $\mathcal{R}$-endomorphism $\mathcal{E}$ of $\mathcal{A}$ lifting E. In particular, we can lift the Frobenius endomorphism F on A to an $\mathcal{R}$-endomorphism $\mathcal{F}$ on $\mathcal{A}$.

For $\mathcal{A}^\dagger$ we define the universal module $\mathcal{D}^1(\mathcal{A}^\dagger)$ of differentials

$$\mathcal{D}^1(\mathcal{A}^\dagger) := (\mathcal{A}^\dagger \, dx + \mathcal{A}^\dagger \, dy)/(\mathcal{A}^\dagger(\frac{\partial g}{\partial x} \, dx + \frac{\partial g}{\partial y} \, dy)). \tag{5}$$

Taking the total differential of the equation $g(x,y) = 0$ gives $\frac{\partial g}{\partial x} \, dx + \frac{\partial g}{\partial y} \, dy = 0$, which accounts for the module $\mathcal{A}^\dagger(\frac{\partial g}{\partial x} \, dx + \frac{\partial g}{\partial y} \, dy)$ in the above definition.

The first cohomology group is then defined as $H^i(A/\mathcal{R}) := \mathcal{D}^1(\mathcal{A}^\dagger)/d(\mathcal{A}^\dagger)$ and $H^1(A/\mathcal{K}) := H^1(A/\mathcal{R}) \otimes_\mathcal{R} \mathcal{K}$ finally defines the first Monsky-Washnitzer cohomology group. Elements of $d(\mathcal{A}^\dagger)$, i.e. differentials of the form $d(l(x,y))$ for $l(x,y) \in \mathcal{A}^\dagger$, are called exact. One can prove that $H^1(A/\mathcal{K})$ is well defined and is in fact a *finite dimensional* vectorspace over $\mathcal{K}$. Furthermore, for a smooth affine curve of genus g, the dimension of $H^1(A/\mathcal{K})$ is $2g + m - 1$, where m is the number of points needed to complete the affine curve to a projective curve.

4   Cohomology of Hyperelliptic Curves in Characteristic 2

Let $\mathbb{F}_q$ be a finite field with $q = 2^n$ elements and consider the smooth affine hyperelliptic curve C of genus g defined by the equation

$$C : y^2 + \overline{h}(x)y = \overline{f}(x),$$

with $\overline{h}(x), \overline{f}(x) \in \mathbb{F}_q[x]$, $\overline{f}(x)$ monic of degree $2g+1$ and $\deg \overline{h}(x) \le g$. Write $\overline{h}(x)$ as $\overline{c} \cdot \prod_{i=0}^s (x - \overline{\theta}_i)^{m_i}$ with $\overline{\theta}_i \in \overline{\mathbb{F}}_q$, $\overline{c} \in \mathbb{F}_q$ the leading coefficient of $\overline{h}(x)$ and define $\overline{H}(x) = \prod_{i=0}^s (x - \overline{\theta}_i) \in \overline{\mathbb{F}}_q[x]$. If $h(x)$ is a constant, we set $\overline{H}(x) = 1$. Without loss of generality we can assume that $\overline{H}(x) \,|\, \overline{f}(x)$. Indeed, the isomorphism defined by $x \mapsto x$ and $y \mapsto y + \sum_{i=0}^s b_i x^i$ transforms the curve in

$$y^2 + h(x)y = f(x) - \sum_{i=0}^s b_i^2 x^{2i} - h(x) \sum_{i=0}^s b_i x^i.$$

The polynomial $\overline{H}(x)$ will divide the right hand side of the above equation if and only if $f(\overline{\theta}_j) = \sum_{i=0}^s b_i^2 \cdot \overline{\theta}_j^{2i}$ for $j = 0, \ldots, s$. This is a system of linear equations in the indeterminates b_i^2 and its determinant is a Vandermonde determinant. Since the $\overline{\theta}_j$ are the zeros of a polynomial defined over $\mathbb{F}_q$, the system of equations is invariant under the q-th power Frobenius automorphism F and it follows that the b_i^2 and therefore the b_i are elements of $\mathbb{F}_q$. We conclude that we can always assume that $\overline{H}(x) \,|\, \overline{f}(x)$.

Let $\pi : C(\overline{\mathbb{F}}_q) \to \mathbb{A}^1(\overline{\mathbb{F}}_q)$ be the projection on the x-axis. It is clear that π ramifies at the points $(\theta_i, 0) \in \overline{\mathbb{F}}_q \times \overline{\mathbb{F}}_q$ for $i = 0, \ldots, s$ where $\overline{H}(\overline{\theta}_i) = 0$. Note that the ordinate of these points is zero, since we assumed that $\overline{H}(x) \,|\, \overline{f}(x)$. Let

C' be the curve obtained from C by removing the ramification points $(\theta_i, 0)$ for $i = 0, \ldots, s$. Then the coordinate ring A of C' is

$$\mathbb{F}_q[x, y, \overline{H}(x)^{-1}]/(y^2 + \overline{h}(x)y - \overline{f}(x)).$$

Note that it is not really necessary to work with the open subset C' instead of with C itself, but it is more efficient to do so. The coordinate ring of C' contains the inverse of $\overline{H}(x)$ which will enable us to choose a particular lift of the Frobenius endomorphism F of A.

Let $\mathcal{K}$ be a degree n unramified extension of $\mathbb{Q}_2$ with valuation ring $\mathcal{R}$ and residue field $\mathcal{R}/2\mathcal{R} = \mathbb{F}_q$. Write $\overline{h}(x) = \overline{c} \cdot \prod_{i=1}^{r} \overline{P}_i(x)^{t_i}$, where the $\overline{P}_i(x)$ are irreducible over $\mathbb{F}_q$. Let $D = \max_i t_i$, then $\overline{h}(x)$ divides $\overline{H}(x)^D$, since we have the identity $\overline{H}(x) = \prod_{i=0}^{r} \overline{P}_i(x)$. Lift $\overline{P}_i(x)$ for $i = 0, \ldots, r$ to any monic polynomial $P_i(x) \in \mathcal{R}[x]$ with $P_i(x) \equiv \overline{P}_i(x)$ mod 2. Define $H(x) = \prod_{i=0}^{r} P_i(x)$ and $h(x) = c \cdot \prod_{i=0}^{r} P_i(x)^{t_i}$, with c any lift of $\overline{c}$ to $\mathcal{R}$. Since $\overline{H}(x)$ divides $\overline{f}(x)$ we can define $\overline{Q}_{\overline{f}}(x) = \overline{f}(x)/\overline{H}(x)$. Let $Q_f(x) \in \mathcal{R}[x]$ be any monic lift of $\overline{Q}_{\overline{f}}(x)$ and finally set $f(x) = H(x) \cdot Q_f(x)$. The result is that we have now constructed a lift $\mathcal{C}$ of the curve C to $\mathcal{R}$ defined by the equation

$$\mathcal{C} : y^2 + h(x)y = f(x).$$

Note that due to the careful construction of $\mathcal{C}$ we have the following properties: $H(x) \mid h(x)$, $H(x) \mid f(x)$ and $h(x) \mid H(x)^D$. Furthermore, let $\pi : \mathcal{C}(\overline{\mathcal{K}}) \to \mathbb{A}^1(\overline{\mathcal{K}})$ be the projection on the x-axis, then it is clear that π ramifies at $(\theta_i, 0)$ for $i = 0, \ldots, s$ where the θ_i are the zeros of $H(x)$.

Let $\mathcal{C}'$ be the curve obtained from $\mathcal{C}$ by deleting the points $(\theta_i, 0)$ for $i = 0, \ldots, s$, then the coordinate ring $\mathcal{A}$ of $\mathcal{C}'$ is

$$\mathcal{R}[x, y, H(x)^{-1}]/(y^2 + h(x)y - f(x)).$$

Let $\mathcal{A}^\dagger$ denote the weak completion of $\mathcal{A}$. Using the equation of the curve, we can represent any element of $\mathcal{A}^\dagger$ as a series $\sum_{i=-\infty}^{\infty} (U_i(x) + V_i(x)y)H(x)^i$, with the degree of $U_i(x)$ and $V_i(x)$ smaller that the degree of $H(x)$ if $\deg H(x) > 0$. If $H(x) = 1$, every element can be written as $\sum_{i=0}^{\infty} (U_i + V_i y)x^i$ with $U_i, V_i \in \mathcal{K}$. The growth condition on the dagger ring implies that there exist real numbers δ and $\epsilon > 0$ such that $\mathrm{ord}_2(U_i(x)) \geq \epsilon \cdot |i| + \delta$ and $\mathrm{ord}_2(V_i(x)) \geq \epsilon \cdot |i+1| + \delta$, where $\mathrm{ord}_2(W(x))$ is defined as $\min_j \mathrm{ord}_2(w_j)$ for $W(x) = \sum w_j x^j \in \mathcal{K}[x]$.

Since $F = \sigma^n$ with σ the 2-nd power Frobenius, it clearly is sufficient to lift σ to an endomorphism Σ of $\mathcal{A}^\dagger$. It is natural to define Σ as the Frobenius substitution on $\mathcal{R}$ and to extend it to $\mathcal{R}[x]$ by mapping x to x^2. Using the equation of the curve we see that $\Sigma(y)$ must satisfy

$$(\Sigma(y))^2 + \Sigma(h(x))\Sigma(y) - \Sigma(f(x)) = 0 \quad \text{and} \quad \Sigma(y) \equiv y^2 \text{ mod } 2.$$

In practice $\Sigma(y)$ is computed as a Laurent series $\sum_{i=-B_L}^{B_U} (S_i(x) + T_i(x)y)H(x)^i$ if $\deg H(x) > 0$ or $\sum_{i=0}^{B_U} (S_i + T_i y)x^i$ if $H = 1$, via the Newton iteration

$$W_{k+1} = W_k - \frac{W_k^2 + \Sigma(h(x)) \cdot W_k - \Sigma(f(x))}{2 \cdot W_k + \Sigma(h(x))} \quad \text{mod } 2^{k+1}. \tag{6}$$

Note that we have to invert $2 \cdot W_k + \Sigma(h(x))$ in the dagger ring $\mathcal{A}^\dagger$. Since $h(x) \mid H(x)^D$, we can define $Q_H(x) = H(x)^D/h(x)$, which immediately leads to $1/h(x) = Q_H(x)/H(x)^D$. We can now compute the inverse of $2 \cdot W_k + \Sigma(h(x))$ as

$$\frac{Q_H(x)^2}{H(x)^{2D} \cdot \left(1 + \dfrac{Q_H(x)^2(2W_k + \Sigma(h(x)) - h(x)^2)}{H(x)^{2D}}\right)}.$$

Note that $\Sigma(h(x)) \equiv h(x)^2 \bmod 2$, which implies that the denominator in the above formula is invertible in $\mathcal{A}^\dagger$. Here we are using the fact that $1/H(x)$ is an element of $\mathcal{A}^\dagger$, which explains why we work with $\mathcal{C}'$ instead of with $\mathcal{C}$.

A detailed analysis of the Newton iteration shows that if we write W_k as $\sum_{i=-L_k}^{U_k}(S_i(x)+T_i(x)y)H(x)^i$ for $\deg H(x) > 0$ or $\sum_{i=0}^{U_k}(S_i+T_iy)x^i$ if $H(x) = 1$, with $\operatorname{ord}_2(S_i(x)) < k$, $\operatorname{ord}_2(T_i(x)) < k$ and $L_k, U_k \in \mathbb{N}$, then we get the following bounds for L_k and U_k:

$$L_k \leq 4kD \quad \text{and} \quad U_k \leq 2k\left(\frac{\deg f(x) - 2\deg h(x)}{\deg H(x)}\right) + 2\frac{\deg h(x)}{\deg H(x)}. \tag{7}$$

In [5] we prove that the first Monsky-Washnitzer cohomology group $H^1(A/\mathcal{K})$ splits into eigenspaces under the hyperelliptic involution: a positive eigenspace $H^1(A/\mathcal{K})^+$ with basis $x^i/H(x)\,dx$ for $i = 0, \ldots, s$ and a negative eigenspace $H^1(A/\mathcal{K})^-$ with basis $x^iy\,dx$ for $i = 0, \ldots, 2g - 1$. Note that the positive eigenspace corresponds to the deleted ramification points $(\theta_i, 0)$ for $i = 0, \ldots, s$ and that only the negative eigenspace $H^1(A/\mathcal{K})^-$ is related to the original curve $\mathcal{C}$.

The final step in the algorithm is to compute the action of the Frobenius operator $\mathcal{F} = \Sigma^n$ on the basis of the first Monsky-Washnitzer cohomology group $H^1(A/\mathcal{K})$. However, since only $H^1(A/\mathcal{K})^-$ corresponds to the original curve $\mathcal{C}$, it is sufficient to compute the action of $\mathcal{F}$ on $H^1(A/\mathcal{K})^-$ to recover the zeta function of $\widetilde{C}$. Let $M_{\mathcal{F}}$ be the matrix through which $\mathcal{F}$ acts on $H^1(A/\mathcal{K})^-$, then we can prove [5] that the characteristic polynomial of $M_{\mathcal{F}}$ is precisely the characteristic polynomial $\chi(T)$ of the Frobenius morphism on the hyperelliptic curve $\widetilde{C}$. Let M be the matrix of Σ on $H^1(A/\mathcal{K})^-$, i.e.

$$\Sigma(x^jy\,dx) \equiv \sum_{i=0}^{2g-1} M(i,j)x^iy\,dx \quad \text{for } j = 0, \ldots, 2g - 1,$$

then one easily verifies that $M_{\mathcal{F}} = M\Sigma(M)\cdots\Sigma^{n-1}(M)$.

The only remaining difficulty in computing M is the reduction of $\Sigma(x^jy\,dx)$ on the basis of $H^1(A/\mathcal{K})^-$. Since $\Sigma(x^jy\,dx) = \Sigma(x)^j\Sigma(y)\,d(\Sigma(x))$, we get the following expansion $\Sigma(x^jy\,dx) = 2x^{2j+1}\sum_{i=-\infty}^{\infty}(S_i(x) + T_i(x)y)H(x)^i\,dx$ if $\deg H(x) > 0$ and $\Sigma(x^jy\,dx) = 2x^{2j+1}\sum_{i=0}^{\infty}(S_i + T_iy)x^i\,dx$ if $H(x) = 1$.

For $i \geq 0$ we can reduce the differential form $T_i(x)H(x)^iy\,dx$ (or $T_ix^iy\,dx$ if $H = 1$), if we know how to reduce the form $x^ky\,dx$ for $k \in \mathbb{N}$. Rewriting the equation of the curve as $(2y + h(x))^2 = 4f(x) + h(x)^2$ and differentiating both sides leads to $(2y + h(x)) \, d(2y + h(x)) = (2f'(x) + h(x)h'(x)) \, dx$. Furthermore, for all $l \geq 1$, we have the following relations

$$x^l(2f'(x) + h(x)h'(x))(2y + h(x)) \, dx = x^l(2y + h(x))^2 \, d(2y + h(x))$$
$$\equiv -\frac{1}{3}(2y + h(x))^3 \, dx^l$$
$$= -\frac{l}{3}x^{l-1}(4f(x) + h(x)^2)(2y + h(x)) \, dx.$$

Since $W(x)h(x) \, dx$ is exact for any polynomial $W(x) \in \mathcal{K}[x]$, we finally obtain that

$$\left[x^l(2f'(x) + h(x)h'(x)) + \frac{l}{3}x^{l-1}(4f(x) + h(x)^2) \right] y \, dx \equiv 0.$$

The polynomial between brackets has degree $2g + l$ and its leading coefficient is $2(2g + 1) + 4l/3 \neq 0$. Note that the formula is also valid for $l = 0$. This means that we can reduce $x^k y \, dx$ for any $k \geq 2g$ by subtracting a suitable multiple of the above differential for $l = k - 2g$.

For $i < 0$ we need an extra trick to reduce the form $T_i(x)H(x)^i y \, dx$. Recall that $Q_f(x) = f(x)/H(x)$ and since the curve is non-singular, we conclude that $\gcd(Q_f(x), H(x)) = 1$. Furthermore, $H(x)$ has no repeated roots which implies $\gcd(H(x), Q_f(x)H'(x)) = 1$. We can partially reduce $T_k(x)/H(x)^k y \, dx$ where $k = -i > 0$, by writing $T_k(x)$ as $A_k(x)H(x) + B_k(x)Q_f(x)H'(x)$, which leads to

$$\frac{T_k(x)}{H(x)^k} y \, dx = \frac{A_k(x)}{H(x)^{k-1}} y \, dx + \frac{B_k(x)Q_f(x)H'(x)}{H(x)^k} y \, dx.$$

The latter differential form can be reduced using the following congruence

$$\frac{B_k(x)}{H(x)^k}(2f'(x) + h(x)h'(x))(2y + h(x)) \, dx \equiv -\frac{1}{3}(2y + h(x))^3 d\left(\frac{B_k(x)}{H(x)^k} \right).$$

Substituting the expressions $h(x) = Q_h(x)H(x)$ and $f(x) = Q_f(x)H(x)$ we get

$$\frac{B_k Q_f H'}{H^k} y \, dx \equiv \frac{B_k(kH'Q_h^2 - 6Q'_f - 3Q_h h') - B'_k(4Q_f + Q_h h')}{(6 - 4k)H^{k-1}} y \, dx + \frac{I}{H} \, dx,$$

where $I(x)/H(x) \, dx$ is some invariant differential. However, we can ignore all invariant differentials since we know that $H^1(A/\mathcal{K})^-$ is stable under Σ.

5   Algorithm and Complexity

Using the formulae devised in the previous section, we describe an algorithm to compute the zeta function of a hyperelliptic curve $\widetilde{C}$ of genus g over $\mathbb{F}_q$ with $q = 2^n$. Theorem 1 implies that it is sufficient to compute the characteristic polynomial $\chi(T)$ of Frobenius and that $\chi(T)$ can be written as

$$\chi(T) = \prod_{i=1}^{2g}(T - \omega_i) = T^{2g} + a_1 T^{2g-1} + \cdots + a_{2g},$$

Algorithm 1 (Hyperelliptic_Zeta_Function)

IN: *Hyperelliptic curve C over $\mathbb{F}_q$ given by equation $y^2 + \overline{h}(x)y = \overline{f}(x)$.*
OUT: *The zeta function $Z(\widetilde{C}/\mathbb{F}_q; T)$.*

---

1. $B = \left\lceil \log_2 \left(\binom{2g}{g} q^{g/2} \right) \right\rceil + 1;\ N - 3 - \lfloor \log_2(2N \deg f + g) \rfloor \geq B;$

2. $h(x), f(x), H(x), D = \texttt{Lift\_Curve}(\overline{h}, \overline{f});$

3. $\alpha_N, \beta_N = \texttt{Lift\_Frobenius\_y}(h, f, H, D, N);$

4. For $i = 0$ To $2g - 1$ Do

   4.1. $Red_i(x) = \texttt{Reduce\_MW\_Cohomology}(2x^{2i+1}\beta_N, h, f, H, N);$

   4.2. For $j = 0$ To $2g - 1$ Do $M[j][i] = \texttt{Coeff}(Red_i, j);$

5. $M_{\mathcal{F}} = M\Sigma(M) \cdots \Sigma^{n-1}(M) \bmod 2^N;$

6. $\chi(T) = \texttt{Characteristic\_Pol}(M_{\mathcal{F}}) \bmod 2^B;$

7. For $i = 0$ To g Do

   7.1. If $\texttt{Coeff}(\chi, 2g - i) > \binom{2g}{i} q^{i/2}$ Then $\texttt{Coeff}(\chi, 2g - i) \mathrel{-}= 2^B;$

   7.2. $\texttt{Coeff}(\chi, i) = q^{g-i}\, \texttt{Coeff}(\chi, 2g - i);$

8. Return $Z(\widetilde{C}/\mathbb{F}_q; T) = \dfrac{T^{2g}\chi(1/T)}{(1 - T)(1 - qT)}.$

---

with $\omega_i \cdot \omega_j = q$ for $i = 1, \ldots, g$, $|\omega_i| = \sqrt{q}$ and $a_i \in \mathbb{Z}$ for $i = 1, \ldots, 2g$. Since $q^{g-i}a_i = a_{2g-i}$ for $i = 0, \ldots, g$, it suffices to compute $a_1, \ldots, a_g$. Moreover, a_i is the sum of $\binom{2g}{i}$ products of i different zeros of the characteristic polynomial of Frobenius, so we can bound the a_i for $i = 1, \ldots, g$ by

$$|a_i| \leq \binom{2g}{i} q^{i/2} \leq \binom{2g}{g} q^{g/2} \leq 2^{2g} q^{g/2}.$$

Hence, to recover all the coefficients $a_1, \ldots, a_{2g}$, we need to compute an approximation of the characteristic polynomial $\chi(T)$ modulo 2^B, with

$$B \geq \left\lceil \log_2 \left(\binom{2g}{g} q^{g/2} \right) \right\rceil + 1.$$

However, it is not sufficient to compute $\Sigma(y) \bmod 2^B$, since the reduction process causes some loss of precision. In [5] we prove that for $i \in \mathbb{Z}$ the valuation of the denominators introduced during the reduction of $T_i(x)H(x)^i y\, dx$ is bounded by $c_1 = 3 + \lfloor \log_2(-i + 1) \rfloor$ for $i < 0$ and $c_2 = 3 + \lfloor \log_2((i + 1) \cdot \deg H + g + 1) \rfloor$ for $i \geq 0$. Combining this with the rate of convergence (7) of the Newton iteration for computing $\Sigma(y)$, we conclude that it is sufficient to compute $\Sigma(y) \bmod 2^N$

Algorithm 2 (Lift_Frobenius_y)

IN: Curve $\mathcal{C} : y^2 + h(x)y = f(x)$ over $\mathcal{K}$, polynomial $H(x) \in \mathcal{R}[x]$ with
   $H|h$ and $H|f$, $D \in \mathbb{N}$ such that $h|H^D$ and precision N.

OUT: Series $\alpha_N, \beta_N \in \mathcal{R}[x, H, H^{-1}]$ with $\Sigma(y) \equiv \alpha_N + \beta_N y \mod 2^N$.

---

1. $B = \lceil \log_2 N \rceil + 1;\ T = N;\ Q_H := H^D$ div h;

2. For $i = B$ Down To 1 Do $P[i] = T;\ T = \lceil T/2 \rceil$;

3. $\alpha \equiv f \mod 2;\ \beta \equiv -h \mod 2;\ \gamma = 1;\ \delta = 0$;

4. For $i = 2$ To B Do

   4.1. $T_A \equiv ((\alpha + \Sigma(h)) \cdot \alpha + \beta^2 \cdot f - \Sigma(f)) \cdot Q_H^2 \cdot H^{-2D} \mod 2^{P[i]}$;

   4.2. $T_B \equiv (2\alpha - h \cdot \beta + \Sigma(h)) \cdot \beta \cdot Q_H^2 \cdot H^{-2D} \mod 2^{P[i]}$;

   4.3. $D_A \equiv 1 + (\Sigma(h) - h^2 + 2\alpha) \cdot Q_H^2 \cdot H^{-2D} \mod 2^{P[i-1]}$;

   4.4. $D_B \equiv 2\beta \cdot Q_H^2 \cdot H^{-2D} \mod 2^{P[i-1]}$;

   4.5. $V_A \equiv D_A \cdot \gamma + D_B \cdot \delta \cdot f - 1 \mod 2^{P[i-1]}$;

   4.6. $V_B \equiv D_A \cdot \delta + D_B \cdot (\gamma - \delta \cdot h) \mod 2^{P[i-1]}$;

   4.7. $\gamma \equiv \gamma - (V_A \cdot \gamma + V_B \cdot \delta \cdot f) \mod 2^{P[i-1]}$;

   4.8. $\delta \equiv \delta - (V_A \cdot \delta + V_B \cdot (\gamma - \delta \cdot h)) \mod 2^{P[i-1]}$;

   4.9. $\alpha \equiv \alpha - (T_A \cdot \gamma + T_B \cdot \delta \cdot f) \mod 2^{P[i]}$;

   4.10. $\beta \equiv \beta - (T_A \cdot \delta + T_B \cdot (\gamma - \delta \cdot h)) \mod 2^{P[i]}$;

5. Return $\alpha_N = \alpha,\ \beta_N = \beta$.

---

where $N \in \mathbb{N}$ satisfies

$$N - 3 - \lfloor \log_2(2N \deg f + g) \rfloor \geq B.$$

The function Hyperelliptic_Zeta_Function given in Algorithm 1 computes the zeta function of a hyperelliptic curve C defined over $\mathbb{F}_q$ where $q = 2^n$. In step 2 we call the subroutine Lift_Curve, which first constructs an isomorphic curve such that $\overline{H}(x) \mid \overline{h}(x)$ and $\overline{H}(x) \mid \overline{f}(x)$ and lifts the curve following the construction described in Section 4. The result of this function is a hyperelliptic curve $\mathcal{C} : y^2 + h(x)y = f(x)$ over $\mathcal{R}$ and a polynomial $H(x)$ such that $H(x) \mid h(x)$, $H(x) \mid f(x)$ and $h(x) \mid H(x)^D$. Since this function is rather straightforward, we have omitted the pseudo-code.

In step 3 we compute $\Sigma(y) \mod 2^N$ using the function Lift_Frobenius_y given in Algorithm 2. This function implements the Newton iteration (6), but has quadratic, instead of linear, convergence. The parameters α_N, β_N are Laurent series in $H(x)$, with coefficients polynomials over $\mathcal{R} \mod 2^N$ of degree smaller than $\deg H(x) > 0$. If $H(x) = 1$, then α_N, β_N are Laurent series in x.

Algorithm 3 (Reduce_MW_Cohomology)

IN: Series $G \in \mathcal{R}[x, H, H^{-1}]$, polynomials $h, f, H \in \mathcal{R}[x]$ and precision N.
OUT: $R \in \mathcal{K}[x]$, with $\deg R < 2g$ such that $Ry \, dx \equiv Gy \, dx$ in $H^1(\overline{A}/K)^-$.

---

1. $D_G = \deg G$; $V_G = \texttt{Valuation}(G)$; $D_T = (D_G + 1) \cdot \deg H$; $T = 0$;

2. For $i = D_G$ Down To 0 Do $T = T \cdot H + \texttt{Coeff}(G, i) \mod 2^N$;

3. For $i = D_T$ Down To $2g$

    3.1. $P \equiv x^{i-2g}(2f' + h \cdot h') + \frac{i-2g}{3} x^{i-2g-1}(4f + h^2) \mod 2^N$;

    3.2. $T \equiv T - (\texttt{Coeff}(T, i) \cdot P) / \texttt{Coeff}(P, i) \mod 2^N$;

4. $Q_f = f$ div H; $Q_h = h$ div H; $P = 0$;

5. For $i = V_G$ To -1

    5.1. $V \equiv P + \texttt{Coeff}(G, i) \mod 2^N$;

    5.2. $P \equiv V$ div $H \mod 2^N$; $V \equiv V - P \cdot H \mod 2^N$;

    5.3. $C, L_A, L_B = \texttt{XGCD}(V \cdot H, V \cdot Q_f \cdot H', N)$;

    5.4. $P \equiv P + L_A + \frac{L_B \cdot (-iQ_h^2 \cdot H' - 3(2Q'_f + Q_h \cdot h')) - L'_B \cdot (4Q_f + Q_h h)}{6 + 4i} \mod 2^N$;

6. **Return** $R \equiv T + P \mod 2^N$.

Note that the function Lift_Frobenius_y is a double Newton iteration: $\alpha + \beta y$ converges to $\Sigma(y)$, whereas $\gamma + \delta y$ is an approximation of the inverse of the denominator in the Newton iteration.

Once we have determined an approximation of $\Sigma(y)$, we compute the action of Σ on the basis of $H^1(A/\mathcal{K})^-$ as $2x^{2i+1}\Sigma(y) \, dx$ for $i = 0, \ldots, 2g - 1$. In step 4 we reduce these forms with the function Reduce_MW_Cohomology given in Algorithm 3. Note that this algorithm is based on the reduction formulae given in Section 4. The result of step 4 of Algorithm 1 is an approximation modulo 2^B of the matrix M through which Σ acts on $H^1(A/\mathcal{K})^-$. In step 5 we compute its norm $M_{\mathcal{F}}$ as $M\Sigma(M) \cdots \Sigma^{n-1}(M)$. Note that since M is not necessarily defined over $\mathcal{R}$, we have to compute this product with slightly increased precision to obtain the correct result. In steps 6 and 7 we recover the characteristic polynomial of Frobenius from the first g coefficients of the characteristic polynomial of $M_{\mathcal{F}}$. Finally, we return the zeta function of the smooth projective hyperelliptic curve $\widetilde{C}$ birational to C in Step 8.

The complexity analysis of the algorithm is similar to Kedlaya's algorithm in [14, Section 5], except that in our case the reduction takes $O(g^{5+\varepsilon}n^{3+\varepsilon})$ time instead of $O(g^{4+\varepsilon}n^{3+\varepsilon})$ time. A detailed complexity analysis can be found in [5], which proves that the zeta function of a genus g hyperelliptic curve C over a finite field $\mathbb{F}_q$ with $q = 2^n$ elements, can be computed deterministically in $O(g^{5+\varepsilon}n^{3+\varepsilon})$ bit operations with space complexity $O(g^3 n^3)$.

6   Implementation and Numerical Results

In this section we present running times of an implementation of Algorithm 1 in the C programming language and give some examples of Jacobians of hyperelliptic curves with almost prime group order.

The basic operations on integers modulo 2^N where $N \leq 256$ were written in assembly language. Elements of $\mathcal{R}$ mod 2^N are represented as polynomials over $\mathbb{Z}/(2^N\mathbb{Z})$ modulo a degree n irreducible polynomial, which we chose to be either a trinomial or a pentanomial. For multiplication of elements in $\mathcal{R}_N := \mathcal{R}$ mod 2^N, polynomials over $\mathcal{R}_N$ and Laurent series over $\mathcal{R}_N[x]$ we used Karatsuba's algorithm. In the near future, we plan to implement Toom's algorithm which will lead to a further speed-up of about 50%.

6.1   Running Times and Memory Usage

Table 1 contains running times and memory usages of our algorithm for genus 2, 3 and 4 hyperelliptic curves over various finite fields of characteristic 2. These results were obtained on an AMD XP 1700+ processor running Linux Redhat 7.1. Note that the fields are chosen such that $g \cdot n$, and therefore the bit size of the group order of the Jacobian, is constant across each row.

Table 1. Running times (s) and memory usage (MB) for genus 2, 3 and 4 hyperelliptic curves over $\mathbb{F}_{2^n}$

Size of Jacobian	Genus 2 curves		Genus 3 curves		Genus 4 curves	
$g \cdot n$	Time (s)	Mem (MB)	Time (s)	Mem (MB)	Time (s)	Mem (MB)
120	30	4.5	38	5.4	35	5.2
144	44	5.7	61	7.3	59	7.2
168	71	8.6	101	11	100	11
192	116	13	143	14	139	13
216	170	16	196	17	185	16

6.2   Hyperelliptic Curve Examples

In this subsection we give two examples of Jacobians of hyperelliptic curves with almost prime group order. The correctness of these results is easily proved by multiplying a random divisor with the given group order and verifying that the result is principal, i.e. is the zero element in the Jacobian $J_{\widetilde{C}}(\mathbb{F}_q)$.

It is clear that both curves are non-supersingular: for the genus 2 curve note that a_2 is odd and for the genus 3 curve this is trivial since there are no hyperelliptic supersingular curves of genus 3 in characteristic 2 [31]. Furthermore, both curves withstand the MOV-FR attack [8, 22].

Genus 2 Hyperelliptic Curve over $\mathbb{F}_{2^{83}}$

Let $\mathbb{F}_{2^{83}}$ be defined as $\mathbb{F}_2[t]/\overline{P}(t)$ with $\overline{P}(t) = t^{83} + t^7 + t^4 + t^2 + 1$ and consider the random hyperelliptic curve C_2 of genus 2 defined by

$$y^2 + \left(\sum_{i=0}^{2} h_i x^i\right) y = x^5 + \sum_{i=0}^{4} f_i x^i,$$

where

$$h_0 = \text{7FF29B08993336B479CD2} \qquad h_1 = \text{32C101713C722F8FB5BC9}$$
$$h_2 = \text{553E16B6A3BC6B2432CA8}$$
$$f_0 = \text{7AD44882C02B9743CD58B} \qquad f_1 = \text{327254FA330B44958262A}$$
$$f_2 = \text{204AB23E12828D061AF04} \qquad f_3 = \text{1C827250FFDEFF93B43BE}$$
$$f_4 = \text{13D80106C0E5571DFD139} \ .$$

The group order of the Jacobian $J_{\widetilde{C}_2}$ of C_2 over $\mathbb{F}_{2^{83}}$ is

$$\#J_{\widetilde{C}_2} = 2 \cdot 46768052394566313810931349196246034325781246483037,$$

where the last factor is prime. The coefficients a_1 and a_2 of the characteristic polynomial of Frobenius $\chi(T) = T^4 + a_1 T^3 + a_2 T^2 + a_3 T + a_4$ are given by

$$a_1 = -4669345964042 \quad \text{and} \quad a_2 = 1898390351338398664676787.$$

Genus 3 Hyperelliptic Curve over $\mathbb{F}_{2^{59}}$

Let $\mathbb{F}_{2^{59}}$ be defined as $\mathbb{F}_2[t]/\overline{P}(t)$ with $\overline{P}(t) = t^{59} + t^7 + t^4 + t^2 + 1$ and consider the random hyperelliptic curve C_3 of genus 3 defined by

$$y^2 + \left(\sum_{i=0}^{3} h_i x^i\right) y = x^7 + \sum_{i=0}^{6} f_i x^i,$$

where

$$h_0 = \text{569121E97EB3821} \qquad h_1 = \text{49F340F25EA38A2}$$
$$h_2 = \text{2DE854D48D56154} \qquad h_3 = \text{0B6372FF7310443}$$
$$f_0 = \text{1104FDBEB454C58} \qquad f_1 = \text{0C426890E5C7481}$$
$$f_2 = \text{34967E2EB7D50C3} \qquad f_3 = \text{1F1728AA28C616C}$$
$$f_4 = \text{1AE177BFE49826A} \qquad f_5 = \text{3895A0E400F7D18}$$
$$f_6 = \text{6DF634A1E2BFA8E} \ .$$

The group order of the Jacobian $J_{\widetilde{C}_3}$ of C_3 over $\mathbb{F}_{2^{59}}$ is

$$\#J_{\widetilde{C}_3} = 2 \cdot 95780971407243394633762332360123160334059170481903949,$$

where the last factor is prime. The coefficients a_1, a_2 and a_3 of the characteristic polynomial of Frobenius $\chi(T) = T^6 + a_1 T^5 + a_2 T^4 + a_3 T^3 + a_4 T^2 + a_5 T + a_6$ are given by

$$a_1 = 620663068,$$
$$a_2 = 848092512078818380,$$
$$a_3 = 341008017371409573053936945 \ .$$

7 Conclusion

We have presented an extension of Kedlaya's algorithm for computing the zeta function of an arbitrary hyperelliptic curve C over a finite field $\mathbb{F}_q$ of characteristic 2. As a byproduct we obtain the group order of the Jacobian $J_{\tilde{C}}(\mathbb{F}_q)$ associated to C which forms the basis of the cryptographic schemes based on hyperelliptic curves. The resulting algorithm runs in $O(g^{5+\varepsilon}n^{3+\varepsilon})$ bit operations and needs $O(g^3n^3)$ storage space for a genus g hyperelliptic curve over $\mathbb{F}_{2^n}$. A first implementation of this algorithm in the C programming language shows that cryptographical sizes are now feasible for any genus g. Computing the order of a 160-bit Jacobian of a hyperelliptic curve of genus 2, 3 or 4 takes less than 100 seconds. In the near future we plan to use the formalism of Monsky-Washnitzer cohomology as a basis for computing the zeta function of any non-singular affine curve over finite fields of small characteristic.

References

1. S. Arita. Algorithms for computations in Jacobians of C_{ab} curve and their application to discrete-log-based public key cryptosystems. In *Proceedings of Conference on The Mathematics of Public Key Cryptography*, Toronto, June 12-17 1999.
2. A.O.L. Atkin. The number of points on an elliptic curve modulo a prime. *Series of e-mails to the* NMBRTHRY *mailing list*, 1992.
3. I.F. Blake, G. Seroussi, and N.P. Smart. *Elliptic curves in cryptography*. London Mathematical Society Lecture Note Series. 265. Cambridge University Press., 1999.
4. J. Denef and F. Vercauteren. An extension of Kedlaya's algorithm to Artin-Schreier curves in characteristic 2. *Algorithmic number theory. 5th international symposium. ANTS-V*, 2002.
5. J. Denef and F. Vercauteren. An extension of Kedlaya's algorithm to hyperelliptic curves in characteristic 2. *Preprint*, 2002.
6. N. Elkies. Elliptic and modular curves over finite fields and related computational issues. *Computational Perspectives on Number Theory*, pages 21–76, 1998.
7. M. Fouquet, P. Gaudry, and R. Harley. On Satoh's algorithm and its implementation. *J. Ramanujan Math. Soc.*, 15:281–318, 2000.
8. G. Frey and H.-G. Rück. A remark concerning m-divisibility and the discrete logarithm in the divisor class group of curves. *Math. Comp.*, 62(206):865–874, April 1994.
9. W. Fulton. *Algebraic curves*. Math. Lec. Note Series. W.A. Benjamin Inc., 1969.
10. S. Galbraith, S. Paulus, and N. Smart. Arithmetic on superelliptic curves. *Math. Comp.*, 71(237):393–405, 2002.
11. P. Gaudry and N. Gürel. An extension of Kedlaya's algorithm for counting points on superelliptic curves. In *Advances in Cryptology - ASIACRYPT 2001*, volume 2248 of *Lecture Notes in Computer Science*, pages 480–494, 2001.
12. P. Gaudry and R. Harley. Counting points on hyperelliptic curves over finite fields. In Wieb Bosma, editor, *Algorithmic number theory. 4th international symposium. ANTS-IV*, volume 1838 of *Lecture Notes in Computer Science*, pages 313–332, 2000.
13. R. Hartshorne. *Algebraic geometry*. Number 52 in Graduate Texts in Mathematics. Springer-Verlag, 1997.

14. K.S. Kedlaya. Counting points on hyperelliptic curves using Monsky-Washnitzer cohomology. *Journal of the Ramanujan Mathematical Society*, 16:323–338, 2001.
15. N. Koblitz. *p-adic Numbers, p-adic Analysis and Zeta-Functions*, volume 58 of *GTM*. Springer-Verlag, 1977.
16. N. Koblitz. Elliptic curve cryptosystems. *Math. Comput.*, 48:203–209, 1987.
17. N. Koblitz. Hyperelliptic cryptosystems. *J. Cryptology*, 1(3):139–150, 1989.
18. A.G.B. Lauder and D. Wan. Counting points on varieties over finite fields of small characteristic. Preprint 2001.
19. A.G.B. Lauder and D. Wan. Computing zeta functions of Artin-Schreier curves over finite fields. *London Mathematical Society JCM*, 5:34–55, 2002.
20. R. Lercier. *Algorithmique des courbes elliptiques dans les corps finis*. PhD thesis, Laboratoire d'Informatique de l'École polytechnique (LIX), 1997. Available at http://ultralix.polytechnique.fr/~lercier.
21. D. Lorenzini. *An Invitation to Arithmetic Geometry*. Number 9 in Graduate Studies in Mathematics. American Mathematical Society, 1996.
22. A. Menezes, T. Okamoto, and S. Vanstone. Reducing elliptic curve logarithms to logarithms in a finite field. In *Proceedings of the twenty third annual ACM Symposium on Theory of Computing, New Orleans, Louisiana, May 6–8, 1991*, pages 80–89, 1991.
23. V. Miller. Uses of elliptic curves in cryptography. *Advances in Cryptology - ASIACRYPT '91, Lecture notes in Computer Science*, 218:460–469, 1993.
24. P. Monsky. Formal cohomology. II: The cohomology sequence of a pair. *Ann. of Math.*, 88:218–238, 1968.
25. P. Monsky. *p-adic analysis and zeta functions*. Lectures in Mathematics, Department of Mathematics Kyoto University. 4. Tokyo, Japan, 1970.
26. P. Monsky. Formal cohomology. III: Fixed point theorems. *Ann. of Math.*, 93:315–343, 1971.
27. P. Monsky and G. Washnitzer. Formal cohomology. I. *Ann. of Math.*, 88:181–217, 1968.
28. J. Pila. Frobenius maps of abelian varieties and finding roots of unity in finite fields. *Math. Comput.*, 55(192):745–763, 1990.
29. T. Satoh. The canonical lift of an ordinary elliptic curve over a finite field and its point counting. *J. Ramanujan Math. Soc.*, 15:247–270, 2000.
30. T. Satoh, B. Skjernaa, and Y. Taguchi. Fast computation of canonical lifts of elliptic curves and its application to point counting. *Preprint*, 2001.
31. J. Scholten and J. Zhu. Hyperelliptic curves in characteristic 2. *International Mathematics Research Notices*, 2002(17):905–917, 2002.
32. R. Schoof. Elliptic curves over finite fields and the computation of square roots mod p. *Math. Comp.*, 44:483–494, 1985.
33. B. Skjernaa. Satoh's algorithm in characteristic 2. *To appear in Math. Comp.*, 2000.
34. M. van der Put. The cohomology of Monsky and Washnitzer. *Mém. Soc. Math. France*, 23:33–60, 1986.
35. F. Vercauteren, B. Preneel, and J. Vandewalle. A memory efficient version of Satoh's algorithm. In *Advances in Cryptology - EUROCRYPT 2001*, volume 2045 of *Lecture Notes in Computer Science*, pages 1–13, 2001.
36. D. Wan. Computing zeta functions over finite fields. *Contemporary Mathematics*, 225:131–141, 1999.
37. A. Weil. *Sur les Courbes Algébriques et les Variétés qui s'en Déduisent*. Hermann, 1948.

Threshold Password-Authenticated Key Exchange

(Extended Abstract)

Philip MacKenzie[1], Thomas Shrimpton[2], and Markus Jakobsson[3]

[1] Bell Laboratories
Lucent Technologies
Murray Hill, NJ 07974 USA
philmac@lucent.com

[2] Dept. of Electrical and Computer Engineering
UC Davis
Davis, CA 95616 USA
teshrim@ucdavis.edu

[3] RSA Laboratories
RSA Security, Inc.
Bedford, MA 01730 USA
mjakobsson@rsasecurity.com

Abstract. In most password-authenticated key exchange systems there is a single server storing password verification data. To provide some resilience against server compromise, this data typically takes the form of a one-way function of the password (and possibly a salt, or other public values), rather than the password itself. However, if the server is compromised, this password verification data can be used to perform an offline dictionary attack on the user's password. In this paper we propose an efficient password-authenticated key exchange system involving a set of servers, in which a certain threshold of servers must participate in the authentication of a user, and in which the compromise of any fewer than that threshold of servers does not allow an attacker to perform an offline dictionary attack. We prove our system is secure in the random oracle model under the Decision Diffie-Hellman assumption against an attacker that may eavesdrop on, insert, delete, or modify messages between the user and servers, and that compromises fewer than that threshold of servers.

1 Introduction

Many real-world systems today rely on password authentication to verify the identity of a user before allowing that user to perform certain functions, such as setting up a virtual private network or downloading secret information. There are many security concerns associated with password authentication, due mainly to the fact that most users' passwords are drawn from a relatively small and easily generated dictionary. Thus if information sufficient to verify a password guess is leaked, the password may be found by performing an offline dictionary attack:

M. Yung (Ed.): CRYPTO 2002, LNCS 2442, pp. 385–400, 2002.
© Springer-Verlag Berlin Heidelberg 2002

one can run through a dictionary of possible passwords, testing each one against the leaked information in order to determine the correct password.

When password authentication is performed over a network, one must be especially careful not to allow any leakage of information to one listening in, or even actively attacking, the network. If one assumes the server's public key is known (or at least can be verified) by the user, then performing password authentication after setting up an anonymous secure channel to the server is generally sufficient to prevent leakage of information, as is done in SSH [28] or on the web using SSL [13]. The problem becomes more difficult if the server's public key cannot be verified by the user. Solutions to this problem have been coined *strong password authentication protocols*, and have the property that (informally) the probability of an active attacker (i.e., one that may eavesdrop on, insert, delete, or modify messages on a network) impersonating a user is only negligibly better than a simple on-line guessing attack, consisting of the attacker iteratively guessing passwords and running the authentication protocol. Strong password authentication protocols were proposed by Jablon [23] and Wu [29], among others. Recently, some protocols were proven secure in the random oracle model[1] (Bellare *et al.* [1], Boyko *et al.* [8] and MacKenzie *et al.* [26]), in the public random string model (Katz *et al.* [25]), and in the standard model (Goldreich and Lindell [21]). However, all of these protocols, even the ones in which the server's public key is known to the user, are vulnerable to server compromise in the sense that compromising the server would allow an attacker to obtain the password verification data on that server (typically some type of one-way function of the password and some public values). This could then be used to perform an offline dictionary attack on the password. To address this issue (without resorting to assumptions like tamper resistance), Ford and Kaliski [17] proposed to distribute the functionality of the server, forcing an attacker to compromise several servers in order to be able to obtain password verification data. Note that the main problem is not just to distribute the password verification data, but to distribute the functionality, i.e., to distribute the password verification data such that it can be used for authentication without ever reconstructing the data on any set of servers smaller than a chosen threshold.

While distributed cryptosystems have been studied extensively (and many proven secure) for other cryptographic operations, such as signatures (e.g., [7, 11, 20, 18]), to our knowledge Ford and Kaliski were the first ones to propose a distributed password-authenticated key exchange system. However, they give no proof of security for their system. Jablon [24] extends the system of Ford and Kaliski, most notably to not require the server's public key to be known to the user, but again does not give a proof of security.

---

[1] In the random oracle model [2], a hash function is modeled as a black box containing an ideal random function. This is not a standard cryptographic assumption. In fact, it is possible for a scheme secure in the random oracle model to be insecure for any real instantiation of the hash function [9]. However, a proof of security in the random oracle model is generally thought to be strong evidence of the practical security of a scheme.

Our contributions. In this paper we propose a completely different distributed password authenticated key exchange system and prove it secure in the random oracle model, assuming the hardness of the Decision Diffie-Hellman (DDH) problem [14] (see [6]). While the system of Ford and Kaliski and the system of Jablon require all servers to perform authentication, our system is a k-out-of-n threshold system (for any $1 \leq k \leq n$), where k servers are required for authentication and the compromise of $k - 1$ servers does not affect the security of the system. This is the first distributed password-authenticated key exchange system proven secure under any standard cryptographic assumption in any model, including the random oracle model. To be specific, we assume the client may store public data, and our security is against an active attacker that may (statically) compromise any number of servers less than the specified threshold.

Technically, we achieve our result by storing a semantically-secure encryption of a function of the password at the servers (instead of simply storing a one-way function of the password), and then leveraging off some known solutions for distributing secret decryption keys, such as Feldman verifiable secret sharing [16]. In other words, we transform the problem of distributing password authentication information to the problem of distributing cryptographic keys. However, once we make this transformation, verifying passwords without leaking information becomes much more difficult, requiring intricate manipulations of ElGamal encryptions [15] and careful use of efficient non-interactive zero-knowledge proofs [5].

We note that a threshold password authentication system does not follow from techniques for general secure multi-party computation (e.g., [22]) since we are working in an asynchronous model, allow concurrent executions of protocols, and assume no authenticated channels. (Note in particular that the *goal* of the protocol is for the client to be authenticated.) The only work on general secure multi-party computation in an asynchronous model, and allowing concurrency, assumes authenticated channels [10].

2   Model

We extend the model of [1] (which builds on [3] and [4], and is also used by [25]). The model of [1] was designed for the problem of authenticated key exchange (ake) between two parties, a client and a server. The goal was for them to engage in a protocol such that after the protocol was completed, they would each hold a session key that is known to nobody but the two of them. Our model is designed for the problem of *distributed authenticated key exchange* (dake) between a client and k servers. The goal is for them to engage in a protocol such that after the protocol is completed, the client would hold k session keys, one being shared with each server, such that the session key shared between the client and a given server is known to nobody but the two of them, even if up to $k - 1$ other servers were to conspire together.

Note that a secure dake protocol allows for secure downloadable credentials, by, e.g., having the servers store an encrypted credentials file with a decryption key stored using a threshold scheme among them, and then having each send a partial decryption of the credentials file to the client, encrypted with the session key it shares with the client. Note that the credentials are secure in a threshold sense: fewer than the given threshold of servers are unable to obtain the credentials. Details are beyond the scope of this paper.

In the following, we will assume some familiarity with the model of [1].

Protocol participants. We have two types of protocol participants: clients and servers. Let $ID \overset{\text{def}}{=} Clients \cup Servers$ be a non-empty set of protocol participants, or *principals*.

We assume *Servers* consists of n servers, denoted $\{S_1, \ldots, S_n\}$, and that these servers are meant to cooperate in authenticating a client[2]. Each client $C \in Clients$ has a secret password π_C, and each server $S \in Servers$ has a vector $\pi_S = [\pi_S[C]]_{C \in Clients}$. Entry $\pi_S[C]$ is the *password record*. Let $Password_C$ be a (possibly small) set from which passwords for client C are selected. We will assume that $\pi_C \overset{R}{\leftarrow} Password_C$ (but our results easily extend to other password distributions). Clients and servers are modeled as probabilistic poly-time algorithms with an input tape and an output tape.

Execution of the protocol. A protocol P is an algorithm that determines how principals behave in response to inputs from their environment. In the real world, each principal is able to execute P multiple times with different partners, and we model this by allowing unlimited number of *instances* of each principal. Instance i of principal $U \in ID$ is denoted Π_i^U.

To describe the security of the protocol, we assume there is an adversary $\mathcal{A}$ that has complete control over the environment (mainly, the network), and thus provides the inputs to instances of principals. We will further assume the network (i.e., $\mathcal{A}$) performs aggregation and broadcast functions[3]. In practice, on a point-to-point network, the protocol implementor would most likely have to implement these functionalities in some way, perhaps using a single intermediate (untrusted) node to aggregate and broadcast messages[4]. Formally, the adversary is a probabilistic algorithm with a distinguished query tape. Queries written to this tape are responded to by principals according to P; the allowed queries are formally defined in [1] and summarized here (with slight modifications for multiple servers):

Send (U, i, M): causes message M to be sent to instance Π_i^U. The instance computes what the protocol says to, state is updated, and the output of the computation is given to $\mathcal{A}$. If this query causes Π_i^U to accept or terminate, this will also be shown to $\mathcal{A}$. To initiate a session between client C and a set of servers, the adversary should send a message containing a set I of k indices of servers in *Servers* to an unused instance of C.

---

[2] Our model could be extended to have multiple sets of servers, but for clarity of presentation we omit this extension.

[3] This is more for notational convenience than anything else. In particular, we make no assumptions about synchronicity or any type of distributed consensus.

[4] Note that since $\mathcal{A}$ controls the network and can deny service at any time, we do not concern ourselves with any denial-of-service attacks that this single intermediate node may facilitate.

Execute $(C, i, ((S_{j_1}, \ell_{j_1}), \ldots, (S_{j_k}, \ell_{j_k})))$: causes P to be executed to completion between Π_i^C (where $C \in Clients$) and $\Pi_{\ell_{j_1}}^{S_{j_1}}, \ldots, \Pi_{\ell_{j_k}}^{S_{j_k}}$, and outputs the transcript of the execution. This query captures the intuition of a passive adversary who simply eavesdrops on the execution of P.

Reveal (C, i, S_j): causes the output of the session key held by Π_i^C corresponding to server S_j, i.e., sk_{C,S_j}^i.

Reveal (S_j, i): causes the output of the session key held by $\Pi_i^{S_j}$, i.e., $sk_{S_j}^i$.

Test (C, i, S_j): causes Π_i^C to flip a bit b. If $b = 1$ the session key sk_{C,S_j}^i is output and if $b = 0$ a string drawn uniformly from the space of session keys is output. A Test query (of either type) may be asked at any time during the execution of P, but may only be asked once.

Test (S_j, i): causes $\Pi_i^{S_j}$ to flip a bit b. If $b = 1$ the session key $sk_{S_j}^i$ is output; otherwise, a string is drawn uniformly from the space of session keys and output. As above, a Test query (of either type) may be asked at any time during the execution of P, but may only be asked once.

The Reveal queries are used to model an adversary who obtains information on session keys in some sessions, and the Test queries are a technical addition to the model that will allow us to determine if an adversary can distinguish a true session key from a random key.

We assume $\mathcal{A}$ may compromise up to $k - 1$ servers, and that the choice of these servers is static. In particular, without loss of generality, we may assume the choice is made before initialization, and we may simply assume the adversary has access to the private keys of the compromised servers.

Partnering. A server instance that accepts holds a partner-id pid, session-id sid, and a session key sk. A client instance that accepts holds a partner-id pid, a session-id sid, and a set of k session keys $(sk_{j_1}, \ldots, sk_{j_k})$. Let sid be the concatenation of all messages (or pre-specified compacted representations of the messages) sent and received by the client instance in its communication with the set of servers. (Note that this excludes messages that are sent only between servers, but not to the client.) Then instances Π_i^C (with $C \in Clients$) holding $(pid, sid, (sk_{j_1}, \ldots, sk_{j_k}))$ for some set $I = \{j_1, \ldots, j_k\}$ and $\Pi_{\ell_j}^{S_j}$ (with $S_j \in Servers$) holding (pid', sid', sk) are said to be *partnered* if $j \in I$, $pid = S_j$, $pid' = C$, $sid = sid'$, and $sk_j = sk$. This is the so-called "matching conversation" approach to defining partnering, as used in [3, 1].

Freshness. A client instance/server pair (Π_i^C, S_j) is *fresh* if (1) S_j is not compromised, (2) there has been no Reveal (C, i, S_j) query, and (3) if $\Pi_i^{S_j}$ is a partner to Π_i^C, there has been no Reveal (S_j, ℓ) query. A server instance $\Pi_i^{S_j}$ is *fresh* if (1) S_j is not compromised, (2) there has been no Reveal (S_j, i) query, and (3) if Π_ℓ^C is the partner to $\Pi_i^{S_j}$, there has been no Reveal (C, ℓ, S_j) query. Intuitively, the adversary should not be able to distinguish random keys from session keys held by fresh instances.

Advantage of the adversary. We now formally define the distributed authenticated key exchange (dake) advantage of the adversary against protocol P. Let $\mathrm{Succ}_P^{\mathrm{dake}}(\mathcal{A})$ be the event that $\mathcal{A}$ makes a single Test query directed to some instance Π_i^U that has terminated and is fresh, and eventually outputs a bit b', where $b' = b$ for the bit b that was selected in the Test query. The dake advantage of $\mathcal{A}$ attacking P is defined to be

$$\mathrm{Adv}_P^{\mathrm{dake}}(\mathcal{A}) \stackrel{\mathrm{def}}{=} 2\Pr\left[\mathrm{Succ}_P^{\mathrm{dake}}(\mathcal{A})\right] - 1.$$

The following fact is easily verified.

Fact 1.
$$\Pr(\mathrm{Succ}_P^{\mathrm{dake}}(\mathcal{A})) = \Pr(\mathrm{Succ}_{P'}^{\mathrm{dake}}(\mathcal{A})) + \epsilon \quad \Longleftrightarrow \quad \mathrm{Adv}_P^{\mathrm{dake}}(\mathcal{A}) = \mathrm{Adv}_{P'}^{\mathrm{dake}}(\mathcal{A}) + 2\epsilon.$$

3   Definitions

Let κ be the cryptographic security parameter. Let $G_q \in \mathcal{G}$ denote a finite (cyclic) group of order q, where $|q| = \kappa$. Let g be a generator of G_q, and assume it is included in the description of G_q.

Notation. We use $(a,b) \times (c,d)$ to mean elementwise multiplication, i.e., (ac, bd). We use $(a,b)^r$ to mean elementwise exponentiation, i.e., (a^r, b^r). For a tuple V, the notation $V[j]$ means the jth element of V.

We denote by Ω the set of all functions H from $\{0,1\}^*$ to $\{0,1\}^\infty$. This set is provided with a probability measure by saying that a random H from Ω assigns to each $x \in \{0,1\}^*$ a sequence of bits each of which is selected uniformly at random. As shown in [2], this sequence of bits may be used to define the output of H in a specific set, and thus we will assume that we can specify that the output of a random oracle H be interpreted as a (random) element of G_q [5]. Access to any public random oracle $H \in \Omega$ is given to all algorithms; specifically, it is given to the protocol P and the adversary $\mathcal{A}$. Assume that secret session keys are drawn from $\{0,1\}^\kappa$.

A function $f : \mathbb{Z} \to [0,1]$ is negligible if for all $\alpha > 0$ there exists an $\kappa_\alpha > 0$ such that for all $\kappa > \kappa_\alpha$, $f(\kappa) < |\kappa|^{-\alpha}$. We say a multi-input function is negligible if it is negligible with respect to each of its inputs.

4   Protocol

In this section we describe our protocol for threshold password-authenticated key exchange. In the next section we prove this protocol is secure under the DDH assumption [6, 14] in the random-oracle model [2].

---

[5] For instance, this can be easily defined when G_q is a q-order subgroup of $\mathbb{Z}_p^*$, where q and p are prime.

4.1   Server Setup

Let there be n servers $\{S_i\}_{i \in \{1,2,\ldots,n\}}$. Let (x,y) be the servers' *global* key pair such that $y = g^x$. The servers share the global secret key x using a (k,n)-threshold Feldman secret sharing protocol [16]. Specifically, a polynomial $f(z) = \sum_{j=0}^{k-1} a_j z^j \bmod q$ is chosen with $a_0 \leftarrow x$ and random coefficients $a_j \xleftarrow{R} \mathbb{Z}_q$ for $j > 0$. Then each server S_i gets a secret share $x_i = f(i)$ and a corresponding public share $y_i = g^{x_i}$, $1 \leq i \leq n$. (In this paper we assume that a trusted dealer generates these shares, but it should be possible to have the servers generate them using a distributed protocol, as in Gennaro *et al.* [19].) In addition, each server S_i independently generates its own *local* key pair (x_i', y_i') such that $y_i' = g^{x_i'}$, $1 \leq i \leq n$. Each server S_i publishes its *local public key* y_i' along with its share of the global public key y_i. Let $H_0, H_1, H_2, H_3, H_4, H_5, H_6 \xleftarrow{R} \Omega$ be random oracles with domain and range defined by the context of their use. Let $h \leftarrow H_0(y)$ and $h' \leftarrow H_1(y)$ be generators for G_q.

Remark 1. We note that in the following protocol the servers are assumed to have stored the $2n+1$ public values y, $\{y_i'\}_{i=1}^n$, and $\{y_i\}_{i=1}^n$. Likewise, the client is assumed to have stored the $n+1$ public values y and $\{y_i'\}_{i=1}^n$. (Alternatively, a trusted certification authority (CA) could certify these values, but we choose to keep our model as simple as possible.)

4.2   Client Setup

A client $C \in Clients$ has a secret password π_C drawn from a set $Password_C$. We assume $Password_C$ can be mapped into $\mathbb{Z}_q$, and for the remainder of the paper, we use passwords as if they were elements of $\mathbb{Z}_q$. C creates an ElGamal ciphertext E_C of the value $g^{(\pi_C)^{-1}}$, using the servers' global public key y. More precisely, he selects $\alpha \xleftarrow{R} \mathbb{Z}_q$ and computes $E_C \leftarrow (y^\alpha g^{(\pi_C)^{-1}}, g^\alpha)$. He sends E_C to each of the servers S_i, $1 \leq i \leq n$, who record (C, E_C) in their database. (Alternatively, a trusted CA could be used, but again we choose to keep our model as simple as possible.)

Remark 2. We assume the the adversary does not observe or participate in either the system or client setup phases. We assume the client saves a copy of E_C locally. It should be clear that since E_C is public information, this is not the same as storing a shared secret key with the client, which would then obviate the need to use a password for authentication. In particular, it should be noted that instead of storing E_C locally, a client alternatively could obtain a certified copy of E_C through interaction with the servers. Details are beyond the scope of the paper.

4.3   Client Login Protocol

A high level description of the protocol is given in Figure 1, and the formal description may be found in the full paper. Our protocol for a client $C \in Clients$ relies on a simulation-sound non-interactive zero-knowledge proof (SS-NIZKP)

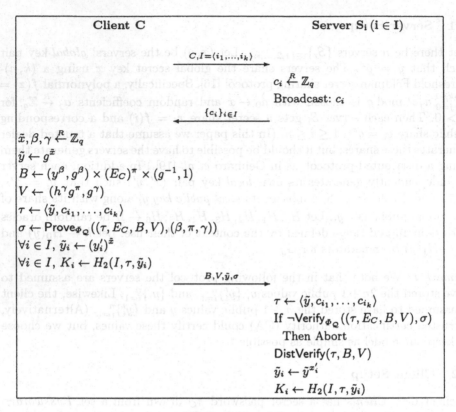

Fig. 1. Protocol P

scheme (see De Santis *et al.* [12] for a definition of an SS-NIZKP scheme) $\mathcal{Q} = (\mathsf{Prove}_{\Phi_{\mathcal{Q}}}, \mathsf{Verify}_{\Phi_{\mathcal{Q}}}, \mathsf{Sim}_{\Phi_{\mathcal{Q}}})$ over a language defined by a predicate $\Phi_{\mathcal{Q}}$ that takes elements of $\{0,1\}^* \times (G_q \times G_q)^3$ and is defined as

$$\Phi_{\mathcal{Q}}(\tau, E_C, B, V) \stackrel{\text{def}}{=}$$
$$\exists \beta, \pi, \gamma : \left(B = \left(y^\beta, g^\beta\right) \times (E_C)^\pi \times (g^{-1}, 1)\right) \text{ and } \left(V = (h^\gamma g^\pi, g^\gamma)\right).$$

The algorithms $\mathsf{Prove}_{\Phi_{\mathcal{Q}}}$, $\mathsf{Verify}_{\Phi_{\mathcal{Q}}}$, and $\mathsf{Sim}_{\Phi_{\mathcal{Q}}}$ use a random oracle H_3. $\mathsf{Prove}_{\Phi_{\mathcal{Q}}}$ may be implemented in a standard way as a three-move honest-verifier proof made non-interactive by using the hash function to generate the verifier's random challenge, and having τ be an extra input to the hash function. Other proofs defined below may be implemented similarly.

Here we discuss Figure 1. The client $C \in$ *Clients* receives a set I of k servers in *Servers* and initiates the protocol with that set, by broadcasting I along with its own identity C. (As stated above, we assume aggregation and broadcast functionalities in the network for the communication between the client and the servers, and among the servers themselves.) In return C receives nonces from the servers in I. The client then "removes" the password from the ciphertext

E_C by raising it to π_C and dividing g out of the first element of the tuple, and reblinds the result to form B. The quantity V is then formed to satisfy the predicate Φ_Q, and an SS-NIZKP σ is created to bind B, V, the session public key $\tilde{y}$, and the nonces from the servers. This SS-NIZKP also forces the client to behave properly, and in particular to allow a simulator in the proof of security to operate correctly. (The idea is similar to the use of a second encryption to achieve (lunchtime) chosen-ciphertext security in [27].) After verifying the SS-NIZKP, if the client has used the password $\pi = \pi_C$, it will be that $B[1] = y^{\beta + \alpha\pi}$ and $B[2] = g^{\beta + \alpha\pi}$. The servers then run DistVerify(τ, B, V) to verify that $\log_g y = \log_{B[2]} B[1]$. Effectively, they are verifying (without decryption) that B is a valid encryption of the plaintext message 1. Each server S_i then computes a session key K_i, which has also been computed by the client.

Efficiency For the following calculations we use the proof constructions of Figures 3 through 7. Recall that there are k servers involved in the execution of the protocol. The protocol requires six rounds, where each round is an exchange of messages among some of the participants. All messages are of length proportional to the size of a group element. The client is involved in only the first three rounds, while the servers are involved in all rounds. The client performs $15 + k$ exponentiations, and each server performs $22 + 34k$ exponentiations.

Remark 3. These costs are obviously much higher than the Ford-Kaliski scheme, but remember that our protocol is the first to achieve *provable* security (in the random oracle model). Also, the costs may be reasonable for practical implementations with k in the range of 2 to 5.

Remark 4. Our protocol does not provide forward security. To achieve forward security, each server S_i would need to generate its Diffie-Hellman values dynamically, instead of simply using y_i'. Then these values would need to be certified somehow by S_i to protect the client against a man-in-the-middle attack. Details are beyond the scope of this paper.

4.4   The DistVerify Protocol

The DistVerify protocol takes three parameters, τ, B, and V, and is run by the servers $\{S_i\}_{i \in I}$ to verify that $\log_g y = \log_{B[2]} B[1]$, i.e., B is an encryption of 1. The parameter V is used in order to allow a proof of security. The protocol is shown in Figure 2, and uses the standard notation for Lagrange coefficients: $\lambda_{j,I} = \prod_{\ell \in I \setminus \{j\}} \frac{-\ell}{j-\ell} \bmod q$. The basic idea of the protocol is as follows. First the servers distributively compute B^r, thus using the (standard) technique of randomizing the quotient $B[1]/(B[2])^x$ if it is not equal to 1. Then they take the second component (i.e., $(B[2])^r$) and distributively compute $((B[2])^r)^x$ using their shared secrets. Finally they verify that $((B[2])^r)^x = (B[1])^r$, implying $B[1] = (B[2])^x$, and hence B is an encryption of 1. DistVerify uses an SS-NIZKP scheme $\mathcal{R} = (\text{Prove}_{\Phi_R}, \text{Verify}_{\Phi_R}, \text{Sim}_{\Phi_R})$ over a language defined by a predicate Φ_R that takes elements of $\mathbb{Z} \times (G_q \times G_q)^6$ and is defined as

$$\boxed{\begin{aligned}
&\textbf{Step 1: } r_i, r_i', \gamma_i, \gamma_i', \gamma_i'' \xleftarrow{R} \mathbb{Z}_q \\
&\qquad B_i \leftarrow B^{r_i} \times (y, g)^{r_i'} \quad V_i \leftarrow (h^{\gamma_i} g^{r_i}, g^{\gamma_i}) \\
&\qquad V_i' \leftarrow (h^{\gamma_i'}(V[1])^{r_i}, g^{\gamma_i'}) \quad V_i'' \leftarrow (h^{\gamma_i''}(V[2])^{r_i}, g^{\gamma_i''}) \\
&\qquad \sigma_i \leftarrow \mathsf{Prove}_{\Phi_{\mathcal{R}}}((i, B, V, B_i, V_i, V_i', V_i''), (r_i, r_i', \gamma_i, \gamma_i', \gamma_i'')) \\
&\qquad \text{Broadcast } (B_i, V_i, V_i', V_i'', \sigma_i) \\[4pt]
&\textbf{Step 2: } \forall j \in I \setminus \{i\} : \text{Receive } (B_j, V_j, V_j', V_j'', \sigma_j) \\
&\qquad \forall j \in I \setminus \{i\} : \text{If } \neg\mathsf{Verify}_{\Phi_{\mathcal{R}}}((j, B, V, B_j, V_j, V_j', V_j''), \sigma_j) \text{ Then Abort} \\
&\qquad (\overline{y}, \overline{g}) \leftarrow \prod_{j \in I} B_j \\
&\qquad \tau' \leftarrow \langle \tau, B, V, B_{i_1}, \ldots, B_{i_k} \rangle \\
&\qquad a_i \leftarrow \lambda_{i,I} x_i \quad \overline{C}_i \leftarrow \overline{g}^{a_i} \quad R_i \leftarrow (h^\zeta (h')^{a_i}, g^\zeta) \\
&\qquad \forall j \in I : C_j \leftarrow (y_j)^{\lambda_{j,I}} \\
&\qquad \Gamma_i \leftarrow \mathsf{Prove}_{\Phi_S}((i, \tau', C_i, R_i), a_i) \\
&\qquad \text{Broadcast } (R_i, \Gamma_i) \\[4pt]
&\textbf{Step 3: } \forall j \in I \setminus \{i\} : \text{Receive } (R_j, \Gamma_j) \\
&\qquad \forall j \in I \setminus \{i\} : \text{If } \neg\mathsf{Verify}_{\Phi_S}((j, \tau', C_j, R_j), \Gamma_j) \text{ Then Abort} \\
&\qquad \Gamma_i' \leftarrow \mathsf{Prove}_{\Phi_{\mathcal{T}}}((i, \tau', \overline{g}, \overline{C}_i, C_i, R_i), a_i) \\
&\qquad \text{Broadcast } (\overline{C}_i, \Gamma_i') \\[4pt]
&\textbf{Step 4: } \forall j \in I \setminus \{i\} : \text{Receive } (\overline{C}_j, \Gamma_j') \\
&\qquad \forall j \in I \setminus \{i\} : \text{If } \neg\mathsf{Verify}_{\Phi_{\mathcal{T}}}((j, \tau', \overline{g}, \overline{C}_j, C_j, R_j), \Gamma_j') \text{ Then Abort} \\
&\qquad \text{If } \Pi_{j \in I} \overline{C}_j \neq \overline{y} \text{ Then Abort}
\end{aligned}}$$

Fig. 2. Protocol DistVerify(τ, B, V) for Server S_i ($i \in I$)

$$\begin{aligned}
\Phi_{\mathcal{R}}(i, B, V, B_i, V_i, V_i', V_i'') &\stackrel{\text{def}}{=} \exists r_i, r_i', \gamma_i, \gamma_i', \gamma_i'' : B_i = B^{r_i} \times (y, g)^{r_i'} \text{ and} \\
&\quad V_i = (h^{\gamma_i} g^{r_i}, g^{\gamma_i}) \text{ and } V_i' = (h^{\gamma_i'}(V[1])^{r_i}, g^{\gamma_i'}) \text{ and} \\
&\quad V_i'' = (h^{\gamma_i''}(V[2])^{r_i}, g^{\gamma_i''}).
\end{aligned}$$

The algorithms $\mathsf{Prove}_{\Phi_{\mathcal{R}}}$, $\mathsf{Verify}_{\Phi_{\mathcal{R}}}$, and $\mathsf{Sim}_{\Phi_{\mathcal{R}}}$ use a random oracle H_4.

DistVerify also uses an SS-NIZKP scheme $S = (\mathsf{Prove}_{\Phi_S}, \mathsf{Verify}_{\Phi_S}, \mathsf{Sim}_{\Phi_S})$ over a language defined by a predicate Φ_S that takes elements of $\mathbb{Z} \times \{0,1\}^* \times G_q \times (G_q \times G_q)$ and is defined as

$$\Phi_S(i, \tau', C_i, R_i) \stackrel{\text{def}}{=} \exists a, \gamma : C_i = g^a \text{ and } R_i = (h^\gamma (h')^a, g^\gamma).$$

The algorithms Prove_{Φ_S}, Verify_{Φ_S}, and Sim_{Φ_S} use a random oracle H_5.

Finally, DistVerify uses an SS-NIZKP scheme $\mathcal{T} = (\mathsf{Prove}_{\Phi_{\mathcal{T}}}, \mathsf{Verify}_{\Phi_{\mathcal{T}}}, \mathsf{Sim}_{\Phi_{\mathcal{T}}})$ over a language defined by a predicate $\Phi_{\mathcal{T}}$ that takes elements of $\mathbb{Z} \times \{0,1\}^* \times G_q \times G_q \times G_q \times (G_q \times G_q)$ and is defined as

$$\Phi_{\mathcal{T}}(i, \tau', \overline{g}, \overline{C}_i, C_i, R_i) \stackrel{\text{def}}{=} \exists a, \gamma : \overline{C}_i = \overline{g}^a \text{ and } C_i = g^a \text{ and } R_i = (h^\gamma (h')^a, g^\gamma).$$

The algorithms $\mathsf{Prove}_{\Phi_{\mathcal{T}}}$, $\mathsf{Verify}_{\Phi_{\mathcal{T}}}$, and $\mathsf{Sim}_{\Phi_{\mathcal{T}}}$ use a random oracle H_6.

$\mu_1, \mu_2, \nu \xleftarrow{R} \mathbb{Z}_q$	$(e, z_1, z_2, z_3) \leftarrow \Gamma_i$
$B' \leftarrow (y^{\mu_1}, g^{\mu_1}) \times (E_C)^{\mu_2}$	
$V' \leftarrow (h^\nu g^{\mu_2}, g^\nu)$	$B' \leftarrow (y^{z_1}, g^{z_1}) \times (E_C)^{z_2} \times (B \times (g, 1))^{-e}$
$e \leftarrow H(\tau, E_C, B, V, B', V')$	$V' \leftarrow (h^{z_3} g^{z_2}, g^{z_3}) \times V^{-e}$
$z_1 \leftarrow \beta e + \mu_1 \bmod q$	
$z_2 \leftarrow \pi e + \mu_2 \bmod q$	Return TRUE if $e = H(\tau, E_C, B, V, B', V')$
$z_3 \leftarrow \gamma e + \nu \bmod q$	
$\sigma \leftarrow (e, z_1, z_2, z_3)$	
Return σ	

Fig. 3. $\mathsf{Prove}_{\Phi_Q}((\tau, E_C, B, V), (\beta, \pi, \gamma))$ and $\mathsf{Verify}_{\Phi_Q}((\tau, E_C, B, V), (e, z_1, z_2, z_3))$

$\mu_1, \mu_2, \nu_1, \nu_2, \nu_3 \xleftarrow{R} \mathbb{Z}_q$
$\tilde{B}_i \leftarrow B^{\mu_1} \times (y^{\mu_2}, g^{\mu_2})$
$\tilde{V}_i \leftarrow (h^{\nu_1} g^{\mu_1}, g^{\nu_1})$
$\tilde{V}_i' \leftarrow (h^{\nu_2} (V[1])^{\mu_1}, g^{\nu_2})$
$\tilde{V}_i'' \leftarrow (h^{\nu_3} (V[2])^{\mu_1}, g^{\nu_3})$
$e \leftarrow H(i, B, V, B_i, V_i, V_i', V_i'', \tilde{B}_i, \tilde{V}_i, \tilde{V}_i', \tilde{V}_i'')$
$z_1 \leftarrow r_i e + \mu_1 \bmod q$
$z_2 \leftarrow r_i' e + \mu_2 \bmod q$
$z_3 \leftarrow \gamma_i e + \nu_1 \bmod q$
$z_4 \leftarrow \gamma_i' e + \nu_2 \bmod q$
$z_5 \leftarrow \gamma_i'' e + \nu_3 \bmod q$
$\sigma \leftarrow (e, z_1, z_2, z_3, z_4, z_5)$
Return σ

Fig. 4. $\mathsf{Prove}_{\Phi_R}((i, B, V, B_i, V_i, V_i', V_i''), (r_i, r_i', \gamma_i, \gamma_i', \gamma_i''))$

5 Security of the Protocol

Here we state the DDH assumption. Following that we prove that the protocol P is secure, based on the DDH assumption.

Decision Diffie-Hellman. Here we formally state the DDH assumption. For full details, see [6]. Let G_q be as in Section 3, with generator g. For two values $X = g^x$ and $Y = g^y$, let $\mathsf{DH}(X, Y) = g^{xy}$. Let $\mathcal{A}$ be an algorithm that on input (X, Y, Z) outputs "1" if it believes that $Z = \mathsf{DH}(X, Y)$, and "0" otherwise. For any $\mathcal{A}$ running in time t

$$\mathsf{Adv}_{G_q}^{\mathsf{DDH}}(\mathcal{A}) \stackrel{\text{def}}{=} \Pr\left[(x, y) \xleftarrow{R} \mathbb{Z}_q; X \leftarrow g^x; Y \leftarrow g^y; Z \leftarrow g^{xy} : \mathcal{A}(X, Y, Z) = 1\right]$$

$$- \Pr\left[(x, y, z) \xleftarrow{R} \mathbb{Z}_q; X \leftarrow g^x; Y \leftarrow g^y; Z \leftarrow g^z : \mathcal{A}(X, Y, Z) = 1\right]$$

$$(e, z_1, z_2, z_3, z_4, z_5) \leftarrow \Gamma_i$$

$$\tilde{B}_i \leftarrow B^{z_1} \times (y^{z_2}, g^{z_2}) \times (B_i)^{-e}$$
$$\tilde{V}_i \leftarrow (h^{z_3} g^{z_1}, g^{z_3}) \times (V_i)^{-e}$$
$$\tilde{V}_i' \leftarrow (h^{z_4} (V[1])^{z_1}, g^{z_4}) \times (V_i')^{-e}$$
$$\tilde{V}_i'' \leftarrow (h^{z_5} (V[2])^{z_1}, g^{z_5}) \times (V_i'')^{-e}$$

Return TRUE if $e = H(i, B, V, B_i, V_i, V_i', V_i'', \tilde{B}_i, \tilde{V}_i, \tilde{V}_i', \tilde{V}_i'')$

Fig. 5. $\mathrm{Verify}_{\Phi_{\mathcal{R}}}((i, B, V, B_i, V_i, V_i', V_i''), (e, z_1, z_2, z_3, z_4, z_5))$

$\mu, \nu \xleftarrow{R} \mathbb{Z}_q$	$(e, z_1, z_2) \leftarrow \Gamma_i$
$W \leftarrow g^\mu$	
$R' \leftarrow (h^\nu (h')^\mu, g^\nu)$	$R' \leftarrow (h^{z_2} (h')^{z_1} (R_i[1])^{-e}, g^{z_2} (R_i[2])^{-e})$
$e \leftarrow H(i, \tau', C_i, R_i, W, R')$	$W \leftarrow g^{z_1} (C_i)^{-e}$
$z_1 \leftarrow ae + \mu \bmod q$	
$z_2 \leftarrow \gamma e + \nu \bmod q$	Return TRUE if $e = H(i, \tau', C_i, R_i, W, R')$
$\Gamma_i \leftarrow (e, z_1, z_2)$	
Return Γ_i	

Fig. 6. $\mathrm{Prove}_{\Phi_S}((i, \tau', C_i, R_i), (a, \gamma))$ and $\mathrm{Verify}_{\Phi_S}((i, \tau', C_i, R_i), \Gamma_i)$

Let $\mathsf{Adv}_{G_q}^{\mathrm{DDH}}(t) = \max_{\mathcal{A}} \left\{ \mathsf{Adv}_{G_q}^{\mathrm{DDH}}(\mathcal{A}) \right\}$, where the maximum is taken over all adversaries of time complexity at most t. The DDH assumption states that for t polynomial in κ, $\mathsf{Adv}_{G_q}^{\mathrm{DDH}}(t)$ is negligible.

5.1 Protocol P

Here we prove that protocol P is secure, in the sense that an adversary attacking the system that compromises fewer than k out of n servers cannot determine session keys with significantly greater advantage than that of an online dictionary attack. Recall that we consider only *static* compromising of servers, i.e., the adversary chooses which servers to compromise before the execution of the system. Let t_{exp} be the time required to perform an exponentiation in G_q.

Theorem 1. *Let P be the protocol described in Figure 1 and Figure 2, using group G_q, and with a password dictionary of size N (that may be mapped into $\mathbb{Z}_q$). Fix an adversary $\mathcal{A}$ that runs in time t, and makes n_{se}, n_{ex}, n_{re} queries of type* Send, Execute, Reveal, *respectively, and n_{ro} queries directly to the random oracles. Then for $t' = O(t + (n_{ro} + k n_{se} + k^2 n_{ex}) t_{exp})$:*

$$\mathsf{Adv}_P^{\mathrm{dake}}(\mathcal{A}) = \frac{n_{se}}{N} + O \left(\mathsf{Adv}_{G_q}^{\mathrm{DDH}}(t') + \frac{n^2 + k n_{ro} n_{se} + n_{ro} n + (n_{se} + k n_{ex})^2}{q} + \right.$$

$$
\begin{array}{l|l}
\mu, \nu \overset{R}{\leftarrow} \mathbb{Z}_q & (e, z_1, z_2) \leftarrow \Gamma_i' \\[4pt]
\overline{W} \leftarrow \overline{g}^\mu & \\[4pt]
W \leftarrow g^\mu & R' \leftarrow (h^{z_2}(h')^{z_1}(R_i[1])^{-e}, g^{z_2}(R_i[2])^{-e}) \\[4pt]
R' \leftarrow (h^\nu(h')^\mu, g^\nu) & \overline{W} \leftarrow \overline{g}^{z_1}(\overline{C}_i)^{-e} \\[4pt]
e \leftarrow H(i, \tau', \overline{g}, \overline{C}_i, C_i, R_i, \overline{W}, W, R') & W \leftarrow g^{z_1}(C_i)^{-e} \\[4pt]
z_1 \leftarrow ae + \mu \bmod q & \\[4pt]
z_2 \leftarrow \gamma e + \nu \bmod q & \text{Return TRUE} \\[4pt]
 & \quad \text{if } e = H(i, \tau', \overline{g}, \overline{C}_i, C_i, R_i, \overline{W}, W, R') \\[4pt]
\Gamma_i' \leftarrow (e, z_1, z_2) & \\[4pt]
\text{Return } \Gamma_i' &
\end{array}
$$

Fig. 7. $\mathsf{Prove}_{\Phi_\mathcal{T}}((i, \tau', \overline{g}, \overline{C}_i, C_i, R_i), (a, \gamma))$ and $\mathsf{Verify}_{\Phi_\mathcal{T}}((i, \tau', \overline{g}, \overline{C}_i, C_i, R_i), \Gamma_i')$

$$
\left. \frac{(n_{\mathrm{se}} + k n_{\mathrm{ex}})(n_{\mathrm{ro}} + n_{\mathrm{se}} + k n_{\mathrm{ex}})}{q^2} \right).
$$

Proof: Our proof will proceed by introducing a series of protocols $P_0, P_1, \ldots, P_7$ related to P, with $P_0 = P$. In P_7, $\mathcal{A}$ will be reduced to simply "guessing" the correct password π_C. We describe these protocols informally in Figure 8. For each i from 1 to 7, we will prove that the difference between the advantage of $\mathcal{A}$ attacking protocols P_{i-1} and P_i is negligible.

We will sketch these proofs here, and leave the details to the full paper.

$P_0 \to P_1$ The probability of a collision of nonces is easily seen to be negligible.

$P_1 \to P_2$ This can be shown using a standard reduction from DDH. On input (X, Y, Z), we plug in random powers of Y for the servers' local public keys, and random powers of X for the clients' $\tilde{y}$ values, and then check H_2 queries for appropriate powers of Z.

$P_2 \to P_3$ This can be shown using a reduction from DDH. On input (X, Y, Z), we plug Y in for $h = H_0(y)$, and we use X and Z to create (randomized) encryptions for all V, V_i, V_i', V_i'', and R_i values. Also, we must factor in the negligible probability of a simulation error in one of the SS-NIZKP proofs,

$P_3 \to P_4$ This can be shown using a reduction from DDH. On input (X, Y, Z), we plug Y in for y, simulate the public shares of the uncompromised servers, and use X and Z to create (randomized) encryptions for all B values. To make sure authentication succeeds for a client that uses this B value, we generate $\overline{C}_i$ values from uncompromised servers in such as way that the product is the $\overline{y}$ value, and we simulate the SS-NIZKP proofs.

The difficulty is now in performing authentication on B values chosen by the adversary, since we do not know the secret shares (the x_i values) for the uncompromised servers. Therefore to perform authentication, we plug a value with a known discrete log in for $h = H_0(y)$ so we can decrypt all V, V_i, V_i', and V_i'' values, and then use these decryptions to aid in computing the correct value of $\overline{g}^x$ (even though we don't know x). Finally, we generate $\overline{C}_i$ values from uncompromised servers in such as way that the product is $\overline{g}^x$, and we simulate the SS-NIZKP proofs.

P_0 The original protocol P.

P_1 The nonces are assumed to be distinct (and thus Reveal queries do not reveal anything that could help in a Test query).

P_2 The Diffie-Hellman key exchange between a client and an uncompromised server is replaced with a perfect key exchange (and thus an adversary that does not succeed in impersonating a client to an uncompromised server does not obtain any information that could help in a Test query).

P_3 Value V from a client, and values V_i, V_i', V_i'', R_i from uncompromised servers, are replaced by random values. The Q-SS-NIZKP σ, and each R-SS-NIZKP σ_i, S-SS-NIZKP Γ_i, and T-SS-NIZKP Γ_i', are constructed using the associated simulators.

P_4 Value B from a client is replaced with a random value, but the $\overline{C}_i$ values from uncompromised servers are changed to force the associated authentication to succeed.

P_5 The adversary succeeds if it ever sends a V value associated with the correct password.

P_6 Abort if the adversary creates a new and valid S-SS-NIZKP proof or T-SS-NIZKP proof associated with an uncompromised server.

P_7 Value E_C for each client is changed to a random value, and on any adversary login attempt for C, the $\overline{C}_i$ values from uncompromised servers are replaced with values generated to form a random $\overline{y}$ (so as to force a failure).

Fig. 8. Informal description of protocols P_0 through P_7

$P_4 \to P_5$ This is straightforward, since this could only increase the probability of the adversary succeeding.

$P_5 \to P_6$ This can be shown using a reduction from DDH. On input (X, Y, Z), we plug Y in for y, simulate the public shares of the uncompromised servers, and let $h' = X$. Given a correct SS-NIZKP for an uncompromised server, we can compute $(h')^x$, where $y = g^x$ (where x is not known). Then we simply check if $Z = (h')^x$.

$P_6 \to P_7$ This can be shown using a reduction from DDH. On input (X, Y, Z), we plug Y in for y, simulate the public shares of the uncompromised servers, and use X and Z to create (randomized) encryptions for all E_C values. This does not affect authentication using B values generated by clients (since these values are random at this point, anyway). The difficulty is in obtaining the right distribution of $\overline{C}_i$ values while authenticating B values chosen by the adversary. To do this we use X and Z in our creation of the B_i values for uncompromised servers, which leaves $\overline{C}_i$ values correct if (X, Y, Z) is a true DH triple, but has the affect of randomizing the $\overline{C}_i$ values if (X, Y, Z) is a random triple. Again, the decryptions of V, V_i, V_i', and V_i'' are used to aid in computing the true $\overline{g}^x$ value (even though we don't know x) when (X, Y, Z) is a true DH triple, or the appropriate random value, when (X, Y, Z) is a random triple.

One can see that in P_2, an adversary that does not succeed in impersonating a client to an uncompromised server gains negligible advantage in determining a real session key from a random session key. The remainder of the protocols are used to show that an adversary gains negligible advantage in impersonating a client over a simple online guessing attack. In particular, in P_7 the password is only used to check V values submitted by the adversary attempting to impersonate a client. The theorem follows. ∎

References

1. M. Bellare, D. Pointcheval, and P. Rogaway. Authenticated key exchange secure against dictionary attacks. In *EUROCRYPT 2000* (LNCS 1807), pp. 139–155, 2000.
2. M. Bellare and P. Rogaway. Random oracles are practical: A paradigm for designing efficient protocols. In 1st *ACM Conference on Computer and Communications Security*, pages 62–73, November 1993.
3. M. Bellare and P. Rogaway. Entity authentication and key distribution. In *CRYPTO '93* (LNCS 773), pp. 232–249, 1993.
4. M. Bellare and P. Rogaway. Provably secure session key distribution—the three party case. In *27th ACM Symposium on the Theory of Computing*, pp. 57–66, 1995.
5. M. Blum, P. Feldman and S. Micali. Non-interactive zero-knowledge and its applications. In *20th ACM Symposium on the Theory of Computing*, pp. 103–112, 1988.
6. D. Boneh. The decision Diffie-Hellman problem. In *Proceedings of the Third Algorithmic Number Theory Symposium* (LNCS 1423), pp. 48–63, 1998.
7. C. Boyd. Digital multisignatures. In H. J. Beker and F. C. Piper, editors, *Cryptography and Coding*, pages 241–246. Clarendon Press, 1986.
8. V. Boyko, P. MacKenzie, and S. Patel. Provably secure password authentication and key exchange using Diffie-Hellman. In *EUROCRYPT 2000* (LNCS 1807), pp. 156–171, 2000.
9. R. Canetti, O. Goldreich, and S. Halevi. The random oracle methodology, revisited. In *30th ACM Symposium on the Theory of Computing*, pp. 209–218, 1998.
10. R. Canetti, Y. Lindell, R. Ostrovsky, and A. Sahai. Universally Composable Twoparty Computation. In *34th ACM Symposium on the Theory of Computing*, 2002.
11. Y. Desmedt and Y. Frankel. Threshold cryptosystems. In *CRYPTO '89* (LNCS 435), pages 307–315, 1989.
12. A. De Santis, G. Di Crescenzo, R. Ostrovsky, G. Persiano and A. Sahai. Robust non-interactive zero knowledge. In *CRYPTO 2001* (LNCS 2139), pp. 566–598, 2001.
13. T. Dierks and C. Allen. The TLS protocol, version 1.0, IETF RFC 2246, January 1999.
14. W. Diffie and M. Hellman. New directions in cryptography. *IEEE Trans. Info. Theory*, 22(6):644–654, 1976.
15. T. ElGamal. A public key cryptosystem and a signature scheme based on discrete logarithm. *IEEE Trans. Info. Theory*, 31:469–472, 1985.
16. P. Feldman. A Practical Scheme for Non-Interactive Verifiable Secret Sharing. In *28th IEEE Symp. on Foundations of Computer Science*, pp. 427-437, 1987
17. W. Ford and B. S. Kaliski, Jr. Server-assisted generation of a strong secret from a password. In *Proceedings of the 5th IEEE International Workshop on Enterprise Security*, 2000.

18. Y. Frankel, P. MacKenzie, and M. Yung. Adaptively-secure distributed threshold public key systems. In *European Symposium on Algorithms* (LNCS 1643), pp. 4–27, 1999.
19. R. Gennaro, S. Jarecki, H. Krawczyk, and T. Rabin. The (in)security of distributed key generation in dlog-based cryptosystems. In *EUROCRYPT '99* (LNCS 1592), pp. 295–310, 1999.
20. R. Gennaro, S. Jarecki, H. Krawczyk, and T. Rabin. Robust threshold DSS signatures. In *EUROCRYPT '96* (LNCS 1070), pages 354–371, 1996.
21. O. Goldreich and Y. Lindell. Session-key generation using human passwords only. In *CRYPTO 2001* (LNCS 2139), pp. 408–432, 2001.
22. O. Goldreich, S. Micali, and A. Wigderson. How to Play any Mental Game – A Completeness Theorem for Protocols with Honest Majority. In *19th ACM Symposium on the Theory of Computing*, pp. 218–229, 1987.
23. D. Jablon. Strong password-only authenticated key exchange. *ACM Computer Communication Review, ACM SIGCOMM*, 26(5):5–20, 1996.
24. D. Jablon. Password authentication using multiple servers. In em RSA Conference 2001, Cryptographers' Track (LNCS 2020), pp. 344–360, 2001.
25. J. Katz, R. Ostrovsky, and M. Yung. Efficient password-authenticated key exchange using human-memorable passwords. In *EUROCRYPT 2001* (LNCS 2045), pp. 475–494, 2001.
26. P. MacKenzie, S. Patel, and R. Swaminathan. Password authenticated key exchange based on RSA. In *ASIACRYPT 2000*, (LNCS 1976), pp. 599–613, 2000.
27. M. Naor and M. Yung. Public-key Cryptosystems Provably Secure against Chosen Ciphertext Attacks. In *22nd ACM Symposium on the Theory of Computing*, pp. 427–437, 1990.
28. SSH Communications Security. http://www.ssh.fi, 2001.
29. T. Wu. The secure remote password protocol. In *Proceedings of the 1998 Internet Society Network and Distributed System Security Symposium*, pp. 97–111, 1998.

A Threshold Pseudorandom Function Construction and Its Applications

Jesper Buus Nielsen

BRICS* Department of Computer Science
University of Aarhus
Ny Munkegade
DK-8000 Arhus C, Denmark
buus@brics.dk

Abstract. We give the first construction of a practical threshold pseudorandom function. The protocol for evaluating the function is efficient enough that it can be used to replace random oracles in some protocols relying on such oracles. In particular, we show how to transform the efficient cryptographically secure Byzantine agreement protocol by Cachin, Kursawe and Shoup for the random oracle model into a cryptographically secure protocol for the complexity theoretic model without loosing efficiency or resilience, thereby constructing an efficient and optimally resilient Byzantine agreement protocol for the complexity theoretic model.

1 Introduction

The notion of pseudorandom function was introduced by Goldreich, Goldwasser and Micali[GGM86] and has found innumerable applications. A pseudorandom function family is a function F taking as input a key K and element x, we write $F_K(x)$, where for a random key the output of F_K cannot be distinguished from uniformly random values if one does not know the key. If one have to require that the input of F_K is uniformly random for the output of F_K to look uniformly random, we say that F is weak pseudorandom.

One immediate application of pseudorandom functions is using them for implementing random oracles: Consider a protocol setting with n parties. A c-threshold random oracle with domain D is an ideal functionality (or trusted party). After c parties have input (evaluate, x), where say $x \in \{0,1\}^*$, the functionality will return a uniformly random value $r_x \xleftarrow{R} D$ to all parties that input (evaluate, x). This functionality defines a uniformly random function from $\{0,1\}^*$ to D. Numerous protocol constructions are known that can be proved secure assuming that a random oracle is available. However, any implementation of such a protocol must also provide an implementation of the oracle. In practice, a hash function is often used to replace a 1-random oracle, but then the implementation is only secure if an adversary can do no better with the hash function than he could with the oracle. This is something that in general cannot

---

* Basic Research in Computer Science,
  Centre of the Danish National Research Foundation

be proved, but must be a belief based on heuristics – in fact, for some protocols, this belief is always wrong [CGH98,Nie02].

In contrast, pseudorandom functions can be used to implement random oracles without loss of security. This can be done by generating K at the beginning of the protocol and letting $r_x = F_K(x)$ when r_x is needed. It is however clearly necessary that no party should know the key of F_K, since the output of a pseudorandom function only looks random to parties who do not have the key. Therefore the key – and hence also ability to evaluate the function – must be distributed among the parties using, for instance, a threshold secret-sharing scheme.

Our Results. In this paper we construct a new pseudorandom function family. The key will be a prime Q, where $P = 2Q + 1$ is also a prime, a random value x from the subgroup Q_P of Z_P^* of order Q, along with $2l$ random values $\{\alpha_{j,b}\}_{j=1,\ldots,l,b=0,1}$ from Z_Q. The function family maps from the set of strings of length at most l to Q_P, and given $\sigma = (\sigma_1, \ldots, \sigma_m) \in \{0,1\}^{\leq l}$, the output will be $x^{\prod_{i=1}^m \alpha_{i,\sigma_i}} \mod P$. We prove this function family secure under the decisional Diffie-Hellman (DDH) assumption.

More importantly, we give a secure n-party protocol for evaluating the function. Our protocol is for the asynchronous model with authenticated public point-to-point channels. This is a very realistic model of communication and can be efficiently implemented[CHH00]. The protocol is statically secure as long as less than $n/3$ parties misbehave. In some applications the protocol can communicate as much as $O(ln^2k)$ bits per evaluation, where k is the security parameter (for each exponent each party sends to each other party k bits). However, in many uses the communication complexity will be $O(n^2k)$ bits.

To demonstrate the applicability of our new threshold pseudorandom function, we describe how to implement efficient Byzantine agreement (BA) in the complexity theoretic model, by replacing the random oracles in the protocol [CKS00] with our threshold pseudorandom function. The resulting protocol has the same resilience as the protocol in [CKS00], namely resilience against a malicious coalition of one third of the parties. It has the same communication complexity of $O(n^2k)$ bits per activation and the same (constant) round complexity up to a small constant factor. As part of the implementation we show how to replace the random oracles in the threshold signature scheme from [Sho00] by our threshold pseudorandom function.

Related Work. The notion of distributed pseudorandom function, which is similar to our threshold pseudorandom function, was introduced by Naor, Pinkas and Reingold in [NPR99]. They do not define distributed pseudorandom functions in a general multiparty computation model – their model is more ad-hoc and implementation-near. Since there are differences between the two notions, we have chosen a different name for our definition.

Until now the most efficient known construction of threshold pseudorandom functions was using general multiparty computation techniques or coin-toss protocols, or were restricted to a (logarithmic) small number of parties because of the way the key was distributed[MS95,BCF00].

In [NPR99] an efficient threshold *weak* pseudorandom function was constructed based on the DDH assumption, and it was left as an interesting open problem to construct an efficient threshold pseudorandom function. Our protocol uses the protocol from [NPR99]. Indeed, our construction contains a general and efficient construction of threshold pseudorandom functions from threshold weak pseudorandom functions. This technique is reminiscent of the construction of pseudorandom functions from pseudorandom generators in [GGM86].

Our pseudorandom function is similar to a function from [NR97], but there are some essential differences, which allows to efficiently distribute our function. Indeed, the construction from [NR97] does not seem to allow an efficient secure distributed protocol.

Organization. In Section 2 we give some preliminary notation and definitions. In Section 3 we describe our pseudorandom function and prove that it is pseudorandom under the DDH assumption. In Section 4 we sketch the framework for secure multiparty computation from [Can01] and define the notions of threshold function family, c-threshold random oracle, threshold trapdoor permutation, threshold signatures and Byzantine agreement in this framework. In Section 5 we construct a threshold pseudorandom function by giving a distributed protocol for our threshold pseudorandom function. In Section 6 we show how to use this threshold function family and the RSA based threshold trapdoor permutation from [Sho00] to implement a threshold signature scheme based on the RSA and DDH assumptions. Finally in Section 7 we show how to use this threshold signature scheme along with our threshold pseudo-random function to implement the BA protocol from [CKS00] in the complexity theoretic model under the RSA and DDH assumptions.

2   Preliminaries

We use ϵ to denote the empty string and for $l \in N$ we use $\{0,1\}^{\leq l}$ to denote the set $\bigcup_{i=0}^{l} \{0,1\}^i$ of all strings of length at most l. For a set S we use $x \xleftarrow{R} S$ to denote the action of sampling an element x (statistically close to) uniformly random from S, and for a probabilistic algorithm we use $a \xleftarrow{R} A$ to denote the action of running A with uniformly random bits and letting a be the output. We use $k \in N$ to denote the security parameter. We will often skip the security parameter if it is implicitly given by the context. If e.g. $S = \{S_k\}_{k \in N}$ is a sequence of sets we will write $x \in S$ to mean $x \in S_k$, where k is the security parameter given by the context.

Trapdoor Commitment Scheme. A trapdoor commitment scheme can be described as follows: first a public key pk is chosen based on a security parameter k, by running a probabilistic polynomial time (PPT) *generator G*. Further more, there is a fixed function commit that the committer C can use to compute a commitment c to s by choosing some random input r, computing $c = \text{commit}_{pk}(s, r)$, and sending c. Opening takes place by sending (s, r); it can then be checked that

$\text{commit}_{pk}(s, r)$ is the value S sent originally. We require that the scheme is perfect hiding and computationally binding. The algorithm for generating pk also outputs a string t, the trapdoor, and there is an efficient algorithm which on input t, pk outputs a commitment c, and then on input any s produces uniformly random r for which $c = \text{commit}_{pk}(s, r)$. In other words, a trapdoor commitment scheme is binding if you know only the public key, but given the trapdoor, you can cheat arbitrarily and undetectable.

Pseudorandom Functions. The following definitions are adaptions of definitions from [Gol01,BDJR97].

Definition 1. *A function family is a sequence $F = \{F_k\}_{k \in N}$ of random variables, so that the random variable F_k assume values which are functions. We say that a function family is a PPT function family if the following two conditions hold:*

Efficient indexing *There exists a PPT algorithm, I, and a mapping from strings to functions, ϕ, so that $\phi(I(1^k))$ and F_k are identically distributed. We denote by f_i the function $\phi(i)$.*

Efficient evaluation *There exists a PPT algorithm, V, so that $V(i, x) = f_i(x)$.*

Let F and G be two function families. We write $F \subset G$ if for all k the functions that receive non-zero probability mass in F_k is a subset of the functions that receive non-zero probability mass in G_k. Consider two sequences $A = \{A_k\}_{k \in N}$ and $B = \{B_k\}_{k \in N}$ of sets. If all A_k and B_k are finite, we use $[A \to B]$ to denote the function family $\{F_k\}_{k \in N}$ where F_k is uniform over the set of all functions from A_k to B_k. We say that a function family F maps from A to B if $F \subset [A \to B]$.

Definition 2. *Let $F \subset [A \to B]$ be a PPT function family. Let $b \in \{0, 1\}$. Let D be a distinguisher that has access to an oracle. Let $\mathcal{O}_f$ be the oracle which on input $s \in A$ outputs $f(s)$, and let $\mathcal{R}_f$ be the oracle which on input gen generates a uniformly random $s \in A$ and outputs $(s, f(s))$. Now consider the following experiments and corresponding advantages.*

$$
\begin{array}{ll}
\underline{\text{proc } \text{Exp}_{F,D}^{\text{vprf-b}}} \equiv & \qquad \underline{\text{proc } \text{Exp}_{F,D}^{\text{prf-b}}} \equiv \\[4pt]
f_0 \xleftarrow{R} [A \to B] & \qquad f_0 \xleftarrow{R} [A \to B] \\[4pt]
f_1 \xleftarrow{R} F & \qquad f_1 \xleftarrow{R} F \\[4pt]
d \leftarrow D^{\mathcal{R}_{f_b}} & \qquad d \leftarrow D^{\mathcal{O}_{f_b}} \\[4pt]
\underline{\text{return } d} & \qquad \underline{\text{return } d}
\end{array}
$$

$$
\mathbf{Adv}_{F,D}^{\text{wprf}} = \Pr[\mathbf{Exp}_{F,D}^{\text{wprf-1}} = 1] - \Pr[\mathbf{Exp}_{F,D}^{\text{wprf-0}} = 1]
$$
$$
\mathbf{Adv}_{F,D}^{\text{prf}} = \Pr[\mathbf{Exp}_{F,D}^{\text{prf-1}} = 1] - \Pr[\mathbf{Exp}_{F,D}^{\text{prf-0}} = 1]
$$

We say that F is a weak pseudorandom function family (WPRF) from A to B (is a pseudorandom function family (PRF) from A to B) if for all PPT distinguishers D the advantage $\mathbf{Adv}_{F,D}^{\text{wprf}}$ ($\mathbf{Adv}_{F,D}^{\text{prf}}$) is negligible.

3 The DDH-Tree Function Family

Definition 3. *The DDH-Tree function family is indexed by values $i = (Q, \{\alpha_{j,b}\}_{j \in \{1,\ldots,l\}, b \in \{0,1\}}, x_\epsilon)$, where Q is a random k-bit prime s.t. $P = 2Q + 1$ is also a prime, l is some polynomial in k, the elements $\alpha_{j,b}$ are random in Z_Q^*, and x_ϵ is random in Q_P (the sub-group of Z_P^* of order Q). For an index i we define a function $f_i : \{0,1\}^{\le l} \to Q_P$, $f_i(\sigma) = x_\epsilon^{\prod_{i=1}^m \alpha_{i,\sigma_i}} \mod P$, where $\sigma = (\sigma_1, \ldots, \sigma_m) \in \{0,1\}^{\le l}$. We will sometimes use the notation x_σ to mean $f_i(\sigma)$, when i is clear from the context. Note that in particular, we have $f_i(\epsilon) = x_\epsilon$.*

We would like the function family to output bit-strings, instead of elements in Q_P. To this end, given an element $y \in Q_P$, let $\lfloor y \rfloor = \min(y, P-y)$. Consider then an index i as above except that Q is a $(k + \delta(k))$-bit prime, where $\log(k)/\delta(k) \in o(1)$. We define the function $g_i : \{0,1\}^{\le l} \to \{0,1\}^k$, $g_i(\sigma) = \lfloor f_i(\sigma) \rfloor \mod 2^k$. The DDH-Tree function family is given by the functions g_i.

Definition 4. *Given $P = 2Q + 1$ and a random generator g of Q_P, the DDH assumption is that the random variable $(g, g^\alpha \mod P, g^\beta \mod P, g^{\alpha\beta} \mod P)$, where $\alpha, \beta \stackrel{R}{\leftarrow} Z_Q$, is computationally indistinguishable from the random variable $(g, g^\alpha \mod P, g^\beta \mod P, g^\gamma \mod P)$, where $\alpha, \beta, \gamma \stackrel{R}{\leftarrow} Z_Q$.*

Theorem 1. *Under the DDH assumption, the DDH-Tree function family is pseudorandom from $\{0,1\}^{\le l}$ to $\{0,1\}^k$.*

Proof: Let $i = (Q, \{\alpha_{j,b}\}_{j \in \{1,\ldots,l\}, b \in \{0,1\}}, x_\epsilon)$ be a random index. Since -1 is not a square in Z_P^*, the map $\lfloor \cdot \rfloor$ is bijective. Since Q is a $(k + \delta(k))$-bit prime, for a uniformly random value $x \in Z_Q$, the value $x \mod 2^k$ is statistically close to uniform in $\{0,1\}^k$. It is therefore enough to prove that the output of f_i for random i cannot be distinguished from uniformly random values from Q_P. For this purpose define for $j \subset \{1,\ldots,l\}$ and $b \in \{0,1\}$ a function $f_{j,b} : Q_P \to Q_P$, $f_{j,b}(x) = x^{\alpha_{j,b}} \mod P$ and a function $g_{j,b}$ which is uniformly random from Q_P to Q_P. Then for $m \in \{0,\ldots,l\}$ let $h_{j,b}^m = f_{j,b}$ if $j \ge m$ and let $h_{j,b}^m = g_{j,b}$ otherwise. Finally let $h_i^m(\sigma) = h_{l,\sigma_l}^m \circ \cdots \circ h_{1,\sigma_1}^m(x_\epsilon)$. Then $f_i = h_i^0$ and h_i^l is statistically close to a uniformly random function from $\{0,1\}^{\le l}$ to Q_P. Only statistically close as collisions will distinguish h_i^l from a uniformly random function: If $|\sigma_1| = |\sigma_2|$ and $h_i^l(\sigma_1) = h_i^l(\sigma_2)$, then for all suffixes σ, $h_i^l(\sigma_1\|\sigma) = h_i^l(\sigma_2\|\sigma)$.

It is therefore enough to prove that h_i^0 and h_i^l cannot be distinguished, which can be done by a hybrids argument. Assume namely that there exists $m \in \{1,\ldots,l\}$ such that the functions h_i^{m-1} and h_i^m can be distinguished by a PPT distinguisher D having black-box access to the functions. We will show that this contradicts the DDH assumption. For this purpose, assume that we have access to a black-box o which returns random values of the form $(x, f_{j,0}(x), f_{j,1}(x))$ if $b = 0$ and returns random values of the form $(x, g_{j,0}(x), g_{j,1}(x))$ if $b = 1$. By a simple application of the DDH assumption it can be seen that no PPT algorithm can guess b with anything but negligible advantage. We reach our contradiction by using D to guess b. To be able to do this we show how to generate values

$\{x_\sigma\}_{\sigma \in \{0,1\}^{\leq l}}$ distributed as those defined by h_i^{m-1+b} given oracle access to o: Pick all the values x_σ for $\sigma \in \{0,1\}^{\leq m-1}$ as uniformly random values with the only restriction that they are consistent with random functions, i.e. if $|\sigma_1| = |\sigma_2|$ and $x_{\sigma_1} = x_{\sigma_2}$, then for all suffixes σ make sure $x_{\sigma_1 \| \sigma} = x_{\sigma_2 \| \sigma}$. To generate a value x_σ where $|\sigma| = m-1$, query o and receive a random evaluation (x, x_1, x_2), where x is uniformly random from Q_P. Then let $x_\sigma = x$, let $x_{\sigma \| 0} = x_1$, and let $x_{\sigma \| 1} = x_2$. Then generate the remaining values x_σ where $|\sigma| > m$ as done in h_i^{m-1} and h_i^m using random exponents. It is straightforward to verify that the values thus defined are distributed as in h_i^{m-1} if $b = 0$ and as in h_i^m if $b = 1$.

To use D to distinguish, run it, and when it queries on $\sigma \in \{0,1\}^{\leq l}$ return x_σ. To make the process efficient, the values x_σ are generated when needed.   ∎

4   The Multiparty Computation Model

We will study our protocol problems in the framework for universally composable asynchronous multiparty computation from [Can01]. Below we sketch the model.

First the real-life execution of the protocol is defined. Here the protocol π is modeled by n interactive Turing machines (ITMs) $P_1, \ldots, P_n$ called the parties of the protocols. Also present in the execution is an adversary $\mathcal{A}$ and an environment $\mathcal{Z}$ modeling the environment in which $\mathcal{A}$ is attacking the protocol. The environment gives inputs to honest parties, receives outputs from honest parties, and can communication with $\mathcal{A}$ at arbitrary points in the execution. Both $\mathcal{A}$ and $\mathcal{Z}$ are PPT ITMs.

Second an ideal process is defined. In the ideal process an ideal functionality $\mathcal{F}$ is present to which all the parties have a secure communication channel. The ideal functionality is an ITM defining the desired input-output behavior of the protocol. Also present is an ideal adversary $\mathcal{S}$, the environment $\mathcal{Z}$, and n so-called dummy parties $\tilde{P}_1, \ldots, \tilde{P}_n$ – all PPT ITMs. The only job of the dummy parties is to take inputs from the environment and send them to the ideal functionality and take messages from the ideal functionality and output them to the environment.

The security of the protocol is then defined by requiring that the protocol emulates the ideal process. We say that the protocol securely realizes the ideal functionality.

The framework also defines the hybrid models, where the execution proceeds as in the real-life execution, but where the parties in addition have access to an ideal functionality. An important property of the framework is that an ideal functionality in a hybrid model can securely be replaced by a sub-protocol securely realizing that ideal functionality.

Below we add a few more details. For a more elaborate treatment of the general framework, see [Can01].

The environment $\mathcal{Z}$ is the driver of the execution. It can either provide a honest party, P_i or $\tilde{P}_i$, with an input or send a message to the adversary. If a party is given an input, that party is then activated. The party can then, in the real-life execution, send a message to another party or give an output to the environment. In the ideal process an activated party just copies its input to the ideal functionality and the ideal functionality is then activated, sending messages to the parties and the adversary according to its program. After the party and/or the ideal functionality stops, the environment is activated again.

If the adversary, $\mathcal{A}$ or $\mathcal{S}$, is activated it can do several things. It can corrupt a honest party, send a message on behalf of a corrupted party, deliver any message sent from one party to another, or communicate with the environment. After a corruption the adversary sends and receives messages on behalf of the corrupted party. We assume a static adversary which corrupts t parties before the execution of the protocol, and then never corrupts again.

The adversary controls the scheduling of the message delivery. In the real-life execution the adversary $\mathcal{A}$ can see the contents of all message and may decide which messages should be delivered and when – it can however not change messages or add messages to a channel. In the ideal process the adversary $\mathcal{S}$ cannot see the contents of the messages as the channels are assumed to be secure. It can only see that a message has been sent and can then decide when the message should be delivered. We will assume that the network is non-blocking. This means that though the adversary is allowed to delay a message for an arbitrary number of activations, any message is eventually delivered if the adversary is activated enough times. If the adversary delivers a message to some party, then this party is activated and the environment resumes control when the party stops.

There is one additional way that the adversary can be activated. An ideal functionality has the opportunity of calling the adversary. This means that the ideal functionality sends a value to the adversary, which then computes a value which is passed back to the ideal functionality. The ideal functionality then proceeds with its actions. Typically, this mechanism is used for modeling some adversarially controlled non-determinism of the ideal functionality. When we specify the functionality for Byzantine agreement, we will use this mechanism to allow the adversary to decide on the result if the honest parties disagree.

At the beginning of the protocol all entities are given as input the security parameter k and random bits. Furthermore the environment is given an auxiliary input z. The environment is then activated and the execution proceeds as described above. At some point the environment stops activating parties and outputs some bit. This bit is taken to be the output of the execution. We use $\text{REAL}_{\pi,\mathcal{A},\mathcal{Z}}(k,z)$ to denote the output of the environment in the real-life execution and use $\text{IDEAL}_{\mathcal{F},\mathcal{S},\mathcal{Z}}(k,z)$ to denote the output of the environment in the ideal process.

We are now ready to state the definition of securely realizing an ideal functionality. For this purpose let $\text{REAL}_{\pi,\mathcal{A},\mathcal{Z}}$ denote the distribution ensemble $\{\text{REAL}_{\pi,\mathcal{A},\mathcal{Z}}(k,z)\}_{k\in N, z\in\{0,1\}^*}$ and let $\text{IDEAL}_{\mathcal{F},\mathcal{S},\mathcal{Z}}$ denote the distribution ensemble $\{\text{IDEAL}_{\mathcal{F},\mathcal{S},\mathcal{Z}}(k,z)\}_{k\in N, z\in\{0,1\}^*}$.

Definition 5. *We say that π t-securely realizes $\mathcal{F}$ if for all real-life adversaries $\mathcal{A}$, which corrupts at most t parties, there exists an ideal-process adversary $\mathcal{S}$ such that for all environments $\mathcal{Z}$ we have that the distribution ensembles $\text{IDEAL}_{\mathcal{F},\mathcal{S},\mathcal{Z}}$ and $\text{REAL}_{\pi,\mathcal{A},\mathcal{Z}}$ are computationally indistinguishable.*

The following theorem (here stated informally) is proven in [Can01].

Theorem 2. *If π t-securely realizes $\mathcal{F}$ in the hybrid model with ideal functionality $\mathcal{G}$ and ρ t-securely realizes $\mathcal{G}$, then π, with the use of $\mathcal{G}$ implemented using ρ, t-securely implements $\mathcal{F}$ in the model without ideal functionality $\mathcal{G}$.*

4.1   Some Functionalities

Definition 6. *Let V be a PPT algorithm outputting an $(n+1)$-tuple of values. A V-preprocessing model is a model equipped with an ideal functionality, which when activated the first time generates $(v_0, v_1, \ldots, v_n) \xleftarrow{R} V$, outputs v_i to party P_i, and outputs v_0 to the adversary.*

We will use the preprocessing model for distributing keys for various cryptographic primitives prior to running the actual protocols.

Definition 7. *A (c, t)-threshold protocol π for function family F is a t-secure realization of the functionality $\mathcal{F}_{F,c}$ described below.*

Init *On the first activation, the functionality generates $f \xleftarrow{R} F$ and outputs $(\texttt{init})$ to all parties.*

Evaluate *If a party P_i inputs (j, x), we say that the permission to evaluate on x is given to P_j by P_i. The message (i, j, x) is output to the adversary and P_j. If at some point a total of c parties have given permission to some party to evaluate on x, then $(x, f(x))$ is given to the adversary. If at some point a total of c parties have given P_j permission to evaluate on x, then $(x, f(x))$ is given to P_j.*

If F is pseudorandom from X to Y we call π a (c, t)-threshold pseudorandom function from X to Y.

Definition 8. *The c-threshold random oracle from X to Y is the ideal functionality $\mathcal{F}_{ro,c,X,Y} = \mathcal{F}_{[X \to Y],c}$ for evaluating a random function from X to Y.*

The following theorem is an easy exercise in using the definitions.

Theorem 3. *A (c, t)-threshold pseudorandom function from X to Y t-securely realizes the c-threshold random oracle from X to Y.*

Definition 9. *Let F be a family of trapdoor permutations. A (c, t)-threshold protocol for F is a protocol t-securely implementing the following functionality $\mathcal{F}_{F,c}$:*

Init *On the first activation, the functionality generates $(f, f^{-1}) \xleftarrow{R} F$ and outputs f to all parties and the adversary.*

Invert *If a party P_i inputs $(\texttt{invert}, j, x)$, we say that the permission to invert on x is given to P_j by P_i. The message $(\texttt{invert}, i, j, x)$ is output to the adversary and P_j. If at some point a total of c parties have given permission to some party to invert on x, then $(\texttt{invert}, f^{-1}(x))$ is given to the adversary. If at some point a total of c parties have given P_j permission to invert on x, then $(\texttt{invert}, f^{-1}(x))$ is given to P_j.*

Definition 10. *The ideal functionality for threshold signatures* $\mathcal{F}_{tsig,c}$ *is given by the following description.*

Init Let $\mathcal{M}$ be the message space. The functionality keeps for each $M \in \mathcal{M}$ a set $\mathrm{see}(M) \subset \{0, 1, \ldots, n\}$. The interpretation of $\mathrm{see}(M) = P$ is that the parties indexed by P (the adversary is index by 0) see a signature on M. Initially $\mathrm{see}(M) = \emptyset$ for all $M \in \mathcal{M}$.

Sign If a party P_i inputs (sign, j, M), we say that the permission to sign M is given by P_i to P_j. The message (sign, i, j, M) is output to the adversary and P_j. If at some point a total of c parties have given some party permission to sign M, then set $\mathrm{see}(M) = \mathrm{see}(M) \cup \{0\}$ and output $(\mathrm{signature}, M)$ to the adversary. If at some point a total of c parties have given P_j permission to sign M, then set $\mathrm{see}(M) = \mathrm{see}(M) \cup \{j\}$ and output $(\mathrm{signature}, M)$ to P_j.

Send If a party P_i or the adversary (P_0) inputs (send, j, M) and $i \in \mathrm{see}(M)$, then set $\mathrm{see}(M) = \mathrm{see}(M) \cup \{j\}$ and output (send, i, M) to P_j and the adversary.

Definition 11. *The ideal functionality for Byzantine agreement* $\mathcal{F}_{ba,t}$ *is given by the following description.*

Vote If a party inputs (vote, vid, b), where $b \in \{0, 1\}$, then $(\mathrm{vote}, vid, i, b)$ is output to the adversary and we say that the party has voted b in voting vid. The adversary is also allowed to vote.

Decide

The result of voting vid is computed using one of the following rules:

- If $n - t$ parties have voted and $t + 1$ of them voted b and the adversary voted b, then the result is b.
- If $n - t$ honest parties have voted b, then the result is b.
- If $n - t$ honest parties have voted, but do not agree, then the adversary is called to decide the result.

When the result of voting vid has been decided to be b, then $(\mathrm{decide}, vid, b)$ is output to all parties and the adversary.

Note, that the three rules for decision are consistent. Especially, if $n - t$ honest parties vote b, then no $t + 1$ parties voted $1 - b$ and therefore the functionality always terminates with decision b.

5   The Threshold DDH-Tree

We now describe our threshold protocol $\pi_{c,\mathrm{DDH\text{-}Tree}}$ for the DDH-Tree function family.

Key Generation The protocol runs in the preprocessing model for the following values.

- For $i = 1, \ldots, n$, a random public key pk_i for a trapdoor commitment scheme.
- $P = 2Q + 1$, where P and Q are random primes and Q is of length $k + \delta(k)$ bits, where $\log(k)/\delta(k) \in o(1)$.

- g, a random generator of Q_P.
- For $j = 1, \ldots, l$ and $b = 0, 1$:
  - $\alpha_{j,b} \in Z_Q^*$, a uniformly random element.
  - $y_{j,b} = g^{\alpha_{j,b}} \bmod P$.
  - $f_{j,b}(X) \in Z_Q[X]$, a uniformly random degree $c - 1$ polynomial for which $f_{j,b}(0) = \alpha_{j,b}$.
  - For $i = 1, \ldots, n$:
    * $\alpha_{j,b,i} = f_{j,b}(i)$.
    * $y_{j,b,i} = g^{\alpha_{j,b,i}} \bmod P$.
- $x_\epsilon \in Q_P$, a uniformly random element.

The values $(\{pk_i\}_{i=1}^n, Q, g, x_\epsilon, \{y_{j,b}\}_{j=1,b=0}^{l,1}, \{y_{j,b,i}\}_{j=1,b=0,i=1}^{l,1,n})$ are output to all parties and the adversary, and the values $\{\alpha_{j,b,i}\}_{j=1,b=0}^{l,1}$ is output to P_i only.

Evaluation On input $(\texttt{evaluate}, \sigma)$, where $\sigma \in \{0,1\}^l$, the party P_i picks the largest possible prefix σ' of σ for which $x_{\sigma'}$ is known. Then for $j = |\sigma'| + 1, \ldots, l$ the party does the following: The party computes the evaluation share $x_{\sigma[1..j],i} = x_{\sigma[1..(j-1)]}^{\alpha_{j,b,i}} \bmod P$ and sends the value to all parties and proves in zero-knowledge (ZK) to all parties that $\log_{x_{\sigma[1..(j-1)]}}(x_{\sigma[1..j],i}) = \log_g(y_{j,b,i})$ (see below for a description of how to do the ZK-proof). When a party has received evaluation shares and accepted ZK-proofs from all $i \in I$, where $|I| = c$, the party computes $x_{\sigma[1..j]} \leftarrow \prod_{i \in I} x_{\sigma[1..j],i}^{\lambda_{i,I}} = x_{\sigma[1..(j-1)]}^{\alpha_{j,b}} \bmod P$, where the $\lambda_{i,I}$ are the appropriate Lagrange coefficients. When x_σ becomes known, output $\lfloor x_\sigma \rfloor \bmod 2^k$.

ZK-Proofs Assume that P_i knows $\alpha \in Z_Q$ and has sent $g, h, A = g^\alpha, B = h^\alpha$ to P_j, where g and h are generators of Q_P, and wants to prove in ZK that $\log_g(A) = \log_h(B)$. This is done as follows.

Commit Message P_i computes $a \leftarrow g^\beta \bmod P$ and $b \leftarrow h^\beta \bmod P$ for uniformly random $\beta \in Z_Q$, and $c \leftarrow \texttt{commit}_{pk_i}((a,b), r_c)$ for appropriate random bits r_c for the commitment scheme, and sends $(\texttt{commit}, c)$ to P_j.

Challenge P_j generates $e \xleftarrow{R} Z_Q$ and sends e to P_i.

Response P_i computes $z \leftarrow \alpha e + \beta \bmod Q$ and sends (a, b, r_c, z) to P_j.

Check P_j checks that $c = \texttt{commit}_{pk_i}((a,b), r_c)$, $A^e a = g^z \bmod P$, and $B^e b = h^z \bmod P$ and if so, accepts the proof.

Theorem 4. *For (c,t) where $0 \le t < c \le n - t$, the protocol $\pi_{c,DDH\text{-}Tree}$ is a (c,t)-threshold pseudorandom function from $\{0,1\}^l$ to $\{0,1\}^k$.*

Proof: In [NPR99] Naor, Pinkas and Reingold described a threshold protocol for the *weak* pseudorandom function $x \mapsto x^\alpha \bmod P$ and analysed it for the case $c = t + 1$. Subsequently in [CKS00] Cachin, Kursawe and Shoup made a generalization of the proof to handle parameters where $0 \le t < c \le n - t$. Our protocol for computing $x_{\sigma[1..(j-1)]}^{\alpha_{j,b}}$ from $x_{\sigma[1..(j-1)]}$ is exactly this protocol, except that we use interactive zero-knowledge proofs, and as detailed below the techniques used in [NPR99,CKS00] generalize straightforwardly to prove our protocol secure. Note, that the theorem specifies input domain $\{0,1\}^l$, and not

$\{0,1\}^{\leq l}$. This is to make it secure to reveal $f_i(\sigma')$ for all prefixes σ' of σ when we compute $f_i(\sigma)$. The only difference between our way of computing x_σ from $x_{\sigma'}$ and the protocol used in [NPR99,CKS00] is that we use interactive ZK-proofs. Here we describe how to generalize the analysis to handle this. We show how to simulate the ZK-protocol to an adversary $\mathcal{A}$ while running as an ideal adversary in the ideal process. Say that P_i is giving a proof to P_j. There are two cases:

Assume first that P_i is honest and we as an ideal adversary must give a proof for (g, A, h, B) that $\log_h(B) = \log_g(A) \bmod P$. The simulators used in [NPR99,CKS00] is such that a honest P_i never has to give such a proof without it actually being true. However, the witness $\alpha = \log_g(A) = \log_h(B) \bmod P$ is not always known, so the simulator cannot just run the protocol. We deal with this as follows: As the commit message c we send a trapdoor commitment, which can be opened arbitrarily. When we receive e we then pick $z \xleftarrow{R} Z_Q$ and compute $a \leftarrow g^z A^{-e} \bmod P$ and $b \leftarrow h^z B^{-e} \bmod P$. Using the trapdoor of the commitment scheme, we then construct random bits r_c such that $c = \mathrm{commit}_{pk_i}((a, b), r_c)$ and send (a, b, r_c, z) to P_j. This conversation is distributed exactly as in the protocol.

Assume then that P_i is corrupted. We can then simply run the protocol, as the code of all other parties than P_i is trivial. All that we have to check is that when we accept (g, A, h, B), then indeed $\log_h(B) = \log_g(A) \bmod P$. That this is the case, except with exponentially small probability, is a well-known result[CP92]. ∎

The theorem specifies input domain $\{0,1\}^l$. If however the oracle is evaluated on consecutive values $\epsilon, 0, 1, 00, 01, 10, 11, 000, \ldots$, or more generally, if it is never evaluated on a prefix of a previous input, then the input domain $\{0,1\}^{\leq l}$ would be secure. In that case the loop in **Evaluation** should just be repeated for $j = |\sigma'| + 1, \ldots, |\sigma|$. For consecutive values the worst-case round complexity would be 3 and the worst-case communication complexity would be about $3n^2 k$ bits.

If the oracle is evaluated on arbitrary values, then an extra overhead of a factor l will be added. If no bound on the length of inputs is known, or to keep l as low as possible, we can use a family of collision resistant hash-functions: If $F : \{0,1\}^l \to \{0,1\}^m$ is a pseudorandom function and $H : \{0,1\}^* \to \{0,1\}^l$ is a collision resistant hash-function, then $F \circ H : \{0,1\}^* \to \{0,1\}^m$ is again a pseudorandom function, which we can distribute by first hashing the input value locally and then running the threshold pseudorandom function on the hash-value. In practice a hash-function with output length at least 160 bits would probably by recommended, and so the round complexity would be 480.

Since the time for one round of network communication probably dominates the time to access even a large database, some of this overhead can by removed by preprocessing: Consider the $2m$ key values $\{\alpha_{i,b}\}_{i=j,\ldots,j+m-1,b=0,1}$. If instead of sharing these values we share the 2^m values $\{\alpha_{j,\sigma} = \prod_{i=1}^m \alpha_{i+j-1,\sigma_i} \bmod Q\}_{\sigma \in \{0,1\}^m}$, then the computation of $x^{\prod_{j=1}^l \alpha_{j,\sigma_i}} \bmod P$ could be speed up by a factor m. By setting $m = 20$ the round complexity could be brought down to 27. The price is a key of about $1Gb$ for $k = 1024$.

6   An RSA and DDH Based Threshold Signature Scheme

In this section we construct a threshold signature protocol $\pi_{\text{tsig},c}$ secure relative to the DDH and RSA assumptions. We will describe the protocol assuming access to a random oracle and an oracle for inverting the RSA function. Using the modular composition theorem, the random oracle can be replaced with our threshold pseudorandom function protocol and the oracle for RSA can be replaced with the protocol given by the following theorem.

Theorem 5. *For the RSA function family with a modulus that is a product of two strong primes and for (c,t), where $n > 3t$ and $t < c \leq n - t$, there exists a statically secure (under the RSA assumption) (c,t)-threshold protocol running in the preprocessing model.*

Proof: In [Sho00] exactly such a protocol is described, which is secure in the model that we consider here, i.e. asynchronous communication and a static adversary. The protocol uses the random oracle model to get non-interactive proofs of equality of discrete logarithms, but we can dispense of the random oracle by doing interactive zero-knowledge proofs. ∎

The round complexity of the protocol from [Sho00], using interactive zero-knowledge proofs, is 3, and the communication complexity is about $3n^2k$ bits.

Our threshold signature protocol $\pi_{\text{tsig},c}$ is given by the following description.

Oracles The protocol runs in the hybrid-model with access to an (n,c)-threshold trapdoor permutation functionality $\mathcal{F}_{F,c}$ and a c-threshold random oracle from $\mathcal{M}$ to $\{0,1\}^k$.
  We denote the value output by the random oracle on input M by $H(M)$.

Sign
  1. If a party P_i gets input (sign, j, M), then P_i inputs $(\text{evaluate}, i, M)$ to the random oracle and sends a message to all other parties instructing them to do the same.
  2. If the party later sees the output $(\text{evaluate}, M, H(M))$ from the random oracle, then the party inputs $(\text{invert}, j, H(M))$ to $\mathcal{F}_{F,c}$.
  3. If a party have received $(\text{evaluate}, M, H(M))$ from the random oracle and $(\text{invert}, f^{-1}(H(M)))$ from $\mathcal{F}_{F,c}$, then the party outputs $(\text{signature}, M)$.

Send
  1. If a party P_i gets input (send, j, M) and receives $(\text{evaluate}, M, H(M))$ from the random oracle and receives $(\text{invert}, f^{-1}(H(M)))$ from $\mathcal{F}_{F,c}$, then P_i sends the message $(\text{send}, M, f^{-1}(H(M)))$ to P_j.
  2. If a party P_j receives $(\text{send}, M, f^{-1}(H(M)))$ from party P_i and receives $(\text{evaluate}, M, H(M))$ from the random oracle, then P_j outputs (send, i, M).

Theorem 6. *For $c < n - t$, the protocol $\pi_{\text{tsig},c}$ t-securely realizes the functionality $\mathcal{F}_{\text{tsig},c}$.*

Proof (sketch): We give a slightly informal proof of the theorem, by arguing correctness (if more than c honest parties give a honest P_j permission to sign M, then P_j will eventually output (signature, M)) and non-forgeability (if at most c honest parties give any other party permission to sign M, then no honest party will ever output (signature, M) or (send, i, M)). Constructing a formal proof by formulating the proof as a simulator is an easy task and is left to the reader.

Correctness: If more than c honest parties give P_j permission to sign M, then because of the non-blocking assumption all these parties will eventually receive $H(M)$ and give P_j permission to invert on $H(M)$. Because of the non-blocking assumption P_j will eventually receive $H(M)$ and $f^{-1}(H(M))$ and will output (signature, M).

Non-forgeability: If at most c honest parties give any other party permission to sign M, then $\mathcal{F}$ will never output $f^{-1}(H(M))$ and especially no honest party will ever output (signature, M). For a honest party to output (send, i, M) the party then has to receive the value $f^{-1}(H(M))$ from (corrupted) P_i. Since $\mathcal{F}$ did not output $f^{-1}(H(M))$ and $H(M)$ is uniformly random, this requires the adversary to break the one-wayness of the trapdoor permutation family F. ∎

To sign a message, one call to the oracle functionality and one call to the threshold permutation functionality is used. Using the implementations of Theorems 4 and 5 instead of the ideal functionalities, the round complexity is 6 and the communication complexity is about $6n^2k$ bits, if consecutive values are signed. The overhead over the protocol in [Sho00] is a factor 6. If arbitrary values are signed, the overhead will be in the order $3l + 3$, where l is the output-length of the hash-function used for hashing the messages.

7   Byzantine Agreement

The protocol $\pi_{\text{ba},t}$ is given by the following description.

Oracles The protocol runs in the hybrid-model with access to an $(n - t, t)$-threshold signature functionality and a $(t+1, t)$-threshold signature functionality. By *input* ($sign_{n-t}$, M) we mean *input* ($sign$, j, M) *to the* $(n - t, t)$-*threshold signature functionality for all* j and we use *input* ($sign_{t+1}$, M) similarly. The protocol also has access to a $(n - t, t)$-threshold random oracle. By *flip the coin* C_r we mean *input* ($evaluate$, r) *to the random oracle and take* C_r *to be the first bit of the output.*

Decide If a party receives a signature on (main, r, o) for some $r \in N$ and some $o \in \{0, 1\}$, then it sends it to all parties and terminates with output (decide, vid, o). In the remaining description we drop the vote id vid.

Initial Round. On input (vote, b_i), party P_i follows the code described below.
Initialize the round counter $r \leftarrow 0$.
Input ($sign_{t+1}$, (pre, 0, b_i)) and wait for $2t + 1$ parties to do the same, i.e. wait until the $(t + 1, t)$-threshold signature functionality has output ($sign$, j, (pre, 0, b_j)) for $b_j \in \{0, 1\}$ for $2t + 1$ different j.

Wait for a signature on some $(\mathtt{pre}, 0, b)$ and input $(\mathtt{sign}_{n-t}, (\mathtt{main}, 0, b))$. Wait for $n - t$ parties to do the same.

Flip coin. Let $r = r + 1$ and flip coin C_r.

Pre-vote.

1. If for some $b \in \{0, 1\}$ a signature on $(\mathtt{pre}, r - 1, b)$ is known, then send the signature to all parties and input $(\mathtt{sign}_{n-t}, (\mathtt{pre}, r, b))$.
2. If a signature on $(\mathtt{main}, r - 1, \perp)$ is known, then send the signature to all parties and input $(\mathtt{sign}_{n-t}, (\mathtt{pre}, r, C_r))$.

Say that a party P_j has pre-voted when values are received verifying that the party did one of the above.

Pre-collect Wait until $n - t$ parties have pre-voted. This either gives a signature on some $(\mathtt{pre}, r, b)$ (if all pre-voted in the same way or $b = C_r$) or (otherwise) gives a signature on $(\mathtt{pre}, r - 1, 1 - C_r)$.

Main-vote.

1. If a signature on $(\mathtt{pre}, r, b)$ is known, then send the signature to all parties and input $(\mathtt{sign}_{n-t}, (\mathtt{main}, r, b))$.
2. If a signature on $(\mathtt{pre}, r - 1, 1 - C_r)$ is known, input $(\mathtt{sign}_{n-t}, (\mathtt{main}, r, \perp))$.

Say that a party P_j has main-voted when values are received verifying that he did one of the above.

Main-collect Wait until $n - t$ parties have main-voted. This either gives a signature on $(\mathtt{main}, r, \perp)$ or gives a signature on some $(\mathtt{pre}, r, b)$.

Go to **Flip coin.**

Theorem 7. *For t where $n > 3t$ the protocol $\pi_{ba,t}$ t-securely realizes $\mathcal{F}_{ba,t}$ in the preprocessing model.*

Proof (sketch): We give an informal proof that the ideal functionality and the protocol behaves consistently. A full proof can be constructed along the lines of the proof in [CKS00].

If at least $n - t$ honest parties agree on the value b, then at most t parties will allow to sign $(\mathtt{pre}, 0, 1 - b)$ and thus it will not be signed. Therefore all honest parties that come through the initial round have seen $n - t$ parties send $(\mathtt{pre}, 0, b)$ and input $(\mathtt{sign}_{n-t}, (\mathtt{main}, 0, b))$, and will thus terminate with decision b. The non-blocking assumption guarantees that all honest parties terminate. This justifies the second decision rule of the ideal BA functionality. The same line of reasoning shows that if less than $t + 1$ parties voted $1 - b$, then all parties that terminate, do so after the initial round with output b. However, if less than $n - t$ *honest* parties participates, termination is not guaranteed. This justifies the first decision rule.

To justify the third decision rule, we must argue that if at least $n - t$ honest parties participate, then all honest parties terminate, and agree. For termination, call a round randomizing if when party number $n - t$ contributed to flipping C_r, for one of the values $b \in \{0, 1\}$ it was impossible for $(\mathtt{pre}, r - 1, b)$ to be signed. We need the fact that out of two consecutive rounds, at least one is randomizing. To see this assume that exactly $n - t$ parties have contributed to

C_r and round r is not randomizing. If any of the contributors had seen a signature on $(\text{pre}, r - 1, b)$, then $n - t$ parties allowed to sign it and thus at least $n - 2t$ honest parties allowed to sign it. This means that at most $2t$ parties would allow to signed $(\text{pre}, r-1, 1-b)$, which will thus never be signed. So, since round r is not randomizing, no party saw a signature on any $(\text{pre}, r-1, b)$. So, the $n - 2t$ honest of them are going to input $(\text{sign}_{n-t}, (\text{pre}, r, C_r))$, and thus $(\text{pre}, r, 1 - C_r))$ will never be signed, which proves that round $r - 1$ is randomizing. Assume then that $n - t$ honest parties have received some input. Since at no point more than $n - t$ parties are waited for, if none of them terminates, arbitrary many rounds will be executed. This means that arbitrary many randomizing rounds are executed, a contradiction as there is a probability of at least $\frac{1}{2}$ that the protocol terminates after a randomizing round (namely if $C_r = 1 - b$, where $(\text{pre}, r - 1, b)$ is never signed). We then argue agreement. For any party to terminate there must be a least r such that $n - t$ parties allowed to sign (main, r, b) for some b. This means that $(\text{main}, r, \perp)$ will never be signed and that (pre, r, b) must have been signed. This means that any honest party that goes through **Main-collect** sees $n - t$ parties go through case one in **Main-vote** and receives a signature on (main, r, b). The parties that do not get through **Main-collect** will receive a copy of (main, r, b) sent be a terminating party.

Finally, there is an implicit fourth decision rule: If less that $n - t - t'$ honest parties, where t' is the number of corrupted parties, participate, none of the other rules apply no matter what the adversary does, and thus the ideal functionality will dead-lock. However, in the protocol less than $n - t$ parties will participate, so no (main, r, b) can get signed, so the protocol behaves accordingly. ∎

If the honest parties do not terminate after the initial round (which costs the signing of two messages), they terminate after an average of four iterations of the main loop as every second round is randomizing. One iteration costs the signing of two messages and one random oracle call. Therefore, the expected round complexity will be at most 72 and the expected communication complexity will be about $72n^2k$ bits. These complexities are about a factor 5 larger than for the protocol in [CKS00].

Acknowledgments

I'm grateful to Ivan Damgård for many stimulating conversations.

References

[BCF00] Ernest F. Brickell, Giovanni Di Crescenzo, and Yair Frankel. Sharing block ciphers. In Ed Dawson, Andrew Clark, and Colin Boyd, editors, *Information Security and Privacy, 5th Australasian Conference, ACISP 2000, Brisbane, Australia, July 10-12, 2000, Proceedings*, pages 457–470. Springer, 2000.

[BDJR97] M. Bellare, A. Desai, E. Jokipii, and P. Rogaway. A concrete security treatment of symmetric encryption. In *38th Annual Symposium on Foundations of Computer Science* [IEE97].

[Can01]   Ran Canetti. Universally composable security: A new paradigm for crypto-graphic protocols. In *42th Annual Symposium on Foundations of Computer Science*. IEEE, 2001.

[CGH98]   Ran Canetti, Oded Goldreich, and Shai Halevi. The random oracle method-ology, revisited (preliminary version). In *Proceedings of the Thirtieth Annual ACM Symposium on the Theory of Computing*, pages 209–218, Dallas, TX, USA, 24–26 May 1998.

[CHH00]   Ran Canetti, Shai Halevi, and Amir Herzberg. Maintaining authenti-cated communication in the presence of break-ins. *Journal of Cryptology*, 13(1):61–106, winter 2000.

[CKS00]   Christian Cachin, Klaus Kursawe, and Victor Shoup. Random oracles in constantinople: Practical asynchronous byzantine agreement using cryp-tography. In *Proceedings of the 19th ACM Symposium on Principles of Distributed Computing (PODC 2000)*, pages 123–132. ACM, July 2000.

[CP92]   D. Chaum and T. P. Pedersen. Wallet databases with observers. In Ernest F. Brickell, editor, *Advances in Cryptology - Crypto '92*, pages 89–105, Berlin, 1992. Springer-Verlag. Lecture Notes in Computer Science Volume 740.

[GGM86]   Oded Goldreich, Shafi Goldwasser, and Silvio Micali. How to construct random functions. *Journal of the ACM*, 33(4):792–807, 1986.

[Gol01]   Oded Goldreich. *The Foundations of Cryptography*, volume 1. Cambridge University Press, 2001.

[IEE97]   IEEE. *38th Annual Symposium on Foundations of Computer Science*, Mi-ami Beach, FL, 19–22 October 1997.

[MS95]   Silvio Micali and Ray Sidney. A simple method for generating and sharing pseudo-random functions, with applications to clipper-like escrow systems. In Don Coppersmith, editor, *Advances in Cryptology - Crypto '95*, pages 185–196, Berlin, 1995. Springer-Verlag. Lecture Notes in Computer Science Volume 963.

[Nie02]   Jesper B. Nielsen. Separating random oracle proofs from complexity theo-retic proofs: the non-committing encryption case. In *Advances in Cryptology - Crypto '02*, 2002.

[NPR99]   Moni Naor, Benny Pinkas, and Omer Reingold. Distributed pseudo-random functions and KDCs. In Jacques Stern, editor, *Advances in Cryptology - EuroCrypt '99*, pages 327–346, Berlin, 1999. Springer-Verlag. Lecture Notes in Computer Science Volume 1592.

[NR97]   Moni Naor and Omer Reingold. Number-theoretic constructions of efficient pseudo-random functions (extended abstract). In *38th Annual Symposium on Foundations of Computer Science* [IEE97], pages 458–467.

[Sho00]   Victor Shoup. Practical threshold signatures. In Bart Preneel, editor, *Advances in Cryptology - EuroCrypt 2000*, pages 207–220, Berlin, 2000. Springer-Verlag. Lecture Notes in Computer Science Volume 1807.

Efficient Computation Modulo a Shared Secret with Application to the Generation of Shared Safe-Prime Products

Joy Algesheimer, Jan Camenisch, and Victor Shoup

IBM Research
Zurich Research Laboratory
CH–8803 Rüschlikon
{jmu,jca,sho}@zurich.ibm.com

Abstract. We present a new protocol for efficient distributed computation modulo a shared secret. We further present a protocol to distributively generate a random shared prime or safe prime that is much more efficient than previously known methods. This allows one to distributively compute shared RSA keys, where the modulus is the product of two safe primes, much more efficiently than was previously known.

Keywords: RSA, safe primes, threshold cryptography, distributed primality test.

1 Introduction

Many distributed protocols, e.g., [14, 17, 19], require that an RSA modulus $N = pq$ is generated during system initialization, together with a public exponent e and shares of the corresponding private exponent. Moreover, many protocols, e.g., [23, 11, 18, 3, 7], even require that N is the product of "safe" primes, i.e., $p = 2p' + 1$ and $q = 2q' + 1$, where p' and q' are themselves prime. While the requirement for safe primes can sometimes be avoided (e.g., [13, 15]), this typically comes at the cost of extra communication, computation, and/or non-standard intractability assumptions.

While the initialization of the system with an RSA modulus N can be accomplished using a "trusted dealer," it would be preferable not to rely on this.

Given a distributed protocol to generate a random (safe) prime, securely shared among the players, it is not too difficult to solve the above problem. One can of course use general multi-party computation techniques of Ben-Or, Goldwasser and Wigderson [5] to generate a random, shared (safe) prime. Indeed, that would work as follows: one starts with a standard algorithm for generating a random (safe) prime, and converts this algorithm into a corresponding Boolean or arithmetic circuit, and then for each gate in this circuit, the players perform a distributed multiplication modulo a small prime t. This protocol is not very practical, especially as the players need to perform a distributed computation for *every gate* in the circuit, and so unlike in the non-distributed prime generation case, they cannot use much more efficient algorithms and practical implementation techniques for working with large integers.

M. Yung (Ed.): CRYPTO 2002, LNCS 2442, pp. 417–432, 2002.
© Springer-Verlag Berlin Heidelberg 2002

In this paper, we present new protocols that allow one to perform arithmetic modulo a secret, shared modulus in a way that is much more efficient than can be done using the general techniques of Ben-Or et al. More specifically, we develop a new protocol to efficiently compute shares of c, where $c \equiv ab \pmod{p}$, given shares of a, b, and p. The shares of a, b, c, and p are integers modulo Q, where Q is a prime whose bit-length is roughly twice that of p, and the cost of this protocol is essentially the cost of performing a small, constant number of distributed multiplications modulo Q. Actually, this is the amortized cost of multiplication modulo p assuming many such multiplications are performed for a fixed p. This protocol, together with several other new supporting protocols, gives us a protocol to generate a random, shared prime, or safe prime, that is much more efficient than the generically derived protocol discussed above. In particular, we obtain a protocol for jointly generating an RSA modulus that is the product of two *safe* primes that is much more efficient in practice than any generic circuit-based protocol (which are the only previously known protocols for this problem), even using the most efficient circuits for integer multiplication, division, etc.

Our protocols work in the so-called "honest-but-curious" model. That is, we assume that all players follow the protocol honestly, but we guarantee that even if a minority of players "pool" their information they cannot learn anything that they were not "supposed" to. Even though we make this restriction, fairly standard techniques can be used to make our protocols robust, while maintaining their practicality. In fact, using "optimistic" techniques for robustness, we can obtain a fully robust protocol for distributively generating an RSA modulus that is not significantly less efficient than our honest-but-curious solution – this is the subject of on-going work.

Related Work. Boneh and Franklin [6] present a protocol for jointly generating an RSA modulus $N = pq$ along with a a public exponent and shares of the corresponding private key. Like us, they also work in the honest-but-curious adversary model. Unlike ours, their protocol is not based on a sub-protocol for generating a random, shared prime. While our protocol for this task is asymptotically more efficient than the protocol of Boneh and Franklin (when the number of players is small), we do not claim that our protocol is in practice more efficient than theirs for typical parameter choices. The relative performance of these protocols in such a practical setting depends on a myriad of implementation details.

Unlike our techniques, those of Boneh and Franklin do not give rise to a protocol for jointly generating an RSA modulus $N = pq$, where p and q are safe primes. Indeed, prior to our work, the only known method for solving this problem was to apply the much less efficient general circuit technique of Ben-Or et al. [5].

As our protocols rely mainly on distributed multiplication over a prime field, rather than over the integers, one can easily make them robust using traditional techniques for verifiable secret sharing modulo a prime, avoiding the somewhat less efficient techniques by Frankel et al. [16] for robust distributed multiplication over the integers. Moreover, using the optimistic approach mentioned above, even further improvements are possible, so that we can get robustness essentially "for free."

2 Model

We consider k players $P_1, \ldots, P_k$ that are mutually connected by secure and authentic channels. Our protocols are secure against a static and honest-but-curious behaving adversary, controlling up to $\tau = \lfloor \frac{k-1}{2} \rfloor$ players. That is, all players follow the protocol honestly but the dishonest players may pool their data and try to derive additional information. We finally assume that no party stops participating prematurely (we use k-out-of-k secret sharing schemes).

However, these assumptions can be relaxed: First, it's possible to force the participants to behave honestly by having them to commit to their inputs, to generate their individual random strings jointly, and to prove (using zero-knowledge proofs) that they followed the protocols correctly. Second, the k-out-of-k secret sharing schemes can easily be converted into $\tau + 1$-out-of-k ones by the 'share back-up' method introduced by Rabin [21]. We do not pursue these possibilities here.

We prove security in the model by Canetti [8]. Here, we describe a simplified version of it for a static adversary in the honest-but-curious model. Such an adversary first chooses the players he wants to corrupt and then gets to see their inputs, their internal state and all the messages they receive. A protocol π is proved secure by specifying the functionality f the protocol should provide in an ideal world where all the parties send their inputs to a trusted third party T who then returns to them the outputs they are to obtain according to f. Let $\pi_i(x_1, \ldots, x_k, \rho)$ denote the output of party P_i when running protocol π on input x_i in the presence of adversary $\mathcal{A}$, where ρ is a security parameter. As $\mathcal{A}$ has honest-but-curious behavior, the output $\pi_i(x_1, \ldots, x_k, \rho)$ does not depend on $\mathcal{A}$.

Definition 1. *A protocol is said to be statistically secure if for any honest-but-curious behaving adversary $\mathcal{A}$ there exists a probabilistic polynomial-time simulator S such that the two ensembles of random variables*

$$\{\mathcal{A}(z), \pi_1(x_1, \ldots, x_k, \rho), \ldots, \pi_k(x_1, \ldots, x_k, \rho)\}_{\rho \in \mathbb{N}; z, x_1, \ldots, x_k \in \{0,1\}^*}$$

and

$$\{S(z), f_1(x_1, \ldots, x_k, \rho), \ldots, f_k(x_1, \ldots, x_k, \rho)\}_{\rho \in \mathbb{N}; z, x_1, \ldots, x_k \in \{0,1\}^*}$$

are statistically indistinguishable.

It can be shown that security in this sense is preserved under non-concurrent, modular composition of protocols [8].

3 Preliminaries

3.1 Notation

Let a be a real number. We denote by $\lfloor a \rfloor$ the largest integer $b \leq a$, by $\lceil a \rceil$ the smallest integer $b \geq a$, and by $\lceil a \rfloor$ the largest integer $b \leq a + 1/2$. We denote by trunc(a) the integer b such that $b = \lceil a \rceil$ if $a < 0$ and $b = \lfloor a \rfloor$ if $a \geq 0$; that is, trunc(a) rounds a towards 0.

Let Q be a positive integer. All modular arithmetic is done centered around 0; to remind the reader of this, we use 'rem' as the operator for modular reduction rather than 'mod', i.e., $c \operatorname{rem} Q$ is $c - \lceil c/Q \rfloor Q$.

Define $\mathcal{Z}_Q$ as the set $\{x \in \mathbb{Z} \mid -Q/2 < x \leq Q/2\}$ (we should emphasize that $\mathcal{Z}_Q$ is properly viewed as a set of integers rather than a ring). We denote an additive sharing of a value $a \in \mathcal{Z}_Q$ over $\mathcal{Z}_Q$ by $\langle a \rangle_1^Q, \ldots, \langle a \rangle_k^Q \in \mathcal{Z}_Q$, i.e., $a = \sum_{j=1}^k \langle a \rangle_j^Q \operatorname{rem} Q$ and by $[a]_1^Q, \ldots, [a]_k^Q \in \mathcal{Z}_Q$ we denote a polynomial sharing (also called Shamir-sharing [22]), i.e., $a = \sum_{j=1}^\tau \lambda_j [a]_j^Q \operatorname{rem} Q$, where λ_j are the Lagrange coefficients. The latter only works if $Q > k$ and if Q is prime.

For $a \in \mathcal{Z}$ we denote by $\langle a \rangle_1^I, \ldots, \langle a \rangle_k^I \in \mathcal{Z}$ an additive sharing of a over the integers, i.e., $a = \sum_{j=1}^k \langle a \rangle_j^I$.

We denote protocols as follows: the term $b := \textsf{PROTOCOLNAME}(a)$ means that the player in consideration runs the protocol $\textsf{PROTOCOLNAME}$ with local input a and gets local output b as the result of the protocol. Finally, $\lg(x)$ denotes the logarithm of x to the base 2.

3.2 Known Primitives

We recall the known secure multi-party protocols for efficient distributed computation with shared secrets that we will use to compose our protocols, and we state the number of bit-operations for which we assume $\lg Q = \Theta(n)$ and that the bit-complexity of a multiplication of two n-bit integers is $O(n^2)$ (which is a reasonable estimate for realistic values of n, e.g., $n = 1024$). The round-complexity of all primitives is $O(1)$ and their communication is $O(kn)$ bits (we consider communication complexity to be the number of bits each player sends on average).

Additive sharing over $\mathcal{Z}_Q$: To share a secret $a \in \mathcal{Z}_Q$ player P_j chooses $\langle a \rangle_i^Q \in_R \mathcal{Z}_Q$ for $i \neq j$, sets $\langle a \rangle_j^Q := a - \sum_{i=1, i \neq j}^k \langle a \rangle_i^Q \operatorname{rem} Q$, and sends $\langle a \rangle_i^Q$ to player P_i. This takes $O(kn)$ bit operations.

Polynomial sharing over $\mathcal{Z}_Q$: To share a secret $a \in \mathcal{Z}_Q$ player P_j chooses coefficients $a_l \in_R \mathcal{Z}_Q$ for $l = 1, \ldots, \tau$, where $\tau = \lfloor (k-1)/2 \rfloor$, and sets $[a]_i^Q := a + \sum_{l=1}^\tau a_l i^l \operatorname{rem} Q$, and sends $[a]_i^Q$ to player P_i. This takes $O(nk^2 \lg k)$ bit operations.

Additive sharing over $\mathbb{Z}$: To share a secret $a \in [-A, A]$ player P_j chooses $\langle a \rangle_i^I \in_R [-A2^\rho, A2^\rho]$ for $i \neq j$, where ρ is a security parameter, and sets $\langle a \rangle_j^I := a - \sum_{i=1, i \neq j}^k \langle a \rangle_i^I$, and sends $\langle a \rangle_i^I$ to player P_i. Note that for any set of $k-1$ players, the distribution of shares of different secrets are statistically indistinguishable for suitably large ρ (e.g., $\rho = 128$). This takes $O(k(\rho + \lg A))$ bit operations.

Distributed computation over $\mathcal{Z}_Q$: Addition and multiplication modulo Q of a constant and a polynomially shared secret is done by having all players locally add or multiply the constant to their shares. Hence $[a]_j^Q + c \operatorname{rem} Q$ is a polynomial share of $a + c \operatorname{rem} Q$ and $c \cdot \langle a \rangle_j^Q \operatorname{rem} Q$ is a polynomial share of

ac rem Q. These operations take $O(n)$ and $O(n^2)$ bit operations, respectively. Addition of two shared secrets is achieved by having the players locally add their shares. Thus $[a]_j^Q + [b]_j^Q$ rem Q is a polynomial share of $a + b$ rem Q and takes $O(\lg Q)$ bit operations.

Multiplication modulo Q of two polynomially shared secrets is done by jointly executing a multiplication protocol due to Ben-Or, Goldwasser and Wigderson [5] or by a more efficient variant due to Gennaro, Rabin and Rabin [20] which requires $O(n^2 k + nk^2 \lg k)$ bit operations for each player. We denote this protocol by $\mathsf{MUL}([a]_j^Q, [b]_j^Q)$.

Joint random sharing over $\mathcal{Z}_Q$: To generate shares of a secret chosen jointly at random from $\mathcal{Z}_Q$, each player chooses a random number $r_i \in_R \mathcal{Z}_Q$ and shares it according to the required type of secret sharing scheme and sends the shares to the respective players. Each player adds up all the shares gotten to obtain a share of a random value. We denote this protocol by $\mathsf{JRS}(\mathcal{Z}_Q)$ in case the players get additive shares and by $\mathsf{JRP}(\mathcal{Z}_Q)$ if they get polynomial shares. The protocols require $O(nk)$ and $O(nk^2 \lg k)$ bit operations per player, respectively.

Joint random sharing of 0: In protocols it is often needed to re-randomized shares obtained from some computation by adding random shares of 0. Such shares can be obtained for any sharing scheme by having each player share 0 according to the required type of secret sharing scheme and sending them to the respective players. Each player adds up all the shares gotten to obtain a share of 0. We denote this protocol by $\mathsf{JRSZ}(\mathcal{Z}_Q)$ in case the players get additive shares over $\mathcal{Z}_Q$ and $\mathsf{JRPZ}(\mathcal{Z}_Q)$ if they get polynomial shares over $\mathcal{Z}_Q$. The protocols require $O(nk)$ and $O(nk^2 \lg k)$ bit operations per player, respectively. In case we want to have additive shares over the integers, it is required to give the range (e.g., $[-A, A]$) from which the players choose the shares they send to the other players. We denote this protocol by $\mathsf{JRIZ}([-A, A])$ and it requires $O(k(\rho + \lg A))$ bit operations per player.

Computing shares of the inverse of a shared secret: This protocol works only for polynomial sharings over $\mathcal{Z}_Q$. Let a be the shared invertible element. Then, using a protocol due to Bar-Ilan and Beaver [4], the players can compute shares of a^{-1} rem Q given shares $[a]_j^Q$. The protocol, denoted by $\mathsf{INV}([a]_j^Q)$, is as follows: first run $[r]_j^Q := \mathsf{JRP}(\mathcal{Z}_Q)$, then compute $[u]_j^Q := \mathsf{MUL}([a]_j^Q, [r]_j^Q)$, reveal $[u]_j^Q$, and reconstruct u. If $u \equiv 0 \pmod{Q}$, the players start over. Otherwise, they each locally compute their share of a^{-1} rem Q as $(u^{-1}$ rem $Q) \cdot [r]_j^Q$ rem Q. This protocol requires an expected number of $O(n^2 k + nk^2 \lg k)$ bit operations per player.

Joint random invertible element sharing: This protocol, denoted $\mathsf{JRP\text{-}INV}(\mathcal{Z}_Q)$, is due to Bar-Ilan and Beaver [4]. The players generate shares of random elements $[r]_j^Q := \mathsf{JRP}(\mathcal{Z}_Q)$ and $[s]_j^Q := \mathsf{JRP}(\mathcal{Z}_Q)$, jointly compute $[u]_j^Q := \mathsf{MUL}([s]_j^Q, [r]_j^Q)$, reveal $[u]_j^Q$ and then reconstruct u. If u is non-zero, they each take $[r]_j^Q$ as their share of a random invertible element. Otherwise, they repeat the protocol. The protocol requires an expected number of $O(nk^2 \lg k + n^2 k)$ bit operations per player.

4　Conversions between Different Sharings

In our protocols, we work with all three secret sharing schemes introduced in the previous section. For this we need methods to convert shares from one sharing scheme into shares of another one. This section reviews the known methods for such transformations and provides a method to transform additive shares over $\mathcal{Z}_Q$ into additive shares over the integers. The latter is apparently new. The section also provides a method to obtain shares of the bits of a shared secret.

4.1　Converting between Integer Shares and $\mathcal{Z}_Q$ Shares

It is well known how to convert additive shares modulo Q into polynomial shares modulo Q and vice versa: If the players hold additive (or polynomial) shares of a value a they re-share those with a polynomial (additive) sharing and send the shares to the respective players, which add up (or interpolate) the received shares to obtain a polynomial (or additive) share of a. We denote the first transformation by $\mathsf{SQ2PQ}(\cdot)$ and the latter by $\mathsf{PQ2SQ}(\cdot)$.

Conversions between shares over the integers into shares over $\mathcal{Z}_Q$ naturally requires that $Q/2$ is bigger than the absolute shared value. If this is the case, an additive sharing $\langle c \rangle_1^I, \ldots, \langle c \rangle_k^I$ over the integers of a secret c with $-2^{n-1} < c < 2^{n-1} < Q/2$ can be converted in an additive sharing over $\mathcal{Z}_Q$ (and thus also a polynomial sharing) by reducing the shares modulo Q, i.e., $\langle c \rangle_i^Q := \langle c \rangle_i^I \operatorname{rem} Q$. We denote this transformation by $\mathsf{SI2SQ}(\cdot)$.

Obtaining additive shares over the integers from additive shares over $\mathcal{Z}_Q$ is not so straightforward. The main problem is that if one considers the additive shares over $\mathcal{Z}_Q$ as additive shares over the integers then one is off by an unknown multiple Q, the multiple being the quotient of the sum of these shares and Q. However, if the shared secret is sufficiently smaller than Q (i.e., ρ bits smaller, where ρ is a security parameter), then the players can reveal the high-order bits of their shares without revealing anything about the secret. Knowledge of these high-order bits is sufficient to compute the quotient. This observation leads to the following protocol.

Let $\langle c \rangle_j^Q \in \mathcal{Z}_Q$ be the share of party P_j and let $-2^{n-1} < c = \sum_i \langle c \rangle_i^Q \operatorname{rem} Q < 2^{n-1}$. If $Q > 2^{\rho+n+\lg k+4}$ holds, the parties can use the following protocol to securely compute additive shares of c over the integers.

Protocol $\mathsf{SQ2SI}(\langle c \rangle_j^Q)$:
Let $t = \rho + n + 2$. Party P_j executes the following steps.

1. Reveal $a_j := \operatorname{trunc}(\frac{\langle c \rangle_j^Q}{2^t})$ to all other parties.
2. Compute $l := \left\lceil \frac{2^t \sum_i a_i}{Q} \right\rceil$.
3. Run $\langle 0 \rangle_j^I := \mathsf{JRIZ}([-Q2^\rho, Q2^\rho])$.
4. If $j \le |l|$ set the output to $\langle c \rangle_j^I := \langle c \rangle_j^Q - Q + \langle 0 \rangle_j^I$ if $l > 0$ and to $\langle c \rangle_j^I := \langle c \rangle_j^Q + Q + \langle 0 \rangle_j^I$ if $l < 0$.
   If $j > |l|$ set the output to $\langle c \rangle_j^I = \langle c \rangle_j^Q + \langle 0 \rangle_j^I$.

Theorem 1. *Let* $\langle c \rangle_1^Q, \ldots, \langle c \rangle_k^Q$ *be a random additive sharing of* $-2^{n-1} \leq c < 2^{n-1}$. *If* $\lg Q > \rho + n + \lg k + 4$, *where* ρ *is a security parameter, then the protocol* $\mathsf{SQ2SI}(\langle c \rangle_j^Q)$ *securely computes additive shares of* c *over the integers.*

Proof. We have to provide a simulator that interacts with the ideal world trusted party T and produces an output indistinguishable from that of the adversary. The trusted party T gets as input the shares $\langle c \rangle_1^Q, \ldots, \langle c \rangle_k^Q$, computes c and re-shares c over the integers by choosing integer shares of 0 the same way as it would be done if the parties ran the protocol $\langle 0 \rangle_i^I := \mathsf{JRIZ}([-Q2^\rho, Q2^\rho])$. Then T sets $\langle c \rangle_1^I := \langle 0 \rangle_1^I + c$ and $\langle c \rangle_i^I := \langle 0 \rangle_i^I$ for $i \neq 1$, and then sends $\langle c \rangle_i^I$ to player P_i. Note that the players' outputs are additive shares of c with the right distribution (i.e., the distribution of any subset of $k - 1$ shares is statistically close to the distribution of the corresponding subset if another value c' was shared).

A simulator is as follows: it forwards the inputs $\langle c \rangle_i^Q$ of the corrupted players to T and obtains the shares $\langle c \rangle_i^I$ for these players from T. It extends the set of shares $\langle c \rangle_i^Q$ of the corrupted players into a full (and random) sharing of any valid c' (e.g., 0). Let $r_1, \ldots, r_n$ be the thereby obtained shares. The simulator then computes $a_i = \mathsf{trunc}(\frac{r_i}{2^t})$ and lets the adversary know the a_i's that the corrupted players would receive in the protocol. Then the simulator computes $l = \left\lceil \frac{2^t \sum_i a_i}{Q} \right\rceil$ and, for every i where Party P_i is corrupted, it sets

$$\langle 0 \rangle_i^I := \begin{cases} \langle c \rangle_i^I - \langle c \rangle_i^Q + Q & \text{if } l > 0, i \leq |l| \\ \langle c \rangle_i^I - \langle c \rangle_i^Q - Q & \text{if } l < 0, i \leq |l| \\ \langle c \rangle_i^I - \langle c \rangle_i^Q & \text{otherwise.} \end{cases}$$

The simulator finally runs the simulator for $\mathsf{JRIZ}([-Q2^\rho, Q2^\rho])$ such that these shares $\langle 0 \rangle_i^I$ are the outputs of the corrupted players. Finally the simulator stops, outputting whatever the adversary outputs.

It remains to show that for this simulator the distributions of the players' and the simulators outputs are statistically indistinguishable from the views and outputs of the players and the adversary when running protocol $\mathsf{SQ2SI}(\langle c \rangle_j^Q)$.

Let us first prove that the players' outputs of protocol $\mathsf{SQ2SI}(\langle c \rangle_j^Q)$ are indeed shares of c. Let $\hat{l} = \left\lceil \frac{\sum_i \langle c \rangle_i^Q}{Q} \right\rceil$. Thus $c = \sum_i \langle c \rangle_i^Q - \hat{l} Q$ fulfills $|c| < 2^{n-1}$ by assumption. Define $b_i = \langle c \rangle_i^Q - a_i 2^t$. Note that $|b_i| < 2^t$. We have to show that $l = \hat{l}$. As $\sum_i a_i 2^t = c + \hat{l} Q - \sum_i b_i$ we have $l = \left\lceil \frac{2^t \sum_i a_i}{Q} \right\rceil = \left\lceil \frac{c}{Q} + \hat{l} - \frac{\sum_i b_i}{Q} \right\rceil$. Because $\hat{l}$ is an integer, we have $l = \hat{l}$ if $|\frac{c}{Q}| < 1/4$ and $|\frac{\sum_i b_i}{Q}| < 1/4$, that is, if $n < \lg Q - 2$ and $2 + t + \lg k = \rho + n + \lg k + 4 < \lg Q$ holds. As $\langle c \rangle_i^Q \in \mathcal{Z}_Q$ we have $|l| < k$ and thus $c = \sum_i \langle c \rangle_i^Q - lQ = \sum_i \langle c \rangle_i^I$. Furthermore it is easy to see that the distribution of the shares output is statistically close to the ones produced by T.

Let us now show that the distribution of the a_i's for different shared values c are statistically indistinguishable. We consider the probability that the a_i's take different values if a different value of c was shared. W.l.o.g., we can assume that $\langle c \rangle_1^Q, \ldots, \langle c \rangle_{k-1}^Q$ are random elements from $\mathcal{Z}_Q$ and that $\langle c \rangle_k^Q = c -$

$\sum_{i=1}^{k-1} \langle c \rangle_i^Q$ rem Q. Clearly, the values $a_1 = \mathrm{trunc}(\frac{\langle c \rangle_1^Q}{2^t}), \ldots, a_{k-1} = \mathrm{trunc}(\frac{\langle c \rangle_{k-1}^Q}{2^t})$ do not depend on the shared value. It remains to consider a_k. We have $\langle c \rangle_k^Q$ rem $Q = a_k 2^t + b_k$ with $b_k < 2^t$. First note that $C = -\sum_{i=1}^{k-1} \langle c \rangle_i$ rem Q is uniformly distributed over $\mathcal{Z}_Q$ and that $Q > 2^t$. If $C > Q - 2^n$ or if C rem $2^t > 2^t - 2^n$ then a_k takes a value that depends on c. These conditions are fulfilled with probability at most $\frac{2^n + 2^n}{2^t + 2^n} < \frac{2^{n+1}}{2^t} = 2^{-t+n+1}$. Therefore, the statistical difference between the distribution of the a_i's for different shared values must be smaller than $2 \cdot 2^{-t+n+1} = 2^{-t+n+2} = 2^{-\rho}$.

As the $\mathsf{JRIZ}([-Q2^\rho, Q2^\rho])$ protocol is secure, the distributions of the outputs in the real world and the outputs of the ideal world with our simulator are statistically indistinguishable.

Combining the above protocols, we can move from polynomial shares over $\mathcal{Z}_Q$ to additive shares over the integers and vice versa. The bit-complexities for these conversions are $O(nk^2 \lg k + n^2 k)$ and $O(nk^2 \lg k)$, respectively. For both, the communication-complexity is $O(kn)$ bits and the round-complexity is $O(1)$.

Moreover, it follows that we can also move from polynomial shares over $\mathcal{Z}_Q$ to polynomial shares over $\mathcal{Z}_{Q'}$ provided Q and Q' are sufficiently large w.r.t. the security parameter and the shared value.

4.2   Computing Shares of the Binary Representation of a Shared Secret

To do a distributed exponentiation with a shared exponent b it is useful when the players are given shares of the bits of b. In the following we assume (w.l.o.g.) that the players hold additive shares of the exponent b over the integers. The idea of the following protocol to obtain shares of b's bits is that each player distributes polynomial shares modulo $\widetilde{Q}$ of the bits of her or his additive share. Then the players perform a (general) multi-party computation to add these bits to obtain shares of the bits of b. This multi-party computation, however, is rather simple. In fact, we need to implement a circuit of size $O(kn)$ and depth $O(\lg k + \lg n)$ (c.f., [10]). Each gate in this circuit requires $O(1)$ invocations of the multiplication protocol $\mathsf{MUL}(\cdot, \cdot)$ over $\mathcal{Z}_{\widetilde{Q}}$, where $\widetilde{Q}$ can be small.

Protocol $\mathsf{I2Q\text{-}BIT}(\langle b \rangle_j^I)$:
Let n to be (an upper-bound on) the number of bits of b. Party P_j runs the following steps.

1. Re-share each bit of the share $\langle b \rangle_j^I$ with a polynomial sharing over $\widetilde{Q}$ and send each share to the respective player. Let $[b_{i,l}]_j^{\widetilde{Q}}$ denote the share held by party P_j of the i-th bit of party P_l's additive share of b.
2. The player use the computation techniques of Ben-Or, Goldwasser and Wigderson [5] on a circuit for adding the k n-bit numbers. This takes $O(\lg k + \lg n)$ steps.
   Let $[b_1]_j^{\widetilde{Q}}$, $i = 1, \ldots, n$, be the shares of the bits of the result. (Recall that it is ensured that b has n-bits.)
3. Output $([b_1]_j^{\widetilde{Q}}, \ldots, [b_n]_j^{\widetilde{Q}})$.

Proving the security of this protocol is straightforward given the security of its sub-protocols and the composition theorem.

Efficiency analysis: computing shares of the bits of b requires $O(nk^3 \lg k \lg \widetilde{Q} + nk^2(\lg \widetilde{Q})^2)$ bit operations per player. This protocol requires only a relatively small $\widetilde{Q}$, e.g., $\rho + 5 + \lg k$ bits. If shares of the bits modulo a larger prime Q are required, it is more efficient to compute shares modulo a small $\widetilde{Q}$ using the above protocol and then convert these shares into ones modulo Q. The following protocol converts polynomial shares of c modulo $\widetilde{Q}$ to polynomial shares of c modulo Q: $\langle c \rangle_j^{\widetilde{Q}} := \mathsf{PQ2SQ}([c]_j^{\widetilde{Q}})$; $\langle c \rangle_j^I := \mathsf{SQ2SI}(\langle c \rangle_j^{\widetilde{Q}})$; $\langle c \rangle_j^Q := \mathsf{SI2SQ}(\langle c \rangle_j^I)$; $[c]_j^Q := \mathsf{SQ2PQ}(\langle c \rangle_j^Q)$. The number of bit operations for this is $O(\gamma nk^3 \lg k + \gamma^2 nk^2 + n^2 k^2 \lg k)$, where $\lg Q = \Theta(n)$ and $\lg \widetilde{Q} = \Theta(\gamma)$, as opposed to $O(n^2 k^3 \lg k + n^3 k^2)$ when using the bigger Q only. This optimization may be quite important in practice as γ may be much smaller than n (e.g., $\gamma = 100$ and $n = 2000$). The communication-complexity for both variants is $O(n^2 k + nk \lg Q)$ bits. and their round-complexity is $O(\lg k + \lg n)$.

4.3 Approximate Truncation

This section presents a truncation protocol, that on input polynomial shares of a and a parameter n outputs polynomial shares of b such that $|b - a/2^n| \le k + 1$.

Protocol $\mathsf{TRUNC}(a, n)$:
Party P_j executes the following steps.

1. Get additive shares of a over the integers: $\langle a \rangle_j^I := \mathsf{SQ2SI}(\mathsf{PQ2SQ}([a]_j^Q))$.
2. Locally compute $\langle b \rangle_j^I := \mathsf{trunc}(\frac{\langle a \rangle_j^I}{2^n})$.
3. Get polynomial shares of b over $\mathcal{Z}_Q$: $[b]_j^Q := \mathsf{SQ2PQ}(\mathsf{SI2SQ}(\langle b \rangle_j^I))$.
4. Output $[b]_j^Q$.

It is easy to see that the protocol is secure and correct, if $\lg Q > \rho + n + \lg k + 4$ holds, where ρ is a security parameter (c.f. requirements of the $\mathsf{SQ2SI}(\cdot)$ protocol). Its bit-complexity is $O(nk^2 \lg k + n^2 k)$, its communication-complexity is $O(kn)$ bits, and its round-complexity is $O(1)$ rounds.

5 Distributed Computation Modulo a Shared Integer

This section provides efficient protocols for distributed computation modulo a shared, secret modulus p. All computations will be done using shares modulo a prime Q whose bit-length is roughly twice that of p. The main building block is an efficient protocol for reducing a shared secret modulo p. This immediately gives us distributed modular addition and multiplication. The section further provides a protocol for efficient modular exponentiation where the exponent is a shared secret as well. As our modular reduction protocol does not compute the smallest residue in absolute value but only one that is bounded by a small multiple of the modulus, the usual approach for comparing two shared secrets no longer works and therefore a new protocol for comparing such 'almost reduced' shared secrets modulo p is also presented.

The idea of our protocol for modular reduction is based on classical algorithmic techniques (c.f. [1]). Recall that $c \operatorname{rem} p = c - \lceil \frac{c}{p} \rfloor p$. Thus the problem reduces to the problem of distributively computing $\lceil \frac{c}{p} \rfloor$.

By interpreting an integer m as the mantissa of a floating point number with a public exponent, we can interpret shares of this integer as shares of the corresponding floating point number. To multiply two such floating point numbers we distributively multiply the mantissas and locally add the exponents. To keep the shared numbers small, we 'round' the product by converting the polynomial shares of the product mantissa modulo Q to additive shares over the integers, by having each party locally right-shift its additive share by ξ bits and add ξ to the exponent, and by converting back to polynomial shares modulo Q. This rounding technique introduces a relative error of $O(k2^{\xi}/m)$.

So we split the problem of distributively computing $\lceil \frac{c}{p} \rfloor$ into the problem of distributively computing a floating point approximation of $1/p$, and of distributively computing $\lceil \frac{c}{p} \rfloor$ using the precomputed shares of $1/p$. The first problem can be solved using Newton iteration and is described in the next subsection. In Section 5.2 we show how to compute a close approximation to $\lceil \frac{c}{p} \rfloor$ if we are given additive shares of a good approximation to $\frac{c}{p}$ over the integers by having each participant locally truncate its share. The resulting (shared) integer s satisfies $|s - \lceil \frac{c}{p} \rfloor| \leq k + 1$. It turns out that this is accurate enough to compute a value congruent to c modulo p that is sufficiently small to allow for on-going computations modulo p (Section 5.3).

5.1 Computation of Shares of an Approximation to $1/p$

Assume each party is given polynomial shares $[p]_i^Q$ of p, with $2^{n-1} < p < 2^n$. This section provides a protocol that allows the parties to compute polynomial shares of an integer $0 < \tilde{p} < 2^{t+2}$ such that $\tilde{p} 2^{-n-t} = 1/p+\epsilon$, where $|\epsilon| < (k+1)2^{-n-t+4}$ and t is a parameter whose choice is discussed below.

As already mentioned we employ Newton iteration for this task with the function $f(x) = 1/x - p/2^n$ which leads to the iteration formula $x_{i+1} := x_i(2 - x_i p/2^n)$ that has quadratic convergence. Using $3/2$ as a start value gives us an initial error of $|2^n/p - 3/2| < 1/2$ and hence we need to do about $\lg t$ iterations to get a t-bit approximation $\tilde{x}$ to $2^n/p$. We set $\tilde{p} = 2^t \tilde{x}$, which is an integer.

Protocol $\mathsf{APPINV}([p]_j^Q)$:
Party P_j executes the following steps.

1. Set $[u_0]_j^Q := u_0 = 3 \cdot 2^{t-1} \operatorname{rem} Q$.
2. For $i = 0$ to $\lceil \lg(t - 3 - \lg(k+1)) \rceil - 1$ run
   (a) Distributively compute $[z_{i+1}]_j^Q := \mathsf{MUL}([p]_j^Q, [u_i]_j^Q)$.
   (b) $[w_{i+1}]_j^Q := \mathsf{TRUNC}([z_{i+1}]_j^Q, n)$.
   (c) Compute $[v_{i+1}]_j^Q := 2^{t+1} \cdot [u_i]_j^Q - \mathsf{MUL}([w_{i+1}]_j^Q, [u_i]_j^Q)$.
   (d) $[u_{i+1}]_j^Q := \mathsf{TRUNC}([v_{i+1}]_j^Q, t)$.
3. Output $[\tilde{p}]_j^Q := [u_{i+1}]_j^Q$.

Theorem 2. *Let ρ be a security parameter and let $Q > 2^{\rho+t+\nu+6+\lg k}$, where $\nu = \max(n,t)$. Then, for any $t > 5+\lg(k+1)$ and any p satisfying $2^{n-1} < p < 2^n$ for some n, the protocol $\mathsf{APPINV}([p]_j^Q)$ securely computes shares of an integer $\tilde{p}$, such that*

$$\left| \frac{2^n}{p} - \frac{\tilde{p}}{2^t} \right| < \frac{k+1}{2^{t-4}} ,$$

with $0 < \tilde{p} < 2^{t+2}$. That is, $\tilde{p}/2^{t+n}$ is an approximation to $1/p$ with relative error $\frac{k+1}{2^{t-4}}$.

Proof. We need to show that the protocol actually computes an approximation to $1/p$. Then the security follows from the security of the sub-protocols for multiplication and truncation of the shares.

Consider how u_{i+1} is computed from u_i in the protocol. Because of the local truncation, we have $2u_i - pu_i^2 2^{-n-t} - (k+1)(1+u_i/2^t) \leq u_{i+1} \leq 2u_i - pu_i^2 2^{-n-t} + (k+1)(1+u_i/2^t)$. As we will see $u_i/2^t < 3$ holds. Thus $|\frac{2^n}{p} - \frac{u_{i+1}}{2^t}| < \frac{2^n}{p} - 2\frac{u_i}{2^t} + \frac{p}{2^n}(\frac{u_i}{2^t})^2 + \frac{(k+1)}{2^t}(1+u_i/2^t) = \frac{p}{2^n}(\frac{2^n}{p} - \frac{u_i}{2^t})^2 + \frac{(k+1)}{2^t}(1+u_i/2^t)$. From this it follows that $|\frac{2^n}{p} - \frac{u_{i+1}}{2^t}| < \epsilon_i^2 + \frac{k+1}{2^{t-2}} =: \epsilon_{i+1}$. As $2^{n-1} < p < 2^n$ and $u_0 = 3 \cdot 2^{t-1}$ we have $\epsilon_0 < 1/2$ and by requiring $k < 2^{t-5} - 1$ we get $\epsilon_1 < 1/2$ and $\epsilon_i = 2^{-2^i} + \frac{k+1}{2^{t-3}} < 1/2$. In particular, we have $\epsilon_i = \frac{k+1}{2^{t-4}}$ for $i = \lceil \lg(t-3-\lg(k+1)) \rceil$.

Consider the size of the integers u_i that are shared during the protocol. As $\epsilon_i < 1/2$ and $1 < 2^n/p < 2$ we have $0 < u_i/2^t < 2+1/2$ and hence $0 < u_i < 2^{t+2}$ for all i and hence $0 < z_i < 2^{n+t+2}$. Similarly, one can show that $0 < v_i < 2^{2t+2}$.

The lower-bound on Q follows from the fact that the $\mathsf{TRUNC}(\cdot,\cdot)$ algorithm must work on the v_i's and the z_i's.

Let us discuss the choice of t: in order for the b most significant bits of $1/p$ and $\tilde{p}/2^{t+n}$ to be equal, t must be chosen bigger than $b+5+\lg(k+1)$. The cost of the protocol is dominated by the $\mathsf{MUL}(\cdot,\cdot)$ protocol and is $O(\lg t(n^2 k + nk^2 \lg k))$ bit-operations per player. Its communication-complexity is $O(kn \lg t)$ bits and its round-complexity is $O(\lg t)$.

5.2 Reduction of a Shared Integer Modulo a Shared p

Assume the players hold polynomial shares modulo Q of the three integers $-2^w < c < 2^w$, $0 < \tilde{p} < 2^{t+2}$, and $2^{n-1} < p < 2^n$, where $\tilde{p}2^{-n-t}$ is an approximation of $1/p$ as computed by the protocol in the previous paragraph. Using the following protocol, the players can compute shares of an integer d such that $d \equiv c \pmod{p}$ and $\lg|d| < \lg(k+1) + w - t + 5$.

As already mentioned this protocol computes d as $c - \lceil c\tilde{p}2^{-n-t} \rceil p$. For distributively computing the product $c\tilde{p}$ the size of Q would need to be about $w+t$ bits. However, as the n least significant bits of c do not significantly affect the computation of the quotient, we can first cut off say $\ell \approx n$ low-order bits, obtaining $\tilde{c}$, and then compute d as $c - \lceil \tilde{c}\tilde{p}2^{-n-t+\ell} \rceil p$ which requires the size of Q to be only about $w + t - \ell$ bits.

Protocol $\mathsf{MOD}([c]_j^Q, [p]_j^Q, [\tilde{p}]_j^Q)$:

Player P_j executes the following steps.

1. $[\tilde{c}]_j^Q := \mathsf{TRUNC}([c]_j^Q, \ell)$.
2. Compute $[\hat{q}]_j^Q := \mathsf{MUL}([\tilde{c}]_j^Q, [\tilde{p}]_j^Q)$.
3. $[q]_j^Q := \mathsf{TRUNC}([\hat{q}]_j^Q, n + t - \ell)$.
4. Compute $[d]_j^Q := [c]_j^Q - \mathsf{MUL}([p]_j^Q, [q]_j^Q)$.

Theorem 3. *Assume* $Q > \max\left(2^{\rho+6+w-\ell+t+2\lg(k+1)}, 2^{\rho+w+4+\lg(k+1)}\right)$. *Then, given shares of three integers* $-2^w < c < 2^w$, $0 < \tilde{p} < 2^{t+2}$, *and* $0 < p < 2^n$, *the above protocol securely computes shares of* $d = (c \operatorname{rem} p) + ip$ *with* $|i| \leq (k+1)(1 + 2^{w+4-n-t} + 2^{\ell-n+2})$, *where* k *is the number of players.*

Proof. Due to the local rounding in the $\mathsf{TRUNC}(\cdot, \cdot)$ protocol in Step 1, we have $c - (k+1)2^\ell \leq \tilde{c}2^\ell \leq c + (k+1)2^\ell$. Due to the local rounding in the $\mathsf{TRUNC}(\cdot, \cdot)$ protocol in Step 3, we have $\operatorname{trunc}(\tilde{c}\tilde{p}2^{-n-t+\ell}) - k \leq q \leq \operatorname{trunc}(\tilde{c}\tilde{p}2^{-n-t+\ell}) + k$. As $\tilde{p}2^{-(n+t)}$ is only an approximation to $1/p$, we have $\operatorname{trunc}(\frac{c}{p} - \frac{c(k+1)}{2^{n-4+t}} - \frac{\tilde{p}(k+1)}{2^{n+t-\ell}}) - k \leq q \leq \operatorname{trunc}(\frac{c}{p} + \frac{c(k+1)}{2^{n-4+t}} + \frac{\tilde{p}(k+1)}{2^{n+t-\ell}}) + k$ and, as $-2^w < c < 2^w$ and $0 < \tilde{p} < 2^{t+2}$, we get $\lceil \frac{c}{p} \rceil - (k+1)(1 + 2^{w+4-n-t} + 2^{\ell-n+2}) \leq q \leq \lceil \frac{c}{p} \rceil + (k+1)(1 + 2^{w+4-n-t} + 2^{\ell-n+2})$. Thus $d = (c \operatorname{rem} p) + ip$ with $|i| < (k+1)(1 + 2^{w+4-n-t} + 2^{\ell-n+2})$.

The bound on Q follows from the requirements of the $\mathsf{SQ2SI}(\cdot)$ in the $\mathsf{TRUNC}(\cdot, \cdot)$ protocol.

The cost of the $\mathsf{MOD}(\cdot, \cdot, \cdot)$ protocol is dominated by the $\mathsf{MUL}(\cdot, \cdot)$ protocol and is $O(n^2 k + nk^2 \lg k)$ bit operations per players. The communication-complexity of the protocol is $O(kn)$ bits and its round-complexity is $O(1)$.

5.3 Computing with a Shared Modulus p

Now, we are ready to discuss "on-going" distributed computation modulo a shared integer. In particular, we discuss how the parameters for the $\mathsf{MOD}(\cdot, \cdot, \cdot)$ and $\mathsf{APPINV}(\cdot)$ protocols must be set such that such computation is possible. Assume that the players hold polynomial shares modulo a prime Q of the integers $0 < \tilde{p} < 2^{t+2}$, and $2^{n-1} < p < 2^n$, where $\tilde{p}2^{-t-n}$ is an approximation of $1/p$ as computed above. Let

$$\ell = n - 2 , \qquad\qquad t = \lceil n + 10 + 2\lg(3(k+1)) \rceil ,$$
$$Q > 2^{\rho+2n+36+6\lg(k+1)} , \text{ and} \qquad v = n + \lg(3(k+1)) + 1 .$$

Then, given polynomial shares modulo a prime Q of an integer $-2^{2v} < c < 2^{2v}$, the players can compute shares of an integer $-2^v < d < 2^v$ as $[d]_j^Q := \mathsf{MOD}([c]_j^Q, [p]_j^Q, [\tilde{p}]_j^Q)$. In particular, given polynomial shares modulo a prime Q of the integers $-2^v < a, b < 2^v$ the players can compute shares of an integer $-2^v < d' < 2^v$ as $[d']_j^Q := \mathsf{MOD}(\mathsf{MUL}([a]_j^Q, [b]_j^Q), [p]_j^Q, [\tilde{p}]_j^Q)$. Thus d and d' can be used as inputs to further modular multiplication computations.

Exponentiation with a Shared Exponent: Assume the players want to compute shares of $c \equiv a^b \pmod{p}$, where a, b, p, $\tilde{p}$ are shared secrets and $\tilde{p}$ is an approximation to $2^{n+t}/p$. This can be done by distributively running the square and multiply algorithm where the fact that $a^{b_i} = (a - 1)b_i + 1$ if $b_i \in \{0, 1\}$ comes in handy. We assume that the players hold shares $([b_1]_j^Q, \ldots, [b_n]_j^Q)$ of the bits of b, where b_1 is the low-order bit of b (as computed, say, by protocol $\mathsf{I2Q\text{-}BIT}(\cdot)$).

Assuming that $|a| < 2^v$ then the following protocol securely computes shares of c such that $|c| < 2^v$ and $c \equiv a^b \pmod{p}$.

Protocol EXPMOD$([a]_j^Q, ([b_1]_j^Q, \ldots, [b_n]_j^Q), [p]_j^Q, [\tilde{p}]_j^Q)$:
Player P_j executes the following steps.

1. Compute $[c_n]_j^Q := \mathsf{MUL}([a]_j^Q - 1 \operatorname{rem} Q, [b_n]_j^Q) + 1 \operatorname{rem} Q$.
2. For $i = n - 1, \ldots, 1$ do
   (a) $[d_i]_j^Q := \mathsf{MUL}([a]_j^Q - 1 \operatorname{rem} Q, [b_i]_j^Q) + 1 \operatorname{rem} Q$.
   (b) $[cs_i]_j^Q := \mathsf{MOD}(\mathsf{MUL}([c_{i+1}]_j^Q, [c_{i+1}]_j^Q), [p]_j^Q, [\tilde{p}]_j^Q)$.
   (c) $[c_i]_j^Q := \mathsf{MOD}(\mathsf{MUL}([cs_i]_j^Q, [d_i]_j^Q), [p]_j^Q, [\tilde{p}]_j^Q)$.
3. Output $[c]_j^Q := [c_1]_j^Q$.

Efficiency analysis: This protocol invokes about $3n$ times $\mathsf{MUL}(\cdot, \cdot)$ and about $2n$ times $\mathsf{MOD}(\cdot, \cdot, \cdot)$ and hence requires $O(n^3 k + n^2 k^2 \lg k))$ bit operations per player. The communication complexity is $O(n^2 k)$ bits and it has $O(n)$ rounds.

Set membership: Assume the players want to establish whether $a \equiv b \pmod{p}$ holds for three shared secrets a, b and p (where p is not necessarily a prime). This can in principle be done by computing shares of $c := a - b \operatorname{rem} p$, (re-)sharing c modulo Q, multiplying it with a jointly generated random invertible element from $\mathcal{Z}_Q$, revealing the result, and checking if it is 0 modulo Q (provided $Q > p$). However, because of the properties of $\mathsf{MOD}(\cdot, \cdot, \cdot)$, we can only compute shares of $c = (a - b \operatorname{rem} p) + ip$ with $|i| < 3(k + 1)$ and therefore the test does not quite work. But as i is relatively small, it is possible to distributively compute the integer $s := \prod_{l=-3(k+1)+1}^{3(k+1)-1}(c - lp)$ which will be zero if $c \equiv 0 \pmod{p}$ and non-zero otherwise. This also holds for s modulo Q because $Q \nmid s$ if $Q > p6(k+1)$ as then $Q > |(c - lp)|$ holds for all $l \in [-3(k + 1), 3(k + 1)]$.

The protocol below is a generalization of what we just described in that it allows the players to check whether a equals one of $b_1, \ldots b_m$ modulo p. Here, first an s_i is computed for each b_i similarly as the s above for b and then it is tested whether $\prod_i s_i \equiv 0 \pmod{Q}$.

Assuming that $a, b_1, \ldots, b_m$ are less than 2^v in absolute value, then the following protocol securely tests if $a \equiv b_i \pmod{p}$ for some i.

Protocol SETMEM$([a]_j^Q, \{[b_1]_j^Q, \ldots, [b_m]_j^Q\}, [p]_j^Q, [\tilde{p}]_j^Q)$:
Player P_j runs the following steps.

1. For all $i = 1, \ldots, m$ compute $[c_i]_j^Q := \mathsf{MOD}([a]_j^Q - [b_i]_j^Q \operatorname{rem} Q, [p]_j^Q, [\tilde{p}]_j^Q)$ (in parallel).

2. For all $i = 1, \ldots, m$ do (in parallel)
   (a) Set $[u_{(i,-3(k+1)+1)}]_j^Q := [c_i]_j^Q - (3(k+1) - 1)[p]_j^Q \operatorname{rem} Q$.
   (b) For $l = -3(k+1) + 2, \ldots, 3(k+1) - 1$ do
       i. Compute $[u_{(i,l)}]_j^Q := \mathsf{MUL}([u_{(i,l-1)}]_j^Q, ([c_i]_j^Q - l[p]_j^Q \operatorname{rem} Q))$.
3. Let $[\tilde{u}_1]_j^Q := [u_{(1,3(k+1)+1)}]_j^Q$.
4. For $i = 2, \ldots, m$ do
   (a) Compute $[\tilde{u}_i]_j^Q := \mathsf{MUL}([\tilde{u}_{i-1}]_j^Q, [u_{(i,3(k+1)+1)}]_j^Q)$.
5. Perform $[r]_j^Q := \mathsf{JRP\text{-}INV}(\mathcal{Z}_Q)$, compute $[z]_j^Q := \mathsf{MUL}([\tilde{u}_m]_j^Q, [r]_j^Q)$ and send $[z]_j^Q$ to all other players.
6. Reconstruct z and output success if $z \equiv 0 \operatorname{rem} Q$ and failure otherwise.

Security of this protocol follows from the security of its sub-protocols, and the fact that if z is non-zero, then it is a random element from $\mathcal{Z}_Q$ and hence no information about a or any of the b_i's is revealed other than that a is different from all the b_i's modulo p.

Note that this protocol includes as a special case the comparison of two almost reduced residues. It requires $O(mk(n^2k + nk^2 \lg k))$ bit operations per player. The communication-complexity $O(mnk^2)$ bits and it takes is $O(k + n)$ rounds. However, it is trivial to get the number of rounds down to $O(\lg k + \lg n)$ by using a "tree multiplication method" in step 2b and 4.

We note that an alternative to the above protocol would be to use the techniques of Ben-Or et al. [5] on a circuit to fully reduce a and b modulo p. As a and b are "almost reduced" modulo p, this circuit is small.

6   Generation of Shared Random Primes and Safe Primes

We shortly discuss how the protocols described in the previous section can be used to distributively generate random primes and safe primes. Clearly, the latter allows to distributively generate random RSA moduli that are the product of two primes or safe primes.

The strategy for generating a random shared prime is the same as the one usually applied in the non-distributed case: choose a random number, do trial division, and then run sufficiently many rounds of some primality test, e.g., the Miller-Rabin test. To generate a random shared safe prime, one can apply the strategy proposed by Cramer and Shoup [12]. To reduce the round complexity, one tests many candidates in parallel. We refer the reader to the full version of this paper [2] for a detailed discussion and complexity analysis.

Many applications require also that the players generate shares of the private exponent. This is much less computationally involved than distributively generating the modulus N. In particular, Boneh and Franklin [6] as well as Catalano et al. [9] present efficient protocols to accomplish this, given additive shares over the integers of the factors of N. Our techniques can in fact be used to improve the latter protocol as well.

Let us compare the computational cost of the method described above of generating a shared prime product to the one by Boneh and Franklin. (We do not consider the improvement on the latter protocol described by Malkin, Wu, and Boneh [24], as most of them apply to our protocol as well.) We first summarize the latter approach. Boneh and Franklin propose to first choose random n-bit strings and to do a distributed trial division of them. When two strings are found that pass this trial division, they are multiplied to obtain N. Then, local trial division is done on N, and finally a special primality test on N is applied that checks whether N is the product of two primes. Thus, from a bird's eyes view, one finds that with this method one needs to test about $(n/\lg n)^2$ candidates as opposed to about $n/\lg n$ with our method.

A more careful analysis assuming $\lg k \ll n$ shows that the expected bit-complexity of their protocol is $O((n/\lg n)^2(n^3 + n^2k + nk^2 \lg k)$ whereas it is $O(n^2/\lg n(k^3 \lg k\gamma + k^2\gamma^2 + nk^2 \lg k + n^2k))$ for ours, where $\gamma \approx 128$ is a security parameter smaller than n. For this analysis we assumed that the bound B for trial division is about $O(n)$. For small number of players k these figures become $O(n^5/(\lg n)^2)$ and $O(n^4/\lg n)$. Round and communication complexities are $O(1)$ rounds and $O(kn^3/(\lg n)^2)$ bits for theirs and $O(n)$ rounds and $O(kn^3/\lg n)$ bits for ours. We note that, in practice, the round-complexities and communication complexities are not relevant as for this kind of application one would run many instances of the protocol in parallel and thereby keep the party with the least computational power constantly busy.

Acknowledgements

We are grateful to Matt Franklin for enlightening discussions that led to a substantially more efficient test for safe-prime products.

References

1. A. V. Aho, J. E. Hopcroft, and J. D. Ullman. *The Design and Analysis of Computer Algorithms.* Addison Wesley, 1974.
2. J. Algesheimer, J. Camenisch, and V. Shoup. Efficient computation modulo a shared secret with application to the generation of shared safe-prime products. http://eprint.iacr.org/2002/029, 2002.
3. G. Ateniese, J. Camenisch, M. Joye, and G. Tsudik. A practical and provably secure coalition-resistant group signature scheme. In *Advances in Cryptology – CRYPTO 2000*, vol. 1880 of *LNCS*, pp. 255–270, 2000.
4. J. Bar-Ilan and D. Beaver. Non-cryptographic fault-tolerant computing in a constant number of rounds of interaction. In *8th ACM PODC*, pp. 201–209, 1989.
5. M. Ben-Or, S. Goldwasser, and A. Wigderson. Completeness theorems for non-cryptographic fault-tolerant distributed computation. In *Proc. 20th STOC*, pp. 1–10, 1988.
6. D. Boneh and M. Franklin. Efficient generation of shared RSA keys. In *Advances in Cryptology – CRYPTO '97*, vol. 1296 of *LNCS*, pp. 425–439, 1997.
7. J. Camenisch and A. Lysyanskaya. Efficient non-transferable anonymous multi-show credential system with optional anonymity revocation. In *Advances in Cryptology – EUROCRYPT 2001*, vol. 2045 of *LNCS*, pp. 93–118, 2001.

8. R. Canetti. Security and composition of multi-party cryptographic protocols. *Journal of Cryptology*, 13(1):143–202, 2000.
9. D. Catalano, R. Gennaro, and S. Halevi. Computing inverses over a shared secret modulus. In *EUROCRYPT 2000*, vol. 1807 of *LNCS*, pp. 190–206, 2000.
10. T. H. Cormen, C. E. Leiserson, and R. L. Rivest. *Introduction to Algorithms*. MIT Press, Cambridge, 1992.
11. R. Cramer and V. Shoup. Signature schemes based on the strong RSA assumption. In *Proc. 6th ACM CCS*, pp. 46–52. ACM press, nov 1999.
12. R. Cramer and V. Shoup. Signature schemes based on the strong RSA assumption. *ACM Transactions on Information and System Security*, 3(3):161–185, 2000.
13. I. Damgård and M. Koprowski. Practical threshold RSA signatures without a trusted dealer. In *EUROCRYPT 2001*, vol. 2045 of *LNCS*, pp. 152–165, 2001.
14. U. Feige, A. Fiat, and A. Shamir. Zero-knowledge proofs of identity. *Journal of Cryptology*, 1:77–94, 1988.
15. P.-A. Fouque and J. Stern. Fully distributed threshold RSA under standard assumptions. In *ASIACRYPT 2001*, vol. 2248 of *LNCS*, pp. 310–330, 2001.
16. Y. Frankel, P. MacKenzie, and M. Yung. Robust efficient distributed RSA key generation. In *Proc. 30th Annual ACM STOC*, pp. 663–672, 1998.
17. M. Franklin and S. Haber. Joint encryption and message-efficient secure computation. In *CRYPTO '93*, vol. 773 of *LNCS*, pp. 266–277, 1994.
18. R. Gennaro, S. Halevi, and T. Rabin. Secure hash-and-sign signatures without the random oracle. In *EUROCRYPT '99*, vol. 1592 of *LNCS*, pp. 123–139, 1999.
19. R. Gennaro, S. Jarecki, H. Krawczyk, and T. Rabin. Robust and efficient sharing of RSA functions. In *Advances in Cryptology – CRYPTO '96*, vol. 1109 of *LNCS*, pp. 157–172, 1996.
20. R. Gennaro, M. O. Rabin, and T. Rabin. Simplified VSS and fast-track multiparty computations with applications to threshold cryptography. In *Proc. 17th ACM PODC*, 1998.
21. T. Rabin. A simplified approach to threshold and proactive RSA. In *Advances in Cryptology – CRYPTO '98*, vol. 1642 of *LNCS*, pp. 89–104, 1998.
22. A. Shamir. How to share a secret. *Communications of the ACM*, 22(11):612–613, Nov. 1979.
23. V. Shoup. Practical threshold signatures. In *Advances in Cryptology: EUROCRYPT 2000*, vol. 1087 of *LNCS*, pp. 207–220, 2000.
24. M. M. T. Wu and D. Boneh. Experimenting with shared generation of RSA keys. In *Proceedings of the Internet Society's 1999 Symposium on Network and Distributed System Security (SNDSS)*, pp. 43–56, 1999.

Hidden Number Problem with the Trace and Bit Security of XTR and LUC

Wen-Ching W. Li[1],[*], Mats Näslund[2],[**], and Igor E. Shparlinski[3],[***]

[1] Department of Mathematics, Penn State University
University Park, PA 16802, USA
wli@math.psu.edu
[2] Ericsson Research
SE-16480 Stockholm, Sweden
mats.naslund@era.ericsson.se
[3] Department of Computing, Macquarie University
Sydney, NSW 2109, Australia
igor@ics.mq.edu.au

Abstract. We consider a certain generalization of the hidden number problem introduced by Boneh and Venkatesan in 1996. Considering the XTR variation of Diffie-Hellman, we apply our results to show security of the $\log^{1/2} p$ most significant bits of the secret, in analogy to the results known for the classical Diffie-Hellman scheme. Our method is based on bounds of exponential sums which were introduced by Deligne in 1977. We proceed to show that the results are also applicable to the LUC scheme. Here, assuming the LUC function is one-way, we can in addition show that each single bit of the argument is a hard-core bit.

1 Introduction

When a new cryptosystem is proposed, some maturity period is normally needed before we see practical deployment. This is natural since some amount of public scrutiny is needed before we feel confident that there are no (serious) attacks possible. We can for instance see this in the case of elliptic curve cryptography where it is not until now, almost twenty years after the introduction by Koblitz and Miller [18, 31], that we start to see commercial use. For this reason, a formal proof of security for a brand new scheme speeds up acceptance.

In 1994, Smith and Skinner [47] proposed a public key scheme, LUC, based on Lucas sequences modulo p. One reason for introducing LUC was a hope that there would not be any sub-exponential attacks on it. Although, this hope has failed, see [2] and the discussion below, LUC still seems to have both better speed and higher security than the classical Diffie-Hellman scheme [20].

While Lucas sequences have a rich mathematical theory, and have seen applications in computer science, for example, primality testing, there has been no formal proof of security relative to for example, discrete logarithms. Indeed,

---

* Supported in part by NSF grant DMS 997-0651.
** Part of work done while visiting Macquarie University.
*** Supported in part by ARC grant A69700294.

shortly after publication, Bleichenbacher et al. showed in [2] that the very parameter settings that made LUC so efficient, also made it possible to reduce the security to discrete logarithms in $\mathbb{F}_{p^2}$, implying sub-exponential attacks. A natural question turns up: does LUC have other (worse) defects, not present in standard discrete logarithm based schemes?

One of the most recently proposed schemes is XTR, invented in 2000 by Lenstra and Verheul [23]. XTR can be thought of as a generalization of LUC to an extension field of degree six rather than two, and is based on initial ideas by Brouwer et al. [6]. The idea is to get a security against attackers corresponding to discrete logarithms in $\mathbb{F}_{p^6}$ while the actual computation and messages exchanged are in $\mathbb{F}_{p^2}$. For XTR, a proof of security exists; breaking XTR is computationally equivalent to computing discrete logarithms in $\mathbb{F}_{p^6}$, see also [23].

Still, even if completely breaking the respective schemes (LUC, XTR) requires finite field discrete logarithm computation, it is not clear what other security properties these schemes have. For instance, one can ask the natural question on "partial breaking", for example, in terms of computing certain bits of the XTR/LUC-secrets (the "logarithms").

The perhaps most interesting application of XTR (and LUC) is the Diffie-Hellman (DH) analogues; the exchanged messages are small, and, according to the recent evaluation [20, 22] of the relative performance of various cryptosystems, XTR and LUC are the fastest non-elliptic curve schemes. Hence, it would be important to establish bit-security results for the DH version; given the exchanged DH messages, can certain parts or bits of the DH secret key be computed? For the conventional DH scheme over $\mathbb{F}_p$, such results were shown by Boneh and Venkatesan [4], as consequences of a generalization of the *hidden number problem* introduced in the same paper [4]: given polynomially many t_i and approximations to $t_i \alpha \pmod{p}$, recover α. Their results roughly state that if one can compute all of the $\lceil \log^{1/2} p \rceil$ most significant bits of the DH secret without errors, then the remaining bits can be found as well.

In fact, the original security proof of the Diffie-Hellman bits in [4] contained a slight gap, corrected in a later work by Shparlinski and González-Vasco [11]. That paper [11] is based on using bounds on certain *exponential sums* to establish approximate uniformity for some distributions over $\mathbb{F}_p$.

For XTR, Shparlinski [45] has recently extended the techniques to show security also for the XTR Diffie-Hellman secret. We here improve these results and also extend the techniques to LUC Diffie-Hellman. Our results follow from bounds on exponential sums, established using algebraic-geometric means.

For prime fields $\mathbb{F}_p$, or cyclic multiplicative groups of prime order, the security of all individual bits (except possibly the the least significant bits, depending on group order) of the discrete logarithms follows from the works [3, 13, 29, 38, 42], showing that computing any single bit with non-negligible advantage (over the trivial $1/2$) implies polynomial time discrete logarithm computations. In this paper we show that the results can be carried over to LUC in a natural way. Further, when the generator has even order, the predictability for the least significant bits also applies.

2   Preliminaries

2.1   Notation

Let p be a prime. To ease notation, we write $\mathbb{F}$ for the prime field $\mathbb{F}_p$, and $\mathbb{K}$ for a degree m extension $\mathbb{F}_{p^m}$ of $\mathbb{F}$. As usual we assume that $\mathbb{F}$ is represented by $\{0, \ldots, p-1\}$. For integers s and $r \geq 1$ denote by $\lfloor s \rfloor_r$ the remainder of s on division by r. We also use $\log z$ to denote the binary logarithm of $z > 0$. Let

$$\mathrm{Tr}(z) = \mathrm{Tr}_{\mathbb{K}/\mathbb{F}}(z) = z + z^p + \ldots + z^{p^{m-1}}$$

be the trace of $z \in \mathbb{K}$ in $\mathbb{F}$, see [28] for basics of the theory of finite fields.

For a prime p and $k \geq 0$ we denote by $\mathrm{MSB}_{k,p}(x)$ any integer u such that

$$\left| \lfloor x \rfloor_p - u \right| \leq p/2^{k+1}. \tag{1}$$

Roughly speaking $\mathrm{MSB}_{k,p}(x)$ gives k most significant bits of x, however this definition is more flexible and suits better our purposes. In particular, in (1), k need not be integer. Also, the notion of most significant bits is tailored to modular residues and does not match the usual definition for integers.

Throughout the paper the implied constants in symbols 'O' depend on m and occasionally, where obvious, may depend on the small positive parameter δ; they all are effective and can be explicitly evaluated.

We now give a short introduction to the XTR and LUC cryptosystem, as well as to the hidden number problem and its use of lattices. The reader familiar with these can proceed directly to Sect. 3.

2.2   The XTR and LUC Cryptosystems

Below, we concentrate only on the details relevant to this work.

Let $m = 2$ so that $\mathbb{K} = \mathbb{F}_{p^2}$ and let $\mathrm{Tr}_{\mathbb{K}/\mathbb{F}}(u) = \mathrm{Tr}(u)$ as above. Let $g \in \mathbb{K}$ be a root of an irreducible quadratic polynomial $f(X) = X^2 - PX + 1 \in \mathbb{F}_p[X]$, thus $P^2 - 4$ is a quadratic non-residue of $\mathbb{F}_p$. It is easy to show that such elements exist. For example, for any root $\vartheta \in \mathbb{K}$ of an arbitrary irreducible quadratic polynomial over $\mathbb{F}$, $g = \vartheta^{p-1}$ is such an element. Note also that f above is the characteristic polynomial of the recurrence $V_n(P) \equiv PV_{n-1}(P) - V_{n-2}(P)$ (mod p), $V_0 = 2, V_1 = P$, and that $V_n(P) = \mathrm{Tr}(g^n)$.

In the LUC variant of Diffie–Hellman (known as LUCDIF) the communicating parties exchange $\mathrm{Tr}\,(g^x)$ and $\mathrm{Tr}\,(g^y)$ (that is, $V_x(P)$ and $V_y(P)$), then, using $V_{xy}(P) = V_x(V_y(P)) = V_y(V_x(P))$, compute the common secret $\mathrm{Tr}\,(g^{xy})$. For details refer to [2, 47].

XTR can be thought of as a generalization of LUC and has been introduced by Lenstra and Verheul, see [6, 23–25, 43, 48] for basic properties and ideas behind XTR. Let $m = 6$ so that $\mathbb{K} = \mathbb{F}_{p^6}$. We also consider the field $\mathbb{L} = \mathbb{F}_{p^2}$, thus we have a tower of extensions $\mathbb{F} \subseteq \mathbb{L} \subseteq \mathbb{K}$. Accordingly, we denote by $\mathrm{Tr}_{\mathbb{K}/\mathbb{L}}(u)$ and $\mathrm{Tr}_{\mathbb{L}/\mathbb{F}}(v)$ the trace of $u \in \mathbb{K}$ in $\mathbb{L}$ and the trace of $v \in \mathbb{L}$ in $\mathbb{F}$. In particular, $\mathrm{Tr}_{\mathbb{L}/\mathbb{F}}\left(\mathrm{Tr}_{\mathbb{K}/\mathbb{L}}(u)\right) = \mathrm{Tr}(u)$ for $u \in \mathbb{K}$.

The idea of XTR is based on the observation that for some specially selected element $g \in \mathbb{K}^*$, *the XTR generator*, of prime order $l > 3$ such that $l | p^2 - p + 1$, one can also here efficiently compute $\mathrm{Tr}_{\mathbb{K}/\mathbb{L}}(g^{xy})$ from the values of x and $\mathrm{Tr}_{\mathbb{K}/\mathbb{L}}(g^y)$ (alternatively from y and $\mathrm{Tr}_{\mathbb{K}/\mathbb{L}}(g^x)$). This reduces the size of the Diffie-Hellman messages to exchange (namely, $\mathrm{Tr}_{\mathbb{K}/\mathbb{L}}(g^x)$ and $\mathrm{Tr}_{\mathbb{K}/\mathbb{L}}(g^y)$ rather than g^x and g^y) to create a common key $\mathrm{Tr}_{\mathbb{K}/\mathbb{L}}(g^{xy})$.

2.3   The Hidden Number Problem

We shall be interested in a variant of the *hidden number problem* introduced by Boneh and Venkatesan [4, 5]. The problem can be stated as follows: recover an unknown $\alpha \in \mathbb{F}$, given approximations to $\lfloor \alpha t \rfloor_p$ for polynomially many known random $t \in \mathbb{F}$.

Let $\mathcal{G}$ be a subgroup of the multiplicative group $\mathbb{K}^*$. Motivated by the application of bit security for XTR and LUC, we consider the following question: recover $\alpha \in \mathbb{K}$, given approximations of $\mathrm{Tr}(\alpha t)$ for polynomially many known random $t \in \mathcal{G}$. Then we apply our results to obtain a statement about the bit security of the XTR and LUC cryptosystems.

For the general hidden number problem, it has turned out that for many applications (as we shall see, including the one at hand) the condition that t is selected uniformly at random is too restrictive. Examples include the earlier bit security results for the Diffie-Hellman, Shamir, and several other cryptosystems [11, 12] and rigorous results on attacks (following the heuristic arguments of [14, 32]) on the DSA(-like) signature schemes [9, 33, 34].

The aforementioned papers [9, 11, 12, 33, 34] have exploited that the method of [4] can be adjusted to the case when t is selected from a sequence which has some uniformity of distribution property. Thus, a central ingredient is bounds on exponential sums; a natural tool to establish such properties.

2.4   Bounds on Exponential Sums

The case when t is selected from a small subgroup of $\mathbb{F}^*$ was studied in [11] to generalize (and correct) the results of [4] on the bit security of the Diffie-Hellman key in $\mathbb{F}_p$. The results of [11] are based on bounds of exponential sums with elements of subgroups of $\mathbb{F}^*$, namely on Theorems 3.4, 5.5 of [19].

Unfortunately, analogues of the bounds of exponential sums of Theorems 3.4, 5.5 of [19] are not known for non-prime fields. However, in [45] an alternative method of [44] has been used, which applies to very small subgroups $\mathcal{G}$ and is based on bounds of [7, 10] for the number of solutions of certain equations in finite fields. Unfortunately it produces much weaker results.

It is remarked in [45] that for subgroups $\mathcal{G}$ of cardinality $|\mathcal{G}| \geq p^{m/2+\delta}$, with any fixed $\delta > 0$, the bound of exponential sums given by Theorem 8.78 in [28] combined with Theorem 8.24 of the same work (see also (3.15) in [19]) can be used. These bounds would provide analogues of the results [4, 11] but, unfortunately, the subgroups $\mathcal{G}$ associated with XTR and LUC fall below this square-root threshold.

Nevertheless, thanks to the special property of these subgroup, we show in Sect. 3 how one can use an approach from [8], developed in [26, 27], to apply the exponential sum technique to studying the subgroups related to XTR and LUC. We get a substantial improvement to the results of [45] on the bit-security for XTR (from αn bits for a constant $0 < \alpha < 1$, to $\log^{1/2} n$ bits), and a new equally strong result for LUC.

2.5   Lattices

As in [4, 5], our results related to the hidden number problem rely on rounding techniques in lattices. We review a few related results and definitions. Let $\{\mathbf{b}_1, \ldots, \mathbf{b}_s\}$ be a set of linearly independent vectors in $\mathbb{R}^s$. The set

$$L = \{\mathbf{z} \ : \ \mathbf{z} = c_1 \mathbf{b}_1 + \ldots + c_s \mathbf{b}_s, \quad c_1, \ldots, c_s \in \mathbb{Z}\}$$

is called an *s-dimensional full rank lattice* with *basis* $\{\mathbf{b}_1, \ldots, \mathbf{b}_s\}$. For a vector $\mathbf{u}$, let $\|\mathbf{u}\|$ denote its *Euclidean norm*.

It has been remarked in [30], and then in [35, 36] that the following statement holds, which is somewhat stronger than that usually used in the literature. It follows from the lattice basis reduction algorithm of [21] and results of [41, 15].

Lemma 1. *There exists a deterministic polynomial time algorithm which, for a given s-dimensional full rank lattice L and $\mathbf{r} \in \mathbb{R}^s$, finds $\mathbf{v} \in L$ with*

$$\|\mathbf{v} - \mathbf{r}\| \leq \exp\left(O\left(\frac{s \log^2 \log s}{\log s}\right)\right) \min\{\|\mathbf{z} - \mathbf{r}\|, \quad \mathbf{z} \in L\}.$$

3   Distribution of Trace

First, we need the following result which, as mentioned, is essentially Theorem 8.78 of [28] (combined with Theorem 8.24 of the same work) or the bound (3.15) of [19]. Let $\mathcal{G}$ be a subgroup of $\mathbb{K} = \mathbb{F}_{p^m}$.

Lemma 2. *For any $\gamma \in \mathbb{K}^*$, we have*

$$\left| \sum_{t \in \mathcal{G}} \exp\left(2\pi i \mathrm{Tr}\left(\gamma t\right)/p\right) \right| \leq p^{m/2}.$$

Lemma 2 is nontrivial when $|\mathcal{G}| \geq p^{m/2+\delta}$. Much less is known when $\mathcal{G}$ has size less than $p^{m/2}$. For prime finite fields, $m = 1$, Theorems 3.4, 5.5 of [19] provide a nontrivial upper bound for $|\mathcal{G}| \geq p^{1/3+\delta}$ for all primes p and for $|\mathcal{G}| \geq p^{\delta}$ for almost all primes p, respectively, which underlies the result of [11]. However these results have not been extended to composite fields (and it seems that such extensions will require some substantially new ideas).

Nevertheless, applying the results of [8, 26, 27] for some special subgroups we obtain non-trivial estimates beyond the $p^{m/2}$-threshold.

For a divisor $s|m$ let $\mathcal{N}_s$ be the set of $z \in \mathbb{K}$ with $\mathrm{Nm}_s(z) = 1$, where

$$\mathrm{Nm}_s(z) = z^{1+p^{m/s}+\ldots+p^{m-m/s}}$$

is the norm of $z \in \mathbb{K} = \mathbb{F}_{p^m}$ in $\mathbb{F}_{p^{m/s}} \subseteq \mathbb{K}$. Thus $|\mathcal{N}_s| = (p^m - 1)/(p^{m/s} - 1)$.

Our results depend on the following estimate conjectured by Deligne in [8] (and proved in the case of the trivial character $\chi = \chi_0$). In the full generality it has been proved by Katz [17] (to be precise Theorem 4.1.1 of [17] and some standard transformations). For the cases relevant to XTR and LUC, we may also refer to the simpler and more explicit statements of [26, 27].

Lemma 3. *For any divisor $s|m$, any $\gamma \in \mathbb{K}^*$, and any multiplicative character χ of $\mathbb{K}$, we have*

$$\left| \sum_{t \in \mathcal{N}_s} \chi(t) \exp\left(2\pi i \mathrm{Tr}\left(\gamma t\right)/p\right) \right| \leq s p^{(m-m/s)/2}.$$

To proceed, recall the following property of the group of characters of an abelian group.

Lemma 4. *Let $\mathcal{H}$ be an abelian group and let $\widehat{\mathcal{H}} = \mathrm{Hom}(\mathcal{H}, \mathbb{C}^*)$ be its dual group. Then for any character χ of $\mathcal{H}$,*

$$\frac{1}{|\mathcal{H}|} \sum_{h \in \mathcal{H}} \chi(h) = \begin{cases} 1, & \text{if } \chi = \chi_0, \\ 0, & \text{if } \chi \neq \chi_0, \end{cases}$$

where $\chi_0 \in \widehat{\mathcal{H}}$ is the trivial character.

Lemma 5. *For any divisor $s|m$, any subgroup $\mathcal{G}$ of $\mathcal{N}_s$ and any $\gamma \in \mathbb{K}^*$,*

$$\left| \sum_{t \in \mathcal{G}} \exp\left(2\pi i \mathrm{Tr}\left(\gamma t\right)/p\right) \right| \leq s p^{(m-m/s)/2}.$$

Proof. Let $\Omega_{\mathcal{G}}$ be the set of all multiplicative characters of $\mathcal{N}_s$, trivial on $\mathcal{G}$. Using Lemma 4, we write

$$\sum_{t \in \mathcal{G}} \exp\left(2\pi i \mathrm{Tr}\left(\gamma t\right)/p\right) = \frac{1}{|\Omega_{\mathcal{G}}|} \sum_{t \in \mathcal{N}_s} \exp\left(2\pi i \mathrm{Tr}\left(\gamma t\right)/p\right) \sum_{\chi \in \Omega_{\mathcal{G}}} \chi(t)$$

$$= \frac{1}{|\Omega_{\mathcal{G}}|} \sum_{\chi \in \Omega_{\mathcal{G}}} \sum_{t \in \mathcal{N}_s} \chi(t) \exp\left(2\pi i \mathrm{Tr}\left(\gamma t\right)/p\right).$$

Applying the inequality of Lemma 3, we obtain the desired estimate.   □

For $\gamma \in \mathbb{K}$ and integers r and h, denote by $N_\gamma(\mathcal{G}, r, h)$ the number of solutions of the congruence

$$\mathrm{Tr}(\gamma t) \equiv r + y \pmod{p}, \qquad t \in \mathcal{G}, \ y = 0, \ldots, h - 1.$$

Using standard relations between the uniformity of distribution and bounds of exponential sums, for example, see Corollary 3.11 of [37], from Lemma 2 and Lemma 5 one immediately derives the following asymptotic formulas for $N_\gamma(\mathcal{G}, r, h)$.

Lemma 6. *For any $\gamma \in \mathbb{K}^*$ and any subgroup $\mathcal{G} \subseteq \mathbb{K}^*$, we have*

$$N_\gamma(\mathcal{G}, r, h) = \frac{h}{p}|\mathcal{G}| + O(p^{m/2}\log p).$$

Lemma 7. *For any divisor $s|m$, any subgroup $\mathcal{G} \subseteq \mathcal{N}_s$, and any $\gamma \in \mathbb{K}^*$,*

$$N_\gamma(\mathcal{G}, r, h) = \frac{h}{p}|\mathcal{G}| + O(p^{(m-m/s)/2}\log p).$$

These bounds can be turned into statements on the form of the shortest vector in certain lattices. Let $\omega_1, \ldots, \omega_m$ be a fixed basis of $\mathbb{K}$ over $\mathbb{F}$. For an integer $k \geq 0$ and $d(\geq 1)$ elements $t_1, \ldots, t_d \in \mathbb{K}$, let $\mathcal{L}_k(t_1, \ldots, t_d)$ be the $d+m$-dimensional lattice generated by the rows of the $(d+m) \times (d+m)$-matrix:

$$\begin{pmatrix} p & \cdots 0 & 0 & \cdots 0 \\ 0 & \cdots 0 & 0 & \cdots 0 \\ \vdots & \ddots & \vdots & \vdots \\ 0 & \cdots p & 0 & \cdots 0 \\ \mathrm{Tr}(\omega_1 t_1) & \cdots \mathrm{Tr}(\omega_1 t_d) & 1/2^{k+1} & \cdots 0 \\ \vdots & \vdots & \vdots & \ddots \vdots \\ \mathrm{Tr}(\omega_m t_1) & \cdots \mathrm{Tr}(\omega_m t_d) & 0 & \cdots 1/2^{k+1} \end{pmatrix}. \tag{2}$$

Lemma 8. *Let p be a sufficiently large prime number and let $\mathcal{G}$ be a subgroup of $\mathbb{K}^*$ with $|\mathcal{G}| \geq p^{m/2+\delta}$ for some fixed $\delta > 0$. Then for*

$$\eta = \left\lceil \log^{1/2} p \right\rceil \quad and \quad d = \left\lceil \frac{m+1}{\eta-3}\log p \right\rceil,$$

the following holds. Let $\alpha = a_1\omega_1 + \ldots + a_m\omega_m$, $a_1, \ldots, a_m \in \mathbb{F}$, be a fixed element of $\mathbb{K}$. Assume that $t_1, \ldots, t_d \in \mathcal{G}$ are chosen uniformly and independently at random. Then with probability exceeding $1 - p^{-1}$ for any $\mathbf{s} = (s_1, \ldots, s_d, 0, \ldots, 0)$ with

$$\left(\sum_{i=1}^{d} (\mathrm{Tr}(\alpha t_i) - s_i)^2 \right)^{1/2} \leq 2^{-\eta} p,$$

all vectors $\mathbf{v} = (v_1, \ldots, v_d, v_{d+1}, \ldots, v_{d+m}) \in \mathcal{L}_k(t_1, \ldots, t_d)$ satisfying

$$\left(\sum_{i=1}^{d} (v_i - s_i)^2 \right)^{1/2} \leq 2^{-\eta} p,$$

are of the form

$$\mathbf{v} = \left(\left\lfloor \sum_{j=1}^{m} b_j \mathrm{Tr}\,(\omega_j t_1) \right\rfloor_p, \ldots, \left\lfloor \sum_{j=1}^{m} b_j \mathrm{Tr}\,(\omega_j t_d) \right\rfloor_p, \frac{b_1}{2^{k+1}}, \ldots, \frac{b_m}{2^{k+1}} \right)$$

with some integers $b_j \equiv a_j \pmod{p}$, $j = 1, \ldots, m$.

Proof. As in [4] we define the modular distance between two integers r and l as

$$\mathrm{dist}_p(r, l) = \min_{b \in \mathbb{Z}} |r - l - bp| = \min \left\{ \lfloor r - l \rfloor_p, p - \lfloor r - l \rfloor_p \right\}.$$

We see from Lemma 6 that for any $\beta \in \mathbb{K}$ with $\beta \neq \alpha$ the probability $P(\beta)$ that

$$\mathrm{dist}_p\left(\mathrm{Tr}(\alpha t), \mathrm{Tr}(\beta t) \right) \leq 2^{-\eta+1} p$$

for $t \in \mathcal{G}$ selected uniformly at random is

$$P(\beta) \leq 2^{-\eta+2} + O(p^{m/2} |\mathcal{G}|^{-1} \log p) \leq 2^{-\eta+2} + O(p^{-\delta} \log p) \leq 2^{-\eta+3},$$

for large enough p. Therefore, with $d = \lceil (m+1)/(\eta - 3) \rceil$, for any $\beta \in \mathbb{K}$,

$$\Pr\left[\forall i \in [1, d] \mid \mathrm{dist}_p\left(\mathrm{Tr}(\alpha t_i), \mathrm{Tr}(\beta t_i) \right) \leq 2^{-\eta+1} p \right] = P(\beta)^d \leq p^{-m-1},$$

where the probability is taken over $t_1, \ldots, t_d \in \mathcal{G}$ chosen uniformly and independently at random. From here, we derive

$$\Pr\left[\forall \beta \in \mathbb{K} \backslash \{\alpha\}, \ \forall i \in [1, d] \mid \mathrm{dist}_p\left(\mathrm{Tr}(\alpha t_i), \mathrm{Tr}(\beta t_i) \right) \leq 2^{-\eta+1} p \right] \leq p^{-1}.$$

The rest of the proof is identical to the proof of Theorem 5 of [4]. Indeed, we fix some $t_1, \ldots, t_d \in \mathcal{G}$ with

$$\min_{\beta \in \mathbb{K} \backslash \{\alpha\}} \max_{i \in [1, d]} \mathrm{dist}_p\left(\mathrm{Tr}(\alpha t_i), \mathrm{Tr}(\beta t_i) \right) > 2^{-\eta+1} p. \tag{3}$$

Let $\mathbf{v} \in \mathcal{L}_k(t_1, \ldots, t_d)$ be a lattice point satisfying

$$\left(\sum_{i=1}^{d} (v_i - s_i)^2 \right)^{1/2} \leq 2^{-\eta} p.$$

Since $\mathbf{v} \in \mathcal{L}_k(t_1, \ldots, t_d)$, there are integers $b_1, \ldots, b_m, z_1, \ldots, z_d$ such that

$$\mathbf{v} = \left(\sum_{j=1}^{m} b_j \mathrm{Tr}\,(\omega_j t_1) - z_1 p, \ldots, \sum_{j=1}^{m} b_j \mathrm{Tr}\,(\omega_j t_d) - z_d p, \frac{b_1}{2^{k+1}}, \ldots, \frac{b_m}{2^{k+1}} \right).$$

If $b_j \equiv a_j \pmod{p}$, $j = 1, \ldots, m$, then for all $i = 1, \ldots, d$ we have

$$\sum_{j=1}^{m} b_j \mathrm{Tr}\,(\omega_j t_i) - z_i p = \left\lfloor \sum_{j=1}^{m} b_j \mathrm{Tr}\,(\omega_j t_i) \right\rfloor_p = \mathrm{Tr}(\alpha t_i),$$

since otherwise there would be $i \in \{1, \ldots, d\}$ such that $|v_i - s_i| > 2^{-\eta} p$.

Now suppose that $b_j \not\equiv a_j \pmod{p}$ for some $j = 1, \ldots, m$. Put $\beta = b_1 \omega_1 + \ldots + b_m \omega_m$. In this case we have

$$\left(\sum_{i=1}^{d} (v_i - s_i)^2 \right)^{1/2} \geq \max_{i \in [1,d]} \operatorname{dist}_p \left(\sum_{j=1}^{m} b_j \operatorname{Tr}(\omega_j t_i), s_i \right)$$

$$\geq \max_{i \in [1,d]} \left(\operatorname{dist}_p \left(\operatorname{Tr}(\alpha t_i), \sum_{j=1}^{m} b_j \operatorname{Tr}(\omega_j t_i) \right) - \operatorname{dist}_p (s_i, \operatorname{Tr}(\alpha t_i)) \right)$$

$$\geq \max_{i \in [1,d]} \left(\operatorname{dist}_p (\operatorname{Tr}(\alpha t_i), \operatorname{Tr}(\beta t_i)) - \operatorname{dist}_p (s_i, \operatorname{Tr}(\alpha t_i)) \right)$$

$$> 2^{-\eta+1} p - 2^{-\eta} p = 2^{-\eta} p$$

that contradicts our assumption. As we have seen, the condition (3) holds with probability exceeding $1 - p^{-1}$ and the result follows. □

Accordingly, using Lemma 7 instead of Lemma 6 we obtain:

Lemma 9. *Let p be a sufficiently large prime number and let s be a divisor of m. Let $\mathcal{G}$ be a subgroup of $\mathcal{N}_s$ with $|\mathcal{G}| \geq |\mathcal{N}_s|^{1/2} p^{\delta}$ for some fixed $\delta > 0$. Then for*

$$\eta = \left\lceil \log^{1/2} p \right\rceil \qquad and \qquad d = \left\lceil \frac{m+1}{\eta - 3} \log p \right\rceil,$$

the following holds. Let $\alpha = a_1 \omega_1 + \ldots + a_m \omega_m$, $a_1, \ldots, a_m \in \mathbb{F}$, be a fixed element of $\mathbb{K}$. Assume that $t_1, \ldots, t_d \in \mathcal{G}$ are chosen uniformly and independently at random. Then with probability exceeding $1 - p^{-1}$ for any $\mathbf{s} = (s_1, \ldots, s_d, 0, \ldots, 0)$ with

$$\left(\sum_{i=1}^{d} (\operatorname{Tr}(\alpha t_i) - s_i)^2 \right)^{1/2} \leq 2^{-\eta} p,$$

all vectors $\mathbf{v} = (v_1, \ldots, v_d, v_{d+1}, \ldots, v_{d+m}) \in \mathcal{L}_k(t_1, \ldots, t_d)$ satisfying

$$\left(\sum_{i=1}^{d} (v_i - s_i)^2 \right)^{1/2} \leq 2^{-\eta} p,$$

are of the form

$$\mathbf{v} = \left(\left\lfloor \sum_{j=1}^{m} b_j \operatorname{Tr}(\omega_j t_1) \right\rfloor_p, \ldots, \left\lfloor \sum_{j=1}^{m} b_j \operatorname{Tr}(\omega_j t_d) \right\rfloor_p, \frac{b_1}{2^{k+1}}, \ldots, \frac{b_m}{2^{k+1}} \right)$$

with some integers $b_j \equiv a_j \pmod{p}$, $j = 1, \ldots, m$.

4   Hidden Number Problem for the Trace

Using Lemma 8 in the same way as Theorem 5 of [4] was used in the proof of Theorem 1 of that paper, we obtain

Theorem 1. *Let p be a sufficiently large prime number and let $\mathcal{G}$ be a subgroup of $\mathbb{K}^*$ with $|\mathcal{G}| \geq p^{m/2+\delta}$ for some fixed $\delta > 0$. Then for $k = \lceil 2\log^{1/2} p \rceil$, $d = \lceil 4(m+1)\log^{1/2} p \rceil$, the following holds. There exists a deterministic polynomial time algorithm $\mathcal{A}$ such that for any $\alpha \in \mathbb{K}$ given $2d$ values $t_i \in \mathcal{G}$ and $s_i = \mathrm{MSB}_{k,p}(\mathrm{Tr}(\alpha t_i))$, $i = 1, \dots, d$, its output satisfies*

$$\Pr_{t_1,\dots,t_d \in \mathcal{G}} [\mathcal{A}(t_1, \dots, t_d; s_1, \dots, s_d) = \alpha] \geq 1 - p^{-1}$$

if $t_1, \dots, t_d$ are chosen uniformly and independently at random from $\mathcal{G}$.

Proof. We follow the arguments in the proof of Theorem 1 in [4], here briefly outlined for completeness. We refer to the first d vectors in the matrix (2) as p-vectors and the remaining m vectors as trace-vectors. Write

$$\alpha = \sum_{j=1}^{m} a_j \omega_j \in \mathbb{K}, \qquad a_1, \dots, a_m \in \mathbb{F}.$$

We consider the vector $\mathbf{s} = (s_1, \dots, s_d, s_{d+1}, \dots, s_{d+m})$ where $s_{d+j} = 0$, for $j = 1, \dots, m$. Multiplying the jth trace-vector of the matrix (2) by a_j and subtracting a certain multiple of the jth p-vector for $j = 1, \dots, m$, we obtain a lattice point

$$\mathbf{u}_\alpha = (u_1, \dots, u_d, a_1/2^{k+1}, \dots, a_m/2^{k+1}) \in \mathcal{L}_k(t_1, \dots, t_d)$$

such that $|u_i - s_i| \leq p2^{-k-1}$, $i = 1, \dots, d+m$, where $u_{d+j} = a_j/2^{k+1}$ for $j = 1, \dots, m$. Therefore,

$$\|\mathbf{u}_\alpha - \mathbf{s}\| \leq (d+m)^{1/2} 2^{-k-1} p.$$

Let $\eta = \lceil \log^{1/2} p \rceil$. By Lemma 1 (with a slightly rougher constant $2^{(d+m)/4}$) in polynomial time we find $\mathbf{v} = (v_1, \dots, v_d, v_{d+1}, \dots, v_{d+m}) \in \mathcal{L}_k(t_1, \dots, t_d)$ such that

$$\|\mathbf{v} - \mathbf{s}\| \leq 2^{o(d+m)} \min\{\|\mathbf{z} - \mathbf{s}\|, \quad \mathbf{z} \in \mathcal{L}_k(t_1, \dots, t_d)\} \leq 2^{-k+o(d)} p \leq 2^{-\eta-1} p,$$

provided that p is large enough. We also have

$$\left(\sum_{i=1}^{d}(u_i - s_i)^2\right)^{1/2} \leq d^{1/2} 2^{-k-1} p \leq 2^{-\eta-1} p.$$

Therefore,

$$\left(\sum_{i=1}^{d}(u_i - v_i)^2\right)^{1/2} \leq 2^{-\eta} p.$$

Applying Lemma 8, we see that $\mathbf{v} = \mathbf{u}_\alpha$ with probability at least $1 - p^{-1}$, and therefore the components $a_1, \ldots, a_m$ of α can be recovered from the last m components of $\mathbf{v} = \mathbf{u}_\alpha$.   □

Accordingly, Lemma 9 implies:

Theorem 2. *Let p be a sufficiently large prime number and let s be a divisor of m, $s|m$. Let $\mathcal{G}$ be a subgroup of of $\mathcal{N}_s$ with $|\mathcal{G}| \geq |\mathcal{N}_s|^{1/2} p^\delta$ for some fixed $\delta > 0$. Then for $k = \lceil 2 \log^{1/2} p \rceil$, $d = \lceil 4(m+1) \log^{1/2} p \rceil$ the following holds. There exists a deterministic polynomial time algorithm $\mathcal{A}$ such that for any $\alpha \in \mathbb{K}$ given $2d$ values $t_i \in \mathcal{G}$ and $s_i = \mathrm{MSB}_{k,p}(\mathrm{Tr}(\alpha t_i))$, $i = 1, \ldots, d$, its output satisfies*

$$\Pr_{t_1,\ldots,t_d \in \mathcal{G}} [\mathcal{A}(t_1, \ldots, t_d; s_1, \ldots, s_d) = \alpha] \geq 1 - p^{-1}$$

if $t_1, \ldots, t_d$ are chosen uniformly and independently at random from $\mathcal{G}$.

5   Bit Security of XTR

From Theorem 24 of [48] (see also [6, 23, 25]), any efficient algorithm to compute $\mathrm{Tr}_{\mathbb{K}/\mathbb{L}}(g^{xy})$ from g^x and g^y can be used to construct an efficient algorithm to compute g^{xy} from the same information. In [43] the same result was obtained with an algorithm which computes $\mathrm{Tr}_{\mathbb{K}/\mathbb{L}}(g^{xy})$ only for a positive proportion of pairs g^x, g^y. Furthermore, the same results hold even for algorithms which compute only $\mathrm{Tr}(g^{xy})$. Any $v \in \mathbb{L}$ can be represented by a pair $(\mathrm{Tr}_{\mathbb{L}/\mathbb{F}}(v), \mathrm{Tr}_{\mathbb{L}/\mathbb{F}}(\vartheta v))$ where ϑ is a root of an irreducible quadratic polynomial over $\mathbb{F}$, so $\mathrm{Tr}(g^{xy})$ is a part of the representation of $\mathrm{Tr}_{\mathbb{K}/\mathbb{L}}(g^{xy})$. In fact the same result holds for $\mathrm{Tr}(\omega g^{xy})$ with any fixed $\omega \in \mathbb{K}^*$.

Thus the above results suggest that breaking XTR is not easier than breaking the classical Diffie–Hellman scheme. Here we obtain one more result of this kind. We would like to show that an oracle computing a certain proportion of bits of $\mathrm{Tr}(g^{xy})$ from $\mathrm{Tr}_{\mathbb{K}/\mathbb{L}}(g^x)$ and $\mathrm{Tr}_{\mathbb{K}/\mathbb{L}}(g^y)$ can be used to break classical Diffie–Hellman. However, we instead prove that an oracle that computes a certain proportion of bits of $\mathrm{Tr}(g^{xy})$ from g^x, g^y can be used to compute g^{xy} from g^x, g^y. This is stronger; the former oracle could be used to simulate the latter so if the latter oracle cannot exist, neither can the former.

Thus, to really benefit from our results, one would need to use the trace over $\mathbb{F}_p$ rather than $\mathbb{F}_{p^2}$ in XTR based systems. As $\mathrm{Tr}(z) = \mathrm{Tr}_{\mathbb{K}/\mathbb{L}}(z) + \mathrm{Tr}_{\mathbb{K}/\mathbb{L}}(z)^p$ and as pth powers are "free" in XTR (due to the specific representation of $\mathbb{F}_{p^2}$) this could easily be done.

For a positive integer k we denote by $\mathcal{XTR}_k$ the oracle such that for any given values of g^x and g^y, it outputs $\mathrm{MSB}_{k,p}(\mathrm{Tr}(g^{xy}))$.

Theorem 3. *Let p be a sufficiently large n-bit prime number. Suppose the XTR generator g has prime order l satisfying $l|p^2 - p + 1$ and $l \geq p^{3/2+\delta}$ for some fixed $\delta > 0$. Then there exists a polynomial time algorithm which, given $U = g^u$ and $V = g^v$, for some $u, v \in \{0, \ldots, l-1\}$, makes $O(\log^{1/2} p)$ calls of the oracle $\mathcal{XTR}_k$ with $k = \lceil 2 \log^{1/2} p \rceil$ and computes g^{uv} correctly with probability at least $1 - p^{-1}$.*

Proof. The case $u = 0$ is trivial. Now assume that $1 \leq u \leq l - 1$. Then $g_u = g^u$ is an element of order l (because l is prime).

Select random $r \in \{0, \ldots, l - 1\}$. Applying the oracle $\mathcal{XTR}_k$ to U and $V_r = g^{v+r} = Vg^r$ we obtain $\mathrm{MSB}_{k,p}\left(\mathrm{Tr}\left(g^{u(v+r)}\right)\right) = \mathrm{MSB}_{k,p}\left(\mathrm{Tr}\left(g^{vu}t\right)\right)$ where $t = g_u^r$.

For $d = O(\log^{1/2} p)$ independent, random $r_1, \ldots, r_d \in [0, l - 1]$, we can now apply Theorem 2 with $\alpha = g^{uv}$, $m = 6$, and the group $\mathcal{G}$ generated by g_u (equal to the group generated by g). Indeed, we see that

$$\mathrm{Nm}_2(g) = g^{1+p^3} = g^{(p+1)(p^2-p+1)} = 1$$

thus $\mathcal{G} \in \mathcal{N}_2$ and from Theorem 2 we obtain the desired result. $\qquad\square$

6   Bit Security of LUC

For a positive integer k we denote by $\mathcal{LUC}_k$ an oracle that for any given values of g^x, g^y, outputs $\mathrm{MSB}_{k,p}\left(\mathrm{Tr}\left(g^{xy}\right)\right)$. In complete analogy with the proof of Theorem 3 we establish the following.

Theorem 4. *Let p be a sufficiently large n-bit prime number. Suppose the LUC generator g has prime order l satisfying $l | p + 1$ and $l \geq p^{1/2+\delta}$ for some fixed $\delta > 0$. Then there exists a polynomial time algorithm which, given $U = g^u$ and $V = g^v$, for some $u, v \in \{0, \ldots, l - 1\}$, makes $O(\log^{1/2} p)$ calls of the oracle $\mathcal{LUC}_k$ with $k = \lceil 2\log^{1/2} p \rceil$ and outputs g^{uv} with probability at least $1 - p^{-\delta/2}$.*

7   Hard-Core Bits of LUC

The security of LUC clearly depends on the hardness of inverting the function

$$x \mapsto f_g(x) = \mathrm{Tr}_{\mathbb{F}_{p^2}/\mathbb{F}_p}(g^x) = \mathrm{Tr}(g^x),$$

where, as above, g is the LUC generator of order $l | p + 1$. That is, it is necessary that this is a *one-way function*, otherwise the LUCDIF scheme would clearly be insecure. Even if this is true, the function could still have other undesirable properties in the form of "leakage", for example, it may be possible to determine individual bits of x. Note that Theorem 4 roughly says that, unless $f_g(x)$ can be efficiently inverted, it cannot be the case that *all of* the $\log^{1/2} p$ most significant bits of x can be simultaneously computed *without errors* from $f_g(x)$. Still, it does not exclude the possibility of computing in polynomial time a single bit of x, with probability, say, $1/2 + \varepsilon(\log p)$ for a non-negligible function $\varepsilon(n)$. (Recall that $\nu(n)$ is negligible if for any $c > 0$, $\nu(n) = o(n^{-c})$.)

As usual let $\{0,1\}^*$ denote the set of all finite binary strings and let G^n, G^* be some subsets of $\{0,1\}^n, \{0,1\}^*$, respectively. Suppose $f : G^* \mapsto \{0,1\}^*$ and $b : G^* \mapsto \{0,1\}$. Then b is called a *hard-core function* for f if for all non-negligible $\varepsilon(n)$ and all probabilistic polynomial time algorithms, for sufficiently large n,

$$\Pr[\mathcal{A}(f(x)) = b(x)] \leq \frac{1}{2} + \varepsilon(n),$$

probability taken over $x \in G^n$ and the random coin tosses required by $\mathcal{A}$.

We are interested in the case that $f(x) = V_x(P) = \text{Tr}(g^x)$ and $b(x) = \text{bit}_i(x)$, the ith bit of x (where $\text{bit}_0(x) = \text{lsb}(x)$ is the least significant bit). A problem is that $V_x(P)$ is not uniquely invertible. Note however that $V_x(P) = V_z(P)$, precisely when $z = l - x$. To make "the ith bit of x" well-defined by $V_x(P)$, we restrict the domain to $\{x \mid x < l/2\}$. (This hurts the group structure, but not our proofs.) Hence, we also assume that the oracle's advantage is averaged over this smaller set.

In Sect. 6, we assumed the existence of an oracle that returned $\text{Tr}(tg^x)$, where the oracle took care of selecting t (and computing the answer). Here we need that we from $\text{Tr}(g^x)$ can ourselves efficiently compute values of form $\text{Tr}(g^{wx+s})$ for w, s of our own choice. Note that by the Diffie-Hellman-like properties of LUC, the case $s = 0$ is easy. We thus start with a certain technical statement. Why traces of this form is of interest will shortly be made clear.

Lemma 10. *Given* $\text{Tr}(g^x)$, *for any set of* $N = (\log p)^{O(1)}$ *triples* $(k_j, r_j, s_j) \in \mathbb{Z}_l^* \times \mathbb{Z}_l \times \mathbb{Z}_l$, *we can in polynomial time compute two sets* $\mathcal{T}_\nu = \{T_{\nu,1}, \ldots T_{\nu,N}\}$, *so that for at least one* $\nu = 1, 2$, $T_{\nu,j} = \text{Tr}(g^{k_j^{-1} r_j x + s_j})$ *for* $j = 1, \ldots, N$.

Proof. As observed in [2], the conjugates of g^x are the roots, H_0, H_1, of $X^2 - \text{Tr}(g^x)X + 1$, and they can be found in polynomial time. The conjugates h_0, h_1 of g are trivial to compute and $H_i = h_{\pi(i)}^x$ for some permutation $\pi \in S_2$. Thus, for any $w = k^{-1}r$ and s,

$$\text{Tr}(g^{wx+s}) = h_{\pi(0)}^s H_0^w + h_{\pi(1)}^s H_1^w.$$

Now, we do not know π, but there are only two possibilities. We thus obtain two candidates $\mathcal{T}_\nu$, $\nu = 1, 2$. $\qquad\Box$

As noted, as with conventional discrete logarithm not all bits are secure.

Theorem 5. *Suppose* $l = 2^s r$, $s > 0$, r *odd. There is a polynomial time algorithm to determine* $\text{bit}_i(x)$ *from* $V_x(P)$ *for every* $i = 0, \ldots, s - 1$.

Proof. For the least significant bit, note that $V_{xl/2}(P) = (-1)^{\text{lsb}(x)} V_0(P)$, a condition that can be easily checked. Using Lemma 10, the rest follows from a straight-forward generalization of the Pohlig-Hellman algorithm [39] to decide $x \pmod{2^s}$. $\qquad\Box$

All other bits are, however, hard:

Theorem 6. *Let* $c > 0$ *be a constant. Except with probability* $o(1)$ *over random choices of* n-*bit prime* $p = 2^s r - 1$, r *odd, and* g *the LUC generator of order* $l = p + 1$, *the following holds. If for some* i, $0 \le i \le n - c\log n$ *there is an algorithm that given* $V_x(P)$ *for random* x, *computes* $\text{bit}_i(x)$ *with probability at least* $1/2 + \varepsilon(n)$, *then there is a probabilistic algorithm that in time polynomial in* $n\varepsilon(n)^{-1}$ *computes* x *with non-negligible probability.*

Proof. In [38, 13], reductions are given from discrete logarithms in $\mathbb{F}_p$ to computing $\mathrm{bit}_i(x)$ from g^x, for any i. The reduction only uses operations transforming $y = g^x$ into values of form $y^{k^{-1}a}g^b = g^{k^{-1}ax+b}$. By Lemma 10, the same transformations can be applied to $\mathrm{Tr}(g^x)$, for invertible k (as will be the case).

The only complication that arise is that we have here restricted the domain of $V_x(P) = \mathrm{Tr}(g^x)$ to $x < l/2$. However, a closer look at [13] we see that in each step of the reduction, a good approximation to the relative magnitude of each $k^{-1}ax + b$ (modulo the order of g) is maintained. Using this information, we simply discard all samples on $k^{-1}ax + b > l/2$. As the samples are uniformly distributed, this increases the complexity by a factor two. Taking into account the two choices for the list of samples from Lemma 10, we loose another (non-critical) factor of two, trying both possibilities.   □

For prime l, $l|p+1$, the same result follows for all i and all large p from [42].

We note that the $k = O(\log n)$ *most significant bits* of x can also be shown to be hard by [13] and a security notion for biased functions from [46]. For such k, from [3, 29] also follows hardness for the "related" function $\mathrm{MSB}_{k,p}$.

7.1   Hard-Core Bits of XTR

Superficially, all details (for example, a generalization of Lemma 10) seem to go through for XTR. However, problems are encountered by the fact that the XTR function, $\mathrm{Tr}_{\mathbb{F}_{p^6}/\mathbb{F}_{p^2}}(g^x)$, is three-to-one, rather than two-to-one, and there seems to be no obvious way to restrict the domain to a set on which: (a) XTR is 1–1, (b) the set has non-negligible density, and, (c) the set is an interval, $[a..b] \subset \mathbb{Z}_l$. These properties seem necessary to apply the techniques of [13].

8   Summary and Open Problems

Establishing security of new cryptosystems is important. We have shown that LUC and XTR share security properties with the more well-established discrete logarithm based systems. The Diffie-Hellman variant enjoys security features as of the original Diffie-Hellman key exchange. To this end, we have seen a new application for exponential sums; we believe there are more such in store. This paper is the most recent in a series studying variants of the hidden number problem, using such exponential sums. One can ask if there is a "general" theorem to be sought, rather than treating each special case. Such generalization seem difficult though: make the group a little smaller (for example, the proposed XTR-extensions to $\mathbb{F}_{p^{2\cdot3\cdot5}}$) and everything breaks down.

For the LUC scheme, we also showed that no non-trivial information about individual bits leak. Though likely to be true, the analogue for XTR is left open.

References

1. M. Ajtai, R. Kumar and D. Sivakumar, *A sieve algorithm for the shortest lattice vector problem*, Proc. 33rd ACM Symp. on Theory of Comput., Crete, Greece, July 6-8, 2001, 601–610.

2. D. Bleichenbacher, W. Bosma and A. K. Lenstra, *Some remarks on Lucas-based Cryptograph*, Lect. Notes in Comp. Sci., Springer-Verlag, **963** (1995), 386–396.
3. M. Blum and S. Micali, *How to Generate Cryptographically Strong Sequences of Pseudo-random Bits*, SIAM J. on Computing, **13**(4), 850–864, 1984.
4. D. Boneh and R. Venkatesan, *Hardness of computing the most significant bits of secret keys in Diffie–Hellman and related schemes*, Lect. Notes in Comp. Sci., Springer-Verlag, **1109** (1996), 129–142.
5. D. Boneh and R. Venkatesan, *Rounding in lattices and its cryptographic applications*, Proc. 8th Annual ACM-SIAM Symp. on Discr. Algorithms, ACM, NY, 1997, 675–681.
6. A. E. Brouwer, R. Pellikaan and E. R. Verheul, *Doing more with fewer bits*, Lect. Notes in Comp. Sci., Springer-Verlag, **1716** (1999), 321–332.
7. R. Canetti, J. B. Friedlander, S. Konyagin, M. Larsen, D. Lieman and I. E. Shparlinski, *On the statistical properties of Diffie–Hellman distributions*, Israel J. Math., **120** (2000), 23–46.
8. P. Deligne, *Cohomologie 'etale (SGA $4\frac{1}{2}$)*, Lect. Notes in Math., Springer-Verlag, **569** (1977).
9. E. El Mahassni, P. Q. Nguyen and I. E. Shparlinski, *The insecurity of Nyberg–Rueppel and other DSA-like signature schemes with partially known nonces*, Lect. Notes in Comp. Sci., Springer-Verlag, **2146** (2001), 97–109.
10. J. B. Friedlander, M. Larsen, D. Lieman and I. E. Shparlinski, *On correlation of binary M-sequences*, Designs, Codes and Cryptography, **16** (1999), 249–256.
11. M. I. González Vasco and I. E. Shparlinski, *On the security of Diffie-Hellman bits*, Proc. Workshop on Cryptography and Computational Number Theory, Singapore 1999, Birkhäuser, 2001, 257–268.
12. M. I. González Vasco and I. E. Shparlinski, *Security of the most significant bits of the Shamir message passing scheme*, Math. Comp., **71** (2002), 333–342.
13. J. Håstad and M. Näslund, *The Security of all RSA and discrete log bits*, Electronic Colloquium on Computational Complexity, Report TR99-037, 1999. (To appear in Jorunal of the ACM)
14. N. A. Howgrave-Graham and N. P. Smart, *Lattice attacks on digital signature schemes*, Designs, Codes and Cryptography, **23** (2001), 283–290.
15. R. Kannan, *Algorithmic geometry of numbers*, Annual Review of Comp. Sci., **2** (1987), 231–267.
16. R. Kannan, *Minkowski's convex body theorem and integer programming*, Math. of Oper. Research, **12** (1987), 231–267.
17. N. M. Katz, *Gauss sums, Kloosterman sums, and monodromy groups*, Ann. of Math. Studies, **116**, Princeton Univ. Press, 1988.
18. N. Koblitz, *Elliptic curve cryptosystems*, Math. Comp., **48**, 203–209, 1987.
19. S. V. Konyagin and I. E. Shparlinski, *Character sums with exponential functions and their applications*, Cambridge Univ. Press, Cambridge, 1999.
20. A. K. Lenstra *Unbelievable security. Matching AES security using public key systems*, Lect. Notes in Comp. Sci., Springer-Verlag, **2248** (2001), 67–86.
21. A. K. Lenstra, H. W. Lenstra and L. Lovász, *Factoring polynomials with rational coefficients*, Mathematische Annalen, **261** (1982), 515–534.
22. A. K. Lenstra and M. Stam, *Speeding up XTR*, Lect. Notes in Comp. Sci., Springer-Verlag, **2248** (2001), pp. 125–143.
23. A. K. Lenstra and E. R. Verheul, *The XTR public key system*, Lect. Notes in Comp. Sci., Springer-Verlag, **1880** (2000), 1–19.
24. A. K. Lenstra and E. R. Verheul, *Key improvements to XTR*, Lect. Notes in Comp. Sci., Springer-Verlag, **1976** (2000), 220–233.

25. A. K. Lenstra and E. R. Verheul, *An overview of the XTR public key system*, Proc. the Conf. on Public Key Cryptography and Computational Number Theory, Warsaw 2000, Walter de Gruyter, 2001, 151–180.
26. W.-C. W. Li, *Character sums and abelian Ramanujan graphs*, J. Number Theory, **41** (1992), 199–217.
27. W.-C. W. Li, *Number theory with applications*, World Scientific, Singapore, 1996.
28. R. Lidl and H. Niederreiter, *Finite fields*, Cambridge University Press, Cambridge, 1997.
29. D. L. Long and A. Wigderson, *The discrete log hides $O(\log n)$ bits*, SIAM J. on Computing, **17**(2):413–420, 1988.
30. D. Micciancio, *On the hardness of the shortest vector problem*, PhD Thesis, MIT, 1998.
31. V. Miller, *Uses of elliptic curves in cryptography*, Lect. Notes in Comp. Sci., Springer-Verlag, **218** (1986), 417–426.
32. P. Q. Nguyen, *The dark side of the Hidden Number Problem: Lattice attacks on DSA*, Proc. Workshop on Cryptography and Computational Number Theory, Singapore 1999, Birkhäuser, 2001, 321–330.
33. P. Q. Nguyen and I. E. Shparlinski, *The insecurity of the Digital Signature Algorithm with partially known nonces*, J. Cryptology, (to appear).
34. P. Q. Nguyen and I. E. Shparlinski, *The insecurity of the elliptic curve Digital Signature Algorithm with partially known nonces*, Designs, Codes and Cryptography, (to appear).
35. P. Q. Nguyen and J. Stern, *Lattice reduction in cryptology: An update*, Lect. Notes in Comp. Sci., Springer-Verlag, **1838** (2000), 85–112.
36. P. Q. Nguyen and J. Stern, *The two faces of lattices in cryptology*, Springer-Verlag, **2146** (2001), 146–180.
37. H. Niederreiter, *Random number generation and Quasi–Monte Carlo methods*, SIAM Press, 1992.
38. R. Peralta, *Simultaneous security of bits in the discrete log*, Lect. Notes in Comp. Sci., Springer-Verlag, **219** (1986), 62–72.
39. S. C. Pohlig and M. Hellman, *An improved algorithm for computing logarithms over* GF(p), IEEE Transactions on Information Theory, **IT-24**(1):106–110, 1978.
40. K. Prachar, *Primzahlverteilung*, Springer-Verlag, 1957.
41. C. P. Schnorr, *A hierarchy of polynomial time basis reduction algorithms*, Theor. Comp. Sci., **53** (1987), 201–224.
42. C. P. Schnorr, *Security of almost all discrete log bits*, Electronic Colloquium on Computational Complexity, Report TR98-033, 1998.
43. I. E. Shparlinski, *Security of polynomial transformations of the Diffie–Hellman key*, Preprint, 2000, 1–8.
44. I. E. Shparlinski, *Sparse polynomial approximation in finite fields*, Proc. 33rd ACM Symp. on Theory of Comput., Crete, Greece, July 6-8, 2001, 209–215.
45. I. E. Shparlinski, *On the generalised hidden number problem and bit security of XTR*, Lect. Notes in Comp. Sci., Springer-Verlag, **2227** (2001), 268–277.
46. A. W. Schrift and A. Shamir, *On the universality of the next bit test*, Lect. Notes in Comp. Sci., Springer-Verlag, **537** (1990), 394–408.
47. P. J. Smith and C. T. Skinner, *A public-key cryptosystem and a digital signature system based on the Lucas function analogue to discrete logarithms*, Lect. Notes in Comp. Sci., Springer-Verlag, **917** (1995), 357–364.
48. E. R. Verheul, *Certificates of recoverability with scalable recovery agent security*, Lect. Notes in Comp. Sci., Springer-Verlag, **1751** (2000), 258–275.

Expanding Pseudorandom Functions; or: From Known-Plaintext Security to Chosen-Plaintext Security

Ivan Damgård and Jesper Buus Nielsen

BRICS* Department of Computer Science
University of Aarhus
Ny Munkegade
DK-8000 Arhus C, Denmark
{ivan,buus}@brics.dk

Abstract. Given any weak pseudorandom function, we present a general and efficient technique transforming such a function to a new weak pseudorandom function with an arbitrary length output. This implies, among other things, an encryption mode for block ciphers. The mode is as efficient as known (and widely used) encryption modes as CBC mode and counter (CTR) mode, but is provably secure against chosen-plaintext attack (CPA) already if the underlying symmetric cipher is secure against known-plaintext attack (KPA). We prove that CBC, CTR and Jutla's integrity aware modes do not have this property. In particular, we prove that when using a KPA secure block cipher, then: CBC mode is KPA secure, but need not be CPA secure, Jutla's modes need not be CPA secure, and CTR mode need not be even KPA secure. The analysis is done in a concrete security framework.

1 Introduction

A block cipher that is secure against known plaintext attacks (KPA) is a natural example of a *weak pseudorandom function*: as long as the key is unknown, and an adversary is given a set of *random* plaintext blocks and corresponding ciphertext blocks (of length, say, k bits each), he cannot distinguish the encryption function from a random function mapping k bits to k bits – or at least he can only do so with a small advantage. In general, a weak pseudorandom function is just a function F that maps a key K and input string x to a string y, we write $y \leftarrow F(K, x)$. It does not have to be invertible and y does not have to be the same length as x. Weak pseudorandomness means that even if an adversary is given $(x_1, F_K(x_1)), \ldots, (x_t, F_K(x_t))$, for uniformly random K and x_i, he cannot distinguish this from $(x_1, R(x_1)), \ldots, (x_t, R(x_t))$ where R is a random function. Pseudorandomness means that distinguishing remains hard, even if the adversary gets to pick the x_i's. An example of a weak pseudorandom function can be

---

* Basic Research in Computer Science,
  Centre of the Danish National Research Foundation.

M. Yung (Ed.): CRYPTO 2002, LNCS 2442, pp. 449–464, 2002.
© Springer-Verlag Berlin Heidelberg 2002

derived in a natural way from the Decisional Diffie-Hellman Assumption (DDH). Here there are some public parameters, a large prime P and an element g of large prime order q such that the DDH assumption holds in the group generated by g. A random element $\alpha \in \mathbf{Z}_q$ functions as the secret key, the domain is the group generated by g, and the function maps h to h^α mod P.

In this paper, we consider the case where we are given any weak pseudorandom function F with output length k. We present a general construction transforming F into a weak pseudorandom function G for which $G_K(x)$ is nk bits long and where n is any positive integer. The construction is efficient in the sense that to evaluate G, we need exactly n evaluations of F. We believe this is the best we can expect, since if nk random looking bits could be generated with fewer evaluations of F, it seems the construction would have to do some sort of pseudorandom generation "on its own". Whereas this construction requires key size logarithmic is n, we give a way to modify the construction such that variable output length can be handled using a constant amount of key material.

Note that G may also be used as a pseudorandom generator that expands K and a randomly chosen input into a random looking string of nk bits. If we apply our construction to the above function based on the DDH assumption, we obtain a pseudorandom generator that outputs k bits for each exponentiation done, where k is the security parameter. This is slightly faster than the generator that follows from Naor and Reingolds DDH-based pseudorandom function [NR97].

Our construction also implies an encryption mode for block ciphers that is secure against chosen plaintext attacks (CPA) already if the block cipher is KPA secure. We call this Pseudorandom Tree (PRT) mode. While an encryption scheme with such properties could be easily constructed using generic methods [BM84,GL89], the result would be very inefficient. In contrast, to encrypt/decrypt a message, PRT requires exactly the same number of block encryptions as the well known CBC and CTR modes and produces ciphertexts of the same lengths as does CBC/CTR. The only overhead compared to CBC/CTR is logarithmic in the length of the message, namely a key-schedule requiring $2 \log_2(n)$) key expansions and block encryptions for a message containing n blocks. This can be precomputed, so that the actual throughput can be exactly the same as in the standard modes. Note that CBC and CTR also provide CPA security, but here we have to assume that the block cipher is also CPA secure. As in CTR mode, our mode allows easy random access decryption: To decrypt an arbitrary block in an n block message, we need at most $\log_2(n)$ block encryptions. Also as in CTR mode, we only use the block cipher in the encryption direction, even when we decrypt – in implementations, this can sometimes save code or chip area.

From a theoretical point of view, it is of course always better to prove security under the weakest possible assumption. But we believe the result is also useful from a more practical point of view: even though a block cipher was designed to be CPA secure, there may always be surprises and hence relying only on the KPA security of the block cipher gives extra protection. Moreover, we prove in the concrete security framework a bound on how CPA security of PRT mode relates to KPA security of the block cipher, i.e., if we assume the block cipher can be broken with at most a certain advantage under a KPA, we obtain a bound on the advantage with which PRT can be broken under a CPA. This bound behaves similarly to the bound that relates CPA security of CBC to CPA security of the block cipher. Now, considering that it may well be reasonable to assume that the block cipher performs better against KPA than against CPA, we see that we may get a better concrete bound on the security of PRT than we can get for CBC.

We mentioned that PRT introduces a slight overhead compared to using CBC or CTR. One might wonder whether it is possible to obtain the "amplification" from KPA to CPA with no overhead at all. As a partial answer, in Section 4, we analyse CBC mode, CTR mode and Jutla's integrity aware modes [Jut01] and show that none of these schemes can guarantee CPA security unless the underlying block cipher is CPA secure itself. Indeed, we prove that when using a KPA secure block cipher, CBC mode and Jutla's modes are KPA secure, but need not be CPA secure and counter mode need not even be KPA secure. The same simple techniques applied there can be applied to most other known modes of operations to demonstrate that they also do not guarantee CPA security unless the underlying block cipher is CPA secure itself (to our knowledge they can be applied to all *efficient* such modes). We therefore leave it as an interesting open question to investigate whether one can amplify from KPA to CPA and make do with a constant amount of overhead.

Finally, we discuss the problem of achieving chosen ciphertext (CCA) security (in addition to CPA). While we can argue that this is possible based only on a KPA secure block cipher, some overhead is introduced, and we do not know how to provide CCA security as efficiently as the CPA secure construction.

Related Work. The first work to relate the security of modes of operations to the underlying block cipher is [BDJR97]. In [BDJR97] notions of CPA security and CCA security of block ciphers and symmetric encryption are developed in a concrete security framework, and the security of three well-known encryption modes, CBC mode and CTR mode (in its deterministic and probabilistic variants), are analysed. In [BDJR97] the modes of operations are analysed under the assumption that the underlying block cipher is CPA secure. In Section 2 we review their framework and make a straightforward extension to cover KPA security.

The first, and to our knowledge only, other paper to consider security enhancing modes of operations is [HN00]. What is investigated there is the possibility to generate a key stream given a one-way function, after which encryption is just Xoring the message and key stream. To be able to base their construction on block-ciphers, they consider the function $f : K \mapsto E_K(P)$ mapping a key to the encryption of a fixed plaintext under that key and assume that it is one-way. This assumption is of course well-motivated: If the encryption function should be secure in any reasonable sense, one should not be able to find the key given a plaintext/ciphertext pair. However, this assumption alone is not enough: The scheme, called Key Feedback Mode (KFB) is a variant of the technique of [BM84,Lev85,GL89], and works by iterating the one-way function, i.e., we compute $f(K), f(f(K)), \ldots$ and extract some number of bits from each value computed. For this to work, one must assume that f stays one-way even when iterated, i.e., from $f^i(K)$ it is hard to find $f^{i-1}(K)$. We believe that this assumption is closely related to our assumption on KPA security, but the assumptions do not seem to be directly comparable. In any case, KFB is significantly less efficient than PRT: in KFB, a key expansion is needed for every block encryption performed, and each block encryption can only contribute significantly less than k bits to the output stream (where k is block size) – from an asymptotic point of view, only a logarithmic number of bits can be produced. In contrast, PRT can use all k bits produced by each block encryption, and needs a logarithmic number of key expansions.

An Intuitive Introduction to the Main Idea. Assume that we have access to a length-preserving and weak pseudorandom function $F.(\cdot)$, where F_K could be, e.g., the encryption function of a KPA secure block cipher. From this we can build a length doubling weak pseudorandom function G_K by letting the key be a pair of keys $K = (K_1, K_2)$ for F, letting the input domain be that of F, and letting the output be $G_{K_1,K_2}(R) = F_{K_1}(R) \| F_{K_2}(R)$. We can repeat this operation as follows. Let $C_1 \| C_2 = G_{K_1,K_2}(R)$ and assume we have another key (K'_1, K'_2) for G. We can then compute $C_3 \| C_4 = G_{K'_1,K'_2}(C_1)$ and $C_5 \| C_6 = G_{K'_1,K'_2}(C_2)$ and let the output (generated from input R) be $C_1 \| C_2 \| C_3 \| C_4 \| C_5 \| C_6$. We then have a function six-doubling its input using six applications of F. Continuing this we will soon have a pseudorandom function with a very large output length. Using essentially the fact that a new key is used for each 'level' in this construction, it can be shown to be weak pseudorandom if F has this property. This construction can also handle variable-length output, if we can find a secure way to produce produce a key for each level in the construction (to an arbitrary depth) from a constant number of keys for F. How to do this using only weak pseudorandomness of F is shown in detail in Section 3. Finally, to use this construction to encrypt a message M, we can simply choose R at random, and let the ciphertext be $R, G_K(R) \oplus M$, where G is the function we construct from the encryption function of the block cipher. Clearly, a CPA attack on this scheme will allow the adversary to get the value of G_K on random inputs – and nothing more than that. Hence weak pseudorandomness of G is sufficient for CPA security.

2   Notions of Security

The following definitions are straightforward extensions of definitions from [BDJR97,Des00] to consider also KPA security. Of the four notions of security considered in [BDJR97] we have chosen real-or-random (ROR) indistinguishability. A symmetric encryption scheme $\mathcal{SE} = (\mathcal{K}, \mathcal{E}, \mathcal{D})$ consists of three randomized algorithms. The key generation algorithm $\mathcal{K}$ returns a key K; we write $K \leftarrow \mathcal{K}$. The encryption algorithm $\mathcal{E}$ takes as input the key K and a plaintext M and returns a ciphertext C; we write $C \leftarrow \mathcal{E}_K(M)$. The decryption algorithm $\mathcal{D}$ takes as input the key K and a string C and returns a unique plaintext M or $\perp$; we write $x \leftarrow \mathcal{D}_K(C)$. We require that $\mathcal{D}_K(\mathcal{E}_K(M)) = M$ for all $M \in \{0,1\}^*$.

Definition 1 (ROR-KPA, ROR-CPA). *Let* $\mathcal{SE} = (\mathcal{K}, \mathcal{E}, \mathcal{D})$ *be a symmetric encryption scheme. Let* $b \in \{0,1\}$. *Let A be an adversary that has access to an oracle. Let* $\mathcal{R}_{K,b}$ *be the oracle which on input* $l \in \mathbb{N}$, *if* $b = 1$, *outputs* $(x, \mathcal{E}_K(x))$ *for uniformly random* $x \in \{0,1\}^l$, *and, if* $b = 0$, *outputs* $(x, \mathcal{E}_K(r))$ *for uniformly random* $x, r \in \{0,1\}^l$. *Let* $\mathcal{O}_{K,b}$ *be the oracle which on input* $x \in \{0,1\}^*$, *if* $b = 1$, *outputs* $\mathcal{E}_K(x)$, *and, if* $b = 0$, *outputs* $\mathcal{E}_K(r)$ *for uniformly random* r *of the same length as* x. *Now consider the following experiments:*

$$
\begin{array}{ll}
\underline{\textbf{proc } \mathbf{Exp}_{\mathcal{SE},A}^{\text{ror-kpa-}b}} \equiv & \qquad \underline{\textbf{proc } \mathbf{Exp}_{\mathcal{SE},A}^{\text{ror-cpa-}b}} \equiv \\
\quad K \leftarrow \mathcal{K} & \qquad\quad K \leftarrow \mathcal{K} \\
\quad d \leftarrow A^{\mathcal{R}_{K,b}} & \qquad\quad d \leftarrow A^{\mathcal{O}_{K,b}} \\
\quad \underline{\textbf{return }} d & \qquad\quad \underline{\textbf{return }} d
\end{array}
$$

We define the **advantage** *of the adversary via*

$$
\mathbf{Adv}_{\mathcal{SE},A}^{\text{ror-kpa}} = \Pr[\mathbf{Exp}_{\mathcal{SE},A}^{\text{ror-kpa-}1} = 1] - \Pr[\mathbf{Exp}_{\mathcal{SE},A}^{\text{ror-kpa-}0} = 1]
$$

$$
\mathbf{Adv}_{\mathcal{SE},A}^{\text{ror-cpa}} = \Pr[\mathbf{Exp}_{\mathcal{SE},A}^{\text{ror-cpa-}1} = 1] - \Pr[\mathbf{Exp}_{\mathcal{SE},A}^{\text{ror-cpa-}0} = 1] .
$$

We define the **advantage function** *of the scheme as follows. For any integers* t, q, μ,

$$
\mathbf{Adv}_{\mathcal{SE}}^{\text{ror-kpa}}(t, q, \mu) = \max_A \left\{ \mathbf{Adv}_{\mathcal{SE},A}^{\text{ror-kpa}} \right\}
$$

$$
\mathbf{Adv}_{\mathcal{SE}}^{\text{ror-cpa}}(t, q, \mu) = \max_A \left\{ \mathbf{Adv}_{\mathcal{SE},A}^{\text{ror-cpa}} \right\}
$$

where the maximum is over all A with "time complexity" l, *making at most q queries to the oracle, these totaling at most μ bits.*

By the "time complexity" we mean the worst case total running time of the experiment with $b = 1$, plus the size of the code of the adversary, in some fixed RAM model of computation. We stress that the total execution time of the experiment includes the time of *all* operations in the experiment, including the time for key generation and the encryptions done by the oracle. For a discussion of this time complexity, see [Des00].

A function family with key-space $\mathcal{K}$, input-length l, and output-length L is a map $F : \mathcal{K} \times \{0,1\}^l \rightarrow \{0,1\}^L$. For each key $K \in \mathcal{K}$ we define a map $F_K : \{0,1\}^l \rightarrow \{0,1\}^L$ by $F_K(x) = F(K, x)$. We write $f \overset{R}{\leftarrow} F$ for the operation $K \overset{R}{\leftarrow} \mathcal{K}; f \leftarrow F_K$. We call F a family of permutations if for all $K \in \mathcal{K}$, F_K is a permutation. We use $\text{Rand}^{l \rightarrow L}$ to denote the family of all functions $\{0,1\}^l \rightarrow \{0,1\}^L$.

If a random function from the function family looks like a random function from $\text{Rand}^{l \rightarrow L}$, we call the family a **pseudorandom function family**. Below we define this notion formally for KPAs. The definitions for CPAs and CCAs can be found in [BDJR97], but will not be used in the present paper.

A variable-length output function family with key-space $\mathcal{K}$ and input-length l is a map $F : \mathcal{K} \times N \times \{0,1\}^l \rightarrow \{0,1\}^*$. For each key $K \in \mathcal{K}$ we define a map $F_K : N \times \{0,1\}^l \rightarrow \{0,1\}^*$ by $F_K(L,x) = F(K,L,x)$. We require that $|F(K,L,x)| = L$ for all inputs. We use VO-Rand^l to denote the probabilistic function generated as follows: On input (L,r) check if a string o_r is defined, if not define it to be the empty string. Then check whether o_r has length at least L, if not append to o_r a uniformly random string from $\{0,1\}^{L-|o_r|}$. Then output the L first bits of o_r. We define what it means for at variable-length output function family to be KPA secure.

Definition 2 (VO-PRF-KPA). *Let F be a variable-length output function family with input-length l. Let D be a distinguisher that has access to an oracle. Let $\mathcal{R}_f$ be the oracle which on input $L \in N$ generates a uniformly random $r \in \{0,1\}^l$ and outputs $(r, f(L,r))$. Now consider the following experiments:*

proc $\mathbf{Exp}_{F,D}^{\text{vo-prf-kpa-}0} \equiv$
$d \leftarrow D^{\mathcal{R}_{\text{VO-Rand}^l}}$
return d

proc $\mathbf{Exp}_{F,D}^{\text{vo-prf-kpa-}1} \equiv$
$f \xleftarrow{R} F$
$d \leftarrow D^{\mathcal{R}_f}$
return d

We define the **advantage** *of the distinguisher via*

$$\mathbf{Adv}_{F,D}^{\text{vo-prf-kpa}} = \Pr[\mathbf{Exp}_{F,D}^{\text{vo-prf-kpa-}1} = 1] - \Pr[\mathbf{Exp}_{F,D}^{\text{vo-prf-kpa-}0} = 1] \ .$$

We define the **advantage** *function of the function family as follows. For any t, q, l,*

$$\mathbf{Adv}_F^{\text{vo-prf-kpa}}(t, q, \mu) = \max_D \left\{ \mathbf{Adv}_{F,D}^{\text{vo-prf-kpa}} \right\} \ .$$

where the maximum is over all D with time complexity t, making at most q queries to the oracle, these totaling at most μ bits..

The notion of KPA security of a fixed output function family with input-length l can easily be derived from the above definition, giving rise to the notions $\mathbf{Adv}_{F,D}^{\text{prf-kpa}}$ and $\mathbf{Adv}_F^{\text{prf-kpa}}(t, q)$ – we skip the explicit mentioning of μ as it is given by q.

3   PRT Mode

PRT mode is a construction of a VO-PRF-KPA secure variable-length output function family from a PRF-KPA secure function family. The encryption will then be done using the variable-length output function family F as

$$\text{VO-PRF-ENC}[F]_K(M) = (r, F_K(|M|, r) \oplus M) \ ,$$

where r is uniformly random in $\{0,1\}^l$. The following theorem relates the ROR-CPA security of VO-PRF-ENC$[F]$ to the VO-PRF-KPA security of F.

Theorem 1. *Suppose F is a variable-length output function family. If F is VO-PRF-KPA secure, then VO-PRF-ENC$[F]$ is ROR-CPA secure[1]. Specifically, for any t, q, μ,*

$$\mathbf{Adv}^{\text{ror-cpa}}_{\text{VO-PRF-ENC}[F]}(t, q, \mu) \leq \mathbf{Adv}^{\text{vo-prf-kpa}}_{F}(t, q, \mu) + \frac{q(q-1)}{2^{l+1}} .$$

Proof: We prove the specific bound. Consider an ROR-CPA distinguisher $\overline{D}$ expecting access to an oracle $\mathcal{O}_{K,b}$ for the VO-PRF-ENC$[F]$ scheme. We construct a distinguisher D having access to a VO-PRF-KPA oracle $\mathcal{R}_f$ for the variable-length output function family F as follows. The distinguisher D runs the code of $\overline{D}$. Each time D request an encryption of message M, request a pair (r, R), where r is uniformly random in $\{0, 1\}^l$ and $R = f(|M|, r)$. Then return $c = (r, M \oplus R)$. When D returns with some value d, return d.

If $b = 1$, then f is a random function from F and the values c are distributed as values from $\mathcal{O}_{K,1}$. If on the other hand $b = 0$, then f is VO-Rand^l, and in that case the values c are distributed as values from $\mathcal{O}_{K,0}$, as long as there are no collisions among the r-values returned by $\mathcal{R}_f$. Since the r values are uniformly random l-bit values and q of them are drawn, the probability of collision are well-known to be upper bounded by $q(q-1)/2^{l+1}$, which proves the theorem. ∎

Security Preserving Operations on KPA Secure PRFs. Before presenting the actual construction, we present some operations on PRFs which preserves KPA security. Assume that we are given PRFs

$$F : \{0, 1\}^{m_1} \to \{0, 1\}^{n_1}, \qquad G : \{0, 1\}^{m_2} \to \{0, 1\}^{n_2}$$

with key-length k_1 resp. k_2. For operations only involving F, we use the notation $m = m_1, n = n_1, k = k_1$. Our first operation makes the output domain larger. It gives the function family

$$F^{\to \beta} : \{0, 1\}^m \to \{0, 1\}^{\beta n}, \qquad F^{\to \beta}_{K_1, \ldots, K_\beta}(R) = F_{K_1}(R) \| \cdots \| F_{K_\beta}(R) ,$$

where we generate a key for $F^{\to \beta}$ by generating β independent keys $K_1, \ldots, K_\beta$ for F. Our second operation is similar, but has shorter key-length. Assume that $k \leq m$, so that an output-block can be used as key, and consider the following function family

---

[1] Actually, we have not assigned a meaning to the claim that VO-PRF-ENC$[F]$ is ROR-CPA secure if F is VO-PRF-KPA secure, as we have no definition of security: In this paper we consider a concrete security framework without a security parameter. If, however, we introduced a security parameter k, then in the asymptotic security framework, all of $t, q, \mu, l,$ and L would be polynomial in k and typically $l = \Theta(k)$. We would then define security by requiring that the advantage of all probabilistic polynomial time (in k) adversaries is negligible (in k). The claim would then follow from the specific bound on $\mathbf{Adv}^{\text{ror-cpa}}_{\text{VO-PRF-ENC}[F]}(t, q, \mu)$ given by the theorem. In the following we will use the term "secure" with this meaning.

$$F^{\to \bar{\beta}} : \{0,1\}^m \to \{0,1\}^{\beta n}, \qquad F^{\to \bar{\beta}}_{K,R_1}(R_2) = F_{K_1}(R_2)\| \cdots \|F_{K_\beta}(R_2) ,$$

where $K_1 = K$ and inductively $K_{i+1} = F_{K_i}(R_1)$. Now consider the operation making both the input and the output domain larger

$$F^{\alpha \to \alpha} : \{0,1\}^{\alpha m} \to \{0,1\}^{\alpha n}, \qquad F^{\alpha \to \alpha}_K(R) = F_K(R_1)\| \cdots \|F_K(R_\alpha) ,$$

where $R = (R_1, \ldots, R_\alpha)$. For completeness we name the following operation

$$F^{\alpha \to \alpha\beta} : \{0,1\}^{\alpha m} \to \{0,1\}^{\alpha\beta n}, \qquad F^{\alpha \to \beta} = (F^{\to \bar{\beta}})^{\alpha \to \alpha} .$$

Finally assume that $n_1 \geq m_2$ and consider the following composition operation

$$G \circ F : \{0,1\}^{m_1} \to \{0,1\}^{n_1+n_2}, \qquad (G \circ F)_{K_1,K_2}(R) = F_{K_1}(R)\|G_{K_2}(F_{K_1}(R)) .$$

We give a short example of the use of these operations. Assume that we are given any KPA secure PRF $F : \{0,1\}^m \to \{0,1\}^m$. From this family, we can using the $\to 2$ operation and construct a KPA secure PRF $G : \{0,1\}^m \to \{0,1\}^{2m}$. From G we can then construct the KPA secure PRF $G^{2\to2} : \{0,1\}^{2m} \to \{0,1\}^{4m}$, and can using the composition operation construct the KPA secure PRF $G^{2\to2} \circ G : \{0,1\}^m \to \{0,1\}^{6m}$. This can be iterated. In Fig. 1 the PRF $G^{8\to8} \circ G^{4\to4} \circ G^{2\to2} \circ G : \{0,1\}^m \to \{0,1\}^{30m}$ is depicted. This construction works even if F is not length preserving. We can always define G by computing $F^{\to\beta}$ for appropriate choice of β and use the first $2m$ bits of the output.

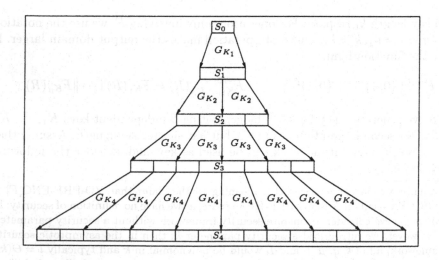

Fig. 1. The structure of the PRF $G^{8\to8} \circ G^{4\to4} \circ G^{2\to2} \circ G$. Note that the output of the function is all levels except the root level, which is the input. The key $K = K_1\|K_2\|K_3\|K_4$ consists of four keys for the PRF G.

Lemma 1. *If F is PRF-KPA secure, then $F^{\rightarrow\beta}$ and $F^{\rightarrow\bar{\beta}}$ are PRF-KPA secure. Specifically, for any t, q,*

$$\mathbf{Adv}_{F\rightarrow\beta}^{\text{prf-kpa}}(t,q) \le \beta\mathbf{Adv}_{F}^{\text{prf-kpa}}(t,q)$$

$$\mathbf{Adv}_{F\rightarrow\bar{\beta}}^{\text{prf-kpa}}(t,q) \le \beta\mathbf{Adv}_{F}^{\text{prf-kpa}}(t,q+1) + \frac{q}{2^m} .$$

Proof: We do the proof for $F^{\rightarrow\bar{\beta}}$. The proof for $F^{\rightarrow\beta}$ is equivalent.

Let $S = S_1\|\cdots\|S_\beta$ be a random function from $\{0,1\}^m$ to $\{0,1\}^{\beta n}$. Let K be a random key for F, let R_1 be uniformly random in $\{0,1\}^m$ and for $i = 1, \ldots, \beta+1$ let

$$H_{S,K,R_1}^i(R_2) = S_1(R_2)\|\cdots\|S_{i-1}(R_2)\|F_{K_i}(R_2)\|\cdots\|F_{K_\beta}(R_2) ,$$

where $K_i = K$ and inductively $K_{j+1} = F_{K_j}(R_1)$. Let D be any distinguisher running in time t using q queries. We want to prove that D cannot distinguish $F^{\rightarrow\bar{\beta}}$ and S with advantage better than $\beta\mathbf{Adv}_{F}^{\text{prf-kpa}}(t, q+1) + \frac{q}{2^m}$. Since the event that R_1 occurs as one of the uniformly random R_2 values is at most $\frac{q}{2^m}$ it is enough to prove that D cannot distinguish $F^{\rightarrow\bar{\beta}}$ and S with advantage better than $\beta\mathbf{Adv}_{F}^{\text{prf-kpa}}(t, q+1)$ when this event does not occur. So, since $H_{S,K,R_1}^1 = F_{K,R_1}^{\rightarrow\bar{\beta}}(R_2)$ and $H_{S,K,R_1}^{\beta+1} = S$ it is in turn enough to prove that for $i = 1, \ldots, \beta$, D cannot distinguish H_{S,K,R_1}^i from H_{S,K,R_1}^{i+1} with better advantage than $\mathbf{Adv}_{F}^{\text{prf-kpa}}(t, q+1)$.

To prove this assume we have access to an oracle $\mathcal{R}_f$ where f is a random function from F or a random function $\{0,1\}^m \rightarrow \{0,1\}^n$. We simulate an oracle to D as follows: First request a generation from $\mathcal{R}_f$ and obtain (R_1, T_1). Let $K_{i+1} = T_1$ and inductively $K_{j+1} = F_{K_j}(R_1)$. Further more maintain a random function $S = S_1\|\cdots\|S_{i-1}$ using a dictionary. When D request a generation, request a generation from $\mathcal{R}_f$ to obtain (R_2, T_2), and return to D the value $(R_2, H_{S,f,R_1}(R_2))$, where

$$H_{S,f,R_1}(R_2) = S_1(R_2)\|\cdots\|S_{i-1}(R_2)\|T_2\|F_{K_{i+1}}(R_2)\|\cdots\|F_{K_\beta}(R_2) .$$

If $f = F_K$, then using the renaming $K_i = K$ we have that $K_{i+1} = F_{K_i}(R_1)$ and $T_2(R_2) = F_{K_i}(R_2)$ and thus $H_{S,f,R_1}(R_2) = H_{S,K,R_1}^i(R_2)$. If f is a random function $R : \{0,1\}^m \rightarrow \{0,1\}^n$, then $K_{i+1} = R(R_1)$ and

$$H_{S,f,R_1}(R_2) = S_1(R_2)\|\cdots\|S_{i-1}(R_2)\|R(R_2)\|F_{K_{i+1}}(R_2)\|\cdots\|F_{K_\beta}(R_2)$$

$$= H_{S,K_{i+1},R_1}^{i+1}(R_2) .$$

Since K_{i+1} is uniformly random and independent of all the other values as long as no R_2 equals R_1, the theorem follows. ∎

Lemma 2. *If $F : \{0,1\}^m \rightarrow \{0,1\}^n$ is PRF-KPA secure, then $F^{\alpha\rightarrow\alpha}$ is PRF-KPA secure. Specifically, for any t, q,*

$$\mathbf{Adv}_{F^{\alpha\rightarrow\alpha}}^{\text{prf-kpa}}(t,q) \le \mathbf{Adv}_{F}^{\text{prf-kpa}}(t,\alpha q) + \frac{q\alpha(q\alpha-1)}{2^{m+1}} .$$

Proof: Assume that D can distinguish $D_1 = (R_1, \ldots, R_\alpha, F_K(R_1), \ldots, F_K(R_\alpha))$ from $D_0 = (R_1, \ldots, R_\alpha, S_1, \ldots, S_\alpha)$ with probability δ when the S_i are uniformly random values. We use D to distinguish values of the form (R_i, T_i) where $T_i = F_K(R_i)$ if $b = 1$ and $T_i = S_i$ if $b = 1$. When D asks for a value, get α values $(R_1, T_1), \ldots, (R_\alpha, T_\alpha)$ and hand $V = (R_1, \cdots, R_\alpha, T_1, \cdots, T_\alpha)$ to D. If $b = 1$, then $V = D_1$ and if $b = 0$, then $V = D_0$ unless there are identical R_i values – if there are identical R_i values, the corresponding S_i values will also be identical, which would typically not occur if $S = S_1 \| \cdots \| S_\alpha$ was a uniformly random αn-bit-string. Since the R_i values are m-bit values and there is generated a total of $q\alpha$ of them, the probability of collisions is bounded by $\frac{q\alpha(q\alpha-1)}{2^{m+1}}$, which proves the theorem.   ∎

Lemma 3. *If F and G are PRF-KPA secure, then $G \circ F$ is PRF-KPA secure. Specifically, for any t, q,*

$$\mathbf{Adv}_{G \circ F}^{\text{prf-kpa}}(t, q) \leq \mathbf{Adv}_F^{\text{prf-kpa}}(t, q) + \mathbf{Adv}_G^{\text{prf-kpa}}(t, q) + \frac{q(q-1)}{2^{m_1+1}} \ .$$

Proof: We must show that one cannot distinguish D_0 and D_1, where $D_1 = (R, F_{K_1}(R), G_{K_2}(F_{K_1}(R)))$, where R is random, and $D_0 = (R, R_1(R), R_2(R))$, where R is random and R_1 and R_2 are random functions. We consider two hybrids: $H_1 = (R, R_1(R), G_{K_2}(R_1(R)))$, where R is random and R_1 is a random function and $H_2 = (R, R_1(R), R_2(R_1(R)))$, where R is random and R_1 and R_2 are random functions. It is easy to see that $H_2 = D_0$ unless identical R_1 values occur. To prove the lemma it is therefore enough to prove that D_1 cannot be distinguished from H_1 with better advantage than $\mathbf{Adv}_F^{\text{prf-kpa}}(t, q)$, and that H_1 cannot be distinguished from H_2 with better advantage than $\mathbf{Adv}_G^{\text{prf-kpa}}(t, q)$. Both claims follows using trivial reductions.   ∎

The PRT Family. As the basic primitive in our PRT construction, we will need a KPA secure PRF $G : \{0,1\}^m \to \{0,1\}^n$ with key length k, where $n \geq \max(2m, k)$. Using Lemmas 1 and 2 such a function can be constructed from any KPA secure PRF F with k-bit keys.

If we use Rijndael with 128 bit keys and 128-bit blocks (call this function family Rin), we can let $G = \text{Rin}^{\to 2}$. Then G has 256-bit keys, 128-bit input, and 256-bit output. Keys will simply consist of two Rijndael keys, and the function will be $\text{Rin}_{K_1, K_2}^{\to 2}(R) = \text{Rin}_{K_1}(R) \| \text{Rin}_{K_2}(R)$. If DES is used, then we can let $G = \text{DES}^{\to 2}$. Then G has 112-bit keys, 64-bit input, and 128-bit output. Since for all families which one would use in practice, the loss of security in going from F to G is minimal, we will in the following express the security of our construction in that of G.

Starting with a random input S_0 for G and keys $K = (K_1, \ldots, K_d)$, we can compute a pseudorandom output of length exponential in d by computing $(G^{2^{d-1} \to 2^{d-1}} \circ \cdots \circ G^{2 \to 2} \circ G)_K(S_0)$. However, we are after a VO-PRF which the construction will not provide for any fixed number of key. This is the reason for

$$\underline{\textbf{proc}} \; \text{PRT}_K(R,l) \; \equiv$$
$$T_0 = K[1..k]$$
$$R_1 = K[(k+1)..(k+m)]$$
$$R_2 = K[(k+m+1)..(k+2m)]$$
$$S_0 = R$$
$$O = \epsilon$$
$$i = 1$$
$$\underline{\textbf{while}} \; |O| < l \; \underline{\textbf{do}}$$
$$\qquad K_i = (G_{T_{i-1}}(R_1))[1..k]$$
$$\qquad T_i = (G_{T_{i-1}}(R_2))[1..k]$$
$$\qquad S_{i+1} = \epsilon$$
$$\qquad j = 1$$
$$\qquad \underline{\textbf{while}} \; j \le |S_{i-1}| - m + 1 \; \underline{\textbf{do}}$$
$$\qquad\qquad S_i = S_i \| G_{K_i}(S_{i-1}[j..(j+m-1)])$$
$$\qquad\qquad j = j + m \; \underline{\textbf{od}}$$
$$\qquad O = O \| S_i$$
$$\qquad i = i + 1 \; \underline{\textbf{od}}$$
$$\underline{\textbf{return}} \; O[1..l]$$

Fig. 2. The VO-PRF $\text{PRT}_K(R,l)$ obtained from a PRF $G : \{0,1\}^m \to \{0,1\}^n$ with key-length k where $n \ge \max(2m,k)$. The domains of the inputs are $K \in \{0,1\}^{k+2m}$ and $R \in \{0,1\}^m$. The first two lines of the outer loop constitutes the key scheduling and can be preprocessed to prepare the functions $G_{K_1},\ldots,G_{K_l}$ to some appropriate depth.

the second requirement that the output of G is at least as long as the key. This allows to schedule an arbitrary number of keys from one key K using Lemma 1: Pick $R \in \{0,1\}^{2m}$ at random and to schedule d keys, compute $G^{\to d}(R)$. This gives a random string of length $dn \ge dk$ and allows to define d random keys. The entire construction is given in more detail in Fig. 2 using pseudo-code.

Theorem 2. *If $G : \{0,1\}^m \to \{0,1\}^n$ is PRF-KPA secure, then $\text{PRT}[G]$ is VO-PRF-KPA secure. Specifically, for any t, q, μ,*

$$\textbf{Adv}^{\text{prf-kpa}}_{\text{PRT}[G]}(t,q,\mu) \le d(\textbf{Adv}^{\text{prf-kpa}}_G(t,ql) + \textbf{Adv}^{\text{prf-kpa}}_G(t,q+1)) + \frac{ql(ql-1)+q}{2^m}$$

where L is the maximal length of a query in bits, $l = \lceil L/m \rceil$, and $d = \log_2(l+1)$ is the maximal depth of any PRT used. Using $L \le \mu$ this easily translates into a bound depending only on t, q and μ.

Proof: A pseudorandom tree of depth d is used to generate between $2^d - 1$ and $2^{d+1} - 2$ blocks. Thus the maximal depth of the pseudorandom trees used in each evaluation is upper bounded by d.

Note that a pseudorandom tree, $\text{PRT}[F]_K(R,l)$, of depth d can be computed by first computing $\overline{K} = G^{\to d}_{K[1..(k+m)]}(K[(k+m+1)..(k+2m)])$ and then computing $O = (\circ^d_{i=1} G^{2^{i-1} \to 2^{i-1}})_{\overline{K}}(R)$ and outputting $O[1..l]$.

If $\overline{K}$ was uniformly random, then by the above observations and Lemmas 2 and 3

$$
\begin{aligned}
\mathbf{Adv}_{\mathrm{PRT}[F]}^{\mathrm{prf\text{-}kpa}}(t,q,\mu) &\leq \sum_{i=0}^{d-1} \mathbf{Adv}_{G^{2^i}\to 2^i}^{\mathrm{prf\text{-}kpa}}(t,q) + \sum_{i=0}^{d-2} \frac{(q(q-1))}{2^{m+1}} \\
&\leq \sum_{i=0}^{d-1} \left(\mathbf{Adv}_{G}^{\mathrm{prf\text{-}kpa}}(t,2^i q) + \frac{q2^i(q2^i-1)}{2^{m+1}} + \frac{q(q-1)}{2^{m+1}} \right) \\
&\leq \sum_{i=0}^{d-1} \mathbf{Adv}_{G}^{\mathrm{prf\text{-}kpa}}(t,2^i q) + \frac{q2^{d-1}(q2^{d-1}-1)}{2^m} \ .
\end{aligned}
$$

The theorem then follows using that $2^{d-1} \leq ql$ and using Lemma 1 in a hybrids argument.   ∎

The theorem tells us that even if G was a perfect PRF, i.e. $\mathbf{Adv}_G^{\mathrm{prf\text{-}kpa}}(t,q) = 0$, the PRF that we build out of it will not necessarily be perfect. Intuitively it is easy to see that this imperfectness is not by failure of our analysis. It is the birthday attack, to which almost all encryption modes must surrender. The intuition is that if at some level in a pseudorandom tree as that in Fig. 1 a collision occurs, then because the next levels are build using functions, sub-trees under collisions will be identical. On the other hand, identical sub-tree will occur very seldom if each bit in the tree is chosen uniformly at random and independent of the other bits. A careful analysis of the probability of find such collisions will allow to prove that the bound in the theorem is fairly sharp.

To see why it is essential to the construction that different keys are used at each level, we refer the reader to Theorem 4 and the discussion following it.

4 Analysis and Comparison of CBC, CTR, Jutla's Modes, and PRT

We will now compare the security of our new encryption mode to that of the well-known encryption modes CBC and CTR, and also the integrity aware modes of Jutla[Jut01] – he proposed two modes, namely IACBC and IAPM, both of which provide both integrity and confidentiality – we will only consider confidentiality in this section, however. We are going to prove the results given by the table below, which gives the "maximal" security that holds in general for various combinations of encryption and attack modes. For instance the entry CBC \ PRF-KPA being equal to ROR-KPA means that CBC-encryption using a KPA-secure PRF family is ROR-KPA secure, and there exists a KPA-secure PRF family G such that CBC[G] is not CPA-secure.

MODE \ ATK$_{\mathrm{impl}}$	PRF-KPA	PRF-CPA
CBC	ROR-KPA	ROR-CPA
CTR	insecure	ROR-CPA
Jutla	ROR-KPA	ROR-CPA
PRT	ROR-CPA	ROR-CPA

The bottom row and the right-most column follows from known results from [BDJR97,Jut01] and Section 3. We now prove the remaining claims in the following theorems. The CBC and CTR encryption modes are given in Fig. 3.

proc $CBC[P]_K(M) \equiv$
  $m \leftarrow \lceil |M|/l \rceil$
  $n \leftarrow ml - |M|$
  $r \overset{R}{\leftarrow} \{0,1\}^n$
  $M \leftarrow M \| r$
  $c_0 \overset{R}{\leftarrow} \{0,1\}^l$
  for $i = 0$ **to** $m - 1$ **do**
    $c_{i+1} \leftarrow P_K(M[il..(il + l - 1)] \oplus c_i)$
  od
  return $(n, c_0 \| c_1 \| \dots \| c_m)$

proc $CTR[F]_K(M) \equiv$
  $m \leftarrow \lceil |M|/L \rceil$
  $n \leftarrow |M| - (m - 1)L$
  $r \overset{R}{\leftarrow} \{0,1\}^l$
  for $i = 1$ **to** m **do**
    $r_i \leftarrow F_K(r + i \bmod 2^l)$
  od
  return $(r, M \oplus (r_1 \| \dots \| r_{m-1} \| r_m[1..n]))$

Fig. 3. $CBC[P]$ mode and $CTR[F]$ mode.

Jutla's IACBC mode is essentially CBC encryption, but where the sequence of blocks coming from the CBC encryption is Xor'ed by a sequence of pseudorandom blocks generated using an independent key. The IAPM mode first Xor's the sequence of plaintext blocks by a pseudorandom sequence of blocks, then encrypts in ECB mode, and finally Xor's the result by the same pseudorandom sequence. Both IACBC and IAPM also generate a checksum that receives special treatment, but this is not relevant for our discussion.

Theorem 3. *Suppose P is a permutation family with length l. If P is PRF-KPA secure, then $CBC[P]$, $IACBC[P]$, and $IAPM[P]$ are ROR-KPA secure. Specifically, for any t, q,*

$$\mathbf{Adv}^{ror\text{-}kpa}_{CBC[P]}(t, q, \mu), \mathbf{Adv}^{ror\text{-}kpa}_{IACBC[P]}(t, q, \mu), \mathbf{Adv}^{ror\text{-}kpa}_{IAPM[P]}(t, q, \mu)$$

$$\leq \mathbf{Adv}^{prf\text{-}kpa}_P(t, \nu) + \frac{\nu(\nu - 1)}{2^{l+1}},$$

where $\nu = \lfloor \mu/l \rfloor + q$.

Proof: Consider an ROR-KPA distinguisher $\overline{D}$ expecting access to an oracle $\mathcal{R}_{K,b}$ for the $CBC[P]$ scheme. We construct a distinguisher D having access to a PRF-KPA oracle $\mathcal{R}_f$ for the permutation family P as follows. The distinguisher D runs the code of $\overline{D}$. Each time D requests an encryption of length m', request $m = \lceil m'/l \rceil$ pairs $(x_i, f(x_i))$ from $\mathcal{R}_f$. Then generate a random l-bit string c_0 and for $i = 1, \dots, m$ let $c_i = f(x_i)$ and let $p_i = x_i \oplus c_{i-1}$. Then output $(M, C) = (p_1 \| \dots \| p_m, (ml - m', c_0 \| c_1 \| \dots \| c_m))$.

In all cases M is uniformly random and C is distributed exactly as a CBC encryption of p using f. So, if $f = P_K$ is a random permutation from P, then (M, C) is distributed exactly as values from $\mathcal{R}_{K,1}$, and if f is a random function, then C is uniformly random and independent of M, unless M has collisions among the blocks, which proves the theorem for CBC.

Since IACBC is clearly no weaker than CBC under any notion of security the result also covers IACBC, and the result for IAPM follows from Lemma 2.   ■

Theorem 4. *For any permutation family P with length l, there exists a permutation family $\overline{P}$ such that $\overline{P}$ is PRF-KPA secure if P is PRF-KPA secure, but neither $CBC[\overline{P}]$ nor $IACBC[\overline{P}]$ are ROR-CPA secure.*

Proof: Given some permutation family P, consider the permutation family $\overline{P}$ given by $\overline{P}_K(x_1, x_2) = (P_K^{-1}(x_2), P_K(x_1))$. A random evaluation of $\overline{P}$ just consists of two random evaluations of P, and so $\overline{P}$ is PRF-KPA secure if P is PRF-KPA secure. To see that $\overline{P}$ is not PRF-CPA secure in CBC mode, ask for an encryption of the all-zero-string of length $4l$ and use that permutations from $\overline{P}$ are their own inverses.

This can be generalized to IACBC mode: In one version of this mode, the pseudorandom sequence S that is Xor'ed to the result of CBC encryption is of form $s_i = e(i)W$, where $e()$ is a public injective map from the integers mod $2^l - 1$ to $GF(2^l)^*$, l is the block length of the cipher, and W is a pseudorandomly generated block. The multiplication is in $GF(2^l)$. Now, it is easy to see that if we request the IACBC encryption of a message with 4 zero-blocks, the first and third block output from the CBC part will be equal. Hence, the Xor of the corresponding blocks in the IACBC encryption will equal $(e(1) + e(3))W$. Since $e()$ is public and $e(1) + e(3) \neq 0$ we can compute W and hence all of S, Xor with the ciphertext and obtain the output from the CBC part. Now we are in a situation equivalent to what we had for CBC. Similar arguments apply to the other suggested variants of IACBC.   ■

Note that the function $\overline{P}$ constructed in the above proof demonstrates that it is essential to the PRT construction that different keys are used at each level. The function family $\overline{P}$ is KPA-PRF secure, but if it was used in a PRT construction with the same key at each level, the tree would be scattered with repeating blocks.

Theorem 5. *For any permutation family P with length l, there exists a permutation family $\overline{P}$ such that $\overline{P}$ is PRF-KPA secure if P is PRF-KPA secure, but $CTR[\overline{P}]$ is not ROR-KPA secure.*

Proof: Given some permutation family P, consider the permutation family $\overline{P}$ given by $\overline{P}_K(x_1, x_2) = (P_K(x_1), P_K(x_2))$. By Lemma 2, $\overline{P}$ is PRF-KPA secure if P is PRF-KPA secure. To see that $\overline{P}$ is not PRF-KPA secure in CTR mode, note that if a permutation from $\overline{P}$ is evaluated on two consecutive elements $x, x + 1$, where $x = x_1 x_2$, then the result will typically be of the form $(P_K(x_1), P_K(x_2)), (P_K(x_1), P_K(x_2 + 1))$, which will not look random.   ■

Theorem 6. *There exists a function family P such that P is PRF-KPA secure, but $IAPM[P]$ is not ROR-CPA secure.*

Proof: In IAPM mode, a ciphertext block is of form $s_i \oplus P_K(p_i \oplus s_i)$ where s_i is a pseudorandom block and p_i is the i'th plaintext block. Suppose that $s_i = e(i)W$

for a random block W as described in the proof of Theorem 4. Then s_i, s_j for $i \neq j$ are related by $s_j = e(j)e(i)^{-1}s_i$. Since $e()$ ranges over all non-zero values in $GF(2^l)$, we can choose i, j such that $\alpha := e(j)e(i)^{-1}$ is a generator of $GF(2^l)^*$. Now, a function P_K in our family is constructed as follows: we will think of it as a mapping from $GF(2^l)$ to itself. We choose two random values as the images of 0 and 1. For every element α^m, where $1 \leq m < 2^l - 1$ is odd we choose a random value as image, whereas we set $P_K(\alpha^{m+1}) = \alpha P_K(\alpha^m)$. Now, P is PRF-KPA secure, because a set of random inputs has to be exponentially large in l in order to contain both of α^m, α^{m+1} for odd m with significant probability. But in a CPA on IAPM, an attacker can choose all $p_i = 0$, which means that P_K receives the sequence of s_i's as inputs. If W is random, then with probability $1/2$, $s_i = \alpha^m$ for an odd m. Hence $s_j = \alpha^{m+1}$, and so we have for ciphertext blocks C_i and C_j that $C_j = s_j + P_K(s_j) = \alpha s_i + \alpha P_K(s_i) = \alpha(s_i + P_K(s_i)) = \alpha C_i$. This correlation allows to distinguish from a random encryption. Other ways to generate the s_i's can be handled in a similar way.   ∎

5   CCA Security

Having constructed CPA secure encryption, we can construct CCA secure encryption using a number of known techniques. One can e.g. do with a KPA secure VO-PRF $G : \{0,1\}^k \to \{0,1\}^*$ acting as key-stream generator and a CPA secure variable-length input PRF (VI-PRF) $MAC : \{0,1\}^* \to \{0,1\}^k$ acting as a MAC. Given a message M one generates a uniformly random input R for G and computes $C = R\|(G_{K_1}(|M|, R) \oplus M)$ and lets the encryption be $E_{K_1,K_2}(M) = (C, MAC_{K_2}(C))$. This scheme can be proven CCA secure using standard techniques, see e.g. [Des00].

We can construct a CPA secure VI-PRF from a KPA secure PRF using known techniques. From any KPA secure PRF F one can build a pseudorandom *generator* by mapping key K and input R for the PRF to $F_K(R)$. Using the technique in Section 3 the PRF can be modified to give this pseudorandom generator expansion factor two. Using the technique in [GGM86] this then allows to build a CPA secure VI-PRF using in the order of l applications of F per evaluation, where l is the length of the message.

To do a CCA secure encryption using PRT mode, one will then need $l/k + \log_2(l/k)$ applications of F for the key-stream, where k is the block-size, and l applications of F for the MACing. This is too large a overhead for the solution to be practical and leaves the open problem of finding an efficient CCA secure encryption scheme relying only on the KPA security of the underlying block cipher.

If one is willing to make the extra assumption that a collision resistant hash-function $H : \{0,1\}^* \to \{0,1\}^h$ is given the above scheme can be made practical. By first hashing the message and then MACing the hash, the result is still a CPA secure VI-PRF. However, now the price for encrypting is (neglecting the price for hashing) $l/k + \log_2(l/k) + h$ applications of F, where h in current practice could be 160.

Ivan Damgård and Jesper Buus Nielsen

6 Conclusion

We have shown how to efficiently enlarge the output-length of a weak pseudorandom function and how to use this for constructing CPA secure encryption from any weak pseudorandom function without essential loss of efficiency compared to known modes as CBC and CTR. We showed that also CCA secure encryption can be based on a KPA secure PRF, and opened the problem of finding an *efficient* CCA secure encryption scheme based on a KPA secure PRF.

References

[BDJR97] M. Bellare, A. Desai, E. Jokipii, and P. Rogaway. A concrete security treatment of symmetric encryption. In *38th Annual Symposium on Foundations of Computer Science* [IEE97].

[BM84] Manuel Blum and Silvio Micali. How to generate cryptographically strong sequences of pseudo-random bits. *SIAM Journal on Computing*, 13(4):850–864, November 1984.

[Des00] Anand Desai. New paradigms for constructing symmetric encryption schemes secure against chosen-ciphertext attack. In Mihir Bellare, editor, *Advances in Cryptology - Crypto 2000*, pages 394–412, Berlin, 2000. Springer-Verlag. Lecture Notes in Computer Science Volume 1880.

[GGM86] Oded Goldreich, Shafi Goldwasser, and Silvio Micali. How to construct random functions. *Journal of the ACM*, 33(4):792–807, 1986.

[GL89] Oded Goldreich and Leonid A. Levin. A hard-core predicate for all one-way functions. In *Proceedings of the Twenty First Annual ACM Symposium on Theory of Computing*, pages 25–32, Seattle, Washington, 15–17 May 1989.

[HN00] Johan Håstad and Mats Näslund. Key feedback mode: a keystream generator with provable security. 2000.

[IEE97] IEEE. *38th Annual Symposium on Foundations of Computer Science*, Miami Beach, FL, 19–22 October 1997.

[Jut01] Charanjit S. Jutla. Encryption modes with almost free message integrity. In *Advances in Cryptology - EuroCrypt 2001*, pages 529–544, Berlin, 2001. Springer-Verlag. Lecture Notes in Computer Science Volume 2045.

[Lev85] Leonid A. Levin. One-way functions and pseudorandom generators. In *Proceedings of the Seventeenth Annual ACM Symposium on Theory of Computing*, pages 363–365, Providence, Rhode Island, 6–8 May 1985.

[NR97] Moni Naor and Omer Reingold. Number-theoretic constructions of efficient pseudo-random functions (extended abstract). In *38th Annual Symposium on Foundations of Computer Science* [IEE97], pages 458–467.

Threshold Ring Signatures and Applications to Ad-hoc Groups

Emmanuel Bresson[1], Jacques Stern[1], and Michael Szydlo[2]

[1] Dépt d'informatique, École normale supérieure, 75230 Paris Cedex 05, France
{Emmanuel.Bresson,Jacques.Stern}@ens.fr
[2] RSA Laboratories, 20 Crosby Drive, Bedford, MA 01730, USA
mszydlo@rsasecurity.com

Abstract. In this paper, we investigate the recent paradigm for group signatures proposed by Rivest *et al.* at Asiacrypt '01. We first improve on their ring signature paradigm by showing that it holds under a strictly weaker assumption, namely the random oracle model rather than the ideal cipher. Then we provide extensions to make ring signatures suitable in practical situations, such as threshold schemes or ad-hoc groups. Finally we propose an efficient scheme for threshold scenarios based on a combinatorial method and provably secure in the random oracle model.

1 Introduction

In many multi-user cryptographic applications, anonymity is required to ensure that information about the user is not revealed. Typical examples are electronic voting [4, 14], digital lotteries [17, 21], or e-cash applications [6, 10]. In these applications releasing private information is highly undesirable and may result in a large financial loss. The concept of group signatures introduced in 1991 [11], allows a registered member of a predefined group to produce anonymous signatures on behalf of the group. However, this anonymity can be revoked by an authority if needed. The extra trapdoor information stored by the authority is used to reveal the identity of the actual signer. This mechanism provides some level of security for non-signing members in case of dispute. Hence, group signatures are only the appropriate tool when members have agreed to cooperate. The distinct but related concept of *ring signature* has recently been formalized by Rivest *et al* [25]. This concept is of particular interest when the members do not agree to cooperate since the scheme requires neither a group manager, nor a setup procedure, nor the action of a non-signing member.

A ring signature specifies a set of possible signers and a proof that is intended to convince any verifier that the author of the signature belongs to this set, while hiding his identity. More precisely, these schemes differ in several ways from classical group signature schemes. First of all, for any message, a signer may add his name to any set of other users he chooses, and produce a ring signature on it which reveals only that the anonymous author (in fact, himself) belongs to this set. This is infeasible with standard group signatures where the possible

M. Yung (Ed.): CRYPTO 2002, LNCS 2442, pp. 465–480, 2002.

signers, by definition, are registered members of the group. In particular, the non-signing members may even be completely unaware that they are involved in such signature. Secondly, the absence of a revocation manager allows *unconditional* anonymity. This is not achievable in typical group signatures for which there *must* be some trap-door information to be used by the authority. Thirdly, ring signatures can be made extremely efficient. While previously proposed group signature schemes heavily use asymmetric computations, such as zero-knowledge proofs, the ring signature scheme proposed in [25] is very fast, and thus very practical to implement.

Using ring signatures in ad-hoc groups. The steadily growing importance of portable devices and mobile applications has spawned new types of groups of interacting parties: ad-hoc groups [2, 22]. The highly dynamic nature of such groups raises new challenges for networking [22]. Ad-hoc networks may be described as networks with minimal infrastructure, lacking fixed routers or stable links. Such ad-hoc networks inherently deal with spontaneous ad-hoc groups; other instances of ad-hoc groups are not dependent on the particular network infrastructure. For example, a group of users who spontaneously decide they wish to communicate sensitive data need a suite of protocols which do not involve any trusted third party or certification of any new public keys. Security goals have to be considered in this new context. Ring signatures are perfectly suited to such a setting, since no setup protocol is required to use them.

Now assume that in order to create a certain signature at least k out of the n parties need to combine their knowledge. Threshold cryptography allows n parties to share the ability to perform a cryptographic operation (e.g., creating a digital signature). We consider the use of ring signatures in this ad-hoc threshold setting, and begin by defining the security notions useful for protocols involving such *ad-hoc groups*. The paradigm of ring signatures provides us with first building block. It gives a recipe for a single member of an ad-hoc group to sign anonymously on behalf of the group. Our flexible construction extends the solution of [25] to solve the threshold problem; in fact it is a solution for all custom signatures on an ad-hoc group.

Contributions. In this paper we significantly improve the ring signature scheme proposed by Rivest *et al.* [25]. We first show that security can be based on strictly weaker complexity assumptions, without sacrificing efficiency. Furthermore, this greatly simplifies the security proof provided in [25].

Next, we show how the ring signature paradigm and the stand alone protocol of [25] extend to a generic building block to make new schemes, in a multi-signer threshold setting. To achieve this aim, we formalize the notion of *ad-hoc group* signature, which is also of independent interest for general multi-party protocols relying on limited public key infrastructure. We define a formal security model to deal with such extended signatures, and we provide an interesting composition theorem which may be used to prove security for classical threshold structures. In case of threshold subgroups, the result remains very efficient for small threshold values and is provably secure, based on RSA, under the strictly weaker assumption of random oracles. In that construction, we use a combinatorial notion we call *fair partition*, which is of independent interest.

Related work. Informal notions of ring signatures were discussed simultaneously with the appearance of group signatures [11, 12] but the concept itself has only been formalized recently in [25].

Many schemes have been proposed for group signatures, offering various additional properties [19, 20, 23] as well as increasing efficiency [3, 8, 9]. The related *Witness hiding* zero knowledge proofs were treated in [13], and an application to group signatures (without a manager) was discussed at the end of this same article. This, and another construction [15] can also be seen as ring signature schemes. However, the scheme by Rivest *et al.* [25] is the most efficient one.

While it is well-known that theoretical very general witness-hiding signature constructions are realizable, they are also widely believed to be completely impractical. In fact, presenting an efficient general signature construction complements the general theory which, for example, tells us that arbitrary statements in $\mathcal{NP}$ can be proven in zero knowledge. Our work combines the known general techniques with some novel constructions and specializes to the case where custom signatures are required in an environment with only a pre-existing PKI. The constructions herein are adaptable to both ad-hoc and traditional threshold structures via a standard procedure which produces custom signatures, complete with a specialization of the security proof and efficiency analysis.

Thus, this work is at the crossroad of three corresponding trends in cryptography research: provable security, custom protocol design, efficiency analysis.

The organization of the rest of the paper is as follows: we review the concept of ring signature in section 2 and explain how improve it in section 3. Next in section 4, we extend the notion, propose a formal security model and finally prove the security of our new scheme in section 5.

2   Overview of Ring Signatures

In this section, we follow the formalization proposed by Rivest, Shamir and Tauman in [25].

2.1   Definitions

One assumes that each user has received (via a PKI or a certificate) a public key P_k, for which the corresponding secret key is denoted S_k. A ring signature scheme consists of the following algorithms.

- **Ring-sign.** A probabilistic algorithm which, on input a message m, the public keys $P_1, \ldots, P_r$ of the r ring members, together with the secret key S_s of a specific member, produces a ring signature σ for the message m.
- **Ring-verify.** A deterministic algorithm which on input (m, σ) (where σ includes the public key of all the possible signers), outputs either "True" or "False".

Properties. A ring signature scheme must satisfy the usual correctness and unforgeability properties: a fairly-generated ring signature should be accepted as valid with respect to the specified ring with overwhelming probability; and it must be infeasible for any user, except with negligible probability, to generate a valid ring signature with respect to a ring he does not belong to.

We also require the signature to be anonymous, in the sense that no verifier should be able to guess the actual signer's identity with probability greater than $1/r + \epsilon$, where r is the size of the ring, and ϵ is negligible.

Note that the size of the ring signature grows linearly with the size of the specified group: this is inherent to the notion, since the ring membership is not known in advance, and therefore, has to be provided as a part of the signature.

Combining Functions. The formal concept of a ring signature can be related to an abstract concept called *combining functions*. We slightly modified the definition given in [25] as follows.

Definition 1 (Combining functions). *A combining function $C_{k,v}(y_1, \ldots, y_n)$ takes as input a key k, an initialization value v, and a list of arbitrary values of the same length ℓ. It outputs a single value $z \in \{0,1\}^\ell$, such that for any k, v, any index s and any fixed values of $\{y_i\}_{i \neq s}$, $C_{k,v}$ is a permutation over $\{0,1\}^\ell$, when seen as a function of y_s. Moreover, this permutation is efficiently computable as well as its inverse.*

The authors of [25] proposed a combining function based on a symmetric encryption scheme E modeled by a (keyed) random permutation

$$z = C_{k,v}(y_1, \ldots, y_n) = E_k\left(y_n \oplus E_k\left(y_{n-1} \oplus E_k\left(\cdots \oplus E_k(y_1 \oplus v)\ldots\right)\right)\right) \quad (1)$$

For any index s, we can easily verify that $C_{k,v}$ is a combining function by rewriting equation (1) as follows:

$$y_s = E_k^{-1}\left(y_{s+1} \oplus \ldots E_k^{-1}\left(y_n \oplus E_k^{-1}(z)\right)\right) \oplus E_k\left(y_{s-1} \oplus \ldots E_k(y_1 \oplus v)\ldots\right)$$

2.2 Ring Signatures by Rivest *et al.* [25]

We denote by ℓ, ℓ_b, ℓ_0 three security parameters. We consider a symmetric encryption scheme E defined over $\{0,1\}^\ell$ using ℓ_0-bit keys and a hash function $\mathcal{H}$ that maps arbitrary strings on ℓ_0-bit strings. In fact, we use $\mathcal{H}$ to define the symmetric key for E. Finally, we assume that each user P_i uses a regular signature scheme built on a trapdoor one-way permutation f_i such as RSA [27] and that the modulus has length $\ell_b < \ell$; typically, we choose $\ell - \ell_b \geq 160$. All these assumptions follow [25].

The scheme proposed by Rivest, Shamir and Tauman is based on combining functions as described above. In that scheme, the inputs y_i to the combining function are computed as $f_i(x_i)$ for some $x_i \in \{0,1\}^{\ell_b}$. A ring signature on a message m consists in a tuple $(v, x_1, \ldots, x_n)$. Setting $z = C_{\mathcal{H}(m),v}(f_1(x_1), \ldots, f_n(x_n))$, the signature is valid iff $z = v$.

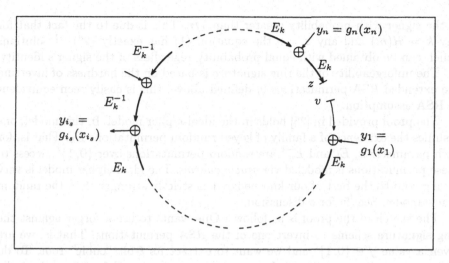

Fig. 1. The ring signature paradigm. The equation is verified if $z = v$. Given all the y_i's, the verifier just goes along the ring and checks $z \stackrel{?}{=} v$. The signer chooses v first, goes clockwise from 1 through $i_s - 1$ and counter-clockwise from n through $i_s + 1$. His trap-door allows him to extract x_{i_s}.

As explained in the previous section, for any message m, any fixed values v and $\{x_i\}_{i \neq s}$, one can efficiently compute the value y_s such that the combining function outputs v. Now using his knowledge of the trapdoor for function f_s, member P_s (the actual signer) is able to compute x_s such that $f_s(x_s) = y_s$. This is illustrated on Figure 1.

However, the RSA moduli n_i involved in the scheme are different. An adaptation has to be made in order to combine them efficiently. One extends the RSA trapdoor permutation $f_i(x) = x^{e_i} \bmod n_i$ in order that all f_i's have identical domain. Briefly speaking, one defines:

$$g_i(x) = \begin{cases} q_i n_i + f_i(r_i) & \text{if } (q_i + 1)n_i \leq 2^\ell \\ x & \text{otherwise} \end{cases}$$

where $x = q_i n_i + r_i$, with $0 \leq r_i < n_i$. That is, the input x is sliced into ℓ_b-bit long parts which then go through the RSA function. The probability that a random input is unchanged becomes negligible as $\ell - \ell_b$ increases. See [25] for more details.

If the trap-door one-way functions are RSA functions with public exponent equal to 3, the scheme is very efficient, requiring only one modular exponentiation (and a linear number of multiplications) for signing, and only two modular multiplications per ring member for verification.

2.3   Assumptions and Security

The authors proved *unconditional* anonymity, in an information-theoretic sense. They showed that even an infinitely powerful adversary cannot guess the identity of the signer with probability greater than $1/r$. This is due to the fact that for any $k = \mathcal{H}(m)$ and any $z = v$ the equation (1) has exactly $(2^b)^{r-1}$ solutions which can be obtained with equal probability regardless of the signer's identity.

The unforgeability of the ring signature is based on the hardness of inverting the extended RSA permutations g_i defined above; this is easily seen equivalent to RSA assumption.

The proof provided in [25] holds in the ideal-cipher model. In this model, one assumes the existence of a family of keyed random permutations E_k. That is, for each parameter k, E_k and E_k^{-1} are random permutations over $\{0,1\}^\ell$; access to these permutations is modeled via *oracle queries*. The ideal-cipher model is very strong, and to the best of our knowledge it is strictly stronger than the random oracle model. See [5] for a discussion.

The sketch of the proof is as follows. One wants to use a forger against the ring signature scheme to invert one of the RSA permutations. That is, we are given a value $y \in \{0,1\}^\ell$ and we want to extract its ℓ-bit "cubic" root. To do so, we try to slip y as a "gap" between two consecutive E functions along the ring. Doing so, the exclusive OR between the input and the output of these E functions is set to y, and then, with non-negligible probability, the forger will have to extract the cubic root of y. Such a slip is feasible at index i only if a query "arriving" to i is asked after the other query "arriving" to or "starting" form i. The proof relies on the following lemma, which proves that this is always the case in a forgery. We refer to it as the *ring lemma*:

Lemma 1 (Ring lemma). *In any forgery output by an adversary, there must be an index between two cyclically consecutive occurrences of E in which the queries were computed in one of the following three ways:*

- *The oracle for the i-th E was queried in the "clockwise" direction and the oracle for the $(i+1)$-st E was queried in the "counterclockwise" direction.*
- *Both E's were queried in the "clockwise" direction, but the i-th E was queried after the $(i+1)$-st E.*
- *Both E's were queried in the "counterclockwise" direction, but the i-th E was queried before the $(i+1)$-st E.*

The proof provided in [25] is based on this lemma; one has to "guess" the index where such a situation will occur as well as the two queries involved. Thus, the guess is correct with probability at least $1/rQ_E^2$, where Q_E is the number of ideal-cipher queries. The concrete security of the scheme is related to the security of inverteing (extended) RSA by this multiplicative factor.

3   Modifications of the Existing Scheme

In this section, we explain how to significantly improve the scheme by Rivest *et al.* by removing the assumption of an ideal-cipher and obtaining at the same time a simplified proof with exactly the same security bound.

Let us first recall the *ring equation* that characterizes the verification of a ring signature:

$$v = C_{k,v}(y_1, \ldots, y_n)$$

$$= E_k\left(y_n \oplus E_k\left(y_{n-1} \oplus E_k\left(\cdots \oplus E_k(y_1 \oplus v)\ldots\right)\right)\right) \quad \text{where } k = \mathcal{H}(m)$$

A Simple Observation. The main idea consists in verifying the ring equation from another starting point than index 1 so that one just needs to go "clockwise". For instance, the signer can put an index of his choice i_0 within the signature, indicating the ring equation should start with $E_k(y_{i_0} \oplus v) \oplus \ldots$. This slight modification allows us to remove the assumption that E is a random permutation; instead, we will only need a hash function, and thus we can simply replace $E_{\mathcal{H}(m)}(x)$ by $\mathcal{H}(m, x)$.

3.1 Modified Algorithms and Simplified Security Proof

We can use this observation to simplify the original scheme. The idea is to have a signer P_s compute successive values along the ring, starting from his *own* index i_s. He chooses a random seed σ, and goes along the ring, hashing $m\|\sigma$, XOR-ing with $g_{i_s+1}(x_{i_s+1})$, concatenate with m and hash, and so on. Of course, we consider index $n+1$ as being 1. We denote the successive values as follows:

$$v_{i_s+1} = \mathcal{H}(m, \sigma), \qquad v_{i_s+2} = \mathcal{H}(m, v_{i_s+1} \oplus g_{i_s+1}(x_{i_s+1})),$$
$$\ldots\ldots \qquad v_{i_s} = \mathcal{H}(m, v_{i_s-1} \oplus g_{i_s-1}(x_{i_s-1}))$$

Just before "closing" the ring, the signer uses his secret key to extract the last input. That is, he computes x_{i_s} such that $v_{i_s} \oplus g_{i_s}(x_{i_s}) = \sigma$. Then, in order to make the signature anonymous, he chooses at random an index i_0, and outputs a modified signature $(i_0, v_{i_0}, x_1, \ldots, x_n)$. The verification is straightforward, the only point is that the verifier starts at index i_0 with value v_{i_0}. The efficiency is unchanged.

The resulting proof of unforgeability is significantly simplified, since we do not use ideal-ciphers anymore, but only a hash function. Indeed, the proof provided in [25] is essentially based on lemma 1. In the original scheme, the difficulty lies in the fact that a forger can go both clockwise and counter-clockwise along the ring. When using hash functions only, the ring lemma becomes trivial: there always exist two cyclically consecutive queries such that the leading one has been asked before the trailing one. Thus, the security bound is unchanged, while the complexity assumptions are weakened.

3.2 How to Simulate a Ring

Another interesting variant in Rivest's scheme [25] is the following; the original scheme defines a ring signature as valid if satisfying $z = v$. However this is purely arbitrary and not necessary *as it* for the security. Indeed, such a condition can be replaced by any other condition fixing the "gap" between z and v, provided that such a "gap" cannot be chosen by a forger. For instance, one can require that $z = v \oplus E_k(0)$ instead of $z = v$.

More precisely, let γ be a publicly known ℓ-bit "gap value" (for instance $\gamma = 0$ or $\gamma = E_k(0)$). We can easily produce ring signatures with only a hashing-oracle as above, but now such that γ appears as a "gap" between z and v, or, more generally, between any two consecutive indices, at any specified position i_γ. The algorithm is modified in that, when arriving at index i_γ, the verifier just replaces v_{i_γ} with $v_{i_\gamma} \oplus \gamma$ before continuing hashing and going along the ring. The figure 2 below illustrates this.

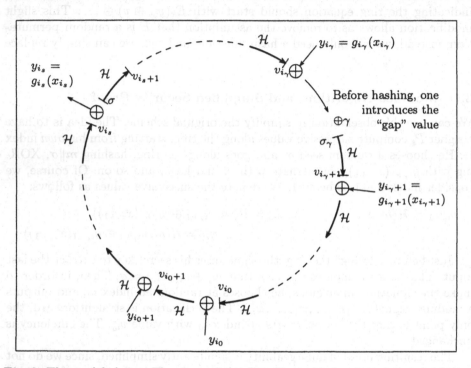

Fig. 2. The modified ring. The signer starts from its own index i_s and a seed value σ and he closes the ring when computing v_{i_s}, using his trapdoor. The verifier starts from i_0 with value v_{i_0} and checks whether the last hash outputs v_{i_0}. A "simulator" starts from i_γ and a seed σ_γ to define γ at closure only, when computing v_{i_γ}. $\mathcal{H}$ is a public hash function.

Doing so, the symmetry of the ring is somewhat broken, but in a crucial manner, since it allows us to *simulate* a ring signature, provided we can choose the value of γ freely. Viewing the gap value as a "challenge" (as in an identification scheme) the ring signature can be simulated if, and only if, the challenge can be chosen. This feature will be in crucial importance in the next section.

Modified Combining Functions. We are going to use a generalized version of the combining functions; Given a "gap" value γ, occurring at index i_γ, and a starting index i_0, we denote this modified function as $C_{v,i_0,m}$:

$C_{v,i_0,m}(i_\gamma, \gamma, y_1, \ldots, y_n) =$
$\mathcal{H}(m, y_{i_0-1} \oplus \mathcal{H}(m, y_{i_0-2} \oplus \ldots \underbrace{\mathcal{H}(m, \gamma \oplus y_\gamma \oplus \mathcal{H}(\cdots \oplus \mathcal{H}(m, y_{i_0} \oplus v) \ldots))}_{v_{i_\gamma+1} \text{ is computed from } \gamma \oplus y_\gamma \oplus v_{i_\gamma}} \ldots))$

The ring equation is verified if $C_{v,i_0,m}(.) = v$. The "gap" value being included in the arguments, this modification generalizes the definition given in [25]; in the original scheme, we have $i_0 = i_\gamma = 1$ and $\gamma = 0$. In the remaining of our paper, and without loss of generality, we assume that i_0 is fixed (the verifier always starts from 1), as well as i_γ (the gap appears between indices n and 1, that is $v_1 = \mathcal{H}(v_n \oplus y_n \oplus \gamma)$). We will omit i_γ in the notations and use $C_{v,i_0,m}(\gamma, y_1, \ldots, y_n)$. Also we (abusively) denote by:

$$y_s = C_{s,\gamma,m}^{-1}(\sigma_s, y_1, \ldots, y_n)$$

the solution that a signer P_s computes using seed σ_s as illustrated in Figure 2. Keep in mind, however, that y_s must not be considered as an argument to this function.

4   Threshold and Ad-hoc Ring Signature Schemes

In this section, we formalize the definition and security requirements for extended ring signatures, which we call *threshold ring signatures*. Traditional "*t*-out-of-*n*" threshold structures may be viewed as a special case of more general access structures, for which any criteria of minimum collaboration may be specified. In this section we also formally define such *ad-hoc groups* and the security requirements of the corresponding *ad-hoc signatures*.

4.1   Preliminaries

Assume that t users want to leak some juicy information, so that any verifier will be convinced that at least t users *among a select group* vouch for its validity. Simply constructing t ring signatures clearly does not prove that the message has been signed by different signers. A threshold ring signature scheme effectively proves that a certain minimum number of users of a certain group must have actually collaborated to produce the signature, while hiding the precise membership of the subgroup. Similarly, an ad-hoc signature might be used to modify the meaning of such a signature by giving certain members increased importance.

Definition 2. *A threshold ring signature scheme consists of two algorithms:*

- **T-ring-sign algorithm.** *On input a message m, a ring of n users including n public keys, and the secret keys of t members, it outputs a (t, n)-ring signature σ on the message m. The value of t as well as the n public keys of all concerned ring members are included in σ.*
- **T-ring-verify algorithm.** *On input a message m and a signature σ, it outputs either "True" or "False".*

We emphasize that there is no key-generation; only existing PKI is used.

The natural formalization for ad-hoc signatures follows that of threshold signatures. In both cases there are some users A_i who want to cooperate, and some others who do not cooperate, say B_i. However for the more general case, not all potential signers have equal standing. The definition of ad-hoc group captures the specification of a particular access structure. That is, it specifies which subsets of potential signers should be allowed to create a valid signature.

Definition 3. *An* **Ad-hoc group**, Σ, *is a list of n users, including n certified public keys, accompanied by a list of subsets S_j of these users, called the* accept-*able subsets. This second list may be optionally replaced with a predicate defining exactly which subsets are acceptable.*

Informally, an ad-hoc signature is one which retains maximal anonymity, yet proves that the signing members all belong to at least one acceptable subset. The signature and verification algorithms are therefore relative to this structure.

Ad-hoc-sign algorithm. On input a message m, a specification of an ad-hoc group Σ (n users with some acceptable subsets), it outputs an ad-hoc-ring signature σ on the message m. The ad-hoc group structure Σ as well as the n public keys of all concerned ring members are included in σ.

Ad-hoc-verify algorithm. On input a message m and a signature σ, it outputs "True" if σ is a valid ad-hoc-ring-signature on m, relative to the user list and acceptable subset list specified in Σ, and "False" otherwise.

4.2   The Boolean Structure of Ad-hoc Signatures

Here we make some remarks on the structure of arbitrary ad-hoc signatures, and draw parallels with threshold ring signatures. The standard threshold ring signature construction contains all of the techniques needed for general ad-hoc signatures, and is presented in detail in Section 5. The observations here serve to indicate exactly how the details of threshold ring signatures apply in the general case. We prefer this exposition, which provides an inclusive proof of security, yet appeals to the reader's intuition of familiar threshold structures.

A threshold ring signature scheme is clearly a special case of a general ad-hoc ring signature; out of all subgroups of the n members, every subgroup consisting of at least t members is an acceptable subset. One approach, though far from the most efficient one, is to list all acceptable subgroups and prove that one contains the cooperating signers. For each subgroup, we form the concatenation of the signatures of all concerned players; obviously such values can be simulated. Then we use a meta-ring mechanism to prove that at least one of these $|\Sigma|$ (concatenation of) values has not been simulated but rather computed using as many private keys as needed. Indeed, such meta-ring shows that at least an acceptable subset has simultaneously received agreement and cooperation from all its members. Thus one can prove for example that player P_1 has signed m, OR P_2 AND P_3 AND P_4 have, OR P_2 AND P_5 have. We observe that any collection of acceptable subsets may be (recursively) described in this way, by using the boolean operations AND, OR.

More complex signatures may be constructed recursively. Because we have shown how to simulate a ring, and the same trivially holds for sequential composition (AND), an ad-hoc signature may serve as a node in a larger, meta-ring. For a general ad-hoc signature, a particular specification of the acceptable subsets in terms of boolean operations corresponds to a particular (nested) composite ring structure.

4.3 Our Security Model

We now derive a security model from [25]. We first focus on standard threshold signatures, and proceed to the general case.

Threshold Security: First of all, we want the signature to be anonymous, in the sense that no information is leaked about the group of actual signers, apart from the fact that they were at least t among the n specified ring members.

Next, we define the unforgeability property as follows. The adversary $\mathcal{A}$ is given access to a signing oracle, which can be queried to threshold-sign any message. Also, $\mathcal{A}$ can corrupt some users in order to obtain their private secret key, and is allowed to do so adaptively. Doing so, we include collusion attacks.

A t-forger is an adversary that is able to sign a message on behalf of t users, having corrupted up to $t - 1$ users, under an adaptive chosen message attack. The scheme is t-**CMA**-secure (*threshold Chosen-Message Attack* -secure) if no t-forger $\mathcal{A}$ can succeed with non-negligible probability. If we denote the number of hash queries by q_H and the number of signing queries by q_S, such a probability is denoted by $\mathrm{Succ}^{cma}_{t,q_H,q_S}(\ell)$, where ℓ is the security parameter.

General Ad-hoc Security: The above security definitions generalize naturally as follows. First, the scheme should be anonymous, in the sense that no information will be leaked about the group of signers except that these actual signers form an acceptable subset. For the unforgeability property, we apply the same adversarial model as above, except that $\mathcal{A}$ can corrupt any number of users, provided that no set of corrupted users contains an acceptable subset.

A Σ-forger against a signature scheme for the ad-hoc group Σ is an adversary that is able to sign a message on behalf of a Σ-acceptable subset, having corrupted no acceptable subset in its entirety, under an adaptive chosen message attack. The scheme is Σ-**CMA**-secure (*ad-hoc group Chosen-Message Attack* - secure) if no Σ-forger can succeed with non-negligible probability. Analogously to above, such a probability is denoted by $\mathrm{Succ}^{cma}_{\Sigma,q_H,q_S}(\ell)$, where ℓ is the security parameter.

5 An Efficient Threshold Ring Signature Scheme

In this section, we describe a methodology to achieve the threshold ring signatures with size $\mathcal{O}(n\log n)$, while remaining essentially efficient in terms of signing/verifying. We not only provide details for signature composition, efficiency tracking, but also present a particularly efficient specification of the acceptable subsets by using fair partitions. Moreover, our solution is provably secure in the random oracle model.

Outline. Consider a ring of r members, and among them two users who want to demonstrate that they have been cooperating to produce a ring signature. The idea is to *split* the group into two disjoint sub-groups, and to show that each of these sub-groups contains one signer. However doing so may compromise perfect anonymity since such split restricts the anonymity of each user to a *sub-ring*. The solution consists in splitting the group many times, in such a way that there always exists a split for which any two users are in two different sub-rings. Then all of these splits are used as *nodes* in a *super-ring*. The super ring proves that at least one split has been solved, that is, two sub-rings has been individually solved. For the other splits, one will have to *simulate* a correct ring signature, for every unsolved sub-ring.

5.1   Fair Partitions of a Ring

Before describing our scheme, we introduce a few notations. Let t be an integer and $\pi = (\pi^1, \ldots, \pi^t)$ denote a partition of $[1, n]$ in t subsets; π defines a partition of the ring $R = (P_1, \ldots, P_n)$ in t sub-rings R^1 through R^t. Finally, the i-th bit of a string x is denoted by $[x]_i$.

Case $t = 2$. Let $\pi = \{\pi^1, \pi^2\}$ be a partition of $[1, n]$ and P_a and P_b two users that want to produce a "2-out-of-n" signature on a message m. If P_a and P_b belongs to distinct sub-rings, for instance, $P_a \in R^1$ and $P_b \in R^2$, then they are able to produce two correct ring signatures, relatively to R^1 and R^2 respectively. In that case, we say that π is a *fair partition* for indices $\{a, b\}$. If is not the case, for instance $\{P_a, P_b\} \subset R^1$, then it is infeasible for P_a and P_b to produce a valid ring signature with respect to R^2.

To ensure anonymity, we need to provide a set Π of partitions such that for any indices a and b in $[1, n]$, there exists a fair partition $\pi \in \Pi$ for $\{a, b\}$. A $(2, n)$-ring signature is a meta-ring over Π, which prove that for at least one partition π, both underlying sub-rings can be solved at the same time. Such a set can be efficiently constructed as stated in the (straightforward) following lemma:

Lemma 2. *For any integer n, there exists a set Π_n of $\lceil \log_2 n \rceil$ partitions satisfying the above requirements.*

General Case. We generalize the previous definitions as follows. Let $\pi = (\pi^1, \ldots, \pi^t)$ a partition of $[1, n]$ in t subsets and $I = \{i_1, \ldots, i_t\}$ a set of t indices in $\{1, n\}$. If all integers in I belongs to t different sub-sets, for instance $i_j \in \pi^j$, we say that π is a *fair partition for I*. Intuitively, a secret is known in every sub-ring R^j defined by π.

Now, to ensure anonymity, we need to provide a set Π of partitions such that there exists a fair partition for any set of cardinality t.

Definition 4. *Let $t < n$ be two integers. We say that a set Π of partitions of $[1, n]$ is a (n, t)-complete partitioning system if for any set I of cardinality t, there exists a fair partition in Π for I:*

$$\forall I \subset [1,n], \ \#(I) = t, \quad \exists \pi = (\pi^1, \ldots, \pi^t) \in \Pi, \quad \forall j \in [1,t], \ \#(I \cap \pi^j) = 1$$

To provide such a complete system, we use the notion of *perfect hash function*. A perfect hash function for a set I is a mapping $h : [1,n] \to [1,t]$ which is 1-1 on I. A (n,t)-family of perfect hash functions, H, is such that that for any I of size t, there exists $h \in H$ which is perfect on I. It is thus clear that defining a partition in t sub-rings for each member of a (n,t)-family makes a (n,t)-complete partitioning system.

The following result has been proven in [1]:

Lemma 3. *There exists a (n,t)-family of perfect hash functions which has size of $2^{O(t)} \log n$. Moreover each of these functions is efficiently computable.*

5.2   Description of the New Scheme

We now describe formally our scheme. We are based on the notion of fair partition, in case of a threshold scenario (which is likely to be used in practice). Consider a (n,t)-complete partitioning system and a set of t signers. If π is fair partition for this set, they can solve all the sub-rings defined by π. For the other partitions, the sub-rings are just simulated and put along a *super-ring*.

We introduce another function G, viewed as a random hash function returning $(t \times \ell)$-bit strings and we denote $p = 2^t \log n$. We assume that for all integers n and $t \le n$, a (n,t)-complete partitioning system is publicly available : $\Pi_n = \{\pi_1, \ldots, \pi_p\}$. Finally, we introduce the straightforward notation $C_{v,i_o,m}(\gamma, y_j, j \in R)$ to properly deal with sub-rings.

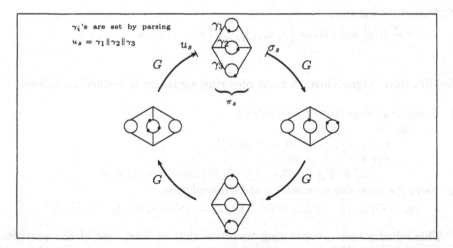

Fig. 3. The ring composition paradigm. Here we have $t = 3$. If π_s is a fair partition w.r.t. three signers, they start from a seed σ_s and compute u_s. Finally they solve the sub-rings in π_s according to the obtained gaps $\gamma_1, \ldots, \gamma_3$. The equation is verified in a straightforward way, starting at any given index.

Signing Algorithm. We denote by $P_{i_1}, \ldots, P_{i_t}$ a subgroup of users that want to sign a message m while proving there were at least t signers among n ring members. The idea is to solve a collection of sub-ring signatures corresponding to a fair partition (each signer belongs to a sub-ring), then to concatenate the results and to apply a ring-like mechanism in order to prove that at least one of the collection of sub-rings has been entirely solved. To do so, they proceed as follows: We denote by π_s a fair partition for $I = \{i_1, \ldots, i_t\}$. We assume for simplicity that for each $j \in [1, t]$, we have $i_j \in \pi_s^j$.

1. Choose random seeds for each sub-ring of each partition.
  For $i = 1, \ldots, p$, Do
    For $k = 1, \ldots, t$, Do $v_i^k \stackrel{R}{\leftarrow} \{0,1\}^\ell$.
2. Simulate rings for all partitions but π_s.
  For $i = 1, \ldots, p$, $i \neq s$, Do
    For $j = 1, \ldots, n$, Do $x_i^j \stackrel{R}{\leftarrow} \{0,1\}^\ell$, and $y_i^j \leftarrow g_j(x_i^j)$.
    For $k = 1, \ldots, t$, Do
      $z_i^k \leftarrow C_{v_i^k, 1, m}(0, y_i^j, j \in \pi_i^k(R))$ and $\gamma_i^k \leftarrow v_i^k \oplus z_i^k$.
3. Compute a "super-ring" with so-obtained gaps.
  $\sigma_s \stackrel{R}{\leftarrow} \{0,1\}^{t\ell}$, and $u_{s+1} \leftarrow G(\sigma_s)$
  For $i = s+2, \ldots, p, 1, \ldots, s$, Do
    $u_i \leftarrow G\big(u_{i-1} \oplus (\gamma_{i-1}^1 \| \ldots \| \gamma_{i-1}^t)\big)$
4. Compute the gap values for sub-rings of π_s by closing the super-ring.
  $(\gamma_s^1 \| \ldots \| \gamma_s^t) \leftarrow u_s \oplus \sigma_s$.
5. Solve the sub-rings for the fair partition π_s.
  For $j \in [1,n] \backslash I$, Do $x_s^j \stackrel{R}{\leftarrow} \{0,1\}^\ell$, and $y_s^j \leftarrow g_j(x_s^j)$.
  For $j \in I$, Do
    $\sigma_k \stackrel{R}{\leftarrow} \{0,1\}^\ell$ for k such that $j \in \pi_s^k$
    $y_s^j \leftarrow C_{j, \gamma_s^k, m}^{-1}(\sigma_k, y_s^j, j \in \pi_s^k(R))$ and $x_s^j \leftarrow g_j^{-1}(y_s^j)$.
6. Output the signature.
  $\nu \stackrel{R}{\leftarrow} [1,p]$ and output $\left(\nu, u_\nu, \bigcup_{1 \leq i \leq p} \left(x_i^1, \ldots, x_i^n, v_i^1, \ldots, v_i^t\right)\right)$.

Verification Algorithm. A t-out-of-n ring signature is verified as follows:

1. Compute all rings starting from index 1.
  For $i = 1, \ldots, p$, Do
    For $j = 1, \ldots, n$, Do $y_i^j \leftarrow g_j(x_i^j)$.
    For $k = 1, \ldots, t$, Do
      $z_i^k \leftarrow C_{v_i^k, 1, m}(0, y_i^j, j \in \pi_i^k(R))$, and $\gamma_i^k \leftarrow v_i^k \oplus z_i^k$.
2. Verify the super-ring from index ν and obtained gaps.
  $u_\nu \stackrel{?}{=} G(\gamma_{\nu-1}^1 \| \ldots \| \gamma_{\nu-1}^t \oplus G(\ldots G(\gamma_\nu^1 \| \ldots \| \gamma_\nu^t \oplus u_\nu) \ldots))$.

This scheme uses a super ring to prove that at least one of the partition is entirely solved. We may seen a ring (either sub-ring or super-ring) as an "OR" connective, while the partitions are used to "AND" the signatures – by concatenating the gaps $\gamma_1 \| \ldots \| \gamma_t$ before embedding them in the super-ring. Thus, our construction can be described as a composition technique.

5.3   Security Result

Theorem 1. *Our scheme is secure in the random oracle model against an adaptive chosen-message attacks involving q_H and q_G hash-queries to $\mathcal{H}$ and G respectively, and q_S signing queries, under the RSA assumption.*

$$\mathsf{Succ}^{\mathsf{cma}}_{t,\tau,q_H,q_G,q_S}(\ell) \leq q_H^2 q_G^2 t^2 \binom{n}{t} \mathsf{Succ}^{\mathsf{ow}}_{RSA}(\ell)$$

Due to lack of space, the proof will appear in the full version [7].

5.4   Discussion

We discuss here the efficiency of our threshold ring signature scheme. The size of the signature grows with both the number of users n and the number of signers t. More precisely, the size of such t-out-of-n signature is: $2^{\mathcal{O}(t)} \lceil \log_2 n \rceil \times (t * \ell + n * \ell) = \mathcal{O}(\ell 2^t n \log n)$. Signing requires t inversions of the g's functions and $\mathcal{O}(2^t n \log n)$ computations in the easy direction. This is clearly a more efficient implementation than the generic solution wherein one lists all the subgroups of cardinality t since this would lead to $\binom{n}{t} = \mathcal{O}(n^t)$ size.

This is also more efficient than the more or less straightforward solution using Shamir secret sharing scheme [28][1]. Indeed, it can be shown that such solution leads to $\mathcal{O}(n^2)$ sized signatures. This is due to additional terms that appear in the security estimates that follow from the reduction. Details will appear in the full version of the paper.

6   Conclusion

This paper addresses the open problem of allowing a subgroup of t members to sign anonymously on behalf of an ad-hoc ring. Our construction thus improves on ring signatures, group signatures, threshold signatures and on the Bellare-Rogaway paradigm for constructing composite protocols. Further work may focus on some new research by R. Canetti on universally composable protocols.

Acknowledgments

The authors thank Moni Naor and Berry Schoenmakers for helpful discussions and the anonymous referees for many extensive, detailed comments.

References

1. N. Alon, R. Yuster and U. Zwick. Color Coding. J. of ACM, (42):844–856.
2. N. Asokan and P. Ginzboorg. Key Agreement in Ad-hoc Networks. Expanded version of a talk given at the Nordsec '99 Workshop.

---

[1] Such solution was first suggested in preliminary version of [26].

3. G. Ateniese, J. Camenisch, M. Joye, G. Tsudik. A practical and provably secure coalition-resistant group signature scheme. In *Crypto '00*, LNCS 1880, pp. 255–270.
4. O. Baudron, P.-A. Fouque, D. Pointcheval, G. Poupard, and J. Stern. Practical multi-candidate election system. In *PODC '01*. ACM, 2001.
5. M. Bellare, D. Pointcheval, and P. Rogaway. Authenticated key exchange secure against dictionary attacks. In *Eurocrypt '00*, LNCS 1807, pp. 139–155.
6. S. Brands. Untraceable off-line cash in wallets with observers. In *Crypto '93*, LNCS 773, pp. 302–318.
7. E. Bresson, J. Stern and M. Szydlo. Threshold Ring Signatures for Ad-Hoc Groups. Full version of this paper, http://www.di.ens.fr/~bresson.
8. J. Camenish and M. Michels. Separability and efficiency for generic group signature schemes. In *Crypto '99*, LNCS 1666, pp. 106–121.
9. J. Camenish and M. Stadler. Efficient group signatures schemes for large groups. In *Crypto '97*, LNCS 1294, pp. 410–424.
10. D. Chaum and T. Pedersen. Wallet databases with observers. In *Crypto '92*, LNCS 740, pp. 89–105.
11. D. Chaum, E. van Heyst. Group signatures. *Eurocrypt '91*, LNCS 547, pp. 257–265.
12. L. Chen and T. Pedersen. New group signature schemes. In *Eurocrypt '94*, LNCS 950, pp. 171–181.
13. R. Cramer, I. Damgård, B. Schoenmakers. Proofs of partial knowledge and simplified design of witness hiding protocols. In *Crypto '94*, LNCS 839, pp. 174–187.
14. R. Cramer, M. Franklin, B. Schoenmakers, and M. Yung. Multi-authority secret-ballot elections with linear work. In *Eurocrypt '96*, LNCS 1070, pp. 72–83.
15. A. De Santis, G. Di Crescenzo, G. Persiano, and M. Yung. On monotone formula closure of SZK. In *FOCS '94*, pp. 454–465.
16. T. ElGamal. A public key cryptosystem and a signature scheme based on discrete logarithms. In *Crypto '84*, LNCS 196, pp. 10–18.
17. D. Goldschlag and S. Stubblebine. Publicly verifiable lotteries: Applications of delaying functions. In *Financial Crypto '98*, LNCS 1465, pp. 214–226.
18. Z. Haas and L. Zhou Securing Ad-Hoc Networks. In *IEEE Networks*, 13(6), 1999.
19. S. Kim, S. Park, and D. Won. Convertible group signatures. In *Asiacrypt '96*, LNCS 1163, pp. 311–321.
20. S. Kim, S. Park, and D. Won. Group signatures for hierarchical multigroups. In *ISW '97*, LNCS 1396, pp. 273–281.
21. E. Kushilevitz and T. Rabin. Fair e-lotteries and e-casinos. In *RSA Conference 2001*, LNCS 2020, pp. 100–109.
22. C. Perkins. Ad-hoc networking. Addison Wesley, 2001.
23. H. Petersen. How to convert any digital signature scheme into a group signature scheme. In *Security Protocols '97*, LNCS 1361, pp. 67–78.
24. D. Pointcheval and J. Stern. Security arguments for digital signatures and blind signatures. *J. of Cryptology*, 13(3):361–396, Aug. 2000.
25. R. Rivest, A. Shamir, and Y. Tauman. How to leak a secret. In *Asiacrypt '01*, LNCS 2248, pp. 552-565.
26. R. Rivest, A. Shamir, Y. Tauman. How to leak a secret. Private Com., Oct. 2001.
27. R. Rivest, A. Shamir, and L. Adleman. A method for obtaining digital signatures and public-key cryptosystems. *Com. of the ACM*, 21(2):120–126, Feb. 1978.
28. A. Shamir. How to share a secret. In *Com. of the ACM*, 22(11):612–613, 1979.

Deniable Ring Authentication*

Moni Naor

Dept. of Computer Science and Applied Math
Weizmann Institute of Science
Rehovot 76100, Israel
naor@wisdom.weizmann.ac.il

Abstract. Digital Signatures enable authenticating messages in a way
that disallows repudiation. While non-repudiation is essential in some
applications, it might be undesirable in others. Two related notions of
authentication are: Deniable Authentication (see Dwork, Naor and Sahai
[25]) and Ring Signatures (see Rivest, Shamir and Tauman [38]). In this
paper we show how to combine these notions and achieve *Deniable Ring
Authentication*: it is possible to convince a verifier that a member of
an *ad hoc* subset of participants (a ring) is authenticating a message m
without revealing which one (source hiding), and the verifier V cannot
convince a third party that message m was indeed authenticated – there
is no 'paper trail' of the conversation, other than what could be produced
by V alone, as in zero-knowledge.

We provide an efficient protocol for deniable ring authentication based
on any strong encryption scheme. That is once an entity has published
a public-key of such an encryption system, it can be drafted to any
such ring. There is no need for any other cryptographic primitive. The
scheme can be extended to yield threshold authentication (e.g. at least
k members of the ring are approving the message) as well.

1 Introduction

An authentication protocol allows a receiver of a message, Bob, to verify that
the message received is indeed the one sent by the sender, Alice. It is one of
the basic issues which cryptography deals with. One of the key insights in the
seminal paper of Diffie and Hellman [22] was the idea that it is possible to make
authentication transferable, i.e. that Bob can convince a third party that Alice
had indeed sent him the message. This involves Alice having a public-key as well
as a secret-key that allows her to produce a digital signature of the message,
verifiable by anyone knowing her public-key, but one that cannot be generated
by anyone not holding her secret-key. This non-repudiation property is essential
to contract signing, e-commerce and a host of other applications. In the last
25 years a lot of effort has been devoted to digital signatures in the research
community, as well as the legal and business one.

However, one question to consider is whether non-repudiation of messages
is *always* desirable. One obvious reason is privacy - one need not be a card

---

* Research supported in part by the RAND/APX grant from the EU Program IST

M. Yung (Ed.): CRYPTO 2002, LNCS 2442, pp. 481–498, 2002.
© Springer-Verlag Berlin Heidelberg 2002

carrying EFF[1] member to appreciate that not everything we ever say should be transferable to anyone else - but this is precisely the case as more and more of our interactions move on-line. Another motivation arises where Bob is *paying* for the authentication (e.g. for checking a piece of software); should he be free to turn and give it away to Charlie? To address these concerns several notions of deniable authentication were developed (see more in Section 1.1 below.) In general an authentication provides (plausible) deniability if the recipient could have generated the authentication all by itself.

A different form of protection to the sender of the messages is hiding its identity or source. This is needed for leaking information - something that can be viewed as an important part of the checks and balances that monitor an open society. Keeping the sender's identity secret while being sure that it is a valid confirmation of the message may sound paradoxical, since the receiver verifies the authenticity of the message with respect to some public information related to the party doing the authentication (e.g. a public key). However a method for doing just that was recently suggested by Rivest, Shamir and Tauman [38]. They proposed the notion of *Ring Signatures* (a generalization of group signatures of Chaum and van Heyst [16]) that allows a member of an *ad hoc* collection of users S (e.g. Crypto'2002 Program Committee members), to prove that a message is authenticated by a member of S. The assumption is that each member of S has published a public signature key of a scheme with certain properties (where RSA and Rabin are examples). The construction given in [38] is very efficient, but its analysis is based on the ideal cipher model (a strengthening of the random oracle one.)

In this work we propose a notion that merges Ring Signatures and Deniable Authentication to form *Deniable Ring Authentication*. Roughly speaking, for a scheme to be Deniable Ring Authentication it should: (i) Enable the sender for any message he wishes and for any ad hoc collection S of users containing the sender to prove (interactively) that a member of S is the one confirming the message. (ii) Be a good authentication scheme, i.e. not allowing forgeries, where the notions of forgeability of Goldwasser, Micali and Rivest [30] are relevant. Ideally an adversary should not be able to make a receiver accept *any* message not sent by a member of S. (iii) The authentication is deniable in the zero-knowledge sense, i.e. the recipient could have simulated the conversation alone and the result would have been indistinguishable. (iv) The authentication should be *source hiding* or preserve the "anonymity in a crowd" of the sender: for any arbitrary subset S of users, any two members of S generate indistinguishable conversations to the recipient. (v) The scheme should not assume that the verifier of the authentication is part of the system and has established a public key. This is needed for two reasons: The PKI may be of a special nature (e.g. high-ranking government officials) and thus there is no reason for the recipient to be part of it. The second reason is that it is difficult to assure the independence of keys in the PKI and there is no reason to assume that the receiver has chosen its key properly (see Footnote 4 for an example.)

---

[1] Electronic Frontier Foundation.

We provide a construction of a Deniable Ring Authentication protocol based sole on the assumption that users have public-keys of some[2] good encryption scheme. The scheme is quite efficient: it requires $|S|$ encryptions by the sender and receiver and a single decryption by the sender. One can view the scheme as evolving from the deniable authentication scheme of Dwork, Naor and Sahai [25] (described in Section 4.) The analysis of the scheme is based on the security of the encryption scheme, without resorting to additional random oracles (or any additional cryptographic primitive.) Note that users have "no choice" about being recruited to the subset S. Once a user has established an encryption key he might be drafted to such a crowd S.

1.1 Related Work

Issues related to deniability and anonymity have been investigated quite extensively from the early days of open scientific investigations of Cryptography[3]. Hence there are quite a few variants of deniability and anonymity protection and we will try to briefly describe them and their relationship to our work.

Group and Ring Signatures: a group signature scheme allows members of a *fixed* group to sign messages on the group's behalf while preserving their anonymity. This anonymity is conditional and a group manager can revoke it. Note that groups here are not ad hoc and the group manager sets up a special type of key assignment protocol. There are quite a few papers on the subject [4, 8, 12], yielding reasonably efficient protocols. A related notion is that of *identity Escrow* allowing proofs of membership in a subset, with the group manager being able to identify and revoke membership [34, 10]. Some of the protocols do support subsets authentication [4] as well as general key choices by the participants, but these all assume special set-up and managers. Ring Signatures, as introduced in [38], support *ad hoc* subset formation and by definition do not require special setup. They rely on a Public-Key Infrastructure (for signatures of certain type in the construction of [38].) Note that some of the protocols for group signatures can actually be used as ring signatures, e.g. [9].

Designated Verifier Proofs were proposed by [31] to enable signatures that convince only the intended recipient, who is assumed to have a public-key. See Footnote 4 for the problems this approach might encounter in our setting.

Deniable Authentication: the work of Dwork, Naor and Sahai [25] on deniable authentication provides a system that addresses the deniability aspects, i.e. that following the protocol there is no paper trail for the authentication of the message. This is the same property we are trying to achieve, and the protocols presented there are our starting point (see Section 4.)

Undeniable signatures are digital signatures in which the recipient cannot transfer the signature without the help of the signer. If forced to either acknowledge or deny a signature, however, the signer cannot deny it if it is authentic (thus the term "invisible" is probably better). They were introduced in 1989 by Chaum and Van Antwerpen [15] and further developed in [17]. A specific and appealing

---

[2] Actually each user can use their favorite encryption scheme.
[3] An early proponent was David Chaum, e.g. [14].

version of them are the *Chameleon Signatures* of Krawczyk and Rabin [35]. The difference in the deniability requirement between this line of work and that of deniable authentication [25] as well as the current paper, is that in our case the authentication is *not* intended for ultimate adjudication by a third party, but rather to assure V – and only V – of the validity of the message.

Contributions of this work: We present a simple and efficient scheme that allows leaking an authenticated secret, without the danger of being traced (Protocol 3 below.) The scheme does not assume any special infrastructure, beyond the one given by standard PKI for encryption. The analysis of the scheme is straightforward and does not resort to random oracles.

We also extend the scheme to be able to authenticate more complex statements than "a member of S is confirming the message", to statements such as "at least k members of S confirm the message" and other access structures (Protocol 4 in Section 6.)

We also deal in Section 7 with the case where the adversary A may have all the secret keys of the authenticator, which is the appropriate model for Identity Based Encryption [41, 5] and the Subset Cover Framework of [37]. Protocol 5 handles this case at the cost of two additional rounds.

There are a number of differences between the properties of our setting and scheme and those of Rivest, Shamir and Tauman [38]: on the negative side (from this paper's point of view), our scheme requires interaction, since the verifier is not assumed to have established a public-key. This requires some mechanism of anonymous routing (e.g. MIX-nets.) Also our scheme involves sending longer messages (proportional to the size of S). On the neutral side, the time complexity of our scheme and that of [38] are roughly comparable (to within multiplicative constants), if one uses an encryption scheme where the encryption process is very efficient, such as RSA with low exponent. On the positive side: (i) Our analysis does not rely on any additional assumptions except the underlying encryption scheme is good (immune to chosen ciphertext attacks.) (ii) Since we only need that the encryption scheme is good, there is no way for an organization that wishes that its members have public-keys to try and fight our system by establishing ones with some weird formats (that deter the [38] scheme, e.g. tree based ones.) (iii) Our deniability guarantees are stronger than in [38]: their deniability is achieved by assuming that the verifier is a member of the system and has established a public key. He is then added to the Ring (and hence could have generated the conversation himself). However this assumes not only that the verifier has a public key, but that this key was properly chosen[4]. (iv) it is not clear how to extend the [38] protocol to handle threshold and other access structures over the ring[5], whereas we do that with no computational penalty in Protocol 4.

---

[4] To see why this issue may be problematic, consider a large corporation A dealing with a small user B. The user B chooses its public-key K_B to be the *same* as K_A, the public key of corporation A. Now suppose that A sends to B a message signed in a ring scheme were the ring consists of A and B. Given that the public keys used are $\{K_A, K_A\}$ this is hardly deniable for A.

[5] But see the recent work [7].

2 Definition of Deniable Ring Authentication

We now summarize the setup and requirements of a deniable ring authentication scheme.

Setup: We assume that participants have published public-keys. The public keys are generated via some key generation process that gives corresponding secret keys. We do not make any particular assumption about this process, except that good participants choose their keys properly, i.e. following the key generation protocol. However bad participants, that are under the control of the adversary, may have chosen them arbitrarily and in particular as a function of the good public keys. A ring S is a any subset of participants. A (good) authenticator P is a member of S. The verifier of a message is an arbitrary party and has not necessarily published a public-key. The only assumption is that both the verifier and the authenticator know the public-keys of *all* members of S. The authenticator P engages with the verifier in an interactive protocol to authenticate a message m. At the end of the interaction the verifier accepts or rejects the authentication.

Given that the protocol is interactive (it must be so, since the verifier has not established any credentials) we must assume that it is possible to route messages anonymously, i.e. that the verifier and prover can exchange message without the adversary being able to trace who is the recipient. How this is achieved is beyond the scope of this paper.

We assume that the adversary $\mathcal{A}$ controls some of the participants of the system. For those participants it chooses (and knows) all the secret bits (we do not deal here with dynamic corruption of good users, though the methods presented seem to be resilient to such attacks as well). The authentication protocol should satisfy:

Completeness: For any subset of participants S and for any good authenticator $P \in S$, for any message m, if the prover and verifier follow the protocol for authenticating the message m (with P using his secret key), then the verifier accepts; this can be relaxed to "accepts with high probability."

Soundness - Existential Unforgeability: Consider an adversary $\mathcal{A}$ trying to forge a message. It may know and choose the secret keys of all bad participants, but the good members choose their public-keys properly. The adversary runs an attack on the protocol as follows: it *adaptively* chooses a sequence of arbitrary messages $m_1, m_2, \ldots$, arbitrary rings $S_1, S_2, \ldots$ and good participants $P_1, P_2, \ldots$ where $P_i \in S_i$, and asks that P_i will authenticate message m_i as part of ring S_i (using the deniable ring authentication protocol) where the verifier is controlled by $\mathcal{A}$. We say that $\mathcal{A}$ successfully attacks the scheme if it can find a ring S of good participants so that a forger C, under control of $\mathcal{A}$ and pretending to be a member of S, succeeds in authenticating to a third party D (running properly the verifier's V protocol) a message $m \notin \{m_i\}_{i=1,2,\ldots}$. The soundness requirement is for all probabilistic polynomial time adversaries $\mathcal{A}$ the probability of success is negligible.

Source Hiding: For any two *good* participants A_1 and A_2, for any subset S containing A_1 and A_2, it is computationally infeasible for any V^* acting as the verifier to distinguish between protocols where A_1 is doing the authentication and A_2 is the one running it (that is the probability it guesses correctly which case it is should be negligibly close to $1/2$.) Note that not all the members of S are necessarily good, but we only protect the anonymity of the good ones.

Zero-Knowledge - Deniability: Consider an adversary $\mathcal{A}$ as above and suppose that a member of S is willing to authenticate any polynomial number of messages. Then for each $\mathcal{A}$ there exists a polynomial-time simulator $\mathcal{Z}$ that outputs an indistinguishable transcript (to everyone but the sender). A possible relaxation is to allow the simulator to depend on ε, the distinguishing advantage (this is known as ε -knowledge.)

Note that Source Hiding and Deniability seem to be related but they are incomparable. In particular, the requirement for Source Hiding should hold even for an online verifier, whereas the requirement for Deniability is only after the fact.

Concurrency: One issue that we have not specified is whether the many various protocols that the adversary may be running are executed concurrently, where timing is under the control of the adversary, or sequentially. This is largely orthogonal to those definitions and we will specify for our main scheme (Protocol 3) for each property whether it withstands concurrent attacks or not.

Big brother: A stronger model for deniability and source hiding is when the adversary $\mathcal{A}$ knows the secret keys of the good players as well as those of the bad ones. This case and its motivating examples is discussed in Section 7.

3 Tools

3.1 Encryption Schemes

Our main tool is encryption schemes. We assume some *good* public-key encryption scheme E. To specify what we mean by good, we have to provide the type of attack that the encryption scheme is assumed to withstand , e.g. known[6] or chosen plaintext, or chosen ciphertext. And we have to specify what breaking the encryption scheme means, where the two leading notions are semantic security and non-malleability. The latter is the relevant notion we will require from E. Roughly speaking, a public key cryptosystem is non-malleable if, seeing an encryption $E(\alpha)$ "does not help" an attacker to generate an encryption $E(\beta)$ such that α and β are related (with certain trivial exceptions). This is formalized and treated at length in [23]. See also [1] and [39].

As for the type of attack, this varies depending on the precise properties we want from our deniable ring authentication process (in particular whether we

---

[6] Not really relevant in public-key encryption.

want to withstand concurrent attacks). We can already gain some properties simply assuming that E is immune to chosen plaintext attack. However for the full strength we require that the scheme be immune against chosen ciphertexts in the post-processing mode, also known as CCA2. This means that the attacker has access to a decryption device and can feed it with ciphertexts of its choice. At some point it gets a challenge ciphertext (so that it should perform some operation on the corresponding plaintext) and still has access to the decryption device, except that now it cannot feed it with the challenge ciphertext. Under such an attack semantic security and non-malleability coincide. (See [23, 1] for background one the subject.)

For a public key K the encryption scheme E_K maps a plaintext into a ciphertext. This mapping must be probabilistic, otherwise the scheme cannot even be semantically secure. Therefore E_K induces for each message m a distribution of ciphertexts. To encrypt m one has to choose a random string ρ and then $C = E_K(m, \rho)$ is a ciphertext of m. Given C and the corresponding private decryption key K^{-1} the decryption process retrieves m, but we do not assume that it retrieves ρ as well (in some schemes the process does retrieve while in others it does not). When we write "generate $C = E_{K_i}(m)$" we mean choose random ρ and let $C = E_{K_i}(m, \rho)$.

A procedure we use quite extensively in our protocols is for the creator of a ciphertext C to prove that C is an encryption of a message m. In order to perform this it is sufficient to produce ρ, the random bits used to generate C and then anyone can verify that $C = E_{K_i}(m, \rho))$. The property we require from E is that it be binding or unique opening. If K was generated properly, then for any ciphertext C there should be a unique message m for which there exists a ρ such that $C = E_K(m, \rho)$ (there could be more than one ρ but no more than one plaintext corresponding to C)[7]. We do not assume any binding in case the key are badly formed (except for Section 6 where this issue arises.) Thus when we write "open ciphertext C" we mean give the plaintext and the random bits ρ used to generate C.

Implementations of the Encryptions Schemes: There are a number of possibilities for encryption scheme meeting the standards outlines above. If one wants to avoid employing random oracles, (which is one of the goals of this paper) then the famed Cramer-Shoup [20] one is the most efficient. It is based on the Decisional Diffie-Hellman problem. One drawback of it is that encryption is as expensive as decryption, i.e. requires (a few) modular exponentiations. Otherwise the system known as OAEP [2] over low exponent RSA or Rabin offers the most efficient implementation (See [42, 27, 3] for the state-of-the-art on the subject.) Using such an encryption in Protocol 3 yields a scheme of complexity comparable (up to multiplicative constants to that of [38]. Another possibility for avoiding

---

[7] This property more or less follows from the non-malleability requirement (without it one has to specify what is the meaning of such a ciphertext) but we added it explicitly to prevent confusion. Note that the complement of this property was used in [13] to obtain deniable *encryption*.

random oracles while maintaining efficiency is is to use interactive encryption, as proposed by Katz [32]. We do not explore it further in this paper, but see [32] for its application for deniable encryption.

3.2 Commitment Schemes

A commitment scheme allows the sender to deposit a hidden value with the receiver, so that the latter has no idea what it is, but at a later point the sender can reveal the hidden value and the receiver can be sure that this is the original one. There are a number of variants on the precise security offered to the two sides. We will be interested in commitment schemes where the sender is offered computational secrecy and the receiver is assured that there is a unique value. More precisely, following the commitment phase the receiver cannot decide (with non-negligible advantage) whether the hidden value is r_1 or r_2. There are quite simply and efficient protocols with these properties (e.g. [36]).

For most of the protocols of the paper we will actually use encryption for the purpose of commitment (this means that it is not secret to the owner of the secret key.) The reason is that we need to obtain *non-malleability* with respect to another encryption, and this is achieved in the easiest way using an encryption scheme which is non-malleable. However, for the big brother setting, where A is assumed to know all the secret keys in the system, this is not good enough and we will need a more involved solution in Protocol 5 in Section 7.

3.3 Zero-Knowledge

We do not apply zero-knowledge protocols as tools, but the deniability requirement means that our protocol should be zero-knowledge[8]. We use the standard tricks of the trade to come up with a simulators $\mathcal{Z}$.

The subject of preserving zero-knowledge for concurrently executed protocols has received much attention recently and in general it is a quite difficult problem. One way to bypass it was proposed in [25] by adding relatively benign timing assumptions. It is possible to use the same techniques to achieve deniability in the presence of concurrent attacks for Protocol 3.

4   Some Background Protocols

In this section we describe two protocols that can be viewed as the precursors of our main protocol. We recommend reading them before Protocol 3. We use the term "prover" for the party doing the authentication or proving the statement "message m is authentic" and "verifier" or V to the receiver or the party doing the verification of the claim. The first protocol simply provides an interactive authenticated protocol. It is based on adding a random secret value to m encrypted under P's public key as a challenge. Note that in all our protocols we

---

[8] This is a relatively rare case where zero-knowledge is needed as an end result and not as a tool in a subprotocol.

assume that the sender and receiver already know what is the candidate message (otherwise an additional preliminary round is needed.)

Protocol 1 Interactive Authentication

The prover has a public key K of an encryption scheme E. The prover wishes to authenticate the message m. The parameter ℓ sufficiently large that $2^{-\ell}$ is negligible. The concatenation of x and y is denoted $x \circ y$.

1. $V \to P$: Choose random $r \in_R \{0,1\}^{\ell}$.
Generate and send the encryption $C = E_K(m \circ r)$ to the prover.
2. $P \to V$: Decrypt C to obtain r.
Verify that the prefix of the plaintext equals m. Send r.
The verifier V accepts if the value P sends in Step 2 equals r.

This protocol was proposed in [23] (see Section 3.5 there) and proved to be existentially unforgeable assuming that the encryption scheme is secure against chosen ciphertext attacks (post-processing, or CCA2). Note that if E is malleable in certain ways then the scheme is not secure, since it is possible to switch the prefix of the message.

Is this scheme deniable? It is deniable against an honest verifier that chooses r at random. However one cannot hope to argue that it remains so against a malicious verifier, since zero-knowledge is impossible to obtain in two rounds (see [29], at least with auxiliary input).

Consider now the following extension that was proposed by Dwork, Naor and Sahai, [25], where the idea is that the verifier should prove knowledge of r before the prover reveals it. For this we use the "opening" of ciphertexts as defined in Section 3.1, by giving away the plaintext and the random bits used to generate it.

Protocol 2 Deniable Authentication

The prover P has a public key K of an encryption scheme E. The message to be authenticated in m, known to both parties.)

1. $V \to P$: Choose $r \in_R \{0,1\}^{\ell}$. Generate and send $C = E_K(m \circ r)$
2. $P \to V$: Decrypt C to obtain r (the suffix of the message).
Generate and send $D = E_K(r)$.
3. $V \to P$: Open C by sending r and ρ,
the random bits used in the encryption in Step 1.
4. $P \to V$: Verify that the prefix of the opened C equals m.
Open D by sending r and σ,
the random bits used in the encryption in Step 2.
V accepts if the value sent in Step 4 equals r and D was opened correctly.

Note that the verification m is only done at Step 4, that is if a bad C was sent, then the prover does not reveal the fact that it detected it at Step 2. The deniability of the scheme is obtained by the possibility that the simulator extract the value of r from any verifier V^*, at least in expected polynomial time, as is common in proofs of zero-knowledge. After r has been extracted it is possible to finish the execution of the protocol. Soundness follows from the fact that the ciphertext $D = E_K(r)$ serves as a non-malleable commitment to r.

5   The Main Scheme

The idea for obtaining a ring authentication protocol from Protocol 2 is to run in parallel a copy of the protocol for each member of S, but using the same r, but otherwise with independent random bits. However there are a few delicate points. In particular if we want to assure source hiding, then it is unsafe for the prover to encrypt the decrypted r using all the K_i's before it verifies the consistency of the Step 1 encryption in all the protocols. Otherwise by using a different r for each encryption key the adversary who may be controlling one member of S may figure out the identity of P. To handle this we let P split r into $r_1, r_2, \ldots r_n$ and encrypt each one separately in Step 2.

Setup: Participants in the system have public keys of an encryption scheme E, as described in Section 3.1. Each good member knows the corresponding secret key. Let the ring be denoted by S and by slight abuse of notation we will also identify S with the set of public keys of its member $\{K_1, K_2, \ldots, K_n\}$. Both P (where we assume $P \in S$) and the verifier V know all the public keys in S.

Protocol 3 Deniable Ring Authentication
for $S = \{K_1, K_2, \ldots, K_n\}$ where P knows the jth decryption key. The message to be authenticated is m.

1. $V \to P$: *Choose random $r \in \{0, 1\}^\ell$. Generate and Send*
    $\langle C_1 = E_{K_1}(m \circ r), C_2 = E_{K_2}(m \circ r), \ldots C_n = E_{K_n}(m \circ r) \rangle$.
2. $P \to V$: *Decrypt C_j to obtain r.*
    Choose random $r_1, r_2, \ldots r_n$ so that $r = r_1 + r_2 + \cdots + r_n$.
    Generate and send $\langle D_1 = E_{K_1}(r_1), \ldots, D_n = E_{K_n}(r_n) \rangle$.
3. $V \to P$: *Open $C_1, C_2 \ldots C_n$ by sending r and $\rho_1, \rho_2, \ldots \rho_n$,*
    the random bits used in the encryption process in Step 1.
4. $P \to V$: *Verify that $C_1, C_2 \ldots C_n$ were properly formed (same m and r).*
    Send $r_1, r_2, \ldots r_n$ and $\sigma_1, \sigma_2, \ldots \sigma_n$,
    the random bits used to generate $D_1, D_2, \ldots D_n$.

V accepts *if $r = r_1 + r_2 + \cdots + r_n$ and $D_1, D_2, \ldots D_n$ were properly formed.*

Complexity: Running the protocol involves on the verifier's side n encryptions and n verifications of encryptions. On the prover's side it involves one decryption, n encryptions and n verifications of encryption. If the underlying encryption scheme is based on low exponents (Rabin or low exponent RSA with OAEP), then this consists of $O(n)$ multiplication and $O(1)$ exponentiations. If the encryption is Diffie-Hellman based (for instance Cramer-Shoup [20]) then $O(n)$ exponentiations are involved. In term of communication, the major burden is sending (both ways) n ciphertexts.

5.1   Functionality and Security of the Scheme

To prove that Protocol 3 is indeed a deniable Ring Authentication Protocol we have to argue that the four requirements, completeness, soundness, source-hiding and deniability are satisfied, as we now sketch. As for completeness it is easy to verify that if both sides follows the protocol than they accept. The only requirement we need from a bad public key K_i is that it will be easy to verify even for bad keys that $C = E_{K_i}(m, \rho)$ which we can assume without loss of generality that holds.

Soundness/Unforgeability: Recall that we may assume that all keys in S are properly formed for this property. The key point of to understanding why the protocol is that $\langle D_1 = E_{K_1}(r_1), D_2 = E_{K_2}(r_2), \ldots D_n = E_{K_n}(r_n) \rangle$ is a non-malleable commitment to $r = r_1 + r_2 + \cdots r_n$, where the non-malleability is with respect to $\langle C_1 = E_{K_1}(m \circ r), C_2 = E_{K_2}(m \circ r), \ldots C_n = E_{K_n}(m \circ r) \rangle$.

For this to hold it is sufficient that E be non-malleable against chosen plain-text attacks (no need for protection against chosen ciphertext attacks, unless we are interested in concurrent attacks .) The fact that P is committed to the value follows from the binding property of E (See Section 3.1.) Once we have established this then soundness follows, as it does for Protocol 2. To handle a concurrent attacks we assume that E is secure against chosen ciphertext secure attacks (post-processing, or CCA2). (We do not know whether this is essential, see Section 8.)

Source Hiding: we claim that the of the key which was used in Step 2 (among well chosen keys in S) is *computationally* hidden during the protocol and *statistically* hidden after protocol, if things went well, i.e. the protocol terminated successfully. This follows from the fact that if at Step 1 all the $\langle C_1 = E_{K_1}(m \circ r), C_2 = E_{K_2}(m \circ r), \ldots C_n = E_{K_n}(m \circ r) \rangle$ are consistent (with the same m and r), at least among the good keys, then the hiding of the source is perfect. Suppose that they are not consistent. Then at step 3 they will be caught (from the binding property of E, and hence Step 4 will not take place.

This property is maintained even when the adversary can schedule concurrent executions. The reason is that witness indistinguishable protocols can be composed concurrently.

Deniability: we can run a simulator 'as usual' and extract r: run the protocol with P using first a random r'. If at Step 3 the verifier opens then rewind to just after Step 1 and run again with the correct r. A few things worth noting: the complications to address, as is usual in proofs of a zero-knowledge property are a V^* that refuses to open. One key point to notice is that the semantic security of E means that it is enough that one key K_i be good and unknown to V^* for $\langle D_1 = E_{K_1}(r_1), D_2 = E_{K_2}(r_2), \ldots D_n = E_{K_n}(r_n) \rangle$ to be a semantically secure commitment scheme to r.

This is the only property that is problematic under concurrent executions. We can appeal to the timing model of [25] and get a variant of this protocol that would work there. However this is beyond the scope of this paper.

6   Extension: Threshold and Other Access Structures

One can view Ring Authentication (both ours and the Rivest Shamir and Tauman one [38]) as a proof system that 1 out of the ring S is confirming the message. In this section we discuss an extension of Rings into proving more general statements, e.g. that (a least) k members out of the ring S are confirming the message, without revealing any information about the subset T of confirmers. In general, we can deal with *any* monotone access structure, provided that it has a good secret sharing scheme (see [43] for bibliography on the subject.)

In this setting we assume that there is a subset $T \subset S$ of members that collude and want to convince the verifier that T satisfies some monotone access structure $\mathcal{M}$. As in the rest of this paper, all this can be ad hoc, i.e. there is no need to fix neither S nor $\mathcal{M}$ in advance (or T of course). We do assume that there is one representative P of T that communicates with the verifier. Note that the members of T need to trust P to the extent that a bad P can make the protocol loose its deniability and source hiding, but the unforgeability.

We adapt an idea suggested by Cramer, Damgård and Shoenmaker [19] and DeSantis et al. [21] for combining zero-knowledge statements via secret sharing. In our context we use this idea by letting the verifier split r according to the secret sharing scheme for $\mathcal{M}$. Only if enough shares are known, then r can be reconstructed, otherwise it remains completely unknown.

We do not have to assume any additional properties from the access structure for $\mathcal{M}$, i.e. the protection could be information theoretic or computational. We assume of course that secret generation and reconstruction are efficient. We also assume that given shares $s_1, s_2, \ldots s_n$ it is possible to verify that they were properly formed, i.e. that for each subset T that satisfies $\mathcal{M}$ the reconstruction algorithm will output the same secret. This is very simple in most, if not all, schemes we are aware of, e.g. Shamir's polynomial based one [40].

Protocol 4 Ring Authentication for Monotone Access Structure $\mathcal{M}$
Ring $S = \{K_1, K_2, \ldots, K_n\}$ where P represents a subset $T \subseteq S$.

1. $V \to P$: *Choose random $r \in \{0,1\}^\ell$.*
    Generate shares $s_1, s_2, \ldots s_n$ of r according to the scheme for $\mathcal{M}$.
    Send $\langle C_1 = E_{K_1}(m \circ s_1), C_2 = E_{K_2}(m \circ s_2), \ldots C_n = E_{K_n}(m \circ s_n) \rangle$.
2. $P \to V$: *P gets from each $j \in T$ the decryption of C_j.*
    P reconstructs r from the shares s_j for $j \in T$.
    Choose random $r_1, r_2, \ldots r_n$ such that $r = r_1 + r_2 + \cdots + r_n$.
    Generate and send $\langle D_1 = E_{K_1}(r_1), \ldots, D_n = E_{K_n}(r_n) \rangle$.
3. $V \to P$: *Open $C_1, C_2 \ldots C_n$ by sending $s_1, s_2, \ldots s_n$ and $\rho_1, \rho_2, \ldots \rho_n$,*
    the random bits used in the encryption in Step 1.
4. $P \to V$: *Verify that $s_1, s_2, \ldots s_n$ yield the same secret for all subsets.*
    Verify that $C_1, C_2 \ldots C_n$ were properly formed
    (same m and corresponding s_i.)
    Send $r_1, \ldots r_n$ and $\sigma_1, \ldots \sigma_n$ (bits used to generate $D_1, \ldots D_n$.)

V **accepts** *if $r = r_1 + r_2 + \cdots + r_n$ and $D_1, \ldots D_n$ were properly formed.*

This extended protocol is not much more complex (computationally as well as to implement) than the original Protocol 3. The additional computation consists simply of the secret sharing generation, reconstruction and verification.

The Completeness and Deniability of Protocol 4 follow from the same principals as Protocol 3. As for unforgeability, we should argue that if the subset $S^* \subset S$ of participants the adversary controls does not satisfy $\mathcal{M}$, then it cannot make the adversary accept a message m (not authenticated by the good participants) with non-negligible probability. One point is the binding property of E. In Protocol 3 the assumption was that all of S consists of good players (otherwise unforgeability is not meaningful), but here some members of S might be under the control of $\mathcal{A}$ (but not enough to satisfy $\mathcal{M}$) and might have chosen their public-keys improperly. For this we either have to assume that E_K is binding even if the key K was not properly chosen, or modify the protocol to and add for each D_i a commitment to r_i as well. The protection against reconstruction by a non qualified subset that the secret sharing scheme $\mathcal{M}$ offers then assures that $\mathcal{A}$ cannot retrieve r following Step 1.

As for source hiding, we should argue that for two sets T_1 and T_2 deciding which one is doing the confirmation is difficult. This follows form the fact that at Step 4 P checks the consistency of $s_1, s_2, \ldots s_n$ and hence revealing $r_1, r_2, \ldots r_n$ will not yield information about T.

7   Deniable Ring Authentication in the Presence of Big Brother

In this Section we deal with the case of where the adversary $\mathcal{A}$ actually knows the secret key of the authenticator. Why is this an interesting case, after all we usually think of the public-key setting as providing users with the freedom of choosing their own keys? There are several possible answers: first, there are settings where users do not choose their own private keys. These include *Identity Based Encryption* [5, 18, 41] where a center provides a key to each users as a function of their identity, and a broadcast encryption type of setting where users receive secret keys of various subsets to which they belong (see in particular [37] and more below). Another answer is that it is desirable to avoid a situation where the distinguisher has an incentive to extract the secret key for j by, say, legal means.

Protocol 3 does not offer anonymity and deniability in case the adversary $\mathcal{A}$ knows the *secret* key of P - it is possible for $\mathcal{A}$ to figure out whether it is j who is authenticating the message by the following active attack: In Step 1 in protocol send $\langle C_1 = E_{K_1}(m \circ r), C_2 = E_{K_2}(m \circ r), \ldots C_j = E_{K_j}(m \circ r'), \ldots C_n = E_{K_n}(m \circ r)\rangle$, that is C_j is the only one with suffix r'. When receiving $D_1, D_2, \ldots D_n$ the adversary can check whether the suffix of decryption of D_j equals r or r'.

There are two possible approaches for correcting this problem, one is for P to make sure that all the C_i's are proper before decrypting any of them. This requires some form of proof of consistency. The other is for P not to commit to r using the E's but rather using a scheme that is secure against everyone.

This requires coming up with non-malleable commitment with respect to the encryptions of Step 1. Both approaches seem viable, but we found a reasonably efficient implementation of only the second one. The major obstacle is to preserve soundness. The idea is simple and is an adaptation of old tricks (e.g. the commitment scheme in [23]): the prover splits r into two parts and will reveal one of them to prove knowledge.

Let W be a commitment scheme with perfect binding and computational protection to the sender, as in, e.g. [36]. We assume that the commitment phase is unidirectional (this even fits the scheme of [36], since it can be sent together with the message.) We assume that commitment to value r involves choosing a random string σ and sending $D = W(r, \sigma)$. The computational protection it offers the sender means that given D which is a commitment to r_1 or r_2 it is hard to distinguish between the two cases. We will have as our security parameter ℓ, we assume is sufficiently large so that $2^{-\ell}$ is negligible.

Protocol 5 Ring Authentication in the presence of big brother
for Ring $S = \{K_1, K_2, \ldots, K_n\}$ where P knows the jth decryption key.
The message to be authenticated is m.

1. $V \to P$: *Choose random $r \in \{0,1\}^\ell$. Generate and Send*
   $\langle C_1 = E_{K_1}(m \circ r), C_2 = E_{K_2}(m \circ r), \ldots C_n = E_{K_n}(m \circ r) \rangle$.
2. $P \to V$: *Decrypt C_j to obtain r.*
   Choose ℓ pairs $(r_1^0, r_1^1), (r_2^0, r_2^1), \ldots (r_\ell^0, r_\ell^1)$ such that $r_i^0 + r_i^1 = r$
   Generate and send ℓ pairs $\langle (D_1^0, D_1^1), (D_2^0, D_2^1), \ldots, (D_\ell^0, D_\ell^1) \rangle$
   where for $1 \leq i \leq \ell$ and $b \in \{0,1\}$ generate $D_i^b = W(r_i^b, \sigma_i^b)$.
3. $V \to P$: *Choose and send ℓ random bits $b_1, b_2, \ldots b_\ell$*
4. $P \to V$: *For $1 \leq i \leq \ell$ open $D_i^{b_i}$ by sending $r_i^{b_i}$ and $\sigma_i^{b_i}$.*
5. $V \to P$: *Verify that the opening are consistent: $\forall 1 \leq i \leq \ell$ $D_i^{b_i} = W(r_i^{b_i}, \sigma_i^{b_i})$*
   Open $C_1, \ldots, C_n$ by sending r and $\rho_1, \rho_2, \ldots \rho_n$,
   the random bits used in the encryption process in Step 1.
6. $P \to V$: *Verify that $C_1, \ldots, C_n$ were properly formed (same m and r).*
   Open the remaining members of $\langle (D_1^0, D_1^1), (D_2^0, D_2^1), \ldots (D_\ell^0, D_\ell^1) \rangle$
   by sending $r_i^{1-b_i}$ and $\sigma_i^{1-b_i}$.

V **accepts** *if the revealed values $(r_1^0, r_1^1), (r_2^0, r_2^1), \ldots (r_\ell^0, r_\ell^1)$ were properly formed and $\forall 1 \leq i \leq \ell$ we have $r_i^0 + r_i^1 = r$.*

In order to prove that soundness/unforgeability still holds, consider Protocol 1. For this protocol we are assured of its soundness in case E is secure against chosen ciphertext attacks in the postprocessing mode (CCA2) [23]. We will need the same assumption here. The key point is that it is possible to extract the value $r_i^0 + r_i^1$ (which should be r) by rewinding the forger to just before Step 3. If the forger has probability δ of succeeding, then with probability at least δ^2 such a value can be extracted and hence a guess for r can be mounted. This (plus the original proof of Protocol 1) are sufficient for proving the soundness of Protocol 5.

As for source hiding, it follows from the semantic security of W. The only problem is when V^* is not following the protocol. But in this case in Step 6 the prover does not open the remaining commitments. Similarly, for deniability, the simulator should extract r from V^*, which can be done by sending commitments to random values in Step 2.

The Subset Cover Environment: An interesting case of using Protocol 5 is in the Subset Cover Environment described in [37]. In this setting there is a collection of subsets of users $U_1, U_2, \ldots$. For each subset U_i in the collection there is an associated public key L_i. Each user u is given secret information enabling it to compute the corresponding secret key of L_i for all subsets U_i such that $u \in U_i$. There are many types of rings S where it is easy express S as the union of not too many subsets from the collection. One such example is where the users correspond to points on a line and the collection $U_1, U_2, \ldots$ to segments[9]. If the ring S consists of a small number of segments, then the number of subsets in the union is small. In such cases using Protocol 5 is very attractive: first even though S may be large, there is no need to perform the encryption in Step 1 for each member of S but rather for each segment. The deniable and source hiding properties assure us that even though the prover did not choose his keys by himself, and they might be known to the adversary, he can still enjoy anonymity.

8   Open Problems and Discussion

There are several specific questions that arise from this work, as well as more general ones:

- In the access scheme of Section 6 is it possible for the members to be mutually untrusting with respect to deniability. The protocol is presented so that P is the one responsible for checking that the shares are proper. Is this necessary (without resorting to a complex multi-part computation).
- What is the communication complexity of ring authentication, in particular is it possible to perform such authentication by sending $o((|S|)$ bits, assuming the identity of the members of S is known, or a the very least without sending $|S|$ encryptions (or signatures). Note that [38] manage to achieve that using random oracles. Also Boaz Barak (personal communication) has pointed out that using Kilian's arguments [33] it is possible in principle to obtain inefficient but succinct protocols, so the remaining question is whether it is possible to do so while maintaing the efficiency of the protocol.
- What is the weakest form of security required from an encryption scheme that is sufficient to be used in our ring deniable authentication scheme. In particular what are the minimum requirements for Protocol 3?
- Is obtaining *source hiding* only an easier task than achieving deniability? Our approach was to take a deniable scheme (Protocol 2 and turn it into a ring scheme, but perhaps aiming directly for source hiding will yield other schemes and in particular 2-round ones (note that Witness Indistinguishability is possible in 2-rounds [24].)

---

[9] The Subtree Difference example of [37] can be adapted to work in this case.

- Is it possible to obtain *source hiding* in the case of shared keys, say in the Subset Cover Framework? Note that shared key authentication implies *deniability*, but running a protocol like 3 is problematic, since proving in zero-knowledge the consistency of shared-key encryptions is difficult.
- Is it possible to use the Fiat-Shamir heuristic[10] and remove the interaction from authentication protocols such as Protocols 1, 2 and 3 and thus get new types of signature schemes from encryptions schemes?

An important social concern that both this work and [38] raise is the implication to PKI. The fact that a user that has established a public key can be 'drafted' to a ring S without his consent might be disturbing to many users. On the other hand we believe in a more positive interpretation of the results. Allowing some degree of anonymity as well as leaking secret has always been important at least in modern societies and this form of protocols allows the re-introduction of it.

In general we find the issue of anonymity and deniability to be at the heart of the open scientific investigation of cryptography. A very natural research program is to find the precise mapping between possible and impossible in this area. It seems that behind every impossibility result lies a small twist (in the model perhaps) that allows the tasks to be performed.

Acknowledgments

I wish to thank Boaz Barak, Cynthia Dwork and Adi Shamir for helpful discussions and the anonymous referees for useful comments.

References

1. M. Bellare, A. Desai, D. Pointcheval and P. Rogaway. *Relations among notions of security for public-key encryption schemes,* Advances in Cryptology – CRYPTO'98, LNCS 1462, Springer, pp. 26–45.
2. M. Bellare and P. Rogaway, *Optimal Asymmetric Encryption,* Advances in Cryptology-Eurocrypt '94, LNCS 950, Springer, 1995, pp. 92–111.
3. Dan Boneh, *Simplified OAEP for the RSA and Rabin Functions,* Advances in Cryptology – CRYPTO 2001, LNCS2139, Springer 2001, pp. 275–291.
4. D. Boneh and M. Franklin, *Anonymous Authentication with Subset Queries,* ACM Conference on Computer and Communications Security 1999, pp. 113–119.
5. D. Boneh and M. Franklin, *Identity-Based Encryption from the Weil Pairing,* Advances in Cryptology – CRYPTO 2001, LNCS 2139, Springer, 2001, pp. 213–229.
6. J. Boyar, D. Chaum, I. Damgård and T. P. Pedersen: *Convertible Undeniable Signatures,* Advances in Cryptology – CRYPTO'90, Springer, 1991, pp. 189–205.
7. E. Bresson, J. Stern and M. Szydlo, *Threshold Ring Signatures for Ad-hoc Groups,* Advances in Cryptology – CRYPTO'2002, (these proceedings).

---

[10] Fiat and Shamir [26] proposed a general method for converting *public* coins zero-knowledge proof systems into signatures. The analysis of the method is based on the random oracle model.

8. J. Camenisch, *Efficient and Generalized Group Signatures*, Advances in Cryptology – EUROCRYPT'97, LNCS 1233, Springer, 1997, pp. 465–479.

9. J. Camenisch and I. Damgård, *Verifiable Encryption, Group Encryption, and Their Applications to Group Signatures and Signature Sharing Schemes*, Advances in Cryptology – Asiacrypt 2000, LNCS 1976, Springer, 2000, pp. 331–345.

10. J. Camenisch and A. Lysyanskaya, *An Identity Escrow Scheme with Appointed Verifiers*, Advances in Cryptology – Crypto 2001, LNCS 2139, Springer, 2001, pp. 388–407.

11. J. Camenisch, M. Michels, *Separability and Efficiency for Generic Group Signature Schemes*, Advances in Cryptology - CRYPTO'99, LNCS 1666, Springer, 1999, pp. 106–121.

12. J. Camenisch and M. Stadler, *Efficient Group Signature Schemes for Large Groups*, Advances in Cryptology – CRYPTO'97, LNCS 1294, Springer, 1997, pp. 410–424.

13. R. Canetti, C. Dwork, M. Naor and R. Ostrovsky, *Deniable Encryption*, Advances in Cryptology – CRYPTO'97, LNCS 1294, Springer, 1997, pp. 90–104.

14. D. Chaum, *Untraceable electronic mail, return addresses, and digital pseudonyms*, Comm. of ACM, vol. 24(2), 1981, pp. 84–88.

15. D. Chaum and H. van Antwerpen, *Undeniable Signatures*, Advances in Cryptology – CRYPTO'89, LNCS 435, Springer, 1990, pp. 212–216.

16. D. Chaum and E. van Heyst, *Group Signatures*, Advances in Cryptology – EUROCRYPT'91, LNCS 541, Springer, 1991, pp. 257–265.

17. D. Chaum and E. van Heyst and B. Pfitzmann, *Cryptographically Strong Undeniable Signatures, Unconditionally Secure for the Signer*, Advances in Cryptology – CRYPTO'91, LNCS 576, Springer, 1992, pp. 470–484.

18. C. Cocks. *An identity based encryption scheme based on quadratic residues*, Cryptography and Coding, LNCS 2260, Springer, 2001, pp. 360–363.

19. R. Cramer, I. Damgård, B. Schoenmakers, *Proofs of Partial Knowledge and Simplified Design of Witness Hiding Protocols*, Advances in Cryptology – CRYPTO'94, LNCS, Springer, 1994, pp. 174–187.

20. R. Cramer and V. Shoup, *A Practical Public Key Cryptosystem Provably Secure against Adaptive Chosen Ciphertext Attack*, Advances in Cryptology – CRYPTO'98, LNCS 1462, Springer, 1998, pp. 13–25.

21. A. De Santis, G. Di Crescenzo, G. Persiano, M. Yung, *On Monotone Formula Closure of SZK*, Proc. 35th IEEE FOCS, 1994, pp. 454–465.

22. W. Diffie, and M.E. Hellman. New Directions in Cryptography. *IEEE Trans. on Info. Theory*, IT-22 (Nov. 1976), pages 644–654.

23. D. Dolev, C. Dwork and M. Naor, *Non-malleable Cryptography*, Siam J. on Computing, vol 30, 2000, pp. 391–437.

24. C. Dwork and M. Naor, *Zaps and Their Applications*, Proc. 41st IEEE Symposium on Foundations of Computer Science, pp. 283–293. Full version: ECCC, Report TR02-001, www.eccc.uni-trier.de/eccc/.

25. C. Dwork, M. Naor and A. Sahai, *Concurrent Zero-Knowledge*, Proc. 30th ACM Symposium on the Theory of Computing, Dallas, 1998, pp. 409–418.

26. A. Fiat and A. Shamir, *How to Prove Yourself: Practical Solutions to Identification and Signature Problems*, Advances in Cryptology – CRYPTO'86, LNCS 263, Springer, 1987, pp. 186–194.

27. E. Fujisaki, T. Okamoto, D. Pointcheval, J. Stern, *RSA-OAEP Is Secure under the RSA Assumption*, Advances in Cryptology – CRYPTO 2001, pp. 260–274.

28. R. Gennaro, H. Krawczyk and T. Rabin, *RSA-Based Undeniable Signatures*, Advances in Cryptology – CRYPTO'97, LNCS 1294, Springer, 1997, pp. 132–149.

29. O. Goldreich and Y. Oren, *Definitions and properties of Zero-Knowledge proof systems*, J. of Cryptology, Vol 7, 1994, pp.1–32.
30. S. Goldwasser, S. Micali and R. Rivest, *A secure digital signature scheme*, SIAM J. on Computing 17, 1988, pp. 281–308.
31. M. Jakobsson, K. Sako and R. Impagliazzo, *Designated Verifier Proofs and Their Applications*, Advances in Cryptology – EUROCRYPT '96, pp. 143–154.
32. J. Katz, *Efficient and Non-Malleable Proofs of Plaintext Knowledge and Applications Cryptology*, ePrint Archive, Report 2002//027, http://eprint.iacr.org/
33. J. Kilian, *A Note on Efficient Zero-Knowledge Proofs and Arguments*, Proc. 24th ACM Symposium on the Theory of Computing, 1992, pp. 723–732.
34. J. Kilian and E. Petrank, *Identity Escrow*, Advances in Cryptology – CRYPTO '98 LNCS 1462, 1998, pp. 169–185.
35. H. Krawczyk and T. Rabin, *Chameleon Hashing Signatures*, Proceedings of Network and Distributed Systems Security Symposium (NDSS) 2000, Internet Society, pp. 143–154.
36. M. Naor. *Bit Commitment Using Pseudo-Randomness*, Journal of Cryptology, vol. 4, 1991, pp. 151–158.
37. D. Naor, M. Naor and J. B. Lotspiech, *Revocation and Tracing Schemes for Stateless Receivers*, Advances in Cryptology – CRYPTO 2001, pp. 41–62. LNCS 2139, Springer, 2001, pp. 205–219. Full version: Cryptology ePrint Archive, Report 2001/059, http://eprint.iacr.org/
38. R. L. Rivest, A. Shamir, and Y. Tauman, *How to Leak A Secret*, Advances in Cryptology – ASIACRYPT 2001, Lecture Notes in Computer Science, Vol. 2248, Springer, pp. 552–565.
39. A. Sahai, *Non-Malleable Non-Interactive Zero Knowledge and Achieving Chosen-Ciphertext Security*, Proc. 40th IEEE Symposium on Foundations of Computer Science, 1999, pp. 543–553.
40. A. Shamir, *How to Share a Secret*, Communications of the ACM 22, 1979, pp. 612–613.
41. A. Shamir, *Identity-Based Cryptosystems and Signature Schemes*, Advances in Cryptology – CRYPTO'84, LNCS 196, Springer, 1985, pp. 47–53.
42. V. Shoup, *OAEP Reconsidered*, Advances in Cryptology – CRYPTO 2001, LNCS, Springer, 2001, pp. 239–259.
43. Bibliography on Secret Sharing Schemes, maintained by D. Stinson and R. Wei. www.cacr.math.uwaterloo.ca/~dstinson/ssbib.html

SiBIR:
Signer-Base Intrusion-Resilient Signatures

Gene Itkis and Leonid Reyzin

Boston University Computer Science Dept.
111 Cummington St.
Boston, MA 02215, USA
{itkis,reyzin}@bu.edu

Abstract. We propose a new notion of *signer-base intrusion-resilient (SiBIR) signatures*, which generalizes and improves upon both forward-secure [And97,BM99] and key-insulated [DKXY02] signature schemes. Specifically, as in the prior notions, time is divided into predefined time periods (e.g., days); each signature includes the number of the time period in which it was generated; while the public key remains the same, the secret keys evolve with time. Also, as in key-insulated schemes, the user has two modules, *signer* and *home base*: the signer generates signatures on his[1] own, and the base is needed only to help update the signer's key from one period to the next.

The main strength of intrusion-resilient schemes, as opposed to prior notions, is that they remain secure even after *arbitrarily many* compromises of *both* modules, as long as the compromises are not simultaneous. Moreover, even if the intruder does compromise both modules simultaneously, she will still be unable to generate any signatures for the previous time periods.

We provide an efficient intrusion-resilient signature scheme, provably secure in the random oracle model based on the strong RSA assumption. We also discuss how such schemes can eliminate the need for certificate revocation in the case of on-line authentication.

1 Introduction

Key exposures appear to be unavoidable. Thus, limiting their impact is extremely important and is the focus of active research. While this issue applies to a wide range of security protocols, here we focus on digital signatures.

1.1 Previous Work

FORWARD SECURITY. Forward-secure signature schemes [And97,BM99] preserve the security of past signatures even after the secret signing key has been exposed: time is divided into predefined time periods, with the signer updating his secret at the end of each time period; the adversary is unable to forge signatures for past periods even if she learns the key for the current one. In this

---

[1] We use masculine pronouns for signer, feminine for adversary, and neuter for base.

M. Yung (Ed.): CRYPTO 2002, LNCS 2442, pp. 499–514, 2002.

model, nothing can be done about the future periods: once the adversary exposes the current secret, she has the same information as the signer.

THRESHOLD AND PROACTIVE SECURITY. An alternative approach explores the multi-party computation paradigm [Yao82,GMW87]: in threshold schemes [DF89], the signing key is somehow shared among a number of signers, and signature generation requires a distributed computation involving some subset of them. The adversary, however, cannot generate valid signatures as long as the number of compromised signers is less than some predetermined security parameter (smaller than the number of signers needed to generate a valid signature). Proactive schemes [OY91,HJJ+97] improve upon this model by allowing multiple corruptions of *all* signers, limiting only the number of *simultaneous* corruptions. Proactive forward-secure signatures considered in [AMN01] combine this with the advantages of forward-security.

KEY-INSULATED SECURITY. The recently proposed model of Dodis, Katz, Xu and Yung [DKXY02] addresses the limitation of forward security: the adversary cannot generate signatures for the future (as well as past) time periods even after learning the current signing key[2]. This is accomplished via the use of two modules: a (possibly mobile) *signer*, and a (generally stationary) *home base*[3]. The signer has the secret signing key, and can generate signatures on its own. At the end of each time period, the signing key expires and the signer needs to *update* his keys by communicating with the home base and performing some local computations (the communication with the base is, in fact, limited to a single message from the base to the signer). Thus, although the signer's keys are vulnerable (because they are frequently accessed, and, moreover, because the signer may be mobile), key exposure is less valuable to the adversary, as it reveals only short-term keys. Perhaps the most compelling application of such a model is the example of a frequently traveling user, whose laptop (or handheld) is the signer, and office computer is the home base. (Alternative approaches with such applications in mind were proposed by [Mic96,Riv98,GPR98,LR00,MR01b,MR01a].) This model enables security that is not possible in ordinary or even forward-secure schemes: even if the signing key is compromised (for up to k time periods, for predetermined security parameter k), the adversary will be unable to forge signatures for *any* other time periods. (Notice that in forward-secure schemes model, signatures for any time period following a compromise are *necessarily* forgeable.)

1.2   Our Results: Intrusion-Resilient Security

Model. We define intrusion-resilient signature schemes to combine benefits of the above three approaches. Namely, while maintaining the efficiency of non-interactive computation of signatures (not provided by threshold and proactive

---

[2] [DKXY02] primarily addresses encryption schemes. Signature schemes are addressed in [DKXY].

[3] The terms *user* and *secure device* are used in [DKXY02]; we find "signer" and "home base" to be more descriptive.

schemes), intrusion-resilient schemes preserve security of past *and* future time periods when *both* signer and base are compromised, though not simultaneously (not preserved by key-insulated and forward-secure schemes), and security of past time periods in the case of simultaneous compromise (not preserved by key-insulated[4] and most proactive schemes).

These points deserve some elaboration. To address potential compromise of the base key, [DKXY02] introduce a stronger version of key-insulated security, which requires that the base cannot generate signatures on its own. However, no security is guaranteed in [DKXY02] if the adversary manages to compromise *both* the base and the signer, even during different time periods. (In fact, the encryption scheme becomes completely insecure in such a case.) This is a serious limitation. If the user's key is compromised even just once, then the prudent thing to do would be to revoke the entire public key and erase the secrets of the home base. Otherwise, a single compromise of the home base would expose not only the future, but also all the past, messages.

In contrast, the salient feature of our new model is the guarantee that a compromise of the home base is *entirely inconsequential* as long as the signer's secret is not exposed at the same time. It thus has the benefits of proactive security. Moreover, our model retains the benefits of forward security even when all the secrets are compromised simultaneously.

Indeed, our intrusion-resilient model appears to provide the *maximum possible* security in the face of corruptions that occur.

Construction. In Section 3 we provide an efficient SiBIR signature scheme we call SiBIR1. Its signing and verifying are as efficient as in the Guillou-Quisquater (ordinary) signatures [GQ88], requiring just two modular exponentiations with short (typically, 128-160 bits) exponents for both signing and verifying. This is as or more efficient than many of the ordinary signatures used in practice today. The construction is based on our forward-secure signature scheme [IR01].

As for that underlying scheme [IR01], our SiBIR1 security proof relies on the strong RSA assumption (see Section 4.1) and is in the random oracle model.

1.3   Towards Obsoletion of Certificate Revocation

On-line authentication is a common application of signatures. For example, a user establishing an authenticated connection to a web site (e.g., over SSL), must verify the web site's signature on a protocol message, as well as the web site's certificate that attests to the authenticity of the web site's public key. If the web site's secret key is compromised, the certificate needs to be *revoked*.

Certificate revocation, however, is a complex logistical problem that results in some of the most cumbersome aspects of public key infrastructures. The most

---

[4] Because the focus of [DKXY02] is on encryption schemes, and no non-trivial forward-secure encryption schemes had been known until very recently [Kat02], it was, in a sense, by necessity that key-insulated notion of [DKXY02] did not provide forward security when all the secrets are compromised.

common, though perhaps not the most efficient, mechanism is to consult a certificate revocation list (CRL), which would most likely be stored at a remote location (certificates usually include a pointer to the corresponding CRL site).

However, if the web site uses our signature scheme, then an exposed secret key would compromise the authenticity of the web site only for a limited time (which could be made less than the time required for the certificate revocation process, which is typically one day). Then the users need not check whether the site's certificate is revoked or not: by the time the revocation information could be updated, the web site would be authentic again, anyway.

Note that forward-secure signatures do not help address this problem: the web site's certificate would still have to be revoked in case of compromise. In contrast, if the web site uses intrusion-resilient signatures, the certificate would have to be revoked only in the unlikely case that the web site and its (presumably, separately protected) home base are compromised *simultaneously*. (We note that short-lived certificates [Mic96,GGM00], key-insulated signatures [DKXY02] and proactive signature [OY91,HJJ+97] can also be used to address certificate revocation; our solution, however, seems to provide the most security if one is interested in abandoning certificate revocation/reissuing entirely and having truly off-line certification authorities.)

2 Intrusion-Resilient Security Model

Our definitions are based on the definitions of key-insulated security [DKXY02], which, in turn, are based on the definitions of forward secure [BM99] and ordinary [GMR88] signatures schemes. Before describing our model formally, we explain its differences from that of [DKXY02].

First, in our model the home base updates its internal state at the end of each time period (in addition to sending the update information to the signer). Second, we also provide for a special refresh procedure (akin to proactivization): if a refresh is run after a compromise of one of the modules but before the compromise of the other, the information the adversary learned during the compromise becomes essentially useless, and the system remains secure (except, in the case of signer compromise, for the current time period). Moreover, because our refresh involves just one message from the home base to the signer, it can be combined with update and thus run at least every time period.

The adversary in our model is allowed the usual adaptive-chosen-message-and-time-period attack, and, additionally, can obtain the secrets from the home base and the signer for time periods of her choice. Furthermore, the adversary can intercept update and refresh messages of her choice between the base and the signer. Like in [DKXY02], if the adversary only compromises the base (in fact, even if the base is continuously monitored by the adversary from the start), she still cannot forge signatures. Also like in [DKXY02], if the adversary compromises the signer, then she can forge signatures only for the periods for which the secrets were obtained (either directly via signer compromise, or by combination of signer compromise and interception of some update and refresh messages).

In contrast to [DKXY02], however, our model tolerates multiple compromises of both base and signer (in arbitrary order), as long as there is a refresh between any compromise of the different modules. Moreover, the scheme still remains forward-secure, even if there is no such refresh between some compromises of the two modules.

We treat all compromises in one definition, as opposed to separately defining security against different kinds of compromises. This allows us to precisely specify the security requirements when different types of compromises (base, signer, update messages) are combined. This is in contrast to the key-insulated definitions of [DKXY02], where compromises of the base key are considered in isolation, and compromises of key update messages are reduced to compromises of pairs of consecutive time periods[5].

The definitions below are given in the standard model, but can easily incorporate random oracles (used in our proofs).

2.1   Functional Definition

We first define the functionality of the various components of the system; the security definition is given in the subsequent section. Recall that the system's secret keys may be modified in two different ways, called *update* and *refresh*. Updates change the secrets from one time period to the next (e.g. from one day to the next), changing also the period number in the signatures. In contrast, refreshes affect only the internal secrets and messages of the system, and are transparent to the verifier.

Thus we use notation $SK_{t.r}$ for secret key SK, where t is the time period (the number of times the key has been updated) and r is the "refresh number" (the number of times the key has been refreshed since the last update). We say $t.r = t'.r'$ when $t = t'$ and $r = r'$. Similarly, we say $t.r < t'.r'$ when either $t < t'$ or $t = t'$ and $r < r'$. We follow the convention of [BM99], which requires key update immediately after key generation in order to obtain the keys for $t = 1$ (this is done merely for notational convenience, in order to make the number of time periods T equal to the number of updates, and need not affect the efficiency of an actual implementation). We also require key refresh immediately after key update in order to obtain keys for $r = 1$ (this is also done for convenience, and need not affect efficiency of an actual implementation; in particular, the update

---

[5] With respect to key update information, [DKXY02] define a scheme as having "secure key updates" if key update information sent for time period i can be computed from the signer's keys for time period i and $i - 1$. We find this requirement to be both too strong and too weak. It is too strong because it is quite possible that, while key update information cannot be *computed* from the signer's keys, it is *no more useful* than the two consecutive signer's keys. It is too weak, because it does not rule out the possibility for the adversary to forge signatures for two consecutive time periods if key update information is compromised. In fact, in [DKXY02], if the number of signer compromises that the scheme resists is limited to c, then number of update information exposures is limited to only $c/2$.

and refresh information that the base sends to the signer can be combined into a single message).

Definition 1. *A* ⟨signer-base⟩ *key-evolving signature scheme is a septuple of probabilistic polynomial-time algorithms* (*Gen, Sign, Ver; US,UB; RB, RS*)[6]:

1. *Gen, the key generation algorithm*[7].
   In: *security parameter(s) (in unary), the total number T of time periods*
   Out: *initial signer key $SKS_{0.0}$, initial home base key $SKB_{0.0}$, and the public key PK.*
2. *Sign, the signing algorithm.*
   In: *current signer key $SKS_{t.r}$, message m*
   Out: *signature (t, sig) on m for time period t*
3. *Ver, the verifying algorithm*
   In: *message m, signature (t, sig) and public key PK*
   Out: *"valid" or "invalid" (as usual, signatures generated by Sign must verify as "valid")*
4. *UB, the base key update algorithm*
   In: *current base key $SKB_{t.r}$*
   Out: *new base key $SKB_{(t+1).0}$ and the key update message SKU_t*
5. *US, the signer update algorithm*
   In: *current signer secret key $SKS_{t.r}$ and the key update message SKU_t*
   Out: *new signer secret key $SKS_{(t+1).0}$*
6. *RB, the base key refresh algorithm*
   In: *current base key $SKB_{t.r}$*
   Out: *new base key $SKB_{t.(r+1)}$ and the corresponding key refresh message $SKR_{t.r}$*
7. *RS, the signer refresh algorithm*
   In: *current signer key $SKS_{t.r}$ and the key refresh message $SKR_{t.r}$*
   Out: *new signer key $SKS_{t.(r+1)}$ (corresponding to the base key $SKB_{t.(r+1)}$)*

Note that this definition implies that messages are processed by the signer in the same order in which they are generated by the base.

DIFFERENCES FROM PRIOR NOTIONS. If only *Gen, Sign, Ver* are used, then $t.r$ and *SKB* can be ignored in these algorithms, and the above functional definition becomes that of an ordinary signature scheme.

---

[6] Intuitively (and quite roughly), the first three correspond to the ordinary signatures; the first four correspond to forward-secure ones, the first five (with some restrictions) correspond to key-insulated ones; and all seven are needed to provide the full power of the intrusion-resilient signatures.

[7] As opposed to the other algorithms below, which are meant to be run by a single module (signer, verifier or base), it may be useful to implement the key generation algorithm as distributed between the signer and the home base modules, in such a way that corruption even during key generation does not fully compromise the scheme. Alternatively, key generation may be run by a trusted third party. For simplicity, we postpone this discussion until Section 3, where we propose a practical intermediate solution.

Relaxing the above restrictions to also allow the use of $\mathcal{US}$ (while setting $SKU_t = 1$ for all t), extends the definition to that of forward-secure signatures (or a "key-evolving" scheme [BM99], to be more precise).

Functional definition of a "key-insulated" signature scheme [DKXY02] is obtained by further relaxing the restrictions to allow the use of $\mathcal{UB}$ as well (and thus removing $SKU_t = 1$ restriction), but restricting $SKB_t = \langle SKB, t \rangle$ for some secret SKB and for every period t (i.e. the base secret does not change).

Finally, our model is obtained by removing the remaining restrictions: allowing the base secret to vary and using $\mathcal{RB}, \mathcal{RS}$.

2.2 Security Definition

In order to formalize security, we need a notation for the number of refreshes in each period: Let $RN(t)$ denote the number of times the keys are refreshed in the period t: i.e., there will be $RN(t)+1$ instances of signer and base keys. Recall that each update is immediately followed by a refresh; thus, keys with refresh index 0 are never actually used. RN is used only for notational convenience: it need not be known; security is defined below in terms of RN (among other parameters).

Now, consider all the keys generated during the entire run of the signature scheme. They can be generated by the following "thought experiment" (we do not need to *actually* run it – it is used just for definitions).

Experiment Generate-Keys(k, T, RN)
  $t \leftarrow 0; r \leftarrow 0$
  $(SKS_{t.r}, SKB_{t.r}, PK) \leftarrow Gen(1^k, T)$
  for $t = 1$ to T
    $(SKB_{t.0}, SKU_{t-1}) \leftarrow \mathcal{UB}(SKB_{(t-1).r})$
    $SKS_{t.0} \leftarrow \mathcal{US}(SKS_{(t-1).r}, SKU_{t-1})$
    for $r = 1$ to $RN(t)$
      $(SKB_{t.r}, SKR_{t.(r-1)}) \leftarrow \mathcal{RB}(SKB_{t.(r-1)})$
      $SKS_{t.r} \leftarrow \mathcal{RS}(SKS_{t.(r-1)}, SKR_{t.(r-1)})$

Let SKS^*, SKB^*, SKU^* and SKR^* be the sets consisting of, respectively, signer and base keys and update and refresh messages, generated during the above experiment. We want these sets to contain all the secrets that can be directly stolen (as opposed to computed) by the adversary. Thus, we omit from these sets the keys $SKS_{t.0}, SKB_{t.0}$ for $0 \leq t \leq T$, SKU_0 and $SKR_{1.0}$, which are never actually stored or sent (because key generation is immediately followed by update, and each update is immediately followed by refresh). Note that $SKR_{t.0}$ for $t > 1$ is used (it is sent together with SKU_{t-1} to the signer), and thus is included into SKR^*.

To define security, let F, the adversary (or "forger"), be a probabilistic polynomial-time oracle Turing machine with the following oracles:

- *Osig*, the signing oracle (constructed using SKS^*), which on input (m, t, r)
  ($1 \leq t \leq T$, $1 \leq r \leq RN(t)$) outputs $Sign(SKS_{t.r}, m)$

- *Osec*, the key exposure oracle (based on the sets SKS^*, SKB^*, SKU^* and SKR^*), which
  1. on input ("s", $t.r$) for $1 \leq t \leq T, 1 \leq r \leq RN(t)$ outputs $SKS_{t.r}$;
  2. on input ("b", $t.r$) for $1 \leq t \leq T, 1 \leq r \leq RN(t)$ outputs $SKB_{t.r}$;
  3. on input ("u", t) for $1 \leq t \leq T - 1$ outputs SKU_t and $SKR_{t+1}.0$; and
  4. on input ("r", $t.r$) for $1 \leq t \leq T, 1 \leq r < RN(t)$, outputs $SKR_{t.r}$.

Queries to *Osec* oracle correspond to intrusions, resulting in the corresponding secrets exposures. Exposure of an update key SKU_{t-1} automatically exposes the subsequent refresh key $SKR_{t.0}$, because they are sent together in one message.

Adversary's queries to *Osig* and *Osec* must have the $t.r$ values within the appropriate bounds. It may be reasonable to require the adversary to "respect erasures" and prohibit a value that should have been erased from being queried (or used for signature computation). However, this restriction is optional, and in fact we do not require it here. Note, that respecting erasures is local to signer and base and is different from requiring any kind of global synchronization. For any set of valid key exposure queries Q, time period $t \geq 1$ and refresh number r, $1 \leq r \leq RN(t)$, we say that key $SKS_{t.r}$ is Q-*exposed*:

[**directly**] if ("s", $t.r$) $\in Q$; or
[**via refresh**] if $r > 1$, ("r", $t.(r-1)$) $\in Q$, and $SKS_{t.(r-1)}$ is Q-exposed; or
[**via update**] if $r = 1$, ("u", $t-1$) $\in Q$, and $SKS_{(t-1).RN(t-1)}$ is Q-exposed.

Replacing SKS with SKB throughout the above definition yields the definition of base key exposure (or more precisely, of $SKB_{t.r}$ being Q-*exposed*). Both definitions are recursive, with direct exposure as the base case.

Clearly, exposure of a signer key $SKS_{t.r}$ for the given t and any r enables the adversary to generate legal signatures for this period t. Similarly, simultaneous exposure of both base and signer keys ($SKB_{t.r}, SKS_{t.r}$, for some t, r) allows the adversary to run the algorithms of definition 1 to generate valid signatures for any messages for all later periods $t' \geq t$.

Thus, we say that the scheme is (t, Q)-*compromised*, if either

- $SKS_{t.r}$ is Q-exposed for some r, $1 \leq r \leq RN(t)$; or
- $SKS_{t'.r}$ and $SKB_{t'.r}$ are both Q-exposed for some $t' < t$.

In other words, a particular time period has been rendered insecure if either the signer was broken into during that time period, or, during a previous time period, the signer and the base were compromised without a refresh in between. Note that update and refresh messages by themselves do not help the adversary in our model – they only help when combined, in unbroken chains, with signer or base keys[8]. If the scheme is (j, Q)-compromised, then clearly adversary, pos-

---

[8] Interestingly, and perhaps counter-intuitively, a secret that has not been exposed might still be deduced by adversary. For example, it may be possible in some implementations to compute $SKB_{t.r}$ from $SKB_{t.r+1}$ and $SKR_{t.r}$ (similarly for update and for the signer). The only case we cannot allow – in order to stay consistent with our definition – is that of adversary computing $SKS_{t.r}$ without Q-exposing some $SKS_{t.r'}$. So, it may be possible to compute $SKS_{t.r}$ from $SKS_{t.r+1}$ and $SKR_{t.r}$, but not from $SKS_{t+1.0}$ and $SKU_{t.r}$ (even if $r = RN(t)$).

sessing the secrets returned by *Osec* in response to queries in Q, can generate signatures for the period t.

The following experiment captures adversary's functionality. Intuitively, adversary succeeds if she generates a valid signature without "cheating": not obtaining this signature from *Osig*, asking only legal queries (e.g. no out of bounds queries), and not compromising the scheme for the given time period. We call this adversary "adaptive" because she is allowed to decide which keys and signatures to query based on previous answers she receives.

Experiment **Run-Adaptive-Adversary**(F, k, T, RN)
> **Generate-Keys**(k, T, RN)
> $(m, j, sig) \leftarrow F^{Osig, Osec}(1^k, T, PK, RN)$
> Let Q be the set of key exposure queries F made to *Osec*;
> if $Ver(m, j, sig) =$"invalid" or (m, j) was queried by F to *Osig* or there
>
>> was an illegal query or the scheme is (j, Q)-compromised
>
> **then return** 0
> **else return** 1

We now define security for the intrusion resilient signature schemes.

Definition 2. *Let* SiBIR$[k, T, RN]$ *be a $\langle$signer-base$\rangle$ key-evolving scheme with security parameter k, number of time periods T, and table RN of T refresh numbers. For adversary F, define adversary success function as*

$$\mathbf{Succ}^{IR}(F, \mathsf{SiBIR}[k, T, RN]) \overset{def}{=} \Pr[\mathbf{Run\text{-}Adaptive\text{-}Adversary}(F, k, T, RN) = 1].$$

Let insecurity function $\mathbf{InSec}^{IR-adaptive}(\mathsf{SiBIR}[k, T, RN], \tau, q_{sig})$ *be the maximum of* $\mathbf{Succ}^{IR}(F, \mathsf{SiBIR}[k, T, RN])$ *over all adaptive adversaries F that run in time at most τ and ask at most q_{sig} signature queries.*

Finally, SiBIR$[k, T, RN]$ *is $(\tau, \epsilon, q_{sig})$-intrusion-resilient if*

$$\mathbf{InSec}^{IR-adaptive}(\mathsf{SiBIR}[k, T, RN], \tau, q_{sig}) < \epsilon.$$

Although the notion of (j, Q)-compromise depends only the *set* Q, it is important how Q is generated by the adversary. Allowing the adversary to decide her queries based on previous answers gives her potentially more power.

While we do not see an attack on our scheme in Sec. 3 by a fully adaptive adversary, the proof for such a strong adversary seems elusive. Instead, we consider two types of slightly restricted adversaries. The first type, which we call *partially adaptive*, is allowed to adaptively choose queries to *Osig*, but not to *Osec*. To be precise, the experiment **Run-Adaptive-Adversary** is modified **Run-Partially-Adaptive-Adversary** by requiring F to output the set Q of *all* her *Osec* queries before she makes any of them. The rest of the definitions remain the same, giving us $\mathbf{InSec}^{IR-partially-adaptive}$ instead of $\mathbf{InSec}^{IR-adaptive}$.

The second type of restricted adversary is *partially synchronous*. She may select all of her queries adaptively, but is not allowed to go back in time "too far." Specifically, upon querying any key in time period t, she is not allowed to query keys in time period $t-2$ (note that the choice of 2 is arbitrary and can be replaced with another constant). This is a reasonable assumption in practice, essentially saying that the base, the network and the signer can be at most one time period apart at any given time. In order to formally define security against such an adversary, we simply expand the definition of "illegal queries" to encompass the above restriction. The rest of the definitions remain the same, giving us $\mathbf{InSec}^{\mathrm{IR-partially-synchronous}}$ instead of $\mathbf{InSec}^{\mathrm{IR-adaptive}}$.

3   Intrusion-Resilient Scheme: Construction

Our scheme, which we call SiBIR1, is based on the [IR01] forward-secure signature scheme (which will call $\mathsf{FSIG}^{\mathsf{IR}}$). In turn, $\mathsf{FSIG}^{\mathsf{IR}}$ is based on the Guillou-Quisquater [GQ88] ordinary signature scheme. In fact, the [IR01] forward-secure scheme can be obtained from our scheme by simply eliminating the base, and setting all the messages that the signer expects equal to 1.

The scheme utilizes two security parameters, l and k. Let $H : \{0,1\}^* \rightarrow \{0,1\}^l$ be a hash function (modeled in the security proof as a random oracle). In the interests of conciseness, we do not present the rationale behind $\mathsf{FSIG}^{\mathsf{IR}}$ here. We do, however, recall how keys are generated and updated in the $\mathsf{FSIG}^{\mathsf{IR}}$ scheme (utilizing our own notation, rather than notation used in [IR01], in the interests of clarity)

KEYS IN $\mathsf{FSIG}^{\mathsf{IR}}$. Both the public and the secret keys contain a modulus $n = p_1 p_2$, where p_1 and p_2 are $(k/2)$-bit *safe* primes: $p_i = 2q_i + 1$ such that q_i are odd primes (such q_i, satisfying $2q_i + 1$ is prime, are known as *Sophie Germain primes*)[9]. The public key also contains a value $v \in Z_n^*$. For each time period t, there is a corresponding $(l+1)$-bit exponent e_t that the signer can easily compute (we require all the exponents to be relatively prime; then they need not be stored in the public key, but rather the appropriate e_t is included in each signature – see [IR01] for further details).

Messages during time period t are signed using the secret value $\widehat{s}_t$, such that $\widehat{s}_t^{e_t} \equiv 1/v \pmod{n}$.

The factorization of n must be erased after key generation in order to achieve forward security. Knowledge of secrets $\widehat{s}_t, \widehat{s}_{t+1}, \ldots, \widehat{s}_T$, is equivalent (by the so-called "Shamir's trick" [Sha83] – see Proposition1 in [IR01]) to knowledge of the root of $1/v$ of degree $e_{[t,T]} \stackrel{def}{=} e_t \cdot e_{t+1} \cdot \ldots \cdot e_T$. Call this root $\widehat{s}_{[t,T]}$: then $\widehat{s}_{[t,T]}^{e_{[t,T]}} \equiv 1/v \pmod{n}$.

At key generation we actually select $\widehat{s}_{[1,T]}$ at random, and compute v as $v \leftarrow 1/\widehat{s}_{[1,T]}^{e_{[1,T]}} \bmod n$.

Subsequently (just before each time period t), $\widehat{s}_{[t,T]}$ is updated as follows: $\widehat{s}_{[t+1,T]} \leftarrow \widehat{s}_{[t,T]}^{e_t} \bmod n$; and the "current" signing secret is computed as $\widehat{s}_t \leftarrow \widehat{s}_{[t,T]}^{e_{[t+1,T]}} \bmod n$.

---

[9] See [CS00] for an excellent discussion on efficient generation of safe primes and short primes.

---

algorithm SiBIR1.$Gen(k, l, T)$
    Generate a modulus n:
        Generate random ($\lceil k/2 \rceil - 1$)-bit primes q_1, q_2 s.t. $p_i = 2q_i + 1$ are both prime
        $n \leftarrow p_1 p_2$
    Generate exponents:
        Generate primes e_i s.t. $2^l(1 + (i-1)/T) \leq e_i < 2^l(1 + i/T)$ for $i = 1, 2, \ldots, T$
        (Some seed $\mathcal{E}$ can be used with H to generate these $e_1, \ldots, e_T$.
        This $\mathcal{E}$ might need to be stored for later regeneration of $e_1, \ldots, e_T$.)
    $s_{[1,T]} \xleftarrow{R} Z_n^*; \ SKS_0 \leftarrow (0, T, n, \emptyset, s_{[1,T]}, \emptyset, \mathcal{E}) \ \% \ \emptyset\text{'s will be filled in in } US$
    $b_{[1,T]} \xleftarrow{R} Z_n^*; \ SKB_0 \leftarrow (0, T, n, b_{[1,T]}, \mathcal{E})$
    $v \leftarrow 1/(s_{[1,T]} b_{[1,T]})^{e_1 \cdots e_T} \bmod n; \ PK \leftarrow (n, v, T)$
    return (SKS_0, SKB_0, PK)

Fig. 1. Key generation. Refresh index on the keys is omitted to simplify notation.

KEY GENERATION AND UPDATE IN SiBIR1. We use essentially the same keys in SiBIR1. However, in order to achieve intrusion-resilience, $\widehat{s}_{[t,T]}$ is never stored explicitly. Rather, it is shared multiplicatively between the signer and the base. The signer stores $s_{[t,T]}$ and the base stores $b_{[t,T]}$, such that $\widehat{s}_{[t,T]} = s_{[t,T]} b_{[t,T]}$. This multiplication is never explicitly performed: instead, the signer computes $s_t = s_{[t,T]}^{e_{[t+1,T]}}$, the base computes $b_t = b_{[t,T]}^{e_{[t+1,T]}}$, and the two values are multiplied together to obtain $\widehat{s}_t$.

Following the conventions that key generation is immediately followed by key update, the first signer secret key contains blanks for $\widehat{s}_0$ and e_0. We note that, in actual implementation, it will be more efficient to combine the first key generation and update.

Also, following the convention that the only "storage" available to the base and the signer is the secret key, we store some values in the secret key that need not really be secret, such as the current time period and the information needed to regenerate the e_t values. The only values that the signer needs to keep secret are $s_{t,T}$ and $\widehat{s}_t$; the only values that the base needs to keep secret are $b_{t,T}$.

Finally, note that key generation and update algorithms do not affect the refresh index, so we omit it in Figures 1 and 2 in order to simplify notation.

DISTRIBUTING KEY GENERATION. Most of the key generation algorithm can be easily split between the signer and the base. Namely, once the shared modulus n is generated and given to both parties (without factoring), the base can generate $b_{[1,T]}$ on its own, and the signer can generate $s_{[1,T]}$ on its own, as well. Both parties can then generate "shares" $b_{[1,T]}^{e_{[1,T]}}$ and $s_{[1,T]}^{e_{[1,T]}}$ that can be combined to compute the public key. The shares themselves can be made public without adversely affecting security. Thus, the amount of cooperation required during key generation is minimal.

The same modulus n can be used by multiple signature schemes. In particular, our signature scheme can be made identity-based if a third party is trusted to take roots modulo n of the identity v.

---

algorithm SiBIR1.$\mathcal{UB}(SKB_t)$
  Let $SKB_t = (t < T, T, n, b_{[t+1,T]}, \mathcal{E})$
  Regenerate $e_{t+1}, \ldots, e_T$ using $\mathcal{E}$
  $b_{t+1} \leftarrow b_{[t+1,T]}^{e_{t+2} \cdots e_T} \bmod n;\ b_{[t+2,T]} \leftarrow b_{[t+1,T]}^{e_{t+1}} \bmod n$
  return $(SKB_{t+1} = (t+1, T, n, b_{[t+2,T]}, \mathcal{E}), SKU_t = b_{t+1})$

---

algorithm SiBIR1.$\mathcal{US}(SKS_t, SKU_t)$
  Let $SKS_t = (t < T, T, n, \widehat{s}_t, s_{[t+1,T]}, e_t, \mathcal{E});\ SKU_t = b_{t+1}$
  Regenerate $e_{t+1}, \ldots, e_T$ using $\mathcal{E}$
  $s_{t+1} \leftarrow s_{[t+1,T]}^{e_{t+2} \cdots e_T} \bmod n;\ s_{[t+2,T]} \leftarrow s_{[t+1,T]}^{e_{t+1}} \bmod n$
  $\widehat{s}_{t+1} \leftarrow s_{t+1} b_{t+1} \bmod n$
  return $SKS_{t+1} = (t+1, T, n, \widehat{s}_{t+1}, s_{[t+2,T]}, e_{t+1}, \mathcal{E})$

Fig. 2. Update algorithms. Refresh index on the keys is omitted to simplify notation.

---

algorithm SiBIR1.$\mathcal{RB}(SKB_{t.r})$
  Let $SKB_{t.r} = (t, T, n, b_{[t+1,T]}, \mathcal{E})$
  $R_{t.r} \overset{R}{\leftarrow} Z_n^*$
  $b_{[t+1,T]} \leftarrow b_{[t+1,T]}/R_{t.r}$
  return $(SKB_{t.r+1} = (t, T, n, b_{[t+1,T]}, \mathcal{E}), SKR_{t.r} = R_{t.r})$

---

algorithm SiBIR1.$\mathcal{RS}(SKS_{t.r}, SKR_{t.r})$
  Let $SKS_{t.r} = (t, T, n, \widehat{s}_t, s_{[t+1,T]}, e_t, \mathcal{E});\ SKR_{t.r} = R_{t.r}$
  $s_{[t+1,T]} \leftarrow s_{[t+1,T]} \cdot R_{t.r}$
  return $SKS_{t.r+1} = (t, T, n, \widehat{s}_t, s_{[t+1,T]}, e_t, \mathcal{E})$

Fig. 3. Key refresh algorithms.

REFRESH.   Because the signer and the base share a single value multiplicatively, the refresh algorithm presented in Figure 3 is quite simple: the base divides its share by a random value, and signer multiplies its share by the same value. Recall that each update is immediately followed by refresh (and, in fact, update and refresh information can be sent by the base to the signer in one message).

SIGNING AND VERIFYING.   Figure 4 describes our signature and verification algorithms. They are exactly the same as in the forward-secure signature scheme of [IR01]. Again, we omit the refresh index on the signer's key for ease of notation.

VARIATIONS.   Our scheme can be easily modified (with no or minimum increase in storage requirement!) to "re-charge" the signer for more than one time period at a time. To enable the signer to compute $\widehat{s}_{t_1}, \widehat{s}_{t_1+1}, \ldots \widehat{s}_{t_2}$, the base simply needs to send the signer $b_{[t_1,t_2]} = b_{[t_1,T]}^{e_{[t_2+1,T]}}$. In fact, it is easy to extend this method to

---

algorithm SiBIR1.$Sign(M, SK_t)$ %same as IR.$Sign$ in [IR01]
Let $SK_t = (t, T, n, \widehat{s}_t, s_{[t+1,T]}, e_t, \mathcal{E})$
$x \xleftarrow{R} Z_n^*$
$y \leftarrow x^{e_t} \bmod n$
$\sigma \leftarrow H(t, e_t, y, M)$
$z \leftarrow x\widehat{s}_t^\sigma \bmod n$
return (z, σ, t, e_t)

---

algorithm SiBIR1.$Ver(M, PK, (z, \sigma, t, e))$ %same as IR.Ver in [IR01]
Let $PK = (n, v, T)$
if $e \geq 2^l(1 + t/T)$ or $e < 2^l$ or e is even **then return** 0
if $z \equiv 0 \pmod{n}$ **then return** 0
$y' \leftarrow z^e v^\sigma \bmod n$
if $\sigma = H(t, e, y', M)$ **then return** 1 **else return** 0

Fig. 4. Signing and verifying algorithms. Refresh index on the keys is omitted to simplify notation.

non-contiguous time periods. This feature may have interesting applications for delegation (including self-delegation).

Another simple modification of our scheme can yield forward-secure threshold and proactive scheme (similar to, but more efficient than, the scheme of [AMN01]). Efficiency for the verifier and for each of the modules participating in the signing will be essentially the same as for the regular Guillou-Quisquater scheme.

4 Security

4.1 Complexity Assumption

We use a variant of the strong RSA assumption (introduced in [BP97] and [FO97], our variant is identical to the one in [IR01]), which postulates that it is hard to compute *any* root of a fixed value modulo a composite integer. More precisely, the strong RSA assumption states that it is intractable, given n that is a product of two primes and a value α in Z_n^*, to find $\beta \in Z_n^*$ and $r > 1$ such that $\beta^r = \alpha$.

In our version, we restrict ourselves to the moduli that are products of so-called "safe" primes (a safe prime is one of the form $2q + 1$, where q itself is a prime). Note that, assuming safe primes are frequent, this restriction does not strengthen the assumption. Second, we upperbound the permissible values or r by 2^{l+1}, where l is a security parameter for our scheme (in an implementation, l will be significantly shorter than the length k of the modulus n).

More formally, let A be an algorithm. Consider the following experiment.

Experiment Break-Strong-RSA(k, l, A)

Randomly choose two primes q_1 and q_2 of length $\lceil k/2 \rceil - 1$ each
such that $2q_1 + 1$ and $2q_2 + 1$ are both prime.

$p_1 \leftarrow 2q_1 + 1; p_2 \leftarrow 2q_2 + 1; n \leftarrow p_1 p_2$

Randomly choose $\alpha \in Z_n^*$.

$(\beta, r) \leftarrow A(n, \alpha)$

If $1 < r \le 2^{l+1}$ and $\beta^r \equiv \alpha$ (mod n) then **return** 1 **else return** 0

Let $\mathbf{Succ}^{\mathrm{SRSA}}(A, k, l) = \Pr[\text{Break-Strong-RSA}(k, l, A) = 1]$. Let the "insecurity function" $\mathbf{InSec}^{\mathrm{SRSA}}(k, l, \tau)$ be the maximum of $\mathbf{Succ}^{\mathrm{SRSA}}(A, k, l)$ over all the adversaries A who run in expected time at most τ. Our assumption is that $\mathbf{InSec}^{\mathrm{SRSA}}(k, l, \tau)$, for τ polynomial in k, is negligible in k. The smaller the value of l, of course, the weaker the assumption.

In fact, for a sufficiently small l, our assumption follows from a variant of the fixed-exponent RSA assumption. Namely, assume that there exists a constant ϵ such that, for every r, the probability of computing, in expected time τ, an r-th root of a random integer modulo a k-bit product of two safe primes, is at most 2^{-k^ϵ}. Then, $\mathbf{InSec}^{\mathrm{SRSA}}(k, l, \tau) < 2^{l+1-k^\epsilon}$, which is negligible if $l = o(k^\epsilon)$.

4.2   Security Proof

Our security proof is more complex that the one of [IR01], although the two proofs are quite similar. Both are based on the forking lemma of [PS96].

Theorem 1. *For any* τ, q_{sig}, *and* q_{hash},

$$\mathbf{InSec}^{\mathrm{IR-partially-adaptive}}(\mathsf{SiBIR1}[k, l, T, RN]; t, q_{\mathrm{sig}}, q_{\mathrm{hash}}) \le$$

$$T\sqrt{(q_{\mathrm{hash}} + 1)\mathbf{InSec}^{\mathrm{SRSA}}(k, l, \tau')} + 2^{-l+1}T(q_{\mathrm{hash}} + 1) + 2^{2-k}q_{\mathrm{sig}}(q_{\mathrm{hash}} + 1),$$

where $\tau' = 4\tau + O(lT(l^2T^2 + k^2))$.

A similar theorem also holds for partially synchronous adversary. The proofs of these theorems will be given in the full version of the paper.

4.3   Active Attacks

Because information flows only from the base to the signer, the adversary's only possible active attack is to send a bad *SKR* or *SKU* value to the signer. An active attacker can thus always prevent signatures from being issued. While our definition does not consider active attacks for the sake of simplicity, in our implementation in Section 3, the active adversary cannot do anything worse that merely sabotage the system. It is easy to show that, in terms of forging new signatures, its powers are no greater than those of a passive attacker who merely obtains *SKR* and *SKU* values.

Acknowledgements

We thank Nenad Dedic, Scott Russell, Yael Tauman and the anonymous referees for helpful comments.

References

[AMN01] Michel Abdalla, Sara Miner, and Chanathip Namprempre. Forward-secure threshold signature schemes. In David Naccache, editor, *Progress in Cryptology - CT-RSA 2001, Lecture Notes in Computer Science* 2020, 2001.

[And97] Ross Anderson. Invited lecture. Fourth Annual Conference on Computer and Communications Security, ACM, 1997.

[BM99]   Mihir Bellare and Sara Miner. A forward-secure digital signature scheme. In Michael Wiener, editor, *Advances in Cryptology - CRYPTO '99*, volume 1666 of *Lecture Notes in Computer Science*, pages 431–448. Springer-Verlag, 15–19 August 1999. Revised version is available from http://www.cs.ucsd.edu/~mihir/.

[BP97]   Niko Barić and Birgit Pfitzmann. Collision-free accumulators and fail-stop signature schemes without trees. In Walter Fumy, editor, *Advances in Cryptology - EUROCRYPT 97*, volume 1233 of *Lecture Notes in Computer Science*, pages 480–494. Springer-Verlag, 11–15 May 1997.

[CS00]   Ronald Cramer and Victor Shoup. Signature schemes based on the strong RSA assumption. *ACM Transactions on Information and System Security*, 3(3):161–185, 2000.

[DF89]   Yvo Desmedt and Yair Frankel. Threshold cryptosystems. In G. Brassard, editor, *Advances in Cryptology - CRYPTO '89*, volume 435 of *Lecture Notes in Computer Science*, pages 307–315. Springer-Verlag, 1990.

[DKXY]   Yevgeniy Dodis, Jonathan Katz, Shouhuai Xu, and Moti Yung. Strong key-insulated signature schemes. Unpublished Manuscript.

[DKXY02] Yevgeniy Dodis, Jonathan Katz, Shouhuai Xu, and Moti Yung. Key-insulated public key cryptosystems. In Lars Knudsen, editor, *Advances in Cryptology - EUROCRYPT 2002*, Lecture Notes in Computer Science. Springer-Verlag, 28 April–2 May 2002.

[FO97]   Eiichiro Fujisaki and Tatsuaki Okamoto. Statistical zero knowledge protocols to prove modular polynomial relations. In Burton S. Kaliski Jr., editor, *Advances in Cryptology - CRYPTO '97*, volume 1294 of *Lecture Notes in Computer Science*, pages 16–30. Springer-Verlag, 17–21 August 1997.

[GGM00] Irene Gassko, Peter Gemmell, and Philip MacKenzie. Efficient and fresh certication, 2000.

[GMR88] Shafi Goldwasser, Silvio Micali, and Ronald L. Rivest. A digital signature scheme secure against adaptive chosen-message attacks. *SIAM Journal on Computing*, 17(2):281–308, April 1988.

[GMW87] Oded Goldreich, Silvio Micali, and Avi Wigderson. How to play any mental game or a completeness theorem for protocols with honest majority. In *Proceedings of the Nineteenth Annual ACM Symposium on Theory of Computing*, pages 218–229, New York City, 25–27 May 1987.

[GPR98]   Oded Goldreich, Birgit Pfitzmann, and Ronald L. Rivest. Self-delegation with controlled propagation – or – what if you lose your laptop. In Hugo Krawczyk, editor, *Advances in Cryptology – CRYPTO '98*, volume 1462 of *Lecture Notes in Computer Science*, pages 153–168. Springer-Verlag, 23–27 August 1998.

[GQ88]   Louis Claude Guillou and Jean-Jacques Quisquater. A "paradoxical" indentity-based signature scheme resulting from zero-knowledge. In Shafi Goldwasser, editor, *Advances in Cryptology – CRYPTO '88*, volume 403 of *Lecture Notes in Computer Science*, pages 216–231. Springer-Verlag, 1990.

[HJJ+97]   Amir Herzberg, Markus Jakobsson, Stanisław Jarecki, Hugo Krawczyk, and Moti Yung. Proactive public key and signature systems. In *Fourth ACM Conference on Computer and Communication Security*, pages 100–110. ACM, April 1–4 1997.

[IR01]   Gene Itkis and Leonid Reyzin. Forward-secure signatures with optimal signing and verifying. In Joe Kilian, editor, *Advances in Cryptology – CRYPTO 2001*, volume 2139 of *Lecture Notes in Computer Science*, pages 332–354. Springer-Verlag, 19–23 August 2001.

[Kat02]   Jonathan Katz. A forward-secure public-key encryption scheme. Cryptology ePrint Archive, Report 2002/60, 2002. http://eprint.iacr.org/.

[LR00]   Anna Lysyanskaya and Ron Rivest. Bepper-based signatures. Presented by Rivest at the CIS seminar at MIT, 27 October 2000.

[Mic96]   Silvio Micali. Efficient certificate revocation. Technical Report MIT/LCS/TM-542b, Massachusetts Institute of Technology, Cambridge, MA, March 1996.

[MR01a]   Philip D. MacKenzie and Michael K. Reiter. Delegation of cryptographic servers for capture-resilient devices. In *Eighth ACM Conference on Computer and Communication Security*, pages 10–19. ACM, November 5–8 2001.

[MR01b]   Philip D. MacKenzie and Michael K. Reiter. Networked cryptographic devices resilient to capture. In *IEEE Symposium on Security and Privacy*, pages 12–25, 2001.

[OY91]   Rafail Ostrovsky and Moti Yung. How to withstand mobile virus attacks. In *10-th Annual ACM Symp. on Principles of Distributed Computing*, 1991.

[PS96]   David Pointcheval and Jacques Stern. Security proofs for signature schemes. In Ueli Maurer, editor, *Advances in Cryptology – EURO-CRYPT 96*, volume 1070 of *Lecture Notes in Computer Science*, pages 387–398. Springer-Verlag, 12–16 May 1996.

[Riv98]   Ronald L. Rivest. Can we eliminate certificate revocation lists? In Rafael Hirschfeld, editor, *Financial Cryptography*, volume 1465 of *Lecture Notes in Computer Science*. Springer-Verlag, 1998.

[Sha83]   Adi Shamir. On the generation of cryptographically strong pseudorandom sequences. *ACM Transactions on Computer Systems*, 1(1):38–44, 1983.

[Yao82]   A.C. Yao. Protocols for secure computations. In *23rd Annual Symposium on Foundations of Computer Science*, pages 160–164, Chicago, Illinois, 3–5 November 1982. IEEE.

Cryptanalysis of Stream Ciphers
with Linear Masking

Don Coppersmith, Shai Halevi, and Charanjit Jutla

IBM T. J. Watson Research Center, NY, USA
{copper,shaih,csjutla}@watson.ibm.com

Abstract. We describe a cryptanalytical technique for distinguishing some stream ciphers from a truly random process. Roughly, the ciphers to which this method applies consist of a "non-linear process" (say, akin to a round function in block ciphers), and a "linear process" such as an LFSR (or even fixed tables). The output of the cipher can be the linear sum of both processes. To attack such ciphers, we look for any property of the "non-linear process" that can be distinguished from random. In addition, we look for a linear combination of the linear process that vanishes. We then consider the same linear combination applied to the cipher's output, and try to find traces of the distinguishing property.

In this report we analyze two specific "distinguishing properties". One is a linear approximation of the non-linear process, which we demonstrate on the stream cipher SNOW. This attack needs roughly 2^{95} words of output, with work-load of about 2^{100}. The other is a "low-diffusion" attack, that we apply to the cipher Scream-0. The latter attack needs only about 2^{43} bytes of output, using roughly 2^{50} space and 2^{80} time.

Keywords: Hypothesis testing, Linear cryptanalysis, Linear masking, Low-Diffusion attacks, Stream ciphers.

1 Introduction

A stream cipher (or pseudorandom generator) is an algorithm that takes a short random string, and expands it into a much longer string, that still "looks random" to adversaries with limited resources. The short input string is called the seed (or key) of the cipher, and the long output string is called the output stream (or key-stream). Although one could get a pseudorandom generator simply by iterating a block cipher (say, in counter mode), it is believed that one could get higher speeds by using a "special purpose" stream cipher.

One approach for designing such fast ciphers, is to use some "non-linear process" that may resemble block cipher design, and to hide this process using linear masking. A plausible rationale behind this design, is that the non-linear process behaves roughly like a block cipher, so we expect its state at two "far away" points in time to be essentially uncorrelated. For close points, on the other hand, it can be argued they are masked by independent parts of the linear process, and so again they should not be correlated.

M. Yung (Ed.): CRYPTO 2002, LNCS 2442, pp. 515–532, 2002.
© Springer-Verlag Berlin Heidelberg 2002

Some examples of ciphers that use this approach include SEAL [18] and Scream [2], where the non-linear process is very much like a block cipher, and the output from each step is obtained by adding together the current state of the non-linear process and some entries from fixed (or slowly modified) secret tables. Other examples are PANAMA [4] and MUGI [21], where the linear process (called buffer) is an LFSR (Linear Feedback Shift Register), which is used as input to the non-linear process, rather than to hide the output. Yet another example is SNOW [5], where the linear LFSR is used both as input to the non-linear finite state machine, and also to hide its output.

In this work we describe a technique that can be used to distinguish such ciphers from random. The basic idea is very simple. We first concentrate on the non-linear process, looking for a characteristic that can be distinguished from random. For example, a linear approximation that has noticeable bias. We then look at the linear process, and find some linear combination of it that vanishes. If we now take the same linear combination of the output stream, then the linear process would vanish, and we are left with a sum of linear approximations, which is itself a linear approximation. As we show below, this technique is not limited to linear approximations. In some sense, it can be used with "any distinguishing characteristic" of the non-linear process. In this report we analyze in details two types of "distinguishing characteristics", and show some examples of its use for specific ciphers.

Perhaps the most obvious use of this technique, is to devise linear attacks (and indeed, many such attacks are known in the literature). This is also the easiest case to analyze. In Section 4 we characterize the statistical distance between the cipher and random as a function of the bias of the original approximation of the non-linear process, and the *weight distribution* of a linear code related to the linear process of the cipher.

Another type of attacks uses the low diffusion in the non-linear process. Namely, some input/output bits of this process depend only on very few other input/output bits. For this type of attacks, we again analyze the statistical distance, as a function of the number of bits in the low-diffusion characteristic. This analysis is harder than for the linear attacks. Indeed, here we do not have a complete characterization of the possible attacks of this sort, but only an analysis for the most basic such attack.

We demonstrate the usefulness of our technique by analyzing two specific ciphers. One is the cipher SNOW [5], for which we demonstrate a linear attack, and the other is the variant Scream-0 of the stream cipher Scream [2], for which we demonstrate a low-diffusion attack.

1.1 Relation to Prior Work

Linear analyses of various types are the most common tool for cryptanalyzing stream ciphers. Much work was done on LFSR-based ciphers, trying to discover the state of the LFSRs using correlation attacks (starting from Meier and Staffelbach [17], see also, e.g., [14, 13]). Golić [9, 10] devised linear models (quite similar to our model of linear attacks) that can be applied in principle to any stream cipher. He then used them to analyze many types of ciphers (including, for example, a linear distinguisher for RC4 [11]). Some examples of linear distinguishers for LFSR based ciphers, very similar to our analysis of SNOW, are [1, 6], among others. Few works used also different cryptanalytical tools. Among them are the distinguishers for SEAL [12, 7] and for RC4 [8].

The main contribution of the current work is in presenting a simple framework for distinguishing attacks. This framework can be applied to many ciphers, and for those ciphers it incorporates linear analysis as a special case, but can be used to devise many other attacks, such as our "low-diffusion attacks". (Also, the attacks on SEAL due to [12] and [7] can be viewed as special cases of this framework.) For linear attacks, we believe that our explicit characterization of the statistical distance (Theorem 1) is new and useful. In addition to the cryptanalytical technique, the explicit formulation of attacks on stream ciphers, as done in Section 3, is a further contribution of this work.

Organization. In Section 2 we briefly review some background material on statistical distance and hypothesis testing. In Section 3 we formally define the framework in which our techniques apply. In Section 4 we describe how these techniques apply to linear attacks, and in Section 5 we show how they apply to low-diffusion attacks.

2   Elements of Statistical Hypothesis Testing

If $\mathcal{D}$ is a distribution over some finite domain X and x is an element of X, then by $\mathcal{D}(x)$ we denote probability mass of x according to $\mathcal{D}$. For notational convenience, we sometimes denote the same probability mass by $\Pr_{\mathcal{D}}[x]$. Similarly, if $S \subseteq X$ then $\mathcal{D}(S) = \Pr_{\mathcal{D}}[S] = \sum_{x \in S} \mathcal{D}(x)$.

Definition 1 (Statistical distance). *Let $\mathcal{D}_1, \mathcal{D}_2$ be two distributions over some finite domain X. The statistical distance between $\mathcal{D}_1, \mathcal{D}_2$, is defined as*

$$|\mathcal{D}_1 - \mathcal{D}_2| \stackrel{\text{def}}{=} \sum_{x \in X} |\mathcal{D}_1(x) - \mathcal{D}_2(x)| = 2 \cdot \max_{S \subseteq X} \mathcal{D}_1(S) - \mathcal{D}_2(S)$$

(We note that the statistical distance is always between 0 and 2.) Below are two useful facts about this measure:

• Denote by $\mathcal{D}^N$ the distribution which is obtained by picking independently N elements $x_1, ..., x_n \in X$ according to $\mathcal{D}$. If $|\mathcal{D}_1 - \mathcal{D}_2| = \epsilon$, then to get $|\mathcal{D}_1^N - \mathcal{D}_2^N| = 1$, the number N needs to be between $\Omega(1/\epsilon)$ and $O(1/\epsilon^2)$. (A proof can be found, for example, in [20, Lemma 3.1.15].) In this work we sometimes make the heuristic assumption that the distributions that we consider are "smooth enough", so that we really need to set $N \approx 1/\epsilon^2$.

• If $\mathcal{D}_1, ..., \mathcal{D}_N$ are distributions over n-bit strings, we denote by $\sum \mathcal{D}_i$ the distribution over the sum (exclusive-or), $\sum_{i=1}^{N} x_i$, where each x_i is chosen according to $\mathcal{D}_i$, independently of all the other x_j's. Denote by $\mathcal{U}$ the uniform distribution over $\{0,1\}^n$. If for all i, $|\mathcal{U} - \mathcal{D}_i| = \epsilon_i$, then $|\mathcal{U} - \sum \mathcal{D}_i| \leq \prod_i \epsilon_i$. (We include a proof of this simple "xor lemma" in the long version of this report [3].) In the analysis in this paper, we sometimes assume that the distributions $\mathcal{D}_i$ are "smooth enough", so that we can use the approximation $|\mathcal{U} - \sum \mathcal{D}_i| \approx \prod_i \epsilon_i$.

Hypothesis testing. We provide a brief overview of (binary) hypothesis testing. This material is covered in many statistics and engineering textbooks (e.g., [16, Ch.5]). In a binary hypothesis testing problem, there are two distributions $\mathcal{D}_1, \mathcal{D}_2$, defined over the same domain X. We are given an element $x \in X$, which was drawn according to either $\mathcal{D}_1$ or $\mathcal{D}_2$, and we need to guess which is the case. A *decision rule* for such hypothesis testing problem is a function $DR : X \to \{1,2\}$, that tells us what should be our guess for each element $x \in X$. Perhaps the simplest notion of success for a decision rule DR, is the statistical advantage that it gives (over a random coin-toss), in the case that the distributions $\mathcal{D}_1, \mathcal{D}_2$ are equally likely a-priori. Namely, adv(DR) = $\frac{1}{2} \left(\Pr_{\mathcal{D}_1}[DR(x) = 1] + \Pr_{\mathcal{D}_2}[DR(x) = 2] \right) - \frac{1}{2}$.

Proposition 1. *For any hypothesis-testing problem $\langle \mathcal{D}_1, \mathcal{D}_2 \rangle$, the decision rule with the largest advantage is the* maximum-likelihood *rule, $ML(x) = 1$ if $\mathcal{D}_1(x) > \mathcal{D}_2(x)$, and 2 otherwise. The advantage of the ML decision rule equals a quarter of the statistical distance, adv(ML) = $\frac{1}{4}|\mathcal{D}_1 - \mathcal{D}_2|$.*

3   Formal Framework

We consider ciphers that are built around two repeating functions (processes). One is a non-linear function $NF(x)$ and the other is a linear function $LF(w)$. The non-linear function NF is usually a permutation on n-bit blocks (typically, $n \approx 100$). The linear function LF is either an LFSR, or just fixed tables of size between a few hundred and a few thousand bits. The state of such a cipher consists of the "non-linear state" x and the "linear state" w. In each step, we apply the function NF to x and the function LF to w, and we may also "mix" these states by xor-ing some bits of w into x and vice versa. The output of the current state is also computed as an xor of bits from x and w. To simplify the presentation of this report, we concentrate on a special case, similar to Scream[1]. In each step i we do the following:

1. Set $w_i := LF(w_{i-1})$
2. Set $y_i := L1(w_i)$, $z_i = L2(w_i)$ // $L1, L2$ are some linear functions
3. Set $x_i := NF(x_{i-1} + y_i) + z_i$ // '+' denotes exclusive-or
4. Output x_i

3.1   The Linear Process

The only property of the linear process that we care about, is that the string $y_1 z_1 y_2 z_2 \ldots$ can be modeled as a random element in some known linear subspace

---

[1] We show how our techniques can handle other variants when we describe the attack on SNOW, but we do not attempt to characterize all the variants where such techniques apply.

of $\{0,1\}^*$. Perhaps the most popular linear process is to view the "linear state" w as the contents of an LFSR. The linear modification function LF clocks the LFSR some fixed number of times (e.g., 32 times), and the functions $L1, L2$ just pick some bits from the LFSR. If we denote the LFSR polynomial by p, then the relevant linear subspace is the subspace orthogonal to $p \cdot Z_2[x]$.

A different approach is taken in Scream. There, the "linear state" resides in some tables, that are "almost fixed". In particular, in Scream, each entry in these tables is used 16 times before it is modified (via the non-linear function NF). For our purposes, we model this scheme by assuming that whenever an entry is modified, it is actually being replaced by a new random value. The masking scheme in Scream can be thought of as a "two-dimensional" scheme, where there are two tables, which are used in lexicographical order[2]. Namely, we have a "row table" $R[\cdot]$ and a "column table" $C[\cdot]$, each with 16 entries of $2n$-bit string. The steps of the cipher are partitioned into batches of 256 steps each. At the beginning of a batch, all the entries in the tables are "chosen at random". Then, in step $i = j + 16k$ in a batch, we set $(y_i|z_i) := R[j] + C[k]$.

3.2 Attacks on Stream Ciphers

We consider an attacker that just watches the output stream and tries to distinguish it from a truly random stream. The relevant parameters in an attack are the amount of text that the attacker must see before it can reliably distinguish the cipher from random, and the time and space complexity of the distinguishing procedure. The attacks that we analyze in this report exploit the fact that for a (small) subset of the bits of x and $NF(x)$, the joint distribution of these bits differs from the uniform distribution by some noticeable amount. Intuitively, such attacks never try to exploit correlations between "far away" points in time. The only correlations that are considered, are the ones between the input and output of a single application of the non-linear function[3].

Formally, we view the non-linear process not as one continuous process, but rather as a sequence of uncorrelated steps. That is, for the purpose of the attack, one can view the non-linear state x at the beginning of each step as a new random value, independent of anything else. Under this view, the attacker sees a collection of pairs $\langle x_j + y_j, NF(x_j) + z_j \rangle$, where the x_j's are chosen uniformly at random and independently of each other, and the y_j, z_j's are taken from the linear process.

One example of attacks that fits in this model are linear attacks. In linear cryptanalysis, the attacker exploits the fact that a one-bit linear combination of $\langle x, NF(x) \rangle$ is more likely to be zero than one (or vice versa). In these attack, it is always assumed that the bias in one step is independent of the bias in all the other steps. Somewhat surprisingly, differential cryptanalysis too fits into this

---

[2] The scheme in Scream is actually slightly different than the one described here, but this difference does not effect the analysis in any significant way.

[3] When only a part of x is used as output, we may be forced to look at a few consecutive applications of NF. This is the case in SNOW, for example.

framework (under our attack model). Since the attacker in our model is not given chosen-input capabilities, it exploits differential properties of the round function by waiting for the difference $x_i + x_j = \Delta$ to happen "by chance", and then using the fact that $NF(x_i) + NF(x_j) = \Delta'$ is more likely than you would expect from a random process. It is clear that this attack too is just as effective against pairs of uncorrelated steps, as when given the output from the real cipher.

We are now ready to define formally what we mean by "an attack on the cipher". The attacks that we consider, observe some (linear combinations of) input and output bits from each step of the cipher, and try to decide if these indeed come from the cipher, or from a random source. This can be framed as a hypothesis testing problem. According to one hypothesis (Random), the observed bits in each step are random and independent. According to the other (Cipher), they are generated by the cipher.

Definition 2 (Attacks on stream ciphers with linear masking). *An attack is specified by a linear function ℓ, and by a decision rule for the following hypothesis-testing problem: The two distributions that we want to distinguish are*

Cipher. *The Cipher distribution is $\mathcal{D}_c = \langle \ell (x_j + y_j, NF(x_j) + z_j) \rangle_{j=1,2,...}$, where the $y_j z_j$'s are chosen at random from the appropriate linear subspace (defined by the linear process of the cipher), and the x_j's are random and independent.*

Random. *Using the same notations, the "random process" distribution is $\mathcal{D}_r \overset{\text{def}}{=} \langle \ell(x_j, x'_j) \rangle_{j=1,2,...}$, where the x_j's and x'_j's are random and independent.*

We call the function ℓ, the distinguishing characteristic *used by attack.*

The amount of text needed for the attack is the smallest number of steps for which the decision rule has a constant advantage (e.g., advantage of $1/4$) in distinguishing the cipher from random. Other relevant parameters of the attack are the time and space complexity of the decision rule. An obvious lower bound on the amount of text is provided by the statistical distance between the Cipher and Random distributions after N steps.

4 Linear Attacks

A linear attack [15] exploits the fact that some linear combination of the input and output bits of the non-linear function is more likely to be zero than one (or vice versa). Namely, we have a (non-trivial) linear function $\ell : \{0,1\}^{2n} \to \{0,1\}$, such that for a randomly selected n bit string x, $\Pr[\ell(x, NF(x)) = 0] = (1+\epsilon)/2$. The function ℓ is called a *linear approximation* (or characteristic) of the non-linear function, and the quantity ϵ is called the *bias* of the approximation.

When trying to exploit one such linear approximation, the attacker observes for each step j of the cipher, a bit $\sigma_j = \ell(x_j + y_j, NF(x_j) + z_j)$. Note that σ_j by itself is likely to be unbiased, but the σ's are correlated. In particular, since the y, z's come from a linear subspace, it is possible to find some linear combination of steps for which they vanish. Let J be a set of steps such that $\sum_{j \in J} y_j = \sum_{j \in J} z_j = 0$. Then we have

$$\sum_{j \in J} \sigma_j = \sum_{j \in J} \ell(x_j, NF(x_j)) + \sum_{j \in J} \ell(y_j, z_j) = \sum_{j \in J} \ell(x_j, NF(x_j))$$

(where the equalities follow since ℓ is linear). Therefore, the bit $\xi_J = \sum_{j \in J} \sigma_j$ has bias of $\epsilon^{|J|}$. If the attacker can observe "sufficiently many" such sets J, it can reliably distinguish the cipher from random.

This section is organized as follows: We first bound the effectiveness of linear attacks in terms of the bias ϵ and the *weight distribution* of some linear subspace. As we explain below, this bound suggests that looking at sets of steps as above is essentially "the only way to exploit linear correlations". Then we show how to devise a linear attack on SNOW, and analyze its effectiveness.

4.1   The Statistical Distance

Recall that we model an attack in which the attacker observes a single bit per step, namely $\sigma_j = \ell(x_j + y_j, NF(x_j) + z_j)$. Below we denote $\tau_j = \ell(x_j, NF(x_j))$ and $\rho_j = \ell(y_j, z_j)$. We can re-write the Cipher and Random distributions as

Cipher. $\mathcal{D}_c \overset{\text{def}}{=} \langle \tau_j + \rho_j \rangle_{j=1,2,\dots}$, where the τ_j's are independent but biased, $\Pr[\tau_j = 0] = (1 + \epsilon)/2$, and the string $\rho_1 \rho_2 \dots$ is chosen at random from the appropriate linear subspace (i.e., the image under ℓ of the linear subspace of the $y_j z_j$'s).

Random. $\mathcal{D}_r \overset{\text{def}}{=} \langle \sigma_j \rangle_{j=1,2,\dots}$, where the σ_j's are independent and unbiased.

Below we analyze the statistical distance between the Cipher and Random distributions, after observing N bits $\sigma_1 \dots \sigma_N$. Denote the linear subspace of the ρ's by $L \subseteq \{0,1\}^N$, and let $L^\perp \subseteq \{0,1\}^N$ be the orthogonal subspace. The *weight distribution* of the space $L^\perp$ plays an important role in our analysis. For $r \in \{0, 1, \dots, N\}$, let $\mathcal{A}_N(r)$ be the set of strings $\chi \in L^\perp$ of Hamming weight r, and let $A_N(r)$ denote the cardinality of $\mathcal{A}_N(r)$. We prove the following theorem:

Theorem 1. *The statistical distance between the Cipher and Random distributions from above, is bounded by* $\sqrt{\sum_{r=1}^{N} A_N(r) \epsilon^{2r}}$.

Proof. Included in the long version [3].

Remark. Heuristically, this bound is nearly tight. In the proof we analyzed the random variable Δ and used the bound $E[|\Delta - E[\Delta]|] \le \sqrt{\text{VAR}[\Delta]}$. One can argue heuristically that as long as the statistical distance is sufficiently small, "Δ should behave much like a Gaussian random variable". If it were a Gaussian, we would have $E[|\Delta|] = \sqrt{\text{VAR}[\Delta]} \cdot \sqrt{2/\pi}$. Thus, we expect the bound from Theorem 1 to be tight up to a constant factor $\sqrt{2/\pi} \approx 0.8$.

4.2   Interpretations of Theorem 1

There are a few ways to view Theorem 1. The obvious way is to use it in order to argue that a certain cipher is resilient to linear attacks. For example, in [2] we use Theorem 1 to deduce a lower-bound on the amount of text needed for any linear attack on Scream-0.

Also, one could notice that the form of Theorem 1 exactly matches the common practice (and intuition) of devising linear attacks. Namely, we always look at sets where the linear process vanishes, and view each such set J as providing "statistical evidence of weight $\epsilon^{2|J|}$" for distinguishing the cipher from random. Linear attacks work by collecting enough of these sets, until the weights sum up of to one. One can therefore view Theorem 1 as asserting that this is indeed the best you can do.

Finally, we could think of devising linear attacks, using the heuristic argument about this bound being tight. However, the way Theorem 1 is stated above, it usually does not imply efficient attacks. For example, when the linear space L has relatively small dimension (as is usually the case with LFSR-based ciphers, where the dimension of L is at most a few hundreds), the statistical distance is likely to approach one for relatively small N. But it is likely that most of the "mass" in the bound of Theorem 1 comes from terms with a large power of ϵ (and therefore very small "weight"). Therefore, if we want to use a small N, we would need to collect very many samples, and this attack is likely to be more expensive than an exhaustive search for the key.

Alternatively, one can try and use an efficient sub-optimal decision rule. For a given bound on the work-load W and the amount of text N, we only consider the first few terms in the power series. That is, we observe the N bits $\sigma = \sigma_1 \ldots \sigma_N$, but only consider the W smallest sets J for which $\chi(J) \in L^{\perp}$. For each such set J, the sum of steps $\sum_{j \in J} \sigma_j$ has bias $\epsilon^{|J|}$, and these can be used to distinguish the cipher from random. If we take all the sets of size at most R, we expect the advantage of such a decision rule to be roughly $\frac{1}{4}\sqrt{\sum_{r=1}^{R} A_N(r)\epsilon^{2r}}$. The simplest form of this attack (which is almost always the most useful), is to consider only the minimum-weight terms. If the minimum-weight of $L^{\perp}$ is r_0, then we need to make N big enough so that $\frac{1}{4}\sqrt{A_N(r_0)} = \epsilon^{-r_0}$.

4.3   The Attack on SNOW

The stream cipher SNOW was submitted to NESSIE in 2000, by Ekdahl and Johansson. A detailed description of SNOW is available from [5]. Here we outline a linear attack on SNOW along the lines above, that can reliably distinguish it from random after observing roughly 2^{95} steps of the cipher, with work-load of roughly 2^{100}.

SNOW consists of a non-linear process (called there a Finite-State Machine, or FSM), and a linear process which is implemented by an LFSR. The LFSR of SNOW consists of sixteen 32-bit words, and the LFSR polynomial, defined over $GF(2^{32})$, is $p(z) = z^{16} + z^{13} + z^7 + \alpha$, where α is a primitive element of $GF(2^{32})$. (The orthogonal subspace $L^{\perp}$ is therefore the space of (bitwise reversal of) polynomials over Z_2 of degree $\leq N$, which are divisible by the LFSR polynomial p.) At a given step j, we denote the content of the LFSR by $L_j[0..15]$, so we have $L_{j+1}[i] = L_j[i-1]$ for $i > 0$ and $L_{j+1}[0] = \alpha \cdot (L_j[15] + L_j[12] + L_j[6])$.

The "FSM state" of SNOW in step j consists of only two 32-bit words, denoted $R1_j, R2_j$. The FSM update function modifies these two values, using one word from the LFSR, and also outputs one word. The output word is then added to another word from the LFSR, to form the step output. We denote the "input word" from the LFSR to the FSM update function by f_j, and the "output word" from the FSM by F_j. The FSM uses a "32×32 S-box" $S[\cdot]$ (which is built internally as an SP-network, from four identical 8×8 boxes and some bit permutation). A complete step of SNOW is described in Figure 1. In this figure, we deviate from the notations in the rest of the paper, and denote exclusive-or by $\oplus$ and integer addition mod 2^{32} by $+$. We also denote 32-bit cyclic rotation to the left by $\lll$.

1. $f_j := L_j[0]$
2. $F_j := (f_j + R1_j) \oplus R2_j$
3. output $F_j \oplus L_j[15]$
4. $R1_{j+1} := R1_j \oplus ((R2_j + F_j) \lll 7)$
5. $R2_{j+1} := S[R1_j]$
6. update the LFSR

Fig. 1. One step of SNOW: $\oplus$ is xor and $+$ is addition mod 2^{32}.

To devise an attack we need to find a good linear approximation of the non-linear FSM process, and low-weight combinations of steps where the $L_j[\cdot]$ values vanish (i.e., low-weight polynomials which are divisible by the LFSR polynomial p). The best linear approximation that we found for the FSM process, uses six bits from two consecutive inputs and outputs, $f_j, f_{j+1}, F_j, F_{j+1}$. Specifically, for each step j, the bit

$$\sigma_j \overset{\text{def}}{=} (f_j)_{15} + (f_j)_{16} + (f_{j+1})_{22} + (f_{j+1})_{23} + (F_j)_{15} + (F_{j+1})_{23}$$

is biased. (Of these six bits, the bits $(f_j)_{15}, (F_j)_{15}$ and $(F_{j+1})_{22}$ are meant to approximate carry bits.) We measured the bias experimentally, and it appears to be at least $2^{-8.3}$.

At first glance, one may hope to find weight-4 polynomials that are divisible by the LFSR polynomial p. After all, p itself has only four non-zero terms. Unfortunately, one of these terms is the element $\alpha \in GF(2^{32})$, whereas we need a low-weight polynomial with 0-1 coefficients. What we can show, however, is the existence of 0-1 polynomials of weight-six that are divisible by p.

Proposition 2. *The polynomial* $q(z) = z^{16 \times 2^{32} - 7} + z^{13 \times 2^{32} - 7} + z^{7 \times 2^{32} - 7} + z^9 + z^6 + 1$ *is divisible by the LFSR polynomial* $p(z) = z^{16} + z^{13} + z^7 + \alpha$.

Proof. Included in the long version [3].

Corollary 1. *For all m, n, the polynomial $q_{m,n}(z) \stackrel{\text{def}}{=} q(z)^{2^m} \cdot z^n$ is divisible by $p(z)$.*

If we take, say, $m = 0, 1, \ldots 58$ and $n = 0, 1, \ldots 2^{94}$, we get about 2^{100} different 0-1 polynomials, all with weight 6 and degree less than $N = 2^{95}$, and all divisible by $p(z)$. Each such polynomial yields a sequence of six steps, $J_{m,n}$, such that the sum of the $L_j[\cdot]$ values in these steps vanishes. Therefore, if we denote the output word of SNOW at step j by S_j, then for all m, n we have,

$$\tau_{m,n} \stackrel{\text{def}}{=} \sum_{j \in J_{m,n}} (S_j)_{15} + (S_{j+1})_{23} = \sum_{j \in J_{m,n}} \sigma_j$$

and therefore each $\tau_{m,n}$ has bias of $2^{-8.3 \times 6} = 2^{-49.8}$. Since we have roughly 2^{100} of them, we can reliably distinguish them from random.

5   Low-Diffusion Attacks

In low-diffusion attacks, the attacker looks for a small set of (linear combinations of) input and output bits of the non-linear function NF, whose values completely determine the values of some other (linear combinations of) input and output bits. The attacker tries to guess the first set of bits, computes the values of the other bits, and uses the computed value to verify the guess against the cipher's output. The complexity of such attacks is exponential in the number of bits that the attacker needs to guess.

We introduce some notations in order to put such attacks in the context of our framework. To simplify the notations, we assume that the guessed bits are always input bits, and the determined bits are always output bits. (Eliminating this assumption is usually quite straightforward.) As usual, let $NF : \{0,1\}^n \to \{0,1\}^n$ be the non-linear function. The attack exploits the fact that some input bits $\ell_{\text{in}}(x)$ are related to some output bits $\ell_{\text{out}}(NF(x))$ via a known deterministic function f. That is, we have $\ell_{\text{out}}(NF(x)) = f(\ell_{\text{in}}(x))$. Here, $\ell_{\text{in}}, \ell_{\text{out}}$ are linear functions, and f is an arbitrary function, all known to the attacker. We denote the output size of $\ell_{\text{in}}, \ell_{\text{out}}$ by m, m', respectively. That is, $\ell_{\text{in}} : \{0,1\}^n \to \{0,1\}^m$, $\ell_{\text{out}} : \{0,1\}^n \to \{0,1\}^{m'}$, and $f : \{0,1\}^m \to \{0,1\}^{m'}$.

In each step j, the attacker observes the bits $\ell_{\text{in}}(x_j + y_j)$ and $\ell_{\text{out}}(NF(x_j) + z_j)$ (where y_j, z_j are from the linear process, as in Section 3.1). Below we denote $u_j = \ell_{\text{in}}(x_j)$, $u'_j = \ell_{\text{out}}(NF(x_j))$, $v_j = \ell_{\text{in}}(y_j)$, $v'_j = \ell_{\text{out}}(z_j)$, and $w_j = u_j + v_j$, $w'_j = u'_j + v'_j$. We can re-write the Cipher and Random distributions as

Cipher. $\mathcal{D}_c \stackrel{\text{def}}{=} \langle (w_j = u_j + v_j, \ w'_j = u'_j + v'_j) \rangle_{j=1,2,\ldots}$, where the u_j's are uniform and independent, $u'_j = f(u_j)$, and the string $v_1 v'_1 v_2 v'_2 \ldots$ is chosen at random from the appropriate linear subspace (i.e., the image under $\ell_{\text{in}}, \ell_{\text{out}}$ of the linear subspace of the y, z's).

Random. $\mathcal{D}_r \stackrel{\text{def}}{=} \langle (w_j, w'_j) \rangle_{j=1,2,\ldots}$, all uniform and independent.

It is not hard to see that there may be enough information there to distinguish these two distributions after only a moderate number of steps of the cipher. Suppose that the dimension of the linear subspace of the v_j's and v_j''s is a, and the attacker observes N steps such that $m'N > a$. Then, the attacker can (in principle) go over all the 2^a possibilities for the v_j's and v_j''s. For each guess, the attacker can compute the u_j's and u_j''s, and verify the guess by checking that $u_j' = f(u_j)$ for all j. This way, the attacker guesses a bits and gets $m'N$ bits of consistency checks. Since $m'N > a$ we expect only the "right guess" to pass the consistency checks.

This attack, however, is clearly not efficient. To devise an efficient attack, we can again concentrate on sets of steps where the linear process vanishes: Suppose that we have a set of steps J, such that $\sum_{j \in J} [v_j, v_j'] = [0, 0]$. Then we get

$$\sum_{j \in J} (w_j, w_j') = \sum_{j \in J} (u_j, u_j') = \sum_{j \in J} (u_j, f(u_j))$$

and the distribution over such pairs may differ from the uniform distribution by a noticeable amount. The distance between this distribution and the uniform one, depends on the specific function f, and on the cardinality of the set J. Below we analyze in details perhaps the simplest cases, where f is a random function. Later we explain how this analysis can be extended for other settings, and in particular for the case of the functions in Scream.

5.1   Analysis for Random Functions

For a given function, $f : \{0,1\}^m \to \{0,1\}^{m'}$, and an integer n, we denote $\mathcal{D}_f^n \stackrel{\text{def}}{=} \left\langle d = \sum_{j=1}^n u_j, \ d' = \sum_{j=1}^n f(u_j) \right\rangle$, where the u_j's are uniform in $\{0,1\}^m$ and independent. We assume that the attacker knows f, and it sees many instances of $\langle d, d' \rangle$. The attacker needs to decide if these instances come from $\mathcal{D}_f^n$ or from the uniform distribution on $\{0,1\}^{m+m'}$. Below we denote the uniform distribution by $\mathcal{R}$. If the function f "does not have any clear structure", it makes sense to analyze it as if it was a random function. Here we prove the following:

Theorem 2. *Let n, m, m' be integers with $n^2 \ll 2^m$* [4]. *For a uniformly selected function $f : \{0,1\}^m \to \{0,1\}^{m'}$, $E_f[\|\mathcal{D}_f^n - \mathcal{R}\|] \le c(n) \cdot 2^{\frac{m' - (n-1)m}{2}}$, where*

$$c(n) = \begin{cases} \sqrt{(2n)! \, / \, (n! \, 2^n)} & \text{if } n \text{ is odd} \\ (1 + o(1)) \sqrt{\frac{(2n)!}{n! \, 2^n} - \left(\frac{n!}{(n/2)! \, 2^{n/2}} \right)^2} & \text{if } n \text{ is even} \end{cases}$$

Proof. Included in the long version [3]. We note that the term $2^{\frac{m' - (n-1)m}{2}}$ is due to the fact that the attacker guesses $(n-1)m$ bits and gets m' bits of consistency check, and the term $c(n)$ is due to the symmetries in the guessed bits. (For example, the vector $\boldsymbol{u} = u_1 \dots u_n$ is equivalent to any permutation of $\boldsymbol{u}$.)

---

[4] It can be shown that the same bounds hold also for larger n's, but assuming $n^2 \ll 2^m$ makes some proofs a bit easier.

How tight is this bound? Here too we can argue heuristically that the random variables in the proof "should behave like Gaussian random variables", and again we expect the ratio between $E[|X - E[X]|]$ and $\sqrt{\mathrm{VAR}[X]}$ to be roughly $\sqrt{2/\pi}$. Therefore, we expect the constant $c(n)$ to be replaced by $\sqrt{2/\pi} \cdot c(n) \approx 0.8c(n)$. Indeed we ran some experiments to measure the statistical distance $|\mathcal{D}_f^n - \mathcal{R}|$, for random functions with $n = 4$ and a few values of m, m'. (Note that $c(4) = (1 + o(1))\sqrt{96} \approx 9.8$ and $\sqrt{2/\pi} \cdot c(4) \approx 7.8$). These experiments are described in the long version of this report [3]. The results confirm that the distance between these distributions is just under $7.8 \cdot 2^{(m'-3m)/2}$.

5.2 Variations and Extensions

Here we briefly discuss a few possible extensions to the analysis from above.

Using different f's for different steps. Instead of using the same f everywhere, we may have different f's for different steps. I.e., in step j we have $\ell_{\mathrm{out}}(NF(x_j)) = f_j(\ell_{\mathrm{in}}(x_j))$, and we assume that the f_j's are random and independent. The distribution that we want to analyze is therefore $\langle d = \sum u_j, d' = \sum f_j(u_j) \rangle$. The analysis from above still works for the most part (as long as $\ell_{\mathrm{in}}, \ell_{\mathrm{out}}$ are the same in all the steps). The main difference is that the factor $c(n)$ is replaced by a smaller one (call it $c'(n)$).

For example, if we use n independent functions, we get $c'(n) = 1$, since all the symmetries in the proof of Theorem 2 disappear. Another example (which is used in the attack on Scream-0) is when we have just two independent functions, $f_1 = f_3 = \cdots$ and $f_2 = f_4 = \cdots$. In this case (and when n is divisible by four), we get $c'(n) = (1 + o(1))\sqrt{\left(\frac{n!}{(n/2)! \, 2^{n/2}}\right)^2 - \left(\frac{(n/2)!}{(n/4)! \, 2^{n/4}}\right)^4}$.

When f is a sum of a few functions. An important special case, is when f is a sum of a few functions. For example, in the functions that are used in the attack on Scream-0, the m-bit input to f can be broken into three *disjoint* parts, each with $m/3$ bits, so that $f(x) = f^1(x^1) + f^2(x^2) + f^3(x^3)$. (Here we have $|x^1| = |x^2| = |x^3| = m/3$ and $x = x^1 x^2 x^3$.) If f^1, f^2, f^3 themselves do not have any clear structure, then we can apply the analysis from above to each of them. That analysis tells us that each of the distributions $\mathcal{D}^i \stackrel{\text{def}}{=} (\sum_j u_j^i, \sum_j f^i(u_j^i))$ is likely to be roughly $c(n) \cdot 2^{(m'-(n-1)m/3)/2}$ away from the uniform distribution.

It is not hard to see that the distribution $\mathcal{D}_f^n$ that we want to analyze can be cast as $\mathcal{D}^1 + \mathcal{D}^2 + \mathcal{D}^3$, so we expect to get $|\mathcal{D}_f^n - \mathcal{R}| \approx \prod |\mathcal{D}^i - \mathcal{R}| \approx \left(c(n) \cdot 2^{(m'-(n-1)m/3)/2}\right)^3 = c(n)^3 2^{(3m'-(n-1)m)/2}$. More generally, suppose we can write f as a sum of r functions over disjoint arguments of the same length. Namely, $f(x) = \sum_{i=1}^r f^i(x^i)$, where $|x^1| = ... = |x^r| = m/r$ and $x = x^1...x^r$. Repeating the argument from above, we get that the expected distance $|\mathcal{D}_f^n - \mathcal{R}|$ is about $c(n)^r 2^{(rm'-(n-1)m)/2}$ (assuming that this is still smaller than one). As before, one could use the "Gaussian heuristics" to argue that for the "actual distance" we should replace $c(n)^r$ by $(c(n) \cdot \sqrt{2/\pi})^r$. (And if we have different functions for different steps, as above, then we would get $(c'(n) \cdot \sqrt{2/\pi})^r$.)

Linear masking over different groups. Another variation is when we do linear masking over different groups. For example, instead of xor-ing the masks, we add them modulo some prime q, or modulo a power of two. Again, the analysis stays more or less the same, but the constants change. If we work modulo a prime $q > n$, we get a constant of $c'(n) = \sqrt{n!}$, since the only symmetry that is left is between all the orderings of $\{u_1, \ldots, u_n\}$. When we work modulo a power of two, the constant will be somewhere between $c'(n)$ and $c(n)$, probably closer to the former.

5.3   Efficiency Considerations

The analysis from above says nothing about the computational cost of distinguishing between $\mathcal{D}_f^n$ and $\mathcal{R}$. It should be noted that in a "real life" attack, the attacker may have access to many different relations (with different values of m, m'), all for the same non-linear function NF. To minimize the amount of needed text, the attacker may choose to work with the relation for which the quantity $(n-1)m - m'$ is minimized. However, the choice of relations is limited by the attacker's computational resources. Indeed, for large values of m, m', computing the maximum-likelihood decision rule may be prohibitively expensive in terms of space and time. Below we review some strategies for computing the maximum-likelihood decision rule.

Using one big table. Perhaps the simplest strategy, is for the attacker to prepare off-line a table of all possible pairs $\langle d, d' \rangle$ with $d \in \{0,1\}^m$, $d' \in \{0,1\}^{m'}$. For each pair $\langle d, d' \rangle$ the table contains the probability of this pair under the distribution $\mathcal{D}_f^n$ (or perhaps just one bit that says whether this probability is more than $2^{-m-m'}$).

Given such a table, the on-line part of the attack is trivial. For each set of steps J, compute $(d, d') = \sum_{j \in J}(w_j, w'_j)$, and look into the table to see if this pair is more likely to come from $\mathcal{D}_f^n$ or from $\mathcal{R}$. After observing roughly $2^{(n-1)m-m'}/c(n)^2$ such sets J, a simple majority vote can be used to determine if this is the cipher or a random process. Thus, the on-line phase is linear in the amount of text that has to be observed, and the space requirement is $2^{m+m'}$.

As for the off-line part (in which the table is computed), the naive way is to go over all possible values of $u_1 \ldots u_n \in \{0,1\}^m$, for each value computing $d = \sum u_i$ and $d' = \sum f(u_i)$ and increasing the corresponding entry $\langle d, d' \rangle$ by one. This takes 2^{mn} time. However, in the (typical) case where $m' \ll (n-1)m$, one can use a much better strategy, whose running time is only $O(\log n(m+m')2^{m+m'})$.

First, we represent the function f by a $2^m \times 2^{m'}$ table, with $F[x, y] = 1$ if $f(x) = y$, and $F[x, y] = 0$ otherwise. Then, we compute the convolution of F with itself, $E \stackrel{\text{def}}{=} F \star F$ [5],

---

[5] Recall that the convolution operator is defined on one-dimensional vectors, not on matrices. Indeed, in this expression we view the table F as a one-dimensional vector, whose indexes are $m + m'$-bits long.

$$E[s,t] \;=\; \sum_{x+x'=s} \sum_{y+y'=t} F[x,y] \cdot F[x',y'] \;=\; |\{x \;:\; f(x) + f(x+s) = t\}|$$

(Note that E represents the distribution $\mathcal{D}_f^2$.) One can use the Walsh-Hadamard transform to perform this step in time $O((m+m')2^{m+m'})$ (see, e.g., [19]). Then, we again use the Walsh-Hadamard transform to compute the convolution of E with itself,

$$
\begin{aligned}
D[d,d'] \stackrel{\text{def}}{=} (E \star E)[d,d'] \;&=\; \sum_{s+s'=d} \sum_{t+t'=d'} E(s,t) \cdot E(s',t') \\
&= |\{\langle x,s,z \rangle : \; f(x) + f(x+s) + f(z) + f(z+s+d) = d'\}| \\
&= |\{\langle x,y,z \rangle : \; f(x) + f(y) + f(z) + f(x+y+z+d) = d'\}|
\end{aligned}
$$

thus getting the distribution $\mathcal{D}_f^4$, etc. After $\log n$ such steps, we get the distribution of $\mathcal{D}_f^n$.

When f is a sum of functions. We can get additional flexibility when f is a sum of functions on disjoint arguments, $f(x) = f^1(x^1) + \cdots + f^r(x^r)$ (with $x = x^1 \ldots x^r$). In this case, one can use the procedure from above to compute the tables $D^i[d,d']$ for the individual f^i's. If all the x^i's are of the same size, then each of the D^i's takes up $2^{m'+(m/r)}$ space, and can be computed in time $O(\log n(m' + (m/r))2^{m'+(m/r)})$. Then, the "global" D table can again be computed using convolutions. Specifically, for any fixed $d = d^1 \ldots d^r$, the $2^{m'}$-vector of entries $D[d, \cdot]$ can be computed as the convolutions of the $2^{m'}$-vectors $D^1[d^1, \cdot]$, $D^2[d^2, \cdot]$, ..., $D^r[d^r, \cdot]$,

$$D[d, \cdot] \;=\; D^1[d^1, \cdot] \star D^2[d^2, \cdot] \star \cdots \star D^r[d^r, \cdot]$$

At first glance, this does not seem to help much: Computing each convolution takes time $O(r \cdot m'2^{m'})$, and we need to repeat this for each $d \in \{0,1\}^m$, so the total time is $O(rm'2^{m+m'})$. However, we can do much better than that.

Instead of storing the vectors $D^i[d^i, \cdot]$ themselves, we store their image under the Walsh-Hadamard transform, $\Delta^i[d^i, \cdot] \stackrel{\text{def}}{=} \mathcal{H}(D^i[d^i, \cdot])$. Then, to compute the vector $D[\langle d^1 \ldots d^r \rangle, \cdot]$, all we need is to multiply (point-wise) the corresponding $\Delta^i[d^i, \cdot]$'s, and then apply the inverse Walsh-Hadamard transform to the result. Thus, once we have the tables $D^i[\cdot, \cdot]$, we need to compute $r \cdot 2^{m/r}$ "forward transforms" (one for each vector $D^i[d^i, \cdot]$), and 2^m inverse transforms (one for each $\langle d^1 \ldots d^r \rangle$). Computing each transform (or inverse) takes $O(m'2^{m'})$ time. Hence, the total time (including the initial computation of the D^i's) is $O\left(\log n(rm' + m)2^{m'+(m/r)} + m'2^{m+m'}\right)$, and the total space that is needed is $O(2^{m+m'})$.

If the amount of text that is needed is less than 2^m, then we can optimize even further. In this case the attacker need not store the entire table D in memory. Instead, it is possible to store only the D^i tables (or rather, the $\Delta^i[\cdot, \cdot]$ vectors), and compute the entries of D during the on-line part, as they are needed. Using this method, the off-line phase takes $O(\log n(rm' + m)2^{m'+(m/r)})$ time and $O(r2^{m'+m/r})$ space to compute and store the vectors $\Delta_i[\cdot, \cdot]$, and the on-line phase takes $O(m'2^{m'})$ time per sample. Thus the total time complexity here is $O(\log n(rm'+m)2^{m'+(m/r)} + Sm'2^{m'})$, where S is the number of samples needed to distinguish $\mathcal{D}$ from $\mathcal{R}$.

5.4   An Attack on Scream-0

The stream cipher Scream (with its variants Scream-0 and Scream-F) was proposed very recently by Coppersmith, Halevi and Jutla. A detailed description of Scream is available in [2]. Below we only give a partial description of Scream-0, which suffices for the purpose of our attack.

Scream-0 maintains a 128-bit "non-linear state" x, two 128-bit "column masks" $c1, c2$ (which are modified every sixteen steps), and a table of sixteen "row masks" $R[0..15]$. It uses a non-linear function NF, somewhat similar to a round of Rijndael. Roughly speaking, the steps of Scream-0 are partitioned to chunks of sixteen steps. A description of one such chunk is found in Figure 2.

1. for $i = 0$ to 15 do
2. $x := NF(x + c1) + c2$
3.   output $x + R[i]$
4.   if i is even, rotate $c1$ by 64 bits
5.   if i is odd, rotate $c1$ by some other amount
6. end-for
7. modify $c1, c2$, and one entry of R, using the function $NF(\cdot)$

Fig. 2. sixteen steps of Scream-0.

Here we outline a low-diffusion attack on the variant Scream-0, along the lines above, that can reliably distinguish it from random after observing merely 2^{43} bytes of output, with memory requirement of about 2^{50} and work-load of about 2^{80}. This attack is described in more details in the long version of [2].

As usual, we need to find a "distinguishing characteristic" of the non-linear function (in this case, a low-diffusion characteristic), and a combination of steps in which the linear process vanishes. The linear process consists of the c_i's and the $R[i]$'s. Since each entry $R[i]$ is used sixteen times before it is modified, we can cancel it out by adding two steps were the same entry is used. Similarly, we can cancel c_2 by adding two steps within the same "chunk" of sixteen steps. However, since $c1$ is rotated after each use, we need to look for two different characteristics of the NF function, such that the pattern of input bits in one characteristic is a rotated version of the pattern in the other.

The best such pair of "distinguishing characteristics" that we found for Scream-0, uses a low-diffusion characteristic for NF in which the input bits pattern is 2-periodic (and the fact that $c1$ is rotated every other step by 64 bits). Specifically, the four input bytes x_0, x_5, x_8, x_{13}, together with two bytes of linear combinations of the output $NF(x)$, yield the two input bytes x_2, x_{10}, and two other bytes of linear combinations of the output $NF(x)$. In terms of the parameters that we used above, we have $m = 48$ input and output bits, which completely determine $m' = 32$ other input and output bits.

To use this relation, we can observe these ten bytes from each of four steps, (i.e., $j, j+1, j+16k, j+1+16k$ for even j and $k < 16$). We can then add them up (with the proper rotation of the input bytes in steps $j+1, j+17$), to cancel both the "row masks" $R[i]$ and the "column masks" $c1, c2$. This gives us the following distribution $\mathcal{D} = \langle u_1 + u_2 + u_3 + u_4, \; f_1(u_1) + f_2(u_2) + f_1(u_3) + f_2(u_4) \rangle$, where the u_i's are modeled as independent, uniformly selected, 48-bit strings, and f_1, f_2 are two known functions $f_j : \{0,1\}^{48} \to \{0,1\}^{32}$. (The reason that we have two different functions is that the order of the input bytes is different between the even and odd steps.) Moreover, each of the two f_j's can be written as a sum of three functions over disjoint parts, $f_j(x) = f_j^1(x^1) + f_j^2(x^2) + f_j^3(x^3)$ where $|x^1| = |x^2| = |x^3| = 16$.

This is one of the "extensions" that were discussed in Section 5.2. Here we have $n = 4$, $m = 48$, $m' = 32$, $r = 3$, and two different functions. Therefore, we expect to get statistical distance of $c'(n)^3 \cdot 2^{(3m' - (n-1)m)/2}$, with

$$c'(n) \approx \sqrt{2/\pi} \cdot \sqrt{\left(\frac{n!}{(n/2)! \; 2^{n/2}} \right)^2 - \left(\frac{(n/2)!}{(n/4)! \; 2^{n/4}} \right)^4}$$

Plugging in the parameters, we have $c'(4) \approx \sqrt{2/\pi} \cdot \sqrt{8}$, and the expected statistical distance is roughly $(16/\pi)^{3/2} \cdot 2^{-24} \approx 2^{-20.5}$. We therefore expect to be able to reliably distinguish $\mathcal{D}$ from random after about 2^{41} samples. Roughly speaking, we can get $8 \cdot \binom{14}{2} \approx 2^{10}$ samples from 256 steps of Scream-0. (We have 8 choices for an even step in a chunk of 16 steps, and we can choose two such chunks from a collection of 14 in which the three row masks in use remain unchanged.) So we need about $2^{31} \cdot 256 = 2^{39}$ steps, or 2^{43} bytes of output.

Also, in Section 5.3 we show how one could efficiently implement the maximum-likelihood decision rule to distinguish $\mathcal{D}$ from $\mathcal{R}$, using Walsh-Hadamard transforms. Plugging the parameters of the attack on Scream-0 into the general techniques that are described there, we have space complexity of $O(r2^{m'+m/r})$, which is about 2^{50}. The time complexity is $O(\log n (rm'+m)2^{m'+(m/r)}+Sm'2^{m'})$, where in our case $S = 2^{41}$, so we need roughly 2^{80} time.

6   Conclusions

In this work we described a general cryptanalytical technique that can be used to attack ciphers that employ a combination of a "non-linear" process and a "linear process". We analyze in details the effectiveness of this technique for two special cases. One is when we exploit linear approximations of the non-linear process, and the other is when we exploit the low diffusion of (one step of) the non-linear process. We also show how these two special cases are useful in attacking the ciphers SNOW [5] and Scream-0 [2].

It remains an interesting open problem to extend the analysis that we have here to more general "distinguishing characteristics" of the non-linear process.

For example, extending the analysis of the low-diffusion attack from Section 5.1 to the case where the functions f is key-dependent (and thus not known to the adversary) may yield an effective attack on Scream [2].

In addition to the cryptanalytical technique, we believe that another contribution of this work is our formulation of attacks on stream ciphers. We believe that explicitly formalizing an attack as considering sequence of uncorrelated steps (as opposed to one continuous process) can be used to shed light on the strength of many ciphers.

References

1. A. Canteaut and E. Filiol. Ciphertext only reconstruction of stream ciphers based on combination generators. In *Fast Software Encryption*, volume 1978 of *Lecture Notes in Computer Science*, pages 165–180. Springer-Verlag, 2000.
2. D. Copersmith, S. Halevi, and C. Jutla. Scream: a software-efficient stream cipher. In *Fast Software Encryption*, Lecture Notes in Computer Science. Springer-Verlag, 2002. to appear. A longer version is available on-line from http://eprint.iacr.org/2002/019/.
3. D. Coppersmith, S. Halevi, and C. Jutla. Cryptanalysis of stream ciphers with linear masking. Available from the ePrint archive, at http://eprint.iacr.org/2002/020/, 2002.
4. J. Daemen and C. S. K. Clapp. Fast hashing and stream encryption with Panama. In S. Vaudenay, editor, *Fast Software Encryption: 5th International Workshop*, volume 1372 of *Lecture Notes in Computer Science*, pages 23–25. Springer-Verlag, 1998.
5. P. Ekdahl and T. Johansson. SNOW – a new stream cipher. Submitted to NESSIE. Available on-line from http://www.it.lth.se/cryptology/snow/.
6. P. Ekdahl and T. Johansson. Distinguishing attacks on SOBER-t16 and t32. In *Fast Software Encryption*, Lecture Notes in Computer Science. Springer-Verlag, 2002. to appear.
7. S. Fluhrer. Cryptanalysis of the SEAL 3.0 pseudorandom function family. In *Proceedings of the Fast Software Encryption Workshop (FSE'01)*, 2001.
8. S. R. Fluhrer and D. A. McGraw. Statistical analysis of the alleged RC4 keystream generator. In *Proceedings of the 7th Annual Workshop on Fast Software Encryption, (FSE'2000)*, volume 1978 of *Lecture Notes in Computer Science*, pages 19–30. Springer-Verlag, 2000.
9. J. D. Golić. Correlation properties of a general binary combiner with memory. *Journal of Cryptology*, 9(2):111–126, 1996.
10. J. D. Golić. Linear models for keystream generators. *IEEE Trans. on Computers*, 45(1):41–49, Jan 1996.
11. J. D. Golić. Linear statistical weakness of alleged RC4 keystream generator. In W. Fumy, editor, *Advances in Cryptology – Eurocrypt'97*, volume 1233 of *Lecture Notes in Computer Science*, pages 226–238. Springer-Verlag, 1997.
12. H. Handschuh and H. Gilbert. χ^2 cryptanalysis of the SEAL encryption algorithm. In *Proceedings of the 4th Workshop on Fast Software Encryption*, volume 1267 of *Lecture Notes in Computer Science*, pages 1–12. Springer-Verlag, 1997.
13. T. Johansson and F. Jönsson. Fast correlation attacks based on turbo code techniques. In *Advances in Cryptology – CRYPTO '99*, volume 1666 of *Lecture Notes in Computer Science*, pages 181–197. Springer-Verlag, 1999.

14. T. Johansson and F. Jönsson. Improved fast correlation attacks on stream ciphers via convolution codes. In *Advances in Cryptology – Eurocrypt '99*, volume 1592 of *Lecture Notes in Computer Science*, pages 347–362. Springer-Verlag, 1999.
15. M. Matsui. Linear cryptanalysis method for DES cipher. In *Advances in Cryptology, EUROCRYPT'93*, volume 765 of *Lecture Notes in Computer Science*, pages 386–397. Springer-Verlag, 1993.
16. R. N. McDonough and A. D. Whalen. *Detection of Signals in Noise*. Academic Press, Inc., 2nd edition, 1995.
17. W. Meier and O. Staffelbach. Fast correlation attacks on stream ciphers. *Journal of Cryptology*, 1(3):159–176, 1989.
18. P. Rogaway and D. Coppersmith. A software optimized encryption algorithm. *Journal of Cryptology*, 11(4):273–287, 1998.
19. D. Sundararajan. *The Discrete Fourier Transform: Theory, Algorithms and Applications*. World Scientific Pub Co., 2001.
20. S. P. Vadhan. *A Study of Statistical Zero-Knowledge Proofs*. PhD thesis, MIT Department of Mathematics, August 1999.
21. D. Watanabe, S. Furuya, H. Yoshida, and B. Preneel. A new keystream generator MUGI. In *Fast Software Encryption*, Lecture Notes in Computer Science. Springer-Verlag, 2002. Description available on-line from http://www.sdl.hitachi.co.jp/crypto/mugi/index-e.html.

The Filter-Combiner Model
for Memoryless Synchronous Stream Ciphers

Palash Sarkar

Cryptology Research Centre
Applied Statistics Unit
Indian Statistical Institute
203, B.T. Road,
Kolkata 700035 India
palash@isical.ac.in

Abstract. We introduce a new model – the Filter-Combiner model –
for memoryless synchronous stream ciphers. The new model combines
the best features of the classical models for memoryless synchronous
stream ciphers – the Nonlinear-Combiner model and the Nonlinear-Filter
model. In particular, we show that the Filter-Combiner model provides
key length optimal resistance to correlation attacks and eliminates weaknesses of the NF model such as the the Anderson leakage and the Inversion Attacks. Further, practical length sequences extracted from the
Filter-Combiner model cannot be distinguished from true random sequences based on linear complexity test. We show how to realise the
Filter-Combiner model using Boolean functions and cellular automata.
In the process we point out an important security advantage of sequences
obtained from cellular automata over sequences obtained from LFSRs.

Keywords: synchronous stream ciphers, linear feedback shift registers,
cellular automata, nonlinear filter model, nonlinear combiner model,
filter-combiner model.

1 Introduction

Stream ciphers are a basic cryptographic primitive. They are used widely for both
defence communications and industrial applications. The underlying principle
behind stream ciphers is the following. Let $m^{(t)}$, $t \geq 0$ be the sequence of message
bits. Let $z^{(t)}$, $t \geq 0$ be a sequence of pseudorandom bits (also called the key
sequence). Then $c^{(t)} = m^{(t)} \oplus z^{(t)}$, $t \geq 0$ is the sequence of cipher bits. Decryption
is done by computing $c^{(t)} \oplus z^{(t)} = m^{(t)}$. The security of the system depends on
the security of the pseudorandom bits $z^{(t)}$.

Stream ciphers are usually classified into two broad categories – synchronous
and asynchronous stream ciphers. In synchronous stream ciphers the key bits do
not depend on the message or cipher bits while in asynchronous stream ciphers
the key bits depend on previous cipher and/or message bits. There are two
classical models of memoryless synchronous stream ciphers – the Nonlinear-
Filter model and the Nonlinear-Combiner model. See [12, 14, 5] for more details
on stream ciphers.

M. Yung (Ed.): CRYPTO 2002, LNCS 2442, pp. 533–548, 2002.

534   Palash Sarkar

Both the standard models are built using Linear Feedback Shift Registers (LFSRs) and Boolean functions. In the Nonlinear-Combiner model exactly one bit sequence is extracted from each LFSR and all the bit sequences are combined using a Boolean function to generate the key sequence. In the Nonlinear-Filter model several bit sequences are generated from a single LFSR and these are then combined using a Boolean function to generate the key sequence.

Here we introduce the Filter-Combiner model for memoryless synchronous stream ciphers. This model is a combination of the Nonlinear-Filter and the Nonlinear-Combiner model. In the Filter-Combiner model there are several Linear Finite State Machines (LFSMs) each of which generate multiple bit sequences. These sequences are combined using a Boolean function to produce the key sequence. We show that the Filter-Combiner model has the following features.

1. Provides key length optimal resistance to correlation attack and hence overcomes the main disadvantage of the Nonlinear-Combiner model.
2. Eliminates weaknesses of the Nonlinear-Filter model which arises due to the fact that multiple sequences are extracted from a single LFSR.
3. Practical sized key sequences extracted from the Filter-Combiner model cannot be distinguised from random strings based on linear complexity tests.

Thus the new model combines the best features of the previous two models. An important part in eliminating LFSR based weaknesses is the realisation of the LFSMs by Cellular Automata (CA). We identify the main problem of using LFSR in the Nonlinear-Filter model and show that this can be eliminated by using an important property of sequences obtained from CA. To the best of our knowledge, this is the first work to identify the important security advantage that can be obtained in replacing LFSR by CA.

We believe that as a consequence of our work, future models for practical stream ciphers will be based on the Filter-Combiner model rather than the Nonlinear-Filter or the Nonlinear-Combiner model.

2   Standard Models

$\mathbb{F}_2$ is the finite field of two elements and $\oplus$ denotes addition over $\mathbb{F}_2$ as well as the vector space $\mathbb{F}_2^l$ over $\mathbb{F}_2$. The common models of generating the key stream are built out of two kinds of primitives – linear finite state machines (LFSMs) and Boolean functions.

An l-bit LFSM $\mathcal{M}$ is a pair $(\mathbb{F}_2^l, M)$, where M is an $l \times l$ matrix. The internal state of $\mathcal{M}$ is described by an l-bit vector. The evolution of $\mathcal{M}$ over discrete time points $t \geq 0$ is described by a sequence of l-bit vectors $S^{(0)}, S^{(1)}, \ldots$, where $S^{(t+1)} = MS^{(t)}$. Thus only the vector $S^{(0)}$ (called the initial condition of $\mathcal{M}$) need to be specified for $\mathcal{M}$ to start operation. For $t \geq 0$, the vector $S^{(t)}$ will be written as $S^{(t)} = (s_1^{(t)}, \ldots, s_l^{(t)})$.

An n-variable Boolean function f is a map $f : \{0,1\}^n \to \{0,1\}$. The weight of a binary string s, denoted by $wt(s)$ is defined to be the number of ones in s.

2.1 Nonlinear-Filter (NF) Model

In this model one LFSM $\mathcal{M} = (\mathbb{F}_2^l, M)$ and one n-variable Boolean function $f(x_1, \ldots, x_n)$ are used. Let $S^{(0)}, S^{(1)}, \ldots$ be the sequence of n-bit vectors generated by $\mathcal{M}$. Then the key stream $z^{(t)}$ is obtained in the following manner.

$$z^{(t)} = f(s_{i_1}^{(t)}, \ldots, s_{i_n}^{(t)}), \qquad \text{for } t \geq 0, \tag{1}$$

where $S^{(t)} = (s_1^{(t)}, \ldots, s_l^{(t)})$, $i_1, \ldots, i_n \in \{1, \ldots, l\}$ and are distinct.

In this case, the secret key of the entire system is the initial condition $S^{(0)}$ of the LFSM giving rise to an l-bit secret key. We will call the nonlinear filter model the NF model.

2.2 Nonlinear-Combiner (NC) Model

In this model n LFSMs $\mathcal{M}_1 = (\mathbb{F}_2^{l_1}, M_1), \ldots, \mathcal{M}_n = (\mathbb{F}_2^{l_n}, M_n)$ and one n-variable Boolean function $f(x_1, \ldots, x_n)$ are used. Let $S_i^{(t)}$, $t \geq 0$ be the sequence of vectors generated by LFSM $\mathcal{M}_i$, $1 \leq i \leq n$. Further, let $S_i^{(t)} = (s_{i,1}^{(t)}, \ldots, s_{i,l_i}^{(t)})$. Then the key stream $z^{(t)}$ is generated in the following manner.

$$z^{(t)} = f(s_{1,1}^{(t)}, \ldots, s_{n,1}^{(t)}), \qquad \text{for } t \geq 0. \tag{2}$$

In this case the secret key of the entire system consists of the intial conditions $S_1^{(0)}, \ldots, S_n^{(0)}$ of all the LFSMs giving rise to an $(l_1 + \ldots + l_n)$-bit secret key. We will call the nonlinear combiner model the NC model.

3 Model Components

3.1 Linear Finite State Machines

Let $\mathcal{M} = (\mathbb{F}_2^l, M)$ be an LFSM generating the sequence of l-bit vectors $\mathcal{S} = S^{(0)}, S^{(1)}, \ldots$, where $S^{(t)} = (s_1^{(t)}, \ldots, s_l^{(t)})$. Let $p(x) = x^l \oplus a_{l-1}x^{l-1} \oplus \ldots \oplus a_1 x \oplus a_0$ be the characteristic polynomial for M. It is known that if $p(x)$ is primitive over $\mathbb{F}_2$, then the sequence $\mathcal{S}$ has period $2^l - 1$ (see [9]). Further, each of the sequences $s_i^{(t)}$, $1 \leq i \leq l$ also has period $2^l - 1$. This is the maximum possible period that can be obtained from a linear machine.

The most popular implementation of an LFSM is by a Linear Feedback Shift Register (LFSR). We will also consider implementation using Cellular Automata (CA). Below we briefly describe both LFSR and CA.

Linear Feedback Shift Register (LFSR). For an LFSR, the matrix M is the companion matrix of $p(x)$ and as a result the following two relations hold.

$$\left.\begin{array}{ll} s_{j+1}^{(t+1)} = s_j^{(t)} & t \geq 0, 1 \leq j < l, \\ s_1^{(t+1)} = \bigoplus_{i=0}^{l-1} a_{l-1+i} s_{i+1}^{(t)}. & \end{array}\right\} \tag{3}$$

Each of the sequences $s_i^{(t)}$, $1 \leq i \leq t$ satisfy a linear recurrence whose character-istic polynomial is $p(x)$ (see [9]). For $i \geq 1$ the sequence $s_{i+1}^{(t)}$ is obtained from the sequence $s_i^{(t)}$ by a single shift in the time domain. We record this as follows.

Fact 1 *The relative shift between two sequences $s_i^{(t)}$ and $s_j^{(t)}$ extracted from a single LFSR is $|i - j|$.*

An LFSR is simple to implement in hardware using an l-bit register and $l_1 = |\{i : a_i = 1, 0 \leq i \leq l - 1\}|$ XOR gates. The initial condition $S^{(0)}$ is loaded into the register to start operation. The next state is determined by (3).

Cellular Automata (CA). In case of CA the matrix M is a tridiagonal ma-trix. If the upper and lower subdiagonal entries of M are all 1 then the CA is called a 90/150 CA. We will only consider 90/150 CA. Let $c_1 \ldots c_l$ be the main diagonal entries of M. Then the following relations hold for the sequence of vectors $S^{(0)}, S^{(1)}, \ldots$.

$$\left.\begin{aligned} s_1^{(t+1)} &= c_1 s_1^{(t)} \oplus s_2^{(t)}, \\ s_i^{(t+1)} &= s_{i-1}^{(t)} \oplus c_i s_i^{(t)} \oplus s_{i+1}^{(t)} \quad \text{for } 2 \leq i \leq l - 1, \\ s_l^{(t+1)} &= s_{l-1}^{(t)} \oplus c_l s_l^{(t)}. \end{aligned}\right\} \quad (4)$$

A CA can be implemented in hardware using an l-bit register and l XOR gates. The initial condition $S^{(0)}$ is loaded into the register for the CA to start operation. The next state of the CA is obtained using (4).

For $1 \leq i < j < l$, the shift between the sequences $s_i^{(t)}$ and $s_j^{(t)}$ depends upon the CA being used. A general algorithm to compute these shifts have been obtained in [16]. Observations suggest that these shifts can be exponential in l. In Section 7, we discuss this point in detail and conclude that this feature is an important security advantage of CA over LFSR.

3.2   Boolean Functions

An n-variable Boolean function $f(x_1, \ldots, x_n)$ can be represented by a unique multivariate polynomial over $\mathbb{F}_2$. Thus $f(x_1, \ldots, x_n)$ can be written as

$$f(x_1, \ldots, x_n) = \bigoplus_{(i_1, \ldots, i_n) \in \mathbb{F}_2^n} g(i_1, \ldots, i_n) x_1^{i_1} \ldots x_n^{i_n} \quad (5)$$

where $g(x_1, \ldots, x_n)$ is another n-variable Boolean function. The representation of f in (5) is called the *algebraic normal form* (ANF) of f. The *degree* of f, $deg(f)$ is defined to be $\max\{wt(i_1 \ldots i_n) : g(i_1, \ldots, i_n) = 1\}$.

The *weight* of an n-variable Boolean function f is denoted by $wt(f)$ and is defined as $wt(f) = |\{(i_1, \ldots, i_n) \in \mathbb{F}_2^n : f(i_1, \ldots, i_n) = 1\}|$. The function f is *balanced* if $wt(f) = 2^{n-1}$. The *distance* between two n-variable functions f and g is denoted by $d(f, g)$ and is defined as $d(f, g) = |\{(i_1, \ldots, i_n) \in \mathbb{F}_2^n :$

$f(i_1, \ldots, i_n) \neq g(i_1, \ldots, i_n)\}|$. The probability that f and g are unequal is given by $Prob[f \neq g] = \frac{d(f,g)}{2^n}$.

The *Walsh transform* of f is an integer valued function $W_f : \{0,1\}^n \to [-2^n, 2^n]$ defined as $W_f(u) = \sum_{x \in \mathbb{F}_2^n} (1)^{f(x) \oplus \langle u, x \rangle}$, where $\langle u, x \rangle = u_1 x_1 \oplus \ldots \oplus u_n x_n$ is the inner product of u and x considered as vectors over $\mathbb{F}_2$.

The notion of correlation immune (CI) functions was introduced by Siegenthaler [18]. A characterization of correlation immunity in terms of Walsh transform was obtained in [21]. We present this characterization as our definition. An n-variable function f is said to be *correlation immune of order m (m-CI)* if $W_f(u) = 0$ for all $1 \leq wt(u) \leq m$. A balanced m-CI function is said to be *m-resilient*.

For $u \in \mathbb{F}_2^n$, let $\lambda_u(x_1, \ldots, x_n)$ be a linear function defined as

$$\lambda_u(x_1, \ldots, x_n) = \langle u, (x_1, \ldots, x_n) \rangle.$$

Then $W_f(u) = 2^n - 2 \times d(f, \lambda_u)$. Let $A_n = \{\lambda_u \oplus b : u \in \mathbb{F}_2^n, b \in \{0,1\}\}$ be the set of all n-variable affine functions. The nonlinearity of f is defined to be $nl(f) = \min_{g \in A_n} d(f, g)$. Equivalently, this can be written as $nl(f) = 2^{n-1} - \frac{1}{2} \max_{u \in \mathbb{F}_2^n} |W_f(u)|$. Any function $g \in A_n$ such that $d(f, g) = nl(g)$ is said to be *a best affine approximation* of f.

4   Correlation Attacks

The currently known most powerful class of attacks on both the NF and the NC model is the class of correlation attacks. We describe the basic idea of a correlation attack with reference to the NC model.

In the NC model, n input bit sequences are combined by an n-variable Boolean function $f(x_1, \ldots, x_n)$ to produce the key sequence $z^{(t)}$. For notational convenience we will denote the n input sequence to f by $x_1^{(t)}, \ldots, x_n^{(t)}$. The input sequence $x_i^{(t)}$ is produced by an LFSM of length l_i. For $t \geq 0$, we have

$$z^{(t)} = f(x_1^{(t)}, \ldots, x_n^{(t)}). \tag{6}$$

Suppose $W_f(u) \neq 0$ for some $u \in \mathbb{F}_2^n$, with $wt(u) = 1$. Let i be such that $u_i = 1$ and for $j \neq i$, $u_j = 0$. In this situation first order correlation attacks are applicable. The function $\lambda_u(x_1, \ldots, x_n)$ is equal to x_i. The idea is to use the bias $\beta_u = |Prob(\lambda_u = f) - \frac{1}{2}| = \frac{|W_f(u)|}{2^{n+1}}$ to estimate the sequence $x_i^{(t)}$ from the sequence $z^{(t)}$ (or even from the cipher sequence $c^{(t)}$). This was originally proposed by Siegenthaler [19]. Recently a great deal of work has been done in this area (see [3, 4]).

If $\beta_u = 0$ for all u with $wt(u) = 1$, then it is not possible to directly estimate any $x_i^{(t)}$ from $z^{(t)}$. In this case a higher order attack can be carried out as follows. Suppose f is m-CI but not $(m + 1)$-CI. Then there exists $u \in \mathbb{F}_2^n$ with $wt(u) = m + 1$ such that $W_f(u) \neq 0$. Let $i_1, \ldots, i_{m+1}$ be such that $u_{i_1} = \ldots = u_{i_{m+1}} = 1$ and $u_j = 0$ for $j \notin \{i_1, \ldots, i_{m+1}\}$. Define β_u as before. Then the bias β_u is used to estimate the sequence $y^{(t)} = x_{i_1}^{(t)} \oplus \ldots \oplus x_{i_{m+1}}^{(t)}$. The individual sequences $x_{i_1}^{(t)}, \ldots, x_{i_{m+1}}^{(t)}$ can be obtained from $y^{(t)}$ by solving a system of linear equations.

Define $L = l_{i_1} + \ldots + l_{i_{m+1}}$. The linear complexity (see Section 8) of the sequence $x_{i_1}^{(t)} \oplus \ldots \oplus x_{i_{m+1}}^{(t)}$ is L (see Lemma 1). Let N be the number of key bits required to successfully carry out the attack. The parameter N depends on the bias β_u and the length L. Most work on correlation attacks present only simulation studies. Recently, some theoretical analysis has been done in [4, 3]. We briefly describe the analysis from [4].

$$N \simeq \frac{1}{4} \cdot (2kt!\ln 2)^{\frac{1}{t}} \cdot \beta_u^{-2} \cdot 2^{\frac{L-k}{t}}, \tag{7}$$

where k and t are algorithm parameters. The attack stores certain parity check relations and consists of a precomputation phase and a decoding phase. The complexity of the precomputation phase is approximately $N^{\lceil (t-1)/2 \rceil}$ and requires $N^{\lfloor (t-1)/2 \rfloor}$ memory. The number of parity check relations that need to be stored is roughly $\frac{N^t}{t!} \cdot 2^{-(L-k)}$ and the decoding complexity is 2^k times the number of parity checks. Thus the attack becomes infeasible if either β_u is sufficiently close to 0 or L is sufficiently large.

4.1   Resistance of the NC Model to Correlation Attacks

For $u \in \{0,1\}^n$, define $l(u) = u_1 l_1 + \cdots + u_n l_n$. For an m-resilient function f define

$$\alpha_f = \min_{W_f(u) \neq 0, wt(u) = m+1} l(u).$$

The lengths of the LFSMs in the NC model are $l_1, \ldots, l_n$ and the secret key length is $l = l_1 + \cdots + l_n$. However, the complexity of a correlation attack depends on the parameter α_f which is less than l. Thus we obtain the following fact.

Fact 2 *The resistance to correlation attack provided by the NC model is suboptimal in the secret key length.*

Remark: *A consequence of this fact is that to obtain a desired level of security one has to choose a significantly longer secret key. This is clearly a major shortcoming of the NC model.*

5   The Filter-Combiner (FC) Model

In this section we present our new model – the Filter-Combiner Model. We will call this model the FC model. We present a formal description of the model.

Components of the model: An n-variable Boolean function $f(x_1, \ldots, x_n)$ and k $(1 < k < n)$ LFSMs $\mathcal{M}_1 = (\mathbb{F}_2^{l_1}, M_1), \ldots, \mathcal{M}_k = (\mathbb{F}_2^{l_k}, M_k)$. The characteristic polynomials of $M_1, \ldots, M_k$ are chosen to be primitive and $l_1, \ldots, l_k$ are chosen to be all distinct.

Keystream generation: LFSM $\mathcal{M}_j$ produces l_j bit sequences. Out of these i_j bit sequences $y_{j,1}, \ldots, y_{j,i_j}$ are chosen, where $i_1 + \ldots + i_k = n$. The key stream $z^{(t)}$ is generated as follows.

$$z^{(t)} = f(y_{1,1}^{(t)}, \ldots, y_{1,i_1}^{(t)}, y_{2,1}^{(t)}, \ldots, y_{2,i_2}^{(t)}, \ldots, y_{k,1}^{(t)}, \ldots, y_{k,i_k}^{(t)}) \quad \text{for } t \geq 0. \quad (8)$$

Constraints on the model: Denote the sequences

$$y_{1,1}^{(t)}, \ldots, y_{1,i_1}^{(t)}, y_{2,1}^{(t)}, \ldots, y_{2,i_2}^{(t)}, \ldots, y_{k,1}^{(t)}, \ldots, y_{k,i_k}^{(t)}$$

by $x_1^{(t)}, \ldots, x_n^{(t)}$, where $x_1, \ldots, x_n$ are the input variables to the function f. For each variable x_i define $FSM(x_i) = j$ such that the sequence $x_i^{(t)}$ is one of the sequences $y_{j,1}^{(t)}, \ldots, y_{j,i_j}^{(t)}$. The following conditions must hold on the model.

1. If $W_f(u) \neq 0$, then $\{FSM(x_{i_1}), \ldots, FSM(x_{i_p})\} = \{1, \ldots, k\}$, where $u_{i_1} = \ldots = u_{i_p} = 1$ and $u_j = 0$ for $j \notin \{i_1, \ldots, i_p\}$.
2. For $1 \leq j \leq k$, i_j bit sequences are extracted from LFSM $\mathcal{M}_j$ where $i_1 + \cdots + i_k = n$. Let $n = qk + r = r(q+1) + (k-r)q$, where $0 \leq r < k$. We require $i_1 = \ldots = i_r = \lceil \frac{n}{k} \rceil$ and $i_{r+1} = \ldots = i_k = \lfloor \frac{n}{k} \rfloor$.
3. $i_j \leq \log_2 l_j$ for $1 \leq j \leq k$.
4. If $FSM(x_i) = FSM(x_j) = p$, then the shift between the sequences $x_i^{(t)}$ and $x_j^{(t)}$ must be in the range $[\frac{2^{l_p}}{i_p} - \epsilon_p, \frac{2^{l_p}}{i_p} + \epsilon_p]$ for some $\epsilon_p \ll 2^{l_p}$.
5. The maximum length of a message that should be encrypted by the system is $\min_{1 \leq j \leq k}(\frac{2^{l_j}}{i_j} - \epsilon_j)$.

Remark: Suppose $x^{(t)}$ and $y^{(t)}$ are obtained from a LFSM of length l having period $2^l - 1$. Further suppose the shift between $x^{(t)}$ and $y^{(t)}$ is s. Since the sequences $x^{(t)}$ and $y^{(t)}$ both have period $2^l - 1$, the backward shift between these two sequences is $2^l - 1 - s$. We would like to have both the forward and backward shifts between $x^{(t)}$ and $y^{(t)}$ to be exponential in l. Hence in Constraint 4 above we require the (forward) shift between $x^{(t)}$ and $y^{(t)}$ to be within a certain range instead of requiring a lower bound on this shift.

Proposition 1. *Let f be m-resilient and suppose Constraint 1 holds. Then $k \leq m + 1$.*

Proof. Suppose $k > m + 1$ and $u \in \mathbb{F}_2^n$ be such that $wt(u) = m + 1$ and $W_f(u) \neq 0$. Let $u_{i_1} = \ldots = u_{i_{m+1}} = 1$ and $u_j = 0$ for $j \notin \{i_1, \ldots, i_{m+1}\}$. Then $|\{FSM(x_{i_1}), \ldots, FSM(x_{i_{m+1}})\}| \leq m + 1 < k$. Hence Constraint 1 is violated. $\square$

Proposition 2. *Suppose Constraint 2 holds. Then Constraint 3 holds if and only if*

$$l_j \geq 2^{\lceil (n/k) \rceil} \quad \text{if } 1 \leq j \leq r$$
$$\geq 2^{\lfloor (n/k) \rfloor} \quad \text{if } r + 1 \leq j \leq k. \quad (9)$$

Proposition 3. *Suppose Constraints 3 and 4 hold. Then the shift between the sequences* $x_i^{(t)}$ *and* $x_j^{(t)}$ *is at least* $2^{l_p - \log_2 \log_2 l_p} - \epsilon_p$, *where* $FSM(x_i) = FSM(x_j) = p$.

Remark: 1. Proposition 3 assures us that the shift between any two sequences obtained from the same LFSM is "exponential" in the length of the LFSM.
2. Constraint 5 guarantees that no bit generated by any LFSM is used more than once.
3. Constraint 4 is to be contrasted with Fact 1 in Section 3.1. An immediate consequence is that Constraint 4 cannot be realised using LFSR. In Section 9.2 we show that Constraint 4 can be achieved using CA.

6   Resistance to Correlation Attacks

In this section we show that the resistance to correlation attacks provided by the FC model is optimal in the key length. This is a direct consequence of Constraint 1 in the design criteria. We first prove the following result.

Lemma 1. *Let* $x_1^{(t)}$ *and* $x_2^{(t)}$ *be two linear recurring sequences having distinct characteristic polynomials* $p_1(x)$ *and* $p_2(x)$ *of degrees* d_1 *and* d_2 *respectively. Assume that* $p_1(x)$ *and* $p_2(x)$ *are both primitive. Then the linear complexity of the sequence* $x^{(t)} = x_1^{(t)} \oplus x_2^{(t)}$ *is* $d_1 + d_2$.

Proof. Let $\alpha_1, \ldots, \alpha_{d_1}$ be the roots of $p_1(x)$ and $\beta_1, \ldots, \beta_{d_2}$ be the roots of $p_2(x)$. We can write (see for example [15])

$$\begin{aligned} x_1^{(t)} &= A_1 \alpha_1^t \oplus \ldots \oplus A_{d_1} \alpha_{d_1}^t, \\ x_2^{(t)} &= B_1 \beta_1^t \oplus \ldots \oplus B_{d_2} \beta_{d_2}^t. \end{aligned} \tag{10}$$

Here $A_1, \ldots, A_{d_1}$ (resp. $B_1, \ldots, B_{d_2}$) are constants determined solely by the initial d_1 (resp. d_2) bits of $x_1^{(t)}$ (resp. $x_2^{(t)}$). Thus we can write

$$x^{(t)} = A_1 \alpha_1^t \oplus \ldots \oplus A_{d_1} \alpha_{d_1}^t \oplus B_1 \beta_1^t \oplus \ldots \oplus B_{d_2} \beta_{d_2}^t \tag{11}$$

The roots $\alpha_1, \ldots, \alpha_{d_1}$ and $\beta_1, \ldots, \beta_{d_2}$ are elements of the field $GF(2^{\text{lcm}(d_1, d_2)})$. Since $p_1(x)$ and $p_2(x)$ are primitive it is not difficult to see that $\{\alpha_1, \ldots, \alpha_{d_1}\} \cap \{\beta_1, \ldots, \beta_{d_2}\} = \emptyset$. Hence using (11) it follows that the linear complexity of $x^{(t)}$ is $d_1 + d_2$ (see [15]).   □

Theorem 1. *The FC model provides key length optimal resistance to correlation attacks.*

Proof: Let $f(x_1, \ldots, x_n)$ be the m-resilient Boolean function which combines the input bit sequences. Let $x_{i_1}, \ldots, x_{i_{m+1}}$ be such that $W_f(u) \neq 0$, where $u_{i_1} = \ldots = u_{i_{m+1}} = 1$ and for $j \notin \{i_1, \ldots, i_{m+1}\}$, $u_j = 0$. Using Constraint 1 this implies $\{FSM(x_{i_1}), \ldots, FSM(x_{i_{m+1}})\} = \{1, \ldots, k\}$.

Let the characteristic polynomials of the LFSMs be $p_1(x), \ldots, p_k(x)$ of degrees $l_1, \ldots, l_k$. Let the roots of the polynomial $p_i(x)$ be $\alpha_{i,j}$, $1 \le j \le l_i$ in the field $GF(2^a)$, where $a = \operatorname{lcm}(l_1, \ldots, l_k)$. Then any bit sequence $y^{(t)}$ obtained from LFSM $\mathcal{M}_i$ can be written as $y^{(t)} = A_1 \alpha_{i,1}^t \oplus \ldots \oplus A_{l_i} \alpha_{i,l_i}^t$, where $A_1, \ldots, A_{l_i}$ are constants dependent on the initial l_i bits of the sequence $y^{(t)}$.

Let $x^{(t)} = x_{i_1} \oplus \ldots \oplus x_{i_{m+1}}$. The characteristics polynomials $p_1(x), \ldots, p_k(x)$ are primitive by model criteria. Hence the roots $\alpha_{i,j}$ are all distinct. Using the fact that $\{FSM(x_{i_1}), \ldots, FSM(x_{i_{m+1}})\} = \{1, \ldots, k\}$, we can write

$$x^{(t)} = \bigoplus_{i=1}^{k} \bigoplus_{j=1}^{l_i} C_{i,j} \alpha_{i,j}. \tag{12}$$

Here $C_{i,j} \in GF(2^a)$ are constants and are completely determined by the bits

$$x_{i_1}^{(1)}, \ldots, x_{i_1}^{(l_1)}, \ldots, x_{i_{m+1}}^{(1)}, \ldots, x_{i_{m+1}}^{l_{m+1}}.$$

Hence the linear complexity of the sequence $x^{(t)}$ is $L = l_1 + \ldots + l_k$. In other words, if we want to obtain $x^{(t)}$ by a linear recurrence, then the degree of the characteristic polynomial of the recurrence is at least L. Thus in any correlation attack, the number of key bits required for a successful attack depends on L. Since the length of the secret key is also L, the resistance to correlation attacks is optimal in the key length.   □

Remark: *Theorem 1 shows that with respect to correlation attacks the FC model is superior to the NC model (see Fact 2 in Section 4.1).*

7 Eliminating Weaknesses of the NF Model

In traditional implementation of the NF model a single LFSR is used to implement the LFSM. This means that more than one sequence is extracted from a single LFSR. Extracting more than one sequence from a single LFSR makes the system vulnerable to certain kinds of attacks.

Anderson Leakage: Suppose an LFSR of length l and an n-variable function $f(x_1, \ldots, x_n)$ is used to implement the NF model. Let the l sequences of the LFSR be $s_i^{(t)}$, for $1 \le i \le t$. Suppose the sequence $x_j^{(t)} = s_{i_j}^{(t)}$ for $1 \le j \le n$. Then the relative shift between two sequences $x_{j_1}^{(t)}$ and $x_{j_2}^{(t)}$ is $|i_{j_1} - i_{j_2}| \le l$. Since the period of any of the sequences $x_j^{(t)}$ is $2^l - 1$, the relative shifts between the sequences are comparatively small. Thus the inputs to the function f are obtained from the same sequence with small shifts. This results in information leakage from the input to the output even if the function f is resilient. No general algorithm is known which can exploit this attack. However, Anderson [1] has provided convincing evidence of the leakage phenomenon.

We use Proposition 3 and Constraint 5 to show that the FC model is resistant to Anderson leakage. Proposition 3 states that the relative shift in the sequences extracted from two tap positions must be "exponential" in the length of the LFSM. Constraint 5 states that the maximum length of the message that should be enciphered by the system is less than the minimum shift between any two sequences obtained from a single LFSM. Thus any bit of an extracted sequence is used at most once to generate the pseudo random key stream. Thus Anderson leakage is not applicable to the FC model.

Inversion Attacks: The idea behind a basic or generalized inversion attack [7, 8] is the following. Suppose the LFSR used is of length l. The attack proceeds as follows.

1. Guess q ($< l$) bits of the initial condition.
2. Extend these q bits to l bits using $(l - q)$ of the known bits of the keystream and the relation among the bits of the LFSR defined by the Boolean function.
3. Use the l-bits to generate a segment of the key and check whether this segment is equal to the segment produced by the secret initial condition. If two are equal, then the l-bits form a possibly correct initial condition.

Step 2 is the most important step in the attack. However, the realisation of this step is crucially dependent on the fact that the n input sequences to the Boolean function satisfy the same linear recurrence, i.e., they are obtained from a single LFSR.

In the FC model, the input sequences to the Boolean function satisfy distinct linear recurrences. There does not seem to be any way of applying the inversion attack even when the input sequences are obtained from only two distinct linear recurrences. In fact, it appears that this is also the reason why the inversion attack has not been applied to the NC model.

Remark: An anonymous referee has provided an example to show that the property of exponential size shifts between the sequences do not necessarily provide resistance to inversion attacks.

We summarize the discussion of this section in the following fact.

Fact 3 *Anderson leakage and Inversion attacks are not applicable to the FC model.*

Remark: *Combining the results of Sections 6 and 7 we see that with respect to the considered attacks the FC model is superior to both the NF and the NC models.*

8   Linear Complexity

Given a bit sequence, a parameter of fundamental importance is its *linear complexity* which is defined to be the length of a minimum length LFSR which can generate the sequence. The linear complexity of a bit sequence generated by an LFSM $\mathcal{M} = (\mathbb{F}_2^l, M)$ is l. Given an arbitrary bit sequence, its linear complexity can be determined using the Berlekamp-Massey algorithm [11]. We record some facts about linear complexity.

Fact 4 *The expected linear complexity of a random string of length L is $\lfloor \frac{L}{2} \rfloor$ (see [12]).*

Fact 5 *In case of the NC model the linear complexity can be determined using a result of Rueppel and Staffelbach [15]. Suppose the lengths of the LFSMs are $l_1, \ldots, l_n$ and the sequences are combined using an n-variable Boolean function $f(x_1, \ldots, x_n)$ whose ANF is $\bigoplus_{(i_1,\ldots,i_n)\in\mathbb{F}_2^n} g(i_1, \ldots, i_n)x_1^{i_1} \ldots x_n^{i_n}$. The linear complexity of $z^{(t)} = x_1^{(t)} \oplus \ldots x_n^{(t)}$ is $\leq \sum_{(i_1,\ldots,i_n)\in\mathbb{F}_2^n} g(i_1, \ldots, i_n)l_1^{i_1} \ldots l_n^{i_n}$ where equality is achieved if the lengths l_i, $1 \leq i \leq n$ are all distinct.*

Fact 5 shows that $(1 + l_1) \ldots (1 + l_n)$ is an upper bound on the maximum possible linear complexity in the NC model. Note that this upper bound is substantially less than the value $2^{l_1 + \cdots + l_n}$.

For the NF model, it is more difficult to compute the linear complexity of the generated entire key sequence. Let l be the secret key length. Rueppel [14] has shown that for a class of Boolean functions it is possible to generate a key sequence of guaranteed linear complexity at least $\binom{l}{\lfloor \frac{l}{2} \rfloor}$. However, the functions in this class do not necessarily satisfy the other requirements of high nonlinearity, high correlation immunity (see [5]).

In case of the FC model, it is difficult to compute the linear complexity of the entire sequence. Instead we conducted several experiments with different set ups. We describe two set ups.

1. System 1 used 3 CA of lengths 15,16 and 17 bits whose characteristic polynomials are primitive. Two sequences were extracted from the first two CA and three sequences were extracted from the third CA satisfying Constraint 4 of the FC model. A 7-variable, resiliency 3, degree 3, nonlinearity 48 function was used to combine the extracted sequences satisfying Constraint 1 of the FC model. The secret key length of the system is 48 bits.

2. System 2 used 3 CA of lengths 16,17 and 18 bits with primitive characteristic polynomials. Two sequences were extracted from the first CA and three sequences each were extracted from the last two CA satisfying Constraint 4 of the FC model. A 8-variable, resiliency 4, degree 3, nonlinearity 96 function was used to combine the extracted sequences satisfying Constraint 1 of the FC model. The secret key length is 51 bits.

In each of the above two cases we generated key sequences of lengths L equal to $2^{10}, 2^{11}, 2^{12}, 2^{13}, 2^{14}, 2^{15}$ from randomly chosen secret keys (initial configurations of the CA involved). The linear complexity was obtained in each case using the Berlekamp-Massey algorithm as described in [12]. In all our experiments we obtained linear complexity very close to $\frac{L}{2}$, which is as expected for a random bit sequence.

The secret key sizes of 48 and 51 bits are not sufficient in practical stream ciphers. In practical situations the secret key length would be at least 128 and the generated key sequence would be at most 2^{30} between two key changes. It would have been better to test the linear complexity of key sequences of length around 2^{30}. However, the Berlekamp-Massey algorithm requires L^2 operations to compute the linear complexity of a key sequence of length L (see [12]). Thus computing the linear complexity of a sequence of length 2^{30} would require around

2^{60} operations. This makes it impractical to run such experiments. On the other hand, our experiments confirm the following fact.

Fact 6 *If the length L of the extracted key sequence of the FC model is small compared to 2^l (where l is the secret key length), then the linear complexity of the sequence cannot be distinguished from the linear complexity of a random string.*

9   Realization of the FC Model

There are four main constraints on the model that must be satisfied to build a particular system. The first concerns the Boolean function and the connection of the Boolean function to the LFSMs. The second to fourth concerns the implementation of the LFSMs. We describe methods for satisfying these constraints.

9.1   Satisfying Constraints 1 and 2

For $u \in \{0,1\}^n$, define $A_u = \{i_1, \ldots, i_p\}$, where $u_{i_1} = \ldots = u_{i_p} = 1$ and $u_j \neq 0$ for $j \notin \{i_1, \ldots, i_p\}$.

Constraint 1 asks the following question. Can we construct an n-variable, m-resilient Boolean function such that the variables $x_1, \ldots, x_n$ can be partitioned into k sets $A_1, \ldots, A_k$, where $A_i \cap A_u \neq \emptyset$ for $1 \leq i \leq k$ and for each $u \in \mathbb{F}_2^n$ with $W_f(u) \neq 0$? We now describe a simple solution to this problem. We begin with the following simple result.

Proposition 4. *Let $f(x_1, \ldots, x_n)$ be a Boolean function of the form*

$$f(x_1, \ldots, x_n) = x_1 \oplus \ldots \oplus x_k \oplus g(x_{k+1}, \ldots, x_n).$$

If $W_f(u) \neq 0$, then $u_1 = \ldots = u_k = 1$.

Proof: Suppose that for some $j \in \{1, \ldots, k\}$, we have $u_j = 0$. Then the variable x_j does not occur in the linear function $l_u(x_1, \ldots, s_n) = \langle u, (x_1, \ldots, x_n) \rangle$. Thus the function

$$f(x_1, \ldots, x_n) \oplus l_u(x_1, \ldots, x_n) = x_j \oplus h(x_1, \ldots, x_{j-1}, x_{j+1}, \ldots, x_n),$$

for some function h. Hence $f \oplus l$ is a balanced function and so $W_f(u) = 0$.   □

We can now describe our construction. Let f be an n-variable, m-resilient function of the form

$$f(x_1, \ldots, x_n) = x_1 \oplus \ldots \oplus x_k \oplus g(x_{k+1}, \ldots, x_n). \tag{13}$$

Construction: Construct the sets A_j $(1 \leq j \leq k)$ as follows.

1. Put element j in A_j.
2. Distribute the elements $k+1, \ldots, n$ to the sets A_j such that $|A_j| = \lceil (n/k) \rceil$ if $1 \leq j \leq r$ and $|A_j| = \lfloor (n/k) \rfloor$ if $r+1 \leq j \leq k$. Note that this can easily be done.

This construction ensures that for any u such that $W_f(u) \neq 0$, we have $A_j \cap A_u \neq \emptyset$ for all $1 \leq j \leq k$. Thus Constraint 1 is satisfied. We extract the input sequences $x_{i_1}^{(t)}, \ldots, x_{i_j}^{(t)}$ from LFSM $\mathcal{M}_i$. By construction, each $|A_j|$ is either $\lfloor (n/k) \rfloor$ or $\lceil (n/k) \rceil$. Hence Constraint 2 is also satisfied.

We briefly comment on the availability of n-variable, m-resilient Boolean functions of the form described in (13). The construction of functions in the form (13) was first described by Siegenthaler [18]. Later work [2, 10] have investigated this construction. Note that a necessary condition is that $k \leq m$. Under this condition it is always possible to get n-variable, m-resilient functions in the form (13). In fact, for certain values of the parameters n and m, it is also possible to get functions in the form (13) which achieve the best possible trade-off among resiliency, degree and nonlinearity (see [17]).

Let us now turn the question around and consider the following problem. We first describe Constraint 1 formally as a decision problem.

Problem: CONS1

Instance: A family $\mathcal{F} = \{A_u \subset \{1, \ldots, n\} : u \in \mathbb{F}_2^n, W_f(u) \neq 0\}$, where f is an n-variable, m-resilient Boolean function and a positive integer k such that $2 \leq k \leq m$.

Question: Is there a k-partition $A_1, \ldots, A_k$ of $\{1, \ldots, n\}$ such that $A_u \cap A_i \neq \emptyset$, for every $A_u \in \mathcal{F}$?

Even though Constraint 1 has been described as a decision problem we are really interested in an actual k-partition $A_1, \ldots, A_k$. If we are able to obtain such a partition, then for each A_i we can assign the variables $x_{j_1}, \ldots, x_{j_i}$ to the FSM i. Solving CONS1 does not seem to be easy in general. We describe a modified version of CONS1 which is easily proved to be NP-complete.

Problem: Generalized Set Splitting (GSS)

Instance: A family $\mathcal{F} = \{T \subset \{1, \ldots, n\} : |T| \geq m\}$, and a positive integer k with $2 \leq k < n$ and $k \leq m$.

Question: Is there a k-partition $A_1, \ldots, A_k$ of $\{1, \ldots, n\}$ such that $A_i \cap T \neq \emptyset$ for $1 \leq i \leq k$ and for each $T \in \mathcal{F}$?

The GSS problem is a generalized version of the set splitting problem (see [6, page 221]) and is easily proved to be NP-complete. This does not prove the CONS1 problem to be NP-complete, since in CONS1 the family $\mathcal{F}$ is obtained from the nonzero points of the Walsh transform of a Boolean function whereas in GSS the family $\mathcal{F}$ is an arbitrary collection. Thus it may be possible to use algebraic properties of the Walsh transform of f to solve CONS1 easily even though GSS is NP-complete. However, the NP-completeness of GSS is very strong evidence of the intractibility of solving CONS1.

Remark: Given an n-variable, m-resilient Boolean function it might not be possible to satisfy Constraint 1, i.e., there might not be a proper partition or it might be computationally intractible to find a proper partition. However, we have shown that for a large class of cryptographically significant functions it is always possible to satisfy Constraint 1. Also there are examples of functions not of the type (13) for which it is possible to satisfy Constraint 1. Further research will throw more light on the set of functions which satisfy Constraint 1.

9.2   Satisfying Constraints 3 and 4

Constraint 4 depends on the properties of the $\mathcal{M}_1, \ldots, \mathcal{M}_k$. Suppose sequences $x_{j_1}^{(t)}, \ldots, x_{j_i}^{(t)}$ are extracted from $\mathcal{M}_i$. We require the relative shift between two sequences $x_{j_k}^{(t)}$ and $x_{j_p}^{(t)}$ to be exponential in l_i.

We consider the use of CA to implement the LFSMs to satisfy Constraint 4. (Note that from Fact 1 in Section 3.1 it follows that LFSRs cannot be used to satisfy Constraint 4.) To do this we need to do the following two things.

1. Given an primitive polynomial $p(x)$ of degree l, we need to construct a 90/150 CA which realizes $\mathcal{M} = (\mathbb{F}_2^l, M)$ such that the characteristic polynomial of M is $p(x)$.

2. Given a 90/150 CA producing l-bit state vectors $S^{(t)} = (s_1^{(t)}, \ldots, s_l^{(t)})$, we need to compute the relative shift between any two sequences $s_i^{(t)}$ and $s_j^{(t)}$.

Based on a result by Mesirov and Sweet [13], an efficient solution to the first problem has been presented in [20]. Further, in [16] an algorithm to solve the second problem has been presented.

Experimental results based on the algorithm of [16] show the following Fact.

Fact 7 *For a 90/150 CA of length l with primitive characteristic polynomial, it is possible to obtain at least p ($\log_2 l \leq p < l$) positions such that the relative shift between any two pair of these p positions is in the range $[\frac{2^l}{p} - \epsilon, \frac{2^l}{p} + \epsilon]$ for some $\epsilon \ll 2^l$.*

Remark: Fact 7 should be contrasted with Fact 1. This underlines the enhanced security features of CA sequences over LFSR sequences.

It immediately follows from Fact 7 that Constraints 3 and 4 can be satisfied using CA. For the purpose of illustration we present a concrete example of a 24-cell 90/150 CA.

Example: Consider a 24 cell CA. Choose $p(x) = x^{24} \oplus x^4 \oplus x^3 \oplus x \oplus 1$ to be the characteristic polynomial of the CA. The polynomial $p(x)$ is primitive (see [12, page 161]). We wish to obtain a 90/150 CA whose characteristic polynomial is $p(x)$. It is enough to obtain the main diagonal entries of the state transition matrix (see Section 3.1). The main diagonal entries can be described by a 24-bit string. Using the algorithm of [20], we obtain this string to be 110100111001001111001011. Let the sequences obtained from the 24 cells of the CA be denoted by $s_1^{(t)}, \ldots, s_{24}^{(t)}$. Define integers $b_1 = 0, b_2, \ldots, b_{24}$, such that $s_1^{(t)} = s_i^{(t+b_i)}$ for all $t \geq 0$ and $1 \leq i \leq 24$. Using the algorithm of [16], we obtain the values $(b_1, \ldots, b_{24})$ to be equal to

$$(0, 11662498, 16777213, 3837988, 12949649, 13910896, 13911015, 959496,$$
$$3720499, 15512414, 9453076, 13780753, 15184694, 2216344, 15313151, 3521236,$$
$$760233, 13711752, 13711633, 12750386, 3638725, 16577950, 11463235, 16577952).$$

We select tap positions $1, 4, 11$ and 20 and extract $4 = \lfloor \log_2(24) \rfloor$ bit sequences from the CA. The tuple $(b_1, b_4, b_{11}, b_{20}) = (0, 3837988, 9453076, 12750386)$ represents the shift of the 4 sequences from the first sequence. The value of 2^{24} is 16777216. We have

1. $b_4 = 2^{22} - a_4$, where $a_4 = 356316$.
2. $b_{11} = 2^{23} + a_{11}$, where $a_{11} = 1064468$.
3. $b_{20} = 2^{23} + 2^{22} + a_{20}$, where $a_{20} = 167474$.

We have $\min\{b_4 - b_1, b_{11} - b_4, b_{20} - b_{11}, 2^{24} - 1 - b_{20}\} = \min\{2^{22} - a_4, 2^{22} + a_{11} + a_4, 2^{22} + a_{20} - a_{11}, 2^{22} - a_{20} - 1\} = 2^{22} - 886994$. Thus the system can encrypt messages of length $2^{22} - 886994 > 2^{21}$. $\qquad\square$

10 Conclusion

In this paper we have introduced new ideas to improve upon the well studied classical models of stream ciphers. An important constituent of our model is the use of cellular automata. We point out an important security advantage of cellular automata over linear feedback shift registers. We believe that our model will form the basic skeleton for designing new practical stream cipher systems.

Acknowledgement

The author wishes to thank Rana Barua and Sounak Mishra for discussions on the set splitting problem. Also comments from the anonymous referees have helped in clearing up certain technical points.

References

1. R. J. Anderson. Searching for the optimum correlation attack. In *Fast Software Encryption – FSE 1994*, pp 137-143.
2. P. Camion, C. Carlet, P. Charpin, and N. Sendrier. On correlation immune functions. In *Advances in Cryptology – CRYPTO'91*, pages 86–100. Springer-Verlag, 1992.
3. A. Canteaut and M. Trabbia. Improved fast correlation attacks using parity checks equations of weight 4 and 5. Advances in Cryptology – EUROCRYPT 2000, Lecture Notes in Computer Science, pp 573-588.
4. V. Chepysov, T. Johansson and B. Smeets. A simple algorithm for fast correlation attacks on stream ciphers, In *Fast Software Encryption – FSE 2000*, Lecture Notes in Computer Science.
5. C. Ding, G. Xiao, and W. Shan. *The Stability Theory of Stream Ciphers*. Number 561 in Lecture Notes in Computer Science. Springer-Verlag, 1991.
6. M.R. Garey and D.S. Johnson. Computers and Intractibility: A Guide to the Theory of NP-completeness. W.H. Freeman, San Francisco, 1979.
7. J. D. Golic. On the Security of Nonlinear Filter Generators. *Fast Software Encryption – Cambridge '96*, D. Gollman, ed., 1996.

8. J. D. Golic, A. Clark and E. Dawson. Generalized inversion attack on nonlinear filter generators. *IEEE Transactions on Computers*, 49(10):1100-1109 (2000).
9. R. Lidl and H. Niederreiter. Introduction to finite fields and their applications. Cambridge University Press, revised edition, 1994.
10. S.Maitra and P. Sarkar. Highly nonlinear resilient functions optimizing Siegenthaler's inequality. *Advances in Cryptology - CRYPTO 1999*, Lecture Notes in Computer Science, pp 198-215.
11. J.L. Massey. Shift register synthesis and BCH decoding. *IEEE Transactions on Information Theory,*, 15(1969), 122-127.
12. A. J. Menezes, P. C. van Oorschot, and S. A. Vanstone. Handbook of Applied Cryptography. CRC Press, 1997.
13. J. P. Mesirov and M. M. Sweet. Continued fraction expansions of rational expressions for built-in self-test. *Journal of Number Theory*, 27, 144-148 (1987).
14. R. A. Rueppel. Analysis and Design of Stream Ciphers Springer-Verlag, 1986.
15. R. A. Rueppel and O. J. Staffelbach. Products of linear recurring sequences with maximum complexity. *IEEE Transactions on Information Theory*, volume IT-33, number 1, pp. 124-131, 1987.
16. P. Sarkar. Computing Shifts in 90/150 Cellular Automata Sequences. CACR Technical Report CORR 2001-46, University of Waterloo, http://www.cacr.math.uwaterloo.ca
17. P. Sarkar and S. Maitra. Nonlinearity bounds and constructions of resilient Boolean functions. In *Advances in Cryptology - CRYPTO 2000*, number 1880 in LNCS, pages 515–532. Springer Verlag, 2000.
18. T. Siegenthaler. Correlation-immunity of nonlinear combining functions for cryptographic applications. *IEEE Transactions on Information Theory*, IT-30(5):776–780, September 1984.
19. T. Siegenthaler. Decrypting a class of stream ciphers using ciphertext only. *IEEE Transactions on Computers*, C-34(1):81–85, January 1985.
20. S. Tezuka and M. Fushimi. A method of designing cellular automata as pseudo random number generators for built-in self-test for VLSI. In *Finite Fields: Theory, Applications and Algorithms*, Contemporary Mathematics, AMS, pages 363–367, 1994.
21. G.-Z. Xiao and J. Massey. A spectral characterization of correlation immune combining functions. *IEEE Transactions on Information Theory*, 34(3):569–571, May 1988.

A Larger Class
of Cryptographic Boolean Functions via a Study
of the Maiorana-McFarland Construction

Claude Carlet

INRIA, Domaine de Voluceau, BP 105 – 78153, Le Chesnay Cedex, France
also member of GREYC-Caen and of University of Paris 8
claude.carlet@inria.fr

Abstract. Thanks to a new upper bound, we study more precisely the nonlinearities of Maiorana-McFarland's resilient functions. We characterize those functions with optimum nonlinearities and we give examples of functions with high nonlinearities. But these functions have a peculiarity which makes them potentially cryptographically weak. We study a natural super-class of Maiorana-McFarland's class whose elements do not have the same drawback and we give examples of such functions achieving high nonlinearities.

Keywords: resilient functions, nonlinearity, stream ciphers.

1   Introduction

The Boolean functions used in stream ciphers are functions from F_2^n to F_2, where n is a positive integer. In practice, n is often small (smaller than or equal to 10), but even for small values of n, searching for the best cryptographic functions by visiting all Boolean functions in n variables is computationally impossible since their number 2^{2^n} is too large (for instance, for $n = 7$, it would need billions of times the age of the universe on a work-station). Thus, we need constructions of Boolean functions satisfying all necessary cryptographic criteria. Before describing the known constructions, we recall what are these cryptographic criteria.

Any Boolean function f in n variables (i.e. any F_2-valued function defined on the set F_2^n of all binary vectors of length n) admits a unique algebraic normal form (A.N.F.):

$$f(x_1, \ldots, x_n) = \sum_{I \subseteq \{1, \ldots, n\}} a_I \prod_{i \in I} x_i,$$

where the additions are computed in F_2, i.e. modulo 2, and where the a_I's are in F_2. We call *algebraic degree* of a Boolean function f and we denote by $d^\circ f$ the degree of its algebraic normal form. The *affine functions* are those functions of degrees at most 1. They are the simplest functions, from cryptographic viewpoint. On the contrary, *cryptographic functions must have high degrees* (cf. [3, 16, 21, 27]).

M. Yung (Ed.): CRYPTO 2002, LNCS 2442, pp. 549–564, 2002.

The *Hamming weight* $w_H(f)$ of a Boolean function f in n variables is the size of its *support* $\{x \in F_2^n; f(x) = 1\}$. The *Hamming distance* $d_H(f,g)$ between two Boolean functions f and g is the Hamming weight of their difference, i.e. of $f + g$ (this sum is computed modulo 2). The *nonlinearity* of f is its minimum distance to all affine functions. We denote by N_f the nonlinearity of f. *Functions used in stream ciphers must have high nonlinearities* to resist the known attacks on these ciphers (correlation and linear attacks) [3]. A Boolean function f is called *bent* if its nonlinearity equals $2^{n-1} - 2^{n/2-1}$, which is the maximum possible value (obviously, n must be even). Then, its distance to every affine function equals $2^{n-1} \pm 2^{n/2-1}$. This property can also be stated in terms of the Walsh (i.e., discrete Fourier, or Hadamard) transform of f defined on F_2^n as $\widehat{f}(u) = \sum_{x \in F_2^n} f(x)(-1)^{x \cdot u}$ (where $x \cdot u$ denotes the usual inner product $x \cdot u = \sum_{i=1}^n x_i u_i$). But it is more easily stated in terms of the Walsh transform of the "sign" function $\chi_f(x) = (-1)^{f(x)}$, equal to $\widehat{\chi_f}(u) = \sum_{x \in F_2^n} (-1)^{f(x)+x \cdot u}$: f is bent if and only if $\widehat{\chi_f}(u)$ has constant magnitude $2^{n/2}$ (cf. [14, 20]). Indeed, the Hamming distances between f and the affine functions $u \cdot x$ and $u \cdot x + 1$ are equal to $2^{n-1} - \frac{1}{2}\widehat{\chi_f}(u)$ and $2^{n-1} + \frac{1}{2}\widehat{\chi_f}(u)$. Thus:

$$N_f = 2^{n-1} - \frac{1}{2} \max_{u \in F_2^n} |\widehat{\chi_f}(u)|. \tag{1}$$

Bent functions have degrees upper bounded by $n/2$. They are characterized by the fact that their derivatives $D_a f(x) = f(x) + f(x+a)$, $a \neq 0$, are all *balanced*, i.e. have weight 2^{n-1}. But *cryptographic functions themselves must be balanced*, so that the systems using them resist statistical attacks [21]. Bent functions are not balanced.

The last (but not least) criterion considered in this paper is resiliency. It plays a central role in stream ciphers: in the standard model of these ciphers (cf. [26]), the outputs of n linear feedback shift registers are the inputs of a Boolean function, called combining function. The output of the function produces the keystream, which is then bitwisely xored with the message to produce the cipher. Some devide-and-conquer attacks exist on this method of encryption (cf. [3, 27]). To resist these attacks, the system must use a combining function whose output distribution probability is unaltered when any m of the inputs are fixed [27], with m as large as possible. This property, called *m-th order correlation-immunity* [26], is characterized by the set of zero values in the Walsh spectrum [30]: f is *m-th order correlation-immune* if and only if $\widehat{\chi_f}(u) = 0$, i.e. $\widehat{f}(u) = 0$, for all $u \in F_2^n$ such that $1 \leq w_H(u) \leq m$, where $w_H(u)$ denotes the Hamming weight of the n-bit vector u, (the number of its nonzero components). Balanced m-th order correlation-immune functions are called *m-resilient* functions. They are characterized by the fact that $\widehat{\chi_f}(u) = 0$ for all $u \in F_2^n$ such that $0 \leq w_H(u) \leq m$.

Siegenthaler's inequality [26] states that any m-th order correlation immune function in n variables has degree at most $n - m$, that any m-resilient function ($0 \leq m < n-1$) has algebraic degree smaller than or equal to $n - m - 1$ and that any $(n-1)$-resilient function has algebraic degree 1. Sarkar and Maitra [23] have shown that the nonlinearity of any m-resilient function ($m \leq n-2$) is divisible by 2^{m+1} and is therefore upper bounded by $2^{n-1} - 2^{m+1}$. If a function achieves this bound (independently obtained by Tarannikov [28] and Zheng and Zhang [31]), then it also achieves Siegenthaler's bound (cf. [28]) and the Fourier spectrum of the function has then three values (such functions are often called "plateaued" or "three-valued"; cf. [2]), these values are 0 and $\pm 2^{m+2}$. More precisely, it has been shown by Carlet and Sarkar [7, 8] that if f is m-resilient and has degree d, then its nonlinearity is divisible by $2^{m+1+\lfloor \frac{n-m-2}{d} \rfloor}$ and can therefore equal $2^{n-1} - 2^{m+1}$ only if $d = n - m - 1$. We shall say that an m-resilient function achieves the best possible nonlinearity if its nonlinearity equals $2^{n-1} - 2^{m+1}$.

If $2^{n-1} - 2^{m+1}$ is greater than the best possible nonlinearity of all balanced functions (and in particular if it is greater than the best possible nonlinearity of all Boolean functions) then the Sarkar-Maitra-Tarannikov-Zheng-Zhang's bound can obviously be improved. In the case n is even, the best possible nonlinearity of all Boolean functions being equal to $2^{n-1} - 2^{n/2-1}$ and the best possible nonlinearity of all balanced functions being smaller than $2^{n-1} - 2^{n/2-1}$, Sarkar and Maitra deduce from their divisibility result that $N_f \leq 2^{n-1} - 2^{n/2-1} - 2^{m+1}$ for every m-resilient function f with $m \leq n/2 - 2$. In the case n is odd, they state that N_f is smaller than or equal to the highest multiple of 2^{m+1} which is less than or equal to the best possible nonlinearity of all Boolean functions, which is smaller than $2^{n-1} - 2^{n/2-1}$ (see [17] for more details). For $m \leq n/2 - 2$, a potentially better upper bound can be given, whatever is the evenness of n: Sarkar-Maitra's divisibility bound shows that $\widehat{\chi_f}(a) = \varphi(a) \cdot 2^{m+2}$ where $\varphi(a)$ is integer-valued. But Parseval's relation $\sum_{a \in F_2^n} \widehat{\chi_f}^2(a) = 2^{2n}$ and the fact that $\widehat{\chi_f}(a)$ is null for every word a of weight $\leq m$ implies $\sum_{a;\, w_H(a) > m} \varphi^2(a) = 2^{2n-2m-4}$ and thus $\max_{a \in F_2^n} |\varphi(a)| \geq \sqrt{\frac{2^{2n-2m-4}}{2^n - \sum_{i=0}^{m} \binom{n}{i}}} = \frac{2^{n-m-2}}{\sqrt{2^n - \sum_{i=0}^{m} \binom{n}{i}}}$. Thus we have $\max_{a \in F_2^n} |\varphi(a)| \geq \left\lceil \frac{2^{n-m-2}}{\sqrt{2^n - \sum_{i=0}^{m} \binom{n}{i}}} \right\rceil$ (where $\lceil \lambda \rceil$ denotes the smallest integer greater than or equal to λ)) and this implies $N_f \leq 2^{n-1} - 2^{m+1} \left\lceil \frac{2^{n-m-2}}{\sqrt{2^n - \sum_{i=0}^{m} \binom{n}{i}}} \right\rceil$.

We shall call "Sarkar et al.'s bounds" all these bounds, in the sequel.

High order resilient functions with high degrees and high nonlinearities are needed for applications in stream ciphers, but designing constructions of Boolean functions meeting these cryptographic criteria is still a crucial challenge nowadays in symmetric cryptography. We observe now some imbalance in the knowledge on cryptographic functions for stream ciphers, after the results recently obtained on the properties of resilient functions [7, 8, 22, 23]. Examples of m-resilient functions achieving the best possible nonlinearities have been obtained for small values of n [19, 22, 23] and for every $m \geq 0.6\, n$ [29] (n being then not limited). But these examples give very limited numbers of functions (they are often defined recursively or obtained after a computer search) and these functions often have cryptographic weaknesses such as linear structures. Designing constructions leading to large numbers of functions would permit to choose in applications cryptographic functions satisfying specific constraints. It would also make more efficient those cryptosystems in which the cryptographic functions themselves would be part of the secret keys.

The paper is organized as follows. At section 2, we study the known constructions of resilient functions and the nonlinearities of the functions they produce. We study the nonlinearities of Maiorana-McFarland's functions more efficiently than the previous papers on this subject could do, thanks to a new upper bound that we introduce. We characterize then those functions which reach Sarkar et al.'s bound and we exhibit functions achieving high nonlinearities. At section 3, we introduce a super-class of Maiorana-McFarland's class. We study the degrees, the nonlinearities and the resiliency orders of its elements and we give examples of functions in this class having good cryptographic parameters.

2   The Known Constructions of Reasonably Large Sets of Cryptographic Functions, and Their Properties

Only one reasonably large class of Boolean functions is known, whose elements can be cryptographically analyzed.

2.1   Maiorana-McFarland's Construction

In [1] is introduced a modification of Maiorana-McFarland's construction of bent functions (cf. [12]) whose elements, viewed as binary vectors of length 2^n, are the concatenations of affine functions[1]: let k and r be integers such that $n \geq r > k \geq 0$; denote $n - r$ by s; let g be any Boolean function on F_2^s and ϕ a mapping from F_2^s to F_2^r such that every element in $\phi(F_2^s)$ has Hamming weight strictly greater than k. Then the function:

$$f_{\phi,g}(x,y) = x \cdot \phi(y) + g(y) = \sum_{i=1}^{r} x_i \phi_i(y) + g(y), \quad x \in F_2^r, \ y \in F_2^s \quad (2)$$

where $\phi_i(y)$ is the ith coordinate of $\phi(y)$, is m-resilient with $m \geq k$. Indeed, for every $a \in F_2^r$ and every $b \in F_2^s$, we have

$$\widehat{\chi_{f_{\phi,g}}}(a,b) = 2^r \sum_{y \in \phi^{-1}(a)} (-1)^{g(y)+b \cdot y}, \quad (3)$$

since every (affine) function $x \mapsto f_{\phi,g}(x,y) + a \cdot x + b \cdot y$ either is constant or is balanced and contributes then for 0 in the sum $\sum_{x \in F_2^r, y \in F_2^s} (-1)^{f_{\phi,g}(x,y)+x \cdot a + y \cdot b}$.

The degree of $f_{\phi,g}$ is $s + 1 = n - r + 1$ if and only if ϕ has degree s (i.e. if at least one of its coordinate functions has degree s), which is possible only if

---

[1] As noted e.g. in [22], concatenations of m-resilient functions produce also, more generally, m-resilient functions. But this observation has not permitted until now to produce larger classes of resilient functions.

$k \leq r - 2$, since if $k = r - 1$ then ϕ is constant. Otherwise, the degree of $f_{\phi,g}$ is at most s. Thus, if $m = k$ then the degree of $f_{\phi,g}$ reachs Siegenthaler's bound $n - m - 1$ if and only if either $m = r - 2$ and ϕ has degree $s = n - m - 2$ or $m = r - 1$ and g has degree $s = n - m - 1$. There are cases where $m > k$ (see below).

The nonlinearity of Maiorana-McFarland's functions could not be determined in the literature in a precise and a general way: the lower bound $N_{f_{\phi,g}} \geq 2^{n-1} - 2^{r-1} \max_{a \in F_2^r} |\phi^{-1}(a)|$ (where $|\phi^{-1}(a)|$ denotes the size of $\phi^{-1}(a)$) obtained in [24] is rather precise, but the upper bound $N_{f_{\phi,g}} \leq 2^{n-1} - 2^{r-1}$ obtained in [9, 10] does not involve the size of $\phi^{-1}(a)$. This upper bound is efficient when ϕ is injective. Notice that in this case, $f_{\phi,g}$ is then exactly k-resilient, where $k + 1$ is the minimum weight of $\phi(y)$, $y \in F_2^s$ and that g plays no role in the nonlinearity of $f_{\phi,g}$ or in its resiliency order. Thanks to these bounds, the nonlinearity of $f_{\phi,g}$ can also be precisely determined when g is null (as noted in [9, 10]) and more generally when g is affine, and also when $\max_{a \in F_2^r} |\phi^{-1}(a)| \leq 2$: according to relation (3), $N_{f_{\phi,g}}$ equals then $2^{n-1} - 2^{r-1} \max_{a \in F_2^r} |\phi^{-1}(a)|$. Notice that, ϕ being chosen, the case g affine is unfortunately not the most interesting one from nonlinearity viewpoint. Indeed, in relation (3), for a given a, the sum $\sum_{y \in \phi^{-1}(a)} (-1)^{g(y) + b \cdot y}$ has maximum magnitude when $g(y) + b \cdot y$ is constant on $\phi^{-1}(a)$ for some b.

In the next proposition, we improve upon the upper bound proved in [9, 10] and we deduce further information on the nonlinearities of Maiorana-McFarland's functions, which shows for instance why Sarkar and Maitra could not find 4-resilient Maiorana McFarland's functions in 10 variables with nonlinearity 480.

Proposition 1. *Let $f_{\phi,g}$ be defined by (2). Then the nonlinearity $N_{f_{\phi,g}}$ of $f_{\phi,g}$ satisfies*

$$2^{n-1} - 2^{r-1} \max_{a \in F_2^r} |\phi^{-1}(a)| \leq N_{f_{\phi,g}} \leq 2^{n-1} - 2^{r-1} \left\lceil \sqrt{\max_{a \in F_2^r} |\phi^{-1}(a)|} \right\rceil. \quad (4)$$

Assume that every element in $\phi(F_2^s)$ has Hamming weight strictly greater than k ($f_{\phi,g}$ is then m-resilient with $m \geq k$). Then $N_{f_{\phi,g}} \leq 2^{n-1} - 2^{r-1} \left\lceil \dfrac{2^{s/2}}{\sqrt{\sum_{i=k+1}^{r} \binom{r}{i}}} \right\rceil$.

Under this hypothesis, if $f_{\phi,g}$ achieves the best possible nonlinearity $2^{n-1} - 2^{k+1}$, then either $r = k + 1$ or $r = k + 2$.

If $r = k + 1$ then ϕ takes constant value $(1, \cdots, 1)$ and $n \leq k + 3$. Either $s = 1$ and $g(y)$ is then any function in one variable or $s = 2$ and g is then any function of the form $y_1 y_2 + l(y)$ where l is affine (thus, f is quadratic, i.e. has degree at most 2).

If $r = k + 2$, then ϕ is injective, $n \leq k + 2 + \log_2(k + 3)$, g is any function in $n - k - 2$ variables and $d^\circ f_{\phi,g} \leq 1 + \log_2(k + 3)$.

Proof: The inequality $N_{f_{\phi,g}} \geq 2^{n-1} - 2^{r-1} \max_{a \in F_2^r} |\phi^{-1}(a)|$ is a direct consequence of relations (1) and (3). Let us prove now the upper bound. The sum

$$\sum_{b \in F_2^s} \left(\sum_{y \in \phi^{-1}(a)} (-1)^{g(y)+b \cdot y} \right)^2 = \sum_{b \in F_2^s} \left(\sum_{y,z \in \phi^{-1}(a)} (-1)^{g(y)+g(z)+b \cdot (y+z)} \right)$$

equals: $2^s |\phi^{-1}(a)|$ (indeed, $\sum_{b \in F_2^s} (-1)^{b \cdot (y+z)}$ is null if $y \neq z$). The maximum of a set of values being always greater than or equal to its mean, we deduce $\max_{b \in F_2^s} |\sum_{y \in \phi^{-1}(a)} (-1)^{g(y)+b \cdot y}| \geq \sqrt{|\phi^{-1}(a)|}$ and thus

$$\max_{a \in F_2^r; b \in F_2^s} |\widehat{\chi_{f_{\phi,g}}}(a,b)| \geq 2^r \left\lceil \sqrt{\max_{a \in F_2^r} |\phi^{-1}(a)|} \right\rceil.$$

Hence, according to relation (1): $N_{f_{\phi,g}} \leq 2^{n-1} - 2^{r-1} \left\lceil \sqrt{\max_{a \in F_2^r} |\phi^{-1}(a)|} \right\rceil$.

If every element in $\phi(F_2^s)$ has Hamming weight strictly greater than k, we have

$$\max_{a \in F_2^r} |\phi^{-1}(a)| \left(\sum_{i=k+1}^r \binom{r}{i} \right) \geq 2^s \text{ and } N_{f_{\phi,g}} \leq 2^{n-1} - 2^{r-1} \left\lceil \frac{2^{s/2}}{\sqrt{\sum_{i=k+1}^r \binom{r}{i}}} \right\rceil.$$

If $N_{f_{\phi,g}} = 2^{n-1} - 2^{k+1}$, then according to (4), we have $\sqrt{\max_{a \in F_2^r} |\phi^{-1}(a)|} \leq 2^{k-r+2}$ and thus $k+1 \leq r \leq k+2$ since $\max_{a \in F_2^r} |\phi^{-1}(a)| \geq 1$. If $r = k+1$, then since every element in $\phi(F_2^s)$ has Hamming weight strictly greater than k, ϕ must take constant value $(1, \cdots, 1)$ and $\max_{a \in F_2^r} |\phi^{-1}(a)|$ is then equal to 2^s. Since $\sqrt{\max_{a \in F_2^r} |\phi^{-1}(a)|} \leq 2^{k-r+2}$, this implies $s \leq 2(k - r + 2) = 2$. Thus, $f_{\phi,g}$ is quadratic and of the form $f(x,y) = \sum_{i=1}^r x_i + g(y)$. Its nonlinearity being equal to $2^{n-1} - 2^{k+1}$, we have $\max_{b \in F_2^s} |\sum_{y \in F_2^s} (-1)^{g(y)+b \cdot y}| = 2$. Thus $s \geq 1$. If $s = 1$ then $f(x,y_1) = \sum_{i=1}^r x_i + g(y_1)$ (if g is constant then f is $(n-2)$-resilient with null nonlinearity and if g is not constant, then f is $(n-1)$-resilient with null nonlinearity). If $s = 2$ then g must be bent, i.e. equal to $y_1 y_2 + l(y)$ where l is affine. If $r = k+2$, then $\max_{a \in F_2^r} |\phi^{-1}(a)| = 1$ and ϕ is injective. Since ϕ is injective and is valued in $\{a \in F_2^r; w_H(a) \geq k+1 = r-1\}$ we deduce $2^s \leq \binom{r}{r-1} + \binom{r}{r} = r+1$ and thus $n - r \leq \log_2(r+1)$. Siegenthaler's inequality completes the proof.   ◇

Examples of Optimum Functions. We give now examples of resilient Maiorana-McFarland's functions with high nonlinearities. The existence of some of these functions have been already shown in the literature. But this was often done by random search while a deterministic construction is provided here. We shall reduce our investigation to m-resilient functions with n even or with n odd and $m > n/2 - 2$, since in the case n is odd and $m \leq n/2 - 2$, we do not know what is the precise bound.

– We first complete Proposition 1 when $\phi(y) = (1, \cdots, 1)$, $\forall y \in F_2^s$. Then $\phi^{-1}(a)$ is empty if $a \neq (1, \cdots, 1)$ and equals F_2^s if $a = (1, \cdots, 1)$, and the function $f_{\phi,g}$ is $(r-1)$-resilient if g is not balanced and it is $(r+k)$-resilient if g is k-resilient. $N_{f_{\phi,g}}$ equals $2^r N_g$ and is at most equal to $2^{n-1} - 2^{r-1+s/2} = 2^{n-1} - 2^{n/2-1+r/2}$. If g is not balanced, the functions $f_{\phi,g}$ achieving Sarkar et al.'s bound have been studied in Proposition 1 for $m = r - 1 > n/2 - 2$. For $m = r - 1 \leq n/2 - 2$ (n even) the only possible cases for which we can obtain functions with nonlinearity

$2^{n-1} - 2^{n/2-1} - 2^r$ are clearly for $r \leq 2$. For $r = 1$, we have $N_g = 2^{n-2} - 2^{n/2-2} - 1$ which is possible for $n = 4$ only. For $r = 2$, the function $f_{\phi,g}$ achieves Sarkar et al.'s bound if and only if $r = n/2 - 1$, i.e. $n = 6$ and g is bent. If g is k-resilient, then if $r + k > n/2 - 2$, $f_{\phi,g}$ achieves Sarkar et al.'s bound if and only if $k > s/2 - 2$ and if g does; and if $r + k \leq n/2 - 2$ then $f_{\phi,g}$ cannot achieve Sarkar et al.'s bound (i.e. N_g cannot equal $2^{s-1} - 2^{n/2-r-1} - 2^{k+1}$) unless, maybe, $r = 1$ and $k = (s-5)/2$.

– We show now that for every even $n \leq 10$, Sarkar et al.'s bound with $m = n/2 - 2$ can be acheived by Maiorana-McFarland's functions. The nonlinearity the function $f_{\phi,g}$ must reach is $2^{n-1} - 2^{n/2}$ (this number is often called the quadratic bound, see next paragraph). Take $r = n/2 + 1$ and $s = n/2 - 1$. For $n \leq 10$, we have $1 + r + \binom{r}{2} \geq 2^s$ and we deduce that there exist injective mappings $\phi : F_2^s \mapsto \{x \in F_2^r; w_H(x) > r - 3 = m\}$. For every such ϕ and for every $g : F_2^s \mapsto F_2$, the function $f_{\phi,g}$ is $(n/2 - 2)$-resilient and its nonlinearity is $2^{n-1} - 2^{r-1} = 2^{n-1} - 2^{n/2}$.

– We describe now a general situation in which Maiorana-McFarland's functions can have high nonlinearities (but do not achieve in general Sarkar et al.'s bound, which is not known to be tight in these ranges except for small values of n). Let r, k and s be three positive integers such that $k \leq r - 1$ and $\sum_{i=k+1}^{r} \binom{r}{i} \geq 2^s$. Set $n = r + s$. Let ϕ be any one-to-one mapping from F_2^s to the set $\{x \in F_2^r; w_H(x) > k\}$ (such mapping ϕ exists thanks to the inequality above). Then for every Boolean function g on F_2^s, the function $f_{\phi,g}$ is a k-resilient function on F_2^n and has nonlinearity $2^{n-1} - 2^{r-1}$. Examples of such situation are the following:

• For any $k > 0$, choose $r = 2k + 1$ and $s = 2k$; then the nonlinearity of $f_{\phi,g}$ equals $2^{n-1} - 2^{2k} = 2^{n-1} - 2^{\frac{n-1}{2}}$ which is known as the best possible nonlinearity of all Boolean functions on F_2^n for odd $n \leq 7$ and the best possible nonlinearity of quadratic functions on F_2^n for every odd n (it is often called the quadratic bound). There exist only few known examples of functions on F_2^n (n odd) with nonlinearities strictly greater than $2^{n-1} - 2^{\frac{n-1}{2}}$ (these examples are known for odd $n \geq 15$, cf. [17]) and of balanced such functions (cf. [15, 22, 25]); here we have an example, for every $n \equiv 1 \mod 4$, of $\frac{n-1}{4}$-resilient functions on F_2^n with nonlinearity equal to $2^{n-1} - 2^{\frac{n-1}{2}}$. This nonlinearity is the best known nonlinearity for k-resilient functions; moreover, for $k = 1, 2$ ($n = 5, 9$) it achieves Sarkar et al.'s bound (this does not imply, in the case $n = 9$, that $f_{\phi,g}$ achieves Siegenthaler's bound because $2^{n-1} - 2^{2k} > 2^{n-1} - 2^{k+1}$; in fact, the maximum possible degree of $f_{\phi,g}$ is $2k + 1$). For $n = 9$, this optimal function can be obtained by Sarkar and Maitra's algorithm A given in [22]. We have here its precise description.

Notice that it is impossible to obtain nonlinearity $2^{n-1} - 2^{\frac{n-1}{2}}$ with a quadratic $\frac{n-1}{4}$-resilient function (or even more generally with a partially-bent function): recall that such function has this nonlinearity if and only if its kernel has dimension 1 (see [5]) and that it can then be $\frac{n-1}{4}$-resilient only if there exists an affine hyperplane with minimum weight strictly greater than $\frac{n-1}{4}$. This is clearly impossible for $n \geq 9$.

- For any $s > 0$, choose $r \geq 2^s - 1$ and set $k = r - 2$. The nonlinearity of $f_{\phi,g}$ equals then $2^{n-1} - 2^{r-1} = 2^{n-1} - 2^{k+1}$ and $f_{\phi,g}$ achieves Sarkar et al.'s bound and Siegenthaler's bound.
- There exist other examples of r, k, s leading to good functions.

Improved Resiliency Orders. The functions satisfying the hypothesis of Proposition 1 are not the only ones in Maiorana-McFarland's class which can achieve Sarkar et al.'s bound. We describe below two other cases. The first one has also been considered by Cusick in [11], but in a more complex way and without looking for the best possible nonlinearity.

Proposition 2. Let $f_{\phi,g}$ be defined by (2).

1. Assume that every element in $\phi(F_2^s)$ has Hamming weight strictly greater than k and that, for every $a \in F_2^r$ of weight $k + 1$, either the set $\phi^{-1}(a)$ is empty or it has an even size and the restriction of g to this set is balanced. Then $f_{\phi,g}$ is m-resilient with $m \geq k + 1$. Under this hypothesis, if $f_{\phi,g}$ achieves the best possible nonlinearity $2^{n-1} - 2^{k+2}$, then $r \leq k + 2$.
If $r = k + 1$ then either $s = 2$ and g and f are affine or $s = 3$ and g is balanced and has nonlinearity 4.
If $r = k + 2$ then $n \leq k + 4 + \log_2(k + 3)$ and $d°f \leq 2 + \log_2(k + 3)$.

2. Assume in addition that:
a. for every $a \in F_2^r$ of weight $k + 1$ and every $i \in \{1, \cdots, s\}$, denoting by H_i the linear hyperplane of equation $y_i = 0$ in F_2^s, either the set $\phi^{-1}(a) \cap H_i$ is empty or it has an even size and the restriction of g to this set is balanced;
b. for every $a \in F_2^r$ of weight $k + 2$, either the set $\phi^{-1}(a)$ is empty or it has an even size and the restriction of g to this set is balanced. Then $f_{\phi,g}$ is m-resilient with $m \geq k + 2$. Under this hypothesis, if $f_{\phi,g}$ achieves the best possible nonlinearity $2^{n-1} - 2^{k+3}$, then $r \leq k + 3$.
If $r = k + 1$, then $3 \leq s \leq 5$ and ϕ takes constant value $(1, \cdots, 1)$. If $s = 3$ then g and f are affine. If $s = 4$, then g has nonlinearity 4. If $s = 5$ then g has nonlinearity 12.
If $r = k + 2$ then $n \leq k + 6 + \log_2(k + 3)$ and $d°f \leq 3 + \log_2(k + 3)$.
If $r = k + 3$ then $n \leq k + 5 + \log_2(\binom{k+3}{2} + k + 3)$ and $d°f \leq 2 + \log_2(\binom{k+3}{2} + k + 3)$.

The proof has to be omitted because of length constraints.

Other Examples of Optimum Functions. Choose again three positive integers r, k and s such that $\sum_{i=k+1}^{r} \binom{r}{i} \geq 2^s$ and a one-to-one mapping ϕ from F_2^s to the set $\{x \in F_2^r; w_H(x) > k\}$. Set $s' = s + 1$ and modify ϕ into a two-to-one mapping $\phi' : F_2^{s'} \mapsto \{x \in F_2^r; w_H(x) > k\}$. For any $x \in F_2^r$ such that $w_H(x) > k$, let $g' : F_2^{s'} \mapsto F_2$ take each value 0 and 1 once on the pair $\phi'^{-1}(x)$. Then, according to Proposition 2, the function $f_{\phi',g'}$ is $(k+1)$-resilient on $F_2^{n'}$ with $n' = s' + r = n + 1$ and its nonlinearity is twice that of $f_{\phi,g}$ for every $g : F_2^s \mapsto F_2$ (thus, $f_{\phi',g'}$ achieves Sarkar et al.'s bound if $f_{\phi,g}$ does).

Another way of modifying $f_{\phi,g}$ into a function with the same number of variables and the same parameters as the function $f_{\phi',g'}$ above would be to take $f'(x, x_{r+1}, y) = f_{\phi,g}(x,y) + x_{r+1}$. But this kind of function having a linear term, it is less suited for cryptographic use (for instance, it has a linear structure, i.e. the derivative $f'(x, x_{r+1}, y) + f'(x, x_{r+1} + 1, y)$ is a constant).

Remark: In the case that every non-empty set $\phi^{-1}(a)$ is an affine set, then more can be said: assume that $\phi^{-1}(a)$ is either the empty set or a flat for every a, that it is empty for every word a of weight $\leq k$, and that, for some positive integer l, the restriction of g to every non-empty set $\phi^{-1}(a)$ such that $w_H(a) = k + i$, $i \leq l$, is $(l - i)$-resilient. Then according to relation 3, $f_{\phi,g}$ is $(k + l)$-resilient. ◇

A drawback of Maiorana-McFarland's functions is that their restrictions obtained by fixing y in their input are affine. Affine functions being cryptographically weak functions, there is a risk that this property be used in attacks. Also, Maiorana-McFarland's functions have high divisibilities of their Fourier spectra, and there is also a risk that this property be used in attacks as it is used in [4] to attack block ciphers. A purpose of this paper is to produce a construction having not this drawback and leading to a larger class of cryptographic functions. Before that, we study the other known constructions.

2.2 Dillon's Construction

In [6] is used an idea of Dillon (cf. [12]) to obtain a construction of resilient functions. Similar observations as for Maiorana-McFarland's construction can be made on the ability of these functions to have nonlinearities near Sarkar et al.'s bound. But this class has few elements.

2.3 Dobbertin's Construction

In [13], Hans Dobbertin studies an interesting method for modifying bent functions into balanced functions with high nonlinearities. Unfortunately:

Proposition 3. *Dobbertin's construction cannot produce m-resilient functions with $m > 0$.*

3   Maiorana-McFarland's Super-class

The functions of the super-class of Maiorana-McFarland's class that we introduce now are concatenations of quadratic functions (i.e. functions of degrees at most 2).

3.1   Quadratic Functions

It is shown in [14] that any quadratic function $f(x)$ is linearly equivalent to a function of the form

$$x_1 x_2 + \cdots + x_{2i-1} x_{2i} + \cdots + x_{2t-1} x_{2t} + l(x) \tag{5}$$

where $2t$ is smaller than or equal to the number of variables and where l is affine. The functions we shall concatenate below are not general quadratic functions, because the parameters of the functions could then not be evaluated, but they have a slightly more general form than (5). They are defined on F_2^{2t} and have the form

$$f(x) = \sum_{i=1}^{t} u_i x_{2i-1} x_{2i} + l(x) = \sum_{i=1}^{t} u_i x_{2i-1} x_{2i} + \sum_{j=1}^{2t} v_i x_i + c, \qquad (6)$$

where $u = (u_1, \cdots, u_t)$ is an element of F_2^t, $v = (v_1, \cdots, v_{2t})$ is an element of F_2^{2t} and c is an element of F_2. We shall need in the sequel to compute sums $\sum_{x \in F_2^{2t}} (-1)^{f(x)}$. We know (and it is a simple matter to check) that if there exists $i = 1, \cdots, t$ such that u_i is null and v_{2i-1} or v_{2i} is not null, then f is balanced and thus $\sum_{x \in F_2^{2t}} (-1)^{f(x)} = 0$. We consider now the case where such an i does not exist. Then we have $f(x) = \sum_{i=1}^{t} u_i (x_{2i-1} + v_{2i})(x_{2i} + v_{2i-1}) + \sum_{i=1}^{t} v_{2i-1} v_{2i} + c$. Changing x_{2i-1} into $x_{2i-1} + v_{2i}$ and x_{2i} into $x_{2i} + v_{2i-1}$ does not change the value of $\sum_{x \in F_2^{2t}} (-1)^{f(x)}$. Hence: $\sum_{x \in F_2^{2t}} (-1)^{f(x)} = \sum_{x \in F_2^{2t}} (-1)^{\sum_{i=1}^{t} u_i x_{2i-1} x_{2i} + \sum_{i=1}^{t} v_{2i-1} v_{2i} + c}$. It is a simple matter to check that $\sum_{x_{2i-1}, x_{2i} \in F_2} (-1)^{u_i x_{2i-1} x_{2i}}$ equals 4 if $u_i = 0$ and equals 2 if $u_i = 1$. Thus $\sum_{x \in F_2^{2t}} (-1)^{f(x)} = 2^{2t - w_H(u)} (-1)^{\sum_{i=1}^{t} v_{2i-1} v_{2i} + c}$. Applying this to the function $f(x) + \sum_{j=1}^{2t} a_j x_j$, we deduce:

Proposition 4. *Let* $u = (u_1, \cdots, u_t) \in F_2^t$, $v = (v_1, \cdots, v_{2t}) \in F_2^{2t}$, $c \in F_2$ *and set* $f(x) = \sum_{i=1}^{t} u_i x_{2i-1} x_{2i} + \sum_{j=1}^{2t} v_j x_j + c$. *Let* a *be any element of* F_2^{2t}. *If there exists* $i = 1, \cdots, t$ *such that* $u_i = 0$ *and* $v_{2i-1} \neq a_{2i-1}$ *or* $v_{2i} \neq a_{2i}$, *then* $\widehat{\chi_f}(a)$ *is null. Otherwise,* $\widehat{\chi_f}(a)$ *equals* $2^{2t - w_H(u)} (-1)^{\sum_{i=1}^{t} (v_{2i-1} + a_{2i-1})(v_{2i} + a_{2i}) + c}$.

3.2   The Maiorana-McFarland's Super-class

Definition 1. *Let* n *and* r *be positive integers such that* $r < n$. *Denote the integer part* $\lfloor \frac{r}{2} \rfloor$ *by* t *and* $n - r$ *by* s. *Let* ψ *be a mapping from* F_2^s *to* F_2^t *and let* $\psi_1, \cdots, \psi_t$ *be its coordinate functions. Let* ϕ *be a mapping from* F_2^s *to* F_2^r *and let* $\phi_1, \cdots, \phi_r$ *be its coordinate functions. Let* g *be a Boolean function on* F_2^s. *The function* $f_{\psi, \phi, g}$ *is defined on* $F_2^n = F_2^r \times F_2^s$ *as*

$$f_{\psi, \phi, g}(x, y) = \sum_{i=1}^{t} x_{2i-1} x_{2i} \psi_i(y) + x \cdot \phi(y) + g(y) =$$

$$\sum_{i=1}^{t} x_{2i-1} x_{2i} \psi_i(y) + \sum_{j=1}^{r} x_i \phi_i(y) + g(y); \quad x \in F_2^r, \; y \in F_2^s.$$

The restrictions of $f_{\psi, \phi, g}$ obtained by fixing y in its input are quadratic functions of the form (6) or their extensions with one linear variable (r odd), and

$f_{\psi,\phi,g}(x, y)$, viewed as a binary vector of length 2^n, equals the concatenation of quadratic functions. Maiorana-McFarland's functions correspond to the case where ψ is the null mapping. As a direct consequence of Proposition 4, we have:

Theorem 1. *Let $f_{\psi,\phi,g}$ be defined as in Definition 1. Then for every $a \in F_2^r$ and every $b \in F_2^s$ we have*

$$\widehat{\chi_{f_{\psi,\phi,g}}}(a, b) = \sum_{y \in E_a} 2^{r - w_H(\psi(y))} (-1)^{\sum_{i=1}^t (\phi_{2i-1}(y) + a_{2i-1})(\phi_{2i}(y) + a_{2i}) + g(y) + y \cdot b},$$

where E_a is the superset of $\phi^{-1}(a)$ equal if r is even to

$$\{y \in F_2^s / \forall i \le t, \psi_i(y) = 0 \Rightarrow (\phi_{2i-1}(y) = a_{2i-1} \text{ and } \phi_{2i}(y) = a_{2i})\},$$

and if r is odd to

$$\left\{ y \in F_2^s / \begin{cases} \forall i \le t, \psi_i(y) = 0 \Rightarrow (\phi_{2i-1}(y) = a_{2i-1} \text{ and } \phi_{2i}(y) = a_{2i}) \\ \phi_r(y) = a_r \end{cases} \right\}.$$

Remark: let y be an element of F_2^s. Denote the weight of $\psi(y)$ by l. Then y belongs to 4^l sets E_a. One of them is $E_{\phi(y)}$. The others correspond to the vectors $a \ne \phi(y)$ such that $a_{2i-1} = \phi_{2i-1}(y)$ and $a_{2i} = \phi_{2i}(y)$ for every index i outside the support of the vector $\psi(y)$.

4   Cryptographic Properties of the Constructed Functions

4.1   Algebraic Degree

Let $f_{\psi,\phi,g}$ be defined as in Definition 1. The degree of $f_{\psi,\phi,g}$ clearly equals $\max(2 + d^\circ \psi_1, \cdots, 2 + d^\circ \psi_t, 1 + d^\circ \phi_1, \cdots, 1 + d^\circ \phi_r, d^\circ g)$. It is upper bounded by $2 + s$.

4.2   Nonlinearity

Theorem 2. *Let $f_{\psi,\phi,g}$ be defined as in Definition 1. Denote by M the maximum weight of $\psi(y)$ for $y \in F_2^s$, and by M' its minimum weight. Then the nonlinearity $N_{f_{\psi,\phi,g}}$ of $f_{\psi,\phi,g}$ satisfies*

$$2^{n-1} - 2^{r - M' - 1} \max_{a \in F_2^r} |E_a| \le 2^{n-1} - \max_{a \in F_2^r} \sum_{y \in E_a} 2^{r - w_H(\psi(y)) - 1} \le N_{f_{\psi,\phi,g}} \le$$

$$2^{n-1} - \max_{a \in F_2^r} \sqrt{\sum_{y \in E_a} 2^{2r - 2w_H(\psi(y)) - 2}} \le 2^{n-1} - 2^{r - M - 1} \max_{a \in F_2^r} \sqrt{|E_a|}$$

where $|E_a|$ denotes the size of the set E_a defined in Theorem 1.

The proof is similar to that of Proposition 1 and is omitted because of length constraints.

We have seen above that the nonlinearity of a Maiorana-McFarland's function $f_{\phi,g}$ can be more easily determined when ϕ is injective. The function is then "three-valued". The nonlinearity of $f_{\psi,\phi,g}$ can similarly be more precisely determined when all the sets E_a have size at most 1, i.e. when the quadratic functions whose concatenation is $f_{\psi,\phi,g}$ have disjoint spectra.

Proposition 5. *Let $f_{\psi,\phi,g}$ be defined as in Definition 1. Every set E_a has at most one element if and only if, for every two distinct elements y and y' of F_2^s, denoting by J_y the set of indices equal to $\{j \leq 2t/ \, \psi_{\lceil \frac{i}{2} \rceil}(y) = 0\}$ if r is even and to $\{j \leq 2t/ \, \psi_{\lceil \frac{i}{2} \rceil}(y) = 0\} \cup \{r\}$ if r is odd, there exists $i \in J_y \cap J_{y'}$ such that $\phi_j(y) \neq \phi_j(y')$.*

Notice that, even in this case, $f_{\psi,\phi,g}$ is not necessarily three-valued: the magnitude of $\widehat{\chi_{f_{\psi,\phi,g}}}$ being bounded between 2^{r-M} and $2^{r-M'}$ where M (resp. M') is the maximum (resp. minimum) weight of $\psi(y)$, $y \in F_2^s$, the function $f_{\psi,\phi,g}$ is three-valued if $M' = M$. We study below a situation in which the hypothesis of Proposition 5 is satisfied.

Corollary 1. *Let $f_{\psi,\phi,g}$ be defined as in Definition 1 and let M be the maximum weight of $\psi(y)$, $y \in F_2^s$. Suppose that ϕ is injective and that for every two distinct elements y and y' of F_2^s, the set $\{i \leq t; \, \phi_{2i-1}(y) \neq \phi_{2i-1}(y') \text{ or } \phi_{2i}(y) \neq \phi_{2i}(y')\}$ has size strictly greater than $2M$ (this condition is satisfied in particular if the set $\phi(F_2^s)$ has minimum Hamming distance strictly greater than $4M$). Then, every set E_a has size at most 1, and $N_{f_{\psi,\phi,g}} = 2^{n-1} - 2^j$ where $r - M - 1 \leq j \leq r - M' - 1$.*

Proof. For every two elements $y \neq y'$ of F_2^s, since $\psi(y)$ and $\psi(y')$ have weights smaller than or equal to M, at most $2M$ indices $i \leq t$ satisfy $\psi_i(y) = 1$ or $\psi_i(y') = 1$. The condition satisfied by ϕ implies that there exists $i \leq t$ such that $\psi_i(y) = \psi_i(y') = 0$ and $\phi_{2i-1}(y) \neq \phi_{2i-1}(y')$ or $\phi_{2i}(y) \neq \phi_{2i}(y')$, and the hypothesis of Proposition 5 is satisfied. Thus every set E_a contains at most one element. Theorem 2 completes the proof.   ◇

4.3   Balancedness and Resiliency

Theorem 3. *Let $f_{\psi,\phi,g}$ be defined as in Definition 1 and let k be non-negative. For every $y \in F_2^s$, denote by I_y the set of indices equal to $\{j \leq 2t/ \, \psi_{\lceil \frac{i}{2} \rceil}(y) = 0 \text{ and } \phi_j(y) = 1\}$ if r is even or if r is odd and $\phi_r(y) = 0$, and to $\{j \leq 2t/ \, \psi_{\lceil \frac{i}{2} \rceil}(y) = 0 \text{ and } \phi_j(y) = 1\} \cup \{r\}$ if r is odd and $\phi_r(y) = 1$. Assume that for every $y \in F_2^s$, I_y has size strictly greater than k. Then $f_{\psi,\phi,g}$ is m-resilient with $m \geq k$.*
In particular, if for every $y \in F_2^s$, the set I_y is not empty, then $f_{\psi,\phi,g}$ is balanced.

Proof: Let $a \in F_2^r$ and $b \in F_2^s$. Assume that (a,b) has weight smaller than or equal to k. Then a has weight smaller than or equal to k. Let y be an element of the set E_a (defined in Theorem 1), then for every index j in I_y, we must have $a_j = 1$. According to the hypothesis on I_y, the word a must then have weight strictly greater than k, a contradiction. We deduce that the set E_a is empty and, thus, that $\widehat{\chi_f}(a,b) = 0$.   ◇

In the case of Maiorana-McFarland's functions, the condition of Theorem 3 reduces to the fact that every element in $\phi(F_2^s)$ has Hamming weight strictly greater than k, since all coordinate functions of ψ are null. Let us translate the condition similarly in the general case.

Corollary 2. *Let $f_{\psi,\phi,g}$ be defined as in Definition 1 and let k be a non-negative integer. Consider the mapping Φ from F_2^s to F_2^r whose jth coordinate function for $j \le 2t$ equals the product of the Boolean functions ϕ_j and $1 + \psi_{\lceil \frac{j}{2} \rceil}$ and whose rth coordinate function equals ϕ_r if r is odd. If the image of every element in F_2^s by Φ has Hamming weight strictly greater than k, then $f_{\psi,\phi,g}$ is m-resilient with $m \ge k$.*

In particular, if the image of every element in F_2^s by Φ is nonzero, then $f_{\psi,\phi,g}$ is balanced.

Proof: For every $y \in F_2^s$, the set I_y introduced in Theorem 3 equals the support of $\Phi(y)$. Thus, it has size strictly greater than k if and only if $\Phi(y)$ has Hamming weight strictly greater than k. Theorem 3 completes the proof.   ◇

Remark:

- If the mapping ϕ satisfies $w_H(\phi(y)) > k$ for every $y \in F_2^s$, then the mapping Φ satisfies $w_H(\Phi(y)) > k - 2M$ for every y, since the vectors $\phi(y)$ and $\Phi(y)$ lie at distance at most $2M$ from each other.
- The results of Theorem 3 and Corollary 2 can be refined the same way as in Proposition 2.

Constructions of Highly Nonlinear Resilient Functions from the Super-class

- Let n be even and $\phi : F_2^{n/2} \mapsto F_2^{n/2} \setminus \{0\}$ be chosen such that every vector different from $(1, \cdots, 1)$ has one reverse image by ϕ and $(1, \cdots, 1)$ has two reverse images by ϕ. For every $g : F_2^{n/2} \mapsto F_2$, the function $f_{\phi,g}$ is then balanced, since ϕ does not take the zero value; but it has nonlinearity $2^{n-1} - 2^r = 2^{n-1} - 2^{n/2}$ only. We shall increase this nonlinearity by considering, instead of $f_{\phi,g}$, a function $f_{\psi,\phi,g}$ where ψ is chosen such that $f_{\psi,\phi,g}$ is still balanced. Choose $\psi(y)$ equal to the zero vector, except at one element u of $\phi^{-1}(1, \cdots, 1)$. If $n/2$ is odd or if it is even and if we choose as value of $\psi(u)$ a vector of F_2^t different from $(1, \cdots, 1)$, then for every $g : F_2^{n/2} \mapsto F_2$, the function $f_{\psi,\phi,g}$ is balanced since $E_{(0,\cdots,0)} = \emptyset$. According to Theorem 1, its nonlinearity equals $2^{n-1} - 2^{n/2-1} - 2^{n/2 - w_H(\psi(u)) - 1}$. So let us choose for $\psi(u)$ a vector of highest possible weight: $\lceil \frac{n}{4} \rceil - 1$. Then $f_{\psi,\phi,g}$ has nonlinearity $2^{n-1} - 2^{n/2-1} - 2^{n/2 - \lceil \frac{n}{4} \rceil} = 2^{n-1} - 2^{n/2-1} - 2^{\lfloor \frac{n}{4} \rfloor}$. If $n/2$ is odd, then $f_{\psi,\phi,g}$ has nonlinearity $2^{n-1} - 2^{n/2-1} - 2^{(n/2-1)/2}$, which is the best known nonlinearity for balanced functions if $n \leq 26$ (this same nonlinearity can be also reached with Dobbertin's method; the other methods do not work here: it can be checked that extending Patterson-Wiedemann's functions [17] or their modifications by Maitra-Sarkar [15] to even numbers of variables gives worse nonlinearities). Notice that this nonlinearity is impossible to exceed with a Maiorana-McFarland's function with $\phi : F_2^{n-r} \mapsto F_2^r \setminus \{0\}$. Indeed, if $r \geq n/2$ then $f_{\phi,g}$ has nonlinearity at most $2^{n-1} - 2^{n/2}$ (since ϕ cannot be injective if $r = n/2$) and if $r < n/2$ then there exists $a \in F_2^r$ such that $|\phi^{-1}(a)| \geq \frac{2^{n-r}}{2^r-1}$ and thus, according to Proposition 1, $N_{f_{\phi,g}} \leq 2^{n-1} - 2^{r-1} \left\lceil \sqrt{\frac{2^{n-r}}{2^r-1}} \right\rceil = 2^{n-1} - 2^{r-1} \left\lceil 2^{n/2-r} \sqrt{\frac{2^r}{2^r-1}} \right\rceil$. We have $\sqrt{\frac{2^r}{2^r-1}} > 1 + 2^{-r-1}$. Thus $N_{f_{\phi,g}} \leq 2^{n-1} - 2^{n/2-1} - 2^{r-1} \left\lceil 2^{n/2-2r-1} + \epsilon \right\rceil$ where $\epsilon > 0$. We checked that $N_{f_{\phi,g}}$ cannot then exceed $2^{n-1} - 2^{n/2-1} - 2^{(n/2-1)/2}$. It seems impossible that $N_{f_{\phi,g}}$ equals $2^{n-1} - 2^{n/2-1} - 2^{(n/2-1)/2}$, but we could not prove it.

– Let n be even and and let k be an integer such that $\sum_{i=0}^{k} \binom{n/2-2}{i} \leq \frac{2^{n/2-2}}{5}$. Then we have $2^{n/2-2} - \sum_{i=k+1}^{n/2-2} \binom{n/2-2}{i} \leq \frac{1}{5} 2^{n/2-2}$, thus $2^{n/2} \leq 5 \sum_{i=k+1}^{n/2-2} \binom{n/2-2}{i}$ and there can exist $\phi : F_2^{n/2} \mapsto \{x \in F_2^{n/2}; w_H(x_3, \cdots, x_{n/2}) > k\}$ such that for every $u \in F_2^{n/2-2}$, at most one element of $F_2^2 \times \{u\}$ has two reverse images by ϕ and the three others have at most one reverse image. For every element $a \in F_2^{n/2}$ which has two reverse images, choose $y \in \phi^{-1}(a)$ and take $\psi(y) = (1, 0, \cdots, 0)$. Take $\psi(y) = (0, \cdots, 0)$ for every other element. Then $f_{\psi,\phi,g}$ is at least k-resilient and has nonlinearity $2^{n-1} - 2^{n/2-1} - 2^{n/2-2}$, while $f_{\phi,g}$ is also at least k-resilient but has nonlinearity $2^{n-1} - 2^{n/2}$.

– A general method: Let $\phi : F_2^s \mapsto F_2^r$ be injective and such that $\phi^{-1}(a) = \emptyset$ for every a of Hamming weight at most k ($f_{\phi,g}$ is then k-resilient for every g; we have seen that such functions can achieve high nonlinearities). Choose a subset I of $\{1, \cdots, t\}$, where $t = \lfloor \frac{r}{2} \rfloor$ and denote its size by M. In our choice of the values taken by ψ, some of the vectors in $\psi(F_2^t)$ will have I as support and the others will be null. To ensure that $f_{\psi,\phi,g}$ is k-resilient, we need that for every $y \in F_2^s$ such that $\psi(y) \neq 0$, the word obtained from $\phi(y)$ by erasing all its coordinates of indices $j \leq 2t$ such that $\lceil \frac{j}{2} \rceil \in I$ has weight strictly greater than k. So we choose a subset U of F_2^{r-2M} of minimum weight at least $k + 1$, we denote by $\tilde{U}$ the set of all $y \in F_2^s$ such that the word $\tilde{\phi}(y)$ obtained from $\phi(y)$ by erasing all these coordinates belongs to U, and we set ψ such that $\psi_i(y) = 1$ if $y \in \tilde{U}$ and $i \in I$ and $\psi_i(y) = 0$ otherwise. Assume that every non-empty set E_a is a flat and that, for every a such that $\phi^{-1}(a) \in \tilde{U}$, the restriction of g to E_a is bent. Then the upper bound of Theorem 2 is achieved. We have $E_a = \emptyset$ for every a of Hamming weight at most k, $|E_a| = 1$ for every a such that $w_H(a) > k$ and $\phi^{-1}(a) \notin \tilde{U}$ and $|E_a| = 2^{2M}$ for every a such that. $w_H(a) > k$ and $\phi^{-1}(a) \in \tilde{U}$. Then $f_{\psi,\phi,g}$ has same resiliency order and nonlinearity as $f_{\phi,g}$.

Acknowledgement

The author thanks one of the anonymous referees for his (her) useful observations.

References

1. P. Camion, C. Carlet, P. Charpin, N. Sendrier, "On correlation-immune functions", *Advances in Cryptology-CRYPTO'91*, Lecture Notes in Computer Science 576, pp. 86-100 (1991).
2. A. Canteaut, C. Carlet, P. Charpin et C. Fontaine. "On cryptographic properties of the cosets of $R(1, m)$". *IEEE Transactions on Information Theory* Vol. 47, no 4, pp. 1494-1513 (2001)
3. A. Canteaut and M. Trabbia. "Improved fast correlation attacks using parity-check equations of weight 4 and 5", *Advanced in Cryptology-EUROCRYPT 2000*. Lecture notes in computer science 1807, pp. 573-588 (2000).
4. A. Canteaut and M. Videau. "Degree of Composition of Highly Nonlinear Functions and Applications to Higher Order Differential Cryptanalysis", *Advances in Cryptology, EUROCRYPT2002*, Lecture Notes in Computer Science 2332, Springer Verlag, pp. 518-533 (2002)
5. C. Carlet. "Partially-bent functions", *Designs Codes and Cryptography*, 3, pp. 135-145 (1993) and *Advances in Cryptology-CRYPTO'92* Lecture Notes in Computer Science 740, pp. 280-291 (1993).
6. C. Carlet. "More correlation-immune and resilient functions over Galois fields and Galois rings". *Advances in Cryptology, EUROCRYPT' 97*, Lecture Notes in Computer Science 1233, Springer Verlag, pp. 422-433 (1997).
7. C. Carlet. "On the coset weight divisibility and nonlinearity of resilient and correlation-immune functions", *Proceedings of SETA'01* (Sequences and their Applications 2001), *Discrete Mathematics and Theoretical Computer Science*, Springer, pp. 131-144 (2001).
8. C. Carlet and P. Sarkar. "Spectral Domain Analysis of Correlation Immune and Resilient Boolean Functions". *Finite fields and Applications* 8, pp. 120-130 (2002).
9. S. Chee, S. Lee, K. Kim and D. Kim. "Correlation immune functions with controlable nonlinearity". *ETRI Journal*, vol 19, no 4, pp. 389-401 (1997).
10. S. Chee, S. Lee, D. Lee and S. H. Sung. "On the correlation immune functions and their nonlinearity" *proceedings of Asiacrypt'96*, LNCS 1163, pp. 232-243 (1997).
11. T. W. Cusick. "On constructing balanced correlation immune functions". *Proceedings of SETA'98* (Sequences and their Applications 1998), *Discrete Mathematics and Theoretical Computer Science*, Springer, pp. 184-190 (1999).
12. J. F. Dillon. Elementary Hadamard Difference sets. Ph. D. Thesis, Univ. of Maryland (1974).
13. H. Dobbertin, " Construction of bent functions and balanced Boolean functions with high nonlinearity", *Fast Software Encryption* (Proceedings of the 1994 Leuven Workshop on Cryptographic Algorithms), Lecture Notes in Computer Science 1008, pp. 61-74 (1995).
14. Mac Williams, F. J. and N. J. Sloane (1977). *The theory of error-correcting codes*, Amsterdam, North Holland.
15. S. Maitra and P. Sarkar. "Modifications of Patterson-Wiedemann functions for cryptographic applications". *IEEE Trans. Inform. Theory*, Vol. 48, pp. 278-284, 2002.

16. W. Meier and O. Staffelbach. " Nonlinearity Criteria for Cryptographic Functions", *Advances in Cryptology*, EUROCRYPT' 89, Lecture Notes in Computer Science 434, Springer Verlag, pp. 549-562 (1990).

17. N.J. Patterson and D.H. Wiedemann. " The covering radius of the $[2^{15}, 16]$ Reed-Muller code is at least 16276". *IEEE Trans. Inform. Theory*, IT-29, pp. 354-356 (1983).

18. N.J. Patterson and D.H. Wiedemann. " Correction to [17]". *IEEE Trans. Inform. Theory*, IT-36(2), pp. 443 (1990).

19. E. Pasalic, S. Maitra, T. Johansson and P. Sarkar. "New constructions of resilient functions and correlation immune Boolean functions achieving upper bound on nonlinearity". Proceedings of the *Workshop on Coding and Cryptography 2001*, pp. 425–434 (2001).

20. O. S. Rothaus. " On bent functions", *J. Comb. Theory*, 20A, 300-305 (1976).

21. R. A. Rueppel *Analysis and design of stream ciphers* Com. and Contr. Eng. Series, Berlin, Heidelberg, NY, London, Paris, Tokyo 1986

22. P. Sarkar and S. Maitra. "Construction of nonlinear Boolean functions with important cryptographic properties". *Advances in Cryptology - EUROCRYPT 2000*, number 1807 in Lecture Notes in Computer Science, Springer Verlag, pp. 485–506 (2000).

23. P. Sarkar and S. Maitra. "Nonlinearity Bounds and Constructions of Resilient Boolean Functions". *CRYPTO 2000, LNCS* Vol. 1880, ed. Mihir Bellare, pp. 515-532 (2000).

24. J. Seberry, X.M. Zhang and Y. Zheng. "On constructions and nonlinearity of correlation immune Boolean functions." *Advances in Cryptology - EUROCRYPT'93*, LNCS 765, pp. 181-199 (1994).

25. J. Seberry, X.M. Zhang and Y. Zheng. "Nonlinearly balanced Boolean functions and their propagation characteristics." *Advances in Cryptology - CRYPTO'93*, pp. 49–60 (1994).

26. T. Siegenthaler. "Correlation-immunity of nonlinear combining functions for cryptographic applications". *IEEE Transactions on Information theory*, V. IT-30, No 5, pp. 776-780 (1984).

27. T. Siegenthaler. "Decrypting a Class of Stream Ciphers Using Ciphertext Only". *IEEE Transactions on Computer*, V. C-34, No 1, pp. 81-85 (1985).

28. Y. V. Tarannikov. " On resilient Boolean functions with maximum possible nonlinearity". *Proceedings of INDOCRYPT 2000*, Lecture Notes in Computer Science 1977, pp. 19-30 (2000).

29. Y. V. Tarannikov. "New constructions of resilient Boolean functions with maximum nonlinearity". *Proceedings of FSE 2001*, to appear in the Lecture Notes in Computer Science Series (2002).

30. Xiao Guo-Zhen and J. L. Massey. "A Spectral Characterization of Correlation-Immune Combining Functions". *IEEE Trans. Inf. Theory*, Vol IT 34, n° 3, pp. 569-571 (1988).

31. Y. Zheng, X.-M. Zhang. " Improved upper bound on the nonlinearity of high order correlation immune functions". *Proceedings of Selected Areas in Cryptography 2000*, Lecture Notes in Computer Science 2012, pp. 262-274 (2001)

Linear VSS and Distributed Commitments Based on Secret Sharing and Pairwise Checks

Serge Fehr[1],* and Ueli Maurer[2]

[1] BRICS**, Department of Computer Science, Aarhus University, Denmark
fehr@brics.dk
[2] Department of Computer Science, ETH Zurich, Switzerland
maurer@inf.ethz.ch

Abstract. We present a general treatment of all non-cryptographic (i.e., information-theoretically secure) linear verifiable-secret-sharing (VSS) and distributed-commitment (DC) schemes, based on an underlying secret sharing scheme, pairwise checks between players, complaints, and accusations of the dealer. VSS and DC are main building blocks for unconditional secure multi-party computation protocols. This general approach covers all known linear VSS and DC schemes. The main theorem states that the security of a scheme is equivalent to a pure linear-algebra condition on the linear mappings (e.g. described as matrices and vectors) describing the scheme. The security of all known schemes follows as corollaries whose proofs are pure linear-algebra arguments, in contrast to some hybrid arguments used in the literature. Our approach is demonstrated for the CDM DC scheme, which we generalize to be secure against mixed adversary settings (some curious and some dishonest players), and for the classical BGW VSS scheme, for which we show that some of the checks between players are superfluous, i.e., the scheme is not optimal. More generally, our approach, establishing the minimal conditions for security (and hence the common denominator of the known schemes), can lead to the design of more efficient VSS and DC schemes for general adversary structures.

1 Introduction

The concept of *secret sharing* was introduced by Shamir [12] as a means to protect a secret simultaneously from exposure and from being lost. It allows a so called *dealer* to share the secret among a set of entities, usually called *players*, in such a way that only certain specified subsets of the players are able to reconstruct the secret (if needed) while smaller subsets have no information about it. While secret sharing only guarantees security against curious players that try to gather information they are not supposed to obtain but otherwise behave honestly, its stronger version *verifiable secret sharing (VSS)*, introduced

---

* Most of this research was carried out while the author was employed at ETH Zurich. Supported by the Swiss National Science Foundation (SNF).
** Basic Research in Computer Science (www.brics.dk), funded by the Danish National Research Foundation.

in [4], is secure in the following sense against dishonest players (which are of course also curious) and a dishonest dealer that behave in an arbitrary manner.

Privacy: If the dealer is honest, then the curious players learn nothing about the secret k.

Correctness: After the secret is shared, there exists a unique value k' that can be reconstructed by the players (no matter how the dishonest players behave), and for an honest dealer k' is equal to the shared secret k.

Reconstruction must work even if the dealer does not cooperate in the reconstruction. If an *efficient* reconstruction of the secret requires the cooperation of the dealer, then such a scheme is called a *distributed commitment (DC)* scheme. In such a scheme a dishonest dealer can prevent the reconstruction by refusing to cooperate, but he cannot achieve that a different secret is reconstructed, not even with the help of the dishonest players. A DC scheme is almost a VSS, except for the efficiency of the reconstruction, since the players could try all possible behaviors of the dealer in the reconstruction.

Linear VSS and DC schemes are a main building block for general secure multi-party protocols. Linearity implies that any linear function on shared values can be computed without interaction by each player (locally) computing the linear function on the corresponding individual shares.

The goal of this paper is a unified treatment of (linear) VSS and DC schemes. We present a very natural and general sharing protocol which converts an arbitrary given linear secret sharing scheme into a DC (or VSS) scheme, provided of course that this is possible at all, by enforcing *pairwise* consistency among the shares of the (honest) players. Namely, by pairwise checking, complaining and accusing, it ensures that pairwise linear dependences among the shares that should hold *do* hold. This seems to be not only a very natural but the only possible approach for the construction of secure DC and VSS schemes in our model (i.e. unconditionally secure and zero error probability), and indeed, all known schemes can be seen as concrete instances of this general approach. Then we state the condition under which such a scheme is a secure DC (or VSS) scheme. This characterization is a predicate in the language of pure linear algebra, depending only on the parameters of the underlying secret sharing scheme and of the sharing protocol.

As a consequence, the security of all known schemes (and possibly even all future ones) follow as corollaries whose proofs are linear-algebra arguments, in contrast to some hybrid arguments used in the literature. This is demonstrated for two schemes, for the CDM DC scheme of [5] and for the classical BGW VSS scheme of [1]. We show how the security of the CDM DC scheme can be proven by a simple linear-algebra argument – even with respect to a mixed adversary which strictly generalizes the results of [5] – and characterize the general-adversary condition under which a secure VSS scheme exists. For the BGW VSS scheme, we show that some of the checks between players are superfluous, i.e., the scheme is not optimal. This also shows that arguing about the security of such schemes becomes conceptually simpler. Finally, our approach, establishing the minimal conditions for security, can lead to the design of linear VSS or DC schemes for general adversary structures which are more efficient than the schemes resulting from generic constructions as for instance that of [5].

The outline of the paper is as follows. In the next section, we introduce the notation we use throughout the paper, describe the communication and adversary model and define VSS and DC schemes. In Section 3, we consider general, i.e. not necessarily linear, secret sharing schemes and investigate what is needed to achieve a unique reconstruction as required by the above correctness property, while in Section 4 we then show how this is reduced to a linear-algebraic property in case of linear schemes. In Section 5 and 6 we then discuss the already mentioned applications to the existing schemes of [5] and [1], and in Section 7 we draw some final conclusions.

2 Preliminaries

2.1 Notation

Throughout the paper, $\mathcal{P}$ stands for the *player set* $\mathcal{P} = \{p_1, \ldots, p_n\}$, and for simplicity we set $p_i = i$. We call a subset Π of the power set $2^{\mathcal{P}}$ of $\mathcal{P}$ a *(monotone) structure* of $\mathcal{P}$ if it is closed under taking subsets, i.e., if $P \in \Pi$ and $P' \subseteq P$ implies $P' \in \Pi$. We call it a *(monotone) anti-structure* if it is closed under taking supersets, i.e., if the complement $\Pi^c := \{P \in 2^{\mathcal{P}} \mid P \notin \Pi\}$ is a structure. Given two structures Π_1 and Π_2, $\Pi_1 \sqcup \Pi_2$ denotes the element-wise union, i.e., the structure

$$\Pi_1 \sqcup \Pi_2 := \{P_1 \cup P_2 \mid P_1 \in \Pi_1, P_2 \in \Pi_2\}.$$

Consider a finite set $\mathcal{K}$ (the set of *secrets*), n finite sets $\mathcal{S}_1, \ldots, \mathcal{S}_n$, where $\mathcal{S}_i$ is the set of possible *shares* for player p_i, and let $\mathcal{S}$ be the Cartesian product $\mathcal{S} = \mathcal{S}_1 \times \cdots \times \mathcal{S}_n$. Elements of $\mathcal{S}$ will sometimes be called a *sharing*.

For two sharings $s = (s_1, \ldots, s_n)$ and $\tilde{s} = (\tilde{s}_1, \ldots, \tilde{s}_n)$, the set $\delta(s, \tilde{s}) \subseteq \mathcal{P}$ is defined as

$$\delta(s, \tilde{s}) := \{i \in \mathcal{P} \mid s_i \neq \tilde{s}_i\}.$$

Note that δ can be treated similar to a metric, as for all $s, s', s'' \in \mathcal{S}$ we have $\delta(s, s) = \emptyset$, $\delta(s, s') = \delta(s', s)$ and $\delta(s, s'') \subseteq \delta(s, s') \cup \delta(s', s'')$.

For a subset $Q = \{i_1, \ldots, i_\ell\} \subseteq \mathcal{P}$, a sharing $s \in \mathcal{S}$ and a subset $U \subseteq \mathcal{S}$ of sharings, pr_Q denotes the projection $\mathrm{pr}_Q : \mathcal{S} \to \mathcal{S}_{i_1} \times \cdots \times \mathcal{S}_{i_\ell}$, and s_Q and U_Q stand for $s_Q = \mathrm{pr}_Q(s)$ and $U_Q = \{\mathrm{pr}_Q(u) \mid u \in U\}$, respectively. Finally, if $\mathcal{S}_1, \ldots, \mathcal{S}_n$ and hence $\mathcal{S}$ are in fact vector spaces, which will be the case in Section 4, then, for a sharing $s \in \mathcal{S}$, the *support* $\mathrm{supp}(s)$ denotes the smallest set $Q \subseteq \mathcal{P}$ with $\mathrm{pr}_{\mathcal{P} \setminus Q}(s) = (0, \ldots, 0)$, in other words $\mathrm{supp}(s) = \delta(s, 0)$, and, for $Q \subseteq \mathcal{P}$ and $U \subseteq \mathcal{S}$, $U|_Q$ denotes the subset $U|_Q = \{u \in U \mid \mathrm{supp}(u) \subseteq Q\}$.

2.2 Model

We consider the *secure-channels model*, as introduced in [1,3], where the set of players (including the dealer) is connected by bilateral synchronous reliable secure channels. Broadcast channels are not assumed to be available, though can be implemented for the cases we consider [2,8] and thus will be treated as given primitives.

Like in previous literature on VSS and secure multi-party computation, we consider a central adversary who can corrupt players, subject to certain constraints, for example an upper bound on the total number of corrupted players. The selection of which player to corrupt can be adaptive, depending on the course of the protocol. The dealer is one of the players that can potentially also be corrupted.

Passive corruption of a player means that the adversary learns the player's entire information, but the player performs the protocol correctly. This models what is often also called "honest but curious" players. Active corruption of a player means that the adversary takes full control and can make the player deviate from the protocol in an arbitrary manner. Such a player is also called dishonest, or simply a cheater. Active corruption is hence strictly stronger than passive corruption. The adversary is characterized by a *privacy structure* $\Delta \subseteq \mathcal{P}$ and an *adversary structure* $\mathcal{A} \subseteq \Delta$ with the intended meaning that the adversary can be tolerated to corrupt any players passively or actively (one variant being the upgrading of a passive corruption to an active corruption), as long as the total set D of corrupted players satisfies $D \in \Delta$ and the subset A of them being actively corrupted satisfies $A \in \mathcal{A}$. In other words, all players in $D \backslash A$ are honest but curious. The complement $\mathcal{H} = \mathcal{A}^c$ is sometimes called the *honest-players structure*.

Finally, we assume that the adversary has unbounded computing power, and we achieve zero error probability.

2.3   Definition of VSS and DC

Let $\mathcal{K}$ be a finite set (as described in Section 2.1), let Δ be a privacy structure, and let $\mathcal{A} \subseteq \Delta$ be an adversary structure (as described in Section 2.2).

Definition 1. *A $(\Delta, \mathcal{A})$-secure verifiable secret sharing (VSS) scheme is a pair* (Share, Rec) *of protocols (phases), the sharing phase, where the dealer shares a secret $k \in \mathcal{K}$, and the reconstruction phase, where the players try to reconstruct k, such that the following two properties hold, even if the players of a set $A \in \mathcal{A}$ are dishonest and behave in an arbitrary manner:*

Privacy: *If the dealer remains honest, then the players of any set $D \in \Delta$ with $A \subseteq D$ learn nothing about the secret k as a result of the sharing phase.*

Correctness: *After the secret is shared, there exists a unique value k' that can be reconstructed by the players, and for an honest dealer this value k' is equal to the shared secret k.*

Reconstruction must work even if the dealer does not cooperate in the reconstruction. If an efficient reconstruction of the secret requires the cooperation of the dealer, then such a scheme is called a distributed commitment (DC) *scheme.*

In a DC scheme, a dishonest dealer can prevent the (efficient) reconstruction by refusing to cooperate correctly, but he cannot achieve that a different secret is reconstructed, not even with the help of the dishonest players. Note that if one would define a default value for the case the dealer refuses to reconstruct, then a cheating dealer would not be committed because he could open a sharing in two different ways: as the real or as the default value.

A VSS or DC scheme is called *linear* if the list of *shares*, i.e., the information given to the players during the sharing phase, is a linear function of the secret and randomly chosen values.

3 General Schemes

Even though our goal is a general treatment of *linear* schemes, we first consider arbitrary, not necessarily linear secret sharing schemes and discuss facts that are independent of the linearity of the scheme. More precisely, we present a sufficient condition on the (possibly not correctly) distributed shares in order to have uniqueness of the shared secret as required by the correctness property of VSS and DC schemes. And then, in the next section, we show how this can be achieved using linear schemes.

Most of the arguments of this section have been used – implicitly or explicitly – in the literature, but typically with respect to some restricted model. This unification not only generalizes arguments that have been used before (to non-linear schemes and to a mixed adversary), it also leads to a better understanding of the security of (linear and general) VSS and DC schemes.

Let $\mathcal{K}$ and $\mathcal{S} = \mathcal{S}_1 \times \cdots \times \mathcal{S}_n$ be defined as defined in Section 2.1.

Definition 2. *A secret sharing scheme is given by a joint conditional probability distribution $P_{S|K} : \mathcal{S} \times \mathcal{K} \longrightarrow [0,1]$. The privacy structure Δ is defined as the structure*

$$\Delta = \{D \subseteq \mathcal{P} \mid P_{S_D|K}(\,\cdot\,,k) = P_{S_D|K}(\,\cdot\,,k') \text{ for all } k,k' \in \mathcal{K}\}\,[1],$$

and the access structure Γ is defined as the anti-structure[2]

$$\Gamma = \{Q \subseteq \mathcal{P} \mid P_{S|K}(s,k), P_{S|K}(s',k') > 0 \wedge s_Q = s'_Q \implies k = k'\}.$$

A sharing $s \in \mathcal{S}$ is called correct *of a secret k if $P_{S|K}(s,k) > 0$, and, by defining the relation* corr $:= \{(s,k) \mid P_{S|K}(s,k) > 0\} \subset \mathcal{S} \times \mathcal{K}$, *is denoted by $(s,k) \in$ corr.*

Typically, a secret sharing scheme $P_{S|K}$ is given in terms of an (efficiently computable) function $f : \mathcal{K} \times \mathcal{R} \to \mathcal{S}$, where $\mathcal{R}$ is some finite set, such that $P_{S|K}(\,\cdot\,,k)$ is the distribution of $f(k,r)$ for a uniformly random chosen $r \in \mathcal{R}$. This is often directly used as the definition of a secret sharing scheme. Note that we do *not* require, as is usually the case in the literature, that the privacy structure Δ be the complement Γ^c of the access structure Γ, but for linear schemes this is the case.

By the definition of Δ and Γ, the following properties are guaranteed.

---

[1] $P_{S_D|K}(\,\cdot\,,k)$ is naturally defined by $P_{S_D|K}(s_D,k) = \sum_{s' \in \mathcal{S}:s'_D=s_D} P_{S|K}(s',k)$.

[2] Note that even though Γ is an anti-structure, it is called access *structure* (and not access anti-structure).

Privacy: For any secret k and for $s = (s_1, \ldots, s_n)$ chosen according to the distribution $P_{S|K}(\cdot, k)$ [3] the shares $s_{i_1}, \ldots, s_{i_k}$ corresponding to a set $D = \{i_1, \ldots, i_k\} \in \Delta$ give no information about the secret k.

Correctness: For any $(s, k) \in$ corr, the shares $s_{j_1}, \ldots, s_{j_\ell}$ corresponding to a set $Q = \{j_1, \ldots, j_\ell\} \in \Gamma$ uniquely define k (and hence k can – at least in principle – be computed from $s_{j_1}, \ldots, s_{j_\ell}$).

Hence, the correctness property guarantees that the secret is uniquely defined by the set of shares even if some are missing, i.e., in this sense the scheme is robust against lost shares. We now investigate what it means to be robust against *incorrect* shares.

Let $\mathcal{A} \subseteq \Delta$ be an adversary structure.

Proposition 1. *The following robustness property is fulfilled if and only if* $\mathcal{P} \notin \Gamma^c \sqcup \mathcal{A} \sqcup \mathcal{A}$.

Robustness: *For any* $(s, k) \in$ corr, *any sharing* $\tilde{s}$ *with* $\delta(s, \tilde{s}) \in \mathcal{A}$ *uniquely defines* k, *in the sense that for any* $\tilde{s} \in S$

$$(s, k), (s', k') \in \text{corr} \land \delta(s, \tilde{s}), \delta(s', \tilde{s}) \in \mathcal{A} \implies k = k'. \tag{1}$$

Namely, by the definition of Γ, (1) holds if and only if for every pair $A_1, A_2 \in \mathcal{A}$ the set $Q = \mathcal{P} \setminus (A_1 \cup A_2)$ is in Γ, which is equivalent to $\mathcal{P} \notin \Gamma^c \sqcup \mathcal{A} \sqcup \mathcal{A}$.

Note that in the literature $\mathcal{A}$ typically coincides with Δ and, as already mentioned, Δ with Γ^c, in which case $\mathcal{P} \notin \Gamma^c \sqcup \mathcal{A} \sqcup \mathcal{A}$ coincides with the Q^3 property of [10] which states that no three sets in $\mathcal{A}$ cover $\mathcal{P}$, which itself generalizes the classical bound $t < n/3$. However, we consider this more general case because it gives deeper insight but also because it makes perfect sense to separate the privacy from the adversary structure, i.e., to consider curious as well as dishonest players as argued in Section 2, and in fact will generalize in Section 5 the DC scheme from [5] to such mixed adversaries.

Robustness guarantees that the secret is uniquely defined by the set of shares even if some might be incorrect. If, as usual, the secret k is shared by a so-called *dealer* by choosing s according to $P_{S|K}(\cdot, k)$ and distributing the shares among the players in $\mathcal{P}$, then this allows the correct reconstruction of the secret even if the players of a set $A \in \mathcal{A}$ are dishonest and do not provide correct shares. However, this is only guaranteed to hold if the dealer is honest and indeed distributes a correct sharing s of k. Hence, it seems that to achieve security against a possibly dishonest dealer, in the sense that a unique secret is defined, the dealer has to be forced to distribute a *correct* sharing s. We will now show in the remainder of this section that this is actually overkill and a weaker condition already suffices.

Definition 3. *A function* $\rho : \Gamma \times S \ni (Q, s) \mapsto \rho^Q(s) \in \mathcal{K}$ *is called a reconstruction function for a secret sharing scheme* $P_{S|K}$ *if, for every* $Q \in \Gamma$, $\rho^Q(s)$ *only depends on* s_Q, *i.e.,* $\rho^Q : S \to \mathcal{K}$ *can be seen as a function* $\rho^Q : S_Q \to \mathcal{K}$,

---

[3] e.g. computed as $s = f(k, r)$ for a random $r \in \mathcal{R}$

and $\rho^Q(s_Q) = k$ for every correct sharing $s \in \mathcal{S}$ of a secret $k \in K$. A (not necessarily correct) sharing $s \in \mathcal{S}$ is called a consistent sharing of a secret k (with respect to ρ) if $\rho^Q(s_Q) = k$ for every $Q \in \Gamma$, and is denoted by $(s, k) \in \text{cons}_\rho$. And, similarly, s_H with $H \in \Gamma$ is called a consistent sharing of a secret k for the players in H (with respect to ρ) if $\rho^Q(s_Q) = k$ for every $Q \in \Gamma$ with $Q \subseteq H$, and is denoted by $(s_H, k) \in \text{cons}_\rho^H$ [4].

It is easy to verify that the access structure Γ coincides with

$$\Gamma_\rho = \{Q \subseteq \mathcal{P} \mid (s, k), (s', k') \in \text{cons}_\rho \wedge s_Q = s'_Q \implies k = k'\}$$

Indeed, if $Q \in \Gamma_\rho$ and $(s, k), (s', k') \in \text{corr}$ with $s_Q = s'_Q$, then $(s, k), (s', k') \in \text{cons}_\rho$ and hence $k = k'$, and therefore $Q \in \Gamma$. On the other hand, if $Q \in \Gamma$ and $(s, k), (s', k') \in \text{cons}_\rho$ with $s_Q = s'_Q$, then, by the properties of ρ, $k = \rho^Q(s_Q) = \rho^Q(s'_Q) = k'$, and therefore $Q \in \Gamma_\rho$.

Hence, arguing as before, we have

Proposition 2. *The following strong robustness property is fulfilled for an arbitrary reconstruction function ρ if and only if $\mathcal{P} \notin \Gamma^c \sqcup \mathcal{A} \sqcup \mathcal{A}$.*

Strong robustness: *For any $(s, k) \in \text{cons}_\rho$, any sharing $\tilde{s}$ with $\delta(s, \tilde{s}) \in \mathcal{A}$ uniquely defines k, in the sense that for any $\tilde{s} \in \mathcal{S}$*

$$(s, k), (s', k') \in \text{cons}_\rho \wedge \delta(s, \tilde{s}), \delta(s', \tilde{s}) \in \mathcal{A} \implies k = k'. \tag{2}$$

Hence, if indeed $\mathcal{P} \notin \Gamma^c \sqcup \mathcal{A} \sqcup \mathcal{A}$, as long as the dealer is partially honest and hands out a *consistent* (but not necessarily correct) sharing s, there is a unique secret k defined, assuming that the shares s_H of an honest-players set $H \in \mathcal{H} = \mathcal{A}^c$ remain unchanged. We finally show that this even holds as long as the dealer hands out a consistent sharing s_H for the players in H.

Proposition 3. *The following very strong robustness property is fulfilled for an arbitrary reconstruction function ρ if and only if $\mathcal{P} \notin \Gamma^c \sqcup \mathcal{A} \sqcup \mathcal{A}$.*

Very strong robustness: *For any honest-players set $H \in \mathcal{H}$ and $(s_H, k) \in \text{cons}_\rho^H$, any sharing $\tilde{s}$ with $s_H = \tilde{s}_H$ uniquely defines k, in the sense that for any $\tilde{s} \in \mathcal{S}$*

$$\left. \begin{array}{l} H \in \mathcal{H} \wedge (s_H, k) \in \text{cons}_\rho^H \wedge s_H = \tilde{s}_H \\ \wedge\ H' \in \mathcal{H} \wedge (s'_{H'}, k') \in \text{cons}_\rho^{H'} \wedge s'_{H'} = \tilde{s}_{H'} \end{array} \right\} \implies k = k'. \tag{3}$$

Indeed, (3) holds if and only if $H \cap H' \in \Gamma$ for all $H, H' \in \mathcal{H}$, which is equivalent to $\mathcal{P} \setminus (A_1 \cup A_2) \in \Gamma$ for all $A_1, A_2 \in \mathcal{A}$, which, as already noticed earlier, is equivalent to $\mathcal{P} \notin \Gamma^c \sqcup \mathcal{A} \sqcup \mathcal{A}$.

---

[4] Clearly, if s is a consistent sharing then, for any $H \in \Gamma$, s_H is a consistent sharing for the players in H; however, if s_H is a consistent sharing for the players in H for some H then, in general, s_H cannot be completed to a consistent sharing s.

4   Linear Schemes

We have seen in the above Section 3 that the uniqueness of the shared secret (that is required by the correctness property of VSS or DC) is guaranteed if (and only if) $\mathcal{P} \notin \Gamma^c \sqcup \mathcal{A} \sqcup \mathcal{A}$ and if the dealer is at least partially honest and hands out a *consistent* sharing to the honest players, or if he can be forced to behave this way. In this section we now concentrate on *linear* schemes, and we present a very natural sharing protocol which enforces some kind of consistency. Namely, by pairwise checking, complaining and accusing, it ensures *pairwise* consistency among the shares (of the honest players). All known DC and VSS schemes can be seen as concrete instances of this general approach. Finally, we give a characterization in the language of linear algebra of when the sharing protocol results in a secure DC (or VSS) scheme. As a consequence, the security of all known schemes follow as corollaries whose proofs are linear-algebra arguments, and, more generally, it becomes conceptually very simple to argue about the security of such schemes, as it involves only pure linear algebra.

From now on, $\mathcal{K}$ is a field, and $\mathcal{S}_1, \ldots, \mathcal{S}_n$ are vector spaces over $\mathcal{K}$ with inner products $\langle \cdot, \cdot \rangle_{\mathcal{S}_1}, \ldots, \langle \cdot, \cdot \rangle_{\mathcal{S}_n}$, respectively, which naturally induce an inner product $\langle \cdot, \cdot \rangle_{\mathcal{S}}$ for the vector space $\mathcal{S} = \mathcal{S}_1 \times \cdots \times \mathcal{S}_n$ by $\langle s, s' \rangle_{\mathcal{S}} = \sum_i \langle s_i, s'_i \rangle_{\mathcal{S}_i}$.

As usual in linear algebra, for a subset $U \subseteq \mathcal{S}$, span(U) denotes the subspace consisting of all linear combinations of vectors in U and the *orthogonal complement* $U^{\perp_{\mathcal{S}}}$ is the subspace defined by $U^{\perp_{\mathcal{S}}} := \{s \in \mathcal{S} \mid \langle s, u \rangle_{\mathcal{S}} = 0 \; \forall u \in U\}$. We also write $s \perp_{\mathcal{S}} U$ instead of $s \in U^{\perp_{\mathcal{S}}}$.

4.1   Secret Sharing

A *linear* secret sharing scheme is given by a pair (M, ε), consisting of a linear map

$$M : \mathcal{V} \longrightarrow \mathcal{S} = \mathcal{S}_1 \times \cdots \times \mathcal{S}_n$$

and a vector $\varepsilon \in \mathcal{V}$, where $\mathcal{V}$ is a vector space over the field $\mathcal{K}$ with inner product $\langle \cdot, \cdot \rangle_{\mathcal{V}}$ and $\mathcal{S}$ is as described above. A secret $k \in \mathcal{K}$ is shared by choosing a random $x \in \mathcal{V}$ such that $\langle \varepsilon, x \rangle_{\mathcal{V}} = k$ and computing s as $s = Mx$.

Consider the special case where $\mathcal{V} = \mathcal{K}^e$ for some e and $\mathcal{S}_i = \mathcal{K}^{d_i}$ for some $d_1, \ldots, d_n$, where every inner product is the respective standard inner product, and where M is a matrix multiplication $M : \mathcal{K}^e \to \mathcal{K}^{\Sigma d_i} = \mathcal{K}^{d_1} \times \ldots \times \mathcal{K}^{d_n}$, $x \mapsto M \cdot x$ [5]. In this case, (M, ε) is called a monotone span program [11]. Clearly, by fixing orthogonal bases of $\mathcal{V}$ and $\mathcal{S}_1, \ldots, \mathcal{S}_n$, respectively, one can always have this simplified and more familiar view. However, as this simplification might (and indeed would in Section 5.1) destroy the naturalness of additional structures in $\mathcal{V}$ or $\mathcal{S}$, we keep this more general view. Nevertheless, because of this reduction, it follows from [11] that the access structure Γ and the privacy structure Δ of a linear secret sharing scheme (M, ε) are given by

$$\Gamma = \{Q \subseteq \mathcal{P} \mid \exists \lambda \in \mathcal{S} : \text{supp}(\lambda) \subseteq Q, M^* \lambda = \varepsilon\}$$

---

[5] We slightly abuse notation and use the same symbol, M, for the matrix $M \in \mathcal{K}^{\Sigma d_i \times e}$ as well as the corresponding linear map $M : \mathcal{K}^e \to \mathcal{K}^{\Sigma d_i} = \mathcal{K}^{d_1} \times \ldots \times \mathcal{K}^{d_n}$.

and $\Delta = \Gamma^c$, respectively, where $M^* : S \to V$ is the conjugate of M (i.e. such that $\langle \lambda, Mx \rangle_S = \langle M^*\lambda, x \rangle_V$ for all $\lambda \in S$ and $x \in V$, and, in the simplified monotone span program view, $M^* = M^T$, the transposed matrix). Furthermore, any subset Λ of

$$\Lambda^{max} = \{\lambda \in S \mid M^*\lambda = \varepsilon\}$$

which is complete in the sense that for every $Q \in \Gamma$ there exists $\lambda \in \Lambda|_Q$, naturally induces a reconstruction function $\rho : \Gamma \times S \to \mathcal{K}$ by

$$\rho^Q(s) = \begin{cases} \langle \lambda, s \rangle_S & \text{if } \langle \lambda, s \rangle_S \text{ is the same for every } \lambda \in \Lambda|_Q \\ 0 & \text{otherwise} \end{cases}$$

Note that $\lambda \in \Lambda^{max}$ fulfills $\langle \lambda, s \rangle_S = \langle \lambda, Mx \rangle_S = \langle M^*\lambda, x \rangle_V = \langle \varepsilon, x \rangle_V = k$ for any correct sharing s of a secret k.

4.2 Verifiable Secret Sharing and Distributed Commitments

Consider a linear secret sharing scheme, given by $M : V \to S = S_1 \times \cdots \times S_n$ and $\varepsilon \in V$, with an access structure Γ. According to Section 3, in order to turn this scheme into a DC scheme or a VSS, secure against the privacy structure $\Delta = \Gamma^c$ and the adversary structure $\mathcal{A} \subseteq \Delta$, it is necessary that $\mathcal{P} \notin \Delta \sqcup \mathcal{A} \sqcup \mathcal{A}$, and additionally, as part of the sharing procedure, it has to be checked that the dealer behaves partially honest and hands out a *consistent* sharing with respect to some reconstruction function ρ to the honest players. However, it seems to be impossible to *directly* check this kind of consistency, i.e. to verify something like $\langle \lambda, s \rangle = \langle \lambda', s \rangle$ for $\lambda \neq \lambda'$, *without violating privacy*. The only thing that can be checked without violating privacy is *pairwise consistency*, i.e. whether (some) pairwise linear dependences $\langle \gamma, s \rangle = 0$ with $\text{supp}(\gamma) = \{i, j\}$ that should hold indeed *do* hold; namely by comparing in private the respective contributions $\langle \gamma_i, s_i \rangle$ and $\langle \gamma_j, s_j \rangle$ (which, up to the sign, are supposed to be equal) of the two involved players. A player *complains* in case of a pairwise inconsistency, but this may be due to the dealer's *or* another player's misbehavior, and he *accuses* (the dealer) if he knows that the dealer misbehaved. In any case, the dealer has to publicly clarify the situation, such that finally the shares of all honest players *are* pairwise consistent *and* privacy is satisfied. This is described in full detail in the protocol Share below. Finally, a simple linear algebra condition is given that is sufficient (and also necessary) in order for the pairwise consistency to imply consistency with respect to a given reconstruction function, and hence in order for the scheme to result in a secure DC respectively VSS.

Consider the set

$$Checks(M) := \{\gamma \in \ker M^* \mid |\text{supp}(\gamma)| \leq 2\} \subseteq S$$

of all possible *(pairwise) checking vectors*, where $\ker M^*$ denotes the kernel $\ker M^* = \{\xi \in S \mid M^*\xi = 0\}$ of M^*. Clearly, for any $\gamma \in Checks(M)$ and any *correct* sharing $s = Mx$, we have

$$\langle \gamma, s \rangle_S = \langle \gamma, Mx \rangle_S = \langle M^*\gamma, x \rangle_V = \langle 0, x \rangle_V = 0.$$

For an arbitrary but fixed subset $C \in Checks(M)$, the following sharing protocol enforces pairwise consistency with respect to the checking vectors $\gamma \in C$ among the players that remain honest during the execution, without revealing any information about the shared secret. The concrete choice of the protocol is somewhat arbitrary, in that it can be modified in different ways without loosing its functionality and without nullifying the upcoming results. For instance, techniques from [9] can be applied to improve the round complexity (at the cost of an increased communication complexity), and some secret sharing schemes M allow "early stopping".

Protocol Share$_{(M,\varepsilon),C}(k)$

1. The dealer chooses a random $x \in V$ such that $\langle \varepsilon, x \rangle = k$, computes $s = Mx$ and sends to every player $p_i \in P$ the corresponding share s_i.
2. For every checking vector $\gamma \in C$, it is as follows checked whether $\langle \gamma, s \rangle_S = 0$:
   If $\text{supp}(\gamma) = \{p_i\}$, then player p_i verifies whether $\langle \gamma_i, s_i \rangle_{S_i} = 0$, and he broadcasts an "accusation" against the dealer if it does not hold.
   If $\text{supp}(\gamma) = \{p_i, p_j\}$ with $p_i < p_j$, then then player p_i sends $c_{ij} = \langle \gamma_i, s_i \rangle_{S_i}$ to p_j who verifies whether $c_{ij} + \langle \gamma_j, s_j \rangle_{S_j} = 0$ and broadcasts a "complaint" if it does not hold. The dealer answers such a complaint by broadcasting $c_{ij} = \langle \gamma_i, s_i \rangle$, and if this value does not coincide with p_i's c_{ij} respectively if it does not fulfill $c_{ij} + \langle \gamma_j, s_j \rangle_{S_j} = 0$, then player p_i respectively p_j broadcasts an "accusation" against the dealer.
3. The following is repeated until there is no further "accusation" or the dealer is declared faulty (which requires at most n rounds). For every "accusation" from a player p_i, the dealer answers by broadcasting p_i's share s_i, and p_i replaces his share by this s_i. If this share contradicts the share of some player p_j, in the sense that $\langle \gamma_i, s_i \rangle_{S_i} + \langle \gamma_j, s_j \rangle_{S_j} \neq 0$ for some $\gamma \in C$ with $\text{supp}(\gamma) = \{p_i, p_j\}$, then p_j broadcasts an "accusation" (if he has not yet done so in an earlier step). If this share s_i contradicts itself, in the sense that $\langle \gamma_i, s_i \rangle \neq 0$ for some $\gamma \in C$ with $\text{supp}(\gamma) = \{p_i\}$, or it contradicts a share s_j that has already been broadcast, then the dealer is publicly declared to be faulty.

It is easy to see that if the dealer remains honest, then no matter what the dishonest players do, nobody learns anything beyond his share, and hence the players of any set $D \in \Delta$ learn nothing about the shared secret. Furthermore, independent of the behavior of the dishonest players, if H denotes the set of players that remain honest during the protocol execution (though some might become curious) then the protocol achieves pairwise consistency among the players in H, i.e., $\langle \gamma, s \rangle_S = 0$ for every $\gamma \in C|_H$, or, in other words,

$$s \perp_S C|_H .$$

In order for the protocol to achieve consistency with respect to a reconstruction function ρ, it must be guaranteed for a complete subset of reconstruction vectors $\Lambda \subseteq \Lambda^{max}$ that $\langle \lambda, s \rangle = \langle \lambda', s \rangle$ for all $\lambda, \lambda' \in \Lambda|_H$, or, in other words, that

$$s \perp_S \{\lambda - \lambda' \mid \lambda, \lambda' \in \Lambda|_H\} .$$

This implies

Proposition 4. *Let* $\rho : \Gamma \times \mathcal{S} \to \mathcal{K}$ *be a reconstruction function induced by a complete subset of reconstruction vectors* $\Lambda \subseteq \Lambda^{max}$ *(as defined Section 4.1). Then, the sharing protocol* $\mathsf{Share}_{(M,\varepsilon),\mathcal{C}}$ *is guaranteed to produce a consistent sharing for the honest players with respect to* ρ *if and only if*

$$\{\lambda - \lambda' \mid \lambda, \lambda' \in \Lambda|_H\} \subseteq \mathrm{span}(\mathcal{C}|_H) \quad \text{for every } H \in \mathcal{H}. \tag{4}$$

Combing this with Proposition 3 leads to

Theorem 1. *Let* (M, ε) *be a linear secret sharing scheme with privacy structure* Δ, $\mathcal{C} \subseteq Checks(M)$ *a subset of checking vectors and* $\mathcal{A} \subseteq \Delta$ *an adversary structure. Then the protocol* $\mathsf{Share}_{(M,\varepsilon),\mathcal{C}}$ *can be completed to a* $(\Delta, \mathcal{A})$-*secure DC scheme* $(\mathsf{Share}_{(M,\varepsilon),\mathcal{C}}, \mathsf{Rec}_{(M,\varepsilon),\mathcal{C}})$ *if (and only if)* $\mathcal{P} \notin \Delta \sqcup \mathcal{A} \sqcup \mathcal{A}$ *and if (4) holds for some complete subset* $\Lambda \subseteq \Lambda^{max}$ *of reconstruction vectors.*
If additionally $\mathcal{C}_{\{i,j\}} \subseteq \mathrm{span}(\mathcal{C}|_{\{i\}\cup Q} \cup \mathcal{C}|_{\{j\}\cup Q})$ *for all* i, j *and every* $Q \notin \Delta$, *then* $\mathsf{Share}_{(M,\varepsilon),\mathcal{C}}$ *can be completed to a* $(\Delta, \mathcal{A})$-*secure VSS scheme.*

Proof. With respect to a not necessarily efficient reconstruction procedure, the claim follows from Proposition 4 and 3 (even without the additional requirement for the VSS case). It remains to show the existence of *efficient* reconstruction procedures: In the DC reconstruction, the dealer publishes the vector x used in Step 1 of the sharing protocol and every player p_i publishes his share s_i, and then the players take $k = \langle \varepsilon, x \rangle_V$ as the reconstructed secret if $\delta(Mx, s) \in \mathcal{A}$ and reject the reconstruction otherwise (as if the dealer had refused to take part at all). In the VSS reconstruction, every player p_i publishes his share s_i, then any share s_i that is pairwise inconsistent (with respect to the checking vectors in $\mathcal{C}$) with the shares of a set $A \notin \mathcal{A}$ is rejected, and the secret is reconstructed from the accepted shares by applying the reconstruction function ρ induced by Λ. Note that the additional requirement for $\mathcal{C}$ implies that all accepted shares are pairwise consistent and hence consistent with respect to ρ. □

To our knowledge, the condition $\mathcal{P} \notin \Delta \sqcup \mathcal{A} \sqcup \mathcal{A}$ for VSS to be possible has not been stated previously in the literature, although the condition for secure multi-party computation has been given in [7]. In the threshold case, this confirms Lemma 1 of [6]: If the total number of (passively) corrupted players is t and if w of them can even be actively corrupted, then VSS is possible if and only if $t + 2w < n$.

The following lemma will be helpful in the next section.

Lemma 1. *Predicate (4) is fulfilled if every pair* $\lambda, \lambda' \in \Lambda$ *fulfills*

$$\lambda - \lambda' \in \mathrm{span}(\mathcal{C}|_{\mathrm{supp}(\lambda) \cup \mathrm{supp}(\lambda')}).$$

Proof. Let $H \in \mathcal{H}$ be arbitrary but fixed. Then, for $\lambda, \lambda' \in \Lambda|_H \subseteq \Lambda$, we have by assumption $\lambda - \lambda' \in \mathrm{span}(\mathcal{C}|_{\mathrm{supp}(\lambda) \cup \mathrm{supp}(\lambda')})$, which is of course contained in $\mathrm{span}(\mathcal{C}|_H)$. □

5   Application I: Proving the Security of the CDM Scheme

We now demonstrate the power of Theorem 1 and prove the security of the CDM DC scheme [5] by proving a pure linear-algebra statement. We only have to show that $\{\lambda - \lambda' \mid \lambda, \lambda' \in \Lambda|_H\} \subseteq \text{span}(\mathcal{C}|_H)$ for every $H \in \mathcal{H}$, or, and that is what we are going to do, that $\lambda - \lambda' \in \text{span}(\mathcal{C}|_{\text{supp}(\lambda) \cup \text{supp}(\lambda')})$ for every pair $\lambda, \lambda' \in \Lambda$. As a by-product, because of our general treatment in Section 3, we generalize the CDM DC scheme to a mixed adversary.

5.1   The CDM Scheme

In [5], a generic construction was presented to convert any linear secret sharing scheme, described by a monotone span program, into a linear DC scheme. As mentioned in Section 4, a monotone span program is given by a matrix $M_o \in \mathcal{K}^{\Sigma d_i \times e}$ and a vector $\varepsilon_o \in \mathcal{K}^e$. The CDM DC scheme works as follows, assuming for simplification that $\varepsilon_o = (1, 0, \ldots, 0)^T$ and $d_1 = \ldots = d_n = 1$. To share (or commit to) a secret k, the dealer chooses a random *symmetric matrix* $X \in \mathcal{K}^{e \times e}$ with k in the upper left corner and sends the share $s_i = M_{oi} \cdot X$ to player p_i, where M_{oi} denotes the i-th row of M_o. Now, every pair p_i, p_j of players verifies whether $M_{oi} \cdot s_j^T = M_{oj} \cdot s_i^T$ and, in case it does not hold, start complaining and accusing as in the protocol from the above section.

It is not hard to see that this scheme is a concrete instance of the class of schemes described in the previous section. Indeed, it coincides with $\text{Share}_{(M,\varepsilon),\mathcal{C}}$ for M, ε and $\mathcal{C}$ as described in the following. M is the linear map

$$M : \mathcal{V} \to \mathcal{S} = \mathcal{K}^{n \times e} = \mathcal{K}^e \times \cdots \times \mathcal{K}^e$$
$$X \mapsto s = M_o \cdot X$$

where $\mathcal{V}$ is the vector space consisting of all symmetric $e \times e$-matrices over $\mathcal{K}$ and $\langle \cdot, \cdot \rangle_{\mathcal{V}}$ is given by

$$\langle a, b \rangle_{\mathcal{V}} = \sum_{1 \leq i, j \leq e} a[i,j] b[i,j]$$

for matrices a and b in $\mathcal{V}$ with entries $a[i,j]$ and $b[i,j]$. Furthermore, $\varepsilon \in \mathcal{V}$ is the matrix with a 1 in the upper left corner and zeros otherwise, and the set $\mathcal{C}$ is given by

$$\mathcal{C} = \{\gamma^{ij} = \mu^{ij} - \mu^{ji} \mid 1 \leq i < j \leq n\} \subseteq Checks(M)$$

where $\mu^{ij} \in \mathcal{S}$ has M_{oj} as i-th row and zero-entries otherwise. In this example the checking vectors $\gamma \in \mathcal{C}$ are in fact matrices.

Note that $\mathcal{S} = \mathcal{S}_1 \times \cdots \times \mathcal{S}_n$ with $\mathcal{S}_i = \mathcal{K}^e$ (and $\langle \cdot, \cdot \rangle_{\mathcal{S}_i}$ the standard inner product) is interpreted as $\mathcal{S} = \mathcal{K}^{n \times e}$. Hence, for any matrix $s \in \mathcal{S}$, s_i is the i-th row of s, and therefore if $s = M_o \cdot X$ then $s_i = M_{oi} \cdot X$. Furthermore, $\langle \varepsilon, X \rangle_{\mathcal{V}} = k$ if and only if the upper left corner of X is k and for a check vector $\gamma^{ij} \in \mathcal{C}$ we have $\langle \gamma^{ij}, s \rangle_{\mathcal{S}} = \langle M_{oj}, s_i \rangle_{\mathcal{S}_i} - \langle M_{oi}, s_j \rangle_{\mathcal{S}_j} = M_{oj} \cdot s_i^T - M_{oi} \cdot s_j^T$. Hence, $\text{Share}_{(M,\varepsilon),\mathcal{C}}$ indeed coincides with the CDM protocol [5].

Finally, note that (as it is also shown in [5]) the access structure Γ of the secret sharing scheme (M, ε) coincides with the access structure Γ_o of the original scheme (M_o, ε_o).

5.2   The Security Proof

If λ_o is a reconstruction vector for the original secret sharing scheme (M_o, ε_o), i.e. $\langle \lambda_o, M_o x \rangle_{\mathcal{K}^n} = k$ for $x \in \mathcal{K}^e$ with k as first entry (such that $\langle \varepsilon_o, x \rangle_{\mathcal{K}^e} = k$), then the matrix $\lambda = [\lambda_o | \ 0 \ ] \in \mathcal{S}$ with λ_o as first column and zero-entries otherwise is a reconstruction vector for M, i.e. $\langle \lambda, M_o X \rangle_{\mathcal{S}} = k$ for $X \in \mathcal{V}$ with k in the upper left corner (such that $\langle \varepsilon, X \rangle_{\mathcal{V}} = k$). Since $\Gamma = \Gamma_o$, the subset $\Lambda \subset \Lambda^{max}$ consisting of such reconstruction vectors $\lambda = [\lambda_o | \ 0 \ ]$ is complete. Furthermore, we will show the following linear-algebraic fact.

Lemma 2. *For every pair* $\lambda, \lambda' \in \Lambda$,

$$\lambda - \lambda' \in \mathrm{span}(\mathcal{C}|_{\mathrm{supp}(\lambda) \cup \mathrm{supp}(\lambda')}).$$

The following corollary now follows directly from Theorem 1 and Lemma 1, generalizing the results of [5] to a *mixed adversary*.

Corollary 1. *The CDM DC scheme based on a linear secret sharing scheme with access structure Γ and corresponding privacy structure $\Delta = \Gamma^c$ is secure with respect to an adversary structure $\mathcal{A} \subseteq \Delta$ if and only if $\mathcal{P} \notin \Delta \sqcup \mathcal{A} \sqcup \mathcal{A}$.*

Proof of Lemma 2: Let $\lambda = [\lambda_o | \ 0 \ ]$ and $\lambda' = [\lambda'_o | \ 0 \ ]$ be reconstruction vectors from $\mathcal{C}$. We have $\sum_i \lambda_o[i] M_{oi} = \lambda_o{}^T \cdot M_o = \varepsilon_o{}^T = (1, 0, \ldots, 0)$ and hence

$$\sum_i \lambda_o[i] \mu^{ji} = \sum_i \lambda_o[i] \begin{pmatrix} 0 \\ \boxed{M_{oi}} \\ 0 \end{pmatrix} = \begin{pmatrix} 0 \\ \boxed{1\,0\cdots0} \\ 0 \end{pmatrix}$$

where the non-zero row is at the j-th position, and hence λ' can be written as

$$\lambda' = [\lambda'_o | \ 0 \ ] = \sum_j \lambda'_o[j] \begin{pmatrix} 0 \\ \boxed{1\,0\cdots0} \\ 0 \end{pmatrix} = \sum_j \lambda'_o[j] \left(\sum_i \lambda_o[i] \mu^{ji} \right) = \sum_{ij} \lambda_o[i] \lambda'_o[j] \mu^{ji}.$$

Similarly $\lambda = \sum_{ij} \lambda_o[i] \lambda'_o[j] \mu^{ij}$ and therefore

$$\lambda - \lambda' = \sum_{ij} \lambda_o[i] \lambda'_o[j] \left(\mu^{ij} - \mu^{ji} \right) = \sum_{ij} \lambda_o[i] \lambda'_o[j] \gamma^{ij} \in \mathrm{span}(\mathcal{C}|_{\mathrm{supp}(\lambda) \cup \mathrm{supp}(\lambda')}),$$

which proves the claim. $\square$

6   Application II: Reducing the Number of Checks in the BGW Scheme

Theorem 1 tells us that as long as the set $\{\lambda - \lambda' \mid \lambda, \lambda' \in \Lambda|_H\}$ is contained in the subspace $\mathrm{span}(\mathcal{C}|_H) \subseteq \mathcal{S}$, where $H \in \mathcal{H}$ collects the honest players, the corresponding scheme is secure. By this it is obvious that if the vectors in $\mathcal{C}|_H$ are not linearly independent, then $\mathcal{C}|_H$ contains more vectors than actually needed. We will now use this simple observation to reduce the number of checks in the (symmetric version of the) BGW VSS scheme [1].

The variation of the scheme of [1] where a *symmetric* bivariate polynomial is used instead of an arbitrary one can be seen as a special case of the CDM scheme, where the matrix M_o is a Van-der-Monde matrix, i.e., $M_{oi} = [1, \alpha_i, \alpha_i^2, \ldots, \alpha_i^t]$ for disjoint $\alpha_1, \ldots, \alpha_n \neq 0$. We have the following fact.

Lemma 3. *Let* $Q^* \in \Gamma = \{Q \subseteq \mathcal{P} \mid |Q| \geq t+1\}$ *and* $H \supseteq Q^*$. *Then*

$$\mathrm{span}(\{\gamma^{ij} \in \mathcal{C}|_H \mid i \in Q^* \text{ or } j \in Q^*\}) = \mathrm{span}(\mathcal{C}|_H).$$

As the proof is purely technical and does not give any new insight, it is moved to the appendix. Similarly, it can be shown using linear algebra that $\mathcal{C}_{\{i,j\}} \subseteq \mathrm{span}(\mathcal{C}|_{\{i\} \cup Q} \cup \mathcal{C}|_{\{j\} \cup Q})$ for all i, j and every Q with $|Q| \geq t+1$. The following corollary follows now from Theorem 1, showing that (the symmetric version of) the classical VSS scheme of [1] is not optimal with respect to the number of required pairwise checks.

Corollary 2. *The symmetric version of the BGW VSS scheme with threshold privacy structure* $\Delta = \{D \subseteq \mathcal{P} \mid |D| \leq t\}$ *is secure with respect to a threshold adversary structure* $\mathcal{A} = \{A \subseteq \mathcal{P} \mid |A| \leq w\}$ *with* $w \leq t$ *if and only if* $n > t+2w$, *even if* $\mathcal{C}$ *is replaced by*

$$\bar{\mathcal{C}} = \{\gamma^{ij} \in \mathcal{C} \mid j > w\}.$$

Proof. Let H be the set of honest players, hence $|H| \geq n-w > t+w$. Clearly, the set $Q^* = \{i \in H \mid i > w\}$ is in Γ and hence $\bar{\mathcal{C}}|_H = \{\gamma^{ij} \in \mathcal{C}|_H \mid i \in Q^* \text{ or } j \in Q^*\}$ fulfills $\mathrm{span}(\bar{\mathcal{C}}|_H) = \mathrm{span}(\mathcal{C}|_H)$. □

Alternatively, this shows that the classical BGW VSS scheme allows "early stopping", as it is also used in the 4-round VSS of [9].

7   Conclusions

We presented a general treatment of all linear VSS and DC schemes based on an underlying linear secret sharing scheme, pairwise checks, complaints and accusations (against the dealer), and we analysed the security of this class of schemes. This class covers all currently known linear schemes, and possibly even all future ones. We reduced the security of these schemes to a pure linear-algebra predicate and showed with two concrete examples that this makes arguing about the security of such schemes conceptually very simple, as no cryptographic reasoning is needed anymore but just pure linear algebra. Furthermore, given a fixed adversary structure (e.g. described by a monotone span program) this might allow the construction of secure schemes which are more efficient than the generic construction of [5].

References

1. M. Ben-Or, S. Goldwasser, and A. Widgerson. Completeness theorems for non-cryptographic fault-tolerant distributed computation. In *20th Annual ACM Symposium on the Theory of Computing*. ACM Press, 1988.
2. P. Berman, J. A. Garay, and K. J. Perry. Towards optimal distributed consensus (extended abstract). In *30th Annual Symposium on Foundations of Computer Science*. IEEE, 1989.
3. D. Chaum, C. Crepeau, and I. Damgård. Multiparty unconditional secure protocols. In *20th Annual ACM Symposium on the Theory of Computing*. ACM Press, 1988.
4. B. Chor, S. Goldwasser, S. Micali, and B. Awerbuch. Verifiable secret sharing and achieving simultaneity in the presence of faults (extended abstract). In *26th Annual Symposium on Foundations of Computer Science*. IEEE, 1985.
5. R. Cramer, I. Damgård, and U. Maurer. General secure multi-party computation from any linear secret-sharing scheme. In *Advances in Cryptology - EUROCRYPT 2000, Lecture Notes in Computer Science*. Springer, 2000.
6. M. Fitzi, M. Hirt, and U. Maurer. Trading correctness for privacy in unconditional multi-party computation. In *Advances in Cryptology – CRYPTO '98, Lecture Notes in Computer Science*. Springer, 1998. Corrected proceedings version.
7. M. Fitzi, M. Hirt, and U. Maurer. General adversaries in unconditional multi-party computation. In *Advances in Cryptology - ASIACRYPT '99, Lecture Notes in Computer Science*. Springer, 1999.
8. M. Fitzi and U. Maurer. Efficient byzantine agreement secure against general adversaries. In *International Symposium on Distributed Computing (DISC), Lecture Notes in Computer Science*. Springer, 1998.
9. R. Gennaro, Y. Ishai, E. Kushilevitz, and T. Rabin. The round complexity of verifiable secret sharing and secure multicast. In *33rd Annual ACM Symposium on the Theory of Computing*. ACM Press, 2001.
10. M. Hirt and U. Maurer. Complete characterization of adversaries tolerable in secure multi-party computation (extended abstract). In *16th ACM Symposium on Principles of Distributed Computing*, 1997. Final version appeared in Journal of Cryptology 2000.
11. M. Karchmer and A. Wigderson. On span programs. In *8th Annual Conference on Structure in Complexity Theory (SCTC '93)*. IEEE, 1993.
12. A. Shamir. How to share a secret. *Communications of the Association for Computing Machinery*, 22(11), 1979.

A   Proof of Lemma 3

Recall that the checking vectors $\gamma^{ij} \in \mathcal{C}$ (which are actually matrices) are of the following form: The i-th row is $M_{\circ j}$, the j-th row is $-M_{\circ i}$, and all remaining entries are zero, i.e., $\gamma_i^{ij} = M_{\circ j}$ and $\gamma_j^{ij} = -M_{\circ i}$ and $\gamma_l^{ij} = 0$ for $l \neq i, j$.

Proof of Lemma 3: We assume without loss of generality that $Q^* = \{p_{n-t}, \ldots, p_n\}$. Consider an arbitrary but fixed checking vector $\gamma^{i_0 j_0}$ with $i_0, j_0 \in H$ and $i_0 < j_0 < n - t$, i.e. $i_0, j_0 \notin Q^*$ (otherwise, nothing needs to be shown). We have to show that $\gamma^{i_0 j_0}$ is contained in $\mathrm{span}(\{\gamma^{ij} \in \mathcal{C}|_H \mid i \in Q^* \text{ or } j \in Q^*\}$. This will be done by the following claim.

Claim: There exists a sequence $\delta^{n-t-1}, \ldots, \delta^n \in \text{span}(\{\gamma^{ij} \in C|_H \, | \, i \in Q^* \text{ or } j \in Q^*\})$ such that for every $n - t - 1 \leq i \leq n$

$$\delta_k^i = \begin{cases} \gamma_k^{i_0 j_0} & \text{if } k \leq i \\ \sum_{k \neq l = i+1}^{n} (\lambda_k^{i_0} \lambda_l^{j_0} - \lambda_l^{i_0} \lambda_k^{j_0}) M_{\circ l} & \text{otherwise} \end{cases}$$

where for $1 \leq i \leq n$ and $n - t \leq l \leq n$ we let $\lambda_l^i \neq 0$ be such that $\sum_{l=n-t}^{n} \lambda_l^i M_{\circ l} = M_{\circ i}$.

Truly, we can set

$$\delta^{n-t-1} = \sum_{l=n-t}^{n} (\lambda_l^{j_0} \gamma^{i_0 l} - \lambda_l^{i_0} \gamma^{j_0 l}) \in \text{span}(\{\gamma^{ij} \in C|_H \, | \, i \in Q^* \text{ or } j \in Q^*\})$$

Then for $k \leq n - t - 1$ we indeed have $\delta_k^{n-t-1} = \gamma_k^{i_0 j_0}$, namely

$$\delta_{i_0}^{n-t-1} = \sum_l \lambda_l^{j_0} M_{\circ l} = M_{\circ j_0} = \gamma_{i_0}^{i_0 j_0}$$

$$\delta_{j_0}^{n-t-1} = -\sum_l \lambda_l^{i_0} M_{\circ l} = -M_{\circ i_0} = \gamma_{j_0}^{i_0 j_0} \quad \text{and}$$

$$\delta_k^{n-t-1} = 0 = \gamma_k^{i_0 j_0} \quad \text{if } k \neq i_0, j_0$$

while for $k > n - t - 1$ we have

$$\delta_k^{n-t-1} = -\lambda_k^{j_0} M_{\circ i_0} + \lambda_k^{i_0} M_{\circ j_0} = -\lambda_k^{j_0}\Big(\sum_{l=n-t}^{n} \lambda_l^{i_0} M_{\circ l}\Big) + \lambda_k^{i_0}\Big(\sum_{l=n-t}^{n} \lambda_l^{j_0} M_{\circ l}\Big)$$

$$= \sum_{l=n-t}^{n} (\lambda_k^{i_0} \lambda_l^{j_0} - \lambda_l^{i_0} \lambda_k^{j_0}) M_{\circ l} = \sum_{\substack{l=n-t \\ l \neq k}}^{n} (\lambda_k^{i_0} \lambda_l^{j_0} - \lambda_l^{i_0} \lambda_k^{j_0}) M_{\circ l}$$

And inductively for $i = n - t - 1, \ldots, n - 1$, given δ^i as demanded, we can set

$$\delta^{i+1} = \delta^i - \sum_{l=i+2}^{n} (\lambda_{i+1}^{i_0} \lambda_l^{j_0} - \lambda_l^{i_0} \lambda_{i+1}^{j_0}) \gamma^{i+1,l} \in \text{span}(\{\gamma^{ij} \in C|_H \, | \, i \in Q^* \text{ or } j \in Q^*\})$$

Then, clearly, for $k < i + 1$ we have $\delta_k^{i+1} = \delta_k^i = \gamma_k^{i_0 j_0}$. Furthermore, we have

$$\delta_{i+1}^{i+1} = \delta_{i+1}^i - \sum_{l=i+2}^{n} (\lambda_{i+1}^{i_0} \lambda_l^{j_0} - \lambda_l^{i_0} \lambda_{i+1}^{j_0}) M_{\circ l} = 0 = \gamma_{i+1}^{i_0 j_0}$$

while for $k > i + 1$

$$\delta_k^{i+1} = \delta_k^i + (\lambda_{i+1}^{i_0} \lambda_k^{j_0} - \lambda_k^{i_0} \lambda_{i+1}^{j_0}) M_{\circ i+1} = \sum_{\substack{l=i+2 \\ l \neq k}}^{n} (\lambda_k^{i_0} \lambda_l^{j_0} - \lambda_l^{i_0} \lambda_k^{j_0}) M_{\circ l}$$

as required. $\square$

Perfect Hiding and Perfect Binding Universally Composable Commitment Schemes with Constant Expansion Factor

Ivan Damgård and Jesper Buus Nielsen

BRICS* Department of Computer Science
University of Aarhus
Ny Munkegade
DK-8000 Arhus C, Denmark
{ivan,buus}@brics.dk

Abstract. Canetti and Fischlin have recently proposed the security notion *universal composability* for commitment schemes and provided two examples. This new notion is very strong. It guarantees that security is maintained even when an unbounded number of copies of the scheme are running concurrently, also it guarantees non-malleability and security against adaptive adversaries. Both proposed schemes use $\Theta(k)$ bits to commit to one bit and can be based on the existence of trapdoor commitments and non-malleable encryption.
We present new universally composable commitment (UCC) schemes based on extractable q one-way homomorphisms. These in turn exist based on the Paillier cryptosystem, the Okamoto-Uchiyama cryptosystem, or the DDH assumption. The schemes are efficient: to commit to k bits, they use a constant number of modular exponentiations and communicates $O(k)$ bits. Furthermore the scheme can be instantiated in either perfectly hiding or perfectly binding versions. These are the first schemes to show that constant expansion factor, perfect hiding, and perfect binding can be obtained for universally composable commitments.
We also show how the schemes can be applied to do efficient zero-knowledge proofs of knowledge that are universally composable.

1 Introduction

The notion of commitment is one of the most fundamental primitives in both theory and practice of modern cryptography. In a commitment scheme, a *committer* chooses an element m from some finite set M, and releases some information about m through a commit protocol to a *receiver*. Later, the committer may release more information to the receiver to open his commitment, so that the receiver learns m. Loosely speaking, the basic properties we want are first that the commitment scheme is *hiding*: a cheating receiver cannot learn m from

---

* Basic Research in Computer Science,
  Centre of the Danish National Research Foundation.

M. Yung (Ed.): CRYPTO 2002, LNCS 2442, pp. 581–596, 2002.
© Springer-Verlag Berlin Heidelberg 2002

the commitment protocol, and second that it is *binding:* a cheating committer cannot change his mind about m, the verifier can check in the opening that the value opened was what the committer had in mind originally. Each of the two properties can be satisfied unconditionally or relative to a complexity assumption. A very large number of commitment schemes are known based on various notions of security and various complexity assumptions.

In [CF01] Canetti and Fischlin proposed a new security measure for commitment schemes called universally composable commitments. This is a very strong notion: it guarantees that security is maintained even when an unbounded number of copies of the scheme are running concurrently and asynchronous. It also guarantees non-malleability and maintains security even if an adversary can decide adaptively to corrupt some of the players and make them cheat. The new security notion is based on the framework for universally composable security in [Can01]. In this framework one specifies desired functionalities by specifying an idealized version of them. An idealized commitment scheme is modeled by assuming a trusted party to which both the committer and the receiver have a secure channel. To commit to m, the committer simply sends m to the trusted party who notifies the receiver that a commitment has been made. To open, the committer asks the trusted party to reveal m to the receiver. Security of a commitment scheme now means that the view of an adversary attacking the scheme can be simulated given access to just the idealized functionality.

It is clearly important for practical applications to have solutions where only the two main players need to be active. However, in [CF01] it is shown that universal composability is so strong a notion that no universally composable commitment scheme for only two players exist. However, if one assumes that a common reference string with a prescribed distribution is available to the players, then two-player solutions do exist and two examples are given in [CF01]. Note that common reference strings are often available in practice, for instance if a public key infrastructure is given.

The commitment scheme(s) from [CF01] uses $\Omega(k)$ bits to commit to one bit, where k is a security parameter, and it guarantees only computational hiding and binding. In fact, as detailed later, one might even get the impression from the construction that perfect hiding, respectively binding cannot be achieved. Here, by perfect, we mean that an unbounded receiver gets zero information about m, respectively an unbounded committer can change his mind about m with probability zero.

Our contribution is a new construction of universally composable commitment schemes, which uses $O(k)$ bits of communication to commit to k bits. The scheme can be set up such that it is perfectly binding, or perfectly hiding, without loosing efficiency[1]. The construction is based on a new primitive which we call a mixed commitment scheme. We give a general construction of mixed

---

[1] [CF01] also contains a scheme which is statistically binding and computationally hiding, the scheme however requires a new setup of the common reference string per commitment and is thus mostly interesting because it demonstrates that statistically binding can be obtained at all.

commitments, based on any family of so called extractable q one-way homomorphisms, and show two efficient implementations of this primitive, one based on the Paillier cryptosystem and one based on the Okamoto-Uchiyama cryptosystem. A third example based on the DDH assumption is less efficient, but still supports perfect hiding or binding. Our commitment protocol has three moves, but the two first messages can be computed independently of the message committed to and thus the latency of a commitment is still one round as in [CF01]. We use a "personalized" version of the common reference string model where each player has a separate piece of the reference string assigned to him. It is an open question if our results can also be obtained with a reference string of size independent of the number of players.

As a final contribution we show that if a mixed commitment scheme comes with protocols in a standard 3-move form for proving in zero-knowledge relations among committed values, the resulting UCC commitment scheme inherits these protocols, such that usage of these is also universally composable. For our concrete schemes, this results in efficient protocols for proving binary Boolean relations among committed values and also (for the version based on Paillier encryption) additive and multiplicative relations modulo N. We discuss how this can be used to construct efficient universally composable zero-knowledge proofs of knowledge for NP, improving the complexity of a corresponding protocol from [CF01].

An Intuitive Explanation of Some Main Ideas. In the simplest type of commitment scheme, both committing and opening are non-interactive, so that committing just consists of running an algorithm commit$_K$, keyed by a public key K, taking as input the message m to be committed to and a uniformly random string r. The committer computes $c \leftarrow$ commit$_K(m, r)$, and sends c to the receiver. To open, the committer sends m and r to the receiver, who checks that $c =$ commit$_K(m, r)$. For this type of scheme, hiding means that given just c the receiver does not learn m and binding means that the committer cannot change his mind by computing m', r', where $c =$ commit(m', r') and $m' \neq m$.

In a *trapdoor scheme* however, to each public key K a piece of trapdoor information t_K is associated which, if known, allows the committer to change his mind. We will call such schemes *equivocable*. One may also construct schemes where a different type of trapdoor information d_K exists, such that given d_K, one can efficiently compute m from commit$_K(m, r)$. We call such schemes *extractable*. Note that equivocable schemes cannot be perfect binding and that extractable schemes cannot be perfect hiding.

As mentioned, the scheme in [CF01] guarantees only computational binding and computational hiding. Actually this is important to the construction: to prove security, we must simulate an adversary's view of the real scheme with access to the idealized functionality only. Now, if the committer is corrupted by the adversary and sends a commitment c, the simulator must find out which message was committed to, and send it to the idealized functionality. The universally composable framework makes very strict demands to the simulation implying that rewinding techniques cannot be used for extracting the message.

A solution is to use an extractable scheme, have the public key K in the reference string, and set things up such that the simulator knows the trapdoor d_k. A similar consideration leads to the conclusion that if instead the receiver is corrupt, the scheme must be equivocable with trapdoor t_K known to the simulator, because the simulator must generate a commitment on behalf of the honest committer before finding out from the idealized functionality which value was actually committed to. So to build universally composable commitments it seems we must have a scheme that is simultaneously extractable *and* equivocable. This is precisely what Canetti's and Fischlin's ingenious construction provides.

In this paper, we propose a different technique for universally composable commitments based on what we call a mixed commitment scheme. A mixed commitment scheme is basically a commitment scheme which on some of the keys is perfectly hiding and equivocable, we call these keys the E-keys, and on some of the keys is perfectly binding and extractable, we call these keys the X-keys. Clearly, no key can be both an X- and an E-key, so if we were to put the entire key in the common reference string, either extractability or equivocability would fail and the simulation could not work. We remedy this by putting only a part of the key, the so-called system key, in the reference string. The rest of the key is set up once per commitment using a two-move protocol. This allows the simulator to force the key used for each commitment to be an E-key or an X-key depending on whether equivocability or extractability is needed.

Our basic construction is neither perfectly binding nor perfectly hiding because the set-up of keys is randomized and is not guaranteed to lead to any particular type of key. However, one may add to the reference string an extra key that is guaranteed to be either an E- or an X-key. Using this in combination with the basic scheme, one can obtain either perfect hiding or perfect binding.

2   Mixed Commitments

We now give a more formal description of mixed commitment schemes. The most important difference to the intuitive discussion above is that the system key N comes with a trapdoor t_N that allows efficient extraction for all X-keys. The E-keys, however, each come with their own trapdoor for equivocability.

Definition 1. *By a* mixed commitment scheme *we mean a commitment scheme* commit_K *with some global system key* N, *which determines the message space* $\mathcal{M}_N$ *and the key space* $\mathcal{K}_N$ *of the commitments. The key space contains two sets, the E-keys and the X-keys, for which the following holds:*

Key generation *One can efficiently generate a system key* N *along with the so-called X-trapdoor* t_N. *One can, given the system key* N, *efficiently generate random keys from* $\mathcal{K}_N$ *and given* t_N *one can sample random X-keys. Given the system key, one can efficiently generate an E-key* K *along with the so-called E-trapdoor* t_K.

Key indistinguishability *Random E-keys and random X-keys are both computationally indistinguishable from random keys from* $\mathcal{K}_N$ *as long as the X-trapdoor* t_N *is not known.*

Equivocability *Given E-key K and E-trapdoor t_K one can generate fake commitments c, distributed exactly as real commitments, which can later be opened arbitrarily, i.e. given a message m one can compute uniformly random r for which $c = \text{commit}_K(m, r)$.*

Extraction *Given a commitment $c = \text{commit}_K(m, r)$, where K is an X-key, one can given the X-trapdoor t_N efficiently compute m.*

Note that the indistinguishability of random E-keys, random X-keys, and random keys from $\mathcal{K}_N$ implies that as long as the X-trapdoor is not known the scheme is computationally hiding for all keys and as long as the E-trapdoor is not known either the scheme is computationally binding for all keys.

For the construction in the next section we will need a few special requirements on the mixed commitment scheme. First of all we will assume that the message space $\mathcal{M}_N$ and the key space $\mathcal{K}_N$ are finite groups in which we can compute efficiently. We denote the group operation by $+$. Second we need that the number of E-keys over the total number of keys is negligible and that the number of X-keys over the total number of keys is negligibly close to 1. Note that this leaves only a negligible fraction which is neither X-keys nor E-keys. We call a mixed commitment scheme with these properties a special mixed commitment scheme.

The last requirement is that the scheme has two 'independent' E-trapdoors per E-key. We ensure this by a transformation. The keys will be of the form (K_1, K_2). We let the E-keys be the pairs of E-keys and let the X-keys be the pairs of X-keys. The message space will be the same. Given a message m we commit as $(\text{commit}_{K_1}(\overline{m}_1), \text{commit}_{K_2}(\overline{m}_2))$, where $\overline{m}_1$ and $\overline{m}_2$ are uniformly random values for which $m = \overline{m}_1 + \overline{m}_2$. If both keys are E-keys and the E-trapdoor of one of them, say K_b, is known a fake commitment is made by committing honestly to random m_{1-b} under K_{1-b} and making a fake commitment c_b under K_b. Then to open to m, open c_b to $m_b = m_{1-b} - m$. Note that the distribution of the result is independent of b – this will become essential later. All requirements for a special mixed commitment scheme are maintained under the transformation.

Special Mixed Commitment Scheme Based on q One-Way Homomorphisms. Our examples of special mixed commitment schemes are all based on q one-way homomorphism generators, as defined in [CD98]. Here we extend the notion to extractable q one-way homomorphisms. In a nutshell, we want to look at an easily computable homomorphism $f : G \to H$ between Abelian groups G, H such that $H/f(G)$ is cyclic and has only large prime factors in its order. And such that random elements in $f(G)$ are computationally indistinguishable from random elements chosen from all of H (which in particular implies that f is hard to invert). However, given also a trapdoor associated with f, it becomes easy to extract information about the status of an element in H.

More formally, a family of extractable q one-way homomorphisms is given by a probabilistic polynomial time (PPT) generator $\mathcal{G}$ which on input 1^k outputs a (description of a) tuple $(G, H, f, g, q, b, b', t)$, where G and H are groups, $f : G \to H$ is an efficiently computable homomorphism, $g \in H \setminus f(G)$, $q, b, b' \in N$, and t is a string called the trapdoor. Let $F = f(G)$. We require that gF generates the factor group H/F and let $\operatorname{ord}(g) = |H/F|$. We require that $\operatorname{ord}(g)$ is superpolynomial in k (e.g. 2^k), that q is a multiple of $\operatorname{ord}(g)$, and that b is a public lower bound on $\operatorname{ord}(g)$, i.e., we require that $2 \leq b \leq \operatorname{ord}(g) \leq q$. We say that a generator has *public order* if $b = \operatorname{ord}(g) = q$. Also b' is superpolynomial in k (e.g. $2^{k/2}$) and it is a public lower bound on the primefactors in $\operatorname{ord}(g)$, i.e., all primefactors in $\operatorname{ord}(g)$ are at least b'. We write operations in G and H multiplicatively and we require that in both groups one can multiply, exponentiate, take inverses, and sample random elements in PPT given (G, H, f, g, q, b, b'). The final central requirements are as follows:

Indistinguishability. Random elements from F are *computationally indistinguishable* from random elements from H given (G, H, f, g, q, b, b').

Extractability. This comes in two flavors. We call the generator *fully extractable* if given $(G, H, f, g, q, b, b', t)$ and $y = g^i f(r)$ one can compute $i \bmod \operatorname{ord}(g)$ in PPT. Note that, given $(G, H, f, g, q, b, b', t)$, one can compute $\operatorname{ord}(g)$ easily. We call a generator *0/1-extractable* if given $(G, H, f, g, q, b, b', t)$ and $y = g^i f(r)$ one can determine whether $i = 0$ in PPT.

q-invertibility Given (G, H, f, g, q, b, b') and $y \in H$, it is easy to compute x such that $y^q = f(x)$. Note that this does not contradict indistinguishability: since q is a multiple of $\operatorname{ord}(g)$, it is always the case that $y^q \in F$.

We give three examples of extractable q one-way homomorphism generators:

Based on Paillier encryption: Let $n = PQ$ be an RSA modulus, where P and Q are $k/2$-bit primes. Let $G = Z_n^*$, let $H = Z_{n^2}^*$, and let $f(r) = r^n \bmod n^2$. Let $g = (n + 1)$, let $b = q = n$, $b' = 2^{k/2-1}$, and let $t = (P, Q)$. Then it follows directly from [Pai99] that relative to the DCRA assumption we have a fully extractable generator with public order.

Based on Okamoto-Uchiyama encryption: Now let $N = Pn = P^2 Q$. Let $G = Z_n^*$, let $H = Z_N^*$, and let $f(r) = r^N \bmod N$. Let $g = (N + 1)$, $q = N$, $b = b' = 2^{k/2-1}$ and $t = (P, Q)$. Then it follows directly from [OU98] that relative to the p-subgroup assumption we have a fully extractable generator.

Based on Diffie-Hellman encryption: Let $\langle \alpha \rangle$ be a group of prime order Q. Let $\beta = \alpha^x$ for uniformly random $x \in Z_Q^*$. Let $G = Z_Q$, let $H = \langle \alpha \rangle \times \langle \alpha \rangle$, and let $f(r) = (\alpha^r, \beta^r)$. Let $g = (1, \beta)$, $b = b' = q = Q$ and $t = x$. Then the scheme is 0/1-extractable: let $(A, B) = g^m f(r) = (\alpha^r, \beta^{r+m})$, then $A^x = B$ iff $m = 0$. Relative to the DDH assumption we have a 0/1-extractable generator with public order.

We now show how to transform an extractable generator into a special mixed commitment scheme. We treat fully extractable and 0/1-extractable generators in parallel, as the differences are minimal.

The key space will be H, the message space will be Z_b for fully extractable schemes and Z_2 for 0/1-extractable schemes. We commit as $\operatorname{commit}_K(m, r) = K^m f(r)$, where r is uniformly random in H. The E-keys will be the set $F = f(G)$ and the E-trapdoor will be $f^{-1}(K)$. By the requirement that $\operatorname{ord}(g)$ is superpolynomial in k, the set of E-keys is a negligible fraction of the keyspace as required. For equivocability, we generate a fake commitment as $c = f(r_c)$ for uniformly random $r_c \in H$. Assume that $K = f(r_K)$ and that we are given $m \in Z_b$. Compute $r = r_K^{-m} r_c$. Then r is uniformly random and $c = \mathrm{commit}_K(m, r)$.

For a fully extractable generator the X-keys will be the elements of form $K = g^i f(r_K)$, where i is invertible in $Z_{\mathrm{ord}(g)}$. By the requirement that $\mathrm{ord}(g)$ only has large primefactors, the X-keys are the entire key-space except for a negligible fraction as required. They can be sampled efficiently given the trapdoor since then $\mathrm{ord}(g)$ is known. Assume that we are given $c = K^m f(r)$ for $m \in Z_b$. Using fully extractability we can from c compute $im \bmod \mathrm{ord}(g)$ and from K we can compute $i \bmod \mathrm{ord}(g)$. Since i is invertible we can then compute $m \bmod \mathrm{ord}(g) = m$. For a 0/1-extractable generator the X-keys will be the elements of the form $K = g^i f(r_K)$, where $i \in Z_{\mathrm{ord}(g)} \setminus \{0\}$. By the 0/1-extractability these keys can be efficiently sampled given t. For extraction, note that $\mathrm{commit}_K(0, r) \in F$ and $\mathrm{commit}_K(1, r) \notin F$ and use the 0/1-extractability of the generator. For the fully extractable construction and the 0/1-extractable construction, the indistinguishability of the key-spaces follows directly from the indistinguishability requirement on the generator. The transformed scheme is given by $\mathrm{commit}_{K_1, K_2}(m, (r_1, r_2, m_1)) = (K_1^{m_1} f(r_1), K_2^{m_2} f(r_2))$, where $m_2 = m - m_1 \bmod q$.

Proofs of Relations. For the mixed commitment schemes we exhibit in this paper, there are efficient protocols for proving in zero-knowledge relations among committed values. As we shall see, it is possible to have the derived universally composable commitment schemes inherit these protocols while maintaining universal composability. In order for this to work, we need the protocols to be non-erasure Σ-protocols.

A non-erasure Σ-protocol for relation R is a protocol for two parties, called the prover P and the verifier V. The prover gets as input $(x, w) \in R$, the verifier gets as input x, and the goal is for the prover to convince the verifier that he knows w such that $(x, w) \in R$, without revealing information about w. We require that it is done using a protocol of the following form. The prover first computes a message $a \leftarrow A(x, w, r_a)$, where r_a is a uniformly random string, and sends a to V. Then V returns a random challenge e of length l. The prover then computes a responds to the challenge $z \leftarrow Z(x, w, r_a, e)$, and sends z to the verifier. The verifier then runs a program B on (x, a, e, z) which outputs $b \in \{0, 1\}$ indicating where to believe that the prover knows a valid witness w or not. Besides the protocol being of this form we furthermore require that the following hold:

Completeness If $(x, w) \in R$, then the verifier always accepts ($b = 1$).

Special honest verifier zero-knowledge There exists a PPT algorithm, the honest verifier simulator hvs, which given instance x (where there exists w such that $(x, w) \in R$) and any challenge e generates $(a, z) \leftarrow \mathrm{hvs}(x, e, r)$, where r is a uniformly random string, such that (x, a, e, z) is distributed identically to a successful conversation where e occurs as challenge.

State construction Given (x, w, a, e, z, r), where $(a, z) = \text{hvs}(x, e, r)$ and $(x, w) \in R$ it should be possible to compute uniformly random r_a for which $a = A(x, w, r_a)$ and $z = Z(x, w, r_a, e)$.

Special soundness There exists a PPT algorithm, which given x, (a, e, z), and (a, e', z'), where $e \neq e'$, $B(x, a, e, z) = 1$, and $B(x, a, e', z') = 1$, outputs w such that $(x, w) \in R$.

In [Dam00] it is shown how to use Σ-protocols in a concurrent setting. This is done by letting the first message be a commitment to a and then letting the third message be (a, r, z), where (a, r) is an opening of the commitment and z is computed as usual. If the commitment scheme used is a trapdoor commitment scheme this will allow for simulation using the honest verifier simulator. In an adaptive non-erasure setting, where an adversary can corrupt parties during the execution, it is also necessary with the State Construction property as the adversary is entitled to see the internal state of a corrupted party.

Proofs of Relations for the Schemes Based on q One-Way Homomorphisms. The basis for the proofs of relations between commitments will be the following proof of knowledge which works for fully extractable generators with public order, so we have $(G, H, f, g, q, b, b', t)$, with $b = \text{ord}(g) = q$. Assume that the prover is given $K \in H$, $m \in \mathbf{Z}_b$ and $r \in G$, and the verifier is given K and $C = K^m f(r)$. To prove knowledge of m, r, we do as follows:

1. The prover sends $\overline{C} = K^{\overline{m}} f(\overline{r})$ for uniformly random $\overline{m} \in \mathbf{Z}_q$ and $\overline{r} \in G$.
2. The verifier sends a uniformly random challenge e from $\mathbf{Z}_{b'}$, where b' is the public bound on the smallest primefactor in $\text{ord}(g)$.
3. The prover replies with $\tilde{m} = em + \overline{m} \bmod q$ and $\tilde{r} = f^{-1}(K^q)^{\tilde{i}} r^e \overline{r}$, where $\tilde{i} = em + \overline{m} \operatorname{div} q$. The verifier accepts iff $K^{\tilde{m}} f(\tilde{r}) = C^e \overline{C}$.

We argue that this is a non-erasure Σ-protocol: The completeness is immediate. For special soundness assume that we have two accepting conversations $(\overline{C}, e, \tilde{m}, \tilde{r})$ and $(\overline{C}, e', \tilde{m}', \tilde{r}')$. By the requirement that b' is smaller than the smallest primefactor of q we can compute α, β s.t. $1 = \alpha q + \beta(e - e')$. By our assumptions, we can compute $r_c = f^{-1}(C^q)$ and $r_K' = f^{-1}(K^q)$. Then compute $n = (\tilde{m} - \tilde{m}')\beta$, $m = n \bmod q$, and $r = (\tilde{r}/\tilde{r}')^\beta r_c^\alpha (r_K')^{n \operatorname{div} q}$. Then $C = K^m f(r)$. For special honest verifier zero-knowledge, given C and e, pick $\tilde{m} \in \mathbf{Z}_q$ and $\tilde{r} \in G$ at random and let $\overline{C} = K^{\tilde{m}} f(\tilde{r}) C^{-e}$. For the state construction, assume that we are then given m, r such that $C = K^m f(r)$. Then let $\overline{m} = \tilde{m} - em \bmod q$, $\tilde{i} = em + \overline{m} \operatorname{div} q$, $\overline{r} = \tilde{r} f^{-1}(K^q)^{\tilde{i}} r^{-e}$. Then all values have the correct distribution.

We extend this scheme to prove relations between committed values. Assume that the prover knows $K_1, m_1, r_1, \dots, K_l, m_l, r_l$ where $\sum_{i=1}^{l} a_i m_i = a_0 \bmod q$ for $a_0, \dots, a_l \in \mathbf{Z}_q$, and assume that the verifier knows K_i and $C_i = K_i^{m_i} f(r_i)$ for $i = 1, \dots, l$ and knows $a_0, \dots, a_l$. The prover proves knowledge as follows: Run a proof of knowledge as that described above for each of the commitments using the same challenge e in them all. Let $\tilde{m}_i$ be the $\tilde{m}$-value of the protocol for C_i.

We furthermore instruct the verifier to check that $\sum_{i=1}^{l} a_i \tilde{m}_i = e a_0$. For special soundness assume that we have accepting conversations for the two challenges $e \neq e'$. Then we can compute $m_i = (\tilde{m}_i - \tilde{m}_i')(e - e')^{-1} \bmod q$ and r_i as above such that $C_i = K_i^{m_i} f(r_i)$. Furthermore $\sum_{i=1}^{l} a_i m_i = (e - e')^{-1} (\sum_{i=1}^{l} a_i \tilde{m}_i - \sum_{i=1}^{l} a_i \tilde{m}_i') = (e - e')^{-1} (e a_0 - e' a_0) = a_0$. The other properties of a non-erasure Σ-protocol follows using similar arguments.

This handles proofs of knowledge for the basic scheme. Recall, however, that in our UCC construction we need a transformed scheme where pairs of basic commitments are used, as described above. So assume, for instance, that we are given transformed commitments $(C_1, C_2), (C_3, C_4), (C_5, C_6)$ and we want to prove that the value committed to by (C_1, C_2) is the sum modulo q of the values committed by (C_3, C_4) and (C_5, C_6). This can be done by using the above protocol to prove knowledge of $m_1, \ldots, m_6$ contained in $C_1, \ldots, C_6$ such that $(m_1 + m_2) - (m_3 + m_4) - (m_5 + m_6) = 0$. All linear relations between transformed commitments can be dealt with in a similar manner.

By extending the proof of multiplicative relations from [CD98] in a manner equivalent to what we did for the additive proof we obtain a non-erasure Σ-protocol for proving multiplicative relations between transformed commitments.

Now for schemes without public order, the Σ-protocols given above do not directly apply because we were assuming that $b = \mathrm{ord}(g) = q$. However, we can modify the basic protocol by setting $m = \overline{m} = \tilde{m} = 0$. This results in a non-erasure Σ-protocol which allows the prover to prove knowledge of r, where $C = f(r)$ is known by the verifier. I.e. the prover can prove that $C \in F$, in other words that C commits to 0. Given $C = K f(r)$ the prover can using the same protocol prove that $C K^{-1} \in F$, i.e. prove that $C \in KF$, in other words that C commits to 1. Using the technique from [CDS94] for monotone logical combination of Σ-protocols we can then combine such proofs. Let $C_1, \ldots, C_l$ be commitments and let $R = \{(b_1^i, \ldots, b_l^i)\}_{i=1}^{a} \subset \{0,1\}^l$ be a Boolean relation. We can then prove that $C_1, \ldots, C_l$ commits to $(m_1, \ldots, m_l) \in R$ by proving $\bigvee_{i=1}^{a} \bigwedge_{i=1}^{l} C_i \in K^{m_i} F$. Let in particular $\mathbf{0} = \{(0,0),(1,1)\}$ and let $\mathbf{1} = \{(0,1),(1,0)\}$. Then proving knowledge of $(m_1, m_2) \in \mathbf{0}$ for transformed commitment $C = (C_1, C_2)$ proves that C commits to 0, similar for $\mathbf{1}$. Then using the relation $\mathbf{And} = \mathbf{0} \times \mathbf{0} \times \mathbf{0} \cup \mathbf{0} \times \mathbf{0} \times \mathbf{1} \cup \mathbf{0} \times \mathbf{1} \times \mathbf{0} \cup \mathbf{1} \times \mathbf{1} \times \mathbf{1}$, we can prove that three transformed commitments C_1, C_2, C_3 commits to bits m_1, m_2, m_3 s.t. $m_1 = m_2 \wedge m_3$. All Boolean relations of arity $O(\log(k))$ can handled in a similar manner. This will work for both fully and 0/1-extractable schemes.

The following theorem summarizes what we have argued:

Theorem 1. *If there exists a fully (0/1) extractable q one-way homomorphism generator, then there exists a special mixed commitment scheme with message space $\mathbf{Z}_b$ $(\mathbf{Z}_2)$ as described above and with proofs of relations of the form $m = f(m_1, m_2, \ldots, m_l)$ where f is a Boolean predicate and $l = O(\log(k))$. If the scheme is with public order $b = \mathrm{ord}(g) = q$ and is fully extractable, we also have proofs of additive and multiplicative relations modulo q.*

3 Universally Composable Commitments

In the framework from [Can01] the security of a protocol is defined by comparing its real-life execution to an ideal evaluation of its desired behavior.

The protocol π is modeled by n interactive Turing Machines $P_1, \ldots, P_n$ called the parties of the protocol. In the real-life execution of π an adversary $\mathcal{A}$ and an environment $\mathcal{Z}$ modeling the environment in which $\mathcal{A}$ is attacking the protocol participates. The environment gives inputs to honest parties, receives outputs from honest parties, and can communication with $\mathcal{A}$ at arbitrary points in the execution. The adversary can see all messages and schedules all message deliveries. The adversary can corrupted parties adaptively. When a party is corrupted, the adversary learns the entire execution history of the corrupted party, including the random bits used, and will from the point of corruption send messages on behalf of the corrupted party. Both $\mathcal{A}$ and $\mathcal{Z}$ are PPT interactive Turing Machines.

Second an ideal evaluation is defined. In the ideal evaluation an ideal functionality $\mathcal{F}$ is present to which all the parties have a secure communication line. The ideal functionality is an interactive Turing Machine defining the desired input-output behavior of the protocol. Also present is an ideal model adversary S, the environment $\mathcal{Z}$, and n so-called dummy parties $\tilde{P}_1, \ldots, \tilde{P}_n$ – all PPT interactive Turing Machines. The only job of the dummy parties is to take inputs from the environment and send them to the ideal functionality and take messages from the ideal functionality and output them to the environment. Again the adversary schedules all message deliveries, but can now not see the contents of the messages. This basically makes the ideal process a trivially secure protocol with the same input-output behavior as the ideal functionality. The framework also defines the hybrid models, where the execution proceeds as in the real-life execution, but where the parties in addition have access to an ideal functionality. An important property of the framework is that these ideal functionalities can securely be replaced with sub-protocols securely realizing the ideal functionality. The real-life model including access to an ideal functionality $\mathcal{F}$ is called the $\mathcal{F}$-hybrid model.

At the beginning of the protocol all parties, the adversary, and the environment is given as input the security parameter k and random bits. Furthermore the environment is given an auxiliary input z. At some point the environment stops activating parties and outputs some bit. This bit is taken to be the output of the execution. We use $\text{REAL}_{\pi,\mathcal{A},\mathcal{Z}}(k,z)$ to denote the output of $\mathcal{Z}$ in the real-life execution and use $\text{IDEAL}_{\mathcal{F},S,\mathcal{Z}}(k,z)$ to denote the output of $\mathcal{Z}$ in the ideal evaluation. Let $\text{REAL}_{\pi,\mathcal{A},\mathcal{Z}}$ denote the distribution ensemble $\{\text{REAL}_{\pi,\mathcal{A},\mathcal{Z}}(k,z)\}_{k\in N, z\in\{0,1\}^*}$ and let $\text{IDEAL}_{\mathcal{F},S,\mathcal{Z}}(k,z)$ denote the distribution ensemble $\{\text{IDEAL}_{\mathcal{F},S,\mathcal{Z}}(k,z)\}_{k\in N, z\in\{0,1\}^*}$.

Definition 2 ([Can01]). *We say that π securely realizes $\mathcal{F}$ if for all real-life adversaries $\mathcal{A}$ there exists an ideal model adversary S such that for all environments $\mathcal{Z}$ we have that $\text{IDEAL}_{\mathcal{F},S,\mathcal{Z}}$ and $\text{REAL}_{\pi,\mathcal{A},\mathcal{Z}}$ are computationally indistinguishable.*

An important fact about the above security notion is that it is maintained even if an unbounded number of copies of the protocol (and other protocols) are carried out concurrently – see [Can01] for a formal statement and proof. In proving the composition theorem it is used essentially that the environment and the adversary can communicate at any point in an execution. The price for this strong security notion, which is called universal composability in [Can01], is that rewinding cannot be used in the simulation.

The Commitment Functionality. We now specify the task that we want to implement as an ideal functionality. We look at a slightly different version of the commitment functionality than the one in [CF01]. The functionality in [CF01] is only for committing to one bit. Here we generalize. The domain of our commitments will be the domain of the special mixed commitment used in the implementation. Therefore the ideal functionality must specify the domain by initially giving a system key N. For technical reasons, in addition, the X-trapdoor of N is revealed to the the ideal model adversary, i.e., the simulator. This is no problem in the ideal model since here the X-trapdoor cannot be used to find committed values – the ideal functionality stores committed values internally and reveals nothing before opening time. The simulator, however, needs the X-trapdoor in order to do the simulation of our implementation. The implementation, on the other hand, will of course keep the X-trapdoor of N hidden from the *real-life* adversary. The ideal functionality for homomorphic commitments is named $\mathcal{F}_{\text{HCOM}}$ and is as follows.

0. Generate a uniformly random system key N along with the X-trapdoor t_N. Send N to all parties and send (N, t_N) to the adversary.
1. Upon receiving $(\text{commit}, sid, cid, P_i, P_j, m)$ from $\tilde{P}_i$, where m is in the domain of system key N, record (cid, P_i, P_j, m) and send the message $(\text{receipt}, sid, cid, P_i, P_j)$ to $\tilde{P}_j$ and the adversary. Ignore subsequent $(\text{commit}, sid, cid, \dots)$ messages. The values sid and cid are a session id and a commitment id.
2. Upon receiving the message $(\text{prove}, sid, cid, P_i, P_j, R, cid_1, \dots, cid_a)$ from $\tilde{P}_i$, where $(cid_1, P_i, P_j, m_1), \dots, (cid_a, P_i, P_j, m_a)$ have been recorded, R is an a-ary relation with a non-erasure Σ-protocol, and $(m_1, m_2, \dots, m_a) \in R$, send the message $(\text{prove}, sid, cid, P_i, P_j, R, cid_1, \dots, cid_a)$ to $\tilde{P}_j$ and the adversary.
3. Upon receiving a message $(\text{open}, sid, cid, P_i, P_j)$ from $\tilde{P}_i$, where (cid, P_i, P_j, m) has been recorded, send the message $(\text{open}, sid, cid, P_i, P_j, m)$ to $\tilde{P}_j$ and the adversary.

It should be noted that a version of the functionality where N and t_N are not specified by the ideal functionality could be used. We could then let the domain of the commitments be a domain contained in (or easy to encode in) the domain of all the system keys.

The Common Reference String Model. As mentioned in the introduction we cannot hope to construct two-party UCC in the plain real-life model. We need a that a common reference string (CRS) with a prescribed distribution is available to the players. It is straightforward to model a CRS as an ideal functionality $\mathcal{F}_{\text{CRS}}$, see e.g. [CF01].

4   UCC with Constant Expansion Factor

Given a special mixed commitment scheme com we construct the following protocol UCC_{com}.

The CRS The CRS is $(N, \overline{K}_1, \ldots, \overline{K}_n)$, where N is a random system key and $\overline{K}_1, \ldots, \overline{K}_n$ are n random E-keys for the system key N, $\overline{K}_i$ for P_i.

Committing

C.1 On input $(\text{commit}, sid, cid, P_i, P_j, m)$ party P_i generates a random commitment key K_1 for system key N and commits to it as $c_1 = \text{commit}_{\overline{K}_i}(K_1, r_1)$, and sends $(\text{com}_1, sid, cid, c_1)$ to P_j [2].

R.1 P_j replies with $(\text{com}_2, sid, cid, K_2)$ for random key K_2.

C.2 P_i computes $K = K_1 + K_2$ and $c_2 = \text{commit}_K(m, r_2)$ for random r_2, and records $(sid, cid, P_j, K, m, r_2)$ and sends the message $(\text{com}_3, sid, cid, K_1, r_1, c_2)$ to P_j.

R.2 P_j checks that $c_1 = \text{commit}_{\overline{K}_i}(K_1, r_1)$, and if so computes $K = K_1 + K_2$, records (sid, cid, P_j, K, c_2), and outputs $(\text{receipt}, sid, cid, P_i, P_j)$.

Opening

C.3 On input $(\text{open}, sid, cid, P_i, P_j)$, P_i sends $(\text{open}, sid, cid, m, r_2)$ to P_j.

R.3 P_j checks that $c_2 = \text{commit}_K(m, r_2)$, and if so outputs $(\text{open}, sid, cid, P_i, P_j, m)$.

Proving Relation

C.4 On input $(\text{prove}, sid, cid, P_i, P_j, R, cid_1, \ldots, cid_a)$, where $(sid, cid_1, P_j, K_1, m_1, r_1)$, $\ldots$, $(sid, cid_a, P_j, K_a, m_a, r_a)$ are recorded commitments, compute the first message, a, of the Σ-protocol from the recorded witnesses and compute $c_3 = \text{commit}_{\overline{K}_i}(a, r_3)$ for random r_3 and send $(\text{prv}_1, sid, cid, R, cid_1, \ldots, cid_a, c_3)$ to P_j.

R.4 P_j generates a random challenge e and sends $(\text{prv}_2, sid, cid, P_j, e)$ to P_i.

C.5 P_i computes the answer z and sends $(\text{prv}_3, sid, cid, a, r_3, z)$ to P_j.

R.5 P_j checks that $c_3 = \text{commit}_{\overline{K}_i}(a, r_3)$ and that (a, e, z) is an accepting conversation. If so P_j outputs $(\text{prove}, sid, cid, P_i, P_j, R, cid_1, \ldots, cid_a)$.

Theorem 2. *If com is a special mixed commitment scheme, then the protocol UCC_{com} securely realizes $\mathcal{F}_{\text{HCOM}}$ in the CRS-hybrid model.*

---

[2] We assume that the key space is a subset of the message space. If this is not the case the message space can be extended to a large enough $\mathcal{M}_N^l$ by committing to l values in the original scheme.

Proof. We construct a simulator S running a real-life adversary A and simulates to it a real-life execution consistent with the values input to and output from the ideal functionality in ideal-world in which S is running. The main requirements is that S given $|m|$ can simulate a commitment to m in such a way that it can later open the commitment to any value of m; That S can extract from the commitments given by A the value committed to; And that S does not rewind A as this is not possible in the model from [Can01].

The simulator S sets up the CRS s.t. the keys $\overline{K}_i$ are E-keys for which S knows the E-trapdoor, and such that the X-trapdoor is known too. When S is simulating a honest party acting as the committing party it use the E-trapdoor to open c_1 to $K_1 = K - K_2$, where K is generated as a random E-key with known E-trapdoor. Then S generates c_2 as an equivocable commitment, which it can later open to the actual committed value once it becomes known. When S is simulating a honest party acting as the receiver in a commitment it simply follows the protocol. Since no trapdoors are known to the adversary, the resulting key K will be random and in particular it will be an X-key except with negligible probability since all but a negligible fraction of the keys are X-keys. So, since S knows the X-trapdoor, it can compute from the c_2 sent by the adversary the value m committed to except with negligible probability.

For the proofs of relations, when A is giving a proof, S simply follows the protocol. The proofs given by honest P_i are simulated by S. Here the non-rewinding simulation technique from [Dam00] applies. If the party P_i is later corrupted, the messages which should have been committed to are learned. Using the E-trapdoor the simulator then opens the commitments in the relation appropriately. Then given the messages and the random bits (the witnesses of the proof), the state construction property of the proof allows to construct a consistent internal state to show to the adversary.

The main technical problem in proving this simulation indistinguishable from the real-life execution is that the X-trapdoor is used by the simulator, so we cannot do reductions to the computational binding of the mixed commitment scheme. We deal with this by defining a hybrid distribution that is generated by the following experiment: Run the simulation except do not use the X-trapdoor. Each time the adversary makes a commitment instead simply use the message 0 as input to the ideal functionality. When the adversary then opens the commitment to m, simply change the ideal-world communication to make it look as if m was the committed value. Up to the time of opening, the entire execution seen from the the environment and A is independent of whether 0 or m was given to the ideal functionality, and hence this hybrid is distributed exactly as the simulation. It is therefore enough to prove this hybrid indistinguishable from the real-life execution, which is possible using standard techniques. $\square$

Perfect Hiding and Perfect Binding. The scheme described above has neither perfect binding nor perfect hiding. Here we construct a version of the commitment scheme with both perfect hiding and perfect binding. The individual commitments are obviously not simultaneously perfect hiding and perfect binding, but it can be chosen at the time of commitment whether a commitment should be perfect binding or perfect hiding and proofs of relations can include both types of commitments. We sketch the scheme and the proof of its security. The details are left to the reader.

In the extended scheme we add to the CRS a random E-key K_E and a random X-key K_X (both for system key N). Then to do a perfect binding commitment to m the committer will in Step C.2 compute $c_2 = \text{commit}_K(m, r_2)$ as before, but will in addition compute $c_3 = \text{commit}_{K_X}(m, r_3)$. To open the commitment the committer will then have to send both a correct opening (m, r_2) of c_2 and a correct opening (m, r_3) of c_3. This is perfect binding as the X-key commitment is perfect binding.

To do a perfect hiding commitment the committer computes a uniformly random message $\overline{m}$ and commits with $c_2 = \text{commit}_K(\overline{m} + m, r_2)$ and $c_3 = \text{commit}_{K_E}(\overline{m}, r_3)$. To open to m the committer must then send a correct opening (m_2, r_2) of c_2 and a correct opening (m_3, r_3) of c_3 for which $m_2 = m_3 + m$. This is perfect hiding because c_3 hides $\overline{m}$ perfectly and $\overline{m} + m$ thus hides m perfectly.

To do the simulation simply let the simulator make the excusable mistake of letting K_E be a random X-key and letting K_X be a random E-key. This mistake will allow to simulate and cannot be detected by the fact that E-keys and X-keys are indistinguishable. For perfect binding commitments both K and K_X will then be E-keys when the simulator does a commitment, which allows to fake. When the adversary does a commitment K will (except with negligible) be an X-key and the simulator can extract m from $\text{commit}_K(m)$. For perfect hiding commitments both K and K_E will (except with negligible probability) be X-keys when the adversary does a commitment, which allows to extract. When the simulator commits, K will be an E-key, which allows to fake an opening by faking $\text{commit}_K(m)$.

For perfect binding commitments the proofs of relations can be used directly for the modified commitments by doing the proof on the $\text{commit}_K(m)$ values. For perfect hiding commitments there is no general transformation that will carry proofs of relations over to the modified system. If however there is a proof of additive relation, then one can publish $\text{commit}_K(m)$ and prove that the sum of the values committed to by $\text{commit}_K(m)$ and $\text{commit}_{K_E}(\overline{m})$ is committed to by $\text{commit}_K(\overline{m} + m)$, and then use the commitment $\text{commit}_K(m)$ when doing the proofs of relations.

5   Efficient Universally Composable Zero-Knowledge Proofs

In [CF01] Canetti and Fischlin showed how universally composable commitments can be used to construct simple zero-knowledge (ZK) protocols which are universally composable. This is a strong security property, which implies concurrent and non-malleable ZK proof of knowledge.

The functionality $\mathcal{F}_{ZK}^R$ for universally composable zero-knowledge (for binary relation R) is as follows.

1. Wait to receive a value $(\text{verifier}, id, P_i, P_j, x)$ from some party P_i. Once such a value is received, send $(\text{verifier}, id, P_i, P_j, x)$ to S, and ignore all subsequent $(\text{verifier}, \ldots)$ values.
2. Upon receipt of a value $(\text{prover}, id, P_j, P_i, x', w)$ from P_j, let $v = 1$ if $x = x'$ and $R(x, w)$ holds, and $v = 0$ otherwise. Send (id, v) to P_i and S, and halt.

Exploiting the Multi-bit Commitment Property. In [CF01] a protocol for Hamiltonian-Cycle (HC) is given and proven to securely realize $\mathcal{F}_{\text{ZK}}^{\text{HC}}$. The protocol is of a common cut-and-choose form. It proceeds in t rounds. In each round the prover commits to l bits $m \in \{0,1\}^l$. Then the verifier sends a bit b as challenge. If $b = 0$ then the prover opens all commitments and if $b = 1$ the prover opens some subset of the challenges. Say that the subset is given by $S \in \{0,1\}^l$, where $S_i = 1$ if commitment number i should be revealed. Then if $b = 0$ the prover should see m and if $b = 1$ the prover should see $(S, m \wedge S)$. The verifier has two predicates V_0 and V_1 for verifying the reply from the prover. If $b = 0$ it verifies that $V_0(m) = 1$ and if $b = 1$ it verifies that $V_1(S, m \wedge S) = 1$. The protocol is such that seeing m or $(S, m \wedge S)$ reveals no knowledge about the witness (Hamiltonian cycle), but if $V_0(m) = 1$ and $V_1(S, m \wedge S) = 1$, then one can compute a witness from m and S. The verifier accepts if it can verify the reply in each of the t rounds. Obviously S should be kept secret when $b = 0$ – otherwise m and S would reveal the witness. This makes it hard to use the multi-bit commitments to commit to the l bits in such a way that just the subset S can be opened later. However, in [KMO89] Kilian, Micali, and Ostrovski presented a general technique for transforming a multi-bit commitment scheme into a multi-bit commitment scheme with the property that individual bits can be open independently. Unfortunately their technique adds one round of interaction. However, we do not need the full generality of their result. This allows us to modify the technique to avoid the extra round of interaction.

We commit by generating a uniformly random pad $m_1 \in \{0,1\}^l$ and committing to the four values $S, m_1, m_2 = m \oplus m_1$, and $m_3 = m_1 \wedge S$ individually using multi-bit commitments. The verifier then challenges uniformly with $b \in \{0, 1, 2\}$. If $b = 0$, then reveal m_1 and m_2 and verify that $V_0(m_1 \oplus m_2) = 1$. If $b = 1$ then reveal S, m_2, m_3 and verify that $V_1(S, m_2 \wedge S \oplus m_3) = 1$. Finally, if $b = 2$, then reveal S, m_1, and m_3 and verify that $m_3 = m_1 \wedge S$. This is still secure as at no time are S and m revealed at the same time. For the soundness, note that if $V_0(m_1 \oplus m_2) = 1$, $V_1(S, m_2 \wedge S \oplus m_3) = 1$, and $m_3 = m_1 \wedge S$, then for $m = m_1 \oplus m_2$ we have that $V_0(m) = 1$ and $V_1(S, m \wedge S) = 1$ and can thus compute a witness. If we increase the number of rounds by a factor $\log_{3/2}(2) < 1.71$ we will get cheating probability no larger than for t rounds with cheating probability $1/2$ in each round. The number of bits committed to in each round is $4l$ for a total of less than $6.84tl$ bits. However, now the bits can be committed to k bits at a time using the multi-bit commitment scheme. Therefore, if we implement the modified protocol using our commitment scheme, we get communication complexity $O((l+k)t)$. This follows because we can commit to $O(l)$ bits by sending $O(l+k)$ bits. This is an improvement by a factor $\theta(\frac{lk}{l+k}) = \theta(\min(l, k))$ over [CF01].

Exploiting Efficient Proofs of Relations. We show how we can use the efficient proofs of relations on committed values to reduce the communication complexity and the round complexity in a different way. This can simply be done by the parties agreeing in a Boolean circuit for the relation. Then the prover commits to the witness and the evaluation of the circuit on the witness and instance bit by bit and proves for each gate in the circuit that the committed values are consistent with the gate. The commitment to the output gate is opened and the prover then takes the revealed value as its output.

This protocol will have no messages of its own. All interaction is done through the ideal commitment functionality $\mathcal{F}_{HCOM}$. Let l be the size of the gate used. This protocol requires $O(l)$ commitments to single bits, each of which require $O(k)$ bits of communication. Then we need to do $O(l)$ proofs of relations, each of which require $O(k)$ bits of communication. This amounts to $O(lk)$ bits of communication, and is an improvement over the $O(lkt)$ bits when using the scheme of [CF01] by a factor $O(t)$.

References

[Can01]   Ran Canetti. Universally composable security: A new paradigm for cryptographic protocols. In *42th Annual Symposium on Foundations of Computer Science*. IEEE, 2001.

[CD98]   Ronald Cramer and Ivan Damgaard. Zero-knowledge proofs for finite field arithmetic, or: Can zero-knowledge be for free. In Hugo Krawczyk, editor, *Advances in Cryptology - Crypto '98*, pages 424–441, Berlin, 1998. Springer-Verlag. Lecture Notes in Computer Science Volume 1462.

[CDS94]   R. Cramer, I. B. Damgård, and B. Schoenmakers. Proofs of partial knowledge and simplified design of witness hiding protocols. In Yvo Desmedt, editor, *Advances in Cryptology - Crypto '94*, pages 174–187, Berlin, 1994. Springer-Verlag. Lecture Notes in Computer Science Volume 839.

[CF01]   Ran Canetti and Marc Fischlin. Universally composable commitments. In J. Kilian, editor, *Advances in Cryptology - Crypto 2001*, pages 19–40, Berlin, 2001. Springer-Verlag. Lecture Notes in Computer Science Volume 2139.

[Dam00]   Ivan Damgård. Efficient concurrent zero-knowledge in the auxiliary string model. In Bart Preneel, editor, *Advances in Cryptology - EuroCrypt 2000*, pages 418–430, Berlin, 2000. Springer-Verlag. Lecture Notes in Computer Science Volume 1807.

[KMO89]   Joe Kilian, Silvio Micali, and Rafail Ostrovsky. Minimum resource zero-knowledge proofs (extended abstract). In *30th Annual Symposium on Foundations of Computer Science*, pages 474–479, Research Triangle Park, North Carolina, 30 October–1 November 1989. IEEE.

[OU98]   Tatsuaki Okamoto and Shigenori Uchiyama. A new public-key cryptosystem as secure as factoring. In K. Nyberg, editor, *Advances in Cryptology - EuroCrypt '98*, pages 308–318, Berlin, 1998. Springer-Verlag. Lecture Notes in Computer Science Volume 1403.

[Pai99]   P. Paillier. Public-key cryptosystems based on composite degree residue classes. In Jacques Stern, editor, *Advances in Cryptology - EuroCrypt '99*, pages 223–238, Berlin, 1999. Springer-Verlag. Lecture Notes in Computer Science Volume 1592.

Unique Signatures and Verifiable Random Functions from the DH-DDH Separation

Anna Lysyanskaya

MIT LCS
200 Technology Square
Cambridge, MA 02139 USA
anna@theory.lcs.mit.edu

Abstract. A *unique* signature scheme has the property that a signature $\sigma_{PK}(m)$ is a (hard-to-compute) function of the public key PK and message m, for all, even adversarially chosen, PK. Unique signatures, introduced by Goldwasser and Ostrovsky, have been shown to be a building block for constructing verifiable random functions. Another useful property of unique signatures is that they are stateless: the signer does not need to update his secret key after an invocation.

The only previously known construction of a unique signature in the plain model was based on the RSA assumption. The only other previously known provably secure constructions of stateless signatures were based on the Strong RSA assumption. Here, we give a construction of a unique signature scheme based on a generalization of the Diffie-Hellman assumption in groups where decisional Diffie-Hellman is easy. Several recent results suggest plausibility of such groups.

We also give a few related constructions of verifiable random functions (VRFs). VRFs, introduced by Micali, Rabin, and Vadhan, are objects that combine the properties of pseudorandom functions (i.e. indistinguishability from random even after querying) with the verifiability property. Prior to our work, VRFs were only known to exist under the RSA assumption.

Keywords: Unique signatures, verifiable random functions, application of groups with DH-DDH separation.

1 Introduction

Signature schemes are one of the most important cryptographic objects. There were invented, together with the entire field of public-key cryptography, by Diffie and Hellman, and Rivest, Shamir and Adleman followed up with the first candidate construction. Goldwasser, Micali and Rivest [GMR88] gave the first signature scheme that is secure even if the adversary is allowed to obtain signatures on messages of its choice. This notion of security for signature schemes is also due to Goldwasser, Micali, and Rivest (GMR).

Since the GMR seminal work, it has become clear that the first requirement from a signature scheme is that it should satisfy the GMR definition of security.

M. Yung (Ed.): CRYPTO 2002, LNCS 2442, pp. 597–612, 2002.
© Springer-Verlag Berlin Heidelberg 2002

However, to be most useful, two additional properties of signature schemes are desirable: (1) that the scheme be secure in the plain model, i.e., without a random oracle or common parameters; and (2) that the scheme be stateless, i.e., not require the signer to update the secret key after each invocation.

The only signature schemes satisfying both of these additional properties are the Strong-RSA-based schemes of Gennaro et al. [GHR99] and of Cramer and Shoup [CS99], and the scheme implied by the verifiable random function due to Micali et al. [MRV99], based on RSA. An open question was to come up with a signature scheme satisfying the two additional properties, such that it would be secure under a different type of assumption. Here, we give such a signature scheme, based on a generalization of the Diffie-Hellman assumption for groups where decisional Diffie-Hellman is easy.

Unique signatures. The signature scheme we propose is a unique signature scheme. Unique signature schemes are GMR-secure signature schemes where the signature is a hard-to-compute function of the public key and the message. They were introduced by Goldwasser and Ostrovsky [GO92][1].

Intuitively, unique signatures are the "right" notion of signatures. This is because if one has verified a signature on a message once, then why should it be necessary to verify the signature on the same message again? Yet, even if indeed the given message has been accepted before, it is a bad idea to just accept it again – what if this time it came from an unauthorized party? Hence, one must verify the signature again, if it happens to be a different signature. If a signature scheme allows the signer to easily (that is to say, more efficiently than the cost of verifying a signature) generate many signatures on the same message, this leads to a simple denial-of-service attack on a verifier who is forced to verify many signatures on the same message. Although this is not a cryptographic attack, it still illustrates that intuitively unique signatures are more desirable.

In the random-oracle model, a realization of unique signatures is well-known. For example, some RSA signatures are unique: $\sigma_{n,e,H}(m) = (H(m))^{1/e} \bmod n$ is a function of (n, e, H, m) so long as e is a prime number greater than n (this is because for a prime e, $e > n$ implies that e is relatively prime to $\phi(n)$). Goldwasser and Ostrovsky give a solution in the common-random-string model. However, in the standard model, the only known construction of unique signatures was the one due to Micali, Rabin, Vadhan [MRV99].

On the negative side, Goldwasser and Ostrovsky have also shown that even in the common random string model, unique signatures require assumptions of the same strength as needed for non-interactive zero knowledge proofs with polynomial-time provers. The weakest assumption known that is required for non-interactive zero knowledge proofs with polynomial-time provers, is existence of trapdoor permutations [FLS99,BY96]. However, we do not know whether this is a necessary assumption; neither do we know whether this assumption is sufficient for constructing unique signatures in the plain model.

Verifiable random functions. Another reason why unique signatures are valuable is that they are closely related to verifiable random functions. Verifiable ran-

---

[1] They call it an *invariant* signature

dom functions (VRFs) were introduced by Micali, Rabin, and Vadhan [MRV99]. They are similar to pseudorandom functions [GGM86], except that they are also verifiable. That is to say, associated with a secret seed SK, there is a public key PK and a function $F_{PK}(\cdot) : \{0,1\}^k \mapsto \{0,1\}^u$ such that (1) $y = F_{PK}(x)$ is efficiently computable given the corresponding SK; (2) a proof $\pi_{PK}(x)$ that this value y corresponds to the public key PK is also efficiently computable given SK; (3) based purely on PK and oracle calls to $F_{PK}(\cdot)$ and the corresponding proof oracle, no adversary can distinguish the value $F_{PK}(x)$ from a random value without explicitly querying for the value x.

VRFs [MRV99] are useful for protocol design. They can be viewed as a commitment to an exponential number of random-looking bits, which can be of use in protocols. For example, using verifiable random functions, one can reduce the number of rounds for resettable zero knowledge proofs to 3 in the bare model [MR01]. Another example application, due to Micali and Rivest [MR02], is a non-interactive lottery system used in micropayments. Here, the lottery organizer holds a public key PK of a VRF. A participant creates his lottery ticket t himself and sends it to the organizer. The organizer computes the value $y = F_{PK}(t)$ on the lottery ticket, and the corresponding proof $\pi = \pi_{PK}(t)$. The value y determines whether the user wins, while the proof π guarantees that the organizer cannot cheat. Since a VRF is hard to predict, the user has no idea how to bias the lottery in his favor.

These objects are not well-studied; in fact, only one construction, based on the RSA assumption, was previously known [MRV99]. Micali, Rabin and Vadhan showed that, for the purposes of constructing a VRF, it is sufficient to construct a unique signature scheme. More precisely, from a unique signature scheme with small (but super-polynomial) message space and security against adversaries that run in super-polynomial time, they constructed a VRF with an arbitrary input size that tolerates a polynomial-time adversary[2] They then gave an RSA-based unique signature scheme for a small message space.

Constructing VRFs from pseudorandom functions (PRFs) is a good problem that we don't know how to solve in the standard model. This is because by its definition, the output of a PRF should be indistinguishable from random. However, if we extend the model to allow interaction, it is possible to commit to the secret seed of a PRF and let that serve as a public key, and then prove that a given output corresponds to the committed seed using a zero-knowledge proof. This solution is unattractive because of the expense of communication rounds. In the so-called *common random string* model where non-interactive zero-knowledge proofs are possible, a construction is possible using commitments and non-interactive zero knowledge [BFM88,BDMP91,FLS99]. However, this is unattractive as well because this model is unusual and non-interactive ZK is expensive. Thus, construction of verifiable random functions from general assumptions remains an interesting open problem.

---

[2] Intuitively, it would seem that the connection ought to be tighter: a unique signature secure against polynomial-time adversaries should imply a secure verifiable random function. This is an interesting open problem, posed by Micali et al.

DH–DDH separation. Recently, Joux and Nguyen [JN01] demonstrated that one can encounter groups in which decisional Diffie-Hellman is easy, and yet computational Diffie-Hellman seems hard. This is an elegant result that, among other things, sheds light on how reasonable it is to assume decisional Diffie-Hellman.

Joux [Jou00] proposed using the DH-DDH separation to a good end, by exhibiting a one-round key exchange protocol for three parties. Subsequently, insight into such groups has proved relevant for the recent construction of identity-based encryption due to Boneh and Franklin [BF01] which resolved a long standing open problem [Sha85]. Other interesting consequences of the study of these groups are a construction of a short signature scheme in the random-oracle model [BLS01] and of a simple credential system [Ver01].

Our results. We give a simple construction of a unique signature scheme based on groups where DH is conjectured hard and DDH is easy. Ours is a tree-like construction. The message space consists of codewords of an error-correcting code that can correct a constant fraction of errors. For n-bit codewords, the depth of the tree is n. The root of the tree is labelled with g, a generator of a group where DH is hard and DDH is easy. The 2^n leaves of the tree correspond all the possible n-bit strings. The 2^i nodes of depth i, $1 \leq i < n$ correspond to all the possible i-bit prefixes of an n-bit string.

A pair of group elements $(A_{i,0} = g^{a_{i,0}}, A_{i,1} = g^{a_{i,1}})$ is associated with each depth of the tree as part of the public key. The label of a node of depth i is derived from the label of its parent by raising the parent's label to the exponent $a_{i,0}$ if this node is its parent's left child, or $a_{i,1}$ if it is the parent's right child.

Computing the signature on each codeword m amounts to computing the labels of the nodes on the path from the root of the tree all the way down to the leaf corresponding to m.

At first glance, this may seem very similar to the Naor-Reingold [NR97] pseudorandom function. However the proof is not immediate. The major difference from the cited result is that here we must give a proof that the function was evaluated correctly. That makes it harder to prove security. For example, proof by a hybrid argument, as done by Naor and Reingold [NR97] is ruled out immediately: there is no way we can answer some of the adversary's queries with truly random bits, since for such bits there will be no proof.

Our proof of security for the unique signature relies on a generalization of the Diffie-Hellman assumption. We call it the "Many-DH" assumption. However, for directly converting this simple US to VRF, we need to make a very strong assumption; however the resulting VRF is very simple. We also suggest a more involved but also more secure construction of a VRF based on the Many-DH assumption. This last construction is closely related to that due to Micali et al. [MRV99].

Outline of the rest of the paper. In Section 2 we introduce our notation. In Section 3 we give definitions of verifiable random functions and unique signatures. In Section 4 we state our complexity assumptions for the unique signatures. In Section 5 we give our unique signature and prove it secure in Section 6. We then provide a simple construction for a VRF under a somewhat stronger assumption, in Section 7. We conclude in Section 8 with a construction of a VRF based on the weaker assumption alone, but whose complexity (both in terms of computation and in terms of conceptual simplicity) is the same as that of Micali et al. [MRV99].

2 Notation

The notation in this paper is based on the Cryptography class taught by Silvio Micali at MIT [Mic].

Let $A(\cdot)$ be an algorithm. $y \leftarrow A(x)$ denotes that y was obtained by running A on input x. In case A is deterministic, then this y is unique; if A is probabilistic, then y is a random variable.

Let b be a boolean function. The notation $(y \leftarrow A(x) : b(y))$ denotes the event that $b(y)$ is true after y was generated by running A on input x.

Finally, the statement such as $\Pr[y \leftarrow A(x); z \leftarrow B(y) : b(z)] = \alpha$ means that the probability that $b(z)$ is TRUE after the value z was obtained by first obtaining y by running algorithm A on input x, and then running algorithm B on input y.

By $A^{O(\cdot)}(\cdot)$, we denote a Turing machine that makes oracle queries to machine O. I.e., this machine will have an additional (read/write-once) query tape, on which it will write its queries in binary; once it is done writing a query, it inserts a special symbol "#". By external means, once the symbol "#" appears on the query tape, an oracle O is invoked on the value just written down, and the oracle's answer appears on the query tape adjacent to the "#" symbol. By $Q = Q(A^{O(\cdot)}(x)) \leftarrow A^{O(\cdot)}(x)$ we denote the contents of the query tape once A terminates, with oracle O and input x. By $(q, a) \in Q$ we denote the event that q was a query issued by A, and a was the answer received from oracle O. Sometimes, we will write, for example, $A^{O(x, \cdot)}$, to denote the fact that the first input to O is fixed, and A's query supplies only the second input to O.

We say that $\nu(k)$ is a negligible function, if for all polynomials $p(k)$, for all sufficiently large k, $\nu(k) < 1/p(k)$.

3 Definitions

3.1 Unique Signatures

Unique signatures are simply secure signatures where the signature is a function as opposed to a distribution.

Definition 1. *A function family* $\sigma_{(\cdot)}(\cdot) : \{0,1\}^k \mapsto \{0,1\}^{\ell(k)}$ *is a unique signature scheme (US) if there exists probabilistic algorithm* G, *efficient deterministic algorithm* Sign, *and probabilistic algorithm* Verify *such that* $G(1^k)$ *generates the key pair* PK, SK, $\text{Sign}(SK, x)$ *computes the value* $\sigma = \sigma_{PK}(x)$ *and* $\text{Verify}(PK, x, \sigma)$ *verifies that* $\sigma = \sigma_{PK}(x)$. *More formally (but assuming,* *for simplicity, that* Verify *is a deterministic algorithm; in case it is not, the adjustment to the definition is straightforward):*

1. *(Uniqueness of* $\sigma_{PK}(m)$*) There do not exist values* $(PK, m, \sigma_1, \sigma_2)$ *such that* $\sigma_1 \neq \sigma_2$ *and* Verify$(PK, m, \sigma_1) = $ Verify$(PK, m, \sigma_2) = 1$.
2. *(Security) For all families of probabilistic polynomial-time oracle Turing machines* $\{A_k^{(\cdot)}\}$*, there exists a negligible function* $\nu(k)$ *such that*

$$\Pr[(PK, SK) \leftarrow G(1^k);$$
$$(Q, x, \sigma) \leftarrow A_k^{\text{Sign}(SK, \cdot)}(1^k); : \text{Verify}(PK, x, \sigma) = 1 \wedge (x, \sigma) \notin Q] \leq \nu(k)$$

On a relaxed definition. Goldwasser and Ostrovsky [GO92] give a relaxed definition. In their definition, even though the signature is unique, the verification procedure may require, as an additional input, a proof that the signature is correct. This proof is output by the signing algorithm together with the signature, and it might not be unique. Here we give the stronger definition because we can satisfy it (Goldwasser and Ostrovsky do not, even though they work in the common random string model). However, note that the relaxed definition is sufficient for constructing VRFs [MRV99].

Unique signatures are stateless. Since a unique signature is a function of the public key and the message, a signature on a given message will be the same whether this was the first message signed by the signer, or the n'th message. As a result, it is easy to see that the signer does not need to remember anything about past transactions, i.e., a unique signature must be stateless.

3.2   Verifiable Random Functions

The definition below is due to Micali et al. [MRV99]. We use somewhat different and more compact notation, however.

The intuition of this definition is that a function is a verifiable random function if it is like a pseudorandom function with a public key and proofs.

Definition 2. *A function family* $F_{(\cdot)}(\cdot) : \{0,1\}^k \mapsto \{0,1\}^{\ell(k)}$ *is a verifiable random function (VRF) if there exist probabilistic algorithm* G*, and deterministic algorithms* Eval*, and* Prove*, and algorithm* Verify *such that:* $G(1^k)$ *generates the key pair* PK, SK*;* Eval(SK, x) *computes the value* $y = F_{PK}(x)$*;* Prove(SK, x) *computes the proof* π *that* $y = F_{PK}(x)$*; and* Verify(PK, x, y, π) *verifies that* $y = F_{PK}(x)$ *using the proof* π*. More formally (but assuming, for simplicity, that* Verify *is a deterministic algorithm; in case it is not, the adjustment to the definition is straightforward):*

1. *(Uniqueness of* $F_{PK}(x)$*) There do not exist values* $(PK, SK, x, y_1, y_2, \pi_1, \pi_2)$ *such that* $y_1 \neq y_2$ *and* Verify$(PK, x, y_1, \pi_1) = $ Verify$(PK, x, y_2, \pi_2) = 1$.
2. *(Computability of* $F_{PK}(x)$*)* $F_{PK}(x) = $ Eval(SK, x).

3. *(Provability of $F_{PK}(x)$)* If $(y, \pi) = \mathrm{Prove}(SK, x)$, then

$$\mathrm{Verify}(PK, x, y, \pi) = 1.$$

4. *(Pseudorandomness of $F_{PK}(x)$)* For all families of probabilistic polynomial-time Turing machines $\{A_k^{(\cdot)}, B_k\}$, there exists a negligible function $\nu(k)$ such that

$$\Pr[(PK, SK) \leftarrow G(1^k);$$

$$(Q_A, x, \mathrm{state}) \leftarrow A_k^{\mathrm{Prove}(SK, \cdot)}(1^k);$$

$$y_0 = \mathrm{Eval}(SK, x);$$

$$y_1 \leftarrow \{0,1\}^{\ell(k)};$$

$$b \leftarrow \{0,1\};$$

$$(Q_B, b') \leftarrow B_k^{\mathrm{Prove}(SK, \cdot)}(\mathrm{state}, y_b) : b = b' \wedge (x, \mathrm{Prove}(SK, x)) \notin Q_A \cup Q_B]$$

$$\leq 1/2 + \nu(k)$$

(The purpose of the state *random variable is so that A_k can save some useful information that B_k will then need.)*
In other words, the only way that an adversary could tell $F_{PK}(x)$ from a random value, for x of its own choice, is by querying it directly.

It is easy to see [MRV99] that given a VRF $F_{(\cdot)} : \{0,1\}^{\ell(k)} \mapsto \{0,1\}$, one can construct a VRF $F'_{(\cdot)} : \{0,1\}^{\ell'(k)} \mapsto \{0,1\}^{m(k)}$, where $\ell'(k) = \ell(k) - \lceil \log m(k) \rceil$, as follows: $F'_S(x_1 \circ \ldots \circ x_{\ell'(k)}) = F_S(x_1 \circ \ldots \circ x_{\ell'(k)} \circ u_0) \circ F_S(x_1 \circ \ldots \circ x_{\ell'(k)} \circ u_1) \circ \ldots \circ F_S(x_1 \circ \ldots \circ x_{\ell'(k)} \circ u_{m(k)})$ where u_i denotes the $\lceil \log m(k) \rceil$-bit representation of the integer i, and "$\circ$" denotes concatenation.

Thus in the sequel we will focus on constructing VRFs with binary outputs.

Unique signatures vs. VRFs. Micali et al. showed how to construct VRFs from unique signatures. The converse, namely construction of a unique signature from a VRF, holds immediately if the proofs in the VRFs are unique. If the proofs are not unique, then no construction satisfying our strong definition of unique signatures in known. However, constructing relaxed unique signatures in the sense of Goldwasser and Ostrovsky (see the end of Section 3.1) is immediate.

4 Assumptions

Let S be the algorithm that, on input 1^k, generates a group $G = (*, q, g)$ with efficiently computable group operation $*$, of prime order q, with generator g. We require that g is written down in binary using $O(\log q)$ bits, and that every element of the group has a unique binary representation.

We require that the decisional Diffie-Hellman problem be easy in G. More precisely, we require that there is an efficient algorithm D for deciding the following language $L_{DDH}(G)$:

$$L_{DDH}(G) = \{(*, q, g, X, Y, Z) \mid \exists x, y \in \mathbb{Z}_q \text{ such that } X = g^x, Y = g^y, Z = g^{xy}\}$$

One the other hand, we will need to make the following assumption which is somewhat stronger than computational Diffie-Hellman. It is essentially like computational Diffie-Hellman, except that instead of taking two inputs, g^x and g^y, and the challenge is computing g^{xy}, we have a logarithmic number of bases $g^{y_1}, \ldots, g^{y_\ell}$, as well as all the products $g^{\prod_{j \in J} y_j}$ for all proper subsets J of the naturals up to and including ℓ, and the challenge is to come up with the value $g^{\prod_{j=1}^{\ell} y_j}$.

Assumption 1 (Many-DH Assumption) *For all $\ell = O(\log k)$, for all probabilistic polynomial-time families of Turing machines $\{A_k\}$,*

$$\Pr[(*, q, g) \leftarrow S(1^k); \{y_i \leftarrow \mathbb{Z}_q \ : \ 1 \leq i \leq \ell\};$$
$$\{z_J = \prod_{j \in J} y_j; Z_J = g^{z_J} \ : \ J \subset [\ell]\};$$
$$Z \leftarrow A_k(*, q, g, \{Z_J \ : \ J \subset [\ell]\}) : Z = g^{\prod y_i}] \leq \nu(k)$$

For an exposition of groups of this flavor, where the decisional Diffie-Hellman problem is easy, and yet generalizations of the computational Diffie-Hellman problem are conjectured hard, we refer the reader, for example, to the recent papers by Joux and Nguyen [JN01], Joux [Jou00] and Boneh and Franklin [BF01]. In the sequel, we will not address the number-theoretic aspects of the subject.

5   Construction of a Unique Signature

Suppose the algorithms S, D, as in Section 4, are given. Let k be the security parameter. Let the message space consist of strings of length n_0. Our only assumption on the size of the message space is that $n_0 = \omega(\log k)$.

Let $C : \{0,1\}^{n_0} \mapsto \{0,1\}^n$ be an error-correcting code of distance cn, where $c > 0$ is a constant. In other words, C is a function such that if $M \neq M'$ are strings of length n_0, then $C(M)$ differs from $C(M')$ in at least cn places. For an overview of error-correcting codes, see the lecture notes of Madhu Sudan [Sud]. Here, we note that since we will not need the decoding operation, only the encoding operation, we can easily achieve $n = O(n_0)$.

We need to construct algorithms G, Sign, Verify as specified in Definition 1.

Algorithm G Run $S(1^k)$ to obtain $G = (*, q, g)$. Choose n pairs of random elements in $\mathbb{Z}_q$: $(a_{1,0}, a_{1,1}), \ldots, (a_{n,0}, a_{n,1})$. Let $A_{i,b} = g^{a_{i,b}}$, for $1 \leq i \leq k$, $b \in \{0,1\}$. Output the following key pair:

$SK =$	$a_{1,0}$	$a_{2,0}$	$\cdots$	$a_{n,0}$
	$a_{1,1}$	$a_{2,1}$	$\cdots$	$a_{n,1}$

$PK =$	$A_{1,0}$	$A_{2,0}$	$\cdots$	$A_{n,0}$
	$A_{1,1}$	$A_{2,1}$	$\cdots$	$A_{n,1}$

Algorithm Sign On input a message M of length n_0, compute the encoding of M using code C: $m = C(M)$. To sign the n-bit codeword $m = m_1 \circ \ldots \circ m_n$, output $\sigma_{PK}(m) = (s_1, \ldots, s_n)$, where $s_0 = g$ and $s_i = (s_{i-1})^{a_{i,m_i}}$ for $1 \leq i \leq n$.

Algorithm Verify Let $s_{m,0} = 1$. Verify that, for all $1 \leq i \leq n$,

$$D(*, q, g, s_{m,i-1}, A_{i,m_i}, s_{m,i}) = ACCEPT$$

Graphically, we view the message space as the leaves of a balanced binary tree of depth n. Each internal node of the tree is assigned a label, as follows: the label of the root is g. The label of a child, denoted l_c is obtained from the label of its parent, denoted l_p, as follows: if the depth of the child is i, and it is the left child, then its label is $l_c = l_p^{a_{i,0}}$, while if it is the right child, its label will be $l_c = l_p^{a_{i,1}}$. The signature on an n-bit message consists of all the labels on the path from the leaf corresponding to this message all the way to the root. To verify correctness of a signature, one uses the algorithm D that solves the DDH for the group over which this is done.

Efficiency. In terms of efficiency, this signature scheme is a factor of $O(n)$ worse than the Cramer-Shoup scheme, in all parameters such as the key and signature lengths and the complexity of relevant operations. This means that it is still rather practical, and yet has two benefits: uniqueness, and security based on a different assumption. In comparison to the unique signature of Micali et al., our construction is preferable as far as efficiency is concerned. This is because our construction is direct, while they first give a unique signature for short messages, then show how to construct a VRF and a unique signature of arbitrary length from that.

Reducing the length of signatures. Boneh and Silverberg [BS02] point out that, if the language $L(*, q, g) = \{g^{y_1}, \ldots, g^{y_n}, g^{\prod_{i=1}^{n} y_i}\}$ is efficiently decidable, then the signature does not need to contain the labels of the intermediate nodes. I.e., to sign a message M, $m = C(M)$, it is sufficient to give $s = g^{\prod_{i=1}^{n} a_{i,m_i}}$. This reduces the length of a signature by a factor of n. However, finding groups where $L(*, q, g)$ is efficiently decidable, and yet the Many-DH Assumption is still reasonable, is an open question [BS02].

6   Proof of Security for the Unique Signature

In this section, we show how to reduce breaking the Many-DH problem to forging a signature of the construction in Section 5.

First, we show the following lemma:

Lemma 1. *Suppose* $\mathtt{Verify}(((*, q, g), \{A_{i,b}\}_{1 \leq i \leq n, b \in \{0,1\}}), m, (s_1, \ldots, s_n)) = 1$. *Then* $s_j = g^{\prod_{i=1}^{j} a_{i,m_i}}$, *where* a_{i,m_i} *denotes the unique value* $\mathbb{Z}_q$ *such that* $g^{a_{i,m_i}} = A_{i,m_i}$.

Proof. We show the lemma by induction on j. For $j = 1$, the verification algorithm will accept only if $s_1 = A_{1,m_i}$.

Let the lemma hold for $j - 1$. The verification algorithm will accept s_j only if $D(*, q, g, s_{j-1}, A_{j,m_j}, s_j) = 1$. But by definition of D, this is the case only if $s_j = g^{\sigma a_{j,m_j}}$, where σ is the unique value in $\mathbb{Z}_q$ such that $g^\sigma = s_{j-1}$. By the induction hypothesis, $\sigma = \prod_{i=1}^{j-1} a_{i,m_i}$. Therefore, $s_j = g^{\sigma a_{j,m_j}} = g^{\prod_{i=1}^{j} a_{i,m_i}}$. $\square$

6.1   Description of the Reduction

In the following, "we" refers to the reduction, and the forger for the signature scheme is "our adversary."

Recall that k is the security parameter, and that $n_0 = \omega(\log k)$, $n = O(n_0)$.

Suppose that our adversary's running time is $t(k)$. Recall that c is the distance of the error-correcting code we use. Let $\ell = \frac{\lceil \log t(k) \rceil + 2}{\log(1/(1-.99c))} + 1$. (Note that $\ell = O(\log k)$). Assume G is as in Section 4.

Input to the reduction: As in Assumption 1, we are given a group $G = (*, q, g)$ and ℓ elements $Y_1, Y_2, \dots, Y_\ell$ of G, where $Y_u = g^{y_u}$. We are also given the values $Z_I = g^{\prod_{u \in I} y_u}$, for $I \subset [\ell]$. The goal is to break Assumption 1, i.e., compute $g^{\prod_{u=1}^{\ell} y_u}$.

Key generation: Pick a random ℓ-bit string $B = b_1 \circ \dots \circ b_\ell$ and a random ℓ-bit subset $J \subset [n]$. (For simpler notation, assume that the indices are ordered such that $j_u < j_{u+1}$ for all $j_u \in J$.)

Set up the public key as follows: $A_{j_u, b_u} = Y_u$. To set up $A_{i,b}$ where $i \notin J$ or $i = j_u \in J$ but $b \neq b_u$, choose value $a_{i,b} \leftarrow \mathbb{Z}_q$ and set $A_{i,b} = g^{a_{i,b}}$. Figure 1 gives an example of what the reduction does in this step.

$g^{a_{1,0}}$	$g^{a_{2,0}}$	$g^{a_{3,0}}$	Y_2	$g^{a_{5,0}}$	$g^{a_{6,0}}$	$g^{a_{7,0}}$	$g^{a_{8,0}}$	$g^{a_{9,0}}$	$g^{a_{10,0}}$
$g^{a_{1,1}}$	$g^{a_{2,1}}$	Y_1	$g^{a_{4,1}}$	$g^{a_{5,1}}$	$g^{a_{6,1}}$	$g^{a_{7,1}}$	$g^{a_{8,1}}$	Y_3	$g^{a_{10,1}}$

$PK =$

Fig. 1. Toy example of a public key produced by the reduction. In this example, $n = 10$, $\ell = 3$. The reduction randomly picked $J = \{3, 4, 9\}$, $B = 101$.

Responding to signature queries: We will respond to at most $2t$ signature queries from the adversary, as follows.

Suppose the adversary's query is M where $C(M) = m = m_1 \circ \dots \circ m_n$. Let $J(m)$ denote the string $m_{j_1} \circ \dots \circ m_{j_\ell}$.

Check if $J(m) = B$. If so, terminate with "Fail." Otherwise, compute the signature as follows: Let $Z_0 = g$. Let $b' = b'(J)$ be an n-bit string such that $b_{j_u} = b_u$ for all $j_u \in J$. By I_u, we denote the ℓ-bit string which has a 1 in position u, and 0's everywhere else.

Compute $(s_1, \ldots, s_n)$ using the following loop:

Initialize $c := 0^\ell$

For $i = 1$ to n

   (1) If $i = j_u \in J$ and $m_i = b_u$, then $c := c \oplus I_u$

   (2) $s_i = Z_c^{\prod_{j \notin J, j \leq i} a_{j,m_j} \prod_{j \in J, j \leq i, m_j \neq b'_j} a_{j,m_j}}$

   (3) Increment i

(end of loop)

Processing the forgery: Now suppose that the adversary comes back with a forged signature on message M' for which it has not queried R. Let $m' = C(M')$. This forgery is $\sigma_{PK}(m') = \{s_{m',i}\}$. This forgery is *good* if $J(m') = B$. If the forgery is good, we obtain the value $g^{\prod_{u=1}^{\ell} y_u}$ by Lemma 1, by simply computing $(s_{m',i})^{1/\prod_{i \notin J} a_i^{m'_i}}$.

6.2 Analysis of the Reduction

Let $t = t(k)$ be the expected running time of the adversary. For the purposes of the analysis, consider the following algorithms:

- Algorithm 1 runs the signing algorithm and responds to the adversary's queries as the true signing oracle would. If the adversary outputs a valid forgery, this algorithm outputs "Success."
- Algorithm 2 runs the signing algorithm but only responds to $2t$ queries. If the adversary issues more queries, output "Fail." Otherwise, if the adversary outputs a valid forgery, this algorithm outputs "Success."
- Algorithm 3 is the same as Algorithm 2, except in one case. Namely, it chooses a random $J \subset [n]$, $|J| = \ell$, $J = \{j_1, \ldots, j_\ell\}$, and outputs "Fail" if it so happens that $J(C(M')) = J(C(M))$ where M' is the adversary's forgery, and M is any previously queried message, and notation $J(m)$ denotes $m_{j_1} \circ m_{j_2} \circ \ldots \circ m_{j_\ell}$.
- Algorithm 4 is just like Algorithm 3 except in one case Namely, it chooses a random ℓ-bit string B and outputs "Fail" whenever the forged message M' is such that $C(M') = m'$ where $J(m') \neq B$.
- Algorithm 5 is just like Algorithm 4 except in one case. Namely, it outputs "Fail" whenever the queried message M is such that $C(M) = m$ where $J(m) = B$.
- Algorithm 6 runs our reduction. It outputs "Success" whenever the reduction succeeds in computing its goal.

By p_i, let us denote the success probability of algorithm i.

Lemma 2. *The success probability of Algorithm 1 is the same as the success probability of the forger.*

Proof. By construction, this algorithm outputs "Success" iff the forger succeeds. $\square$

Lemma 3. $p_2 \geq p_1/2$.

Proof. Suppose a successful forgery requires t queries on the average. Then by the Markov inequality, $2t$ queries are sufficient at least half the time. □

Lemma 4. $p_3 \geq p_2/2$.

Proof. Recall that $m' = C(M')$ is a codeword of an error-correcting code of distance cn. That implies that for all M, m' differs from $m = C(M)$ on at least cn locations. So, if we picked ℓ locations at random with replacement, $\Pr_J[J(m) = J(m')] \leq (1 - c)^{-\ell}$. Since we are picking without replacement, $\Pr_J[J(m) = J(m')] \leq \prod_{i=1}^{\ell}(n - cn - i)/n \leq (1 - c + \epsilon)^{\ell}$ for any constant $\epsilon > 0$. Let $\epsilon = .01c$ for simplicity. Let Q denote the set of messages queried by the adversary. By the union bound

$$\Pr_J[\exists M \in Q \text{ such that } J(C(M)) = J(m')] \leq 2t(1 - c + \epsilon)^{\ell} < 1/2$$

if $4t < (1/(1 - c + \epsilon))^{\ell}$. Taking the logarithm on both sides of the equation, we get the condition $\log t + 2 < \ell \log(1/(1 - .99c))$, and so since we have set $\ell > \frac{\log t + 2}{\log(1/(1 - .99c))}$, this is satisfied. □

Lemma 5. $p_4 \geq p_3/2^{\ell}$.

Proof. Note that Algorithm 3 and Algorithm 4 can be run with the same adversary, but Algorithm 4 may output "Fail" while Algorithm 3 outputs "Success." Consider the case when both of them output "Success." In this case, (1) $J(C(M')) \neq J(C(M))$ for all previously queried M, and (2) $J(C(M)) = B$. Note that, given that (1) is true, the probability of (2) is exactly $2^{-\ell}$. □

Lemma 6. $p_5 = p_4$.

Proof. Note that the only difference between the two algorithms is that Algorithm 5 will sometimes output "Fail" sooner. Algorithm 4 will continue answering queries, but, if Algorithm 5 has output "Fail," then $J(C(M')) \neq J(C(M)) = B$, and so Algorithm 4 will output "Fail" as well. □

Lemma 7. $p_6 = p_5$.

Proof. First, note that whether we are running Algorithm 5 or Algorithm 6, the view of the adversary is the same. Namely: (1) the public key is identically distributed; (2) both algorithms respond to at most $2t$ signature queries; (3) both algorithms pick J and B uniformly at random and refuse to respond to a signature query M if $J(C(M)) = B$. Therefore, the probability that the adversary comes up with a forgery in the two cases is the same. Now, note that the probability that the forgery is good is also the same: in both cases, the forgery is good if $J(C(M')) = B$. □

Putting these together, we have:

Lemma 8. *If the forger's success probability is p, and expected running time is t, then the success probability of the reduction is $p/2^{\ell+2} = p/O(t)$.*

In turn, this implies the following theorem:

Theorem 1. *Under Assumption 1, the construction presented in Section 5 is a unique signature.*

7   A Simple Construction of a VRF

Consider the following, rather strong, complexity assumption:

Assumption 2 *(Very-Many-DH-Very-Hard Assumption) There exists a constant $\epsilon > 0$ such that for all probabilistic polynomial-time families of Turing machines $\{A_k\}$ with running time $O(2^{k^\epsilon})$,*

$$\Pr[(*, q, g) \leftarrow S(1^k); \{y_i \leftarrow \mathbb{Z}_q \; : \; 1 \leq i \leq k^\epsilon\};$$
$$Z \leftarrow A_k^{\mathcal{O}[*,q,g,\{y_i\}](\cdot)}(*, q, g, \{g^{y_i} \; : \; 1 \leq i \leq k^\epsilon\}) : Z = g^{\prod y_i}] \leq poly(k) * 2^{-2k^\epsilon}$$

where $\mathcal{O}[*, q, g, \{y_i\}](\cdot)$, on input an k^ϵ-bit string I, outputs $g^{\prod_{i=1}^{k^\epsilon} y_i^{I_i}}$ iff I is not an all-1 string.

Definition 3 (VUF [MRV99]). *A verifiable unpredictable function (VUF) $(G, \mathsf{Eval}, \mathsf{Prove}, \mathsf{Verify})$ with input length $a(k)$, output length $b(k)$, and security $s(k)$ is defined as a VRF except that requirement 4 of Definition 2 is replaced with the following: Let $T(\cdot, \cdot)$ be any oracle algorithm that runs in time $s(k)$ when its first input is 1^k. Then:*

$$\Pr[(PK, SK) \leftarrow G(1^k);$$
$$(Q, x, y) \leftarrow T^{\mathsf{Prove}(SK, \cdot)}(1^k) : y = \mathsf{Eval}(SK, x) \land$$
$$(x, \mathsf{Prove}(SK, x)) \notin Q] \leq 1/s(k)$$

Lemma 9. *The unique signature in Section 5 is a VUF with security 2^{k^ϵ} under Assumption 2.*

Proof. (Sketch) Following the proof in Section 6, let $\ell = k^\epsilon$. Then, by Lemma 8, using an adversary that succeeds in breaking the unique signature in 2^{k^ϵ} steps with probability 2^{-k^ϵ} corresponds to computing $g^{\prod y_i}$ in time 2^{k^ϵ} with probability $\Omega(2^{-2k\epsilon})$, which contradicts the assumption.   □

The following proposition completes the picture:

Proposition 1 ([MRV99]). *If there is a VUF $(G, \mathsf{Eval}, \mathsf{Prove}, \mathsf{Verify})$ with input length $a(k)$, output length $b(k)$, and security $s(k)$, then, for any $a'(k) \leq a(k)$, there is a VRF $(G', \mathsf{Eval}', \mathsf{Prove}', \mathsf{Verify}')$ with input length $a'(k)$, output length $b'(k) = 1$, and security $s'(k) = s(k)^{1/3}/(poly(k) \cdot 2^{a'(k)})$. Namely, the following is a VRF with security $s'(k)$:*

- $G'(1^k) = (G(1^k), r)$, where r is a $b(k)$-bit string chosen at random.
- $\text{Eval}'(SK, x) = \text{Eval}(SK, x).r$, where "." denotes the inner product.
- $\text{Verify}'(SK, x) = (\text{Eval}(SK, x), \text{Verify}(SK, x))$.

Corollary 1. We have obtained a VRF with input length k, output length 1, and security $poly(k)$.

A natural open problem is to give better security guarantee to a VRF obtained from a unique signature in this fashion, or to show evidence of impossibility.

8 Complicated but More Secure VRF

Here, we construct a VRF and a unique signature based on the weaker assumption alone. This is the same as the Micali et al. [MRV99] construction, except that the underlying verifiable unpredictable function is different.

Proposition 2 ([MRV99]). If there is a VRF with input length $a(k)$, output length 1, and security $s(k)$, then there is a VRF with unrestricted input length, output length 1, and security at least $\min(s(k)^{1/5}, 2^{a(k)/5})$.

The proof of the proposition gives the VRF construction, which we omit here.
Now let us restate Assumption 1 to make explicit use of security:

Assumption 3 (s(k)-Many-DH assumption) For some $s(k) > poly(k)$, for all $\{A_k\}$, if $\{A_k\}$ is a family of probabilistic Turing machines with running time $t(k) = O(s(k))$, then for all $\ell > \log t(k)$,

$$\Pr[(*, q, g) \leftarrow S(1^k); \{y_i \leftarrow \mathbb{Z}_q \; : \; 1 \leq i \leq \ell\};$$
$$\{z_J = \prod_{j \in J} y_j; Z_J = g^{z_J} \; : \; J \subset [\ell]\};$$
$$Z \leftarrow A(*, q, g, \{Z_J \; : \; J \subset [\ell]\}) : Z = g^{\prod y_i}] < 1/s^2(k)$$

We will construct a VUF with input length $\Omega(\log s'(k))$, and security $s'(k) = 2^{\omega(\log k)}$ based on this assumption. By Propositions 1 and 2, this implies a VRF with unrestricted input length and security $\min(s'(k)^{1/5}, 2^{a(k)/5}) = 2^{\omega(\log k)}$.

This is the same as our unique signature construction, only here the public key and the depth of the tree are smaller. The proof works in exactly the same way as in the previous construction.

Theorem 2. The following construction is a VUF: set $a_{i,b}$ at random from $\mathbb{Z}_q$, for $1 \leq i \leq \ell$, $\ell > \log s(k)$, $b \in \{0, 1\}$. Let $PK = \{g^{a_{i,b}} \; : \; 1 \leq i \leq \ell, b \in \{0, 1\}\}$, $SK = \{a_{i,b}\}$. If J is a binary string of length ℓ, then let $F_{PK}(g^{\prod_{i=1}^{\ell} a_{i,J_i}})$. The verification is as in the construction described in Section 5. Its security is at least $s(k)$ under Assumption 3.

Proof. (Sketch) This proof is essentially the same as in Section 6, with $n = \ell = \log s(k)$; except this case is simpler as there is no code. Here, too, we hope that the adversary's forgery will be good, and by Lemma 8, if $s(k)$ is the size of the message space, then $1/s(k)$ of the forgeries will be good. By contrapositive, in $O(s(k))$ running time, the probability of a forgery is $1/s(k)$. Therefore, the probability of a good forgery is $1/s^2(k)$. This contradicts Assumption 3.   □

Combining the above, we get a VRF with unrestricted input length, output length 1, and security $\min(s'(k)^{1/5}, 2^{a(k)/5}) = \min(s(k)^{1/10}, 2^{\log(s(k))/5}) = \Theta(s(k)^{1/10}) = 2^{\omega(\log k)}$.

Acknowledgements

I am grateful to Dan Boneh for pointing out an error in a previous version of this paper. I thank Ron Rivest, Silvio Micali, Yevgeniy Dodis, and Leo Reyzin for helpful discussions. This research was supported by an NSF graduate research fellowship, Lucent Technologies GRPW, and NTT grant #6762700.

References

[BDMP91] Manuel Blum, Alfredo De Santis, Silvio Micali, and Guiseppe Persiano. Non-interactive zero-knowledge. *SIAM Journal of Computing*, 20(6):1084–1118, 1991.

[BF01] Dan Boneh and Matthew Franklin. Identity-based encryption from the Weil pairing. In Joe Kilian, editor, *Advances in Cryptology – CRYPTO 2001*, volume 2139 of *Lecture Notes in Computer Science*, pages 213–229. Springer Verlag, 2001.

[BFM88] Manuel Blum, Paul Feldman, and Silvio Micali. Non-interactive zero-knowledge and its applications (extended abstract). In *Proceedings of the Twentieth Annual ACM Symposium on Theory of Computing*, pages 103–112, Chicago, Illinois, 2–4 May 1988.

[BLS01] Dan Boneh, Ben Lynn, and Hovav Shacham. Short signatures from the Weil pairing. In Colin Boyd, editor, *Advances in Cryptology – ASIACRYPT 2001*, volume 2248 of *Lecture Notes in Computer Science*, pages 514–532. Springer Verlag, 2001.

[BM84] Manuel Blum and Silvio Micali. How to generate cryptographically strong sequences of pseudo-random bits. *SIAM Journal on Computing*, 13(4):850–863, November 1984.

[BS02] Dan Boneh and Alice Silverberg. Applications of multilinear forms to cryptography. Manuscript obtained by personal communication, 2002.

[BY96] Mihir Bellare and Moti Yung. Certifying permutations: Non-interactive zero-knowledge based on any trapdoor permutation. *Journal of Cryptology*, 9(1):149–166, 1996.

[CS99] Ronald Cramer and Victor Shoup. Signature schemes based on the strong RSA assumption. In *Proc. 6th ACM Conference on Computer and Communications Security*, pages 46–52. ACM press, nov 1999.

[FLS99]   Uriel Feige, Dror Lapidot, and Adi Shamir. Multiple noninteractive zero knowledge proofs under general assumptions. *SIAM Journal on Computing*, 29(1):1–28, 1999.

[GGM86]   Oded Goldreich, Shafi Goldwasser, and Silvio Micali. How to construct random functions. *Journal of the ACM*, 33(4):792–807, October 1986.

[GHR99]   Rosario Gennaro, Shai Halevi, and Tal Rabin. Secure hash-and-sign signatures without the random oracle. In Jacques Stern, editor, *Advances in Cryptology - EUROCRYPT '99*, volume 1592 of *Lecture Notes in Computer Science*, pages 123–139. Springer Verlag, 1999.

[GMR88]   Shafi Goldwasser, Silvio Micali, and Ronald Rivest. A digital signature scheme secure against adaptive chosen-message attacks. *SIAM Journal on Computing*, 17(2):281–308, April 1988.

[GO92]   Shafi Goldwasser and Rafail Ostrovsky. Invariant signatures and noninteractive zero-knowledge proofs are equivalent. In Ernest F. Brickell, editor, *Advances in Cryptology - CRYPTO '92*, pages 228–244. Springer-Verlag, 1992. Lecture Notes in Computer Science No. 740.

[JN01]   Antoine Joux and Kim Nguyen. Separating decision Diffie-Hellman from Diffie-Hellman in cryptographic groups. Manuscript. Available from http://eprint.iacr.org, 2001.

[Jou00]   Antoine Joux. A one-round protocol for tripartite Diffie-Hellman. In *Proceedings of the ANTS-IV conference*, volume 1838 of *Lecture Notes in Computer Science*, pages 385–394. Springer-Verlag, 2000.

[Mic]   Silvio Micali. 6.875: Introduction to cryptography. MIT course taught in Fall 1997.

[MR01]   Silvio Micali and Leonid Reyzin. Soundness in the public-key model. In Joe Kilian, editor, *Advances in Cryptology - CRYPTO 2001*, volume 2139 of *Lecture Notes in Computer Science*, pages 542–565. Springer Verlag, 2001.

[MR02]   Silvio Micali and Ronald L. Rivest. Micropayments revisited. In Bart Preneel, editor, *Proceedings of the Cryptographer's Track at the RSA Conference*, volume 2271 of *Lecture Notes in Computer Science*, pages 149–163. Springer Verlag, 2002.

[MRV99]   Silvio Micali, Michael Rabin, and Salil Vadhan. Verifiable random functions. In *Proc. 40th IEEE Symposium on Foundations of Computer Science (FOCS)*, pages 120–130. IEEE Computer Society Press, 1999.

[NR97]   Moni Naor and Omer Reingold. Number-theoretic constructions of efficient pseudo-random functions. In *Proc. 38th IEEE Symposium on Foundations of Computer Science (FOCS)*, 1997.

[Sha85]   Adi Shamir. Identity-based cryptosystems and signature schemes. In George Robert Blakley and David Chaum, editors, *Advances in Cryptology - CRYPTO '84*, volume 196 of *Lecture Notes in Computer Science*, pages 47–53. Springer Verlag, 1985.

[Sud]   Madhu Sudan. Algorithmic introduction to coding theory. MIT course taught in Fall 2001. Lecture notes available from http://theory.lcs.mit.edu/~madhu/FT01/.

[Ver01]   Eric Verheul. Self-blindable credential certificates from the weil pairing. In Colin Boyd, editor, *Advances in Cryptology - ASIACRYPT 2001*, volume 2248 of *Lecture Notes in Computer Science*, pages 533–551. Springer Verlag, 2001.

Security Proof
for Partial-Domain Hash Signature Schemes

Jean-Sébastien Coron

Gemplus Card International
34 rue Guynemer
Issy-les-Moulineaux, F-92447, France
jean-sebastien.coron@gemplus.com

Abstract. We study the security of partial-domain hash signature schemes, in which the output size of the hash function is only a fraction of the modulus size. We show that for $e = 2$ (Rabin), partial-domain hash signature schemes are provably secure in the random oracle model, if the output size of the hash function is larger than $2/3$ of the modulus size. This provides a security proof for a variant of the signature standards ISO 9796-2 and PKCS#1 v1.5, in which a larger digest size is used.

Keywords: Signature Schemes, Provable Security, Random Oracle Model.

1 Introduction

A common practice for signing with RSA or Rabin consists in first hashing the message m, then padding the hash value with some predetermined or message-dependent block, and eventually raising the result $\mu(m)$ to the private exponent d. This is commonly referred to as the "hash-and-sign" paradigm:

$$s = \mu(m)^d \mod N$$

For digital signature schemes, the strongest security notion was defined by Goldwasser, Micali and Rivest in [8], as *existential unforgeability under an adaptive chosen message attack*. This notion captures the property that an attacker cannot produce a valid signature, even after obtaining the signature of (polynomially many) messages of his choice.

The random oracle model, introduced by Bellare and Rogaway in [2], is a theoretical framework allowing to prove the security of hash-and-sign signature schemes. In this model, the hash function is seen as an oracle which outputs a random value for each new query. Bellare and Rogaway defined in [3] the Full Domain Hash (FDH) signature scheme, in which the output size of the hash function is the same as the modulus size. FDH is provably secure in the random oracle model assuming that inverting RSA is hard. Actually, a security proof in the random oracle model does not necessarily imply that the scheme is secure in the real world (see [4]). Nevertheless, it seems to be a good engineering principle to design a scheme so that it is provably secure in the random oracle model.

M. Yung (Ed.): CRYPTO 2002, LNCS 2442, pp. 613–626, 2002.
© Springer-Verlag Berlin Heidelberg 2002

Many encryption and signature schemes were proven to be secure in the random oracle model

Other hash-and-sign signature schemes include the widely used signature standards PKCS#1 v1.5 and ISO 9796-2. In these standards, the digest size is only a fraction of the modulus size. As opposed to FDH, no security proof is known for those standards. Moreover, it was shown in [5] that ISO 9796-2 was insecure if the size of the hash function was too small, and the standard was subsequently revised.

In this paper, we study the security of partial-domain hash signature schemes, in which the hash size is only a fraction of the modulus size. We show that for $e = 2$, partial-domain hash signature schemes are provably secure in the random oracle model, assuming that factoring is hard, if the size of the hash function is larger than 2/3 of the modulus size. The proof is based on a modification of Vallée's generator of small random squares [16]. This provides a security proof for a variant of PKCS#1 v1.5 and ISO 9796-2 signatures, in which the digest size is larger than 2/3 of the size of the modulus.

2 Definitions

In this section we briefly present some notations and definitions used throughout the paper. We start by recalling the definition of a signature scheme.

Definition 1 (Signature Scheme). *A signature scheme* (Gen, Sign, Verify) *is defined as follows:*

- *The key generation algorithm* Gen *is a probabilistic algorithm which given* 1^k, *outputs a pair of matching public and private keys,* (pk, sk).
- *The signing algorithm* Sign *takes the message* M *to be signed, the private key* sk, *and returns a signature* $x = \text{Sign}_{sk}(M)$. *The signing algorithm may be probabilistic.*
- *The verification algorithm* Verify *takes a message* M, *a candidate signature* x' *and* pk. *It returns a bit* $\text{Verify}_{pk}(M, x')$, *equal to one if the signature is accepted, and zero otherwise. We require that if* $x \leftarrow \text{Sign}_{sk}(M)$, *then* $\text{Verify}_{pk}(M, x) = 1$.

In the previously introduced *existential unforgeability under an adaptive chosen message attack* scenario, the forger can dynamically obtain signatures of messages of his choice and attempt to output a valid forgery. A *valid forgery* is a message/signature pair (M, x) such that $\text{Verify}_{pk}(M, x) = 1$ whereas the signature of M was never requested by the forger. Moreover, in the random oracle model, the attacker cannot evaluate the hash function by himself; instead, he queries an oracle which outputs a random value for each new query.

RSA [14] is undoubtedly the most widely used cryptosystem today:

Definition 2 (RSA). *The RSA cryptosystem is a family of trapdoor permutations, specified by:*

– The RSA generator $\mathcal{RSA}$, which on input 1^k, randomly selects two distinct $k/2$-bit primes p and q and computes the modulus $N = p \cdot q$. It picks an encryption exponent $e \in \mathbb{Z}^*_{\phi(N)}$ and computes the corresponding decryption exponent d such that $e \cdot d = 1 \mod \phi(N)$. The generator returns (N, e, d).
– The encryption function $f : \mathbb{Z}^*_N \to \mathbb{Z}^*_N$ defined by $f(x) = x^e \mod N$.
– The decryption function $f^{-1} : \mathbb{Z}^*_N \to \mathbb{Z}^*_N$ defined by $f^{-1}(y) = y^d \mod N$.

An inverting algorithm $\mathcal{I}$ for RSA gets as input (N, e, y) and tries to find $y^d \mod N$. Its success probability is the probability to output $y^d \mod N$ when (N, e, d) are obtained by running $\mathcal{RSA}(1^k)$ and y is set to $x^e \mod N$ for some x chosen at random in $\mathbb{Z}^*_N$.

The Full-Domain-Hash scheme (FDH) [3] was the first practical and provably secure signature scheme based on RSA. It is defined as follows: the key generation algorithm, on input 1^k, runs $\mathcal{RSA}(1^k)$ to obtain (N, e, d). It outputs (pk, sk), where the public key pk is (N, e) and the private key sk is (N, d). The signing and verifying algorithms use a hash function $H : \{0, 1\}^* \to \mathbb{Z}^*_N$ which maps bit strings of arbitrary length to the set of invertible integers modulo N.

$\text{SignFDH}_{N,d}(M)$	$\text{VerifyFDH}_{N,e}(M, x)$
$y \leftarrow H(M)$	$y \leftarrow x^e \mod N$
return $y^d \mod N$	if $y = H(M)$ then return 1 else return 0.

The following theorem [6] proves the security of FDH in the random oracle model, assuming that inverting RSA is hard. It provides a better security bound than [3].

Theorem 1. *Assume that there is no algorithm which inverts RSA with probability greater than ε within time t. Then the success probability of a FDH forger making at most q_{hash} hash queries and q_{sig} signature queries within running time t' is less than ε', where*

$$\varepsilon' = 4 \cdot q_{sig} \cdot \varepsilon$$
$$t' = t - (q_{hash} + q_{sig} + 1) \cdot \mathcal{O}(k^3)$$

We say that a hash-and-sign signature scheme is a *partial-domain hash signature scheme* if the encoding function $\mu(m)$ can be written as:

$$\mu(m) = \gamma \cdot H(m) + f(m) \tag{1}$$

where γ is a constant, H a hash function and f some function of m. A typical example of a partial-domain hash signature scheme is the ISO 9796-2 standard with full message recovery [11]:

$$\mu(m) = \text{4A}_{16} \| m \| H(m) \| \text{BC}_{16}$$

The main result of this paper is to show that for $e = 2$, partial-domain hash signature schemes are provably secure, if the hash size is larger than 2/3 of the modulus size. In the following, we recall the Rabin-Williams signature scheme [12]. It uses a padding function $\mu(m)$ such that for all m, $\mu(m) = 6 \mod 16$.

- Key generation: on input 1^k, generate two $k/2$-bit primes p and q such that $p = 3 \mod 8$ and $q = 7 \mod 8$. The public key is $N = p \cdot q$ and the private key is $d = (N - p - q + 5)/8$.
- Signature generation: compute the Jacobi symbol

$$J = \left(\frac{\mu(m)}{N}\right)$$

The signature of m is $s = \min(\sigma, N - \sigma)$, where:

$$\sigma = \begin{cases} \mu(m)^d \mod N & \text{if } J = 1 \\ (\mu(m)/2)^d \mod N & \text{otherwise} \end{cases}$$

- Signature verification: compute $\omega = s^2 \mod N$ and check that:

$$\mu(m) \stackrel{?}{=} \begin{cases} \omega & \text{if } \omega = 6 \mod 8 \\ 2 \cdot \omega & \text{if } \omega = 3 \mod 8 \\ N - \omega & \text{if } \omega = 7 \mod 8 \\ 2 \cdot (N - \omega) & \text{if } \omega = 2 \mod 8 \end{cases}$$

3   Security of Partial-Domain Hash Signature Schemes

To prove the security of a signature scheme against chosen message attacks, one must be able to answer the signature queries of the attacker. In FDH's security proof, when answering a hash query, one generates a random $r \in \mathbb{Z}_N$ and answers $H(m) = r^e \mod N$ so that the signature r of m is known. Similarly, for partial-domain hash signature schemes, we should be able to generate a random r such that:

$$\mu(m) = \gamma \cdot H(m) + f(m) = r^e \mod N$$

with $H(m)$ being uniformly distributed in the output space of the hash function. For example, if we take $\mu(m) = H(m)$ where $0 \leq H(m) \leq N^\beta$ and $\beta < 1$, one should be able to generate a random r such that $r^e \mod N$ is uniformly distributed between 0 and N^β.

Up to our knowledge, no such algorithm is known for $e \geq 3$. For $e = 2$, Vallée constructed in [16] a random generator where the size of $r^2 \mod N$ is less than 2/3 of the size of the modulus. [16] used this generator to obtain proven complexity bounds for the quadratic sieve factoring algorithm. Vallée's generator has a quasi-uniform distribution; a distribution is said to be *quasi-uniform* if there is a constant ℓ such that for all x, the probability to generate x lies between $1/\ell$ and ℓ times the probability to generate x under the uniform distribution. However, quasi-uniformity is not sufficient here, as we must simulate a random oracle and therefore our simulation should be indistinguishable from the uniform distribution.

Our contribution is to modify Vallée's generator in order to generate random squares in any interval of size $N^{2/3+\epsilon}$, with a distribution which is statistically indistinguishable from the uniform distribution. From this generator we will derive a security proof for partial-domain hash signatures, in which the digest size is at least $2/3$ of the modulus size.

Remark: for Paillier's trapdoor permutation [13] with parameter $g = 1 + N$, it is easy to show that half-domain hash is provably secure in the random oracle model, assuming that inverting RSA with $e = N$ is hard.

4   Generating Random Squares in a Given Interval

4.1   Notations

We identify $\mathbb{Z}_N$, the ring of integers modulo N with the set of integers between 0 and $N - 1$. We denote by $\mathbb{Z}_N^+$ the set of integers between 0 and $(N-1)/2$. We denote by Q the squaring operation over $\mathbb{Z}_N$:

$$Q(x) = x^2 \mod N$$

Given positive integers a and h such that $a + h < N$, let B be the set:

$$B = \{x \in \mathbb{Z}_N^+ \mid a \leq Q(x) \leq a + h\}$$

Our goal is to generate integers $x \in B$ with a distribution statistically indistinguishable from the uniform distribution. The *statistical distance* between two distributions X and Y is defined as the function:

$$\delta = \frac{1}{2} \sum_{\alpha} |\Pr[X = \alpha] - \Pr[Y = \alpha]|$$

We say that two ensembles $X = \{X_n\}_{n \in \mathbb{N}}$ and $Y = \{Y_n\}_{n \in \mathbb{N}}$ are *statistically indistinguishable* if their statistical distance δ_n is a negligible function of n.

4.2   Description of B

In this section, we recall Vallée's description of the set B. We denote by b the cardinality of B. The following lemma, which proof can be derived from equation (6) in [16], shows that b is close to $h/2$.

Lemma 1. *Let N be a ℓ-bit RSA modulus. We have for $\ell \geq 64$:*

$$\left| b - \frac{h}{2} \right| \leq 4 \cdot \ell \cdot 2^{\ell/2}$$

In the following, we assume that the bit size of N is greater than 64. As in [16], we introduce Farey sequences [9]:

Definition 3 (Farey sequence). *The Farey sequence $\mathcal{F}_k$ of order k is the ascending sequence of irreducible fractions between 0 and 1 whose denominators do not exceed k. Thus p/q belongs to $\mathcal{F}_k$ if $0 \leq p \leq q \leq k$ and $\gcd(p, q) = 1$.*

The characteristic property of Farey sequences is expressed by the following theorem [9]:

Theorem 2. *If p/q and p'/q' are two successive terms of $\mathcal{F}_k$, then $q \cdot p' - p \cdot q' = 1$*

Given $p/q \in \mathcal{F}_k$, we define the *Farey interval* $I(p,q)$ as the interval of center $pN/(2q)$ and radius $N/(2kq)$. Given the terms p'/q' and p''/q'' of $\mathcal{F}_k$ which precede and follow p/q, we let $J(p,q)$ be the interval:

$$J(p,q) = \left[\frac{N(p+p')}{2(q+q')}, \frac{N(p+p'')}{2(q+q'')} \right]$$

If $p/q = 0/1$, then p/q has no predecessor and we take $p'/q' = 0/1$. Similarly, if $p/q = 1/1$, we take $p''/q'' = 1/1$. The set of intervals $J(p,q)$ forms a partition of $\mathbb{Z}_N^+$. The following lemma [16] shows that intervals $I(p,q)$ and $J(p,q)$ are closely related.

Lemma 2. *$I(p,q)$ contains $J(p,q)$ and its length is at most twice the length of $J(p,q)$.*

Given $p/q \in \mathcal{F}_k$ with $p/q \neq 0/1$, let x_0 be the integer nearest to the rational $pN/2q$:

$$x_0 - \frac{pN}{2q} = u_0 \quad \text{with } |u_0| \leq \frac{1}{2}$$

Let $L(x_0)$ be the lattice spanned by the two vectors $(1, 2x_0)$ and $(0, N)$. Let $\mathcal{P}_1$ and $\mathcal{P}_2$ be the two parabolas of equations:

$$\mathcal{P}_1 : \omega + u^2 + x_0^2 = a + h \quad \text{and} \quad \mathcal{P}_2 : \omega + u^2 + x_0^2 = a$$

Let P be the domain of lattice points comprised between the two parabolas:

$$P = \{(u, \omega) \in L(x_0) \mid a \leq \omega + u^2 + x_0^2 \leq a + h\}$$

The following lemma, which proof is straightforward, shows that the elements of B arise from the intersection of the lattice $L(x_0)$ and the domain comprised between the two parabolas (see figure 1).

Lemma 3. *$x = x_0 + u$ belongs to B iff there exists a unique ω such that the point (u, ω) belongs to P.*

We let $B(p,q)$ be the set of integers in $B \cap J(p,q)$. From Lemma 3 the integers in $B(p,q)$ arise from the domain of lattice points:

$$P(p,q) = \{(u, \omega) \in P \mid x_0 + u \in J(p,q)\}$$

From Lemma 2, the set $P(p,q)$ is included inside the set of lattice points:

$$Q(p,q) = \{(u, \omega) \in P \mid x_0 + u \in I(p,q)\}$$

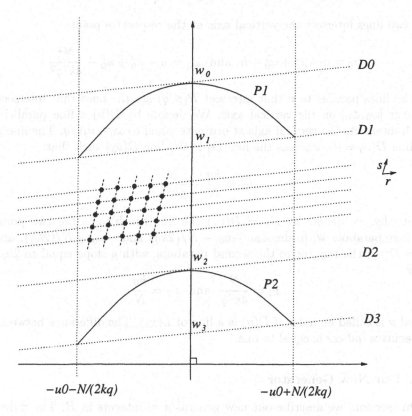

Fig. 1. The intersection between the lattice $L(x_0)$ and the domain between the two parabolas $\mathcal{P}_1$ and $\mathcal{P}_2$

whose abcissae u are comprised between $-u_0 - N/(2kq)$ and $-u_0 + N/(2kq)$. In the following, we describe the domain $Q(p, q)$, using the following two short vectors of $L(x_0)$ (see figure 1):

$$r = q(1, 2x_0) - p(0, N) = (q, 2qu_0) \qquad (2)$$

$$s = q'(1, 2x_0) - p'(0, N) = (q', 2q'u_0 + N/q) \qquad (3)$$

where p'/q' is the term of $\mathcal{F}_k$ which precedes p/q.

We consider the lines of the lattice parallel to vector r which intersect the domain $Q(p, q)$. These lines have a slope equal to $2u_0$. The first extremal position of these lines is the tangent D_0 to the first parabola:

$$D_0 : \; \omega - (-u_0^2 - x_0^2 + a + h) = 2u_0(u + u_0)$$

The second extremal position joins the two points of the second parabola with abscissae $-u_0 - N/(2kq)$ and $-u_0 + N/(2kq)$. This line D_3 has also a slope equal to $2u_0$ and satisfies the equation:

$$\omega + (u_0 + \frac{N}{2kq})^2 - a + x_0^2 = 2u_0(u + u_0 + \frac{N}{2kq})$$

The two lines intersect the vertical axis at the respective points:

$$\omega_0 = a - x_0^2 + u_0^2 + h \quad \text{and} \quad \omega_3 = a - x_0^2 + u_0^2 - \frac{N^2}{4k^2q^2}$$

All the lines parallel to r that intersect $P(p,q)$ are the ones that intersect the segment $[\omega_3, \omega_0]$ on the vertical axis. We denote by $D(\nu)$ a line parallel to r which intersects the vertical axis at ordinate equal to $\omega_0 - \nu N/q$. The line D_0 is the line $D(\nu_0 = 0)$, whereas the line D_3 is the line $D(\nu_3)$ such that:

$$\nu_3 = \frac{hq}{N} + \frac{N}{4k^2q} \tag{4}$$

Eventually, we denote by $D_1 = D(\nu_1)$ the line which joins the two points of the first parabola with abcissae $-u_0 - N/(2kq)$ and $-u_0 + N/(2kq)$, and by $D_2 = D(\nu_2)$ the tangent to the second parabola, with a slope equal to $2u_0$. We have:

$$\nu_1 = \frac{N}{4k^2q} \quad \text{and} \quad \nu_2 = \frac{hq}{N} \tag{5}$$

A real ν is called an index if $D(\nu)$ is a line of $L(x_0)$. The difference between two consecutive indices is equal to one.

4.3   Our New Generator

In this section, we describe our new generator of integers in B. The difference with Vallée's generator is that we use different parameters for k and h, and we do not generate all the integers in B; instead we avoid a negligible subset of B.

First, we describe a generator $\mathcal{G}(p,q)$ of integers in $B(p,q)$, and we show that its distribution is statistically indistinguishable from the uniform distribution. We assume that $N \leq 2 \cdot k \cdot q \cdot \sqrt{h}$, which gives $\nu_1 \leq \nu_2$. Therefore the line D_1 is above the line D_2 (see figure 1). We restrict ourselves to the integers in $B(p,q)$ such that the corresponding points $(u, \omega) \in P(p,q)$ lie on $D(\nu)$ with $\nu_1 \leq \nu \leq \nu_2$. These points are the points on $D(\nu)$ whose abscissae u are such that $x_0 + u \in J(p,q)$.

Generator $\mathcal{G}(p,q)$ of integers in $B(p,q)$:

1. Generate a random index ν uniformly distributed between ν_1 and ν_2.
2. Generate a point $(u, \omega) \in P(p,q)$ on $D(\nu)$ such that $x_0 + u \in J(p,q)$, with the uniform distribution.
3. Output $x_0 + u$.

The following lemma shows that under some conditions on k, h and q, the cardinality $b(p,q)$ of $B(p,q)$ is close to $h \cdot j(p,q)/N$, where $j(p,q)$ is the number of integers in the interval $J(p,q)$. Moreover, under the same conditions, the distribution induced by $\mathcal{G}(p,q)$ is statistically indistinguishable from the uniform distribution in $B(p,q)$. The proof is given in appendix A.

Lemma 4. *Let $\alpha > 0$ and $k = N^{\frac{1}{3}-\alpha}$. Assume that $k \geq 6$, $N^{\alpha} \geq 3$ and $N^{\frac{2}{3}+13\cdot\alpha} \leq h < N$. Then for all $p/q \in \mathcal{F}_k$ such that $N^{1/3-4\alpha} \leq q \leq k$, we have:*

$$\left| b(p,q) - \frac{h \cdot j(p,q)}{N} \right| \leq \frac{4h \cdot j(p,q)}{N} N^{-3\alpha} \tag{6}$$

Moreover, $\mathcal{G}(p,q)$ generates elements in $B(p,q)$ with a distribution whose distance $\delta_{\mathcal{G}}$ from the uniform distribution is at most $7 \cdot N^{-3\alpha}$.

Now we construct a generator $\mathcal{V}$ of $p/q \in \mathcal{F}_k$ such that the probability to generate p/q is close to $b(p,q)/b$. It only generates $p/q \in \mathcal{F}_k$ such that $q \geq N^{1/3-4\alpha}$, so that from the previous lemma, $b(p,q)$ is nearly proportional to the number of integers in $J(p,q)$, and the distribution induced by $\mathcal{G}(p,q)$ is close to the uniform distribution.

Generator $\mathcal{V}$ of $p/q \in \mathcal{F}_k$

1. Generate a random integer $x \in \mathbb{Z}_N^+$ with the uniform distribution.
2. Determine which interval $J(p,q)$ contains x.
3. If $q \geq N^{1/3-4\alpha}$ then output $p/q \in \mathcal{F}_k$, otherwise output $\perp$.

Lemma 5. *Let denote by $\mathcal{D}$ the distribution induced by choosing $p/q \in \mathcal{F}_k$ with probability $b(p,q)/b$. Under the conditions of lemma 4, the statistical distance $\delta_{\mathcal{V}}$ between $\mathcal{D}$ and the distribution induced by $\mathcal{V}$ is at most $9 \cdot N^{-3\alpha}$.*

Proof. See appendix B.

Eventually, our generator $\mathcal{G}$ of elements in B combines the two generators $\mathcal{V}$ and $\mathcal{G}(p,q)$:

Generator $\mathcal{G}$ of $x \in B$

1. Generate y using $\mathcal{V}$.
2. If $y = \perp$, then output $\perp$.
3. Otherwise, $y = p/q$ and generate $x \in B(p,q)$ using $\mathcal{G}(p,q)$. Output x.

The following theorem, whose proof is given in appendix C, shows that the distribution induced by $\mathcal{G}$ is statistically indistinguishable from the uniform distribution in B.

Theorem 3. *For any $\varepsilon > 0$, letting $h = N^{\frac{2}{3}+\varepsilon}$ and $\alpha = \varepsilon/13$. If $N^{\alpha} \geq 3$, then the distance δ between the distribution induced by $\mathcal{G}$ and the uniform distribution in B is at most $16 \cdot N^{-3\cdot\varepsilon/13}$. The running time of $\mathcal{G}$ is $\mathcal{O}(\log^3 N)$.*

5   A Security Proof
for Partial-Domain Hash Signature Schemes

In this section, using the previous generator $\mathcal{G}$ of random squares, we show that partial-domain hash signature schemes are provably secure in the random oracle model, for $e = 2$, assuming that factoring is hard, if the size of the hash function is larger than 2/3 of the modulus size. Moreover, we restrict ourselves to small constants γ in (1), e.g. $\gamma = 16$ or $\gamma = 256$. This is the case for all the signature standards of the next section. We denote by k_0 the hash function's digest size. The proof is similar to the proof of theorem 1 and is given in the full version of this paper [7].

Theorem 4. *Let S be the Rabin-Williams partial-domain hash signature scheme with constant γ and hash size k_0 bits. Assume that there is no algorithm which factors a RSA modulus with probability greater than ε within time t. Then the success probability of a forger against S making at most q_{hash} hash queries and q_{sig} signature queries within time t' is upper bounded by ε', where:*

$$\varepsilon' = 8 \cdot q_{sig} \cdot \varepsilon + 32 \cdot (q_{hash} + q_{sig} + 1) \cdot k_1 \cdot \gamma \cdot 2^{-\frac{3}{13} \cdot k_1} \tag{7}$$

$$t' = t - k_1 \cdot \gamma \cdot (q_{hash} + q_{sig} + 1) \cdot \mathcal{O}(k^3) \tag{8}$$

and $k_1 = k_0 - \frac{2}{3}k$.

6   Application to Signature Standards

6.1   PKCS#1 v1.5 and SSL-3.02

The signature scheme PKCS#1 v1.5 [15] is a partial-domain hash signature scheme, with:

$$\mu(m) = 0001_{16} \| FFFF_{16} \ldots FFFF_{16} \| 00_{16} \| c_{\text{SHA}} \| H(m)$$

where c_{SHA} is a constant and $H(m) = \text{SHA}(m)$, or

$$\mu(m) = 0001_{16} \| FFFF_{16} \ldots FFFF_{16} \| 00_{16} \| c_{\text{MD5}} \| H(m)$$

where c_{MD5} is a constant and $H(m) = \text{MD5}(m)$.

The standard PKCS#1 v1.5 was not designed to work with Rabin ($e = 2$). However, one can replace the last nibble of $H(m)$ by 6 and obtain a padding scheme which is compatible with the Rabin-Williams signature scheme. The standard is then provably secure if the size of the hash-function is larger than 2/3 of the size of the modulus. This is much larger than the 128 or 160 bits which are recommended in the standard. The same analysis applies for the SSL-3.02 padding scheme [10].

6.2   ISO 9796-2 and ANSI x9.31

The ISO 9796-2 encoding scheme [11] is defined as follows:

$$\mu(m) = 6A_{16} \| m[1] \| H(m) \| BC_{16}$$

where $m[1]$ is the leftmost part of the message, or:

$$\mu(m) = 4\mathtt{A_{16}}\|m\|H(m)\|\mathtt{BC_{16}}$$

[11] describes an application of ISO 9796-2 with the Rabin-Williams signature scheme. Note that since $\mu(m) = 12 \mod 16$ instead of $\mu(m) = 6 \mod 16$, there is a slight change in the verification process. However, the same security bound applies: the scheme is provably secure if the size of the hash-function is larger than 2/3 of the size of the modulus. The same analysis applies for the ANSI x9.31 padding scheme [1].

7   Conclusion

We have shown that for Rabin, partial-domain hash signature schemes are provably secure in the random oracle, assuming that factoring is hard, if the size of the hash function is larger than 2/3 of the modulus size. Unfortunately, this is much larger than the size which is recommended in the standards PKCS#1 v1.5 and ISO 9796-2. An open problem is to obtain a smaller bound for the digest size, and to extend this result to RSA signatures.

Acknowledgements

I wish to thanks the anonymous referees for their helpful comments.

References

1. ANSI X9.31, *Digital signatures using reversible public-key cryptography for the financial services industry (rDSA)*, 1998.
2. M. Bellare and P. Rogaway, *Random oracles are practical : a paradigm for designing efficient protocols*. Proceedings of the First Annual Conference on Computer and Communications Security, ACM, 1993.
3. M. Bellare and P. Rogaway, *The exact security of digital signatures - How to sign with RSA and Rabin*. Proceedings of Eurocrypt'96, LNCS vol. 1070, Springer-Verlag, 1996, pp. 399-416.
4. R. Canetti, O. Goldreich and S. Halevi, *The random oracle methodology, revisited*, STOC' 98, ACM, 1998.
5. J.S. Coron, D. Naccache and J.P. Stern, *On the security of RSA Padding*, Proceedings of Crypto'99, LNCS vol. 1666, Springer-Verlag, 1999, pp. 1-18.
6. J.S. Coron, *On the exact security of Full Domain Hash*, Proceedings of Crypto 2000, LNCS vol. 1880, Springer-Verlag, 2000, pp. 229-235.
7. J.S. Coron, *Security proof for partial-domain hash signature schemes*. Full version of this paper. Cryptology ePrint Archive, http://eprint.iacr.org.
8. S. Goldwasser, S. Micali and R. Rivest, *A digital signature scheme secure against adaptive chosen-message attacks*, SIAM Journal of computing, 17(2):281-308, april 1988.
9. G.H. Hardy and E.M. Wright, *An introduction to the theory of numbers*, Oxford science publications, fifth edition.
10. K. Hickman, *The SSL Protocol*, December 1995. Available electronically at : http://www.netscape.com/newsref/std/ssl.html

11. ISO/IEC 9796-2, *Information technology - Security techniques - Digital signature scheme giving message recovery, Part 2* : Mechanisms using a hash-function, 1997.
12. A.J. Menezes, P. C. van Oorschot and S.A. Vanstone, *Handbook of Applied Cryptography*, CRC press, 1996.
13. P. Paillier, *Public-key cryptosystems based on composite degree residuosity classes*, proceedings of Eurocrypt'99, LNCS 1592, pp. 223-238, 1999.
14. R. Rivest, A. Shamir and L. Adleman, *A method for obtaining digital signatures and public key cryptosystems*, CACM 21, 1978.
15. RSA Laboratories, PKCS #1 : *RSA cryptography specifications*, version 1.5, November 1993 and version 2.0, September 1998.
16. B. Vallée, *Generation of elements with small modular squares and provably fast integer factoring algorithms*, Mathematics of Computation, vol. 56, number 194, april 1991, pp. 823-849.

A   Proof of Lemma 4

From the conditions of lemma 4, we obtain:

$$\frac{hq}{N} \geq N^{9\alpha} \quad \text{and} \quad \frac{N}{k^2 q} \leq N^{6\alpha} \tag{9}$$

which gives $N \leq 2 \cdot k \cdot q \cdot \sqrt{h}$ and then $\nu_1 < \nu_2$.

Recall that $j(p,q)$ denotes the number of integers in interval $J(p,q)$. From lemma 2 the length of $J(p,q)$ is at least $N/(2kq)$ and therefore, $j(p,q) \geq N/(2kq) - 1$, which gives using $k \geq 6$:

$$\frac{j(p,q)}{q} \geq \frac{N^{3\alpha}}{3} \tag{10}$$

Let us denote by $n(\nu)$ the number of points of $P(p,q)$ on a line $D(\nu)$. The distance between the abcissae of two consecutive points of $P(p,q)$ on a line $D(\nu)$ is equal to q. Therefore, for all indices ν, we have $n(\nu) \leq \lfloor j(p,q)/q \rfloor + 1$. Moreover, for $\nu_1 \leq \nu \leq \nu_2$, $n(\nu)$ is either $\lfloor j(p,q)/q \rfloor$ or $\lfloor j(p,q)/q \rfloor + 1$. This gives the following bound for $b(p,q)$:

$$(\nu_2 - \nu_1 - 1) \cdot \left(\frac{j(p,q)}{q} - 1 \right) \leq b(p,q) \leq (\nu_3 + 1) \cdot \left(\frac{j(p,q)}{q} + 1 \right)$$

which gives using (4), (5), (9), (10) and $N^\alpha \geq 3$:

$$\left| b(p,q) - \frac{h \cdot j(p,q)}{N} \right| \leq \frac{4h \cdot j(p,q)}{N} N^{-3\alpha} \tag{11}$$

Let n' be the number of indices ν such that $\nu_1 \leq \nu \leq \nu_2$. We have $n' = \lfloor \nu_2 - \nu_1 \rfloor$ or $n' = \lfloor \nu_2 - \nu_1 \rfloor + 1$. The probability that $\mathcal{G}(p,q)$ generates an element $x \in B(p,q)$ corresponding to a point of index ν is given by:

$$\Pr[x] = P(\nu) = \frac{1}{n' \cdot n(\nu)}$$

for $\nu_1 \leq \nu \leq \nu_2$ and $P(\nu) = 0$ otherwise. The number of integers $x \in B(p,q)$ such that $\Pr[x] = 0$ is then at most:

$$(\nu_1 + \nu_3 - \nu_2 + 2) \cdot \left(\frac{j(p,q)}{q} + 1 \right) \leq N^{6\alpha} \cdot \frac{j(p,q)}{q} \tag{12}$$

For all $\nu_1 \leq \nu \leq \nu_2$, we have using (4), (5), (9), (10), (11) and $N^\alpha \geq 3$:

$$\left| P(\nu) - \frac{1}{b(p,q)} \right| \leq 10 \cdot \frac{N}{h \cdot j(p,q)} \cdot N^{-3\alpha} \tag{13}$$

Eventually, the statistical distance from the uniform distribution is:

$$\delta_G = \frac{1}{2} \sum_{x \in B(p,q)} \left| \Pr[x] - \frac{1}{b(p,q)} \right|$$

and we obtain using (11), (12) and (13):

$$\delta_G \leq 7 \cdot N^{-3\alpha}$$

B Proof of Lemma 5

Let us denote $q_m = N^{1/3 - 4\alpha}$. For $q \geq q_m$, the probability to generate $p/q \in \mathcal{F}_k$ using $\mathcal{V}$ is $j(p,q)/|\mathbb{Z}_N^+|$. Moreover, using lemma 2, the probability that $\mathcal{V}$ generates $\perp$ is at most:

$$\Pr[\perp] = \sum_{\mathcal{F}_k | q < q_m} \frac{2 \cdot j(p,q)}{N+1} \leq 3\frac{q_m}{k} \leq 3 \cdot N^{-3\alpha} \tag{14}$$

Consequently, the statistical distance δ_V between $\mathcal{D}$ and the distribution induced by $\mathcal{V}$ is at most:

$$\delta_V = \frac{1}{2} \sum_{\mathcal{F}_k | q \geq q_m} \left| \frac{2 \cdot j(p,q)}{N+1} - \frac{b(p,q)}{b} \right| + \frac{1}{2} \Pr[\perp] + \frac{1}{2} \sum_{\mathcal{F}_k | q < q_m} \frac{b(p,q)}{b} \tag{15}$$

Let ℓ be the size of N in bits. From lemma 1, we obtain for $\ell \geq 64$:

$$\left| b - \frac{h}{2} \right| \leq 4 \cdot \ell \cdot 2^{\ell/2} \leq \frac{1}{2} \cdot N^{2/3} \leq \frac{h}{2} \cdot N^{-3\alpha} \tag{16}$$

For $q \geq q_m$, we obtain from Lemma 4 and (16):

$$\left| \frac{b(p,q)}{b} - \frac{2 \cdot j(p,q)}{N+1} \right| \leq \frac{12 \cdot j(p,q)}{N+1} \cdot N^{-3\alpha} \tag{17}$$

This gives:

$$\sum_{\mathcal{F}_k | q < q_m} \frac{b(p,q)}{b} = 1 - \sum_{\mathcal{F}_k | q \geq q_m} \frac{b(p,q)}{b} \leq 1 - (1 - 6 \cdot N^{-3\alpha}) \cdot \sum_{\mathcal{F}_k | q \geq q_m} \frac{2 \cdot j(p,q)}{N+1}$$

From (14) and using:

$$\sum_{\mathcal{F}_k} \frac{2 \cdot j(p,q)}{N+1} = 1$$

we obtain:

$$\sum_{\mathcal{F}_k \mid q < q_m} \frac{b(p,q)}{b} \leq 9 \cdot N^{-3\alpha} \qquad (18)$$

From equation (15) and inequalities (14), (17) and (18), we obtain:

$$\delta_V \leq 9 \cdot N^{-3\alpha}$$

C   Proof of Theorem 3

The generator $\mathcal{G}$ combines the generators $\mathcal{V}$ and $\mathcal{G}(p,q)$. Moreover, $\mathcal{V}$ generates $p/q \in \mathcal{F}_k$ such that the statistical distance δ_G of the distribution induced by $\mathcal{G}(p,q)$ from the uniform distribution in $B(p,q)$ is at most $7 \cdot N^{-3\alpha}$. Therefore the statistical distance δ of $\mathcal{G}$ from the uniform distribution in B is at most:

$$\delta \leq \delta_V + \delta_G \leq 16 \cdot N^{-3\varepsilon/13}$$

Author Index

Lecture Notes in Computer Science

For information about Vols. 1–2341
please contact your bookseller or Springer-Verlag

Vol. 2383: M.S. Lew, N. Sebe, J.P. Eakins (Eds.), Image and Video Retrieval. Proceedings, 2002. XII, 388 pages. 2002.

Vol. 2384: L. Batten, J. Seberry (Eds.), Information Security and Privacy. Proceedings, 2002. XII, 514 pages. 2002.

Vol. 2385: J. Calmet, B. Benhamou, O. Caprotti, L. Henocque, V. Sorge (Eds.), Artificial Intelligence, Automated Reasoning, and Symbolic Computation. Proceedings, 2002. XI, 343 pages. 2002. (Subseries LNAI).

Vol. 2386: E.A. Boiten, B. Möller (Eds.), Mathematics of Program Construction. Proceedings, 2002. X, 263 pages. 2002.

Vol. 2387: O.H. Ibarra, L. Zhang (Eds.), Computing and Combinatorics. Proceedings, 2002. XIII, 606 pages. 2002.

Vol. 2388: S.-W. Lee, A. Verri (Eds.), Pattern Recognition with Support Vector Machines. Proceedings, 2002. XI, 420 pages. 2002.

Vol. 2389: E. Ranchhod, N.J. Mamede (Eds.), Advances in Natural Language Processing. Proceedings, 2002. XII, 275 pages. 2002. (Subseries LNAI).

Vol. 2391: L.-H. Eriksson, P.A. Lindsay (Eds.), FME 2002: Formal Methods – Getting IT Right. Proceedings, 2002. XI, 625 pages. 2002.

Vol. 2392: A. Voronkov (Ed.), Automated Deduction – CADE-18. Proceedings, 2002. XII, 534 pages. 2002. (Subseries LNAI).

Vol. 2393: U. Priss, D. Corbett, G. Angelova (Eds.), Conceptual Structures: Integration and Interfaces. Proceedings, 2002. XI, 397 pages. 2002. (Subseries LNAI).

Vol. 2395: G. Barthe, P. Dybjer, L. Pinto, J. Saraiva (Eds.), Applied Semantics. IX, 537 pages. 2002.

Vol. 2396: T. Caelli, A. Amin, R.P.W. Duin, M. Kamel, D. de Ridder (Eds.), Structural, Syntactic, and Statistical Pattern Recognition. Proceedings, 2002. XVI, 863 pages. 2002.

Vol. 2398: K. Miesenberger, J. Klaus, W. Zagler (Eds.), Computers Helping People with Special Needs. Proceedings, 2002. XXII, 794 pages. 2002.

Vol. 2399: H. Hermanns, R. Segala (Eds.), Process Algebra and Probabilistic Methods. Proceedings, 2002. X, 215 pages. 2002.

Vol. 2401: P.J. Stuckey (Ed.), Logic Programming. Proceedings, 2002. XI, 486 pages. 2002.

Vol. 2402: W. Chang (Ed.), Advanced Internet Services and Applications. Proceedings, 2002. XI, 307 pages. 2002.

Vol. 2403: Mark d'Inverno, M. Luck, M. Fisher, C. Preist (Eds.), Foundations and Applications of Multi-Agent Systems. Proceedings, 1996-2000. X, 261 pages. 2002. (Subseries LNAI).

Vol. 2404: E. Brinksma, K.G. Larsen (Eds.), Computer Aided Verification. Proceedings, 2002. XIII, 626 pages. 2002.

Vol. 2405: B. Eaglestone, S. North, A. Poulovassilis (Eds.), Advances in Databases. Proceedings, 2002. XII, 199 pages. 2002.

Vol. 2406: C. Peters, M. Braschler, J. Gonzalo, M. Kluck (Eds.), Evaluation of Cross-Language Information Retrieval Systems. Proceedings, 2001. X, 601 pages. 2002.

Vol. 2407: A.C. Kakas, F. Sadri (Eds.), Computational Logic: Logic Programming and Beyond. Part I. XII, 678 pages. 2002. (Subseries LNAI).

Vol. 2408: A.C. Kakas, F. Sadri (Eds.), Computational Logic: Logic Programming and Beyond. Part II. XII, 628 pages. 2002. (Subseries LNAI).

Vol. 2409: D.M. Mount, C. Stein (Eds.), Algorithm Engineering and Experiments. Proceedings, 2002. VIII, 207 pages. 2002.

Vol. 2410: V.A. Carreño, C.A. Muñoz, S. Tahar (Eds.), Theorem Proving in Higher Order Logics. Proceedings, 2002. X, 349 pages. 2002.

Vol. 2412: H. Yin, N. Allinson, R. Freeman, J. Keane, S. Hubbard (Eds.), Intelligent Data Engineering and Automated Learning – IDEAL 2002. Proceedings, 2002. XV, 597 pages. 2002.

Vol. 2413: K. Kuwabara, J. Lee (Eds.), Intelligent Agents and Multi-Agent Systems. Proceedings, 2002. X, 221 pages. 2002. (Subseries LNAI).

Vol. 2414: F. Mattern, M. Naghshineh (Eds.), Pervasive Computing. Proceedings, 2002. XI, 298 pages. 2002.

Vol. 2415: J. Dorronsoro (Ed.), Artificial Neural Networks – ICANN 2002. Proceedings, 2002. XXVIII, 1382 pages. 2002.

Vol. 2417: M. Ishizuka, A. Sattar (Eds.), PRICAI 2002: Trends in Artificial Intelligence. Proceedings, 2002. XX, 623 pages. 2002. (Subseries LNAI).

Vol. 2418: D. Wells, L. Williams (Eds.), Extreme Programming and Agile Methods – XP/Agile Universe 2002. Proceedings, 2002. XII, 292 pages. 2002.

Vol. 2419: X. Meng, J. Su, Y. Wang (Eds.), Advances in Web-Age Information Management. Proceedings, 2002. XV, 446 pages. 2002.

Vol. 2420: K. Diks, W. Rytter (Eds.), Mathematical Foundations of Computer Science 2002. Proceedings, 2002. XII, 652 pages. 2002.

Vol. 2421: L. Brim, P. Jančar, M. Křetínský, A. Kučera (Eds.), CONCUR 2002 – Concurrency Theory. Proceedings, 2002. XII, 611 pages. 2002.

Vol. 2423: D. Lopresti, J. Hu, R. Kashi (Eds.), Document Analysis Systems V. Proceedings, 2002. XIII, 570 pages. 2002.

Vol. 2430: T. Elomaa, H. Mannila, H. Toivonen (Eds.), Machine Learning: ECML 2002. Proceedings, 2002. XIII, 532 pages. 2002. (Subseries LNAI).

Vol. 2431: T. Elomaa, H. Mannila, H. Toivonen (Eds.), Principles of Data Mining and Knowledge Discovery. Proceedings, 2002. XIV, 514 pages. 2002. (Subseries LNAI).

Vol. 2436: J. Fong, R.C.T. Cheung, H.V. Leong, Q. Li (Eds.), Advances in Web-Based Learning. Proceedings, 2002. XIII, 434 pages. 2002.

Vol. 2440: J.M. Haake, J.A. Pino (Eds.), Groupware – CRIWG 2002. Proceedings, 2002. XII, 285 pages. 2002.

Vol. 2442: M. Yung (Ed.), Advances in Cryptology – CRYPTO 2002. Proceedings, 2002. XIV, 627 pages. 2002.

Vol. 2444: A. Buchmann, F. Casati, L. Fiege, M.-C. Hsu, M.-C. Shan (Eds.), Technologies for E-Services. Proceedings, 2002. X, 171 pages. 2002.